Die Grundlehren der mathematischen Wissenschaften

in Einzeldarstellungen
mit besonderer Berücksichtigung
der Anwendungsgebiete

Band 211

W. W. Comfort S. Negrepontis

The Theory of Ultrafilters

Springer-Verlag
New York Heidelberg Berlin 1974

W. Wistar Comfort

Department of Mathematics, Wesleyan University, Middletown, Connecticut, U.S.A.

Stylianos Negrepontis

Department of Mathematics, Athens University, Athens, Greece and Department of Mathematics, McGill University, Montréal, Québec, Canada

AMS Subject Classification (1970)

Primary 02H20, 02H25, 02K25, 02K35, 04A20, 04A30, 12L10, 54C45, 54D40

Secondary 05C05, 06A40, 08A05, 54A10, 54A25, 54C50, 54D35, 54G05

ISBN 0-387-06604-7 Springer-Verlag New York Heidelberg Berlin
ISBN 3-540-06604-7 Springer-Verlag Berlin Heidelberg New York

 Library of Congress Catalog Card Number 74-3673. Printed in Germany. Typesetting and printing: Zechnersche Buchdruckerei, Speyer. Bookbinding: Konrad Triltsch, Würzburg.

Introduction

An ultrafilter is a truth-value assignment to the family of subsets of a set, and a method of convergence to infinity. From the first (logical) property arises its connection with two-valued logic and model theory; from the second (convergence) property arises its connection with topology and set theory. Both these descriptions of an ultrafilter are connected with compactness. The model-theoretic property finds its expression in the construction of the ultraproduct and the compactness type of theorem of Łoś (implying the compactness theorem of first-order logic); and the convergence property leads to the process of completion by the *adjunction* of an ideal element for every ultrafilter—*i.e.*, to the Stone-Čech compactification process (implying the Tychonoff theorem on the compactness of products).

Since these are two ways of describing the same mathematical object, it is reasonable to expect that a study of ultrafilters from these points of view will yield results and methods which can be fruitfully crossbred. This unifying aspect is indeed what we have attempted to emphasize in the present work. As examples, we mention the Baire category type of proof of the existence (in fact, the abundance) of good and minimal ultrafilters that results from an examination of the topological properties of spaces of ultrafilters; the existence of a topological embedding of the Stone space of the α^+-homogeneous-universal Boolean algebra of cardinality α^+ into the space of uniform ultrafilters on α, proved by making use of a good ultrafilter; and the deeper, subtle similarities that exist (below the superficial differences) between the homogeneous-universal Boolean algebras and the corresponding spaces of uniform ultrafilters.

Our study of ultrafilters eventually concentrates on the study of those that are defined by certain extremal properties. Thus the good ultrafilters (and some related ones) are used to produce saturated ultraproducts and isomorphisms of ultrapowers of elementarily equivalent structures; and the minimal ultrafilters are used to analyze the structure and the topology of the set of ultrafilters. But it is only partly because they are useful and subject to successful investigation that we study these special classes. We recall C. L. Siegel's observation (made in the context of Celestial Mechanics): since, according to Leibniz, ours is the best of all possible worlds, its universal laws must be expressed by extremal properties; and of the widespread incidence throughout Mathematics of the phenomenon of Banach: the extremal and even the "pathological" are typical (in the Baire category sense). It seems to us that there is an affinity of our study, leading naturally as it does to extremal properties, with these general philosophical statements.

April 19, 1974

W. Wistar Comfort
Stylianos Negrepontis

Acknowledgements

We thank the following institutions and organizations, which have supported one or both of us with facilities or funds during part of the period 1967—1974: Athens University, McGill University, National Research Council (Canada), National Science Foundation (U.S.A.), Summer Research Institute of Canadian Mathematical Congress (at Queen's University, Kingston and at Carleton University, Ottawa), University of Washington, Wesleyan University.

We thank Mrs. Esther Massa of Montréal for her excellent and patient typing of several versions of the manuscript.

Table of Contents

§ 1. Set Theory

This section is devoted to a self-contained development of those elements of set theory that are used in this work.

The set theory within which we work is the Zermelo-Fraenkel set theory together with the axiom of choice.

The system of axioms of the theory and the basic definitions are stated in informal language. The ordinals are defined in such a way that an ordinal is the set of smaller ordinals. The maximal principle and the well-ordering principle are proved (using the axiom of choice). A cardinal is defined as an ordinal that is not in a one-to-one correspondence with any smaller ordinal. The second half of the section is concerned mostly with cardinal arithmetic.

The fundamental undefined symbol of set theory is the (binary) *membership* predicate symbol $\in$. Together with the logical symbols of equality $=$, negation $\neg$, conjunction $\wedge$, existential quantification $\exists$ (and the derived symbols $\vee$, $\rightarrow$, and $\forall$) and variable symbols $a, b, \cdots, A, B, \cdots, x, y, \cdots$, etc. (denoting *sets*), they form the symbols of the formal *language of set theory*. The formulas of this language, which include the atomic formulas "$a=b$" and "$a \in b$" (read a is an *element* of b) and whose rules of formation are described in § 11 below, are the set-theoretic *properties*, and they determine *classes of sets*. Thus the formula/property $\phi(x)$ *defines* the class, denoted by $\{x: \phi(x)\}$, of all sets a such that $\phi(a)$ is satisfied. Classes are denoted by $\mathbb{A}, \mathbb{B}, \ldots$.

We now state the first axiom of set theory.

(I) *The axiom of extensionality.* Two sets are equal if they have the same elements.

Thus a set a is the class $\{b: b \in a\}$. A class that is not a set is a *proper* class. We extend by definition the applicability of $\in$ as follows: if $\phi(a)$ holds for some set a, we write $a \in \{x: \phi(x)\}$ and say that a is an element of $\{x: \phi(x)\}$. Two classes $\mathbb{A}$ and $\mathbb{B}$ are defined to be *equal* (denoted by $\mathbb{A} = \mathbb{B}$) if they have the same elements.

We introduce the following definitions of classes.

$$\mathbb{V} = \{a\colon a=a\}, \text{ the } universal \text{ class;}$$
$$\emptyset = \{a\colon a\neq a\}, \text{ the } empty \text{ class;}$$
$$\{a,b\} = \{c\colon c=a \vee c=b\}, \text{ the } unordered\ pair \text{ of } a,b;$$
$$\{a\} = \{a,a\}, \text{ the } singleton \text{ of } a;$$
$$\bigcup \mathbb{A} = \{a\colon \exists b(b\in\mathbb{A} \wedge a\in b)\}, \text{ the } union \text{ class of } \mathbb{A};$$
$$\bigcap \mathbb{A} = \{a\colon \forall b(b\in\mathbb{A} \to a\in b)\}, \text{ the } intersection \text{ class of } \mathbb{A}.$$

We further define the *Boolean operations* of *class inclusion, union, intersection*, and *difference*, as follows.

$\mathbb{A}\subset\mathbb{B}$ means $a\in\mathbb{A}\to a\in\mathbb{B}$;
$\mathbb{A}\cup\mathbb{B}$ is the class $\{a\colon a\in\mathbb{A}\vee a\in\mathbb{B}\}$;
$\mathbb{A}\cap\mathbb{B}$ is the class $\{a\colon a\in\mathbb{A}\wedge a\in\mathbb{B}\}$; and
$\mathbb{A}\setminus\mathbb{B}$ is the class $\{a\colon a\in\mathbb{A}\wedge a\notin\mathbb{B}\}$,

respectively.

We note that the axiom of extensionality expresses the anti-symmetric (Cantor-Bernstein) property of inclusion for sets.

With these definitions we state the next four axioms.

(II) *The existence of a set.* The empty class $\emptyset$ is a set.

(III) *The axiom of unordered pairs.* If a and b are sets, then $\{a,b\}$ is a set.

(IV) *The union axiom.* If a is a set, then $\cup a$ is a set.

$(\mathrm{V})_{\mathbb{A}}$ *The axiom schema of comprehension.* If a is a set and $\mathbb{A}$ is a class, then $a\cap\mathbb{A}$ is a set.

Note that, in particular, if $\mathbb{A}$ is a class included in the set a, then $\mathbb{A}$ is a set. We define

$$\mathscr{P}(\mathbb{A}) = \{a\colon a\subset\mathbb{A}\}, \quad \text{the } power\ class \text{ of } \mathbb{A}.$$

(VI) *The power set axiom.* If a is a set, then $\mathscr{P}(a)$ is a set.

The *ordered pair* of a,b, denoted by $\langle a,b\rangle$, is the set $\{\{a\},\{a,b\}\}$. The *Cartesian product* $\mathbb{A}\times\mathbb{B}$ of two classes is the class $\mathbb{A}\times\mathbb{B} = \{\langle a,b\rangle\colon a\in\mathbb{A}\wedge b\in\mathbb{B}\}$. A *(binary) relation* between $\mathbb{A}$ and $\mathbb{B}$ is a subclass of $\mathbb{A}\times\mathbb{B}$. A *function* f is a relation defined in the usual way, and we write $f\colon\mathbb{A}\to\mathbb{B}$, where $\mathbb{A}$ is the *domain* of f and $\mathbb{B}$ the *range* of f. If $\mathbb{C}\subset\mathbb{A}$ and $\mathbb{D}\subset\mathbb{B}$, then $f[\mathbb{C}]$ denotes the *direct image of* $\mathbb{C}$ *under* f, $f|\mathbb{C}$ the *restriction* of the function f to $\mathbb{C}$, and $f^{-1}(\mathbb{D})$ the *inverse image of* $\mathbb{D}$ *under* f.

$(\mathrm{VII})_{\mathbb{A}}$ *The axiom schema of replacement.* If $\mathbb{A}$ is a function and a is a set, then the image class $\mathbb{A}[a]$ is a set.

(VIII) *The axiom of infinity.* The class

$$\{a\colon \emptyset\in a\wedge\forall b(b\in a\to b\cup\{b\}\in a)\}$$

is not empty.

The system of axioms (I)—(VIII) is the *Zermelo-Fraenkel set theory*, denoted ZF.

The *Cartesian product* of a set S of sets, denoted by $\prod_{a\in S} a$, is the set of functions $f: S \to \bigcup S$ such that $f(a) \in a$ for $a \in S$.

(IX) *The axiom of choice.* The Cartesian product of a non-empty set of non-empty sets is non-empty.

The system of axioms (I)—(IX) is the *Zermelo-Fraenkel set theory with the axiom of choice*, denoted ZFC.

Of these axioms, (I) establishes the fundamental connection between $\in$ and $=$; (II) and (VIII) are the only axioms that assert the existence of some set; (IX) asserts the existence of a non-empty set if certain sets are non-empty; and the remaining axioms state that certain defined classes are sets.

Definitions. Let A be a set and $\leq$ a binary relation on A. Then

(i) $\leq$ is *reflexive* if $a \leq a$ for $a \in A$;

(ii) $\leq$ is *transitive* if $a \leq b$ and $b \leq c$ implies $a \leq c$ for $a, b, c \in A$;

(iii) $\leq$ is *anti-symmetric* if $a \leq b$ and $b \leq a$ implies $a = b$ for $a, b \in A$; and

(iv) $\leq$ is *connected* if $a \leq b$ or $b \leq a$ for $a, b \in A$.

A binary relation that is reflexive and transitive is a *pre-order*. A pre-order that is anti-symmetric is a *partial order*. And a partial order that is connected is a *linear order*.

If $\leq$ is a pre-order, or a partial order, or a linear order on A, then $\langle A, \leq \rangle$ is a *pre-ordered* set, or a *partially ordered* set, or a *linearly ordered* set, respectively.

If $\leq$ is a pre-order on A then for $a, b \in A$ we write $a < b$ if $a \leq b$ holds and $b \leq a$ fails. Thus if $\leq$ is a partial order then $a < b$ if and only if $a \leq b$ and $a \neq b$.

Set inclusion $\subset$ is a partial order. For sets a and b we write $a \subsetneqq b$ (and we say that a is a *proper* subset of b) if $a \subset b$ and $a \neq b$.

Definitions. Let $\langle A, \leq \rangle$ be a partially ordered set, $a \in A$ and $C \subset A$.

The set of *predecessors* of a is the set

$$P(a) = \{b \in A : b < a\}.$$

The element a is the *greatest* element of C (denoted by $\max C$) if $a \in C$ and $c \leq a$ for $c \in C$; a is an *upper bound* of C if $c \leq a$ for $c \in C$; a is the (unique) *least upper bound* (or *supremum*) of C if a is an upper bound of C such that if c is any upper bound of C then $a \leq c$; a is a *maximal* element of A if $a \leq c$ implies $a = c$ for $c \in A$. Dually, we define the *least* element (denoted by $\min C$), a *lower* bound, the *greatest lower bound* (or, *infimum*), and a *minimal* element.

If C is linearly ordered with the induced order then C is a *chain* of $\langle A, \leq \rangle$.

If any two distinct elements of C are *non-comparable* (*i.e.*, if $c \leq c'$ and $c' \leq c$ both fail for $c, c' \in C$ and $c \neq c'$), then C is an *anti-chain* of $\langle A, \leq \rangle$.

If $\langle B, \leq \rangle$ is a partially ordered set then a function

$$f: \langle A, \leq \rangle \to \langle B, \leq \rangle$$

is an *embedding (of partially ordered sets)* if f is one to one and

$$a \leq a' \quad \text{if and only if } f(a) \leq f(a') \quad \text{for } a, a' \in A.$$

If in addition f is onto then f is called an *isomorphism (of partially ordered sets)*, or an *order isomorphism.*

A linearly ordered set $\langle A, \leq \rangle$ is *well-ordered* if every non empty subset of A has a least element, and *anti-well-ordered* if every non-empty subset of A has a greatest element.

The following simple result expresses a fundamental property of well-ordered sets.

1.1. Theorem *(Principle of proof by transfinite induction). Let $\langle A, \leq \rangle$ be a well-ordered set and $B \subset A$. Assume that for $a \in A$,*

$$\textit{if } P(a) \subset B \quad \textit{then } a \in B.$$

Then $B = A$.

Proof. Suppose that $A \neq B$, and let a be the least element of $A \setminus B$ (in the well-ordering of $\leq$). Thus $P(a) \subset B$, so by the assumption we have $a \in B$, a contradiction. Thus $A = B$.

1.2. Corollary. *Let $\langle A, \leq \rangle$ be a well-ordered set and let $f: \langle A, \leq \rangle \to \langle A, \leq \rangle$ be an embedding (of partially ordered sets). Then f is non-decreasing (i. e., $a \leq f(a)$ for $a \in A$).*

Proof. We set

$$B = \{a \in A : a \leq f(a)\},$$

and we prove that B satisfies the condition of Theorem 1.1. Let $a \in A$ and $P(a) \subset B$, and suppose that $a \notin B$. Since $f(a) < a$ we have $f(a) \in P(a) \subset B$, and since f is an embedding we have $f(f(a)) < f(a)$. But $f(a) \in B$ and hence $f(a) \leq f(f(a))$, a contradiction. Hence $a \in B$. From Theorem 1.1. it follows that $B = A$, so that f is non-decreasing.

1.3. Corollary. *Let $\langle A, \leq \rangle$ and $\langle B, \leq \rangle$ be order isomorphic well-ordered sets. Then there is a unique order-isomorphism from $\langle A, \leq \rangle$ onto $\langle B, \leq \rangle$.*

Proof. Let f and g be order isomorphisms from $\langle A, \leq \rangle$ onto $\langle B, \leq \rangle$. Then $g^{-1} \circ f$ is an embedding of $\langle A, \leq \rangle$ onto $\langle A, \leq \rangle$, so

$a \le (g^{-1} \circ f)(a)$ for $a \in A$ by Corollary 1.2. Thus, since g is an embedding, $g(a) \le f(a)$ for $a \in A$. With the roles of f and g reversed we have $f(a) \le g(a)$ for $a \in A$, and thus $f = g$.

Let $\langle A, \le \rangle$ be a linearly ordered set, and $B \subset A$. Then
B is an *initial segment* of A if there is $a \in A$ such that $B = P(a)$, and
B is an *interval* of A if $b_0 \le a \le b_1$ (with $b_0, b_1 \in B$ and $a \in A$) implies $a \in B$.

1.4. Corollary. *A well-ordered set is not order-isomorphic to an initial segment of itself.*

Proof. Let $\langle A, \le \rangle$ be a well-ordered set and let $B = P(a)$ for some $a \in A$. If

$$f: A \to B \subset A$$

is an embedding of ordered sets then $f(a) < a$, in contradiction to Corollary 1.2.

Ordinals

Definition. A set ξ is a *(von Neumann) ordinal (number)* if it satisfies the following conditions:

(i) if $\zeta \in \xi$ then $\zeta \subset \xi$;
(ii) if $\zeta, \eta \in \xi$, then either $\zeta = \eta$, or $\zeta \in \eta$, or $\eta \in \zeta$; and
(iii) if A is a non-empty subset of ξ, then there is $\zeta \in A$ such that $\zeta \cap A = \emptyset$.

Ordinal variables are denoted by $\xi, \zeta, \eta, \rho, i$, and j.

1.5. Lemma. *Let ξ be an ordinal. Then there are no sets ξ', ξ'', ξ''' such that*

$$\xi' \in \xi''' \in \xi'' \in \xi' \in \xi.$$

In particular, $\xi \notin \xi$; and there are no sets ξ', ξ'' such that $\xi' \in \xi'' \in \xi' \in \xi$.

Proof. Suppose there are such sets ξ', ξ'', ξ'''. We set $A = \{\xi', \xi'', \xi'''\}$. Since $\xi' \in \xi$ we have $\xi' \subset \xi$ and thus $\xi'' \in \xi$. Likewise we conclude that $\xi''' \in \xi$, so $A \subset \xi$. We prove that the defining condition (iii) of an ordinal fails for A and ξ. Clearly A is a non-empty subset of ξ and

$$\xi'' \in A \cap \xi', \ \xi''' \in A \cap \xi'', \quad \text{and } \xi' \in A \cap \xi'''.$$

Thus $A \cap \zeta \neq \emptyset$ for $\zeta \in A$.

1.6. Lemma. *If ξ is an ordinal and $\zeta \in \xi$, then ζ is an ordinal.*

Proof. We verify that ζ satisfies the three defining conditions of an ordinal.

(i) Let $\eta\in\zeta$; we want to prove that $\eta\subset\zeta$. Let $\rho\in\eta$; thus $\rho\in\eta\in\zeta\in\xi$. Since ξ is an ordinal, it is clear that $\rho\in\xi$. Then $\rho\in\xi$ and $\zeta\in\xi$, so since ξ is an ordinal

$$\text{either } \rho=\zeta, \quad \text{or } \rho\in\zeta, \quad \text{or } \zeta\in\rho.$$

If $\rho=\zeta$, then $\zeta\in\eta\in\zeta\in\xi$, contradicting Lemma 1.5; if $\zeta\in\rho$, then $\zeta\in\rho\in\eta\in\zeta\in\xi$, contradicting Lemma 1.5. Thus $\rho\in\zeta$, and hence $\eta\subset\zeta$.

(ii) Let $\rho\in\zeta$ and $\eta\in\zeta$. Then $\rho\in\zeta\in\xi$ and $\eta\in\zeta\in\xi$, and since ξ is an ordinal we have $\rho\in\xi$ and $\eta\in\xi$. Hence, since ξ is an ordinal,

$$\text{either } \rho=\eta, \quad \text{or } \rho\in\eta, \quad \text{or } \eta\in\rho.$$

(iii) Let A be a non-empty subset of ζ. Since $\zeta\in\xi$ and ξ is an ordinal we have $\zeta\subset\xi$, so that $A\subset\xi$. Then since ξ is an ordinal there is $\eta\in A$ such that $\eta\cap A=\emptyset$.

The proof is complete.

1.7. Lemma. *Let ξ and ζ be ordinals. Then*

$$\xi\in\zeta \quad \textit{if and only if} \quad \xi\subsetneqq\zeta.$$

Proof. If $\xi\in\zeta$ then $\xi\subset\zeta$ since ζ is an ordinal. If $\xi=\zeta$ then $\xi\in\xi$, which contradicts Lemma 1.5. Thus $\xi\subsetneqq\zeta$.

Let $\xi\subsetneqq\zeta$. Then $\zeta\setminus\xi$ is a non-empty subset of ζ, and since ζ is an ordinal there is $\eta\in\zeta\setminus\xi$ such that

$$\eta\cap(\zeta\setminus\xi)=\emptyset.$$

We claim that $\eta=\xi$. Since $\eta\in\zeta$ we have $\eta\subset\zeta$ and hence $\eta\subset\xi$.

If $\eta\neq\xi$ then $\xi\setminus\eta$ is a non-empty subset of ξ, and since ξ is an ordinal there is $\rho\in\xi\setminus\eta$ such that

$$\rho\cap(\xi\setminus\eta)=\emptyset.$$

Then $\rho\subset\eta$ (because $\rho\subset\xi$ and $\rho\cap(\xi\setminus\eta)=\emptyset$). Since $\eta\in\zeta\setminus\xi\subset\zeta$ and $\rho\in\xi\setminus\eta\subset\zeta$ and ζ is an ordinal,

$$\text{either } \eta=\rho, \quad \text{or } \eta\in\rho, \quad \text{or } \rho\in\eta.$$

If $\eta=\rho$, then $\eta=\rho\in\zeta\setminus\xi\cap\xi\setminus\eta=\emptyset$; if $\eta\in\rho$, then $\eta\in\rho\cap(\zeta\setminus\xi)=\emptyset$; and if $\rho\in\eta$ then $\rho\in\eta\cap(\xi\setminus\eta)=\emptyset$. Thus $\eta=\xi$ and the claim is proved.

Hence $\xi=\eta\in\zeta\setminus\xi\subset\zeta$ and we have $\xi\in\zeta$.

This completes the proof of the lemma.

1.8. Corollary. *If ξ is an ordinal, then $\langle\xi,\subset\rangle$ is a well-ordered set.*

Proof. We note that $\subset$ is a well-defined partial order on ξ, since if $\zeta\in\xi$ then $\zeta\subset\xi$. Let $\zeta,\eta\in\xi$. Then either $\zeta=\eta$, or $\zeta\in\eta$, or $\eta\in\zeta$. If $\zeta\in\eta$, then since η is an ordinal by Lemma 1.6, it follows that $\zeta\subsetneqq\eta$, and if $\eta\in\zeta$ then $\eta\subsetneqq\zeta$. Thus $\langle\xi,\subset\rangle$ is a linearly ordered set.

We prove that $\langle\xi,\subset\rangle$ is a well-ordered set. Let A be a non-empty subset of ξ. Since ξ is an ordinal there is $\zeta\in A$ such that $\zeta\cap A=\emptyset$. Now let $\eta\in A$. Since ξ is an ordinal

$$\text{either } \eta=\zeta, \quad \text{or } \eta\in\zeta, \quad \text{or } \zeta\in\eta.$$

If $\eta\in\zeta$, then $\eta\in\zeta\cap A=\emptyset$. Thus either $\eta=\zeta$ or $\zeta\in\eta$. In this latter case $\zeta\subsetneqq\eta$, so ζ is the least element of A (in the ordering induced by $\subset$). Hence $\langle\xi,\subset\rangle$ is a well-ordered set.

1.9. Lemma. *If ξ and ζ are ordinals, then $\xi\cap\zeta$ is an ordinal.*

Proof. We verify the three defining conditions of an ordinal.

(i) Let $\eta\in\xi\cap\zeta$. Then $\eta\subset\xi$ and $\eta\subset\zeta$, so $\eta\subset\xi\cap\zeta$.

(ii) Let $\eta,\rho\in\xi\cap\zeta$. Then $\eta,\rho\in\xi$, and hence either $\eta=\rho$, or $\eta\in\rho$, or $\rho\in\eta$.

(iii) Let A be a non-empty subset of $\xi\cap\zeta$. Then A is a non-empty subset of ξ, and hence there is $\eta\in A$ such that $\eta\cap A=\emptyset$.

Thus $\xi\cap\zeta$ is an ordinal.

1.10. Lemma. *Let ξ and ζ be ordinals. Then*

(a) *either $\xi\subset\zeta$ or $\zeta\subset\xi$; and*

(b) *if $\xi\neq\zeta$, then $\langle\xi,\subset\rangle$ and $\langle\zeta,\subset\rangle$ are not order-isomorphic.*

Proof. (a) Suppose $\xi\not\subset\zeta$ and $\zeta\not\subset\xi$. Then $\xi\cap\zeta\subsetneqq\xi$ and $\xi\cap\zeta\subsetneqq\zeta$. By Lemma 1.9, $\xi\cap\zeta$ is an ordinal, so by Lemma 1.7 $\xi\cap\zeta\in\xi$ and $\xi\cap\zeta\in\zeta$. Hence $\xi\cap\zeta\in\xi\cap\zeta$, in contradiction to Lemma 1.5.

(b) By part (a), either $\xi\subset\zeta$ or $\zeta\subset\xi$. With no loss of generality we assume that $\xi\subsetneqq\zeta$. We note that $\xi=\{\eta\in\zeta:\eta\subsetneqq\xi\}$ by Lemma 1.7. Thus $\langle\xi,\subset\rangle$ is an initial segment of $\langle\zeta,\subset\rangle$. By Corollary 1.4, $\langle\xi,\subset\rangle$ and $\langle\zeta,\subset\rangle$ are not order-isomorphic.

Definition. For ordinals ξ,ζ we define

$$\xi\leq\zeta \quad \text{if } \xi\subset\zeta.$$

Thus, the three statements

$$\xi<\zeta, \qquad \xi\subsetneqq\zeta, \qquad \xi\in\zeta$$

are equivalent for ordinals; and an ordinal is the set of smaller ordinals.

1.11. Lemma. (a) *If ξ is an ordinal then $\xi\cup\{\xi\}$ is an ordinal and $\xi\cup\{\xi\}$ is the immediate successor of ξ (in the sense that if η is an ordinal and $\xi<\eta$, then $\xi\cup\{\xi\}\leq\eta$).*

(b) *If A is a set of ordinals then $\bigcup A$ is an ordinal and $\bigcup A$ is the supremum of A (in the sense that if η is an ordinal such that $\zeta\leq\eta$ for $\zeta\in A$, then $\bigcup A\leq\eta$).*

Proof. (a) Of the three defining properties of an ordinal, we verify only (iii) for $\xi \cup \{\xi\}$.

Let A be a non-empty subset of $\xi \cup \{\xi\}$, and let $\zeta \in A$. If $\zeta \cap A = \emptyset$ there is nothing to prove, so let $\zeta \cap A \neq \emptyset$; then since ζ is an ordinal, there is $\eta \in \zeta \cap A$ such that $\eta \cap \zeta \cap A = \emptyset$. But $\eta \in \zeta$ and hence $\eta \subset \zeta$, so $\eta \cap A = (\eta \cap \zeta) \cap A = \emptyset$.

The simple proof that $\xi \cup \{\xi\}$ is the immediate successor of ξ is left to the reader.

(b) (i) Let $\xi \in \bigcup A$ and let $\zeta \in \xi$. Then $\xi \in \eta$ for some ordinal $\eta \in A$, and from $\zeta \in \xi \in \eta$ we have $\zeta \in \eta$ and then $\zeta \in \bigcup A$.

(ii) Let $\xi, \zeta \in \bigcup A$. Then $\xi \in \eta$, $\zeta \in \rho$ for some $\eta, \rho \in A$. By Lemma 1.10 (a), either $\eta \subset \rho$ or $\rho \subset \eta$; say $\rho \subset \eta$. Then $\xi, \zeta \in \eta$ and since η is an ordinal, either $\xi = \zeta$ or $\xi \in \zeta$ or $\zeta \in \xi$.

(iii) We let B be a non-empty subset of $\bigcup A$ and we prove that there is $\xi \in B$ such that $\xi \cap B = \emptyset$. Let $\zeta \in B$. If $\zeta \cap B = \emptyset$ we set $\xi = \zeta$. If $\zeta \cap B \neq \emptyset$ then, since ζ is an ordinal, there is $\xi \in \zeta \cap B$ such that $\xi \cap \zeta \cap B = \emptyset$; since $\xi \in \zeta$ we have $\xi \subset \zeta$, so that $\xi \cap B = (\xi \cap \zeta) \cap B = \emptyset$.

The simple proof that $\bigcup A$ is the supremum of A is left to the reader.

Sometimes we denote the ordinal $\bigcup A$ by $\sup A$.

A consequence of this lemma that will be often used in the sequel is this: if A is a set, then there is a set t (even a least ordinal t) such that $t \notin A$. In fact, let $B = \{a \in A : a \text{ is an ordinal}\}$, and set $t = \bigcup B$ if $\bigcup B \notin A$ and $t = \bigcup B \cup \{\bigcup B\}$ otherwise.

Thus the class $\mathbb{O}n$ of ordinals is (not a set but) a proper class.

An ordinal ξ is a *non-limit ordinal* if $\xi = \zeta \cup \{\zeta\}$ for some ordinal ζ; otherwise ξ is a *limit ordinal.* In particular, the empty set $\emptyset$ (also denoted 0 and called *zero*) is a limit ordinal.

1.12. Lemma. *There is a non-zero limit ordinal.*

Proof. From the axiom of infinity there is a set a such that $0 \in a$ and if $b \in a$ then $b \cup \{b\} \in a$. We set $b = a \cap \mathbb{O}n$ and $\xi = \bigcup b$. Since b is a set of ordinals it follows from Lemma 1.11 (b) that ξ is an ordinal. Also $0 < \xi$ since $0 \in a$ and $0 \in \mathbb{O}n$. Suppose that there is ζ such that $\xi = \zeta \cup \{\zeta\}$; then $\zeta \in \xi$ and so $\zeta \in \eta$ for some $\eta \in b$. We have $\xi \leq \eta < \eta \cup \{\eta\} \in a \cap \mathbb{O}n = b$, and thus $\eta \cup \{\eta\} \leq \xi$, a contradiction. It follows that ξ is a limit ordinal.

Definition. ω is the least non-zero limit ordinal. It is easy to verify that ω satisfies the usual Peano axioms (not given here).

The elements of ω are called *natural numbers* or *finite ordinals* and are denoted by n, k, m, l, i, and j.

We set $1 = 0 \cup \{0\}$, $2 = 1 \cup \{1\}$, $3 = 2 \cup \{2\}$; if ξ is an ordinal we set $\xi + 1 = \xi \cup \{\xi\}$.

The next theorem establishes that there are "sufficiently many" von Neumann ordinals.

1.13. Theorem. *Let $\langle A, \leq \rangle$ be a well-ordered set. Then there is a unique ordinal ξ such that $\langle A, \leq \rangle$ and ξ are order-isomorphic.*

This ordinal ξ is called the *order-type* of $\langle A, \leq \rangle$.

Proof. The uniqueness of ξ follows from Lemma 1.10 (b). Let t be a set such that $t \notin A$, and set $B = A \cup \{t\}$. We extend the well-ordering of A by letting $t > a$ for $a \in A$. Clearly $\langle B, \leq \rangle$ is a well-ordered set. Let $C = \{x \in B : P(x)$ is order-isomorphic to an ordinal$\}$, and let $x \in B$ be such that $P(x) \subset C$. For every $y \in P(x)$ there is, by the definition of C and by Lemma 1.3, an ordinal ξ_y and a unique order-isomorphism f_y such that $f_y : P(y) \to \xi_y$. Then $\bigcup \{f_y : y \in P(x)\}$ is an order-isomorphism between $\bigcup \{P(y) : y \in P(x)\}$ and the ordinal $\bigcup \{\xi_y : y \in P(x)\}$. Hence $P(x)$ is order-isomorphic either to the ordinal $\bigcup \{\xi_y : y \in P(x)\}$ or to its immediate successor, and thus $x \in C$. By (the principle of proof by) transfinite induction (Theorem 1.1), we conclude that $C = B$. Thus $t \in C$ and hence $P(t) = A$ is order-isomorphic to an ordinal.

Definition. Let ξ_0 and ξ_1 be ordinals. We set

$$A = (\xi_0 \times \{0\}) \cup (\xi_1 \times \{1\})$$

and define a binary relation $\preccurlyeq$ on A as follows: if $\langle \eta, i \rangle, \langle \zeta, i \rangle \in \xi_i \times \{i\}$ for some $i < 2$ then $\langle \eta, i \rangle \preccurlyeq \langle \zeta, i \rangle$ if $\eta \leq \zeta$; and if $\langle \eta, 0 \rangle \in \xi_0 \times \{0\}$ and $\langle \zeta, 1 \rangle \in \xi_1 \times \{1\}$ then $\langle \eta, 0 \rangle \prec \langle \zeta, 1 \rangle$.

It is clear that $\langle A, \preccurlyeq \rangle$ is a well-ordered set; the unique ordinal isomorphic to $\langle A, \preccurlyeq \rangle$ is the *ordinal sum* of ξ_0 and ξ_1, denoted by $\xi_0 + \xi_1$. The notation $\xi + 1$ is unambiguous.

With these definitions, it is easy to prove that for every ordinal ξ there is a unique limit ordinal ζ and a unique finite ordinal n such that

$$\xi = \zeta + n.$$

If n is even, then ξ is an *even* ordinal; otherwise ξ is an *odd* ordinal.

1.14. Theorem *(Principle of definition by transfinite recursion). Let $h(A)$ be a set defined for every set A, and let ξ be an ordinal. Then there is a unique function*

$$f : \xi + 1 \to \mathbb{V}$$

such that $f(\eta) = h(f | \eta)$ for $\eta \leq \xi$.

Proof. (a) We first prove the uniqueness using transfinite induction. Let f and g be functions from $\xi + 1$ to $\mathbb{V}$ such that $f(\eta) = h(f | \eta)$ and $g(\eta) = h(g | \eta)$ for $\eta \leq \xi$. Let $A = \{\zeta \leq \xi : f(\zeta) = g(\zeta)\}$, let $\eta \leq \xi$ and suppose that $P(\eta)\subset A$. Then $f|\eta=g|\eta$ and therefore $f(\eta)=h(f|\eta)=h(g|\eta)=h(\eta)$, *i. e.*, $\eta\in A$. By transfinite induction $A=\xi+1$ and hence $f=g$.

(b) We prove the existence of f, using transfinite induction. Set

$$A=\{\zeta\leq\xi\colon \text{there is } f_\zeta\colon\zeta+1\to\mathbb{V} \text{ such that } f_\zeta(\eta)=h(f_\zeta|\eta) \text{ for } \eta\leq\zeta\},$$

let $\eta\leq\xi$ and suppose that $P(\eta)\subset A$. The uniqueness part of the theorem, already established, implies that

$$f_\zeta|\zeta'=f_{\zeta'} \quad \text{for } \zeta'<\zeta<\eta.$$

We set

$$f_\eta=\bigcup_{\zeta<\eta}f_\zeta\cup\{\langle\eta,h(\bigcup_{\zeta<\eta}f_\zeta)\rangle\}.$$

Then $f_\eta(\zeta)=h(f_\eta|\zeta)$ for $\zeta\leq\eta$, so $\eta\in A$. By transfinite induction we conclude that $A=\xi+1$. We set

$$f=f_\xi,$$

and the proof is complete.

In using this result, we do not give the explicit definition of the function h, as this tends to be awkward and artificial.

1.15. Theorem. (a) *If $\langle A,\leq\rangle$ is a non-empty partially ordered set such that every chain of $\langle A,\leq\rangle$ has an upper bound, then $\langle A,\leq\rangle$ has a maximal element. (The maximal principle; or Zorn's lemma.)*

(b) *Every set can be well-ordered. (The well-ordering principle of Zermelo.)*

Proof. (a) Let $\langle A,\leq\rangle$ be a partially ordered set such that every chain has an upper bound. Let t be a set such that $t\notin A$, and extend the partial order of $\langle A,\leq\rangle$ to $A\cup\{t\}$ by the condition that $a<t$ for $a\in A$.

By the axiom of choice there is a function $f\colon\mathscr{P}(A)\to A\cup\{t\}$ such that $f(B)\in B$ for $B\neq\emptyset$ and $f(\emptyset)=t$. We define $g\colon\mathscr{P}(A)\to\mathscr{P}(A)$ by

$$g(B)=\{a\in A\colon a>b \text{ for } b\in B\}.$$

We note that $g(\emptyset)=A$. We set

$$\mathscr{C}=\{R\subset A\times A\colon R \text{ is a well-ordering of a subset } \operatorname{dom}R \text{ of } A\}.$$

For $R\in\mathscr{C}$ we let ξ_R be the unique ordinal such that $\langle\operatorname{dom}R,R\rangle$ and ξ_R are order-isomorphic, and we set

$$\xi=\sup\{\xi_R\colon R\in\mathscr{C}\}+1.$$

We define $\{a_\zeta\colon\zeta<\xi\}$ by transfinite recursion. Set $a_0=f(g(\emptyset))$. Let $\zeta<\xi$, and suppose that a_η has been defined for $\eta<\zeta$. If $a_\eta=t$ for some $\eta<\zeta$, set $a_\zeta=t$. If $\{a_\eta\colon\eta<\zeta\}\subset A$, set

$$a_\zeta=f(g(\{a_\eta\colon\eta<\zeta\})).$$

Thus we have defined $\{a_\zeta : \zeta < \xi\}$. If $a_\zeta \in A$ for $\zeta < \xi$ it is clear that

$$a_\zeta < a_{\zeta'} \quad \text{for } \zeta < \zeta' < \xi;$$

thus $\{a_\zeta : \zeta < \xi\}$ is a well-ordered subset of $\langle A, \leq \rangle$ isomorphic to $\langle \xi, \subset \rangle$, contradicting the choice of ξ. Let η be the first ordinal $\zeta < \xi$ such that $a_\zeta = t$. Let a be an upper bound of the chain $\{a_\zeta : \zeta < \eta\}$. It is clear that a is a maximal element of $\langle A, \leq \rangle$.

(b) Let A be a set. We set

$$\mathscr{C} = \{R \subset A \times A : R \text{ is a well-ordering of a subset } \operatorname{dom} R \text{ of } A\}.$$

A partial order $\preccurlyeq$ is defined on $\mathscr{C}$ by

$$R \prec S$$

if and only if

$$\operatorname{dom} R \subset \operatorname{dom} S$$

and

$$\langle \operatorname{dom} R, R \rangle \quad \text{is an initial segment of } \langle \operatorname{dom} S, S \rangle.$$

It is easy to verify that $\langle \mathscr{C}, \preccurlyeq \rangle$ satisfies the maximal principle, and hence, by part (a), there is a maximal element R. If $\operatorname{dom} R \neq A$, let $a \in A \setminus \operatorname{dom} R$, and define $\operatorname{dom} S = \operatorname{dom} R \cup \{a\}$, and $S = R \cup \{\langle b, a \rangle : b \in \operatorname{dom} R\}$. It is clear that

$$R \prec S,$$

contradicting the maximality of R.

1.16. Theorem. *A linearly ordered set $\langle A, \leq \rangle$ is well-ordered if and only if there is no infinite decreasing sequence of elements in A.*

Proof. The direct implication is clear.

Suppose for the converse that $\langle A, \leq \rangle$ is not well-ordered. Then there is a non-empty subset B of A with no least element. We proceed recursively. Let $b_0 \in B$ and suppose that $b_0, \cdots, b_{n-1} \in B$ have been defined. Clearly $B \setminus \{b_0, \cdots, b_{n-1}\}$ is non-empty and has no least element. We choose $b_n \in B$ such that

$$b_n < b_{n-1}.$$

Then $b_0 > b_1 > \cdots > b_{n-1} > b_n > \cdots$ is an infinite decreasing sequence in A.

Cardinal Arithmetic

Definition. Let ξ be an ordinal. We say that ξ is a *cardinal (number)* if for every ordinal $\zeta < \xi$ there is no one-to-one correspondence between ζ and ξ.

Cardinal variables are denoted by $\alpha, \beta, \gamma, \delta, \kappa, \lambda$, and μ.

We note that the finite ordinals and ω are cardinals, and that $\omega+1$ is not a cardinal.

If α and β are cardinals, then since they are ordinals we have immediately the *Cantor-Bernstein* theorem on the property of antisymmetry for cardinal inclusion:

$$\text{if } \alpha \leq \beta \quad \text{and } \beta \leq \alpha \quad \text{then } \alpha = \beta.$$

Furthermore, for any two cardinals α and β,

$$\alpha \leq \beta \quad \text{or } \beta \leq \alpha.$$

If A is a set then, by the well-ordering principle of Zermelo, A can be well-ordered and thus there is a one-to-one correspondence between A and an ordinal.

Definition. The *cardinality* of a set A, denoted by $|A|$, is the least ordinal α such that there is a one-to-one correspondence between A and α.

It is clear that $|A|$ is a cardinal.

A set is *finite* if its cardinality is less than ω, *infinite* otherwise. A set is *countable* if its cardinality is at most ω.

It is clear that for sets A and B the following are equivalent.

(i) $|A| \leq |B|$;
(ii) there is a one-to-one function from A into B;
(iii) there is a function from B onto A.

Definitions. Let $\{\alpha_i : i \in I\}$ be a set of cardinals. The *cardinal sum* $\sum_{i \in I} \alpha_i$ and *cardinal product* $\prod_{i \in I} \alpha_i$ are

$$\sum_{i \in I} \alpha_i = \left| \bigcup_{i \in I} \alpha_i \times \{i\} \right|$$

and

$$\prod_{i \in I} \alpha_i = \left| \prod_{i \in I} \alpha_i \right|,$$

respectively.

The sum of two cardinals α_0 and α_1 is denoted by $\alpha_0 + \alpha_1$. For cardinals α_0 and α_1 the notation $\alpha_0 + \alpha_1$ is ambiguous in that it denotes both an ordinal sum and a cardinal sum; the meaning in every case will be clear from the context.

Note that we use the same notation for the Cartesian product $\prod_{i \in I} \alpha_i$ of the set $\{\alpha_i : i \in I\}$ of cardinals, and for its cardinality; the meaning in every case will be clear from the context. (Note, however, that $\alpha_0 \cdot \alpha_1$ denotes the cardinal product and $\alpha_0 \times \alpha_1$ the Cartesian product of the cardinal numbers α_0 and α_1.)

In particular, we have defined α^β and 2^β. It is clear that

$$|\mathscr{P}(\alpha)| = 2^\alpha.$$

1.17. Theorem. $\alpha^2=\alpha$ *for every infinite cardinal* α.

Proof. We define a well-ordering $\preccurlyeq$ on $\alpha\times\alpha$ as follows:

$$\begin{aligned}&\langle\xi_0,\xi_1\rangle\preccurlyeq\langle\zeta_0,\zeta_1\rangle \quad \text{if and only if}\\ &\max(\xi_0,\xi_1)\le\max(\zeta_0,\zeta_1), \quad \text{or}\\ &\max(\xi_0,\xi_1)=\max(\zeta_0,\zeta_1) \quad \text{and } \xi_0\le\zeta_0, \quad \text{or}\\ &\max(\xi_0,\xi_1)=\max(\zeta_0,\zeta_1) \quad \text{and } \xi_0=\zeta_0 \quad \text{and } \xi_1\le\zeta_1.\end{aligned}$$

It is easy to verify that $\preccurlyeq$ is indeed a well-ordering of $\alpha\times\alpha$ such that

$$\xi\times\xi \quad \text{is an initial segment of } \alpha\times\alpha \quad \text{for } \xi<\alpha.$$

We now prove that for any infinite cardinal α, $\langle\alpha\times\alpha,\preccurlyeq\rangle$ is order-isomorphic with α. Suppose this is false, and (using the fact that the set of cardinals less than or equal to a given cardinal is well-ordered) let β be the least infinite cardinal such that $\langle\beta\times\beta,\preccurlyeq\rangle$ is not order-isomorphic with β. Since both $\langle\beta\times\beta,\preccurlyeq\rangle$ and β are well-ordered sets, it is clear that β is order-isomorphic to an initial segment of $\langle\beta\times\beta,\preccurlyeq\rangle$, hence to a subset of $\xi\times\xi$ for some (infinite ordinal) $\xi<\beta$. Since $\xi<\beta$ we have $|\xi|<\beta$, and by the choice of β we have $|\xi\times\xi|=|\xi|$. Thus $\beta\le|\xi\times\xi|=|\xi|$, a contradiction proving the theorem.

1.18. Corollary. *If* α *and* β *are infinite cardinals, then*

$$\max(\alpha,\beta)=\alpha+\beta=\alpha\cdot\beta.$$

Proof. With no loss of generality we assume that $\alpha\ge\beta$, so that $\max(\alpha,\beta)=\alpha$. Then $\alpha\le\alpha+\beta\le\alpha+\alpha=\alpha$ and $\alpha\le\alpha\cdot\beta\le\alpha\cdot\alpha=\alpha$, and the result follows from the Cantor-Bernstein theorem.

The next theorem is proved by a classical diagonal argument.

1.19. Theorem *(J. König). Let I be a non-empty set, and let* $\alpha_i<\beta_i$ *for* $i\in I$. *Then*

$$\textstyle\sum_{i\in I}\alpha_i<\prod_{i\in I}\beta_i.$$

Proof. We first prove that $\sum_{i\in I}\alpha_i\le\prod_{i\in I}\beta_i$. Since $\alpha_i<\beta_i$ we have $\beta_i\backslash\alpha_i\neq\emptyset$, so there is $\xi_i\in\beta_i\backslash\alpha_i$ (*e.g.*, let $\xi_i=\alpha_i$) for $i\in I$. We set

$$T_j=\{p=\langle p_i: i\in I\rangle\in\textstyle\prod_{i\in I}\beta_i: p_i=\xi_i \text{ for all } i\neq j, \text{ and } p_j\in\alpha_j\}$$

for $j\in I$. Then

$$\begin{aligned}|T_j|&=\alpha_j \quad \text{for } j\in I, \quad \text{and}\\ T_i\cap T_j&=\emptyset \quad \text{for } i,j\in I \quad \text{and } i\neq j.\end{aligned}$$

Hence $\left|\bigcup_{i\in I}T_i\right|=\sum_{i\in I}\alpha_i$, and since $\bigcup_{i\in I}T_i\subset\prod_{i\in I}\beta_i$, it follows that

$$\textstyle\sum_{i\in I}\alpha_i\le\prod_{i\in I}\beta_i.$$

Next suppose that $\sum_{i\in I}\alpha_i=\prod_{i\in I}\beta_i$. Then there is a family $\{S_i: i\in I\}$

of subsets of $\prod_{i\in I}\beta_i$ such that

$$|S_i|=\alpha_i \quad \text{for } i\in I,$$
$$S_i\cap S_j=\emptyset \quad \text{for } i,j\in I \quad \text{and } i\neq j, \quad \text{and}$$
$$\bigcup_{i\in I} S_i=\prod_{i\in I}\beta_i.$$

We set $P_i=\pi_i[S_i]\subset\beta_i$, where $\pi_i:\prod_{j\in I}\beta_j\to\beta_i$ denotes the projection function for $i\in I$. Then $|P_i|\leq|S_i|=\alpha_i$ and hence $\beta_i\setminus P_i\neq\emptyset$; thus there is $\xi_i\in\beta_i\setminus P_i$ for $i\in I$; we set $p=\langle\xi_i\colon i\in I\rangle$, and we have $p\in\prod_{i\in I}\beta_i\setminus\bigcup_{i\in I}S_i$. This contradiction proves the theorem.

The next two corollaries to König's theorem are very useful. We first need some definitions.

Definitions. Let ξ be a limit ordinal. The *cofinality* of ξ, denoted by $\mathrm{cf}(\xi)$, is the least ordinal ζ for which there is a function $f\colon\zeta\to\xi$ such that

$$f \text{ is } \textit{order-preserving} \ (i.e., f(\eta)\leq f(\eta') \text{ for } \eta<\eta'<\zeta),$$

and

$$f \text{ is } \textit{unbounded} \ (i.e., \sup_{\eta<\zeta} f(\eta)=\xi).$$

An infinite cardinal α is *regular* if $\alpha=\mathrm{cf}(\alpha)$, *singular* otherwise.

It is not difficult to verify that

$$\mathrm{cf}(\xi)\leq\xi,$$
$$\mathrm{cf}(\xi) \text{ is a cardinal, and}$$
$$\mathrm{cf}(\mathrm{cf}(\xi))=\mathrm{cf}(\xi)$$

for limit ordinals ξ.

We note also that if α and β are cardinals and α is infinite then the following are equivalent:

$$\beta<\mathrm{cf}(\alpha);$$
$$\text{if } \alpha_\xi<\alpha \text{ for } \xi<\beta, \text{ then } \textstyle\sum_{\xi<\beta}\alpha_\xi<\alpha.$$

1.20. Corollary. *If α is an infinite cardinal, then $\alpha<\alpha^{\mathrm{cf}(\alpha)}$.*

Proof. Let $\alpha=\sum_{\xi<\mathrm{cf}(\alpha)}\alpha_\xi$, with $\alpha_\xi<\alpha$ for $\xi<\mathrm{cf}(\alpha)$. Then

$$\alpha=\sum_{\xi<\mathrm{cf}(\alpha)}\alpha_\xi<\prod_{\xi<\mathrm{cf}(\alpha)}\alpha=\alpha^{\mathrm{cf}(\alpha)}$$

by J. König's theorem (1.19).

1.21. Corollary. *If α is an infinite cardinal, then $\alpha<\mathrm{cf}(2^\alpha)$.*

Proof. If $\mathrm{cf}(2^\alpha)\leq\alpha$ then from Corollary 1.20 we have

$$2^\alpha<(2^\alpha)^{\mathrm{cf}(2^\alpha)}\leq(2^\alpha)^\alpha,$$

a contradiction.

It follows from König's theorem (and from Corollary 1.21 for infinite cardinals) that

$$\alpha < 2^\alpha \quad \text{for all cardinals } \alpha.$$

Thus the class $\mathbb{C}$ of cardinals is (not a set but) a proper class.

Definitions. Let α be a cardinal. Then α^+ is the least cardinal β such that $\alpha < \beta$.

Such a cardinal α^+ exists, because $\alpha < 2^\alpha$ and the set of cardinals less than or equal to 2^α is well-ordered.

For ordinals ξ we have

$$|\xi| = \alpha \quad \text{if and only if } \alpha \leq \xi < \alpha^+.$$

A cardinal α is a *non-limit cardinal* if $\alpha = \beta^+$ for some (cardinal) β; and α is a *limit cardinal* otherwise. We note that

0 is a limit cardinal;
every infinite cardinal is a limit ordinal; and
every infinite non-limit cardinal is regular.

Let α and κ be cardinals. We define

$$\begin{aligned}
\alpha^{\underline{\kappa}} &= \sum \{\alpha^\lambda : (\lambda \text{ is a cardinal and) } \lambda < \kappa\}; \\
[\alpha]^\kappa &= \{A \in \mathscr{P}(\alpha) : |A| = \kappa\}; \quad \text{and} \\
\mathscr{P}_\kappa(\alpha) &= \{A \in \mathscr{P}(\alpha) : |A| < \kappa\}.
\end{aligned}$$

1.22. Lemma. *Let α and κ be cardinals.*

(a) *If $\alpha \geq \omega$, then $\alpha^{\underline{\omega}} = \alpha$;*
(b) *if $\omega + \kappa \leq \alpha$, then $|[\alpha]^\kappa| = \alpha^\kappa$;*
(c) *if $\omega + \kappa \leq \alpha^+$, then $|\mathscr{P}_\kappa(\alpha)| = \alpha^{\underline{\kappa}}$;*
(d) *if $\omega + \kappa \leq \alpha$, then*
$|\{f \in \alpha^A : A \in [\alpha]^\kappa$ and f is one-to-one$\}| = \alpha^\kappa$;
(e) *if $\omega + \kappa \leq \alpha^+$, then*
$|\{f \in \alpha^A : A \in \mathscr{P}_\kappa(\alpha)$ and f is one-to-one$\}| = \alpha^{\underline{\kappa}}$;
(f) *if $2 \leq \alpha$, then $\sum_{\xi < \kappa} \alpha^{|\xi|} = \alpha^{\underline{\kappa}}$; and*
(g) *if $2 \leq \alpha$, then $\kappa \leq \alpha^{\underline{\kappa}}$.*

Proof. (a) From Theorem 1.17 and (its) Corollary 1.18 we have

$$\alpha^{\underline{\omega}} = \sum_{n<\omega} \alpha^n = \sum_{n<\omega} \alpha = \alpha \cdot \omega = \alpha.$$

(b) For $A \in [\alpha]^\kappa$ there is, by the axiom of choice, a function f_A from κ onto A. We define $\phi: [\alpha]^\kappa \to \alpha^\kappa$ (where α^κ denotes the Cartesian product) by $\phi(A) = f_A$. Clearly ϕ is one-to-one, and therefore $|[\alpha]^\kappa| \leq \alpha^\kappa$.

Recalling that a function is a set of ordered pairs, we have $\alpha^\kappa \subset [\kappa \times \alpha]^\kappa$ so that $\alpha^\kappa \leq |[\kappa \times \alpha]^\kappa| = |[\alpha]^\kappa|$.

(c) The set $\mathscr{P}_\kappa(\alpha)$ is the disjoint union of the family $\{[\alpha]^\lambda : \lambda < \kappa\}$. Hence

$$|\mathscr{P}_\kappa(\alpha)| = \sum_{\lambda<\kappa} \alpha^\lambda = \alpha^{\underline{\kappa}}$$

by part (b).

(d) Set $E = \{f \in \alpha^A : |A| = \kappa$ and f is one to one$\}$, for $A \in [\alpha]^\kappa$ set $E_A = \{f \in \alpha^A : f$ is one-to-one$\}$, and let $e_A : A \to \alpha$ be the inclusion function. Then

$$|E_A| \leq \alpha^\kappa \quad \text{for } A \in [\alpha]^\kappa,$$

and $E = \bigcup\{E_A : A \in [\alpha]^\kappa\}$; hence

$$|E| \leq \sum \{|E_A| : A \in [\alpha]^\kappa\} \leq \alpha^\kappa \cdot \alpha^\kappa = \alpha^\kappa$$

by part (b). Now let $\phi : [\alpha]^\kappa \to E$ be given by $\phi(A) = e_A$. Then ϕ is one-to-one, so by part (b) again we have $|E| \geq |[\alpha]^\kappa| = \alpha^\kappa$.

(e) follows easily from part (d).

(f) For cardinals β we have $1 \leq |\beta^+ \setminus \beta| \leq \beta^+ \leq 2^\beta$, and $|\xi| = \beta$ for $\xi \in \beta^+ \setminus \beta$. Thus

$$\sum_{\xi \in \beta^+ \setminus \beta} \alpha^{|\xi|} = \sum_{\xi \in \beta^+ \setminus \beta} \alpha^\beta = \alpha^\beta \cdot |\beta^+ \setminus \beta| = \alpha^\beta,$$

so

$$\sum_{\xi<\kappa} \alpha^{|\xi|} = \sum_{\beta<\kappa} \Big(\sum_{\xi \in \beta^+ \setminus \beta} \alpha^{|\xi|}\Big) = \sum_{\beta<\kappa} \alpha^\beta = \alpha^{\underline{\kappa}}.$$

(g) From part (f) we have

$$\kappa \leq \sum_{\xi<\kappa} 2^{|\xi|} = 2^{\underline{\kappa}} \leq \alpha^{\underline{\kappa}}.$$

Definitions. A cardinal α is a *strong limit* cardinal if $2^\beta < \alpha$ for $\beta < \alpha$.

The *beth* cardinals are defined by transfinite recursion as follows.

$\beth_0 = \omega$;

if ξ is a non-zero limit ordinal, then $\beth_\xi = \sum_{\eta<\xi} \beth_\eta$; and

if ξ is an ordinal, then $\beth_{\xi+1} = 2^{\beth_\xi}$.

It is clear by transfinite induction that $\xi \leq \beth_\xi$ for all ordinals ξ.

The following simple lemma relates the two definitions just given.

1.23. Lemma. *An infinite cardinal α is a strong limit cardinal if and only if*

$$\alpha = \beth_\xi \quad \textit{for some limit ordinal } \xi.$$

Proof. Let α be a strong limit cardinal, and let ξ be the least ordinal such that $\alpha \leq \beth_\xi$. Then $\xi \leq \alpha$ since $\alpha \leq \beth_\alpha$. If $\xi = \eta + 1$ then $\beth_\eta < \alpha$, and then $\beth_\xi = \beth_{\eta+1} = 2^{\beth_\eta} < \alpha$ since α is a strong limit cardinal. Thus ξ is a limit ordinal. Further

$$\beth_\xi = \sum_{\eta<\xi} \beth_\eta \leq |\xi| \cdot \alpha \leq \alpha,$$

since $\beth_\eta < \alpha$ for $\eta < \xi$. Thus $\alpha = \beth_\xi$ and ξ is a limit ordinal.

Conversely, let ξ be a limit ordinal such that $\alpha=\beth_\xi$. If $\beta<\alpha$ there is $\eta<\xi$ such that $\beta<\beth_\eta$, and therefore $2^\beta\le 2^{\beth_\eta}=\beth_{\eta+1}<\beth_\xi=\alpha$. Thus α is a strong limit cardinal.

In order to compute the cofinality of the beths (and hence of the strong limit cardinals) we need the following result.

1.24. Lemma *(Tarski's recursion formula). Let α and β be cardinals with α a limit cardinal and $\beta>0$ and let ξ be a limit ordinal such that $\beta<\mathrm{cf}(\xi)$. Let $\{\alpha_\eta:\eta<\xi\}$ be a set of cardinals such that*

$$\begin{aligned}&\alpha_\zeta<\alpha_\eta \quad \text{for } \zeta<\eta<\xi,\\ &\alpha_\eta<\alpha \quad \text{for } \eta<\xi, \quad \text{and}\\ &\textstyle\sum_{\eta<\xi}\alpha_\eta=\alpha.\end{aligned}$$

Then $\alpha^\beta=\sum_{\eta<\xi}\alpha_\eta^\beta$.

Proof. We note that $\sup\{\alpha_\eta:\eta<\xi\}=\alpha$, since if there is $\gamma<\alpha$ such that $\alpha_\eta<\gamma$ for $\eta<\xi$ then $\xi\le\gamma$ and $\sum_{\eta<\xi}\alpha_\eta\le\gamma\cdot|\xi|<\alpha$.

Let $f\in\alpha^\beta$. Then $f(\zeta)<\alpha$ so there is $\eta(\zeta)<\xi$ such that $f(\zeta)<\alpha_{\eta(\zeta)}$ for $\zeta<\beta$. We set $\eta=\sup_{\zeta<\beta}\eta(\zeta)$. Then $\eta<\xi$ because $\beta<\mathrm{cf}(\xi)$, and hence $f\in\alpha_\eta^\beta$. It follows that $\alpha^\beta\subset\bigcup_{\eta<\xi}\alpha_\eta^\beta$, so $\alpha^\beta\le\sum_{\eta<\xi}\alpha_\eta^\beta$.

For the reverse inequality we note that $\xi\le\alpha$, so that

$$\textstyle\sum_{\eta<\xi}\alpha_\eta^\beta\le\alpha^\beta\cdot|\xi|=\alpha^\beta.$$

1.25. Corollary. *If ξ is a non-zero limit ordinal, then* $\mathrm{cf}(\beth_\xi)=\mathrm{cf}(\xi)$.

Proof. Since $\beth_\xi=\sum_{\eta<\xi}\beth_\eta$ we have $\mathrm{cf}(\beth_\xi)\le\mathrm{cf}(\xi)$. If $\beta<\mathrm{cf}(\xi)$ then from Lemma 1.24 we have

$$\begin{aligned}\beth_\xi^\beta&=\textstyle\sum_{\eta<\xi}\beth_\eta^\beta=\sum_{\eta<\xi}\beth_{\eta+1}^\beta=\sum_{\eta<\xi}(2^{\beth_\eta})^\beta\\ &=\textstyle\sum_{\beta\le\eta<\xi}(2^{\beth_\eta})^\beta=\sum_{\beta\le\eta<\xi}2^{\beth_\eta}=\sum_{\beta\le\eta<\xi}\beth_{\eta+1}\\ &=\textstyle\sum_{\beta\le\eta<\xi}\beth_\eta=\beth_\xi.\end{aligned}$$

Thus $\beta<\mathrm{cf}(\beth_\xi)$ and we have $\mathrm{cf}(\xi)\le\mathrm{cf}(\beth_\xi)$.

1.26. Lemma. *If $\alpha\ge\omega$ and $\kappa\ge 2$, then $(\kappa^{\underset{\smile}{\alpha}})^{\underset{\smile}{\mathrm{cf}(\alpha)}}=\kappa^{\underset{\smile}{\alpha}}$.*

Proof. It is sufficient to prove that

$$(\kappa^{\underset{\smile}{\alpha}})^\beta=\kappa^{\underset{\smile}{\alpha}} \quad \text{for } \beta<\mathrm{cf}(\alpha),$$

for then

$$(\kappa^{\underset{\smile}{\alpha}})^{\underset{\smile}{\mathrm{cf}(\alpha)}}=\textstyle\sum_{\beta<\mathrm{cf}(\alpha)}(\kappa^{\underset{\smile}{\alpha}})^\beta=\sum_{\beta<\mathrm{cf}(\alpha)}\kappa^{\underset{\smile}{\alpha}}=\kappa^{\underset{\smile}{\alpha}}\cdot\mathrm{cf}(\alpha)=\kappa^{\underset{\smile}{\alpha}},$$

using Lemma 1.22 (g).

Let $\beta<\mathrm{cf}(\alpha)$, and let $f\in(\kappa^{\underset{\smile}{\alpha}})^\beta$. Then

$$f(\zeta)\in\kappa^{\underset{\smile}{\alpha}}=\textstyle\sum_{\gamma<\alpha}\kappa^\gamma$$

for $\zeta<\beta$, and hence there is $\gamma(\zeta)<\alpha$ such that $f(\zeta)\in\kappa^{\gamma(\zeta)}$ for $\zeta<\beta$. Set $\gamma=\sup_{\zeta<\beta}\gamma(\zeta)$. Then $\gamma<\alpha$ since $\beta<\mathrm{cf}(\alpha)$ and we have $f(\zeta)\in\kappa^{\gamma(\zeta)}\subset\kappa^{\gamma}$ for $\zeta<\beta$; thus $f\in(\kappa^{\gamma})^{\beta}$. It follows that

$$(\kappa^{\underline{\alpha}})^{\beta}\leq\sum_{\gamma<\alpha}(\kappa^{\gamma})^{\beta}=\sum_{\gamma<\alpha}\kappa^{\gamma}=\kappa^{\underline{\alpha}},$$

as required.

It follows from the lemma that if α is an infinite regular cardinal then $(2^{\underline{\alpha}})^{\underline{\alpha}}=2^{\underline{\alpha}}=\alpha^{\underline{\alpha}}$ (for $\alpha^{\underline{\alpha}}\leq(2^{\underline{\alpha}})^{\underline{\alpha}}=2^{\underline{\alpha}}\leq\alpha^{\underline{\alpha}}$).

The next theorem provides useful characterizations of the cardinals α such that either $\alpha=\alpha^{\underline{\alpha}}$ or $\alpha=2^{\underline{\alpha}}$.

1.27. Theorem. *Let α be an infinite cardinal. Then*

(a) $\alpha=\alpha^{\underline{\alpha}}$ *if and only if* $\alpha=2^{\underline{\alpha}}$ *and α is a regular cardinal; and*

(b) $\alpha=2^{\underline{\alpha}}$ *if and only if either* $\alpha=\alpha^{\underline{\alpha}}$ *or α is a strong limit cardinal.*

Thus there are cardinals α with arbitrarily large cofinality such that $\alpha=2^{\underline{\alpha}}$.

Proof. (a) Let $\alpha=\alpha^{\underline{\alpha}}$. Then $\alpha\leq 2^{\underline{\alpha}}\leq\alpha^{\underline{\alpha}}=\alpha$, and hence $\alpha=2^{\underline{\alpha}}$. Further $\alpha\leq\mathrm{cf}(\alpha)$ by Corollary 1.20, *i. e.*, α is a regular cardinal.

Conversely if $\alpha=2^{\underline{\alpha}}$ and α is regular then using the remark after Lemma 1.26 we have

$$\alpha^{\underline{\alpha}}=(2^{\underline{\alpha}})^{\underline{\alpha}}=2^{\underline{\alpha}}=\alpha.$$

(b) If either $\alpha=\alpha^{\underline{\alpha}}$ or α is a strong limit cardinal, then clearly $\alpha=2^{\underline{\alpha}}$.

Conversely, let $\alpha=2^{\underline{\alpha}}$. If α is regular then $\alpha=\alpha^{\underline{\alpha}}$ by part (a). We suppose that α is singular and we prove that α is a strong limit cardinal. If not then there is a cardinal β such that $\mathrm{cf}(\alpha)\leq\beta<\alpha$ and $\alpha\leq 2^{\beta}$. Then $2^{\beta}=\alpha$ and using Corollary 1.20 we have

$$\alpha<\alpha^{\mathrm{cf}(\alpha)}\leq\alpha^{\beta}=(2^{\beta})^{\beta}=2^{\beta}=\alpha.$$

The last statement of the theorem follows by Lemma 1.23 and Corollary 1.25.

The *continuum hypothesis* (CH) is the statement $\omega^{+}=2^{\omega}$; the *generalized continuum hypothesis* (GCH) is the statement $\alpha^{+}=2^{\alpha}$ for every infinite cardinal α.

A cardinal α is *(weakly) inaccessible* if α is an infinite, regular, limit cardinal; and α is *strongly inaccessible* if α is an infinite, regular, strong limit cardinal.

We note that ω is a strongly inaccessible cardinal; and that if α is strongly inaccessible then α is (weakly) inaccessible and $\alpha=\alpha^{\underline{\alpha}}$.

The corollary below follows immediately from Theorem 1.26 and the definitions just given.

1.28. Corollary. *Assume the generalized continuum hypothesis, and let* α *be an infinite cardinal. Then*

(a) *the statements*

$\alpha = \alpha^{\underline{\alpha}}$,
α *is regular,*
α *is a non limit cardinal or* α *is (strongly) inaccessible*

are equivalent; and

(b) $\alpha = 2^{\underline{\alpha}}$.

Notes for § 1

Set theory is the creation of Cantor [62]. The first axiomatization of set theory was given by Zermelo [08]. His axiom of replacement, stated in terms of the essentially undefined concept of *definite property*, was clarified by Skolem [22] in terms of formulas in the first order language of set theory. The axiom of replacement was also formulated in approximately the same way by Mirimanoff [17] and Fraenkel [22]. For an exhaustive list of statements equivalent to the axiom of choice (cf. Theorem 1.15) the reader is referred to Rubin and Rubin [63].

Cantor [78] (p. 257) conjectured the continuum hypothesis $\omega^+ = 2^\omega$ and tried desperately but without success to establish it; in Cantor [83] (p. 574, lines 28—34) the segment $2^{2^\omega} = (2^\omega)^+ = \omega^{++}$ of the generalized continuum hypothesis is conjectured. An exhaustive list of statements equivalent to the continuum hypothesis and of consequences of the continuum hypothesis is given in Sierpiński [56]. The consistency of GCH with ZFC was obtained by Gödel [38], [39] using the axiom of constructibility, proved in detail in Gödel [40], and explained in elementary terms in Gödel [47]. The independence of GCH from ZFC was obtained by Cohen in [63], [64], using the method of forcing, and proved in detail in Cohen [66].

The first authoritative texts on set theory were the works of Schoenflies [13] and Hausdorff [14]. Fraenkel [53], a book containing some 130 pages of bibliographical references, is an exhaustive treatment of the early developments with careful attention to historical details. Basic information on the elements of set theory and the arithmetic of cardinal and ordinal numbers is given in Kuratowski and Mostowski [68] and Bachmann [67]. Jech [71 b] contains many recent consistency results on set theory, covering work of Cohen, Solovay, Vopěnka, and others.

The ordinals were introduced as in 1.5—1.11 by von Neumann [23], [28]. Theorem 1.17 is related to Cauchy's diagonal argument; the proof given here is adapted from Jech [71b] (Theorem 6).

Theorem 1.19 was proved by J. König [04] for countable index sets I and by Jourdain and by Zermelo [08] in general. For regular cardinals König's theorem was shown by Easton [64], [70] to be the best possible (provable) result providing information on the cardinality of power sets. Corollary 1.21 and the remark immediately following it is the celebrated second diagonal argument of Cantor [90]. Manifestations of this technique abound in this work: cf. Theorem 12.21 on the cardinalities of ultraproducts, a descendant of du Bois-Reymond's orders of infinity (Hardy [10]); the disjoint refinement lemma (7.5); Lemma 12.6; Lemma 9.11; etc.

Results on cardinal arithmetic (of which 1.22—1.26 is a part) were obtained and systematized by Cantor, Hausdorff (*e.g.* [04], [14]), Tarski, and others. The notation $\alpha^{\underline{\kappa}}$ and most of Lemma 1.22 is due to Tarski [30b] (Lemmes 5—9). The notation $[\alpha]^\kappa$ was introduced by Erdős and his school in their study of partition calculus, *e.g.* in Erdős-Rado [56] (p. 429). The cofinality of limit ordinals is introduced in Tarski [25] (p. 2). Lemma 1.24 and Corollary 1.25 are given in Tarski [25], Théorème 7 (1) and Théorème II, respectively.

Lemma 1.26 is proved in Tarski [30b] (Lemme 8).

Inaccessible cardinals were first considered and given their name by Kuratowski (cf. Sierpiński and Tarski [30]); strongly inaccessible cardinals were introduced by Sierpiński-Tarski [30], and were further studied by Tarski [38a] and others. For more references on large cardinals the reader is referred to the Notes for § 8.

§ 2. Topology and Boolean Algebras

The material of this section is divided into four sub-sections dealing with Topology, the finitary properties of Boolean algebras, the duality of Stone between Boolean algebras and compact totally disconnected spaces, and the completion of a Boolean algebra and the (essentially dual) Gleason space of a compact space. Three spaces are involved: The Stone-Čech compactification of a space, the Stone space of a Boolean algebra, and the Stone space of the (complete) Boolean algebra of regular-open sets of a space. The definitions use either the topological version of an ultrafilter, namely a z-ultrafilter on a space, or the algebraic version of an ultrafilter, namely an ultrafilter of a Boolean algebra; these are (non-comparable) generalizations of the set-theoretic version of an ultrafilter, namely the ultrafilter on a cardinal number.

Topology

In this sub-section we are occupied mostly but not exclusively with the definition (as a set of z-ultrafilters with an appropriate topology) and the characteristic properties of the Stone-Čech compactification of a space. We also fix topological notation and mention a few other things.

$\mathbb{R}$ denotes the ordered field of real numbers with its usual topology. For $a, b \in \mathbb{R}$ with $a \leq b$ we set

$$[a,b] = \{p \in \mathbb{R} : a \leq p \leq b\}.$$

A *space* is a completely regular, Hausdorff topological space. Thus for a space X we have

(i) X satisfies the *Hausdorff separation property* (*i.e.*, if $p, q \in X$ and $p \neq q$ then there are open subsets U and V of X such that $p \in U$, $q \in V$, and $U \cap V = \emptyset$); and

(ii) if F is a closed subset of X and $p \in X \setminus F$, then there is a continuous function $f: X \to [0,1]$ such that $f(p) = 0$ and $f(q) = 1$ for $q \in F$.

We recall from general topology that the product (in the usual product topology) of a set of spaces is a space; that a subset of a space (in the induced or relative topology) is a space; and that $\mathbb{R}$ is a (metric) space.

A *compact* space is a space X such that every open cover of X has a finite subcover. We note that a topological space with the Hausdorff separation property in which every open cover has a finite subcover is a (compact) space. A *σ-compact* space is a space that is equal to the union of a countable family of compact subspaces. And a *normal* space is a space X such that if A and B are closed subsets of X and $A \cap B = \emptyset$, then there are open subsets U and V of X such that $A \subset U$, $B \subset V$, and $U \cap V = \emptyset$.

The implications X compact $\Rightarrow X$ σ-compact $\Rightarrow X$ normal hold for spaces X. The space $[0,1]$ is compact.

A non-empty family of sets has the *finite intersection property* if the intersection of every non-empty finite subfamily is non-empty.

A space is compact if and only if every family of closed subsets of the space with the finite intersection property has non-empty intersection.

If A is a subset of a space X we denote by $\mathrm{cl}_X A$ (or $\mathrm{cl}\, A$, or $\bar{A}$) the closure of A in X, by $\mathrm{int}_X A$ (or $\mathrm{int}\, A$) the interior of A in X, and by $Bd_X A$ (or $Bd\, A$) the boundary of A. If $p \in \mathrm{int}\, A$ then we say that A is a *neighborhood* of p in X. A subset A of $\mathbb{R}$ is *bounded* if there is $a \in \mathbb{R}$ such that $A \subset [-a, a]$.

Notation. Let X be a space. Then $C(X) = \{f \in \mathbb{R}^X : f \text{ is continuous}\}$, and $C^*(X) = \{f \in C(X) : f[X] \text{ is bounded}\}$.

Definitions. Let X be a space and let A and B be subsets of X.

(a) A is *C-embedded* in X if for $f \in C(A)$ there is $g \in C(X)$ such that $f \subset g$;

(b) A is *C*-embedded* in X if for $f \in C^*(A)$ there is $g \in C^*(X)$ such that $f \subset g$;

(c) A and B are *completely separated (in X)* if there is a continuous function $f: X \to [0,1]$ such that $f(p) = 0$ for $p \in A$ and $f(p) = 1$ for $p \in B$.

We recall from general topology the following basic results.

Urysohn's lemma. *Disjoint closed subsets of a normal space are completely separated.*

Tietze's extension theorem. *Every closed subset of a normal space is C-embedded (and hence C*-embedded).*

A *zero-set* of X is a set equal to $f^{-1}(\{0\})$ for some $f \in C(X)$. The family of all zero-sets of X is denoted $\mathscr{Z}(X)$. A *cozero-set* of X is the complement of a zero-set of X.

We note that if A and B are subsets of a space X then A and B are completely separated in X if and only if A and B are contained in disjoint zero-sets. Indeed, if f is a continuous function from X to $[0,1]$

such that $f(p)=0$ for $p\in A$ and $f(p)=1$ for $p\in B$, then $f^{-1}(\{0\})$ and $f^{-1}(\{1\})$ are disjoint zero-sets containing A and B respectively; and conversely if $f^{-1}(\{0\})$ and $g^{-1}(\{0\})$ are disjoint sets containing A and B respectively with $f,g\in C(X)$, then the (continuous) function $h:X\to\mathbb{R}$ defined by

$$h(p)=f(p)^2/(f(p)^2+g(p)^2)$$

separates A and B.

2.1. Lemma. *Let X be a space.*

(a) *If $\{A_n:n<\omega\}$ is a sequence of zero-sets of X then $\bigcap_{n<\omega}A_n$ is a zero-set of X.*

(b) *A zero-set of X is a G_δ; a cozero-set of X is an F_σ.*

(c) *If X is compact then every cozero-set of X is σ-compact.*

Proof. (a) Let $f_n\in C(X)$ with $A_n=f_n^{-1}(\{0\})$ and with $0\le f_n(x)\le 1/2^n$ for $n<\omega$ and $x\in X$, and set $f=\sum_{n<\omega}f_n$. Then $f\in C(X)$ and $\bigcap_{n<\omega}A_n=f^{-1}(\{0\})$.

(b) Let $f\in C(X)$ and set

$$G_n=\{x\in X:|f(x)|<1/n\}\quad\text{for } 1\le n<\omega.$$

Then G_n is open in X for $1\le n<\omega$, and $f^{-1}(\{0\})=\bigcap_{1\le n<\omega}G_n$.

This proves the first part of (b). The second part, from which (c) follows, is proved by taking complements.

Definition. A (proper) *z-filter* $\mathscr{F}$ on a space X is a non-empty family of non-empty zero-sets of X such that

$$\begin{array}{l}\text{if } A,B\in\mathscr{F}\quad\text{then } A\cap B\in\mathscr{F},\quad\text{and}\\ \text{if } A\in\mathscr{F},\ A\subset B,\ \text{ and } B\in\mathscr{Z}(X),\ \text{ then } B\in\mathscr{F}.\end{array}$$

The improper z-filter on X is $\mathscr{Z}(X)$; it is not (considered to be) a z-filter.

A *z-ultrafilter* on X is a z-filter on X not properly contained in any other z-filter on X.

A trivial application of Zorn's lemma establishes the following fundamental result.

2.2. Theorem. *Let X be a space and let $\mathscr{F}\subset\mathscr{Z}(X)$. If $\mathscr{F}$ has the finite intersection property then there is a z-ultrafilter p on X such that $\mathscr{F}\subset p$.*

A consequence of Theorem 2.2 is that every z-filter on X is contained in a z-ultrafilter on X.

2.3. Lemma. *Let p and q be z-ultrafilters on a space X.*

(a) *If $B\in\mathscr{Z}(X)$ and $A\cap B\neq\emptyset$ for $A\in p$ then $B\in p$;*

(b) *if $A,B\in\mathscr{Z}(X)$ and $A\cup B\in p$ then either $A\in p$ or $B\in p$; and*

(c) *if $p\neq q$ then there are $A\in p$ and $B\in q$ such that $A\cap B=\emptyset$.*

Proof. (a) The family $\mathscr{F}=\{A\cap B: A\in p\}$ has the finite intersection property, so there is by Theorem 2.2 a z-ultrafilter r on X such that $\mathscr{F}\subset r$. It is clear that $B\in r$ and that $p\subset r$; it follows that $p=r$ and $B\in p$.

(b) Let $A,B\in\mathscr{Z}(X)$ and assume that $A\notin p$, $B\notin p$. From (a) there are $C,D\in p$ such that $A\cap C=B\cap D=\emptyset$. Then $(A\cup B)\cap(C\cap D)=\emptyset$, and thus $A\cup B\notin p$.

(c) It is clear that there is $B\in q$ such that $B\notin p$. From (a) there is $A\in p$ such that $A\cap B=\emptyset$.

A z-ultrafilter on a space X is an element of $\mathscr{P}(\mathscr{P}(X))$ and hence the class of all z-ultrafilters on X is indeed a set (of cardinality at most $2^{2^{|X|}}$).

Definition. The *Stone-Čech compactification* $\boldsymbol{\beta}X$ of a space X is the set of all z-ultrafilters on X with the topology determined by the subbase

$$\mathscr{B}=\{\{p\in\boldsymbol{\beta}X: A\notin p\}: A\in\mathscr{Z}(X)\}.$$

We note that if $x\in X$ then the family $\{A\in\mathscr{Z}(X): x\in A\}$ is a z-filter on X. In fact this family is a z-ultrafilter, for if $B\in\mathscr{Z}(X)$ and $x\in X\backslash B$ then (since X is a space) there is $A\in\mathscr{Z}(X)$ such that $x\in A$ and $A\cap B=\emptyset$. Thus the function e defined on X by the rule

$$e(x)=\{A\in\mathscr{Z}(X): x\in A\}$$

is a one-to-one function from X into $\boldsymbol{\beta}X$. This function is called the *canonical embedding* of X into $\boldsymbol{\beta}X$.

2.4. Lemma. *Let X be a space. Then*

(a) *$\mathscr{B}$ is a base for $\boldsymbol{\beta}X$;*
(b) *$\boldsymbol{\beta}X$ is a compact space;*
(c) $e[X\backslash A]=e[X]\cap\{p\in\boldsymbol{\beta}X: A\notin p\}$ *for* $A\in\mathscr{Z}(X)$;
(d) *e is a topological embedding of X into $\boldsymbol{\beta}X$;*
(e) $\mathrm{cl}_{\boldsymbol{\beta}X}e[A]=\{p\in\boldsymbol{\beta}X: A\in p\}$ *for* $A\in\mathscr{Z}(X)$;
(f) *$e[X]$ is dense in $\boldsymbol{\beta}X$;*
(g) $p=\{A\in\mathscr{Z}(X): p\in\mathrm{cl}_{\boldsymbol{\beta}X}e[A]\}$ *for* $p\in\boldsymbol{\beta}X$; *and*
(h) $\mathrm{cl}_{\boldsymbol{\beta}X}e[A]\cap\mathrm{cl}_{\boldsymbol{\beta}X}e[B]=\mathrm{cl}_{\boldsymbol{\beta}X}e[A\cap B]$ *for* $A,B\in\mathscr{Z}(X)$.

Proof. (a) Let $A_0, A_1\in\mathscr{Z}(X)$ and set

$$B_i=\{p\in\beta X: A_i\notin p\} \quad \text{for } i<2.$$

For $p\in\boldsymbol{\beta}X$ we have

$$A_0\cup A_1\notin p \quad \text{if and only if } A_0\notin p \quad \text{and } A_1\notin p$$

by Lemma 2.3(b), and hence

$$B_0\cap B_1=\{p\in\boldsymbol{\beta}X: A_0\cup A_1\notin p\}\in\mathscr{B}.$$

(b) We first prove that $\boldsymbol{\beta} X$ satisfies the Hausdorff separation property. If $p, q \in \boldsymbol{\beta} X$ and $p \neq q$ then by Lemma 2.3 (c) there are $A, B \in \mathscr{Z}(X)$ such that $A \in p$, $B \in q$, and $A \cap B = \emptyset$. Since disjoint zero-sets of X are completely separated there are cozero-sets C and D of X and zero-sets E and F of X such that

$$A \subset C \subset E, \quad B \subset D \subset F, \quad \text{and } E \cap F = \emptyset.$$

Then $\{t \in \boldsymbol{\beta} X : X \setminus C \notin t\}$ and $\{t \in \boldsymbol{\beta} X : X \setminus D \notin t\}$ are neighborhoods of p and q respectively, and

$$\begin{aligned} &\{t \in \boldsymbol{\beta} X : X \setminus C \notin t\} \cap \{t \in \boldsymbol{\beta} X : X \setminus D \notin t\} \\ &\quad \subset \{t \in \boldsymbol{\beta} X : E \in t\} \cap \{t \in \boldsymbol{\beta} X : F \in t\} = \emptyset . \end{aligned}$$

To prove that $\boldsymbol{\beta} X$ is compact it is enough, according to (a), to show that if $\{A_i : i \in I\} \subset \mathscr{Z}(X)$ and

$$\{\{p \in \boldsymbol{\beta} X : A_i \in p\} : i \in I\}$$

has the finite intersection property then

$$\bigcap_{i \in I} \{p \in \boldsymbol{\beta} X : A_i \in p\} \neq \emptyset .$$

There is, by Theorem 2.2, $q \in \boldsymbol{\beta} X$ such that $\{A_i : i \in I\} \subset q$, and we have

$$q \in \bigcap_{A \in q} \{p \in \boldsymbol{\beta} X : A \in p\} \subset \bigcap_{i \in I} \{p \in \boldsymbol{\beta} X : A_i \in p\} .$$

(c) is clear from the definition of e. It implies that e is a one-to-one function which takes a base for X onto a base for $e[X]$.

(d) follows immediately.

(e) Let $A \in \mathscr{Z}(X)$. Then $\{p \in \boldsymbol{\beta} X : A \in p\}$ is closed in $\boldsymbol{\beta} X$ and $e[A] \subset \{p \in \boldsymbol{\beta} X : A \in p\}$, so that

$$\mathrm{cl}_{\boldsymbol{\beta} X}\, e[A] \subset \{p \in \boldsymbol{\beta} X : A \in p\} .$$

For the reverse containment let $p \in \boldsymbol{\beta} X$ with $A \in p$ and suppose that $p \notin \mathrm{cl}_{\boldsymbol{\beta} X}\, e[A]$. Then by part (a) there is $B \in \mathscr{Z}(X)$ such that $B \notin p$ (*i.e.*, $p \in \{t \in \boldsymbol{\beta} X : B \notin t\}$) and

$$e[A] \cap \{t \in \boldsymbol{\beta} X : B \notin t\} = \emptyset .$$

Then $e[A] \subset e[X] \cap \{t \in \boldsymbol{\beta} X : B \in t\} = e[B]$ by part (c), and we have $A \subset B$ and hence the contradiction $B \in p$.

(f) From (e) we have

$$\mathrm{cl}_{\boldsymbol{\beta} X}\, e[X] = \{p \in \boldsymbol{\beta} X : X \in p\} = \boldsymbol{\beta} X .$$

(g) is formally equivalent to (e).

(h) follows from (e).

Via the canonical embedding $e: X \to \beta X$, we usually (for convenience) identify X with the subspace $e[X]$ of βX. Thus an element $p \in X$ is identified with the z-ultrafilter $\{A \in \mathscr{Z}(X): p \in A\}$.

2.5. Lemma. *Let X be a dense subset of Y and let f be a continuous function from X to a compact space K.*

(a) *If for $p \in Y$ there is a continuous function $f_p: X \cup \{p\} \to K$ such that $f \subset f_p$, then there is a continuous function $g: Y \to K$ such that $f \subset g$.*

(b) *If $p \in Y$ and there is no continuous function $g: X \cup \{p\} \to K$ such that $f \subset g$, then there are $A, B \in \mathscr{Z}(K)$ such that*

$$A \cap B = \emptyset, \quad \textit{and} \quad p \in \mathrm{cl}_Y f^{-1}(A) \cap \mathrm{cl}_Y f^{-1}(B).$$

Proof. (a) Set $g = \bigcup\{f_p: p \in Y\}$. To verify the continuity of g at $p \in Y$ let V and W be neighborhoods of $g(p)$ such that $\mathrm{cl}\, W \subset V$, and let U be a neighborhood of p in Y such that $f(q) \in W$ for $q \in U \cap X$. If $q \in U$ then $q \in \mathrm{cl}_Y(U \cap X)$ and thus

$$g(q) = f_q(q) \in \mathrm{cl}\, f_q[U \cap X] = \mathrm{cl}\, f[U \cap X] \subset \mathrm{cl}\, W \subset V.$$

(b) The family

$$\mathscr{F} = \{\mathrm{cl}_K f[U \cap X]: U \text{ is a neighborhood of } p \text{ in } Y\}$$

is a family of closed subsets of K with the finite intersection property (since X is dense in Y); hence $\bigcap \mathscr{F} \neq \emptyset$. If $|\bigcap \mathscr{F}| = 1$, say $\bigcap \mathscr{F} = \{s\}$, then the function $f \cup \{\langle p, s \rangle\}$ is a continuous extension of f. Hence there are $s, t \in \bigcap \mathscr{F}$ with $s \neq t$, and it is easy to see that if $A, B \in \mathscr{Z}(K)$ with $s \in A$ and $t \in B$ and $A \cap B = \emptyset$, then

$$p \in \mathrm{cl}_Y f^{-1}(A) \cap \mathrm{cl}_Y f^{-1}(B).$$

2.6. Theorem. *Let X be a dense subset of a compact space Y. The following statements are equivalent.*

(a) *For every $f: X \to K$ with f continuous and K compact there is a continuous function $g: Y \to K$ such that $f \subset g$.*

(b) *X is C^*-embedded in Y.*

(c) *If $A, B \in \mathscr{Z}(X)$ and $A \cap B = \emptyset$, then $\mathrm{cl}_Y A \cap \mathrm{cl}_Y B = \emptyset$.*

(d) *If $A, B \in \mathscr{Z}(X)$ then $\mathrm{cl}_Y A \cap \mathrm{cl}_Y B = \mathrm{cl}_Y(A \cap B)$.*

(e) *The family $\{A \in \mathscr{Z}(X): p \in \mathrm{cl}_Y A\}$ is a z-filter on X for $p \in Y$.*

Furthermore the space βX satisfies these conditions, and for every space Y which satisfies these conditions there is a (unique) homeomorphism h from βX to Y such that $h(p) = p$ for $p \in X$.

Proof. (a)$\Rightarrow$(b) is trivial.

(b)$\Rightarrow$(c). Let $A, B \in \mathscr{Z}(X)$ and $A \cap B = \emptyset$. We have noted earlier that there is a continuous function $f: X \to [0,1]$ such that $A \subset f^{-1}(\{0\})$

and $B \subset f^{-1}(\{1\})$. From (b) there is a continuous $g: Y \to [0,1]$ such that $f \subset g$. Then $g(p)=0$ for $p \in \mathrm{cl}_Y A$ and $g(p)=1$ for $p \in \mathrm{cl}_Y B$, and (c) follows.

(c)⇒(d). It is clear that $\mathrm{cl}_Y(A \cap B) \subset \mathrm{cl}_Y A \cap \mathrm{cl}_Y B$. If $p \in Y$ and $p \notin \mathrm{cl}_Y(A \cap B)$ then there is $f \in C(Y)$ such that $f(p)=0$ and $f(q)=1$ for $q \in (A \cap B)$. Set $C=\{q \in Y: f(q) \leq 1/2\}$. Then C is a neighborhood of p in Y, and $A \cap B \cap C = \emptyset$. Since $A \cap C \in \mathscr{Z}(X)$ and $B \cap C \in \mathscr{Z}(X)$ we have

$$\mathrm{cl}_Y(A \cap C) \cap \mathrm{cl}_Y(B \cap C) = \emptyset$$

by (c), and hence $p \notin \mathrm{cl}_Y A \cap \mathrm{cl}_Y B$.

(d)⇒(e). Set $\tilde{p} = \{A \in \mathscr{Z}(X): p \in \mathrm{cl}_Y A\}$ for $p \in Y$. It is clear that $\tilde{p} \neq \emptyset$ and $\emptyset \notin \tilde{p}$, and also that if $A, B \in \mathscr{Z}(X)$ with $A \in \tilde{p}$ and $A \subset B$, then $B \in \tilde{p}$. From (d) it follows that the intersection of any two elements of $\tilde{p}$ is an element of $\tilde{p}$. Thus $\tilde{p}$ is a z-filter on X.

(e)⇒(a). If condition (a) fails then by Lemma 2.5(a) there are a compact space K, $p \in Y$, and a continuous function $f: X \to K$ such that f is not contained in a continuous function from $X \cup \{p\}$ to K. Then by Lemma 2.5(b) there are $C, D \in \mathscr{Z}(K)$ such that $C \cap D = \emptyset$ and $p \in \mathrm{cl}_Y f^{-1}(C) \cap \mathrm{cl}_Y f^{-1}(D)$. Since

$$f^{-1}(C), f^{-1}(D) \in \mathscr{Z}(X) \quad \text{and } f^{-1}(C) \cap f^{-1}(D) = \emptyset,$$

the family $\{A \in \mathscr{Z}(X): p \in \mathrm{cl}_Y A\}$ is not a z-filter of X.

According to Lemma 2.4, $\boldsymbol{\beta} X$ is a compact space containing X as a dense subset; further for $p \in \boldsymbol{\beta} X$ we have

$$\{A \in \mathscr{Z}(X): p \in \mathrm{cl}_{\boldsymbol{\beta} X} A\} = p$$

by Lemma 2.4(g), so $\boldsymbol{\beta} X$ satisfies condition (e) (because the elements of $\boldsymbol{\beta} X$ are z-ultrafilters on X) and hence all conditions (a)—(e) of the present theorem.

Suppose now that Y is a compact space containing X densely and satisfying conditions (a)—(e). It follows from condition (a) that there is a continuous function $h: \boldsymbol{\beta} X \to Y$ such that $h(p)=p$ for $p \in X$ and there is a continuous function $k: Y \to \boldsymbol{\beta} X$ such that $k(p)=p$ for $p \in X$. Then $k \circ h(p) = p$ for $p \in X$, and since $k \circ h$ is continuous on $\boldsymbol{\beta} X$ and X is dense in $\boldsymbol{\beta} X$ we have $k \circ h(p)=p$ for $p \in \boldsymbol{\beta} X$. Thus h is a homeomorphism from $\boldsymbol{\beta} X$ onto Y.

The proof is complete.

If $f: X \to K$ is continuous and K is compact then the unique continuous function $\bar{f}: \boldsymbol{\beta} X \to K$ such that $f \subset \bar{f}$ is called the *Stone extension* of f. We consistently use the bar above a continuous function into a compact space to denote its Stone extension. (In case of conceivable ambiguity we specify the compact space we consider.)

2.7. Corollary *(Tychonoff product theorem). If $\{X_i : i \in I\}$ is a family of compact spaces then $\prod_{i\in I} X_i$ is compact.*

Proof. Let $X = \prod_{i\in I} X_i$, let $\pi_i : X \to X_i$ denote the projection function, and let $\bar{\pi}_i$ be the Stone extension of π_i for $i \in I$. Then the function $f : \boldsymbol{\beta} X \to X$ defined by $f(x) = \langle \bar{\pi}_i(x) : i \in I \rangle$ is continuous and onto, and thus X is compact.

2.8. Corollary *(Urysohn's extension theorem). If X is a subset of a space Y then X is C^*-embedded in Y if and only if every two subsets of X which are completely separated in X are completely separated in Y.*

Proof. ($\Rightarrow$) Let A and B be completely separated subsets of X. Then there is $f \in C^*(X)$ such that $f(p) = 0$ for $p \in A$ and $f(p) = 1$ for $p \in B$, and since X is C^*-embedded in Y there is $g \in C^*(Y)$ such that $f \subset g$. Thus A and B are completely separated in Y.

($\Leftarrow$) Set $K = \mathrm{cl}_{\boldsymbol{\beta} Y} X$. Since K is C^*-embedded in $\boldsymbol{\beta} Y$ (by Tietze's extension theorem) it suffices to show that X is C^*-embedded in K, for then X is C^*-embedded in $\boldsymbol{\beta} Y$ and *a fortiori* in the intermediate space Y. If $A, B \in \mathscr{Z}(X)$ with $A \cap B = \emptyset$ then as was remarked earlier A and B are completely separated in X; then by hypothesis A and B are completely separated in Y, hence (by the implication ($\Rightarrow$) above applied to Y and $\boldsymbol{\beta} Y$ in place of X and K) in $\boldsymbol{\beta} Y$. Thus $\mathrm{cl}_K A \cap \mathrm{cl}_K B = \emptyset$, so X is C^*-embedded in K by Theorem 2.6(c)$\Rightarrow$(b).

2.9. Lemma. (a) *If X is dense in Y and X is locally compact, then X is open in Y.*

(b) *X is locally compact if and only if $\boldsymbol{\beta} X \setminus X$ is compact.*

Proof. (a) Let $p \in X$ and let U be a compact neighborhood of p in X. There is an open subset V of Y such that $X \cap V = \mathrm{int}_X U$, and

$$V \subset \mathrm{cl}_Y V = \mathrm{cl}_Y(X \cap V) = \mathrm{cl}_Y(\mathrm{int}_X U) \subset U$$

(the last inclusion since U is compact). It follows that $V \subset X$. From the fact that every element of X has a neighborhood in X which is open in Y, it follows that X is open in Y.

(b) follows directly from (a).

A family $\mathscr{S}$ of functions $f : X \to Y_f$ whose domain is a space X is an *evaluation family for* X. Given such a family we define the *evaluation function* (relative to $\mathscr{S}$)

$$e : X \to \prod_{f\in\mathscr{S}} Y_f$$

by $e(p)_f = f(p)$ for $p \in X$ and $f \in \mathscr{S}$. We say that $\mathscr{S}$ *separates elements* of X if for every $p, q \in X$ with $p \neq q$ there is $f \in \mathscr{S}$ such that $f(p) \neq f(q)$; and $\mathscr{S}$ is a *completely regular family* if for $p \in X$ and every closed subset F of X such that $p \notin F$ there is $f \in \mathscr{S}$ such that $f(p) \notin \mathrm{cl}_{Y_f} f[F]$.

2.10. Lemma (*Embedding lemma*). *Let X be a space and let $\mathcal{S}$ be an evaluation family for X.*

(a) *If the elements of $\mathcal{S}$ are continuous then e is continuous.*

(b) *If the elements of $\mathcal{S}$ are continuous and $\mathcal{S}$ is a completely regular family, then e is a topological embedding.*

(c) *If X is compact and the elements of $\mathcal{S}$ are continuous and $\mathcal{S}$ separates elements of X, then e is a topological embedding.*

Proof. Set $Y=\prod_{f\in\mathcal{S}} Y_f$ and let $\pi_f: Y\to Y_f$ denote the projection function for $f\in\mathcal{S}$.

(a) e is continuous since $\pi_f\circ e$ is continuous for $f\in\mathcal{S}$.

(b) e is continuous by (a). If $p,q\in X$ and $p\neq q$ and f separates p from $\{q\}$, then $e(p)_f\neq e(q)_f$; hence e is one-to-one. Finally if U is open in X and $p\in U$ then there is $f\in\mathcal{S}$ such that $f(p)\notin \mathrm{cl}_{Y_f} f[X\setminus U]$, and hence

$$e(p)\in e[X]\cap\pi_f^{-1}(Y_f\setminus \mathrm{cl}_{Y_f} f[X\setminus U])\subset e[U].$$

Thus $e[U]$ is the union of open subsets of $e[X]$. We conclude that e is an open function.

(c) The function e is continuous and one-to-one; since X is compact, e is a topological embedding.

The *weight* $\mathbf{w}(X)$ of a space X is the least cardinal α such that there is a base for X of cardinality α.

2.11. Corollary. *For every space X there is an embedding of X into $[0,1]^{\mathbf{w}(X)}$.*

Proof. If $\mathbf{w}(X)<\omega$ then $|X|<\omega$ and the statement is obvious, so we assume $\mathbf{w}(X)\geq\omega$. Let $\mathcal{B}$ be a base for X such that $|\mathcal{B}|=\mathbf{w}(X)$. If $\langle U,V\rangle\in\mathcal{B}\times\mathcal{B}$ and U and $X\setminus V$ are completely separated then we choose a continuous function $f_{U,V}: X\to[0,1]$ such that $f_{U,V}(p)=0$ for $p\in U$ and $f_{U,V}(p)=1$ for $f\in X\setminus V$; if $\langle U,V\rangle\in\mathcal{B}\times\mathcal{B}$ and U and $X\setminus V$ are not completely separated we set $f_{U,V}(p)=0$ for $p\in X$. Let

$$e: X\to[0,1]^{\mathcal{B}\times\mathcal{B}}$$

be the evaluation function determined by the evaluation family $\mathcal{S}=\{f_{U,V}:\langle U,V\rangle\in\mathcal{B}\times\mathcal{B}\}$. It is easy to verify, using the fact that for $p\in X$ and every closed subset F of X such that $p\notin F$ there is $\langle U,V\rangle\in\mathcal{B}\times\mathcal{B}$ such that

$$p\in U\subset V\subset X\setminus F$$

and U and $X\setminus V$ are completely separated, that $\mathcal{S}$ is a completely regular family. Hence e is a topological embedding, and since $|\mathcal{B}\times\mathcal{B}|=\mathbf{w}(X)$ the result follows.

2.12. Lemma. *Let X be a space and $\mathscr{B}$ a base for X. Then there is a subfamily $\mathscr{A}$ of $\mathscr{B}$ such that*

$$\mathscr{A} \text{ is a base for } X \text{ and}$$
$$|\mathscr{A}| = \boldsymbol{w}(X).$$

Proof. If $\boldsymbol{w}(X)$ is finite the conclusion is obvious. Now let $\boldsymbol{w}(X) \geq \omega$ and let $\mathscr{C}$ be a base for X such that $|\mathscr{C}| = \boldsymbol{w}(X)$. If $C_0, C_1 \in \mathscr{C}$ and there is $B \in \mathscr{B}$ such that $C_0 \subset B \subset C_1$ we choose $B_{C_0, C_1} \in \mathscr{B}$ such that $C_0 \subset B_{C_0, C_1} \subset C_1$; otherwise for $C_0, C_1 \in \mathscr{C}$ we let B_{C_0, C_1} be an arbitrary element of $\mathscr{B}$. It is easy to verify that the family

$$\mathscr{A} = \{B_{C_0, C_1} : C_0, C_1 \in \mathscr{C}\}$$

satisfies the conclusion of the lemma.

P_α-points. An element p of a space X is a *P_α-point* of X if the intersection of any family of less than α neighborhoods of p is a (not necessarily open) neighborhood of p. A space X is a *P_α-space* if every element of X is a P_α-point.

It is easy to verify that X is a P_α-space if and only if the intersection of every family of less than α open subsets of X is an open subset of X.

Every space is a P_ω-space.

A P_{ω^+}-point is a *P-point*, and a P_{ω^+}-space is a *P-space.*

Remark. If α is an infinite singular cardinal then

$$\text{a } P_\alpha\text{-point is a } P_{\alpha^+}\text{-point, and}$$
$$\text{a } P_\alpha\text{-space is a } P_{\alpha^+}\text{-space.}$$

The second statement follows from the first, so we verify the first only. Let $\{\alpha_\xi : \xi < \mathrm{cf}(\alpha)\}$ be a set of cardinals such that

$$\alpha_\xi < \alpha \quad \text{for } \xi < \mathrm{cf}(\alpha),$$
$$\alpha_\xi < \alpha_\eta \quad \text{for } \xi < \eta < \mathrm{cf}(\alpha), \quad \text{and}$$
$$\textstyle\sum_{\xi < \mathrm{cf}(\alpha)} \alpha_\xi = \alpha.$$

Let p be a P_α-point of X and let $\{U_\zeta : \zeta < \alpha\}$ be a family of neighborhoods of p. We note that

$$\textstyle\bigcap_{\zeta < \alpha_\xi} U_\zeta$$

is a neighborhood of p for $\xi < \mathrm{cf}(\alpha)$, since $\alpha_\xi < \alpha$ and p is a P_α-point. Then

$$\textstyle\bigcap_{\xi < \mathrm{cf}(\alpha)} \bigcap_{\zeta < \alpha_\xi} U_\zeta = \bigcap_{\zeta < \alpha} U_\zeta$$

is a neighborhood of p, since $\mathrm{cf}(\alpha) < \alpha$ and p is a P_α-point.

If $\alpha \geq \omega$ and X is a set and $\mathscr{C} \subset \mathscr{P}(X)$ we set

$$\mathscr{C}^{(\alpha)} = \{\bigcap_{i \in I} C_i : C_i \in \mathscr{C} \text{ for } i \in I \text{ and } |I| < \alpha\};$$

if X is a space we say that $\mathscr{C}$ is an *α-subbase* for X if $\mathscr{C}^{(\alpha)}$ is a subbase for X. We note that if $\mathscr{C}$ is an α-subbase for X then since $(\mathscr{C}^{(\alpha)})^{(\omega)} = C^{(\alpha)}$ the family $\mathscr{C}^{(\alpha)}$ is a base for X. An ω-subbase is a subbase.

For a space X we denote by $\mathscr{T}(X)$ the topology of X.

Definition. Let $\alpha \geq \omega$ and X a space. Then X_α is the space with underlying set equal to the underlying set of X and with the smallest topology $\mathscr{T}(X_\alpha)$ such that

$$\mathscr{T}(X) \subset \mathscr{T}(X_\alpha) \quad \text{and}$$
$$X_\alpha \text{ is a } P_\alpha\text{-space.}$$

Since the discrete topology on the set X is a P_α-space topology and the intersection of P_α-space topologies on the set X is a P_α-space topology, the topology $\mathscr{T}(X_\alpha)$ is well-defined. For every space X we have $X_\omega = X$ and hence X_ω is indeed a space. To verify that X_α is a space for $\alpha > \omega$ we note first that X_α satisfies the Hausdorff separation axiom since $\mathscr{T}(X) \subset \mathscr{T}(X_\alpha)$. If α is regular then

$\mathscr{Z}(X) \subset \mathscr{T}(X)^{(\alpha)}$ (by Lemma 2.1(b)),
$\mathscr{T}(X)$ is an α-subbase for X_α, and
every element of $\mathscr{T}(X)^{(\alpha)}$ is equal to the union of elements of $\mathscr{Z}(X)^{(\alpha)}$;

it follows that $\mathscr{Z}(X)^{(\alpha)}$ is a base for X_α and that the elements of $\mathscr{Z}(X)^{(\alpha)}$ are open-and-closed in X_α. Similarly X_α has a base of open-and-closed sets when $\alpha > \omega$ and α is singular, since in this case α^+ is regular and $X_\alpha = X_{\alpha^+}$. Thus for $\alpha > \omega$ and for every closed subset F of X_α and $p \in X_\alpha \backslash F$ there is a continuous function $f: X_\alpha \to \{0,1\}$ such that $f(p) = 0$ and $f(q) = 1$ for $q \in F$.

We note that $X = X_\alpha$ if and only if X is a P_α-space, and that $(X_\alpha)_\alpha = X_\alpha$ for every space X and $\alpha \geq \omega$. We note also for regular $\alpha \geq \omega$ that every base for X is an α-subbase for X_α; hence we have

$$\mathbf{w}(X_\alpha) \leq (\mathbf{w}(X))^{\underline{\alpha}} \quad \text{for every space } X \text{ and regular } \alpha \geq \omega.$$

The order topology. Let $\langle A, \leq \rangle$ be a linearly ordered set. The *order topology on A* is the topology whose subbase is the family

$$\{\{p \in A : p < a\} : a \in A\} \cup \{\{p \in A : a < p\} : a \in A\}.$$

Of course the (usual) topology on $\mathbb{R}$ is the order topology determined by its (usual) order.

Finitary Properties of Boolean Algebras

We develop here the usual finitary properties of Boolean algebras, involving the concepts of subalgebra, homomorphism, and filter.

Definition. A *Boolean algebra* $\mathfrak{B}$ consists of a non-empty set B and three functions (operations)

$$\wedge, \vee : B \times B \to B \quad \text{and}$$
$$' : B \to B$$

such that

$$\begin{aligned} &a \wedge b = b \wedge a, && a \vee b = b \vee a, \\ &(a \wedge b) \wedge c = a \wedge (b \wedge c), && (a \vee b) \vee c = a \vee (b \vee c), \\ &(a \vee b) \wedge b = b, && (a \wedge b) \vee b = b, \\ &a \wedge (b \vee c) = (a \wedge b) \vee (a \wedge c), && a \vee (b \wedge c) = (a \vee b) \wedge (a \vee c), \end{aligned}$$

and

$$(a \vee a') \wedge b = b, \qquad (a \wedge a') \vee b = b$$

for $a, b, c \in B$.

The *elements* of $\mathfrak{B}$ are the elements of B and we write $a \in \mathfrak{B}$ for $a \in B$.

It is easy to show that

$$a \wedge a' = b \wedge b' \quad \text{and} \quad a \vee a' = b \vee b'$$

for $a, b \in \mathfrak{B}$. We define

$$\mathbb{0} = a \wedge a' \quad \text{and} \quad \mathbb{1} = a \vee a';$$

if necessary we write $\mathbb{0}_{\mathfrak{B}}$ and $\mathbb{1}_{\mathfrak{B}}$ in place of $\mathbb{0}$ and $\mathbb{1}$.

A Boolean algebra $\mathfrak{B}$ is *improper* if $\mathbb{0}_{\mathfrak{B}} = \mathbb{1}_{\mathfrak{B}}$, *i.e.*, if $B = \{\mathbb{0}\}$; $\mathfrak{B}$ is *proper* if $|B| > 1$.

If $a, b \in \mathfrak{B}$ we set

$$a \setminus b = a \wedge b';$$

thus in particular, $\mathbb{1} \setminus b = b'$.

Two elements a and b of a Boolean algebra are *disjoint* if $a \wedge b = \mathbb{0}$.

If $\mathfrak{B}$ is a Boolean algebra, $n < \omega$ and

$$F = \{b_0, b_1, \ldots, b_n\} \in \mathscr{P}^*_\omega(B),$$

then as a consequence of the defining conditions above both

$$b_0 \wedge b_1 \wedge \cdots \wedge b_n \quad \text{and} \quad b_0 \vee b_1 \vee \cdots \vee b_n$$

are unambiguously defined elements of $\mathfrak{B}$. We write

$$\bigwedge F, \quad \text{or } \bigwedge_{b\in F} b, \quad \text{or } \bigwedge_{k\le n} b_k$$

for the first of these, and

$$\bigvee F, \quad \text{or } \bigvee_{b\in F} b, \quad \text{or } \bigvee_{k\le n} b_k$$

for the latter.

We set $\bigwedge \emptyset = \mathbb{1}$ and $\bigvee \emptyset = \mathbb{0}$.

The *laws of de Morgan*, proved by induction from the defining properties of a Boolean algebra, are the following:

If $b\in\mathfrak{B}$, $n<\omega$ and $b_k\in\mathfrak{B}$ for $k\le n$, then

$$b\setminus\bigvee_{k\le n} b_k = \bigwedge_{k\le n}(b\setminus b_k), \quad \text{and}$$
$$b\setminus\bigwedge_{k\le n} b_k = \bigvee_{k\le n}(b\setminus b_k).$$

The following lemma is proved from these definitions by finite induction.

2.13. Lemma. *Let $\mathfrak{B}$ be a Boolean algebra, let $n<\omega$, $r_k<\omega$ for $k\le n$ and $b_{k,i}\in\mathfrak{B}$ for $k\le n$ and $i\le r_k$. Set*

$$K = \prod_{k\le n}(r_k+1).$$

Then

$$\bigwedge_{k\le n}\bigvee_{i\le r_k} b_{k,i} = \bigvee_{a\in K}\bigwedge_{k\le n} b_{k,a_k}; \quad \textit{and}$$
$$\bigvee_{k\le n}\bigwedge_{i\le r_k} b_{k,i} = \bigwedge_{a\in K}\bigvee_{k\le n} b_{k,a_k}.$$

Definition. Let $\mathfrak{B}$ be a Boolean algebra. A *Boolean subalgebra of* $\mathfrak{B}$ is a Boolean algebra $\mathfrak{A}$ such that

$A\subset B$, and
the functions $\wedge$, $\vee$ and $'$ of $\mathfrak{A}$ are the restrictions of the corresponding functions of $\mathfrak{B}$.

We note that if $\mathfrak{B}$ is a Boolean algebra and A is a subset of $\mathfrak{B}$ such that

$$a\vee b\in A \quad \text{and } a'\in A \quad \text{for } a,b\in A,$$

then A is a Boolean subalgebra of $\mathfrak{B}$.

It is clear that the intersection of a set of Boolean subalgebras of a Boolean algebra $\mathfrak{B}$ is a Boolean subalgebra of $\mathfrak{B}$. If $C\subset\mathfrak{B}$ then the intersection of the set of all Boolean subalgebras of $\mathfrak{B}$ containing C is the *Boolean subalgebra of* $\mathfrak{B}$ *generated by* C. This is denoted by $\langle\!\langle C\rangle\!\rangle_{\mathfrak{B}}$ or, simply, by $\langle\!\langle C\rangle\!\rangle$.

If $b\in\mathfrak{B}$ we set $1b=b$ and $-1b=b'$.

2.14. Lemma. *Let* $\mathfrak{B}$ *be a Boolean algebra.*

(a) *If* $\emptyset \neq C \subset B$ *then*

$$\langle\langle C\rangle\rangle = \{\bigvee_{k\leq n} \bigwedge_{i\leq r_k} \varepsilon_{k,i} c_{k,i} : n<\omega, r_k<\omega \text{ for } k\leq n,$$
$$c_{k,i} \in C, \varepsilon_{k,i} \in \{-1,1\} \text{ for } k\leq n, i\leq r_k\}.$$

(b) *If* $\mathfrak{A}$ *is a subalgebra of* $\mathfrak{B}$ *and* $b \in \mathfrak{B}$ *then*

$$\langle\langle \mathfrak{A} \cup \{b\}\rangle\rangle = \{(a_0 \wedge b) \vee (a_1 \wedge b') : a_0, a_1 \in \mathfrak{A}\}.$$

Proof. (a) Let A be the set on the right-hand side of the equality to be proved. It is clear that $C \subset A \subset \langle\langle C\rangle\rangle$. Thus it is enough to prove that A is a subalgebra of $\mathfrak{B}$. It is clear that if $a, b \in A$ then $a \vee b \in A$. Let $b = \bigvee_{k\leq n} \bigwedge_{i\leq r_k} \varepsilon_{k,i} c_{k,i} \in A$. We set

$$K = \prod_{k\leq n} (r_k + 1)$$

and we note, using de Morgan's laws and Lemma 2.13, that

$$b' = \bigwedge_{k\leq n} \bigvee_{i\leq r_k} -\varepsilon_{k,i} c_{k,i} = \bigvee_{a\in K} \bigwedge_{k\leq n} -\varepsilon_{k,a_k} c_{k,a_k} \in A.$$

(b) This is the special case $C = \mathfrak{A} \cup \{b\}$ of (a).

Definition. Let $\mathfrak{A}$ and $\mathfrak{B}$ be Boolean algebras. A function $\phi: \mathfrak{A} \to \mathfrak{B}$ is a *(Boolean algebra) homomorphism* if

$$\phi(a \wedge b) = \phi(a) \wedge \phi(b),$$
$$\phi(a \vee b) = \phi(a) \vee \phi(b), \quad \text{and}$$
$$\phi(a') = (\phi(a))'$$

for $a, b \in \mathfrak{A}$. If the homomorphism ϕ is a one-to-one function, then ϕ is a *(Boolean algebra) embedding* of $\mathfrak{A}$ into $\mathfrak{B}$. If ϕ is an embedding and is an onto function, then ϕ is an *isomorphism* from $\mathfrak{A}$ to $\mathfrak{B}$.

We make some remarks concerning these definitions.

A function $\phi: \mathfrak{A} \to \mathfrak{B}$ is a homomorphism if

$$\phi(a \wedge b) = \phi(a) \wedge \phi(b), \quad \text{and}$$
$$\phi(a') = (\phi(a))'$$

for $a, b \in \mathfrak{A}$.

If $\phi: \mathfrak{A} \to \mathfrak{B}$ is a homomorphism, then

$$\phi(\mathbb{0}) = \mathbb{0} \quad \text{and} \quad \phi(\mathbb{1}) = \mathbb{1}.$$

If $\mathfrak{A}$ is a Boolean subalgebra of $\mathfrak{B}$ then the inclusion function from $\mathfrak{A}$ to $\mathfrak{B}$ is an embedding of $\mathfrak{A}$ into $\mathfrak{B}$.

Definition. Let $\mathfrak{B}$ be a Boolean algebra. A binary relation $\leq$ is defined on $\mathfrak{B}$ by the rule

$$a \leq b \quad \text{if } a \wedge b = a.$$

It follows from the defining conditions of a Boolean algebra that $\leq$ is a partial order on $\mathfrak{B}$ such that $a \vee b$ and $a \wedge b$ are the supremum and infimum respectively of $\{a,b\}$.

Clearly $\mathbb{0} \leq a \leq \mathbb{1}$ for $a \in \mathfrak{B}$.

If $\phi: \mathfrak{A} \to \mathfrak{B}$ is a Boolean algebra homomorphism and $a_0 \leq a_1$ in $\mathfrak{A}$, then $\phi(a_0) \leq \phi(a_1)$ in $\mathfrak{B}$.

2.15. Lemma. *Let $\mathfrak{B}$ be a Boolean algebra generated by a non-empty subset C, let $\mathfrak{A}$ be a Boolean algebra and let $f: C \to \mathfrak{A}$ be such that*

$$\text{if } \bigwedge_{k \leq n} \varepsilon_k c_k = \mathbb{0} \quad \text{then } \bigwedge_{k \leq n} \varepsilon_k f(c_k) = \mathbb{0}$$

for $n < \omega$ and $c_k \in C$, $\varepsilon_k \in \{-1, 1\}$ for $k \leq n$. Then there is a unique homomorphism

$$\phi: \mathfrak{B} \to \mathfrak{A}$$

such that $f \subset \phi$.

Proof. We claim that if

$$\bigvee_{k \leq n} \bigwedge_{i \leq r_k} \varepsilon_{k,i} c_{k,i} = \bigvee_{\bar{k} \leq \bar{n}} \bigwedge_{\bar{i} \leq \bar{r}_{\bar{k}}} \bar{\varepsilon}_{\bar{k},\bar{i}} \bar{c}_{\bar{k},\bar{i}}$$

then

$$\bigvee_{k \leq n} \bigwedge_{i \leq r_k} \varepsilon_{k,i} f(c_{k,i}) = \bigvee_{\bar{k} \leq \bar{n}} \bigwedge_{\bar{i} \leq \bar{r}_{\bar{k}}} \bar{\varepsilon}_{\bar{k},\bar{i}} f(\bar{c}_{\bar{k},\bar{i}})$$

for $n, \bar{n} < \omega$, $r_k < \omega$ for $k \leq n$, $\bar{r}_{\bar{k}} < \omega$ for $\bar{k} \leq \bar{n}$,

$$c_{k,i} \in C, \quad \varepsilon_{k,i} \in \{-1, 1\} \quad \text{for } k \leq n,\ i \leq r_k, \quad \text{and}$$
$$\bar{c}_{\bar{k},\bar{i}} \in C, \quad \bar{\varepsilon}_{\bar{k},\bar{i}} \in \{-1, 1\} \quad \text{for } \bar{k} \leq \bar{n},\ \bar{i} \leq \bar{r}_{\bar{k}}.$$

We set $K = \prod_{\bar{k} \leq \bar{n}} (\bar{r}_{\bar{k}} + 1)$. Using Lemma 2.13 and de Morgan's laws we have

$$\begin{aligned} \mathbb{0} &= \bigvee_{k \leq n} \bigwedge_{i \leq r_k} \varepsilon_{k,i} c_{k,i} \setminus \bigvee_{\bar{k} \leq \bar{n}} \bigwedge_{\bar{i} \leq \bar{r}_{\bar{k}}} \bar{\varepsilon}_{\bar{k},\bar{i}} \bar{c}_{\bar{k},\bar{i}} \\ &= \bigvee_{k \leq n} \bigwedge_{i \leq r_k} \varepsilon_{k,i} c_{k,i} \setminus \bigwedge_{a \in K} \bigvee_{\bar{k} \leq \bar{n}} \bar{\varepsilon}_{\bar{k},a_{\bar{k}}} \bar{c}_{\bar{k},a_{\bar{k}}} \\ &= \bigvee_{a \in K} \bigvee_{k \leq n} \bigwedge_{\bar{k} \leq \bar{n}} \bigwedge_{\bar{i} \leq \bar{r}_{\bar{k}}} (\varepsilon_{k,i} c_{k,i} \setminus \bar{\varepsilon}_{\bar{k},a_{\bar{k}}} \bar{c}_{\bar{k},a_{\bar{k}}}). \end{aligned}$$

Thus

$$\bigwedge_{\bar{k} \leq \bar{n}} \bigwedge_{\bar{i} \leq \bar{r}_{\bar{k}}} (\varepsilon_{k,i} c_{k,i} \wedge -\bar{\varepsilon}_{\bar{k},a_{\bar{k}}} \bar{c}_{\bar{k},a_{\bar{k}}}) = \mathbb{0}$$

for $a \in K$ and $k \leq n$. The assumption on f clearly implies that

$$\bigwedge_{\bar{k} \leq \bar{n}} \bigwedge_{\bar{i} \leq \bar{r}_{\bar{k}}} (\varepsilon_{k,i} f(c_{k,i}) \wedge -\bar{\varepsilon}_{\bar{k},a_{\bar{k}}} f(\bar{c}_{\bar{k},a_{\bar{k}}})) = \mathbb{0}$$

for $a \in K$ and $k \leq n$, from which it follows (using as above de Morgan's laws and Lemma 2.13) that

$$\bigvee_{k \leq n} \bigwedge_{i \leq r_k} \varepsilon_{k,i} f(c_{k,i}) \leq \bigvee_{\bar{k} \leq \bar{n}} \bigwedge_{\bar{i} \leq \bar{r}_{\bar{k}}} \bar{\varepsilon}_{\bar{k},\bar{i}} f(\bar{c}_{\bar{k},\bar{i}}).$$

A symmetric argument proves the reverse inequality and establishes the claim.

Now let $b \in \mathfrak{B}$. By Lemma 2.14(a) there are $n < \omega$, $r_k < \omega$ for $k \leq n$, and $c_{k,i} \in C$, $\varepsilon_{k,i} \in \{-1, 1\}$ for $k \leq n$, $i \leq r_k$ such that

$$b = \bigvee_{k \leq n} \bigwedge_{i \leq r_k} \varepsilon_{k,i} c_{k,i} .$$

We define

$$\phi(b) = \bigvee_{k \leq n} \bigwedge_{i \leq r_k} \varepsilon_{k,i} f(c_{k,i}) .$$

The above claim proves that the function

$$\phi: \mathfrak{B} \to \mathfrak{A}$$

is well-defined. It is easy to verify that ϕ is the unique homomorphism such that $f \subset \phi$.

2.16. Lemma. *Let $\mathfrak{A}$ and $\mathfrak{B}$ be Boolean algebras generated by non-empty sets C and D, respectively, and let $f: C \to D$ be a one-to-one onto function such that*

$$\bigwedge_{k \leq n} \varepsilon_k c_k = \mathbb{0} \quad \textit{if and only if} \quad \bigwedge_{k \leq n} \varepsilon_k f(c_k) = \mathbb{0}$$

for $n < \omega$ and $c_k \in C$, $\varepsilon_k \in \{-1, 1\}$ for $k \leq n$. Then there is a unique isomorphism

$$\phi: \mathfrak{A} \to \mathfrak{B}$$

such that $f \subset \phi$.

Proof. By Lemma 2.15 there is a homomorphism $\phi: \mathfrak{A} \to \mathfrak{B}$ such that $f \subset \phi$ and there is a homomorphism $\psi: \mathfrak{B} \to \mathfrak{A}$ such that $f^{-1} \subset \psi$. Since $(\psi \circ \phi)(c) = c$ for $c \in C$, the function $\psi \circ \phi$ is the identity on $\mathfrak{A}$ by Lemma 2.14(a). Similarly $\phi \circ \psi$ is the identity on $\mathfrak{B}$, so $\psi = \phi^{-1}$ and ϕ is an isomorphism.

The statement that ϕ is unique follows from the corresponding statement in Lemma 2.15.

Definition. Let A be a set. A *field of sets $\mathscr{S}$ on A* is a family of subsets of A such that

$$a \cap b,\ a \cup b,\ A \backslash a \in \mathscr{S} \quad \text{for } a, b \in \mathscr{S}.$$

With the set-theoretic operations $\cap$ and $\cup$ for $\wedge$ and $\vee$ and with $A \backslash a$ for a', a field of sets $\mathscr{S}$ on A is a Boolean algebra such that $\mathbb{0} = \emptyset$, $\mathbb{1} = A$, and $a \leq b$ if and only if $a \subset b$ for $a, b \in \mathscr{S}$. It is clear that $\mathscr{S}$ is a proper Boolean algebra if and only if $A \neq \emptyset$. For a cardinal number α, the family $\mathscr{P}(\alpha)$ of all subsets of α is a field of sets.

For a space X we denote by $\mathfrak{B}(X)$ the set of open-and-closed subsets of X.

It is clear that $\mathfrak{B}(X)$ is a field of sets (on the underlying set of X); and that $\mathfrak{B}(\alpha) = \mathscr{P}(\alpha)$ for a cardinal α (with the discrete topology).

We denote by $\mathbb{2}$ the field of sets $\mathscr{P}(1)$; thus $\mathbb{2}$ is the unique (up to isomorphism) Boolean algebra $\mathfrak{B}$ such that $|\mathfrak{B}| = 2$.

Definitions. Let $\mathfrak{B}$ be a Boolean algebra and $\emptyset \neq \mathscr{F} \subset \mathfrak{B}$.

(a) $\mathscr{F}$ has the *finite intersection property* if

$$\bigwedge_{k \leq n} b_k \neq \mathbb{0} \quad \text{for } n<\omega, \quad b_k \in \mathscr{F} \quad \text{for } k \leq n.$$

(b) $\mathscr{F}$ is a *filter of* $\mathfrak{B}$ if

(i) $a \wedge b \in \mathscr{F}$ for $a, b \in \mathscr{F}$, and

(ii) if $a \in \mathscr{F}$, $b \in \mathfrak{B}$ and $a \leq b$, then $b \in \mathscr{F}$.

A filter $\mathscr{F}$ is *proper* if $\mathscr{F} \neq \mathfrak{B}$ (*i.e.*, if $\mathbb{0} \notin \mathscr{F}$); the *improper* filter of $\mathfrak{B}$ is $\mathfrak{B}$.

(c) The filter of $\mathfrak{B}$ *generated by* $\mathscr{F}$, denoted $\langle \mathscr{F} \rangle$, is the intersection of the family of all filters of $\mathfrak{B}$ that contain $\mathscr{F}$. It is clear that

$\langle \mathscr{F} \rangle = \{b \in \mathfrak{B}$: there are $n<\omega$ and $b_k \in \mathscr{F}$ for $k \leq n$ such that $b_0 \wedge \cdots \wedge b_n \leq b\}$; and

$\langle \mathscr{F} \rangle$ is a proper filter if and only if $\mathscr{F}$ has the finite intersection property.

A filter of $\mathfrak{B}$ is *principal* if there is $a \in \mathfrak{B}$ such that the filter is generated by $\{a\}$.

(d) A *filter on (a cardinal)* α is a filter of $\mathscr{P}(\alpha)$. We note that a proper filter on α is a z-filter on α (with the discrete topology).

Filters and homomorphisms. (a) If

$$\phi: \mathfrak{A} \to \mathfrak{B}$$

is a Boolean algebra homomorphism we define

$$\mathscr{F} = \{a \in \mathfrak{A}: \phi(a) = \mathbb{1}\}.$$

It is easy to verify that $\mathscr{F}$ is a filter of $\mathfrak{B}$; it is called the *filter associated with* ϕ.

(b) Let $\mathfrak{B}$ be a Boolean algebra and let $\mathscr{F}$ be a filter of $\mathfrak{B}$. For $a, b \in \mathfrak{B}$ we set

$$a \underset{\mathscr{F}}{=} b \text{ if there is } c \in \mathscr{F} \text{ such that } a \wedge c = b \wedge c.$$

We verify that $\underset{\mathscr{F}}{=}$ is an equivalence relation on $\mathfrak{B}$. Indeed if $a, b, c \in \mathfrak{B}$ then

(i) $a \underset{\mathscr{F}}{=} a$ since $\mathbb{1} \in \mathscr{F}$;

(ii) $a \underset{\mathscr{F}}{=} b$ implies $b \underset{\mathscr{F}}{=} a$; and

(iii) if $a \underset{\mathscr{F}}{=} b$ and $b \underset{\mathscr{F}}{=} c$ then there are $d, e \in \mathscr{F}$ such that $a \wedge d = b \wedge d$ and $b \wedge e = c \wedge e$, and since $d \wedge e \in \mathscr{F}$ and $a \wedge (d \wedge e) = c \wedge (d \wedge e)$ we have $a \underset{\mathscr{F}}{=} c$.

We denote by $\mathfrak{B}/\mathscr{F}$ the set of $\underset{\mathscr{F}}{=}$-equivalence classes of $\mathfrak{B}$, and by $a/\mathscr{F}$ the $\underset{\mathscr{F}}{=}$-equivalence class of a (for $a \in \mathfrak{B}$).

If $a_0, b_0, a_1, b_1 \in \mathfrak{B}$ and $a_0 \underset{\mathscr{F}}{\equiv} b_0$ and $a_1 \underset{\mathscr{F}}{\equiv} b_1$ then

$$a_0 \wedge a_1 = b_0 \wedge b_1, \quad a_0 \vee a_1 = b_0 \vee b_1, \quad \text{and } a_0' = b_0'.$$

For there are $c_0, c_1 \in \mathscr{F}$ such that

$$a_0 \wedge c_0 = b_0 \wedge c_0 \quad \text{and } a_1 \wedge c_1 = b_1 \wedge c_1,$$

and then $c_0 \wedge c_1 \in \mathscr{F}$ and we have

$$\begin{aligned} (a_0 \wedge a_1) \wedge (c_0 \wedge c_1) &= (b_0 \wedge b_1) \wedge (c_0 \wedge c_1), \\ (a_0 \vee a_1) \wedge (c_0 \wedge c_1) &= (b_0 \vee b_1) \wedge (c_0 \wedge c_1), \quad \text{and} \\ a_0' \wedge c_0 = c_0 \backslash (a_0 \wedge c_0) &= c_0 \backslash (b_0 \wedge c_0) = b_0' \wedge c_0. \end{aligned}$$

It follows that the set $\mathfrak{B}/\mathscr{F}$, together with the (well-defined) operations given by

$$\begin{aligned} a_0/\mathscr{F} \wedge a_1/\mathscr{F} &= (a_0 \wedge a_1)/\mathscr{F}, \\ a_0/\mathscr{F} \vee a_0/\mathscr{F} &= (a_0 \vee a_1)/\mathscr{F} \quad \text{and} \\ (a_0/\mathscr{F})' &= a_0'/\mathscr{F}, \end{aligned}$$

is a Boolean algebra. This is called the *quotient Boolean algebra of* $\mathfrak{B}$ *modulo* $\mathscr{F}$. The function

$$\pi_{\mathscr{F}}: \mathfrak{B} \to \mathfrak{B}/\mathscr{F},$$

which is clearly a Boolean homomorphism of $\mathfrak{B}$ onto $\mathfrak{B}/\mathscr{F}$, is called the *canonical homomorphism associated with* $\mathscr{F}$.

The definitions and verifications above establish the following statements.

2.17. Lemma. (a) *Let* $\mathfrak{B}$ *be a Boolean algebra and let* $\mathscr{F}$ *and* $\mathscr{G}$ *be filters of* $\mathfrak{B}$. *Then* $\mathscr{F} \subset \mathscr{G}$ *if and only if there is a homomorphism* ψ *from* $\mathfrak{B}/\mathscr{F}$ *onto* $\mathfrak{B}/\mathscr{G}$ *such that* $\psi \circ \pi_{\mathscr{F}} = \pi_{\mathscr{G}}$.

(b) *Let* $\mathfrak{A}$ *and* $\mathfrak{B}$ *be Boolean algebras and let* ϕ *be a homomorphism of* $\mathfrak{A}$ *onto* $\mathfrak{B}$ *with associated filter* $\mathscr{F}$. *Then there is a unique isomorphism* $\psi: \mathfrak{A}/\mathscr{F} \to \mathfrak{B}$ *such that* $\psi \circ \pi_{\mathscr{F}} = \phi$.

Stone's Duality

The principal purpose of this sub-section is to establish the duality of Stone between Boolean algebras and compact totally disconnected spaces (as described in Theorem 2.21(d), Corollary 2.25, and 2.29 below). However, for the part of this duality leading to Corollary 2.25 (*i.e.*, for the part dealing with $\mathsf{S} \circ \mathfrak{B}$ rather than $\mathfrak{B} \circ \mathsf{S}$) we take the somewhat more general point of view in considering (arbitrary) spaces with a base of open-and-closed sets for their topology; this leads to Theorem 2.27, a two-valued analogue of Theorem 2.6.

Definition. Let $\mathfrak{B}$ be a Boolean algebra. An *ultrafilter* of $\mathfrak{B}$ is a proper filter not properly contained in any proper filter of $\mathfrak{B}$.

The fundamental fact concerning ultrafilters is the following statement, proved by a trivial application of Zorn's lemma.

2.18. Theorem. *Let $\mathfrak{B}$ be a Boolean algebra and let $\mathscr{F} \subset \mathfrak{B}$. If $\mathscr{F}$ has the finite intersection property then there is an ultrafilter p of $\mathfrak{B}$ such that $\mathscr{F} \subset p$.*

In particular *every proper filter of $\mathfrak{B}$ is contained in an ultrafilter of $\mathfrak{B}$.*

2.19. Theorem. *Let $\mathscr{F}$ be a proper filter of a Boolean algebra $\mathfrak{B}$. The following statements are equivalent.*

(a) *$\mathscr{F}$ is an ultrafilter;*

(b) *if $a, b \in \mathfrak{B}$ and $a \vee b \in \mathscr{F}$, then $a \in \mathscr{F}$ or $b \in \mathscr{F}$;*

(c) *if $a \in \mathfrak{B}$, then $a \in \mathscr{F}$ or $a' \in \mathscr{F}$;*

(d) $\mathfrak{B}/\mathscr{F}$ is (isomorphic to) $\mathbb{2}$.

Proof. (a)$\Rightarrow$(b). If $a \vee b \in \mathscr{F}$ and $a \notin \mathscr{F}$ then $\langle \mathscr{F} \cup \{a\} \rangle = \mathfrak{B}$ (since $\mathscr{F}$ is an ultrafilter) and there is $c \in \mathscr{F}$ such that $\mathbb{0} = a \wedge c$. We have

$$b \geq b \wedge c = \mathbb{0} \vee (b \wedge c) = (a \wedge c) \vee (b \wedge c) = (a \vee b) \wedge c \in \mathscr{F},$$

and hence $b \in \mathscr{F}$.

(b)$\Rightarrow$(c). For $a \in \mathfrak{B}$ we have $a \vee a' = \mathbb{1} \in \mathscr{F}$, and hence $a \in \mathscr{F}$ or $a' \in \mathscr{F}$ by (b).

(c)$\Rightarrow$(d). Let $\pi: \mathfrak{B} \to \mathfrak{B}/\mathscr{F}$ be the canonical homomorphism. Since $\mathscr{F}$ is a proper filter the Boolean algebra $\mathfrak{B}/\mathscr{F}$ is proper. If $a \in \mathfrak{B}$ and $\pi(a) \neq \mathbb{1}$ then $a \notin \mathscr{F}$ and hence $a' \in \mathscr{F}$ by (c); it follows that

$$\pi(a) = \pi(a'') = (\pi(a'))' = \mathbb{1}' = \mathbb{0}.$$

Hence $\mathfrak{B}/\mathscr{F}$ is (isomorphic to) $\mathbb{2}$.

(d)$\Rightarrow$(a). Let p be an ultrafilter of $\mathfrak{B}$ such that $\mathscr{F} \subset p$ and suppose there is $a \in p \backslash \mathscr{F}$. Then $\pi_{\mathscr{F}}(a) \neq \mathbb{1}$, so that $\pi_{\mathscr{F}}(a) = \mathbb{0}$ by (d) and hence $\pi_{\mathscr{F}}(a') = \mathbb{0}' = \mathbb{1}$. Then $a' \in \mathscr{F} \subset p$ and

$$\mathbb{0} = a \wedge a' \in p,$$

contradicting the fact that p is a proper filter.

The proof of the theorem is complete.

For a Boolean algebra $\mathfrak{B}$ we denote by $\mathsf{S}(\mathfrak{B})$ the set of all ultrafilters of $\mathfrak{B}$.

It is clear that $|\mathsf{S}(\mathfrak{B})| \leq 2^{|\mathfrak{B}|}$, and that $\mathsf{S}(\mathfrak{B}) = \emptyset$ if and only if $\mathfrak{B}$ is the improper Boolean algebra.

2.20. Lemma. *Let* $\mathfrak{B}$ *be a Boolean algebra and let* $\emptyset \neq T \subset \mathsf{S}(\mathfrak{B})$. *Define* $\phi: \mathfrak{B} \to \mathscr{P}(T)$ *by* $\phi(a) = \{p \in T : a \in p\}$. *Then*

(a) ϕ *is a Boolean algebra homomorphism; and*

(b) *if for every* $a \in \mathfrak{B}$ *such that* $a \neq \mathbb{0}$ *there is* $p \in T$ *such that* $a \in p$, *then* ϕ *is an embedding.*

Proof. (a) If $a, b \in \mathfrak{B}$ then

$$\begin{aligned}\phi(a \wedge b) &= \{p \in T : a \wedge b \in p\} = \{p \in T : a \in p\} \cap \{p \in T : b \in p\} \\ &= \phi(a) \cap \phi(b)\end{aligned}$$

and

$$\phi(a') = \{p \in T : a' \in p\} = \{p \in T : a \notin p\} = \mathscr{P}(T) \backslash \phi(a)$$

by Theorem 2.19.

(b) If $a, b \in \mathfrak{B}$ and $a \neq b$ then $(a \backslash b) \vee (b \backslash a) \neq \mathbb{0}$. There is $p \in T$ such that $(a \backslash b) \vee (b \backslash a) \in p$, and then from

$$p \in (\phi(a) \backslash \phi(b)) \cup (\phi(b) \backslash \phi(a))$$

it follows that $\phi(a) \neq \phi(b)$.

For a Boolean algebra $\mathfrak{B}$ we denote by $\psi_{\mathfrak{B}}$, or simply by ψ, the function from $\mathfrak{B}$ to $\mathscr{P}(\mathsf{S}(\mathfrak{B}))$ defined by

$$\psi(a) = \{p \in \mathsf{S}(\mathfrak{B}) : a \in p\}.$$

The *Stone topology* on $\mathsf{S}(\mathfrak{B})$ is the topology determined by the subbase $\psi[\mathfrak{B}]$. The set $\mathsf{S}(\mathfrak{B})$ with the Stone topology is called the *Stone space of (the Boolean algebra)* $\mathfrak{B}$; we denote this space by $\mathsf{S}(\mathfrak{B})$.

A space X is *totally disconnected* if for $p, q \in X$ and $p \neq q$ there is an open-and-closed subset of X containing p and not containing q.

2.21. Theorem. *Let* $\mathfrak{B}$ *be a Boolean algebra.*

(a) $\psi[\mathfrak{B}]$ *is a base for* $\mathsf{S}(\mathfrak{B})$;

(b) $\psi[\mathfrak{B}] \subset \mathfrak{B}(\mathsf{S}(\mathfrak{B}))$;

(c) $\mathsf{S}(\mathfrak{B})$ *is a compact, totally disconnected space; and*

(d) ψ *is a Boolean algebra isomorphism from* $\mathfrak{B}$ *onto* $\mathfrak{B}(\mathsf{S}(\mathfrak{B}))$.

Proof. We note that if $\mathfrak{B}$ is the improper Boolean algebra then $\mathsf{S}(\mathfrak{B}) = \emptyset$, ψ is an isomorphism from $\mathfrak{B}$ onto $\mathfrak{B}(\mathsf{S}(\mathfrak{B})) = \{\emptyset\}$, and the statements are obvious. We assume that $\mathfrak{B}$ is a proper Boolean algebra.

(a) We have $\psi(\mathbb{1}) = \mathsf{S}(\mathfrak{B})$, and

$$\psi(a) \cap \psi(b) = \psi(a \wedge b)$$

for $a, b \in \mathfrak{B}$ by Lemma 2.20.

(b) From the definition of the topology of $\mathsf{S}(\mathfrak{B})$, every element of $\psi[\mathfrak{B}]$ is open. Then from Lemma 2.20(a) it follows that

$$\mathsf{S}(\mathfrak{B}) \backslash \psi(a) = \psi(a') \in \mathsf{S}(\mathfrak{B})$$

for $a \in \mathfrak{B}$, so that $\psi[\mathfrak{B}] \subset \mathfrak{B}(\mathsf{S}(\mathfrak{B}))$.

(c) If $p, q \in \mathsf{S}(\mathfrak{B})$ and $p \neq q$ then $p \backslash q \neq \emptyset$ and there is $a \in p \backslash q$. It follows from (b) that $\psi(a)$ is an open-and-closed subset of $\mathsf{S}(\mathfrak{B})$ containing p and not containing q. Thus $\mathsf{S}(\mathfrak{B})$ satisfies the Hausdorff separation property.

Now let $A \subset \mathfrak{B}$ and suppose that $\{\psi(a): a \in A\}$ is a cover of $\mathsf{S}(\mathfrak{B})$ with no finite subcover. Then the family

$$\mathscr{F} = \{a' : a \in A\}$$

has the finite intersection property, and hence there is $p \in \mathsf{S}(\mathfrak{B})$ such that $\mathscr{F} \subset p$. Since $\{\psi(a): a \in A\}$ is a cover of $\mathsf{S}(\mathfrak{B})$ there is $a \in A$ such that $p \in \psi(a)$, *i.e.*, $a \in p$. It follows that

$$\mathbb{0} = a \wedge a' \in p,$$

a contradiction. Thus $\mathsf{S}(\mathfrak{B})$ is compact.

It is now clear, using the two paragraphs above and (b), that $\mathsf{S}(\mathfrak{B})$ is a compact, totally disconnected space.

(d) It follows from (b) and Lemma 2.20(b) that ψ is a Boolean algebra embedding of $\mathfrak{B}$ into $\mathfrak{B}(\mathsf{S}(\mathfrak{B}))$. We verify that $\psi[\mathfrak{B}] = \mathfrak{B}(\mathsf{S}(\mathfrak{B}))$. If $A \in \mathfrak{B}(\mathsf{S}(\mathfrak{B}))$ then by (a) there is $\{a_i : i \in I\} \subset \mathfrak{B}$ such that $A = \bigcup\{\psi(a_i): i \in I\}$; since A is compact there are $n < \omega$ and $i_k \in I$ for $k \leq n$ such that

$$A = \psi(a_0) \cup \cdots \cup \psi(a_n) = \psi(a_0 \vee \cdots \vee a_n) \in \psi[\mathfrak{B}].$$

The proof of the theorem is complete.

If X is a totally disconnected space and $p \in X$ then it is easily verified that

$$\{a \in \mathfrak{B}(X): p \in a\} \in \mathsf{S}(\mathfrak{B}(X)).$$

We define a function

$$h: X \to \mathsf{S}(\mathfrak{B}(X))$$

by $h(p) = \{a \in \mathfrak{B}(X): p \in a\}$.

2.22. Lemma. *Let X be a totally disconnected space and set $\psi = \psi_{\mathfrak{B}(X)}$. Then*

(a) *$h[X] \cap \psi(A) = h[A]$ for $A \in \mathfrak{B}(X)$;*
(b) *h is a continuous function;*
(c) *$\mathrm{cl}_{\mathsf{S}(\mathfrak{B}(X))} h[A] = \psi(A)$ for $A \in \mathfrak{B}(X)$;*
(d) *$h[X]$ is dense in $\mathsf{S}(\mathfrak{B}(X))$;*
(e) *$p = \{A \in \mathfrak{B}(X): p \in \mathrm{cl}_{\mathsf{S}(\mathfrak{B}(X))} h[A]\}$ for $p \in \mathsf{S}(\mathfrak{B}(X))$;*
(f) *$\mathrm{cl}_{\mathsf{S}(\mathfrak{B}(X))} h[A] \cap \mathrm{cl}_{\mathsf{S}(\mathfrak{B}(X))} h[B] = \mathrm{cl}_{\mathsf{S}(\mathfrak{B}(X))} h[A \cap B]$ for $A, B \in \mathfrak{B}(X)$; and*
(g) *if $\mathfrak{B}(X)$ is a base for X then h is a topological embedding.*

Proof. (a) is clear from the definitions of ψ and h.

(b) The family $\{h[X]\cap\psi(A): A\in\mathfrak{B}(X)\}$ is a base for $h[X]$, and from part (a) and the fact that h is one-to-one we have

$$h^{-1}(h[X]\cap\psi(A))=A$$

for $A\in\mathfrak{B}(X)$.

(c) For $A\in\mathfrak{B}(X)$ we have $\psi(A)\in\mathfrak{B}(\mathsf{S}(\mathfrak{B}(X)))$ by Theorem 2.21(b), and thus $\mathrm{cl}_{\mathsf{S}(\mathfrak{B}(X))}h[A]\subset\psi(A)$ by part (a). For the reverse inclusion let

$$p\in\mathsf{S}(\mathfrak{B}(X))\backslash\mathrm{cl}_{\mathsf{S}(\mathfrak{B}(X))}h[A].$$

By Theorem 2.21(a) there is $B\in\mathfrak{B}(X)$ such that $h[A]\subset\psi(B)$ and $p\notin\psi(B)$. We have $h[A]\subset h[B]$ by part (a), so from the fact that h is one-to-one it follows that $A\subset B$ and hence $\psi(A)\subset\psi(B)$. Thus $p\notin\psi(A)$.

(d) From part (c) we have

$$\mathrm{cl}_{\mathsf{S}(\mathfrak{B}(X))}h[X]=\psi(X)=\mathsf{S}(\mathfrak{B}(X)).$$

(e) is equivalent to (c).

(f) follows from (e).

(g) Since $\{\psi(A)\cap h[X]: A\in\mathfrak{B}(X)\}$ is a base for $h[X]$ and h is a one-to-one function, the statement follows from part (a).

Definition. A space X is *zero-dimensional* if for $A,B\in\mathscr{Z}(X)$ with $A\cap B=\emptyset$ there is $C\in\mathfrak{B}(X)$ such that $A\subset C$ and $B\cap C=\emptyset$.

2.23. Lemma. *Let X be a totally disconnected space and let $\overline{h}:\boldsymbol{\beta}X\to\mathsf{S}(\mathfrak{B}(X))$ be the Stone extension of (the continuous function) h. Then*

(a) $\overline{h}(p)=p\cap\mathfrak{B}(X)$ *for* $p\in\boldsymbol{\beta}X$;

(b) *if X is zero-dimensional then $\overline{h}$ is a homeomorphism from $\boldsymbol{\beta}X$ onto $\mathsf{S}(\mathfrak{B}(X))$; and*

(c) *if X is a P_{ω^+}-space (in particular, if X is discrete) then $\boldsymbol{\beta}X=\mathsf{S}(\mathfrak{B}(X))$ and $\overline{h}$ is the identity function on $\boldsymbol{\beta}X$.*

Proof. (a) It is clear that if $p\in\boldsymbol{\beta}X$ then $p\cap\mathfrak{B}(X)\in\mathsf{S}(\mathfrak{B}(X))$. Further if $A\in p\cap\mathfrak{B}(X)$ then $p\in\mathrm{cl}_{\boldsymbol{\beta}X}A$ and from the continuity of $\overline{h}$ we have

$$\overline{h}(p)\in\mathrm{cl}_{\mathsf{S}(\mathfrak{B}(X))}\overline{h}[A]=\mathrm{cl}_{\mathsf{S}(\mathfrak{B}(X))}h[A]=\psi(A)$$

by Lemma 2.22(c); hence $A\in\overline{h}(p)$. Thus $p\cap\mathfrak{B}(X)\subset\overline{h}(p)$, and since $p\cap\mathfrak{B}(X)$ is an ultrafilter of $\mathfrak{B}(X)$ it follows that $p\cap\mathfrak{B}(X)=\overline{h}(p)$.

(b) Since $\boldsymbol{\beta}X$ is compact we have $\overline{h}[\boldsymbol{\beta}X]=\mathsf{S}(\mathfrak{B}(X))$ by Lemma 2.22 and it is enough to prove that $\overline{h}$ is one-to-one. If $p,q\in\boldsymbol{\beta}X$ and $p\neq q$ then by Lemma 2.3(c) there are $A,B\in\mathscr{Z}(X)$ such that $A\in p$, $B\in q$, and $A\cap B=\emptyset$. Since X is zero-dimensional there are $C,D\in\mathfrak{B}(X)$ such that $A\subset C$, $B\subset D$, and $C\cap D=\emptyset$, and from part (a) we have $\overline{h}(p)\neq\overline{h}(q)$.

(c) It follows from Lemma 2.1(b) that if X is a P_{ω^+}-space then the elements of $\mathscr{Z}(X)$ are open-and-closed. Thus X is zero-dimensional and $\mathfrak{B}(X)=\mathscr{Z}(X)$, and from part (a) we have

$$p=p\cap\mathfrak{B}(X)=\bar{h}(p)$$

for $p\in\boldsymbol{\beta}X$.

2.24. Lemma. *If X is a compact space then the following statements are equivalent.*

(a) *X is zero-dimensional;*

(b) *$\mathfrak{B}(X)$ is a base for X; and*

(c) *X is totally disconnected.*

Furthermore if X is an infinite totally disconnected compact space then $\mathbf{w}(X)=|\mathfrak{B}(X)|$.

Proof. (a)$\Rightarrow$(b) is clear.

(b)$\Rightarrow$(c) is clear.

(c)$\Rightarrow$(a). Let $A,B\in\mathscr{Z}(X)$ with $A\cap B=\emptyset$. From (c) there is $A_{p,q}\in\mathfrak{B}(X)$ such that $p\in A_{p,q}$ and $q\notin A_{p,q}$ for $p\in A$, $q\in B$. The family $\{A_{p,q}: p\in A\}$ is an open cover of the compact set A, and thus there are $n(q)<\omega$ and $p_0,\dots,p_{n(q)}\in A$ such that

$$A\subset A_{p_0,q}\cup\cdots\cup A_{p_{n(q)},q}\quad\text{for } q\in B.$$

We set $B_q=X\setminus(A_{p_0,q}\cup\cdots\cup A_{p_{n(q)},q})$, so that $q\in B_q\in\mathfrak{B}(X)$ for $q\in B$. The family $\{B_q: q\in B\}$ is an open cover of B, and thus there are $n<\omega$ and $q_0,\dots,q_n\in B$ such that

$$B\subset B_{q_0}\cup\cdots\cup B_{q_n}.$$

We set $C=B_{q_0}\cup\cdots\cup B_{q_n}$. Then $C\in\mathfrak{B}(X)$, $B\subset C$ and $A\cap C=\emptyset$.

Assume now that X is infinite, compact and totally disconnected. It follows from (b) that $\mathbf{w}(X)\le|\mathfrak{B}(X)|$. For the reverse inequality let $\mathscr{C}$ be a base for X with $|\mathscr{C}|=\mathbf{w}(X)$. For $B\in\mathfrak{B}(X)$ the compactness of B implies that there is $\phi(B)\in\mathscr{P}_\omega(\mathscr{C})$ whose union is B. The resulting function

$$\phi:\mathfrak{B}(X)\to\mathscr{P}_\omega(\mathscr{C})$$

is one-to-one, and since $\mathscr{C}$ is infinite we have $|\mathfrak{B}(X)|\le|\mathscr{P}_\omega(\mathscr{C})|=|\mathscr{C}|=\mathbf{w}(X)$.

2.25. Corollary. *If X is a compact, totally disconnected space then h is a homeomorphism from X onto* $\mathsf{S}(\mathfrak{B}(X))$.

Proof. This follows from Lemmas 2.23(b) and 2.24.

2.26. Corollary. *Let X be a space. Then X is zero-dimensional if and only if $\boldsymbol{\beta}X$ is zero-dimensional.*

Proof. If X is zero-dimensional and $A,B\in\mathscr{Z}(\boldsymbol{\beta}X)$ and $A\cap B=\emptyset$ then there is a continuous function $f:\boldsymbol{\beta}X\to[0,1]$ such that $f(p)=0$ for $p\in A$ and $f(p)=1$ for $p\in B$. Set

$$E=\{p\in X:f(p)\leq 1/3\} \quad\text{and}\quad F=\{p\in X:f(p)\geq 2/3\}.$$

Since $E,F\in\mathscr{Z}(X)$ and $E\cap F=\emptyset$ there is $C\in\mathfrak{B}(X)$ such that $E\subset C$ and $F\cap C=\emptyset$. It is clear that $\mathrm{cl}_{\beta X}C\in\mathfrak{B}(\boldsymbol{\beta}X)$ and that $A\subset\mathrm{cl}_{\beta X}C$ and $B\cap\mathrm{cl}_{\beta X}C=\emptyset$.

(Alternatively we note that $\boldsymbol{\beta}X$ is homeomorphic to $\mathsf{S}(\mathfrak{B}(X))$ by Lemma 2.23(b), so that $\boldsymbol{\beta}X$ is zero-dimensional by Theorem 2.21(c) and Lemma 2.24.)

Now let $\boldsymbol{\beta}X$ be zero-dimensional and let $A,B\in\mathscr{Z}(X)$ and $A\cap B=\emptyset$. Then $\mathrm{cl}_{\beta X}A\cap\mathrm{cl}_{\beta X}B=\emptyset$ by Lemma 2.4 and from Urysohn's lemma there are $C,D\in\mathscr{Z}(\boldsymbol{\beta}X)$ such that $\mathrm{cl}_{\beta X}A\subset C$, $\mathrm{cl}_{\beta X}B\subset D$, and $C\cap D=\emptyset$. Since $\boldsymbol{\beta}X$ is zero-dimensional there is $E\in\mathfrak{B}(\boldsymbol{\beta}X)$ such that $C\subset E$ and $D\cap E=\emptyset$. Then $E\cap X\in\mathfrak{B}(X)$, $A\subset E\cap X$, and $B\cap(E\cap X)=\emptyset$.

2.27. Theorem. *Let X be a dense subset of a compact totally disconnected space Y. The following statements are equivalent.*

(a) *For every $f:X\to K$ with f continuous and K compact and totally disconnected there is a continuous function $g:Y\to K$ such that $f\subset g$.*

(b) *For every continuous function $f:X\to\{0,1\}$ there is a continuous function $g:Y\to\{0,1\}$ such that $f\subset g$.*

(c) *If $A,B\in\mathfrak{B}(X)$ and $A\cap B=\emptyset$, then $\mathrm{cl}_Y A\cap\mathrm{cl}_Y B=\emptyset$.*

(d) *If $A,B\in\mathfrak{B}(X)$ then $\mathrm{cl}_Y A\cap\mathrm{cl}_Y B=\mathrm{cl}_Y(A\cap B)$.*

(e) *The family $\{A\in\mathfrak{B}(X):p\in\mathrm{cl}_Y A\}$ is a proper filter of $\mathfrak{B}(X)$ for $p\in Y$.*

Furthermore the space $\mathsf{S}(\mathfrak{B}(X))$ satisfies these conditions, and for every space Y which satisfies these conditions there is a (unique) homeomorphism k from $\mathsf{S}(\mathfrak{B}(X))$ to Y such that $k(p)=p$ for $p\in X$.

Proof. (a)$\Rightarrow$(b) is trivial.

(b)$\Rightarrow$(c). Let $A,B\in\mathfrak{B}(X)$ and $A\cap B=\emptyset$. Let $f:X\to\{0,1\}$ be defined by $f(p)=0$ if $p\in A$ and $f(p)=1$ if $p\in X\setminus A$. From (b) there is a continuous function $g:Y\to\{0,1\}$ such that $f\subset g$; clearly $g(p)=0$ for $p\in\mathrm{cl}_Y A$ and $g(p)=1$ for $p\in\mathrm{cl}_Y B$.

(c)$\Rightarrow$(d). It is clear that $\mathrm{cl}_Y(A\cap B)\subset\mathrm{cl}_Y A\cap\mathrm{cl}_Y B$. If $p\in Y$ and $p\notin\mathrm{cl}_Y(A\cap B)$ then there is $C\in\mathfrak{B}(Y)$ such that $p\in C$ and $A\cap B\cap C=\emptyset$. Then $\mathrm{cl}_Y(A\cap C)\cap\mathrm{cl}_Y(B\cap C)=\emptyset$ by (c), and hence $p\notin\mathrm{cl}_Y A\cap\mathrm{cl}_Y B$.

(d)$\Rightarrow$(e). Set $\tilde{p}=\{A\in\mathfrak{B}(X):p\in\mathrm{cl}_Y A\}$ for $p\in Y$. It is clear that $p\neq\emptyset$ and $\emptyset\notin\tilde{p}$, and also that if $A,B\in\mathfrak{B}(X)$ with $A\in\tilde{p}$ and $A\subset B$, then $B\in\tilde{p}$. From (d) it follows that the intersection of any two elements of $\tilde{p}$ is an element of $\tilde{p}$. Thus $\tilde{p}$ is a proper filter of $\mathfrak{B}(X)$.

(e)$\Rightarrow$(a). If condition (a) fails then by Lemma 2.5(a) there are a compact totally disconnected space K, $p \in Y$, and a continuous function $f: X \to K$ such that f is not contained in a continuous function defined on $X \cup \{p\}$. Then by Lemmas 2.5(b) and 2.24 there are $C, D \in \mathfrak{B}(K)$ such that $C \cap D = \emptyset$ and $p \in \mathrm{cl}_Y f^{-1}(C) \cap \mathrm{cl}_Y f^{-1}(D)$. Since

$$f^{-1}(C), f^{-1}(D) \in \mathfrak{B}(X) \quad \text{and} \quad f^{-1}(C) \cap f^{-1}(D) = \emptyset,$$

the family $\{A \in \mathfrak{B}(X) : p \in \mathrm{cl}_Y A\}$ is not a filter of $\mathfrak{B}(X)$.

According to Lemma 2.22, $\mathsf{S}(\mathfrak{B}(X))$ is a compact, totally disconnected space containing (the homeomorphic image $h[X]$ of) X as a dense subset; further for $p \in \mathsf{S}(\mathfrak{B}(X))$ we have

$$\{A \in \mathfrak{B}(X) : p \in \mathrm{cl}_{\mathsf{S}(\mathfrak{B}(X))} h[A]\} = p$$

by Lemma 2.22(e), so $\mathsf{S}(\mathfrak{B}(X))$ satisfies condition (e) (for the space $h[X]$ rather than X) and hence all conditions (a)—(e) of the present theorem.

Identifying p with $h(p)$ for $p \in X$, suppose now that Y is a compact totally disconnected space containing X densely and satisfying conditions (a)—(e). It follows from condition (a) that there is a continuous function $k: \mathsf{S}(\mathfrak{B}(X)) \to Y$ such that $k(p) = p$ for $p \in X$ and there is a continuous function $g: Y \to \mathsf{S}(\mathfrak{B}(X))$ such that $g(p) = p$ for $p \in X$. Then $g \circ k(p) = p$ for $p \in X$, and since $g \circ k$ is continuous on $\mathsf{S}(\mathfrak{B}(X))$ and X is dense in $\mathsf{S}(\mathfrak{B}(X))$ we have $g \circ k(p) = p$ for $p \in \mathsf{S}(\mathfrak{B}(X))$. Thus k is a homeomorphism from $\mathsf{S}(\mathfrak{B}(X))$ onto Y.

The proof is complete.

2.28. Theorem. *Let X be a locally compact, zero-dimensional space, set*

$$\mathscr{F}(X) = \{A \in \mathfrak{B}(X) : X \setminus A \text{ is compact}\},$$

and for $A \in \mathfrak{B}(X)$ set $\hat{A} = (\mathrm{cl}_{\beta X} A) \cap (\beta X \setminus X)$.

(a) *If $A \in \mathfrak{B}(X)$ then $\hat{A} = \beta X \setminus X$ if and only if $A \in \mathscr{F}(X)$.*

(b) *If $C \in \mathfrak{B}(\beta X \setminus X)$ then there is $A \in \mathfrak{B}(X)$ such that $\hat{A} = C$.*

(c) *If $A, B \in \mathfrak{B}(X)$ then*

(i) *$(A \cap B)\hat{} = \hat{A} \cap \hat{B}$;*

(ii) *$\hat{A} \subset \hat{B}$ if and only if $A \setminus B$ is compact; and*

(iii) *$\hat{A} = \hat{B}$ if and only if there is $F \in \mathscr{F}(X)$ such that $A \cap F = B \cap F$.*

Thus the function $A \to \hat{A}$ is a Boolean algebra homomorphism from $\mathfrak{B}(X)$ onto $\mathfrak{B}(\beta X \setminus X)$ with associated filter $\mathscr{F}(X)$, and $\mathfrak{B}(\beta X \setminus X)$ is isomorphic to $\mathfrak{B}(X)/\mathscr{F}(X)$.

Proof. (a) For $A \in \mathfrak{B}(X)$ we have $\beta X \setminus X = \hat{A} \cup (X \setminus A)\hat{}$ and $\hat{A} \cap (X \setminus \hat{A}) = \emptyset$. Thus $\hat{A} = \beta X \setminus X$ if and only if $\mathrm{cl}_{\beta X}(X \setminus A) \subset X$; statement (a) follows.

(b) Since $\beta X \setminus X$ is a closed subset of βX by Lemma 2.9(b), the sets C and $(\beta X \setminus X) \setminus C$ are completely separated in βX by Urysohn's

Lemma. Since βX is zero-dimensional (Corollary 2.26) there is $B \in \mathfrak{B}(\beta X)$ such that $C \subset B$ and $((\beta X \setminus X) \setminus C) \cap B = \emptyset$. Set $A = B \cap X$. Then $A \in \mathfrak{B}(X)$ and $\mathrm{cl}_{\beta X} A = B$. Since $C \subset \hat{A}$ and $((\beta X \setminus X) \setminus C) \cap \hat{A} = \emptyset$ we have $\hat{A} = C$.

(c) (i) follows from Lemma 2.4 and Theorem 2.6.

(c) (ii) We note that if $C \in \mathfrak{B}(X)$ then $\hat{C} = \emptyset$ if and only if C is compact. If $A, B \in \mathfrak{B}(X)$ then $A \setminus B \in \mathfrak{B}(X)$ and

$$(A \setminus B)\hat{} = (A \cap (X \setminus B))\hat{} = \hat{A} \cap (X \setminus B)\hat{} = \hat{A} \setminus \hat{B}.$$

Thus $\hat{A} \subset \hat{B}$ if and only if $A \setminus B$ is compact.

(c)(iii) If $\hat{A} = \hat{B}$ then $(A \setminus B) \cup (B \setminus A)$ is compact by (c)(ii) and we set $F = X \setminus ((A \setminus B) \cup (B \setminus A))$. Then $F \in \mathscr{F}$, and $A \cap F = B \cap F$. Conversely if there is $F \in \mathscr{F}$ such that $A \cap F = B \cap F$, then

$$\hat{A} = \hat{A} \cap \hat{F} = (A \cap F)\hat{} = (B \cap F)\hat{} = \hat{B} \cap \hat{F} = \hat{B}$$

by parts (a) and (c)(i).

Definitions. (a) Let $\phi: \mathfrak{A} \to \mathfrak{B}$ be a Boolean algebra homomorphism. The function $\mathsf{S}(\phi)$ is defined on $\mathsf{S}(\mathfrak{B})$ by

$$\mathsf{S}(\phi)(p) = \{a \in \mathfrak{A} : \phi(a) \in p\}.$$

(b) Let $f: X \to Y$ be a continuous function between compact, totally disconnected spaces. The function $\mathfrak{B}(f)$ is defined on $\mathfrak{B}(Y)$ by

$$\mathfrak{B}(f)(B) = f^{-1}(B).$$

It is clear that

$$\mathsf{S}(\phi): \mathsf{S}(\mathfrak{B}) \to \mathsf{S}(\mathfrak{A}) \quad \text{and}$$
$$\mathfrak{B}(f): \mathfrak{B}(Y) \to \mathfrak{B}(X).$$

In the following theorem we use the notation just introduced, as well as the notation h and ψ to denote respectively the homeomorphism and the Boolean algebra isomorphism

$$h: X \to \mathsf{S}(\mathfrak{B}(X)) \quad \text{and} \quad \psi: \mathfrak{B} \to \mathfrak{B}(\mathsf{S}(\mathfrak{B})).$$

2.29. Theorem. *Let $\phi: \mathfrak{A} \to \mathfrak{B}$ be a Boolean algebra homomorphism and let $f: X \to Y$ be a continuous function between compact, totally disconnected spaces. Then*

(a) *$\mathsf{S}(\phi)$ is a continuous function;*

(b) *if ϕ is an embedding, then $\mathsf{S}(\phi)$ is an onto function; if ϕ is an onto function, then $\mathsf{S}(\phi)$ is one-to-one; and if $\mathfrak{A} = \mathfrak{B}$ and ϕ is the identity function, then $\mathsf{S}(\varphi)$ is the identity function.*

(c) *$\mathfrak{B}(f)$ is a Boolean algebra homomorphism;*

(d) *if f is a one-to-one function, then $\mathfrak{B}(f)$ is an onto function; if f is an onto function, then $\mathfrak{B}(f)$ is a one-to-one function; and if $X = Y$ and f is the identity function, then $\mathfrak{B}(f)$ is the identity function.*

(e) *the diagrams*

$$\begin{array}{ccc} \mathfrak{A} & \xrightarrow{\ \phi\ } & \mathfrak{B} \\ \downarrow{\scriptstyle\psi} & & \downarrow{\scriptstyle\psi} \\ \mathfrak{B}(\mathsf{S}(\mathfrak{A})) & \xrightarrow{\mathfrak{B}(\mathsf{S}(\phi))} & \mathfrak{B}(\mathsf{S}(\mathfrak{B})) \end{array} \quad \textit{and} \quad \begin{array}{ccc} X & \xrightarrow{\ f\ } & Y \\ \downarrow{\scriptstyle h} & & \downarrow{\scriptstyle h} \\ \mathsf{S}(\mathfrak{B}(X)) & \xrightarrow{\mathsf{S}(\mathfrak{B}(f))} & \mathsf{S}(\mathfrak{B}(Y)) \end{array}$$

commute;

(f) *if in addition* $\chi:\mathfrak{B}\to\mathfrak{C}$ *is a Boolean algebra homomorphism and* $g:Y\to Z$ *is a continuous function between compact, totally disconnected spaces, then*

$$\mathsf{S}(\chi\circ\phi)=\mathsf{S}(\phi)\circ\mathsf{S}(\chi) \quad \textit{and} \quad \mathfrak{B}(g\circ f)=\mathfrak{B}(f)\circ\mathfrak{B}(g).$$

Proof. (a) For every $a\in\mathfrak{A}$ we have

$$\begin{aligned}(\mathsf{S}(\phi))^{-1}(\psi(a)) &= \{p\in\mathsf{S}(\mathfrak{B}):\mathsf{S}(\phi)(p)\in\psi(a)\}=\{p\in\mathsf{S}(\mathfrak{B}):a\in\mathsf{S}(\phi)(p)\}\\ &=\{p\in\mathsf{S}(\mathfrak{B}):\phi(a)\in p\}=\psi(\phi(a))\in\mathfrak{B}(\mathsf{S}(\mathfrak{B})).\end{aligned}$$

The result follows since $\psi[\mathfrak{A}]$ is a subbase for the topology of $\mathsf{S}(\mathfrak{A})$.

(b) If ϕ is an embedding and $p\in\mathsf{S}(\mathfrak{A})$ then $\phi[p]=\{\phi(a):a\in p\}$ has the finite intersection property. Thus there is an ultrafilter q of $\mathfrak{B}$ such that $\phi[p]\subset q$ and then

$$p=\{a\in\mathfrak{A}:\phi(a)\in\phi[p]\}\subset\{a\in\mathfrak{A}:\phi(a)\in q\}=\mathsf{S}(\phi)(q).$$

Since $\{a\in\mathfrak{A}:\phi(a)\in q\}$ is a filter containing p, we have $p=\mathsf{S}(\phi)(q)$, and thus $\mathsf{S}(\phi)$ is an onto function.

If $p,p'\in\mathsf{S}(\mathfrak{B})$ and $p\neq p'$ there is $b\in\mathfrak{B}$ such that $b\in p$ and $b'\in p'$. If ϕ is an onto function there is $a\in\mathfrak{A}$ such that $\phi(a)=b$, and since $a\in\mathsf{S}(\phi)(p)$ and $a'\in\mathsf{S}(\phi)(p')$ we have $\mathsf{S}(\phi)(p)\neq\mathsf{S}(\phi)(p')$.

The last statement follows from the definition of $\mathsf{S}(\phi)$.

(c) This statement is easily verified.

(d) Let f be a one-to-one function and let $A\in\mathfrak{B}(X)$. Since f is a homeomorphism from X onto $f[X]$ there is an open subset G of Y such that $G\cap f[X]=f[A]$. For $p\in A$ there is $B_p\in\mathfrak{B}(Y)$ such that $f(p)\in B_p\subset G$. The family $\{B_p:p\in A\}$ is an open cover of the compact set $f[A]$ and thus there are $n<\omega$ and $p_k\in A$ for $k\leq n$ such that

$$f[A]=\bigcup_{k\leq n}B_{p_k}\cap f[X].$$

We set $B=\bigcup_{k\leq n}B_{p_k}$. Then $B\in\mathfrak{B}(Y)$ and $f[A]=B\cap f[X]$, so

$$\mathfrak{B}(f)(B)=f^{-1}(B)=A$$

since f is one-to-one.

Let f be an onto function and let $B \in \mathfrak{B}(Y)$ be such that $\mathfrak{B}(f)(B) = f^{-1}(B) = \emptyset$. Since f is an onto function it follows that $B = \emptyset$, and thus $\mathfrak{B}(f)$ is one-to-one.

The last statement follows from the definition of $\mathfrak{B}(f)$.

(e) As in the proof of part (a) we have

$$\mathfrak{B}(\mathsf{S}(\phi))(\psi(a)) = (\mathsf{S}(\phi))^{-1}(\psi(a)) = \psi(\phi(a))$$

for $a \in \mathfrak{A}$; and from the definition of the function h we have

$$\begin{aligned}\mathsf{S}(\mathfrak{B}(f))(h(p)) &= \{B \in \mathfrak{B}(Y) : \mathfrak{B}(f)(B) \in h(p)\} \\ &= \{B \in \mathfrak{B}(Y) : p \in \mathfrak{B}(f)(B)\} \\ &= \{B \in \mathfrak{B}(Y) : f(p) \in B\} = h(f(p))\end{aligned}$$

for $p \in X$.

(f) We have

$$\begin{aligned}\mathsf{S}(\chi \circ \phi)(p) &= (\chi \circ \phi)^{-1}(p) = \phi^{-1}(\chi^{-1}(p)) \\ &= (\mathsf{S}(\phi) \circ \mathsf{S}(\chi))(p)\end{aligned}$$

for $p \in \mathsf{S}(\mathfrak{C})$, and

$$\begin{aligned}\mathfrak{B}(g \circ f)(C) &= (g \circ f)^{-1}(C) = f^{-1}(g^{-1}(C)) \\ &= (\mathfrak{B}(f) \circ \mathfrak{B}(g))(C)\end{aligned}$$

for $C \in \mathfrak{B}(Z)$.

The proof is complete.

Theorem 2.21 (d), Corollary 2.25, and Theorem 2.29 describe *Stone's duality between Boolean algebras and compact totally disconnected spaces.* It allows us to identify a Boolean algebra $\mathfrak{B}$ with the field of sets $\mathfrak{B}(\mathsf{S}(\mathfrak{B}))$ and a compact totally disconnected space X with $\mathsf{S}(\mathfrak{B}(X))$. As a consequence of this identification, the Boolean algebra operations are replaced by the corresponding field of sets operations. We think of Stone's duality as expressing primarily the process of *geometrization* of the algebraic notions of Boolean algebra and Boolean algebra homomorphism.

In the language of categories this duality can be expressed as follows: Let $\mathbb{K}$ be the *category* of compact totally disconnected spaces and continuous functions, and denote by $\mathbb{K}^{\mathrm{opp}}$ the *opposite* category of $\mathbb{K}$ and let $\mathbb{B}$ be the category of Boolean algebras and Boolean algebra homomorphisms; then $\mathfrak{B}: \mathbb{K}^{\mathrm{opp}} \to \mathbb{B}$ and $\mathsf{S}: \mathbb{B} \to \mathbb{K}^{\mathrm{opp}}$ are *functors* such that $\mathfrak{B} \circ \mathsf{S}$ and $\mathsf{S} \circ \mathfrak{B}$ are *naturally equivalent* to the identity functors, and hence $\mathbb{B}$ and $\mathbb{K}^{\mathrm{opp}}$ are *equivalent* categories. We are not concerned in this work with categories, and thus we do not define or amplify these category-theoretic notions.

The Completion of a Boolean Algebra and the Gleason Space of a Compact Space

In this subsection we introduce the infinitary properties of a Boolean algebra that lead to the study of complete Boolean algebras and the definition and characteristic property (in terms of denseness) of the completion of a Boolean algebra. A simple generalization of the dual result leads to the definition and characteristic property (in terms of irreducible functions) of the Gleason space of an (arbitrary) compact space.

If $\mathfrak{B}$ is a Boolean algebra and $A \subset \mathfrak{B}$ then the *greatest lower bound of A in $\mathfrak{B}$*, denoted $\bigwedge A$, is that element b of $\mathfrak{B}$ such that

(i) $b \leq a$ for $a \in A$, and

(ii) if $c \in \mathfrak{B}$ and $c \leq a$ for $a \in A$, then $c \leq b$.

The *least upper bound of A in $\mathfrak{B}$*, denoted $\bigvee A$, is defined analogously.

It is clear that if $A \subset \mathfrak{B}$ then at most one element b of $\mathfrak{B}$ satisfies (i) and (ii). Hence $\bigwedge A$, if it exists, is uniquely defined in $\mathfrak{B}$.

We note also that if $\bigwedge A$ exists and $b \in \mathfrak{B}$ then $\bigwedge \{a \vee b : a \in A\}$ exists in $\mathfrak{B}$ and is $(\bigwedge A) \vee b$. For clearly $(\bigwedge A) \vee b \leq a \vee b$ for $a \in A$, and if $c \in \mathfrak{B}$ and $c \leq a \vee b$ for $a \in A$ then $b' \wedge c \leq a$ for $a \in A$ and hence $b' \wedge c \leq \bigwedge A$ and $c \leq (\bigwedge A) \vee b$.

Definitions. Let $\mathfrak{B}$ be a Boolean algebra.

(a) $\mathfrak{B}$ is *complete* if $\bigwedge A$ exists for $A \subset \mathfrak{B}$.

(b) $\mathfrak{B}$ is *injective* (in the category of Boolean algebras and Boolean algebra homomorphisms) if for every Boolean algebra $\mathfrak{C}$ and every homomorphism ϕ from a Boolean subalgebra $\mathfrak{A}$ of $\mathfrak{C}$ into $\mathfrak{B}$ there is a homomorphism $\psi : \mathfrak{C} \to \mathfrak{B}$ such that $\phi \subset \psi$.

(c) Let $\mathfrak{B}$ be a Boolean subalgebra of $\mathfrak{C}$. Then $\mathfrak{B}$ is a *retract* of $\mathfrak{C}$ if there is a homomorphism $\phi : \mathfrak{C} \to \mathfrak{B}$ such that $\phi(b) = b$ for $b \in \mathfrak{B}$.

The following result is proved by an argument of Hahn-Banach type.

2.30. Theorem *(Sikorski). Every complete Boolean algebra is injective.*

Proof. Let $\mathfrak{B}$ be a complete Boolean algebra and let ϕ be a homomorphism into $\mathfrak{B}$ from a subalgebra $\mathfrak{A}$ of a Boolean algebra $\mathfrak{C}$. We must prove that there is a homomorphism $\psi : \mathfrak{C} \to \mathfrak{B}$ such that $\phi \subset \psi$. We set

$$\mathscr{S} = \{\langle \mathfrak{D}, \psi \rangle : \mathfrak{A} \subset \mathfrak{D} \subset \mathfrak{C},\ \psi : \mathfrak{D} \to \mathfrak{B},\ \phi \subset \psi \text{ and } \psi \text{ is a homomorphism}\}.$$

We define a binary relation $\leq$ on $\mathscr{S}$ by the rule

$$\langle \mathfrak{D}_0, \psi_0 \rangle \leq \langle \mathfrak{D}_1, \psi_1 \rangle \quad \text{if } \mathfrak{D}_0 \subset \mathfrak{D}_1 \quad \text{and } \psi_0 \subset \psi_1 .$$

It is clear that $\langle \mathscr{S}, \leq \rangle$ is a partially ordered set satisfying the condition for the application of Zorn's lemma. Let $\langle \mathfrak{D}, \psi \rangle$ be a maximal element of $\langle \mathscr{S}, \leq \rangle$. It is enough to prove that $\mathfrak{D} = \mathfrak{C}$.

If $\mathfrak{D} \neq \mathfrak{C}$, then there is $c \in \mathfrak{C} \setminus \mathfrak{D}$ and we define

$$b_0 = \bigvee \{\psi(d) : d \in \mathfrak{D},\ d \leq c\} \quad \text{and}$$
$$b_1 = \bigwedge \{\psi(d) : d \in \mathfrak{D},\ c \leq d\}$$

and we note that since $\psi(d_0) \leq \psi(d_1)$ whenever $d_0 \leq c \leq d_1$ we have $b_0 \leq b_1$. We choose $b \in \mathfrak{B}$ such that $b_0 \leq b \leq b_1$; we note that if $d \in \mathfrak{D}$ and $d \wedge c = \mathbb{0}$ then $\psi(d) \wedge b = \mathbb{0}$; and if $d \in \mathfrak{D}$ and $d \wedge c' = \mathbb{0}$ then $\psi(d) \wedge b' = \mathbb{0}$. It follows that

$$\text{if } \bigwedge_{k \leq n} \varepsilon_k d_k \wedge \varepsilon_{n+1} c = \mathbb{0} \quad \text{then } \bigwedge_{k \leq n} \varepsilon_k \psi(d_k) \wedge \varepsilon_{n+1} b = \mathbb{0}$$

for $n < \omega$, $\varepsilon_k \in \{-1, 1\}$ for $k \leq n+1$, and $d_k \in \mathfrak{D}$ for $k \leq n$. Hence by Lemma 2.15 there is a homomorphism χ: $\langle\!\langle \mathfrak{D} \cup \{c\} \rangle\!\rangle \to \mathfrak{B}$ such that $\psi \subset \chi$. This contradicts the maximility of $\langle \mathfrak{D}, \psi \rangle$ in $\mathscr{S}$. The proof is complete.

2.31. Lemma. *Every retract of a complete Boolean algebra is complete.*

Proof. Let $\mathfrak{B}$ be a Boolean subalgebra of a complete Boolean algebra $\mathfrak{C}$, let $\phi: \mathfrak{C} \to \mathfrak{B}$ be a homomorphism such that $\phi(b) = b$ for $b \in \mathfrak{B}$, and let $A \subset \mathfrak{B}$. It is enough to prove that if $c = \bigwedge_{\mathfrak{C}} A$, then $\phi(c) = \bigwedge_{\mathfrak{B}} A$. Indeed $\phi(c) \leq \phi(a) = a$ for $a \in A$; and if $b \in \mathfrak{B}$ and $b \leq a$ for $a \in A$ then $b \leq c$ and hence $b = \phi(b) \leq \phi(c)$.

Definitions. (a) A space X is *extremally disconnected* if $\operatorname{cl}_X U$ is open for every open subset U of X.

(b) A compact space X is *projective* (in the category of compact spaces and continuous functions) if for every two compact spaces Y and Z, and for every two continuous functions $f: Y \to Z$ and $g: X \to Z$ with $f[Y] = Z$, there is a continuous function $h: X \to Y$ such that $f \circ h = g$.

(c) A subspace X of a space Y is a *retract* of Y if there is a continuous function $r: Y \to X$ such that $r(p) = p$ for $p \in X$.

2.32. Lemma. *A space X is extremally disconnected if and only if $\operatorname{cl}_X U \cap \operatorname{cl}_X V = \emptyset$ for every two disjoint open subsets U, V of X.*

Proof. If X is extremally disconnected and U and V are disjoint open subsets of X then $\operatorname{cl}_X U$ is an open subset of X disjoint from V. Hence $\operatorname{cl}_X U \cap \operatorname{cl}_X V = \emptyset$.

Conversely if U is an open subset of X and every two disjoint open sets have disjoint closures then

$$\operatorname{cl}_X U \cap \operatorname{cl}_X (X \setminus \operatorname{cl}_X U) = \emptyset .$$

Since $\operatorname{cl}_X U$ and $\operatorname{cl}_X (X \setminus \operatorname{cl}_X U)$ are disjoint, closed subsets of X whose union is X, each of these sets is open.

2.33. Theorem. *For every Boolean algebra $\mathfrak{B}$ the following statements are equivalent:*

(a) *$\mathfrak{B}$ is complete;*

(b) *$\mathfrak{B}$ is injective (in the category of Boolean algebras and Boolean algebra homomorphisms);*

(c) *$\mathfrak{B}$ is (isomorphic to) a retract of $\mathscr{P}(\alpha)$ for some cardinal α;*

(d) *$\mathsf{S}(\mathfrak{B})$ is extremally disconnected;*

(e) *$\mathsf{S}(\mathfrak{B})$ is projective (in the category of compact spaces and continuous functions); and*

(f) *$\mathsf{S}(\mathfrak{B})$ is (homeomorphic to) a retract of $\boldsymbol{\beta}(\alpha)$ for some cardinal α.*

Furthermore if K is a projective compact space then K is extremally disconnected and is (homeomorphic to) a space $\mathsf{S}(\mathfrak{B})$ with $\mathfrak{B}$ as above.

Proof. The equivalences will be established by proving implications as follows:

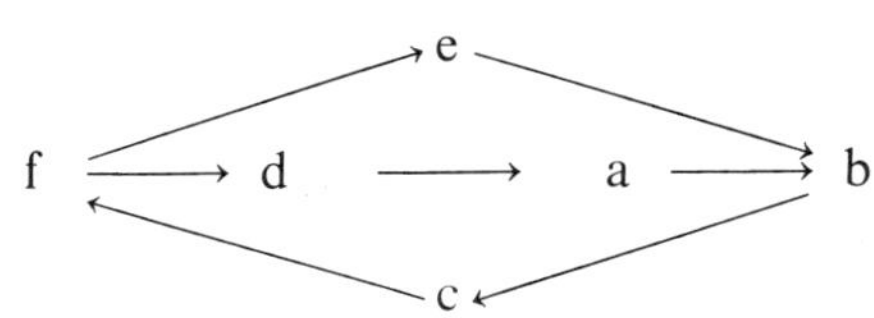

(a)$\Rightarrow$(b). This is Theorem 2.30.

(b)$\Rightarrow$(c). Let $\alpha = |\mathsf{S}(\mathfrak{B})|$. Since $\mathfrak{B}(\mathsf{S}(\mathfrak{B}))$ is injective and is a Boolean subalgebra of $\mathscr{P}(\alpha)$ the identity function $\mathrm{id}_{\mathfrak{B}(\mathsf{S}(\mathfrak{B}))}$ is contained in a homomorphism

$$\phi : \mathscr{P}(\alpha) \to \mathfrak{B}(\mathsf{S}(\mathfrak{B})).$$

Hence $\mathfrak{B}(\mathsf{S}(\mathfrak{B}))$ is a retract of $\mathscr{P}(\alpha)$.

(c)$\Rightarrow$(f). If $e : \mathfrak{B} \to \mathscr{P}(\alpha)$ is the inclusion function and $\phi : \mathscr{P}(\alpha) \to \mathfrak{B}$ is a retraction, then from Lemma 2.23(c) and (parts (a), (b) and (f) of) Theorem 2.29 it follows that

$$\mathsf{S}(e) : \mathsf{S}(\mathscr{P}(\alpha)) \to \mathsf{S}(\mathfrak{B})$$

is a retraction of $\boldsymbol{\beta}(\alpha)$ onto $\mathsf{S}(\mathfrak{B})$.

(f)$\Rightarrow$(d). It is clear that $\boldsymbol{\beta}(\alpha)$ is extremally disconnected. If $r : \boldsymbol{\beta}(\alpha) \to \mathsf{S}(\mathfrak{B})$ is a retraction and U and V are disjoint open subsets of $\mathsf{S}(\mathfrak{B})$ then since $U \subset r^{-1}(U)$ and $V \subset r^{-1}(V)$ we have

$$\mathrm{cl}_{\mathsf{S}(\mathfrak{B})} U \cap \mathrm{cl}_{\mathsf{S}(\mathfrak{B})} V = \mathrm{cl}_{\boldsymbol{\beta}(\alpha)} U \cap \mathrm{cl}_{\boldsymbol{\beta}(\alpha)} V \subset \mathrm{cl}_{\boldsymbol{\beta}(\alpha)} r^{-1}(U) \cap \mathrm{cl}_{\boldsymbol{\beta}(\alpha)} r^{-1}(V) = \emptyset .$$

It follows from Lemma 2.32 that $\mathsf{S}(\mathfrak{B})$ is extremally disconnected.

(d)$\Rightarrow$(a). It is easy to verify that if $\{U_i : i \in I\} \subset \mathfrak{B}(\mathsf{S}(\mathfrak{B}))$ then

$$\mathrm{cl}_{\mathsf{S}(\mathfrak{B})}\left(\bigcup_{i \in I} U_i\right) = \bigvee \{U_i : i \in I\} .$$

Thus $\mathfrak{B}(\mathsf{S}(\mathfrak{B}))$ is complete.

(f)⇒(e). Let r be a retraction of $\boldsymbol{\beta}(\alpha)$ onto $\mathsf{S}(\mathfrak{B})$ and let Y and Z be compact spaces and $f: Y \to Z$ and $g: \mathsf{S}(\mathfrak{B}) \to Z$ continuous functions with $f[Y]=Z$. For $\xi<\alpha$ let $k(\xi) \in f^{-1}(g(r(\xi)))$ (and denote by $\bar{k}: \boldsymbol{\beta}(\alpha) \to Y$ the Stone extension of k). We set $h=\bar{k} \mid \mathsf{S}(\mathfrak{B})$. Since $f \circ \bar{k}=g \circ r$ we have

$$(f \circ h)(p)=(f \circ \bar{k})(p)=(g \circ r)(p)=g(p)$$

for $p \in \mathsf{S}(\mathfrak{B})$.

(e)⇒(b). Let $\mathfrak{A}$ be a Boolean subalgebra of $\mathfrak{C}$ and let $\chi: \mathfrak{A} \to \mathfrak{C}$ and $\phi: \mathfrak{A} \to \mathfrak{B}$ denote respectively the inclusion function and a Boolean algebra homomorphism. Since $\mathsf{S}(\mathfrak{B})$ is projective there is a continuous function $h: \mathsf{S}(\mathfrak{B}) \to \mathsf{S}(\mathfrak{C})$ such that $\mathsf{S}(\chi) \circ h=\mathsf{S}(\phi)$. From Stone's duality we have $\mathfrak{B}(h): \mathfrak{C} \to \mathfrak{B}$ and

$$\phi=\mathfrak{B}(\mathsf{S}(\phi))=\mathfrak{B}(\mathsf{S}(\chi) \circ h)=\mathfrak{B}(h) \circ \mathfrak{B}(\mathsf{S}(\chi))=\mathfrak{B}(h) \circ \chi .$$

The proof that the statements (a)–(f) are equivalent is complete.

Now let K be a compact space which is projective (in the category of compact spaces and continuous functions), set $\alpha=|K|$ and let f be a continuous function from $\boldsymbol{\beta}(\alpha)$ onto K. Since K is projective there is a continuous function $h: K \to \boldsymbol{\beta}(\alpha)$ such that $f \circ h=\mathrm{id}_K$, and since h is one-to-one it is an embedding. Thus K is totally disconnected by Lemma 2.24, K is homeomorphic to $\mathsf{S}(\mathfrak{B}(K))$ by Corollary 2.25, and K is extremally disconnected by the equivalences just proved.

Definition. A subset U of a space X is *regular-open* if $U=\operatorname{int}_X \operatorname{cl}_X U$.

It is easy to see that a subset U of X is regular-open if and only if there is an open subset V of X such that $U=\operatorname{int}_X \operatorname{cl}_X V$.

We denote by $\mathfrak{R}(X)$ the set of regular-open subsets of a space X. If $U, V \in \mathfrak{R}(X)$ we define

$$\begin{aligned} U \wedge V &= U \cap V, \\ U \vee V &= \operatorname{int}_X \operatorname{cl}_X (U \cup V), \quad \text{and} \\ U' &= \operatorname{int}_X (X \setminus U) . \end{aligned}$$

It is to these definitions of $\wedge$, $\vee$ and $'$ that we refer in the next theorem.

2.34. Lemma. *If U and V are open subsets of a space X then*

$$\operatorname{int}_X \operatorname{cl}_X (U \cap V)=(\operatorname{int}_X \operatorname{cl}_X U) \cap (\operatorname{int}_X \operatorname{cl}_X V) .$$

Proof. It is clear that

$$\operatorname{int}_X \operatorname{cl}_X (U \cap V) \subset (\operatorname{int}_X \operatorname{cl}_X U) \cap (\operatorname{int}_X \operatorname{cl}_X V) .$$

For the reverse inclusion we note first that for every open subset W of X we have

$$(*)\qquad W\cap(\operatorname{int}_X \operatorname{cl}_X V)\subset \operatorname{int}_X \operatorname{cl}_X (W\cap V).$$

We take $W=\operatorname{int}_X \operatorname{cl}_X U$ in (*) and apply (*) (with U and V in place of V and W respectively) so that

$$\begin{aligned}(\operatorname{int}_X \operatorname{cl}_X U)\cap(\operatorname{int}_X \operatorname{cl}_X V)&\subset \operatorname{int}_X \operatorname{cl}_X[(\operatorname{int}_X \operatorname{cl}_X U)\cap V]\\ &\subset \operatorname{int}_X \operatorname{cl}_X[\operatorname{int}_X \operatorname{cl}_X (U\cap V)]=\operatorname{int}_X \operatorname{cl}_X (U\cap V).\end{aligned}$$

Definition. A subset C of a Boolean algebra $\mathfrak{B}$ is *dense* in $\mathfrak{B}$ if for $\mathbb{0}\neq b\in\mathfrak{B}$ there is $c\in C$ such that $\mathbb{0}\neq c\leq b$.

2.35. Theorem. *Let X be a space. Then*

(a) *$\mathfrak{R}(X)$ is a complete Boolean algebra such that $\bigvee\mathscr{A}=\operatorname{int}_X \operatorname{cl}_X(\bigcup\mathscr{A})$ for $\mathscr{A}\subset\mathfrak{R}(X)$;*

(b) *$\mathfrak{B}(X)$ is a Boolean subalgebra of $\mathfrak{R}(X)$; and*

(c) *if X is compact and totally disconnected then $\mathfrak{B}(X)$ is dense in $\mathfrak{R}(X)$.*

Proof. (a) It follows from Lemma 2.34 and the remarks which precede it that if $U, V\in\mathfrak{R}(X)$ then the three sets $U\wedge V$, $U\vee V$ and U' are elements of $\mathfrak{R}(X)$. The verification that $\mathfrak{R}(X)$ satisfies the definition of a Boolean algebra is tedious but routine. We note for example, using Lemma 2.34, that if $U, V, W\in\mathfrak{R}(X)$ then

$$\begin{aligned}U\wedge(V\vee W)&=U\cap[\operatorname{int}\operatorname{cl}(V\cup W)]=\operatorname{int}\operatorname{cl}U\cap[\operatorname{int}\operatorname{cl}(V\cup W)]\\ &=\operatorname{int}\operatorname{cl}[U\cap(V\cup W)]=\operatorname{int}\operatorname{cl}[(U\cap V)\cup(U\cap W)]\\ &=(U\wedge V)\vee(U\wedge W)\quad\text{and}\\ U\vee(V\wedge W)&=\operatorname{int}\operatorname{cl}[U\cup(V\cap W)]=\operatorname{int}\operatorname{cl}[(U\cup V)\cap(U\cup W)]\\ &=\operatorname{int}\operatorname{cl}(U\cup V)\cap\operatorname{int}\operatorname{cl}(U\cup W)\\ &=(U\vee V)\wedge(U\vee W).\end{aligned}$$

We leave to the reader the remaining verifications required in (a).

(b) Since every open-and-closed subset of X is regular-open, we have *as sets* the inclusion $\mathfrak{B}(X)\subset\mathfrak{R}(X)$. Let $\wedge_{\mathfrak{B}}$ and $\wedge_{\mathfrak{R}}$ denote the $\wedge$ function in $\mathfrak{B}(X)$ and $\mathfrak{R}(X)$ respectively, and let $\vee_{\mathfrak{B}}$, $\vee_{\mathfrak{R}}$, ${}'^{\mathfrak{B}}$ and ${}'^{\mathfrak{R}}$ be defined analogously. If $U, V\in\mathfrak{B}(X)$ then

$$\begin{aligned}&U\wedge_{\mathfrak{B}}V=U\cap V=U\wedge_{\mathfrak{R}}V,\\ &U\vee_{\mathfrak{B}}V=U\cup V=\operatorname{int}\operatorname{cl}(U\cup V)=U\vee_{\mathfrak{R}}V,\quad\text{and}\\ &\quad U'^{\mathfrak{B}}=X\setminus U=\operatorname{int}\operatorname{cl}(X\setminus U)=U'^{\mathfrak{R}};\end{aligned}$$

hence $\mathfrak{B}(X)$ is a Boolean subalgebra of $\mathfrak{R}(X)$.

(c) follows from Lemma 2.24.

Definition. Let $\mathfrak{B}$ be a Boolean algebra. The *completion* of $\mathfrak{B}$ (denoted $\hat{\mathfrak{B}}$) is the Boolean algebra $\mathfrak{R}(\mathsf{S}(\mathfrak{B}))$. The *canonical embedding* $\phi: \mathfrak{B} \to \mathfrak{R}(\mathsf{S}(\mathfrak{B}))$ is the composition

$$\mathfrak{B} \xrightarrow{\psi} \mathfrak{B}(\mathsf{S}(\mathfrak{B})) \subset \mathfrak{R}(\mathsf{S}(\mathfrak{B})).$$

2.36. Theorem. *Let $\mathfrak{B}$ be a Boolean algebra.*

(a) *$\hat{\mathfrak{B}}$ is a complete Boolean algebra and ϕ is an embedding of $\mathfrak{B}$ into $\hat{\mathfrak{B}}$ such that $\phi[\mathfrak{B}]$ is a dense subalgebra of $\hat{\mathfrak{B}}$.*

(b) *If $\mathfrak{C}$ is a complete Boolean algebra and ϕ' is an embedding of $\mathfrak{B}$ into $\mathfrak{C}$ such that $\phi'[\mathfrak{B}]$ is a dense subalgebra of $\mathfrak{C}$ then there is a Boolean isomorphism χ from $\hat{\mathfrak{B}}$ onto $\mathfrak{C}$ such that $\chi \circ \phi = \phi'$.*

(c) *$\mathfrak{B}$ is a complete Boolean algebra if and only if $\phi[\mathfrak{B}] = \hat{\mathfrak{B}}$.*

Proof. (a) The space $\mathsf{S}(\mathfrak{B})$ is compact and totally disconnected by Theorem 2.21 (c). Thus part (a) follows from Theorem 2.35.

(b) If $\mathfrak{C}$ and ϕ' satisfy the conditions of (b) then since by Theorem 2.30 $\mathfrak{C}$ is injective (in the category of Boolean algebras and Boolean algebra homomorphisms) and $\phi' \circ \phi^{-1}$ is a homomorphism from $\phi[\mathfrak{B}]$, a subalgebra of $\hat{\mathfrak{B}}$, into $\mathfrak{C}$, there is a homomorphism $\chi: \hat{\mathfrak{B}} \to \mathfrak{C}$ such that $\phi' \circ \phi^{-1} \subset \chi$. If $\hat{b} \in \hat{\mathfrak{B}}$ and $\hat{b} \neq \mathbb{0}$ then since $\phi[\mathfrak{B}]$ is dense in $\hat{\mathfrak{B}}$ there is $b \in \mathfrak{B}$ such that $\mathbb{0} \neq \phi(b) \leq \hat{b}$; it follows that

$$\mathbb{0} \neq \phi'(b) = \chi(\phi(b)) \leq \chi(\hat{b}).$$

Thus $\chi: \hat{\mathfrak{B}} \to \mathfrak{C}$ is an embedding. If $\mathbb{0} \neq c \in \mathfrak{C}$ then since $\phi'[\mathfrak{B}]$ is dense in $\mathfrak{C}$ there is $A \subset \mathfrak{B}$ such that $\bigvee \{\phi'(a): a \in A\} = c$; it follows easily that

$$\chi(\bigvee \{\phi(a): a \in A\}) = c.$$

Thus $\chi: \hat{\mathfrak{B}} \to \mathfrak{C}$ is an onto function.

(c) If $\mathfrak{B}$ is complete then by part (b) there is an isomorphism $\chi: \hat{\mathfrak{B}} \to \mathfrak{B}$ such that

$$\chi(\phi(b)) = b \quad \text{for } b \in \mathfrak{B}.$$

For $\hat{b} \in \hat{\mathfrak{B}}$ we have $\chi(\phi(\chi(\hat{b}))) = \chi(\hat{b})$ and therefore

$$\hat{b} = \phi(\chi(\hat{b})) \quad \text{for } \hat{b} \in \hat{\mathfrak{B}}.$$

The converse follows from part (a). The proof is complete.

2.37. Lemma. *Let X be a compact totally disconnected space and let f be a continuous function from X onto the (compact) space K. Then there are a compact totally disconnected space Y such that $\mathbf{w}(Y) = \mathbf{w}(K)$, and continuous functions g and h from X onto Y and from Y onto K respectively, such that $f = h \circ g$.*

Proof. Set $\alpha = w(K)$. If $\alpha < \omega$ the statement is obvious, so we assume that $\alpha \geq \omega$. There is by Lemma 2.12 a base $\{C_\xi : \xi < \alpha\}$ for K such that

$$C_\xi \text{ is a cozero-set of } K \text{ for } \xi < \alpha;$$

since $f^{-1}(C_\xi)$ is a cozero-set of X, there is by Lemmas 2.24 and 2.1 (c) a family $\{A_{\xi,n} : \xi < \alpha,\ n < \omega\}$ of subsets of X such that

$$\begin{aligned} &A_{\xi,n} \in \mathfrak{B}(X) \quad \text{for } \xi < \alpha,\ n < \omega \quad \text{and} \\ &f^{-1}(C_\xi) = \bigcup_{n<\omega} A_{\xi,n} \quad \text{for } \xi < \alpha. \end{aligned}$$

Let $\mathfrak{B}$ denote the Boolean subalgebra of $\mathfrak{B}(X)$ generated by $\{A_{\xi,n} : \xi < \alpha,\ n < \omega\}$, let e be the inclusion function of $\mathfrak{B}$ into $\mathfrak{B}(X)$, and set

$$Y = \mathsf{S}(\mathfrak{B}) \quad \text{and} \quad g = \mathsf{S}(e).$$

Since $w(Y) = |\mathfrak{B}| = \alpha$ and g is a continuous function from X onto Y by Lemma 2.24 and Stone's duality, it remains only to show that there is a continuous function h from Y onto K such that $f = h \circ g$.

We claim first that if $t \in Y$ then the function f is constant on the set $g^{-1}(\{t\})$. Indeed if $p, q \in g^{-1}(\{t\})$ and $f(p) \neq f(q)$ then there are $\xi, \zeta < \alpha$ such that

$$f(p) \in C_\xi, \qquad f(q) \in C_\zeta \quad \text{and} \quad C_\xi \cap C_\zeta = \emptyset$$

and hence there are $m, n < \omega$ such that

$$p \in A_{\xi,n}, \qquad q \in A_{\zeta,m}, \quad \text{and} \quad A_{\xi,n} \cap A_{\zeta,m} = \emptyset.$$

Then since

$$t = g(p) = \mathsf{S}(e)(p) = \{B \in \mathfrak{B} : e(B) \in p\} = \{B \in \mathfrak{B} : p \in B\}$$

we have $A_{\xi,n} \in t$ and similarly $A_{\zeta,m} \in t$; we now have the contradiction

$$\emptyset = A_{\xi,n} \cap A_{\zeta,m} \in t$$

and the claim is established.

We define $h: Y \to K$ by the condition

$$\{h(t)\} = f[g^{-1}(\{t\})] \quad \text{for } t \in Y.$$

Then $h \circ g = f$, $h[Y] = K$, and h is continuous because $g: X \to Y$ is a quotient function.

A continuous function $f: X \to Y$ is said to be *irreducible* (onto Y) if f is onto Y and there is no proper closed subset A of X such that $f|A$ is onto Y. It follows easily from Zorn's lemma that if X and Y are compact spaces and f is a continuous function from X onto Y, then there is a closed subspace F of X such that $f|F$ is irreducible.

2.38. Corollary. *For every compact space K there are a compact totally disconnected space Y and a continuous function* $h\colon Y \to K$ *such that*

$$w(Y) = w(K),$$
$$h[Y] = K, \quad \textit{and}$$
$$h \textit{ is irreducible.}$$

Proof. Set $\alpha = |K|$. There is a continuous function from $\boldsymbol{\beta}(\alpha)$ onto K, so by the remark following Lemma 2.37 there are a closed subset X of $\boldsymbol{\beta}(\alpha)$ and an irreducible continuous function f from X onto K. Let Y, g and h be the space and functions given by Lemma 2.37. Clearly g is an irreducible function from X onto Y, so it follows from the fact that f is irreducible that if A is a proper closed subspace of Y then

$$h[A] = (h \circ g)[g^{-1}(A)] = f[g^{-1}(A)] \subsetneqq K;$$

hence h is irreducible.

2.39. Lemma. *Let X and Y be compact totally disconnected spaces and let h be a continuous function from X onto Y. Then h is irreducible if and only if* $\mathfrak{B}(h)[\mathfrak{B}(Y)]$ *is a dense Boolean subalgebra of* $\mathfrak{B}(X)$.

Proof. We note that $\mathfrak{B}(h)$ is an embedding of $\mathfrak{B}(Y)$ into $\mathfrak{B}(X)$ by Theorem 2.29 (d).

If h is irreducible then for $\emptyset \neq A \in \mathfrak{B}(X)$ there is $B \in \mathfrak{B}(Y)$ such that

$$\emptyset \neq B \subset Y \setminus h[X \setminus A]$$

and hence $\emptyset \neq \mathfrak{B}(h)(B) \subset A$; thus $\mathfrak{B}(h)[\mathfrak{B}(Y)]$ is dense in $\mathfrak{B}(X)$.

Conversely if $\mathfrak{B}(h)[\mathfrak{B}(Y)]$ is dense in $\mathfrak{B}(X)$ then for $\emptyset \neq A \in \mathfrak{B}(X)$ there is $B \in \mathfrak{B}(Y)$ such that

$$\emptyset \neq \mathfrak{B}(h)(B) \subset A$$

and hence $\emptyset \neq B \subset Y \setminus h[X \setminus A]$; thus h is irreducible.

Definition. Let X be a compact space and $q \in X$. Then

$$\mathscr{F}_q = \{A \in \mathfrak{R}(X) : q \in A\}.$$

If $U, V \in \mathfrak{R}(X)$ then $U \cap V \in \mathfrak{R}(X)$. Hence $\mathscr{F}_q$ is a filter of $\mathfrak{R}(X)$ for $q \in X$. It is clear also that

$$\bigcap \{\mathrm{cl}_X A : A \in \mathscr{F}_q\} = \{q\}$$

for $q \in X$.

2.40. Lemma. *Let X be a compact space and let* $p \in \mathsf{S}(\mathfrak{R}(X))$. *Then there is a unique* $q \in X$ *such that* $\mathscr{F}_q \subset p$.

Proof. Since $\{\mathrm{cl}_X A : A \in p\}$ has the finite intersection property there is $q \in \bigcap\{\mathrm{cl}_X A : A \in p\}$. Then $\mathscr{F}_q \cup p$ has the finite intersection property and since p is an ultrafilter of $\mathfrak{R}(X)$ we have $\mathscr{F}_q \subset p$. If $q' \in X$ and $q \neq q'$ then $\mathscr{F}_q \cup \mathscr{F}_{q'}$ does not have the finite intersection property; hence the relation

$$\mathscr{F}_q \cup \mathscr{F}_{q'} \subset p$$

is impossible.

Lemma 2.40 shows that the function π below is well-defined.

Definition. Let X be a compact space. The *Gleason space* of X (sometimes denoted $\mathsf{G}(X)$) is the space $\mathsf{S}(\mathfrak{R}(X))$. The *canonical projection* $\pi : \mathsf{G}(X) \to X$ is defined by the rule

$$\{\pi(p)\} = \bigcap\{\mathrm{cl}_X A : A \in p\}.$$

2.41. Theorem. *Let X be a compact space.*

(a) *$\mathsf{G}(X)$ is a compact extremally disconnected space and π is a continuous irreducible function from $\mathsf{G}(X)$ onto X.*

(b) *If G is a compact extremally disconnected space and π' is a continuous irreducible function from G onto X then there is a homeomorphism k from $\mathsf{G}(X)$ onto G such that $\pi' \circ k = \pi$.*

(c) *X is extremally disconnected if and only if π is a homeomorphism.*

Proof. (a) $\mathfrak{R}(X)$ is a complete Boolean algebra by Theorem 2.35(a); hence $\mathsf{G}(X)$ is extremally disconnected by Theorem 2.33.

For $q \in X$ the family $\mathscr{F}_q$ is a filter of $\mathfrak{R}(X)$; hence there is $p \in \mathsf{S}(\mathfrak{R}(X))$ such that $\mathscr{F}_q \subset p$. It follows from Lemma 2.40 that $\pi(p) = q$. Thus π is an onto function.

Since $\psi[\mathfrak{R}(X)]$ is a base for $\mathsf{G}(X)$ by Theorem 2.21(a) (where

$$\psi(A) = \{p \in \mathsf{G}(X) : A \in p\}),$$

to prove that π is continuous it is enough to verify that

(*) if $A \in \mathfrak{R}(X)$, then $\pi[\psi(A)] \subset \mathrm{cl}_X A$.

Let $q \in \pi[\psi(A)]$; there is $p \in \psi(A)$ such that $q = \pi(p)$, and since $A \in p$ we have

$$q = \pi(p) \in \bigcap\{\mathrm{cl}_X B : B \in p\} \subset \mathrm{cl}_X A;$$

statement (*) follows. If π is not irreducible there are $A \in \mathfrak{R}(X)$, $q \in X$ and $p \in \mathsf{G}(X)$ such that $q \in A$, $p \in \mathsf{G}(X) \backslash \psi(A)$, and $\pi(p) = q$. Since $\psi(A)$ is an open-and-closed subset of $\mathsf{G}(X)$ there is $B \in \mathfrak{R}(X)$ such that $p \in \psi(B)$ and $\psi(B) \cap \psi(A) = \emptyset$. It follows easily that $A \cap B = \emptyset$, hence (since A is open in X) that $A \cap \mathrm{cl}_X B = \emptyset$. From (*) we have

$$q = \pi(p) \in A \cap \pi[\psi(B)] \subset A \cap \mathrm{cl}_X B,$$

a contradiction completing the proof of (a).

(b) By Corollary 2.38 there are a compact totally disconnected space Y and a continuous, irreducible function h from Y onto X, and since $\mathsf{G}(X)$ and G are projective (by Theorem 2.33) there are continuous functions g and g' from $\mathsf{G}(X)$ and G respectively to Y such that $\pi = h \circ g$ and $\pi' = h \circ g'$. It is clear that g and g' are irreducible functions (from $\mathsf{G}(X)$ and G respectively onto Y). Again because $\mathsf{G}(X)$ is projective there is a continuous function $k: \mathsf{G}(X) \to G$ such that $g = g' \circ k$, and again it is clear that k is an irreducible function onto G; further, we have

$$\pi = h \circ g = h \circ g' \circ k = \pi' \circ k .$$

Since k is irreducible the algebra $\mathfrak{B}(k)[\mathfrak{B}(G)]$ is dense in $\mathfrak{B}(\mathsf{G}(X))$ by Lemma 2.39, and hence

$$\mathfrak{B}(k)[\mathfrak{B}(G)] = \mathfrak{B}(\mathsf{G}(X))$$

by Theorem 2.36(c). Thus k is a homeomorphism (from $\mathsf{G}(X)$ onto G) by Theorem 2.29(b).

(c) follows from parts (a) and (b).

It follows from Theorem 2.41 that a continuous irreducible function between compact extremally disconnected spaces is a homeomorphism. We note a consequence of this statement: *If f is a continuous irreducible function from a compact space X onto a (compact) extremally disconnected space Y, then f is a homeomorphism.* For if $\pi: \mathsf{G}(X) \to X$ is the canonical projection then $f \circ \pi: \mathsf{G}(X) \to Y$ is irreducible and hence a homeomorphism; thus π and f are homeomorphisms.

Definition. Let X be a space.

A *pseudobase* for X is a family $\mathscr{C}$ of non-empty open subsets of X such that for every non-empty open subset V of X there is $C \in \mathscr{C}$ such that $C \subset V$. The smallest cardinal number that is the cardinal of a pseudobase for X is the *pseudoweight* of X and is denoted $\boldsymbol{pw}(X)$.

A subset D of X is *dense in X* if $V \cap D \neq \emptyset$ for every non-empty open subset V of X. The smallest cardinal number that is the cardinal of a dense subset of X is the *density character* of X and is denoted $\boldsymbol{d}(X)$.

2.42. Theorem. *Let X be a compact space. Then*

(a) $\boldsymbol{pw}(X) = \boldsymbol{pw}(\mathsf{G}(X))$ *and*

(b) $\boldsymbol{d}(X) = \boldsymbol{d}(\mathsf{G}(X))$.

Proof. By Theorem 2.41(a) the canonical projection $\pi: \mathsf{G}(X) \to X$ is a continuous irreducible function from $\mathsf{G}(X)$ onto X.

(a) If $\mathscr{C}$ is a pseudobase for $\mathsf{G}(X)$ then $\{\mathrm{int}_X \pi[C]: C \in \mathscr{C}\}$ is a pseudobase for X. If $\mathscr{D}$ is a pseudobase for X then $\{\pi^{-1}(D): D \in \mathscr{D}\}$ is a pseudobase for $\mathsf{G}(X)$.

(b) If D is dense in $\mathsf{G}(X)$ then $\pi[D]$ is dense in X. If E is dense in X and D is a subset of $\mathsf{G}(X)$ such that ($|D|=|E|$ and) $\pi[D]=E$ then D is dense in $\mathsf{G}(X)$.

Notes for § 2

The concepts of C-embedding, C^*-embedding and zero-sets were systematically studied in Hewitt [48]. For information relevant to the subsection on topology the reader is referred to Gillman-Jerison [60].

Tychonoff [30] proved that every space can be embedded in a product of closed intervals. Čech [37] noted that the compactification resulting from Tychonoff's embedding satisfies the C^*-embedding property of Theorem 2.6 which is characteristic of the Stone-Čech compactification. Stone [37a] (Theorems 78, 79, 88) independently defined the Stone-Čech compactification of a space X as the maximal ideal space of the ring $C(X)$. Additional information on the various equivalent definitions of the Stone-Čech compactification is given in Gillman-Jerison [60] (see especially Theorem 6.5 and Notes), which we follow (with some simplifications) in Theorem 2.6. Thus the Stone-Čech compactification *is* the set of all z-ultrafilters on X with a suitable topology.

Lemma 2.5 (a) is from Bourbaki-Dieudonné [39]; see also McDowell [58] (Proposition 3.2). Corollaries 2.7 and 2.11 are in Tychonoff [30], and the formulation of Urysohn's extension theorem (Corollary 2.8) follows Gillman-Jerison [60] (Theorem 1.17).

Our treatment of Boolean algebras generally follows Sikorski [64], to which the reader is referred for an extensive bibliography. In particular the (redundant) set of axioms defining a Boolean algebra is from Sikorski [64] (page 3).

The concept of an ultrafilter on a set was apparently enunciated first by F. Riesz [09], in an address that unfortunately did not receive at the time the attention it deserved.

Ulam [29] in measure-theoretic language proved the existence of a non-principal ultrafilter on ω by a transfinite induction involving a well-ordering of all subsets of ω. Theorem 2.18 for the field of all subsets of a set is due to Tarski [30a], who was also motivated by abstract measure-theoretic considerations (see Theorem 7.1 below), and to Stone [36] (Theorems 63 and 64) in general. The significant applications of filters and ultrafilters to topology were made independently by Cartan [37a], [37b] and Stone [36], [37a].

The fundamental duality of Stone (Theorem 2.21 (d), Corollary 2.25 and Theorem 2.29) is developed in Stone [34], [36], [37a] (§§ 1, 2, especially Theorems 1, 2, 5).

In contrast to the case for z-filters, the improper filter of a Boolean algebra is considered to be a filter. This slight asymmetry in our conventions is due to our wish on the one hand to avoid improper filters whenever possible (*e.g.*, in the case of z-filters), and on the other to avoid introducing unnecessary difficulties in the statement of Stone's duality.

Lemma 2.23 (b) is related to Pierce [61] (Theorem 1.7.3).

Theorem 2.30 is due to Sikorski [48]. (We note that Theorem 2.18 is a special case, $\mathfrak{B}=\mathbb{2}$.) Its method of proof follows the classical theorem of Hahn [27] and Banach [29] (cf. also Baer [40]). This similarity in the method of proof is not accidental, as is seen from the characterizations of "injective" normed spaces given by Goodner [50], Nachbin [50] and Kelley [52] in terms of extremally disconnected spaces (introduced and studied by Stone [37 b] (§ 4) and Hewitt [43]).

In Theorem 2.33, the equivalences (a)⇔(d), (b)⇔(e) and (c)⇔(f) are easily proved, (d)⇔(e) is due to Gleason [58], and (e)⇔(f) to Rainwater [59].

The existence and uniqueness of the completion of a Boolean algebra (Theorem 2.36) are given by MacNeille [37] (9.8—9.10) and Sikorski [50] (Theorem 3.6) and [64] (Theorem 35.2). The Gleason space is defined (directly, by a method essentially dual to that of Theorem 2.41) by Gleason [58]; another definition, together with an extensive bibliography, is given by Hager [71].

Lemma 2.37 and its Corollary 2.38 is from Negrepontis [69 a] (Lemmas 2.1 and 2.5). The proof of the (existence and) uniqueness of the Gleason space (Theorem 2.41), derived here from the (existence and) uniqueness of the completion of Boolean algebras (Theorem 2.36), Stone's duality and Lemma 2.37, follows the suggestion of Negrepontis [69 a] (§ 4). For a discussion of the relation between Sikorski's theorem, Gleason's theorem, and other related results, see also Halmos [59].

We thank Professor M. H. Stone for historical remarks on the origin and applications of filters (letter to the authors, December, 1972).

§ 3. Intersection Systems and Families of Large Oscillation

Several key concepts of set-theoretic and topological type are introduced in this section. A combinatorial theorem of Erdős-Rado on intersection systems based on Dedekind's pigeon-hole principle characterizes those pairs of cardinals α, κ (with α regular and $\omega \leq \kappa < \alpha$) for which every family of α sets, each with fewer than κ elements, has a subfamily with α elements every two of which have the same intersection (Theorem 3.2). This result gives estimates for the Souslin number of certain product spaces with various topologies (3.8—3.11). The Erdős-Tarski theorem on the regularity of the Souslin number is Theorem 3.5. Some basic results of the Erdős-Rado partition calculus, developed below in § 8, are applied here (Theorems 3.13 and 3.14) to estimate the Souslin number of certain spaces.

Apart from their intrinsic interest, however, these results are significant to us only as they relate to the vital concept of a family of large oscillation (Theorem 3.16), studied in all essential generality by Hausdorff in 1936. Though simple applications of the existence of such families have been known for some time, it was only recently that Kunen discovered the deeper applications to the existence of ultrafilters with refined properties without any use of the generalized continuum hypothesis.

The long list of applications of the existence of such a family of large oscillation in connection with ultrafilter "constructions" include: the computation of the number of ultrafilters (Theorem 7.3 and its Corollaries 7.4 and 7.8); the proof of existence of many (uniform) ultrafilters with large local character (Corollary 7.15); the proof of the existence of many good ultrafilters not comparable in the Rudin-Keisler order (Theorem 10.4); Theorem 10.9 and its Corollaries; the application given in Corollary 12.4; the exploration of the fundamental saturation properties of ultraproducts given by Shelah's theorem (Theorem 13.19).

Various topological consequences and refinements of Theorem 3.16 are given at the end of the section.

Intersection Systems and the Souslin Number

Definition. Let α and κ be cardinals. Then α is *strongly κ-inaccessible* if $\beta^\lambda < \alpha$ whenever $\beta < \alpha$ and $\lambda < \kappa$.

We note that if $\alpha > 2$ and α is strongly κ-inaccessible then $\kappa \le \alpha$; an infinite regular cardinal α is strongly inaccessible if and only if α is strongly α-inaccessible; and every infinite cardinal is strongly ω-inaccessible.

3.1. Lemma. (a) *If $\omega \le \alpha$ and $\kappa \le \mathrm{cf}(\alpha)$ then α is strongly κ-inaccessible if and only if the following condition is satisfied:*

$$\text{if } \lambda < \kappa \quad \text{and } \beta_i < \alpha \quad \text{for } i < \lambda \quad \text{then } \textstyle\prod_{i<\lambda} \beta_i < \alpha .$$

(b) *If $\omega \le \alpha$ then α^+ is strongly κ-inaccessible if and only if $\alpha = \alpha^{\underline{\kappa}}$.*

Proof. (a) Let α be strongly κ-inaccessible, and let $\lambda < \kappa$ and $\beta_i < \alpha$ for $i < \lambda$. We set $\beta = \sum_{i<\lambda} \beta_i$ and from the fact that $\lambda < \kappa \le \mathrm{cf}(\alpha)$ we have $\beta < \alpha$. Thus

$$\textstyle\prod_{i<\lambda} \beta_i \le \beta^\lambda < \alpha .$$

For the converse we apply condition (a) with $\beta_i = \beta$ for $i < \lambda$.

(b) If α^+ is strongly κ-inaccessible and $\lambda < \kappa$ then $\alpha^\lambda < \alpha^+$, so $\alpha^\lambda = \alpha$ and we have $\alpha^{\underline{\kappa}} = \sum_{\lambda<\kappa} \alpha^\lambda \le \alpha \cdot \kappa = \alpha$. If $\alpha = \alpha^{\underline{\kappa}}$ and $\lambda < \kappa$ and $\beta < \alpha^+$ then $\beta^\lambda \le \alpha^\lambda \le \alpha^{\underline{\kappa}} = \alpha$.

One of the two central results of this section is the following theorem of Erdős and Rado from the theory of "infinitary combinatorics".

3.2. Theorem *(Erdős-Rado). Let α be an infinite regular cardinal and let $\kappa < \alpha$. The following are equivalent.*

(a) *α is strongly κ-inaccessible;*

(b) *for every function $f: \alpha \to \mathscr{P}_\kappa(\alpha)$ there are a set $J \in \mathscr{P}_\kappa(\alpha)$ and $A \in \mathscr{P}(\alpha)$ such that*

$$|A| = \alpha \quad \text{and}$$
$$f(\xi) \cap f(\xi') = J \quad \text{for } \xi < \xi' < \alpha \quad \text{and } \xi, \xi' \in A .$$

Proof. (a)$\Rightarrow$(b). We give two proofs of this implication.

First proof. Since α is a regular cardinal there is an ordinal $\rho < \kappa$ such that

$$|\{\xi < \alpha : f(\xi) \text{ is order-isomorphic to } \rho\}| = \alpha .$$

Without any loss of generality we assume that $f(\xi)$ is order-isomorphic to ρ for all $\xi < \alpha$. Thus there is a unique order-isomorphism $e_\xi : \rho \to f(\xi)$ for $\xi < \alpha$. Set

$$\eta_i = \sup\{e_\xi(i) : \xi < \alpha\} \quad \text{for } i < \rho, \quad \text{and}$$
$$\eta = \sup\{\eta_i : i < \rho\} .$$

It is clear that $\eta_i \leq \eta_j$ for $i<j<\rho$. There are two possibilities to consider.

Case 1. $\eta<\alpha$.
Thus $f(\xi)\subset\eta+1$ for $\xi<\alpha$ and hence

$$f:\alpha\to[\eta+1]^{|\rho|}.$$

Since $|\eta+1|<\alpha$ and $|\rho|<\kappa$ and α is strongly inaccessible, we have $|[\eta+1]^{|\rho|}|=|\eta+1|^{|\rho|}<\alpha$. Since α is a regular cardinal it follows that there are a set $J\in[\eta+1]^{|\rho|}$ and $A\in\mathscr{P}(\alpha)$ such that

$$|A|=\alpha \quad \text{and}$$
$$|\{\xi<\alpha: f(\xi)=J\}|=\alpha.$$

Then certainly the requirements of the theorem are satisfied.

Case 2. $\eta=\alpha$.
Since $\kappa<\alpha=\mathrm{cf}(\alpha)$ there is $i<\rho$ such that $\eta_i=\alpha$. We set

$$\bar{i}=\min\{i<\rho:\eta_i=\alpha\}.$$

Thus $\sup\{e_\xi(\bar{i}):\xi<\alpha\}=\alpha$ and there is a one-to-one, order-preserving function $\phi:\alpha\to\alpha$ such that

$$e_{\phi(\xi)}(\bar{i})<e_{\phi(\xi')}(\bar{i}) \quad \text{for } \xi<\xi'<\alpha \quad \text{and}$$
$$\sup\{e_{\phi(\xi)}(\bar{i}):\xi<\alpha\}=\alpha.$$

Without any loss of generality and without changing $\bar{i}$ we assume that ϕ is the identity function on α. Thus

(*) $$e_\xi(\bar{i})<e_{\xi'}(\bar{i}) \quad \text{for } \xi<\xi'<\alpha.$$

We define

$$\bar{\eta}=\sup\{\eta_i: i<\bar{i}\}$$

(so that $\bar{\eta}=0$ if $\bar{i}=0$); since $\bar{i}<\rho<\alpha$ and $\eta_i<\alpha$ for $i<\bar{i}$ and α is regular it follows that $\bar{\eta}<\alpha$. We claim that there are $J\subset\bar{\eta}+1$ and $B\in\mathscr{P}(\alpha)$ such that

$$|B|=\alpha \quad \text{and}$$
$$f(\xi)\cap(\bar{\eta}+1)=J \quad \text{for } \xi\in B.$$

Indeed the function $\xi\to f(\xi)\cap(\bar{\eta}+1)$ is a function from α into $\mathscr{P}_{|\rho|^+}(\bar{\eta}+1)$; furthermore $|\mathscr{P}_{|\rho|^+}(\bar{\eta}+1)|<\alpha$ since $|\bar{\eta}+1|<\alpha$ and $|\rho|<\kappa$ and α is strongly κ-inaccessible. Since α is regular there is $J\subset\bar{\eta}+1$ such that

$$|\{\xi<\alpha: f(\xi)\cap(\bar{\eta}+1)=J\}|=\alpha,$$

proving the claim. Without any loss of generality and without changing $\bar{i}$ (because of (*)), we assume that

(**) $$f(\xi)\cap(\bar{\eta}+1)=J \quad \text{for } \xi<\alpha.$$

We now define a set of ordinals $\{\psi(\zeta):\zeta<\alpha\}$ less than α such that

(i) $\psi(0)=0$, and

(ii) if $\xi\in\bigcup_{\zeta'<\zeta}f(\psi(\zeta'))$ then $\xi<e_{\psi(\zeta)}(\bar{i})$ for $\zeta<\alpha$.

We proceed recursively. We set $\psi(0)=0$; if $0<\zeta<\alpha$ and if $\psi(\zeta)$ is defined for $\zeta'<\zeta$ then we note that

$$|\bigcup_{\zeta'<\zeta}f(\psi(\zeta'))|\leq\sum_{\zeta'<\zeta}|f(\psi(\zeta'))|\leq|\rho|\cdot|\zeta|<\alpha,$$

and since α is regular and $\sup\{e_{\xi}(\bar{i}):\xi<\alpha\}=\alpha$ there is an ordinal $\psi(\zeta)<\alpha$ such that

$$\sup\{\xi<\alpha:\xi\in\bigcup_{\zeta'<\zeta}f(\psi(\zeta'))\}<e_{\psi(\zeta)}(\bar{i}).$$

This completes the recursive definition. We set

$$A=\{\psi(\zeta):\zeta<\alpha\}.$$

We note that if $\xi\in f(\psi(\zeta))$ and $\bar{\eta}<\xi$ then $e_{\psi(\zeta)}(i)<\xi$ for $i<\bar{i}$, and hence $e_{\psi(\zeta)}(\bar{i})\leq\xi$ for $\zeta<\alpha$, *i.e.*, $e_{\psi(\zeta)}(\bar{i})$ is the smallest element of $f(\psi(\zeta))$ in $\alpha\backslash(\bar{\eta}+1)$. It follows from (ii) that

$$f(\psi(\zeta'))\cap f(\psi(\zeta))\cap(\alpha\backslash(\bar{\eta}+1))=\emptyset\quad\text{for }\zeta'<\zeta<\alpha,$$

which together with (**) implies (b).

Second Proof. For $A\in\mathscr{P}(\alpha)$ we define

$$F(A)=\bigcup_{\xi\in A}f(\xi).$$

Using transfinite recursion and Zorn's lemma we define a family $\{A_i:i<\kappa\}$ of sets such that

(i) $A_i\subset\alpha\backslash\bigcup_{j<i}A_j$;

(ii) if $\xi,\xi'\in A_i$ and $\xi\neq\xi'$, then $f(\xi)\cap f(\xi')\subset F(\bigcup_{j<i}A_j)$; and

(iii) A_i is maximal with respect to (i) and (ii)

for $i<\kappa$.

We note in particular that $\{f(\xi):\xi\in A_0\}$ is a maximal pairwise disjoint subfamily of $f[\alpha]$. We note also that if $i<\kappa$ and $\xi\in\alpha\backslash\bigcup_{j\leq i}A_j$ then from (iii) there is $p_{\xi,i}$ such that

$$p_{\xi,i}\in f(\xi)\cap F(A_i)\quad\text{and}$$
$$p_{\xi,i}\notin F(\bigcup_{j<i}A_j).$$

There are two possibilities to consider.

Case 1. $|A_i|<\alpha$ for $i<\kappa$. Since α is regular and $\kappa<\alpha$ there is $\xi\in\alpha\backslash\bigcup_{i<\kappa}A_i$. Then $\xi\in\alpha\backslash\bigcup_{j\leq i}A_j$ for $i<\kappa$ and it is clear that

$$p_{\xi,j}\neq p_{\xi,i}\quad\text{for }j<i<\kappa.$$

Thus $\kappa=|\{p_{\xi,i}:i<\kappa\}|\leq|f(\xi)|$, a contradiction.

Case 2. There is $i<\kappa$ such that $|A_i|=\alpha$. We define

$$\bar{\imath}=\min\{i<\kappa:|A_i|=\alpha\}, \quad \text{and}$$
$$A_{\bar{\imath},\lambda}=\{\xi\in A_{\bar{\imath}}:|f(\xi)\cap F(\bigcup_{j<\bar{\imath}}A_j)|=\lambda\} \quad \text{for } \lambda<\kappa\,.$$

Since α is regular there is $\lambda<\kappa$ such that $|A_{\bar{\imath},\lambda}|=\alpha$. We note that $|A_j|<\alpha$ for $j<\bar{\imath}(<\kappa)$, and since α is regular we have $|\bigcup_{j<\bar{\imath}}A_j|<\alpha$. Hence

$$|F(\bigcup_{j<\bar{\imath}}A_j)|=|\bigcup_{\xi\in\bigcup_{j<\bar{\imath}}A_j}f(\xi)|\leq\sum_{\xi\in\bigcup_{j<\bar{\imath}}A_j}|f(\xi)|<\alpha\,.$$

Since α is strongly κ-inaccessible we have

$$(*) \qquad |\mathscr{P}_{\lambda^+}(F(\bigcup_{j<\bar{\imath}}A_j))|=|F(\bigcup_{j<\bar{\imath}}A_j)|^{\lambda}<\alpha\,.$$

The function $\xi\to f(\xi)\cap F(\bigcup_{j<\bar{\imath}}A_j)$ for $\xi\in A_{\bar{\imath},\lambda}$ is a function from $A_{\bar{\imath},\lambda}$ into $\mathscr{P}_{\lambda^+}(F(\bigcup_{j<\bar{\imath}}A_j))$; from (*) and the regularity of α it follows that there are a subset A of $A_{\bar{\imath},\lambda}$ and a subset J of $F(\bigcup_{j<\bar{\imath}}A_j)$ such that

$$|A|=\alpha \quad \text{and}$$
$$f(\xi)\cap F(\bigcup_{j<\bar{\imath}}A_j)=J \quad \text{for } \xi\in A\,.$$

Now let $\xi,\xi'\in A$ and $\xi\neq\xi'$; from (ii) we have $f(\xi)\cap f(\xi')\subset F(\bigcup_{j<\bar{\imath}}A_j)$ and hence

$$f(\xi)\cap f(\xi')=(f(\xi)\cap F(\bigcup_{j<\bar{\imath}}A_j))\cap(f(\xi')\cap F(\bigcup_{j<\bar{\imath}}A_j))=J\cap J=J\,.$$

(b)$\Rightarrow$(a). Suppose that there are cardinals $\beta<\alpha$ and $\lambda<\kappa$ such that $\alpha\leq\beta^{\lambda}$. Let $\mathscr{S}\subset\beta^{\lambda}$ be such that $|\mathscr{S}|=\alpha$; we note that $\mathscr{S}\subset\mathscr{P}_{\lambda^+}(\lambda\times\beta)$. Let

$$f:\alpha\to\mathscr{P}_{\lambda^+}(\lambda\times\beta)\subset\mathscr{P}_{\kappa}(\alpha)$$

be a one-to-one function such that $f[\alpha]=\mathscr{S}$. From condition (b) there are $A\in\mathscr{P}(\alpha)$ and $J\in\mathscr{P}_{\kappa}(\alpha)$ such that

$$|A|=\alpha\,, \quad \text{and}$$
$$f(\xi)\cap f(\xi')=J \quad \text{for } \xi<\xi'<\alpha \quad \text{and } \xi,\xi'\in A\,.$$

Since $|A|>\beta$, clearly for $i<\lambda$ there are $\xi,\xi'\in A$ with $\xi<\xi'<\alpha$ and $\phi(i)<\beta$ such that

$$f(\xi)(i)=f(\xi')(i)\,.$$

Then $\langle i,\phi(i)\rangle\in f(\xi)\cap f(\xi')=J$ for all $i<\lambda$ and $\xi,\xi'\in A$ with $\xi<\xi'<\alpha$ and we have $A=\{\phi\}$, a contradiction. The proof of the theorem is complete.

A rewording of condition (b) of the Erdős-Rado Theorem 3.2 is as follows:

for every family $\{A_\xi : \xi < \alpha\}$ of sets such that $|A_\xi| < \kappa$ for $\xi < \alpha$ there are a set J and a subset A of α such that

$$|A| = \alpha \quad \text{and}$$
$$A_\xi \cap A_{\xi'} = J \quad \text{for } \xi < \xi' < \alpha \quad \text{and } \xi, \xi' \in A.$$

The Erdős-Rado theorem just proved is closely related to the Souslin number of topological products, which we now define and study.

Definitions. Let X be a space.

A *cellular family* in X is a family of pairwise disjoint non-empty open subsets of X.

The *Souslin number* of X, denoted by $\boldsymbol{S}(X)$, is the smallest cardinal α such that no cellular family in X has α elements.

By an application of Zorn's lemma every cellular family in a space X is included in a maximal cellular family in X.

It is clear that if Y is a dense subset of the space X then $\boldsymbol{S}(X) = \boldsymbol{S}(Y)$.

3.3. Lemma. (a) *If X is a space then $\boldsymbol{S}(X) \le \boldsymbol{d}(X)^+$;*

(b) *If X is an infinite space then $\omega^+ \le \boldsymbol{S}(X)$.*

Proof. (a) Let D be a dense subset of X such that $|D| = \boldsymbol{d}(X)$ and let $\mathscr{C}$ be a cellular family in X. For every $C \in \mathscr{C}$ let $p_C \in C \cap D$. The function $C \to p_C$ is one-to-one from $\mathscr{C}$ into D. Thus $|\mathscr{C}| \le \boldsymbol{d}(X)$ for every cellular family $\mathscr{C}$ in X, and hence $\boldsymbol{S}(X) \le \boldsymbol{d}(X)^+$.

(b) We note that if Y is an infinite space then there is a non-empty open subset U of Y such that $|Y \setminus \mathrm{cl}_Y U| \ge \omega$.

We proceed recursively to define a sequence $\{U_n : n < \omega\}$ such that U_n is a non-empty open subset of $X \setminus \bigcup_{k<n} \mathrm{cl}_X U_k$ and $|X \setminus \bigcup_{k \le n} \mathrm{cl}_X U_k| \ge \omega$ for $n < \omega$.

Then $\{U_n : n < \omega\}$ is a faithfully indexed cellular family in X, and thus $\omega^+ \le \boldsymbol{S}(X)$.

If the Souslin number of a space X is at most ω^+ then we say that X has the *countable chain condition.*

We say that a space X is *Souslin-homogeneous* if $\boldsymbol{S}(U) = \boldsymbol{S}(X)$ for every open non-empty subspace U of X.

The usefulness of this concept lies with the following result, which contains the main idea for the proof of the Erdős-Tarski theorem (3.5).

3.4. Lemma. *If $\boldsymbol{S}(X)$ is a limit cardinal for some space X then there is an open subset U of X such that*

$$\boldsymbol{S}(U) = \boldsymbol{S}(X) \quad \textit{and}$$
$$U \textit{ is Souslin-homogeneous.}$$

Proof. Let X be a space for which the statement fails. Then for every open non-empty subset U of X there is an open non-empty subset V of U such that $\mathbf{S}(V)<\mathbf{S}(X)$. Thus there is a faithfully indexed maximal cellular family $\{V_i: i\in I\}$ in X such that $\mathbf{S}(V_i)<\mathbf{S}(X)$ for $i\in I$. We define $\alpha=\sum\{\mathbf{S}(V_i): i\in I\}$ and we claim that $\alpha<\mathbf{S}(X)$.

Suppose on the contrary that $\alpha=\mathbf{S}(X)$, so that α is a limit cardinal. Let $\{\alpha_\xi: \xi<\mathrm{cf}(\alpha)\}$ be a family of cardinals such that

$$\begin{aligned}&\alpha_\xi<\alpha_\zeta \quad \text{for } \xi<\zeta<\mathrm{cf}(\alpha),\\ &\alpha_\xi<\alpha \quad \text{for } \xi<\mathrm{cf}(\alpha), \quad \text{and}\\ &\textstyle\sum_{\xi<\mathrm{cf}(\alpha)}\alpha_\xi=\alpha.\end{aligned}$$

Let $i(\xi)\in I$ be such that $\mathbf{S}(V_{i(\xi)})\geq\alpha_\xi^+$ and let $\mathscr{C}_\xi$ be a cellular family in $V_{i(\xi)}$ such that $|\mathscr{C}_\xi|=\alpha_\xi$ for $\xi<\mathrm{cf}(\alpha)$. Then the family

$$\mathscr{C}=\textstyle\bigcup_{\xi<\mathrm{cf}(\alpha)}\mathscr{C}_\xi$$

is a cellular family in X such that $|\mathscr{C}|=\alpha=\mathbf{S}(X)$, a contradiction proving the claim.

Let $\beta=(|I|\cdot\alpha)^+$; since $|I|<\mathbf{S}(X)$ and $\alpha<\mathbf{S}(X)$ and $\mathbf{S}(X)$ is a limit cardinal it follows that $\beta<\mathbf{S}(X)$. Let $\{W_\xi: \xi<\beta\}$ be a faithfully indexed cellular family in X. We define

$$A_i=\{\xi<\beta: V_i\cap W_\xi\neq\emptyset\}$$

for $i\in I$ and we note that the maximality of the family $\{V_i: i\in I\}$ implies that $\bigcup_{i\in I}V_i$ is dense in X and hence that $\bigcup_{i\in I}A_i=\beta$. Furthermore the family $\{V_i\cap W_\xi: \xi\in A_i\}$ is a cellular family in V_i, so we have $|A_i|<\mathbf{S}(V_i)<\alpha$ for $i\in I$. Then

$$\beta=|\textstyle\bigcup_{i\in I}A_i|\leq\sum_{i\in I}|A_i|=|I|\cdot\alpha,$$

contradicting the definition of β.

3.5. Theorem *(Erdős-Tarski). If X is an infinite space then $\mathbf{S}(X)$ is an uncountable regular cardinal.*

Proof. Let $\alpha=\mathbf{S}(X)$; it follows from Lemma 3.3(b) that $\omega^+\leq\alpha$. Assume now that α is a singular cardinal; then α is a limit cardinal and it follows from Lemma 3.4 that there is a Souslin-homogeneous, non-empty open subset U of X such that $\mathbf{S}(U)=\alpha$. Since $\mathrm{cf}(\alpha)<\alpha$ there is a faithfully indexed cellular family $\{V_\xi: \xi<\mathrm{cf}(\alpha)\}$ in U. Let $\{\alpha_\xi: \xi<\mathrm{cf}(\alpha)\}$ be a family of cardinals such that

$$\begin{aligned}&\alpha_\xi<\alpha \quad \text{for } \xi<\mathrm{cf}(\alpha) \quad \text{and}\\ &\textstyle\sum_{\xi<\mathrm{cf}(\alpha)}\alpha_\xi=\alpha.\end{aligned}$$

Since U is Souslin-homogeneous we have $\boldsymbol{S}(V_\xi)=\alpha$ and so there is a cellular family $\mathscr{C}_\xi$ in V_ξ such that $|\mathscr{C}_\xi|=\alpha_\xi$ for $\xi<\mathrm{cf}(\alpha)$. Then the family

$$\mathscr{C}=\bigcup_{\xi<\mathrm{cf}(\alpha)}\mathscr{C}_\xi$$

is a cellular family in X such that $|\mathscr{C}|=\alpha=S(X)$, a contradiction proving the theorem.

We include at this point the particularly simple description of the Souslin number of a metric space.

3.6. Theorem. *If X is a metric space then $\boldsymbol{S}(X)=\boldsymbol{d}(X)^+$.*

Proof. Let ρ be the metric for X and set $S(p,r)=\{q\in X:\rho(p,q)<r\}$ for positive real numbers r.

Let D_n be a subset of X such that

(i) if $p,q\in D_n$ and $p\neq q$ then $\rho(p,q)\geq 1/n$, and

(ii) D_n is maximal with respect to (i)

for $0<n<\omega$,

and set $D=\bigcup_{0<n<\omega}D_n$. It is clear that D is dense in X and thus $\boldsymbol{d}(X)\leq|D|$. We define

$$\mathscr{C}_n=\{S(p,1/(2n)):p\in D_n\}$$

for $0<n<\omega$; then $\mathscr{C}_n$ is a cellular family in X with $|\mathscr{C}_n|=|D_n|$ and from Lemma 3.3 (a) we have $|D_n|<\boldsymbol{S}(X)\leq\boldsymbol{d}(X)^+$ for $0<n<\omega$.

If the theorem fails we have ($\boldsymbol{d}(X)\geq\omega$ and)

$$|D_n|<\boldsymbol{S}(X)\leq\boldsymbol{d}(X)\leq|D|\leq\sum_{0<n<\omega}|D_n|;$$

hence $\boldsymbol{S}(X)=\boldsymbol{d}(X)=\sum_{0<n<\omega}|D_n|$ and $\mathrm{cf}(\boldsymbol{S}(X))=\omega$, contradicting Theorem 3.5. The proof is complete.

Notation. Let $\{X_i:i\in I\}$ be a family of sets and $X=\prod_{i\in I}X_i$. If $J\subset I$ and $J\neq\emptyset$ we set

$$X_J=\prod_{i\in J}X_i,$$

and we denote by π_J the projection function from X onto X_J given by $\pi_J(\langle p_i:i\in I\rangle)=\langle p_i:i\in J\rangle$. In particular $X=X_I$. If $p\in X$ we sometimes write p_J for $\pi_J(p)$, though we write (as usual) π_i and p_i for $\pi_{\{i\}}$ and $p_{\{i\}}$.

A subset A of X such that

$$A=\prod_{i\in I}A_i$$

where $A_i\subset X_i$ for $i\in I$ is called a *(generalized) rectangle* in X and the set

$$R(A)=\{i\in I:A_i\neq X_i\}$$

is called the *restriction set* of A.

Definition. Let $\{X_i : i \in I\}$ be a family of spaces, let X be the *set* $\prod_{i \in I} X_i$ and let κ be an infinite cardinal. The *κ-product topology* on X is the topology with base (called the *canonical* base) equal to the family of all subsets U of X such that

$$U = \prod_{i \in I} U_i,$$
$$U_i \text{ is open in } X_i \text{ for } i \in I, \text{ and}$$
$$|R(U)| < \kappa.$$

It is simple to verify that the canonical base for the κ-product topology just defined is indeed a base (for a topology). The product set $X = \prod_{i \in I} X_i$ with the κ-product topology is denoted by $X_{(\kappa)}$ or $(\prod_{i \in I} X_i)_{(\kappa)}$.

The ω-product topology is the usual product topology; we write simply X and $\prod_{i \in I} X_i$ for $X_{(\omega)}$ and $(\prod_{i \in I} X_i)_{(\omega)}$, respectively.

In discussing the κ-product topology below we use, among the various bases for it that exist, only its canonical base. Thus the expression "U is basic in $(\prod_{i \in I} X_i)_{(\kappa)}$" means that U is an element of the canonical base for the κ-product topology.

We recall that if X is a space and $\kappa \geq \omega$ then X_κ denotes the set X with the P_κ-space topology generated by (*i.e.*, with the topology whose κ-subbase is) the topology of X.

The following simple fact will be used in the sequel without explicit mention.

If κ is an infinite regular cardinal and $\{X_i : i \in I\}$ is a family of P_κ-spaces (in particular, discrete spaces) then $(\prod_{i \in I} X_i)_{(\kappa)}$ is a P_κ-space and in fact $(\prod_{i \in I} X_i)_{(\kappa)} = (\prod_{i \in I} X_i)_\kappa$. On the other hand if κ is singular then $(2^\kappa)_\kappa$ is a P_{κ^+}-space and therefore discrete, and $(2^\kappa)_{(\kappa)}$ is not discrete.

3.7. Lemma. *Let $\kappa \geq \omega$, $\{X_i : i \in I\}$ a family of spaces, U and V non-empty basic open subsets of $(\prod_{i \in I} X_i)_{(\kappa)}$, and let J be a set such that $R(U) \cap R(V) \subset J \subset I$. Then the following are equivalent.*

(a) *$U \cap V = \emptyset$;*

(b) *there is $i \in J$ such that $U_i \cap V_i = \emptyset$; and*

(c) *$J \neq \emptyset$ and $\pi_J[U] \cap \pi_J[V] = \emptyset$.*

Proof. (a)⇒(b). Suppose that (b) fails and let p be an element of $(\prod_{i \in I} X_i)_{(\kappa)}$ such that

$$\begin{aligned} p_i &\in U_i \cap V_i && \text{if } i \in J, \\ p_i &\in U_i && \text{if } i \in R(U) \setminus J, \quad \text{and} \\ p_i &\in V_i && \text{if } i \in R(V) \setminus J. \end{aligned}$$

Then $p \in U \cap V$ and thus (a) fails.

(b)⇒(c) is obvious.

(c)⇒(a). If $p \in U \cap V$ and $J \neq \emptyset$ then $p_J \in \pi_J[U] \cap \pi_J[V]$.

3.8. Theorem. *Let $\omega \leq \kappa < \alpha$ with α regular and strongly κ-inaccessible and let $\{X_i : i \in I\}$ be a family of non-empty spaces. Then $\boldsymbol{S}((\prod_{i\in I} X_i)_{(\kappa)}) \leq \alpha$ if and only if $\boldsymbol{S}((X_J)_{(\kappa)}) \leq \alpha$ for $J \in \mathscr{P}_\kappa^*(I)$.*

Proof. The necessity of the latter condition is clear, for if for some $J \in \mathscr{P}_\kappa^*(I)$ the family $\mathscr{C}$ is a cellular family in $(X_J)_{(\kappa)}$ then $\{\pi_J^{-1}(U) : U \in \mathscr{C}\}$ is a cellular family in $(X_I)_{(\kappa)}$.

For the converse suppose that $\{U^\xi : \xi < \alpha\}$ is a faithfully indexed cellular family of basic sets in $(X_I)_{(\kappa)}$. According to the Erdős-Rado Theorem 3.2 applied to the family $\{R(U^\xi) : \xi < \alpha\}$, there are a set J and a subset A of α with $|A| = \alpha$ such that

$$R(U^\xi) \cap R(U^{\xi'}) = J$$

if ξ and ξ' are distinct elements of A. Lemma 3.7 then implies that $J \neq \emptyset$ and that $\{\pi_J[U^\xi] : \xi \in A\}$ is a faithfully indexed cellular family in $(X_J)_{(\kappa)}$.

3.9. Theorem. *If $\omega \leq \kappa < \alpha$ and α is regular, then the following are equivalent.*

(a) *α is strongly κ-inaccessible;*

(b) *if $\{X_i : i \in I\}$ is a family of spaces with $\boldsymbol{d}(X_i) < \alpha$ for $i \in I$ then $\boldsymbol{S}((\prod_{i\in I} X_i)_{(\kappa)}) \leq \alpha$.*

Proof. (a)⇒(b). Let $J \subset I$ with $|J| < \kappa$ and for $i \in J$ let D_i be dense in X_i with $|D_i| < \alpha$. We set $D = \prod_{i \in J} D_i$ and we note that D is dense in $(X_J)_{(\kappa)}$ and (using Lemma 3.1(a)) that $|D| < \alpha$ by (a). Thus $\boldsymbol{S}((X_J)_{(\kappa)}) \leq \alpha$ by Lemma 3.3(a), so Theorem 3.8 implies (b).

(b)⇒(a). Let $\beta < \alpha$ and $\lambda < \kappa$. Because $\lambda < \kappa$ it follows that the space $(\beta^\lambda)_{(\kappa)}$ is discrete. For any discrete space X we have $\boldsymbol{S}(X) = |X|^+$, so from condition (b) it follows that $|\beta^\lambda|^+ \leq \alpha$, *i.e.*, $\beta^\lambda < \alpha$.

3.10. Corollary. *Let $\omega \leq \kappa < \alpha$ with α regular and strongly κ-inaccessible and κ regular, and let $\{X_i : i \in I\}$ be a family of spaces with $\boldsymbol{w}(X_i) < \alpha$ for $i \in I$. Then each family of pairwise disjoint, non-empty sets, each of which is the intersection of fewer than κ open subsets of $\prod_{i \in I} X_i$, has fewer than α elements.*

Proof. We note that

$$\boldsymbol{d}((X_i)_\kappa) \leq \boldsymbol{w}((X_i)_\kappa) \leq \boldsymbol{w}(X_i)^{\underline{\kappa}} < \alpha \quad \text{for } i \in I\,,$$

using the fact that α is strongly κ-inaccessible and that $\omega \leq \kappa < \mathrm{cf}(\alpha) = \alpha$. It is easy to verify that

$$(\textstyle\prod_{i\in I} X_i)_\kappa = (\prod_{i\in I} (X_i)_\kappa)_\kappa = (\prod_{i\in I} (X_i)_\kappa)_{(\kappa)}$$

and, using Theorem 3.9, that

$$\boldsymbol{S}((\textstyle\prod_{i\in I} (X_i)_\kappa)_{(\kappa)}) \leq \alpha\,.$$

Then $\boldsymbol{S}((\prod_{i\in I} X_i)_\kappa)\leq\alpha$ and the required statement follows.

The assumption that α is regular cannot be omitted in the statement of Theorem 3.9, even if $\kappa=\omega$. Indeed let α be a singular cardinal and let $\{\alpha_\xi:\xi<\mathrm{cf}(\alpha)\}$ be a family of cardinals such that

$$\alpha_\xi<\alpha \quad \text{for } \xi<\mathrm{cf}(\alpha), \quad \text{and}$$

$$\textstyle\sum_{\xi<\mathrm{cf}(\alpha)}\alpha_\xi=\alpha\,.$$

Then

$$\alpha_\zeta^+=\boldsymbol{S}(\alpha_\zeta)\leq\boldsymbol{S}(\textstyle\prod_{\xi<\mathrm{cf}(\alpha)}\alpha_\xi) \quad \text{for } \zeta<\mathrm{cf}(\alpha)$$

and therefore $\boldsymbol{S}(\prod_{\xi<\mathrm{cf}(\alpha)}\alpha_\xi)\geq\alpha$. Since α is a singular cardinal, the Erdős-Tarski Theorem 3.5 implies that $\boldsymbol{S}(\prod_{\xi<\mathrm{cf}(\alpha)}\alpha_\xi)\geq\alpha^+$. We note that it follows from Theorem 3.9 (with α^+ and ω replacing α and κ, respectively) that in fact $\boldsymbol{S}(\prod_{\xi<\mathrm{cf}(\alpha)}\alpha_\xi)=\alpha^+$.

3.11. Theorem. *Let $\{X_i:i\in I\}$ be a set of non-empty spaces, set $X=\prod_{i\in I}X_i$, and define*

$$\alpha=\sup\{S(X_F):F\in\mathscr{P}_\omega^*(I)\}\,.$$

Then

$$\begin{aligned}\boldsymbol{S}(X)&=\alpha \quad \textit{if either } \alpha<\omega \textit{ or } \alpha \textit{ is uncountable and regular}\\ &=\alpha^- \quad \textit{otherwise}.\end{aligned}$$

Proof. We observe first that $\boldsymbol{S}(X)\geq\alpha$. For if $\beta<\alpha$ then there is for some element F of $\mathscr{P}_\omega^*(I)$ a cellular family $\mathscr{C}$ in X_F with $|\mathscr{C}|=\beta$; then $\{\pi_F^{-1}(U):U\in\mathscr{C}\}$ is a cellular family with β elements in X, so that $\boldsymbol{S}(X)>\beta$.

Now let $\alpha<\omega$. Then there is for some $F_0\in\mathscr{P}_\omega^*(I)$ a cellular family $\mathscr{C}$ in X_{F_0} such that $\alpha=|\mathscr{C}|^+$, and we have $|X_i|=1$ for $i\in I\backslash F_0$; indeed otherwise for some $i\in I\backslash F_0$ there are disjoint non-empty open subsets V_0 and V_1 of X_i; then

$$\{U\times V_0:U\in\mathscr{C}\}\cup\{U\times V_1:U\in\mathscr{C}\}$$

is a cellular family in $X_{F_0\cup\{i\}}$ with at least α elements, a contradiction. Thus X is homeomorphic with X_{F_0} and $\boldsymbol{S}(X)=\boldsymbol{S}(X_{F_0})=\alpha$.

Suppose next that $\alpha\geq\omega$ and let β be the least uncountable, regular cardinal such that $\beta\geq\alpha$. Since $\boldsymbol{S}(X_F)\leq\beta$ for $F\in\mathscr{P}_\omega^*(I)$ we have

$$(*) \qquad \omega\leq\alpha\leq\boldsymbol{S}(X)\leq\beta\,,$$

the last inequality from Theorem 3.8. Now if α is regular and uncountable then $\beta=\alpha$ and we have $\boldsymbol{S}(X)=\alpha$; otherwise $\beta=\alpha^+$, the equality $\boldsymbol{S}(X)=\alpha$ cannot occur by Theorem 3.5, and from (*) we have $\boldsymbol{S}(X)=\alpha^+$, as required.

We remark, using the example that follows Theorem 3.9 above, that the case $\boldsymbol{S}(X)=\alpha^+$ in Theorem 3.11 may in fact occur.

3.12. Corollary. *If α is an uncountable regular cardinal then there is a space whose Souslin number is equal to α. In fact*

$$\boldsymbol{S}((\prod_{\omega \le \beta < \alpha} \beta)^I) = \alpha$$

for all non-empty sets I.

Proof. It suffices to prove the second statement, which follows from Theorem 3.11.

The following two theorems have similar proofs based on a combinatorial theorem of Erdős-Rado from the partition calculus (see Theorem 8.6 and its Corollary 8.10 below). The first gives an upper bound for the Souslin number of a product space that depends only on the Souslin numbers of its coordinate spaces, rather than (as in Theorem 3.8) on the Souslin numbers of its "small" sub-products; the second gives an upper bound for the Souslin number of the P_{α^+}-space associated with a compact space X for certain cardinals α.

3.13. Theorem. *If $\alpha \ge \omega$ and $\{X_i : i \in I\}$ is a family of spaces such that $\boldsymbol{S}(X_i) \le \alpha^+$ for $i \in I$, then $\boldsymbol{S}((\prod_{i \in I} X_i)_{(\alpha^+)}) \le (2^\alpha)^+$.*

Proof. Suppose that there is a faithfully indexed cellular family $\{U^\xi : \xi < (2^\alpha)^+\}$ of basic subsets of $(\prod_{i \in I} X_i)_{(\alpha^+)}$, and using Lemma 3.7 let $i(\xi, \xi') \in I$ be such that

$$U^\xi_{i(\xi,\xi')} \cap U^{\xi'}_{i(\xi,\xi')} = \emptyset$$

for $\{\xi, \xi'\} \in [(2^\alpha)^+]^2$. We define

$$I(\xi) = \{i(\xi, \xi') : \xi' < (2^\alpha)^+ \text{ and } \xi' \ne \xi\}$$

for $\xi < (2^\alpha)^+$. Since

$$i(\xi, \xi') \in R(U^\xi) \cap R(U^{\xi'})$$

for $\{\xi, \xi'\} \in [(2^\alpha)^+]^2$ we have

$$|I(\xi)| \le |R(U^\xi)| \le \alpha$$

for $\xi < (2^\alpha)^+$. Let $\{i_{\xi,\eta} : \eta < \alpha\}$ be a well-ordering of $I(\xi)$ for $\xi < (2^\alpha)^+$ and set

$$P_{\eta,\eta'} = \{\{\xi, \xi'\} \in [(2^\alpha)^+]^2 : \xi < \xi' \text{ and } i_{\xi,\eta} = i_{\xi',\eta'}\}$$

for $\langle \eta, \eta' \rangle \in \alpha \times \alpha$. Since

$$[(2^\alpha)^+]^2 = \bigcup_{\langle \eta, \eta' \rangle \in \alpha \times \alpha} P_{\eta,\eta'}$$

it follows from the special case $(2^\alpha)^+ \to (\alpha^+)^2_\alpha$ of the Erdős-Rado Theorem (8.10 below) that there are a subset A of $(2^\alpha)^+$ and $\langle \bar\eta, \bar\eta' \rangle \in \alpha \times \alpha$ such that

$$|A| = \alpha^+, \quad \text{and}$$
$$[A]^2 \subset P_{\bar\eta,\bar\eta'}.$$

Thus there is $\bar{i}\in I$ such that if $\{\xi,\xi'\}\in[A]^2$ and $\xi<\xi'$ then

$$i(\xi,\xi')=i_{\xi,\bar{\eta}}=i_{\xi',\bar{\eta}'}=\bar{i}$$

and hence

$$U_{\bar{i}}^{\xi}\cap U_{\bar{i}}^{\xi'}=\emptyset\,.$$

Thus $\alpha^+=|A|<\boldsymbol{S}(X_{\bar{i}})$, a contradiction completing the proof.

3.14. Theorem. *If $\alpha\geq\omega$ and X is a compact space such that $\boldsymbol{S}(X)\leq\alpha^+$, then $\boldsymbol{S}(X_{\alpha^+})\leq(2^\alpha)^+$.*

Proof. From the definition of the space X_{α^+} and the fact that X is (locally) compact it follows that if U is a non-empty open subset of X_{α^+} then there are a set V and a family $\{V_\eta:\eta<\alpha\}$ of compact subsets of X such that

$$\emptyset\neq V=\bigcap_{\eta<\alpha}\operatorname{int}_X V_\eta\subset\bigcap_{\eta<\alpha}V_\eta\subset U\,,\quad\text{and}$$
$$\{V_\eta:\eta<\alpha\}\ \text{is closed under finite intersections.}$$

Hence if the conclusion of the theorem fails there is a faithfully indexed cellular family $\{V(\xi):\xi<(2^\alpha)^+\}$ of X_{α^+} and for $\xi<(2^\alpha)^+$ a family $\{V_\eta(\xi):\eta<\alpha\}$ of compact subsets of X, closed under finite intersections, such that

$$V(\xi)=\bigcap_{\eta<\alpha}\operatorname{int}_X V_\eta(\xi)\subset\bigcap_{\eta<\alpha}V_\eta(\xi)\,.$$

A standard compactness argument shows that if $\xi<\xi'<(2^\alpha)^+$, then, since $V(\xi)\cap V(\xi')=\emptyset$, there are $\eta,\eta'<\alpha$ such that $V_\eta(\xi)\cap V_{\eta'}(\xi')=\emptyset$. We set

$$P_{\eta,\eta'}=\{\{\xi,\xi'\}\in[(2^\alpha)^+]^2:\xi<\xi'\ \text{and}\ V_\eta(\xi)\cap V_{\eta'}(\xi')=\emptyset\}$$

for $\langle\eta,\eta'\rangle\in\alpha\times\alpha$. Since

$$[(2^\alpha)^+]^2=\bigcup_{\langle\eta,\eta'\rangle\in\alpha\times\alpha}P_{\eta,\eta'}$$

it follows from the special case $(2^\alpha)^+\to(\alpha^+)^2_\alpha$ of the Erdős-Rado Theorem (8.10 below) that there are a subset A of $(2^\alpha)^+$ and $\langle\bar{\eta},\bar{\eta}'\rangle\in\alpha\times\alpha$ such that

$$|A|=\alpha^+\,,\quad\text{and}$$
$$[A]^2\subset P_{\bar{\eta},\bar{\eta}'}\,.$$

We set $U(\xi)=V_{\bar{\eta}}(\xi)\cap V_{\bar{\eta}'}(\xi)$ for $\xi\in A$. Then

$$\operatorname{int}_X U(\xi)\neq\emptyset\,,\quad\text{and}$$
$$U(\xi)\cap U(\xi')=\emptyset$$

for $\xi,\xi'\in A$ with $\xi\neq\xi'$.
Thus $\alpha^+=|A|<\boldsymbol{S}(X)$, a contradiction completing the proof.

We remark that from every partition relation (as in § 8) $\alpha\to(\kappa)^2_\lambda$ with $\lambda\geq\omega$ there are consequences analogous to 3.13 and 3.14. Specifically we have

(a) *if* $\{X_i : i \in I\}$ *is a family of spaces such that* $S(X_i) \leq \kappa$ *for* $i \in I$ *then* $S((\prod_{i \in I} X_i)_{(\lambda^+)}) \leq \alpha$; *and*

(b) *if* X *is a compact space such that* $S(X) \leq \kappa$, *then* $S(X_{\lambda^+}) \leq \alpha$.

3.15. The problem of Souslin. We discuss the relation that exists between the Souslin number of the product of two spaces (having the countable chain condition) and the classical problem of Souslin about linearly ordered sets. It follows from the undecidability of the problem of Souslin that it is not possible to determine the Souslin number of the product (of even two spaces) from the Souslin numbers of its coordinate spaces.

Let us consider a linearly ordered set $\langle A, \leq \rangle$ such that

(i) A does not have a first or a last element;

(ii) A has no jumps and no gaps, *i. e.*, A with the order topology is connected; and

(iii) every family of pairwise disjoint non-empty open intervals of A is countable, *i. e.*, A with the order topology has the countable chain condition.

The *problem of Souslin* [20] is whether A is similar (*i. e.*, order-isomorphic) to the ordered set $\mathbb{R}$ of real numbers. In view of the simple fact that any separable ordered set with properties (i) and (ii) is similar to $\mathbb{R}$, the problem of Souslin is equivalent to this question: Is there an ordered set A satisfying (i), (ii), (iii) and also

(iv) A does not contain a countable dense set?

A set satisfying (i), (ii), (iii) and (iv) is called a *Souslin line.* Kurepa [50], [52] has established the following connection between the problem of Souslin and the Souslin number of finite products: If A is a Souslin line, then $S(A \times A) > \omega^+$.

Tennenbaum [68] and Jech [67] have proved the consistency of the existence of a Souslin line with ZFC. (In fact, Tennenbaum proved the consistency even with ZFC+GCH or, alternatively, with ZFC+$\neg$CH.) Together with Kurepa's result, these statements imply the consistency of the existence of a space L for which $S(L) = \omega^+$ and $S(L \times L) > \omega^+$. (We recall from Theorem 3.13 that $S(L \times L) \leq (2^\omega)^+$.)

The consistency of the non-existence of a Souslin line with ZFC can be proved with the aid of Martin's axiom. Martin's axiom has been introduced and studied by Solovay and Tennenbaum [71], Martin and Solovay [70] and Kunen [68]. To state it, we introduce the following definitions. Let $\langle P, \leq \rangle$ be a non-empty partially ordered set. Two elements a and b of P are *compatible* if there is $c \in P$ such that $c \leq a$ and $c \leq b$; a subset D of P is *dense* if for $a \in P$ there is $b \in D$ such that $b \leq a$; if D is a dense subset of P, then a subset G of P is called a *D-generic filter* on P if the following conditions are satisfied:

if $a \in G$ and $a \leq b$ then $b \in G$,
if $a, b \in G$ then there is $c \in G$ such that $c \leq a$ and $c \leq b$, and $D \cap G \neq \emptyset$. The partially ordered set P has the *countable chain condition* if every subset of P of pairwise incompatible elements is countable.

With these definitions, which are motivated by the forcing conditions of Cohen [66], *Martin's axiom* is the following statement: If $\langle P, \leq \rangle$ is a partially ordered set with the countable chain condition and if $\mathscr{D}$ is a family of dense subsets of P such that $|\mathscr{D}| < 2^{\omega}$, then there is a subset of P which is a D-generic filter for $D \in \mathscr{D}$.

The fundamental consistency theorem of Solovay-Tennenbaum [71] (Theorem 7.11) states that if there is a model of ZFC for which $\kappa = \kappa^{\underline{\kappa}}$ for an uncountable cardinal κ then there is a model of ZFC for which Martin's axiom holds and $2^{\omega} = \kappa$. The Marczewski and Šanin lemma (*i. e.*, the Erdős-Rado Theorem 3.2 for $\kappa = \omega$) plays a central rôle in the Solovay-Tennenbaum consistency proof.

In connection with the Souslin number of (finite) products, we are interested in the following consequence of Martin's axiom.

Lemma. *Assume Martin's axiom and* $2^{\omega} > \omega^{+}$, *and let X be a space with the countable chain condition. Then every uncountable family of open subsets of X contains an uncountable subfamily with the finite intersection property.*

It follows from this lemma that if Martin's axiom and $2^{\omega} > \omega^{+}$ are assumed, then $\boldsymbol{S}(X \times Y) \leq \omega^{+}$ when $\boldsymbol{S}(X) \leq \omega^{+}$ and $\boldsymbol{S}(Y) \leq \omega^{+}$. (For if $\{U_{\xi} \times V_{\xi} : \xi < \omega^{+}\}$ is a faithfully indexed cellular family in $X \times Y$, we choose an uncountable subset I of ω^{+}, and then (using the lemma again) an uncountable subset J of I, for which both $\{U_{\xi} : \xi \in I\}$ and $\{V_{\xi} : \xi \in J\}$ have the finite intersection property; then the family $\{U_{\xi} \times V_{\xi} : \xi \in J\}$ has the finite intersection property.) From the Solovay-Tennenbaum consistency theorem it follows that it is consistent with ZFC to assume that every finite product of spaces with the countable chain condition is another such space; in this case, by Kurepa's result, no Souslin line exists.

Families of Large Oscillation

Definition. Let α, β and κ be cardinals and let $\mathscr{S} \subset \beta^{\alpha}$. The family $\mathscr{S}$ is a family of *κ-large oscillation* if the following condition is satisfied: if $\lambda < \kappa$, $\{f_{\zeta} : \zeta < \lambda\}$ are distinct elements of $\mathscr{S}$, and $\{\eta_{\zeta} : \zeta < \lambda\}$ are elements of β, then there is $\xi < \alpha$ such that $f_{\zeta}(\xi) = \eta_{\zeta}$ for $\zeta < \lambda$.

The following theorem, which is the second main result of this section, is used for the exact computation of the number of ultrafilters on a cardinal (Theorem 7.3 and its corollary) and as a method of avoiding the use of the generalized continuum hypothesis in proofs of the existence of certain types of ultrafilters (cf. Theorem 10.4 and Theorem 13.19).

3.16. Theorem. *Let $\omega \leq \kappa \leq \alpha$. Then the following statements are equivalent.*

(a) $\alpha = \alpha^{\underline{\kappa}}$;

(b) *if* $\{A_\xi : \xi < \alpha^+\}$ *is a family of sets with* $|A_\xi| < \kappa$ *for* $\xi < \alpha^+$, *then there are a set J and a subset A of* α^+ *with* $|A| = \alpha^+$ *such that* $A_\xi \cap A_{\xi'} = J$ *for* $\xi < \xi' < \alpha$ *and* $\xi, \xi' \in A$;

(c) *if* X_i *is a space such that* $\boldsymbol{d}(X_i) \leq \alpha$ *for* $i \in I$ *then* $\boldsymbol{S}((\prod_{i\in I} X_i)_{(\kappa)}) \leq \alpha^+$;

(d) *there is* $\mathscr{S} \subset \alpha^\alpha$ *such that* $|\mathscr{S}| = 2^\alpha$ *and* $\mathscr{S}$ *is of* κ*-large oscillation;*

(e) *there is* $\mathscr{S} \subset \alpha^\alpha$ *such that* $|\mathscr{S}| = \alpha^+$ *and* $\mathscr{S}$ *is of* κ*-large oscillation;*

(f) $\boldsymbol{d}((\alpha^{(2^\alpha)})_{(\kappa)}) = \alpha$; *and*

(g) $\boldsymbol{d}((\alpha^{(\alpha^+)})_{(\kappa)}) = \alpha$.

Proof. From Lemma 3.1(b), $\alpha = \alpha^{\underline{\kappa}}$ if and only if α^+ is strongly κ-inaccessible. Thus (a)⇔(b) follows from Theorem 3.2 and (a)⇔(c) follows from Theorem 3.9.

(a)⇒(d). We define

$$\mathscr{F} = \{\langle F, G, s\rangle : F \in \mathscr{P}_\kappa(\alpha),\ G \in \mathscr{P}_\kappa(\mathscr{P}(F)), \text{ and } s \in \alpha^G\}.$$

If $F \in \mathscr{P}(\alpha)$ and $|F| = \lambda < \kappa$ then

$$|\mathscr{P}_\kappa(\mathscr{P}(F))| \leq (2^\lambda)^{\underline{\kappa}} \leq (\alpha^\lambda)^{\underline{\kappa}} \leq \alpha^{\underline{\kappa}} = \alpha,$$

and if $G \in \mathscr{P}_\kappa(\mathscr{P}(F))$ then $|\alpha^G| \leq \alpha^{\underline{\kappa}} = \alpha$. Thus $|\mathscr{F}| = \alpha$. It is sufficient (by identifying $\mathscr{F}$ with α) to find $\mathscr{S} \subset \alpha^{\mathscr{F}}$ such that $|\mathscr{S}| = 2^\alpha$ and $\mathscr{S}$ is of κ-large oscillation. We define

$$f : \mathscr{P}(\alpha) \to \alpha^{\mathscr{F}}$$

by the following rule:

$$\begin{aligned} f_A(\langle F, G, s\rangle) &= s(A \cap F) \quad \text{if } A \cap F \in G \\ &= 0 \qquad\qquad\ \ \text{if } A \cap F \notin G \end{aligned}$$

for $A \in \mathscr{P}(\alpha)$ and $\langle F, G, s\rangle \in \mathscr{F}$.

We set $\mathscr{S} = \{f_A : A \in \mathscr{P}(\alpha)\}$ and we prove that $\mathscr{S}$ is a family of κ-large oscillation of cardinality 2^α. Let $\lambda < \kappa$, $\{A_\zeta : \zeta < \lambda\}$ a family of distinct subsets of α and $\{\eta_\zeta : \zeta < \lambda\}$ a set of elements of α. For $\zeta < \zeta' < \lambda$ we choose

$$a_{\zeta,\zeta'} \in (A_\zeta \backslash A_{\zeta'}) \cup (A_{\zeta'} \backslash A_\zeta)$$

and we set

$$F = \{a_{\zeta,\zeta'} : \zeta < \zeta' < \lambda\}.$$

We define

$$G=\{A_\zeta\cap F:\zeta<\lambda\},$$

and a function s from G to α by the rule

$$s(A_\zeta\cap F)=\eta_\zeta.$$

The function s is well-defined because $A_\zeta\cap F\neq A_{\zeta'}\cap F$ for $\zeta<\zeta'<\lambda$. Then $\langle F,G,s\rangle\in\mathscr{F}$, and

$$f_{A_\zeta}(\langle F,G,s\rangle)=s(A_\zeta\cap F)=\eta_\zeta \quad \text{for } \zeta<\lambda.$$

The proof of (a)⇒(d) is complete.

If $\mathscr{S}\subset\alpha^\alpha$ and $\mathscr{S}$ is a family of κ-large oscillation then the set $\{p(\xi):\xi<\alpha\}$, where $p(\xi)_f=f(\xi)$ for $f\in\mathscr{S}$ and $\xi<\alpha$, is clearly dense in $(\alpha^{\mathscr{S}})_{(\kappa)}$. Thus (d)⇒(f) and (e)⇒(g).

If β is a cardinal and $\{p(\xi):\xi<\alpha\}$ is dense in $(\alpha^\beta)_{(\kappa)}$ then the family $\mathscr{S}=\{f_\eta:\eta<\beta\}$, where $f_\eta(\xi)=p(\xi)_\eta$ for $\xi<\alpha$ and $\eta<\beta$, is of κ-large oscillation. Thus (f)⇒(d) and (g)⇒(e).

(d)⇒(e) follows from the fact that any subfamily of a family of κ-large oscillation is of κ-large oscillation.

(g)⇒(c). Let $\{X_i:i\in I\}$ be a family of spaces such that $\boldsymbol{d}(X_i)\le\alpha$ for $i\in I$ and $\boldsymbol{S}((\prod_{i\in I}X_i)_{(\kappa)})>\alpha^+$. Then there is a faithfully indexed cellular family $\{U^\xi:\xi<\alpha^+\}$ of basic open subsets of $(\prod_{i\in I}X_i)_{(\kappa)}$. We set

$$J=\bigcup\{R(U^\xi):\xi<\alpha^+\}$$

and we note that $|J|\le\alpha^+$ and that $\{\pi_J[U^\xi]:\xi<\alpha^+\}$ is a faithfully indexed cellular family in $(X_J)_{(\kappa)}$. Since $\boldsymbol{d}(X_i)\le\alpha$ for $i\in J$ it is clear that there is a continuous function from $(\alpha^{(\alpha^+)})_{(\kappa)}$ onto a dense subset of $(X_J)_{(\kappa)}$. From (g) we have $\boldsymbol{d}((X_J)_{(\kappa)})\le\alpha$, which is from Lemma 3.3(a) a contradiction to the inequality $\boldsymbol{S}((X_J)_{(\kappa)})>\alpha^+$.

The proof of the theorem is complete.

3.17. Corollary. *Let $\omega\le\kappa\le\alpha$, $2\le\beta\le\alpha$, and $\alpha=\alpha^{\underline{\kappa}}$. Then there is $\mathscr{S}\subset\beta^\alpha$ such that $|\mathscr{S}|=2^\alpha$ and $\mathscr{S}$ is of κ-large oscillation.*

Proof. From Theorem 3.16 there is $\mathscr{S}'\subset\alpha^\alpha$ such that $|\mathscr{S}'|=2^\alpha$ and $\mathscr{S}'$ is of κ-large oscillation. Let $\pi:\alpha\to\beta$ be a function onto β and set

$$\mathscr{S}=\{\pi\circ f:f\in\mathscr{S}'\}.$$

3.18. Corollary. *If $\omega\le\kappa\le\alpha=\alpha^{\underline{\kappa}}$ and if $\{X_i:i\in I\}$ is a family of spaces such that $\boldsymbol{d}(X_i)\le\alpha$ for $i\in I$ and $|I|\le2^\alpha$, then $\boldsymbol{d}((\prod_{i\in I}X_i)_{(\kappa)})\le\alpha$.*

Proof. This follows easily from the implication (a)⇒(f) of Theorem 3.16.

Remark. We note that the proof of Corollary 3.18 depends only on the implications (a)⇒(d)⇒(f) of Theorem 3.16, which do not use the Erdős-Rado theorem ((a)⇔(b) of 3.16). It is possible to derive 3.16(c) from Corollary 3.18, thus obtaining a proof of 3.16(c) without the Erdős-Rado theorem. This is, historically, the way 3.16(c) was first proved for $\kappa=\omega$. Here is the direct argument. Let $\{U^\xi:\xi<\alpha^+\}$ be a faithfully indexed, cellular family in $(\prod_{i\in I}X_i)_{(\kappa)}$ and set

$$J=\bigcup\{R(U^\xi):\xi<\alpha^+\}.$$

Since $|J|\leq\kappa\cdot\alpha^+=\alpha^+\leq 2^\alpha$ we have $\boldsymbol{d}((X_J)_{(\kappa)})\leq\alpha$ by Corollary 3.18. But $\{\pi_J(U^\xi):\xi<\alpha^+\}$ is by Lemma 3.7 a faithfully indexed, cellular family in $(X_J)_{(\kappa)}$, so that $\boldsymbol{S}((X_J)_{(\kappa)})>\alpha^+$. The inequalities

$$\boldsymbol{d}((X_J)_{(\kappa)})\leq\alpha<\alpha^+<\boldsymbol{S}((X_J)_{(\kappa)})$$

contradict Lemma 3.3(a).

3.19. Lemma. *If* $\alpha\geq\omega$ *and* $X=\prod_{i\in I}X_i$ *with* $|X_i|\geq 2$ *for* $i\in I$ *and* $|I|>2^\alpha$, *then* $\boldsymbol{d}(X)>\alpha$.

Proof. Let U_i and V_i be disjoint non-empty open subsets of X_i for $i\in I$, and let D be dense in X and $|D|=\boldsymbol{d}(X)$. The function

$$i\to D\cap\pi_i^{-1}(U_i)$$

is one-to-one from I into $\mathscr{P}(D)$, for if i and j are different elements of I and $p\in D\cap\pi_i^{-1}(U_i)\cap\pi_j^{-1}(V_j)$, then

$$p\in D\cap\pi_i^{-1}(U_i) \quad\text{and}\quad p\notin D\cap\pi_j^{-1}(U_j).$$

It follows that $|\mathscr{P}(D)|\geq|I|>2^\alpha$, so that $\boldsymbol{d}(X)=|D|>\alpha$.

Definition. For an infinite cardinal α the *logarithm of* α (denoted by $\log\alpha$) is the least cardinal β such that $\alpha\leq 2^\beta$.

3.20. Theorem. *If* $\{X_i:i\in I\}$ *is a family of spaces such that* $|X_i|\geq 2$ *for* $i\in I$ *and* $|I|\geq\omega$, *then*

$$\boldsymbol{d}((\prod_{i\in I}X_i))=\max\{\log|I|,\,\sup\{\boldsymbol{d}(X_i):i\in I\}\}.$$

Proof. Set $X=\prod_{i\in I}X_i$ and

$$\beta=\max\{\log|I|,\,\sup\{\boldsymbol{d}(X_i):i\in I\}\},$$

so that $\boldsymbol{d}(X_i)\leq\beta$ for $i\in I$ and $|I|\leq 2^\beta$; then $\boldsymbol{d}(X)\leq\beta$ by Corollary 3.18.

It is clear that $\boldsymbol{d}(X_i)\leq\boldsymbol{d}(X)$ for $i\in I$. We prove $\log|I|\leq\boldsymbol{d}(X)$. If not then $2^{\boldsymbol{d}(X)}<|I|$, and from Lemma 3.19 with $\boldsymbol{d}(X)$ replacing α we get $\boldsymbol{d}(X)<\boldsymbol{d}(X)$, a contradiction.

Remark. We consider the following conditions where $\omega \leq \kappa \leq \alpha$.

(a′) $2^{\underline{\kappa}} \leq \alpha$;

(c′) $\boldsymbol{S}((2^I)_{(\kappa)}) \leq \alpha^+$ for all sets I;

(d′) there is $\mathscr{S} \subset 2^\alpha$ such that $|\mathscr{S}| = 2^\alpha$ and $\mathscr{S}$ is of κ-large oscillation;

(e′) there is $\mathscr{S} \subset 2^\alpha$ such that $|\mathscr{S}| = \alpha^+$ and $\mathscr{S}$ is of κ-large oscillation;

(f′) $\boldsymbol{d}((2^{(2^\alpha)})_{(\kappa)}) \leq \alpha$; and

(g′) $\boldsymbol{d}((2^{(\alpha^+)})_{(\kappa)}) \leq \alpha$.

Each of these conditions follows from its analogue in Theorem 3.16. (For (d)⇒(d′) and (e)⇒(e′) see the proof of Corollary 3.17.) Thus they all hold for $\kappa = \omega$. With the methods of Theorem 3.15 the following implications are proved:

$$(d') \Leftrightarrow (f') \Rightarrow (e') \Leftrightarrow (g') \Rightarrow (c') \Leftrightarrow (a').$$

Only the implication (a′)⇒(c′) requires comment.

On the contrary suppose that $\boldsymbol{S}((2^I)_{(\kappa)}) > \alpha^+$ for some I, so that there is a family $\{U^\xi : \xi < \alpha^+\}$ of pairwise disjoint non-empty basic open subsets of $(2^I)_{(\kappa)}$. Then $|R(U^\xi)| < \kappa$ for $\xi < \alpha^+$ and since α^+ is regular there is $\lambda < \kappa$ such that

$$|\{\xi < \alpha : |R(U^\xi)| = \lambda\}| = \alpha^+ .$$

Thus $\boldsymbol{S}((2^I)_{(\lambda^+)}) > \alpha^+$. On the other hand from the implication (a)⇒(c) of Theorem 3.16 (with 2^λ and λ^+ replacing α and κ, respectively) we have

$$\boldsymbol{S}((2^I)_{(\lambda^+)}) \leq (2^\lambda)^+ \leq (2^{\underline{\kappa}})^+ \leq \alpha^+ ,$$

a contradiction.

Notes for § 3

Pairs of cardinals $\langle \kappa, \alpha \rangle$ with α strongly κ-inaccessible were introduced by Fuhrken [65] (with different terminology) in connection with a generalization of Łoś' ultraproduct theorem (11.5).

The special case $\kappa = \omega$, $\alpha = \omega^+$ of Theorem 3.2 was proved implicitly by Marczewski-Szpilrajn [41] and used to show that the product of second-countable spaces has no uncountable cellular family, and by Marczewski [47] and Bockstein [48]; Mazur [52] gives an explicit statement and proof of this special case and uses it to prove that under certain conditions functions sequentially continuous on a product are continuous. Šanin [46a] had proved Theorem 3.2 for $\kappa = \omega$ and general α and used this result to prove in Šanin [46b], [48] theorems about calibres of product spaces.

Erdős-Rado [60] prove Theorem 3.2 for non-limit cardinals α. Short proofs of this case were also given by Michael [62] and Davies [67]. The general form of Theorem 3.2 is from Erdős and Rado [69] (Lemma 1). The second proof of Theorem 3.2 is theirs; the first proof, essentially the same as an unpublished proof of Davies referred to by Erdős-Rado [69] (page 467), is from Comfort and Negrepontis [72a] (Theorem 2.1). This proof is related to arguments given by Całczynska-Karlowicz [64], Marek [64] and Mostowski [69] (Theorem 13.3.1).

The Marczewski-Šanin precursor of Theorem 3.2 was rediscovered by Solovay and Tennenbaum [71] (Theorem 6.3) in the form of a result which in topological language reads as follows. *Let* $\{X_\xi : \xi < \alpha\}$ *be an inverse system of compact, extremally disconnected spaces such that* $\boldsymbol{S}(X_\xi) \leq \omega^+$ *for* $\xi < \alpha$ *and the connecting functions* $f_{\xi\zeta}: X_\xi \to X_\zeta$ *are continuous, surjective, and "complete" in the sense that* $f_{\xi\zeta}^{-1}(\mathrm{cl}\, U) = \mathrm{cl}\, f_{\xi\zeta}^{-1}(U)$ *for every open subset* U *of* X_ζ*; then* $\boldsymbol{S}(\varprojlim_\xi X_\xi) \leq \omega^+$.

Theorem 3.5 has been proved independently by several authors, in some cases with a view to the application of Boolean-valued models in set theory. Cf. Vopěnka [65]; Bukovský [68] (introduction to § 2), where the theorem is derived from some related results of Pierce [58a]; Štěpánek and Vopěnka [67], where Theorem 3.6 is also proved. A generalized treatment of Theorem 3.5 appears in Mostowski [69] (Chapter XIII, § 3). The first proof of Theorem 3.6 is apparently due to Haratomi [31].

The treatment of 3.8—3.11 follows Comfort and Negrepontis [72a]. Corollary 3.10 has been given by Engelking and Karłowicz [65] (Theorem 6) (see also Engelking [65] (Theorem 7)) in the case that $\alpha = \mathfrak{m}^+$, $\kappa = \mathfrak{n}^+$, and $\mathfrak{m}^{\mathfrak{n}} = \mathfrak{m}$.

Other topological consequences of Theorem 3.2, not considered in this work and dealing mostly with the possibility of extending continuous functions from dense subsets of product spaces, are given in Corson [59], Corson and Isbell [60], Ross and Stone [64], Ulmer [70], Noble and Ulmer [72], and Comfort and Negrepontis [72b], [*].

The first statement of Corollary 3.12 is proved by a different method in Erdős-Tarski [43]. The given proof of Corollary 3.12 (together with a proof of Theorem 3.5) is by Efimov [68].

Theorem 3.13 for the usual product topology is due to Kurepa [62]. Our proof is from Hedrlín [66] and Juhász [71] (Theorem 4.6). Theorem 3.14 (and other topological applications of the partition calculus of Erdős-Rado) are proved in Juhász [71] (Theorem 2.11).

On questions of priority regarding the consistency of the existence of a Souslin line, cf. the remarks in the review of Jech [67] by Morley [68b]. The consistency of the existence of a generalized (to higher cardinals) Souslin line with ZFC has been proved by Hrbáček [67] using an extension of the method of Jech [67].

For further discussion of Souslin's problem (3.15) see the expository articles of M. E. Rudin [69] and Jech [71a]. The consequence of Martin's axiom given in 3.15 is stated by Juhász [70] and attributed there to Kunen and independently to Solovay and Rowbottom.

The implication (a′)⇒(d′) in the remark following Theorem 3.20 (closely related to the vital implication (a)⇒(d) of Theorem 3.16) is proved by Fichtenholz-Kantorovitch [35] (Lemmes II, IV) for $\kappa=\omega$ and for $\alpha=\omega$ or $\alpha=2^{\omega}$; by Hausdorff [36a] for general α and $\kappa=\omega$ in his fundamental and original paper; and by Tarski [39] (Hilfsatz 3.16) for general α and κ, using preliminary results from Tarski [30b] (Lemme 58).

The implication (a)⇒(d) of Theorem 3.16 is proved by Engleking-Karłowicz [65] (Theorem 3) for $\kappa=\omega$, using the result of Hausdorff [36a]; by Ketonen for $\kappa=\omega$ in 1969 (and used by Kunen [72] to prove the existence of good ultrafilters without GCH (Theorem 10.4)). Theorem 3.16 is from Comfort-Negrepontis [72a] (Theorem 3.1); see also Ketonen [71] (Chapter 0).

For $\kappa=\omega$ Corollary 3.18 is due to Pondiczery [44] and Hewitt [46]; Lemma 3.19 and Theorem 3.20 are in Pondiczery [44].

The implication (a′)⇒(c′) in the final Remark is from a letter by S. Shelah to the authors (December, 1971).

§ 4. The General Theory of Jónsson Classes

The theory of Jónsson classes is the setting of proper generality for the application of the classical principle of back-and-forth induction of Cantor and Hausdorff. Cantor conceived of this principle and used it to prove that the set of rational numbers is described as an ordered set by the following properties: it is countable, without a least or a greatest element, and densely ordered. Hausdorff generalized this result in his theory of η_α-sets (studied in § 5 below). Erdős-Gillman-Henriksen proved similar results for the theory of real-closed fields.

Jónsson found the list of conditions ((I)—(VI)) that an abstract class of relational structures must satisfy in order that it possess unique α-homogeneous-universal structures of powers $\alpha = \alpha^{\underline{\alpha}}$.

The inability to prove in ZFC the existence of many cardinals α such that $\alpha = \alpha^{\underline{\alpha}}$ (unless the generalized continuum hypothesis or large cardinal axioms are assumed) led to the somewhat cumbersome generalization by Morley-Vaught of the notion of "homogeneous-universal" structures to "special" structures; their existence and uniqueness is proved for powers $\alpha = 2^{\underline{\alpha}}$ (Theorem 4.10).

The close relation of these results to model theory is developed in § 11, where it is shown that an elementary type is an (elementary) Jónsson class (Theorem 11.11) whose (elementarily) α-homogeneous-universal and (more generally) special structures are characterized in terms of saturation (Theorem 11.18).

The last part of this section outlines a generalized approach to the theory of Jónsson classes, which dispenses with the (superficial) use of relational structures.

Definitions. A *(similarity) type* τ is a function $\tau:\rho\to\omega$ where ρ is an ordinal.

A *structure* (or, a *relational system*) of *type* τ, denoted $\mathfrak{A} = \langle A, R_i^{\mathfrak{A}}\rangle_{i<\rho}$, consists of a non-empty set A and

a *constant* of A (*i. e.*, an element $R_i^{\mathfrak{A}}$ of A) if $i<\rho$ and $\tau(i)=0$, and
a $\tau(i)$-ary *relation* on A (*i. e.*, a subset $R_i^{\mathfrak{A}}$ of $\mathfrak{A}^{\tau(i)}$) if $i<\rho$ and $\tau(i)>0$.

The ordinal ρ is the *index* of $\mathfrak{A}$, and the set A is the *universe* of $\mathfrak{A}$. The universes of structures $\mathfrak{A}, \mathfrak{B}, \mathfrak{C}, \ldots$ are denoted by $A, B, C, \ldots$.

The *elements* of $\mathfrak{A}$ are the elements of A. We often write $a \in \mathfrak{A}$ instead of $a \in A$. The *cardinality* of $\mathfrak{A}$, denoted by $|\mathfrak{A}|$, is the cardinality $|A|$ of A.

Two systems $\mathfrak{A}$ and $\mathfrak{B}$ are called *similar* if they are of the same type. Let $\mathfrak{A} = \langle A, R_i^{\mathfrak{A}} \rangle_{i<\rho}$ and $\mathfrak{B} = \langle B, R_i^{\mathfrak{B}} \rangle_{i<\rho}$ be two (similar) systems of type τ. Then $\mathfrak{B}$ is a *subsystem* of $\mathfrak{A}$ (denoted by $\mathfrak{B} \subset \mathfrak{A}$) if

$$B \subset A,$$
$$R_i^{\mathfrak{B}} = R_i^{\mathfrak{A}} \quad \text{if } i<\rho \quad \text{and } \tau(i)=0, \quad \text{and}$$
$$R_i^{\mathfrak{B}} = R_i^{\mathfrak{A}} \cap B^{\tau(i)} \quad \text{if } i<\rho \quad \text{and } \tau(i)>0.$$

If A and B are sets and $f: B \to A$ is a (one-to-one) function, then for $1 \leq n < \omega$ the function $f^{(n)}: B^n \to A^n$ is defined by

$$f^{\text{ⁿ}}(\langle b_1, b_2, \ldots, b_n \rangle) = \langle f(b_1), f(b_2), \ldots, f(b_n) \rangle.$$

Let $\mathfrak{A} = \langle A, R_i^{\mathfrak{A}} \rangle_{i<\rho}$ and $\mathfrak{B} = \langle B, R_i^{\mathfrak{B}} \rangle_{i<\rho}$ be two (similar) structures of type τ. A function $f: B \to A$ is an *embedding* of $\mathfrak{B}$ into $\mathfrak{A}$ (also denoted by $f: \mathfrak{B} \to \mathfrak{A}$) if

$$f \text{ is one-to-one, and}$$
$$\langle f[B], f^{(\tau(i))}[R_i^{\mathfrak{B}}] \rangle_{i<\rho} \subset \mathfrak{A}$$

(where $f^{\tau(i)}[R_i^{\mathfrak{B}}]$ denotes $R_i^{\mathfrak{A}}$ if $i<\rho$ and $\tau(i)=0$). The structure $\mathfrak{B}$ is the *domain* of f, and the structure $\langle f[B], f^{\tau(i)}[R_i^{\mathfrak{B}}] \rangle_{i<\rho}$ is the *image* of f and is denoted by $f[\mathfrak{B}]$. If $f: \mathfrak{B} \to \mathfrak{A}$ is an embedding, and if in addition $f: B \to A$ is an onto function, then f is an *isomorphism* (*from* $\mathfrak{B}$ *to* $\mathfrak{A}$, or *between* $\mathfrak{B}$ *and* $\mathfrak{A}$); this is denoted by $f: \mathfrak{B} \cong \mathfrak{A}$ (and we write $\mathfrak{B} \cong \mathfrak{A}$). An *automorphism* on $\mathfrak{A}$ is an isomorphism from $\mathfrak{A}$ onto $\mathfrak{A}$.

Let ζ be a non-zero ordinal and let $\{\mathfrak{A}_\xi : \xi < \zeta\}$ be a family of (similar) relational systems of type $\tau \in \omega^\rho$. Then $\{\mathfrak{A}_\xi : \xi < \zeta\}$ *is a chain (with length ζ)* if $\mathfrak{A}_\eta \subset \mathfrak{A}_\xi$ for $\eta < \xi < \zeta$. If $\mathfrak{A}_\xi = \langle A_\xi, R_i^{\mathfrak{A}_\xi} \rangle_{i<\rho}$ for $\xi < \zeta$ and $\{\mathfrak{A}_\xi : \xi < \zeta\}$ is a chain, then the *union* of the chain $\{\mathfrak{A}_\xi : \xi < \zeta\}$, denoted $\bigcup_{\xi<\zeta} \mathfrak{A}_\xi$, is the relational system

$$\langle \bigcup_{\xi<\zeta} A_\xi, \bigcup_{\xi<\zeta} R_i^{\mathfrak{A}_\xi} \rangle_{i<\rho};$$

this is also a system of type τ. A chain $\{\mathfrak{A}_\xi : \xi < \zeta\}$ is *continuous* if $\mathfrak{A}_\xi = \bigcup_{\eta<\xi} \mathfrak{A}_\eta$ for all non-zero limit ordinals $\xi < \zeta$.

We note that if $\zeta = \xi + 1$, then $\bigcup_{\eta<\zeta} \mathfrak{A}_\eta = \mathfrak{A}_\xi$.

Let $\mathbb{K}$ be a (possibly proper) class of similar relational systems and let α be any cardinal number. By $\mathbb{K}_\alpha$ we denote the class of all members of $\mathbb{K}$ of cardinality less than α, *i.e.*, $\mathbb{K}_\alpha = \{\mathfrak{A} : \mathfrak{A} \in \mathbb{K}, |\mathfrak{A}| < \alpha\}$. We set $\mathbb{P}^{\mathbb{K}}(\mathfrak{A}) = \{\mathfrak{B} : \mathfrak{B} \subset \mathfrak{A} \text{ and } \mathfrak{B} \in \mathbb{K}\}$; these are the $\mathbb{K}$-*subsystems* of $\mathfrak{A}$. We also set $\mathbb{P}_\alpha^{\mathbb{K}}(\mathfrak{A}) = \{\mathfrak{B} : \mathfrak{B} \in \mathbb{P}^{\mathbb{K}}(\mathfrak{A}) \text{ and } |\mathfrak{B}| < \alpha\}$.

If $\mathbb{C}$ is a subclass of $\mathbb{K}$, we denote by $\mathbb{I}(\mathbb{C})$ the class of isomorphism equivalence classes of the elements of $\mathbb{C}$; *e.g.* $\mathbb{I}(\mathbb{K})$, $\mathbb{I}(\mathbb{K}_\alpha)$, $\mathbb{I}(\mathbb{P}_\alpha^{\mathbb{K}}(\mathfrak{A}))$.

Definition. Let α be an infinite cardinal. A *Jónsson class* (*with* $(\mathrm{VI})_\alpha$) is a class $\mathbb{K}$ of similar relational systems satisfying the following conditions.

(I) $\mathbb{K}$ contains systems of arbitrarily large cardinality.

(II) If $\mathfrak{A} \in \mathbb{K}$ and $\mathfrak{B}$ is isomorphic to $\mathfrak{A}$ then $\mathfrak{B} \in \mathbb{K}$.

(III) The *embedding property*. If $\mathfrak{A}$ and $\mathfrak{B}$ are in $\mathbb{K}$ then there are $\mathfrak{C} \in \mathbb{K}$ and embeddings $f:\mathfrak{A} \to \mathfrak{C}$ and $g:\mathfrak{B} \to \mathfrak{C}$.

(IV) The *amalgamation property*. If $\mathfrak{A}, \mathfrak{B}$ and $\mathfrak{C}$ are in $\mathbb{K}$, and $f:\mathfrak{C} \to \mathfrak{A}$ and $g:\mathfrak{C} \to \mathfrak{B}$ are embeddings, then there are $\mathfrak{D} \in \mathbb{K}$ and embeddings $f_1:\mathfrak{A} \to \mathfrak{D}$ and $g_1:\mathfrak{B} \to \mathfrak{D}$ such that the diagram

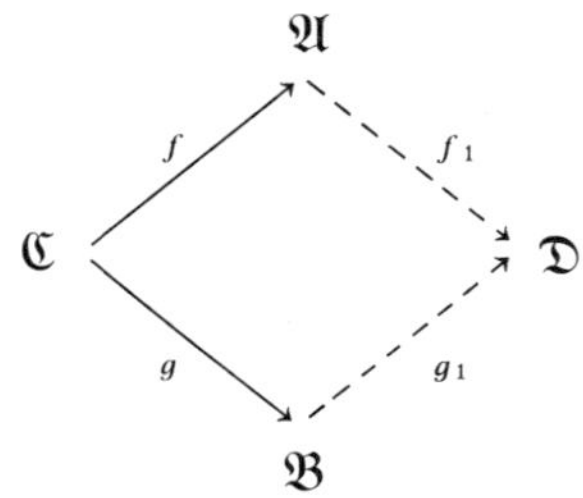

is *commutative (i. e., $f_1 \circ f = g_1 \circ g$)*.

(V) The *compactness property*. The union of any chain of structures, each of which is in $\mathbb{K}$, is in $\mathbb{K}$.

$(\mathrm{VI})_\alpha$ The *downward Löwenheim-Skolem property for the cardinal* α. If $\mathfrak{A} \in \mathbb{K}$ and $B \in \mathscr{P}_\alpha(A)$ then there is $\mathfrak{C} \in \mathbb{K}$ such that $B \subset C$ and $\mathfrak{C} \in \mathbb{P}_\alpha^{\mathbb{K}}(\mathfrak{A})$.

Reference to (some of the) properties (I)—$(\mathrm{VI})_\alpha$ for a class of structures is always according to these definitions.

The following two lemmas are frequently used in this section.

4.1. Lemma. *Let* $\alpha \geq \omega$. *If* $\mathbb{K}$ *satisfies* (V) *and* $(\mathrm{VI})_\alpha$, *and if* $\mathfrak{A} \in \mathbb{K}$ *and* $|A| = \alpha$, *then* $\mathfrak{A}$ *is the union of a continuous chain of elements of* $\mathbb{K}_\alpha$ *of length* $\mathrm{cf}(\alpha)$.

Proof. Let $\{B_\xi : \xi < \mathrm{cf}(\alpha)\}$ be a family of subsets of A such that

$$|B_\xi| < \alpha \quad \text{for } \xi < \mathrm{cf}(\alpha) \quad \text{and}$$
$$\bigcup_{\xi < \mathrm{cf}(\alpha)} B_\xi = A .$$

For $\xi < \mathrm{cf}(\alpha)$ we define structures $\mathfrak{A}_\xi$ in $\mathbb{K}$ such that

(i) $B_\xi \subset A_{\xi+1}$ for $\xi < \mathrm{cf}(\alpha)$,

(ii) $\mathfrak{A}_\xi \subset \mathfrak{A}_\zeta$ for $\xi < \zeta < \mathrm{cf}(\alpha)$,

(iii) $\mathfrak{A}_\xi = \bigcup_{\zeta < \xi} \mathfrak{A}_\zeta$ for non-zero limit ordinals $\xi < \mathrm{cf}(\alpha)$, and

(iv) $|\mathfrak{A}_\xi| < \alpha$ for $\xi < \mathrm{cf}(\alpha)$.

We proceed by transfinite recursion. Let $\mathfrak{A}_0 \in \mathbb{P}_\alpha^{\mathbb{K}}(\mathfrak{A})$ be chosen arbitrarily; the existence of such a structure $\mathfrak{A}_0$ follows from $(\mathrm{VI})_\alpha$. If ξ is a non-zero limit ordinal and $\xi < \mathrm{cf}(\alpha)$, we define $\mathfrak{A}_\xi$ by condition (iii). Then $\mathfrak{A}_\xi \in \mathbb{K}$ by (V), and $|\mathfrak{A}_\xi| < \alpha$ since $\xi < \mathrm{cf}(\alpha)$ and $|\mathfrak{A}_\zeta| < \alpha$ for $\zeta < \xi$.

Now let $\xi < \mathrm{cf}(\alpha)$. We define $\mathfrak{A}_{\xi+1}$. We note that $|A_\xi \cup B_\xi| < \alpha$, so by $(\mathrm{VI})_\alpha$ there is $\mathfrak{A}_{\xi+1} \in \mathbb{P}_\alpha^{\mathbb{K}}(\mathfrak{A})$ such that $A_\xi \cup B_\xi \subset A_{\xi+1}$. Clearly then $\mathfrak{A}_\xi \subset \mathfrak{A}_{\xi+1}$.

This completes the recursive definition of the chain $\{\mathfrak{A}_\xi : \xi < \mathrm{cf}(\alpha)\}$. It is clear that this is a continuous chain, and that $\bigcup_{\xi < \mathrm{cf}(\alpha)} \mathfrak{A}_\xi = \mathfrak{A}$.

4.2. Lemma. *Let* $\alpha \geq \omega$. *If* $\mathbb{K}$ *satisfies* (V) *and* $(\mathrm{VI})_\alpha$, *then* $\mathbb{K}$ *satisfies* $(\mathrm{VI})_\beta$ *for all cardinals* $\beta > \alpha$.

Proof. (a) We first prove that if $\mathbb{K}$ satisfies $(\mathrm{VI})_\alpha$ and (V) then $\mathbb{K}$ satisfies $(\mathrm{VI})_{\alpha^+}$. Let $\mathfrak{A} \in \mathbb{K}$ and let $B \subset A$ with $|B| < \alpha^+$. If $|B| < \alpha$, then from $(\mathrm{VI})_\alpha$ there is a structure $\mathfrak{C} \in \mathbb{P}_\alpha^{\mathbb{K}}(\mathfrak{A})$ such that $B \subset C$. We thus assume that $|B| = \alpha$. Let $\{B_\xi : \xi < \mathrm{cf}(\alpha)\}$ be a family of subsets of B such that

$$|B_\xi| < \alpha \quad \text{for } \xi < \mathrm{cf}(\alpha) \quad \text{and}$$
$$\bigcup_{\xi < \mathrm{cf}(\alpha)} B_\xi = B.$$

For $\xi < \mathrm{cf}(\alpha)$ we define structures $\mathfrak{A}_\xi \in \mathbb{K}$ such that

(i) $B_\xi \subset A_{\xi+1}$ for $\xi < \mathrm{cf}(\alpha)$,

(ii) $\mathfrak{A}_\xi \subset \mathfrak{A}_\zeta$ for $\xi < \zeta < \mathrm{cf}(\alpha)$, and

(iii) $|\mathfrak{A}_\xi| < \mathrm{cf}(\alpha)$ for $\xi < \mathrm{cf}(\alpha)$.

We proceed by transfinite recursion. Let $\mathfrak{A}_0 \in \mathbb{P}_\alpha^{\mathbb{K}}(\mathfrak{A})$ be chosen arbitrarily; the existence of such a structure $\mathfrak{A}_0$ follows from $(\mathrm{VI})_\alpha$. If ξ is a non-zero limit ordinal and $\xi < \mathrm{cf}(\alpha)$, we set $\mathfrak{A}_\xi = \bigcup_{\zeta < \xi} \mathfrak{A}_\zeta$. Then $\mathfrak{A}_\xi \in \mathbb{K}$ by (V), and $|\mathfrak{A}_\xi| < \alpha$ since $\xi < \mathrm{cf}(\alpha)$.

Now for $\xi < \mathrm{cf}(\alpha)$ we define $\mathfrak{A}_{\xi+1}$. We have $|A_\xi \cup B_\xi| < \alpha$, so by $(\mathrm{VI})_\alpha$ there is $\mathfrak{A}_{\xi+1} \in \mathbb{P}_\alpha^{\mathbb{K}}(A)$ such that $A_\xi \cup B_\xi \subset A_{\xi+1}$. Clearly then $\mathfrak{A}_\xi \subset \mathfrak{A}_{\xi+1}$.

This completes the recursive definition of the chain $\{\mathfrak{A}_\xi : \xi < \mathrm{cf}(\alpha)\}$. We define

$$\mathfrak{C} = \bigcup_{\xi < \mathrm{cf}(\alpha)} \mathfrak{A}_\xi,$$

and it is clear that $B \subset C$ and $\mathfrak{C} \in \mathbb{P}_{\alpha^+}^{\mathbb{K}}(\mathfrak{A})$. Thus $\mathbb{K}$ satisfies $(\mathrm{VI})_{\alpha^+}$.

(b) We next prove that if β is a limit cardinal and if $\mathbb{K}$ satisfies $(\mathrm{VI})_\gamma$ for all γ such that $\alpha \leq \gamma < \beta$, then $\mathbb{K}$ satisfies $(\mathrm{VI})_\beta$. Let $\mathfrak{A} \in \mathbb{K}$ and let $B \subset A$ with $|B| = \gamma < \beta$. Since β is a limit cardinal we have $|B| < \gamma^+ < \beta$. By assumption, $(\mathrm{VI})_{\gamma^+}$ is satisfied in $\mathbb{K}$. Thus there is $\mathfrak{C} \in \mathbb{P}_{\gamma^+}^{\mathbb{K}}(\mathfrak{A})$ with $B \subset C$, so $(\mathrm{VI})_\beta$ is satisfied in $\mathbb{K}$.

(c) We complete the proof of the lemma by employing transfinite induction. Let β be the smallest cardinal greater than α for which $(\mathrm{VI})_\beta$

fails. If $\beta=\gamma^+$, we use (a); if β is a limit cardinal, we use (b). This completes the proof of the lemma.

Definitions. Let α be an infinite cardinal, let $\mathbb{K}$ be a Jónsson class, and let $\mathfrak{A}\in\mathbb{K}$.

(a) $\mathfrak{A}$ is *α-universal in* $\mathbb{K}$ if for every $\mathfrak{B}\in\mathbb{K}$ with $|\mathfrak{B}|\leq\alpha$ there is an embedding of $\mathfrak{B}$ into $\mathfrak{A}$.

(b) $\mathfrak{A}$ is *α-homogeneous in* $\mathbb{K}$ if for every $\mathfrak{B}\in\mathbb{K}$ with $|\mathfrak{B}|<\alpha$, and for every pair of embeddings f,g of $\mathfrak{B}$ into $\mathfrak{A}$, there is an automorphism h of $\mathfrak{A}$ such that $h\circ f=g$.

(c) If $\mathfrak{A}$ is both α-homogeneous and α-universal in $\mathbb{K}$, then $\mathfrak{A}$ is *α-homogeneous-universal in* $\mathbb{K}$.

(d) $\mathfrak{A}$ is *special (of cardinality α) in* $\mathbb{K}$ if $|\mathfrak{A}|=\alpha$ and $\mathfrak{A}$ is equal to the union of a chain $\{\mathfrak{A}_\beta:\beta<\alpha\}$ such that

$$\mathfrak{A}_\beta\in\mathbb{K},\quad\text{and}$$
$$\mathfrak{A}_\beta \text{ is } \beta^+\text{-homogeneous-universal}$$

for $\beta<\alpha$. The chain $\{\mathfrak{A}_\beta:\beta<\alpha\}$ is called a *specializing chain for* $\mathfrak{A}$.

Mention of the class $\mathbb{K}$ in these definitions is omitted if there is no danger of confusion.

We note that a specializing chain for a special system of cardinality α is indexed by the set of all *cardinals* less than α.

We note that if a system $\mathfrak{A}$ is α-homogeneous-universal in $\mathbb{K}$ and of cardinality α, then $\mathfrak{A}$ is special in $\mathbb{K}$. Indeed, a specializing chain for $\mathfrak{A}$ is given by $\mathfrak{A}_\beta=\mathfrak{A}$ for $\beta<\alpha$.

4.3. Lemma. *Let* $\mathbb{K}$ *satisfy* (II) *and* (V), *let* $\mathfrak{A}\in\mathbb{K}$ *and let* $f:\mathfrak{A}\to\mathfrak{A}$ *be an embedding. Then there are* $\mathfrak{B}\in\mathbb{K}$ *and* $g:\mathfrak{B}\to\mathfrak{B}$ *such that*

$$\mathfrak{A}\subset\mathfrak{B},$$
$$|\mathfrak{A}|=|\mathfrak{B}|,$$
$$f\subset g,\quad\textit{and}$$
$$g \textit{ is an automorphism of } \mathfrak{B}.$$

Proof. If $|\mathfrak{A}|$ is finite, then f itself is an automorphism; in this case we take $\mathfrak{B}=\mathfrak{A}$ and $g=f$. Thus we assume that $|\mathfrak{A}|$ is infinite.

Let $\mathfrak{A}_0=f[\mathfrak{A}]$ and $\mathfrak{A}_1=\mathfrak{A}$, and let f_0 be the isomorphism f from $\mathfrak{A}_1$ onto $\mathfrak{A}_0$ obtained by restricting the range of f. We define A_2 to be the disjoint union of A_1 with a set which is in a one-to-one correspondence with the set $A\setminus f[A]$ and which contains no elements of A. Let $f_1:A_2\to A$ be the one-to-one function determined by this one-to-one correspondence on $A_2\setminus A_1$, and by the requirement that $f_1|A_1=f_0$. Let $\mathfrak{A}_2$ be the relational system defined on A_2 and determined by the requirement that f_1 be an isomorphism from $\mathfrak{A}_2$ to $\mathfrak{A}_1$; then $\mathfrak{A}_2 \in \mathbb{K}$ by (II). Thus we have the following commutative diagram:

$$\begin{array}{ccc} \mathfrak{A}_1 & \xrightarrow[\cong]{f_0} & \mathfrak{A}_0 \\ \cap & & \cap \\ \mathfrak{A}_2 & \xrightarrow[\cong]{f_1} & \mathfrak{A}_1 \end{array}$$

Now f_1 can be considered as an embedding

$$f_1 : \mathfrak{A}_2 \to \mathfrak{A}_2$$

and the same process as above can be repeated with $\mathfrak{A}_2, \mathfrak{A}_1$ and f_1 replacing $\mathfrak{A}_1, \mathfrak{A}_0$ and f_0, respectively. By recursion we obtain structures $\mathfrak{A}_n \in \mathbb{K}$, and isomorphisms $f_n : \mathfrak{A}_{n+1} \stackrel{\cong}{\to} \mathfrak{A}_n$, such that

$$\mathfrak{A}_n \subset \mathfrak{A}_{n+1}, \quad \text{and}$$
$$f_n \subset f_{n+1}$$

for $n < \omega$. We set

$$\mathfrak{B} = \bigcup_{n<\omega} \mathfrak{A}_n \quad \text{and}$$
$$g = \bigcup_{n<\omega} f_n .$$

Then $\mathfrak{B} \in \mathbb{K}$ by (V), and it is clear that g is an automorphism of $\mathfrak{B}$.

4.4. Lemma. *Let α be an infinite cardinal and let $\mathbb{K}$ be a Jónsson class with* $(\mathrm{VI})_{\alpha^+}$. *Let $\mathfrak{A}$ be in $\mathbb{K}$ with $|\mathfrak{A}| = \alpha$, and let $\mathscr{S}$ be a family of embeddings of $\mathbb{K}$-subsystems of $\mathfrak{A}$ (i.e., of elements of $\mathbb{P}^{\mathbb{K}}(\mathfrak{A})$) into $\mathfrak{A}$, with $|\mathscr{S}| \leq \alpha$. Then there is $\mathfrak{B} \in \mathbb{K}$ such that*

$$\mathfrak{A} \subset \mathfrak{B},$$
$$|\mathfrak{B}| = \alpha, \quad \textit{and}$$

every element of $\mathscr{S}$ can be extended to an automorphism of $\mathfrak{B}$.

Proof. (a) We first prove the lemma for the case $|\mathscr{S}| = 1$, say $\mathscr{S} = \{f_0\}$ where $f_0 : \mathfrak{A}_0 \to \mathfrak{A}$ for some $\mathfrak{A}_0 \in \mathbb{P}^{\mathbb{K}}(\mathfrak{A})$. We set $\mathfrak{A}_1 = \mathfrak{A}$. By the amalgamation property (IV) for the following diagram

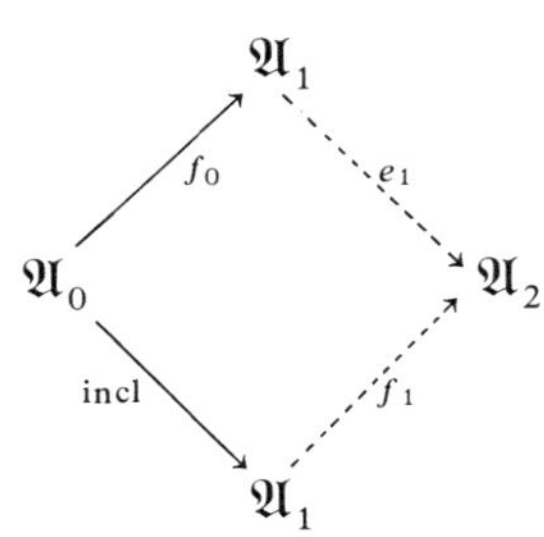

there are $\mathfrak{A}_2 \in \mathbb{K}$ and embeddings e_1 and f_1, as indicated, making the diagram commutative. We assume without loss of generality by $(\mathrm{VI})_{\alpha^+}$ that $|\mathfrak{A}_2| = \alpha$, and by (II) that $\mathfrak{A}_1 \subset \mathfrak{A}_2$. Then $f_0 \subset f_1$, and we have the following commutative diagram

$$\begin{array}{ccc} \mathfrak{A}_0 & \xrightarrow{f_0} & \mathfrak{A}_1 \\ \cap & & \cap \\ \mathfrak{A}_1 & \xrightarrow{f_1} & \mathfrak{A}_2 . \end{array}$$

We proceed by an argument similar to the one above, thus obtaining by recursion two sequences $\{\mathfrak{A}_n : n < \omega\}$ and $\{f_n : n < \omega\}$ such that

$$\mathfrak{A}_n \in \mathbb{K},$$
$$f_n : \mathfrak{A}_n \to \mathfrak{A}_{n+1},$$
$$\mathfrak{A}_n \subset \mathfrak{A}_{n+1}, \quad \text{and}$$
$$f_n \subset f_{n+1}$$

for $n < \omega$. We set $\mathfrak{B} = \bigcup_{n<\omega} \mathfrak{A}_n$ and $g = \bigcup_{n<\omega} f_n$. (We note, using (V), that $\mathfrak{B} \in \mathbb{K}$.) Thus $g : \mathfrak{B} \to \mathfrak{B}$ is an embedding. From Lemma 4.3, with $\mathfrak{B}$ and g replacing $\mathfrak{A}$ and f respectively, we obtain the required automorphism.

(b) Let $|\mathscr{S}| \leq \alpha$ and let $\langle f_\xi : \xi < \alpha \rangle$ be a well-ordering of $\mathscr{S}$ such that every element of $\mathscr{S}$ appears α times. We define systems $\mathfrak{A}_\xi \in \mathbb{K}$ and embeddings $g(f,\xi)$ of $\mathbb{K}$-subsystems of $\mathfrak{A}_\xi$ for $\xi < \alpha$, $f \in \mathscr{S}$, such that

(i) $\mathfrak{A}_0 = \mathfrak{A}$ and $g(f,0) = f$ for $f \in \mathscr{S}$;

(ii) $\mathfrak{A}_\xi \subset \mathfrak{A}_\eta$ and $g(f,\xi) \subset g(f,\eta)$ for $\xi < \eta < \alpha$ and $f \in \mathscr{S}$;

(iii) $|\mathfrak{A}_\xi| = \alpha$ for $\xi < \alpha$;

(iv) if ξ is a non-zero limit ordinal and $\xi < \alpha$, then $\mathfrak{A}_\xi = \bigcup_{\eta<\xi} \mathfrak{A}_\eta$ and $g(f,\xi) = \bigcup_{\eta<\xi} g(f,\eta)$; and,

(v) if $f_\xi = f$ then $g(f,\xi+1)$ is an automorphism of $\mathfrak{A}_{\xi+1}$ for $\xi < \alpha$.

We proceed by transfinite recursion. We define $\mathfrak{A}_0$ and $g(f,0)$ for $f \in \mathscr{S}$ by (i), and $\mathfrak{A}_\xi$ and $g(f,\xi)$ for non-zero limit ordinals $\xi < \alpha$ by condition (iv). We now let $\xi < \alpha$ and we define $\mathfrak{A}_{\xi+1}$ and $g(f,\xi+1)$ for $f \in \mathscr{S}$. If $f \neq f_\xi$ we set $\mathfrak{A}_{\xi+1} = \mathfrak{A}_\xi$ and $g(f,\xi+1) = g(f,\xi)$. If $f = f_\xi$ then by part (a) above there are $\mathfrak{A}_{\xi+1} \in \mathbb{K}$ and an automorphism $g(f,\xi+1)$ of $\mathfrak{A}_{\xi+1}$ such that

$$\mathfrak{A}_\xi \subset \mathfrak{A}_{\xi+1},$$
$$|\mathfrak{A}_{\xi+1}| = \alpha, \quad \text{and}$$
$$g(f,\xi) \subset g(f,\xi+1).$$

This completes the recursive definition. We set

$$\mathfrak{B} = \bigcup_{\xi<\alpha} \mathfrak{A}_\xi, \quad \text{and}$$
$$g(f) = \bigcup_{\xi<\alpha} g(f,\xi) \quad \text{for } f \in \mathscr{S}.$$

It follows from (V), (i) and (ii) that $\mathfrak{B}\in\mathbb{K}$, and from (i) and (iii) that $|\mathfrak{B}|=\alpha$. We have $\mathfrak{A}\subset\mathfrak{B}$ and $f\subset g(f)$ by (i) and (ii), and $g(f)$ is an automorphism of $\mathfrak{B}$ since f appears α times in the well-ordering of $\mathscr{S}$ (so that $\{\xi<\alpha:f_\xi=f\}$ is cofinal in α). The proof is complete.

4.5. Theorem. *Let* $\omega\leq\kappa\leq\alpha=\alpha^{\underline{\kappa}}$, *let* $\mathbb{K}$ *be a Jónsson class with* $(\mathrm{VI})_{\alpha^+}$, *and let* $\mathfrak{A}$ *be a structure in* $\mathbb{K}$ *with* $|\mathfrak{A}|=\alpha$. *Then there is a structure* $\mathfrak{B}\in\mathbb{K}$ *such that*

$$\mathfrak{A}\subset\mathfrak{B},$$
$$|\mathfrak{B}|=\alpha, \quad \textit{and}$$
$$\mathfrak{B} \textit{ is } \kappa\textit{-homogeneous}.$$

Proof. For a structure $\mathfrak{C}\in\mathbb{K}$, we let $\mathscr{E}^{\mathbb{K}}_\kappa(\mathfrak{C})$ be the set of all embeddings $f:\mathfrak{D}\to\mathfrak{C}$, where $\mathfrak{D}\in\mathbb{P}^{\mathbb{K}}_\kappa(\mathfrak{C})$. We define structures $\mathfrak{A}_\xi\in\mathbb{K}$ for $\xi<\mathrm{cf}(\alpha)$ and automorphisms $g(f,\xi)$ of $\mathfrak{A}_\xi$ for $f\in\mathscr{E}^{\mathbb{K}}_\kappa(\mathfrak{A}_\zeta)$ and $\zeta<\xi<\mathrm{cf}(\alpha)$, such that

(i) $\mathfrak{A}_0=\mathfrak{A}$;

(ii) $\mathfrak{A}_\xi\subset\mathfrak{A}_\eta$ for $\xi<\eta<\mathrm{cf}(\alpha)$;

(iii) $|\mathfrak{A}_\xi|=\alpha$ for $\xi<\mathrm{cf}(\alpha)$;

(iv) $\mathfrak{A}_\xi=\bigcup_{\eta<\xi}\mathfrak{A}_\eta$ for non-zero limit ordinals $\xi<\mathrm{cf}(\alpha)$; and

(v) if $f\in\mathscr{E}^{\mathbb{K}}_\kappa(\mathfrak{A}_\zeta)$ and $\zeta<\xi<\mathrm{cf}(\alpha)$, then $g(f,\xi)$ is an automorphism of $\mathfrak{A}_\xi$, such that

$$f\subset g(f,\xi) \quad \text{and}$$
$$g(f,\xi)\subset g(f,\eta) \quad \text{for } \zeta<\xi<\eta<\mathrm{cf}(\alpha).$$

We proceed by transfinite recursion. We set $\mathfrak{A}_0=\mathfrak{A}$; and if ξ is a non-zero limit ordinal and $\xi<\mathrm{cf}(\alpha)$ we define

$$\mathfrak{A}_\xi=\bigcup_{\eta<\xi}\mathfrak{A}_\eta, \quad \text{and}$$
$$g(f,\xi)=\bigcup_{\zeta<\eta<\xi} g(f,\eta) \quad \text{for } f\in\mathscr{E}^{\mathbb{K}}_\kappa(\mathfrak{A}_\zeta),\ \zeta<\xi.$$

We now let $\xi<\mathrm{cf}(\alpha)$ and we define $\mathfrak{A}_{\xi+1}$. We note that the cardinality of the set

$$\mathscr{S}_\xi=\mathscr{E}^{\mathbb{K}}_\kappa(\mathfrak{A}_\xi)\cup\{g(f,\xi):f\in\mathscr{E}^{\mathbb{K}}_\kappa(\mathfrak{A}_\zeta),\ \zeta<\xi\}$$

is at most $\alpha^{\underline{\kappa}}=\alpha$ (Lemma 1.22). By Lemma 4.4 there is a structure $\mathfrak{A}_{\xi+1}\in\mathbb{K}$ such that

$$|\mathfrak{A}_{\xi+1}|=\alpha,$$
$$\mathfrak{A}_\xi\subset\mathfrak{A}_{\xi+1}, \quad \text{and}$$

every element h of $\mathscr{S}_\xi$ has an extension to an automorphism of $\mathfrak{A}_{\xi+1}$, denoted by $g(h,\xi+1)$.

This completes the recursive definitions. We set

$$\mathfrak{B}=\bigcup_{\xi<\mathrm{cf}(\alpha)}\mathfrak{A}_\xi .$$

Then $\mathfrak{B}\in\mathbb{K}$ by (ii) and (V), $\mathfrak{A}\subset\mathfrak{B}$ by (i), and $|\mathfrak{B}|=\alpha$ by (iii). To prove that $\mathfrak{B}$ is κ-homogeneous let $f:\mathfrak{C}\to\mathfrak{B}$ be an embedding, where $\mathfrak{C}\in\mathbb{P}_\kappa^{\mathbb{K}}(\mathfrak{B})$; since $|C\cup f[C]|<\kappa\le\mathrm{cf}(\alpha)$ there is $\xi<\mathrm{cf}(\alpha)$ such that $C\cup f[C]\subset A_\xi$, and then

$$\bigcup_{\xi<\eta<\mathrm{cf}(\alpha)}g(f,\eta)$$

is an automorphism of $\mathfrak{B}$ containing f.

4.6. Theorem. *(Existence of homogeneous-universal structures). Let $\omega\le\kappa\le\alpha=\alpha^{\underline{\kappa}}$, let $\mathbb{K}$ be a Jónsson class with $(\mathrm{VI})_\kappa$ such that $|\mathbb{I}(\mathbb{K}_\kappa)|\le\alpha$, and let $\mathfrak{A}$ be a structure in $\mathbb{K}$ with $|\mathfrak{A}|\le\alpha$. Then there is a structure $\mathfrak{B}\in\mathbb{K}$ such that*

$$\mathfrak{A}\subset\mathfrak{B},$$
$$|\mathfrak{B}|=\alpha, \quad \textit{and}$$
$$\mathfrak{B} \textit{ is } \kappa\textit{-homogeneous-universal}.$$

Proof. Let $\langle\mathfrak{A}_\xi:\xi<\alpha\rangle$ be a well-ordering of a set of elements of $\mathbb{K}_\kappa$ which contains one representative from every isomorphism class of $\mathbb{K}_\kappa$.

We define structures $\mathfrak{B}_\xi\in\mathbb{K}$ for $\xi<\alpha$ such that

(i) $\mathfrak{B}_0=\mathfrak{A}$;
(ii) $\mathfrak{B}_\xi\subset\mathfrak{B}_\eta$ for $\xi<\eta<\alpha$;
(iii) $|\mathfrak{B}_\xi|\le\alpha$ for $\xi<\alpha$;
(iv) $\mathfrak{A}_\xi\subset\mathfrak{B}_{\xi+1}$ for $\xi<\alpha$; and,
(v) if ξ is a non-zero limit ordinal and $\xi<\alpha$, then $\mathfrak{B}_\xi=\bigcup_{\eta<\xi}\mathfrak{B}_\eta$.

We proceed by transfinite recursion. We define $\mathfrak{B}_\xi$ for limit ordinals $\xi<\alpha$ by (i) and (v). Now for $\xi<\alpha$ we define $\mathfrak{B}_{\xi+1}$. By Lemma 4.2 the class $\mathbb{K}$ has $(\mathrm{VI})_{\alpha^+}$. Thus by (III) there is $\mathfrak{B}_{\xi+1}\in\mathbb{K}$ such that

$$|\mathfrak{B}_{\xi+1}|\le\alpha, \quad \text{and}$$
$$\mathfrak{B}_\xi\subset\mathfrak{B}_{\xi+1}, \quad \mathfrak{A}_\xi\subset\mathfrak{B}_{\xi+1}.$$

This completes the recursive definition, and we set

$$\mathfrak{C}=\bigcup_{\xi<\alpha}\mathfrak{B}_\xi .$$

Then $\mathfrak{C}\in\mathbb{K}$ by (ii) and (V); $|\mathfrak{C}|\le\alpha$ by (iii); $\mathfrak{A}\subset\mathfrak{C}$ by (i); and from (v), the system $\mathfrak{C}$ contains a structure from every isomorphism class of $\mathbb{K}_\kappa$. From (I), (III) and $(\mathrm{VI})_{\alpha^+}$ there is a structure $\mathfrak{C}_0\in\mathbb{K}$ such that

$$|\mathfrak{C}_0|=\alpha \quad \text{and}$$
$$\mathfrak{C}\subset\mathfrak{C}_0,$$

and by Theorem 4.5 there is a structure $\mathfrak{B} \in \mathbb{K}$ such that

$$\mathfrak{C}_0 \subset \mathfrak{B}\,,$$
$$|\mathfrak{B}| = \alpha\,, \quad \text{and}$$
$$\mathfrak{B} \text{ is } \kappa\text{-homogeneous.}$$

It remains to show that $\mathfrak{B}$ is κ-universal.

Let $\mathfrak{D} \in \mathbb{K}_{\kappa^-}$ and use Lemma 4.1 (which applies because $\mathbb{K}$ has $(\mathrm{VI})_\kappa$) to express $\mathfrak{D}$ in the form

$$\mathfrak{D} = \bigcup_{\xi < \mathrm{cf}(\kappa)} \mathfrak{D}_\xi$$

where $\{\mathfrak{D}_\xi : \xi < \mathrm{cf}(\kappa)\}$ is a continuous chain of elements of $\mathbb{K}_\kappa$. For every $\xi < \mathrm{cf}(\kappa)$ let f_ξ be an embedding of $\mathfrak{D}_\xi$ into $\mathfrak{B}$. We define embeddings $g_\xi : \mathfrak{D}_\xi \to \mathfrak{B}$ for $\xi < \alpha$ such that

$$g_0 = f_0 \quad \text{and}$$
$$g_\xi \subset g_\zeta \quad \text{for } \xi < \zeta < \mathrm{cf}(\kappa)\,.$$

We proceed by transfinite recursion. We set $g_0 = f_0$. If ξ is non-zero limit ordinal and $\xi < \mathrm{cf}(\kappa)$, we set $g_\xi = \bigcup_{\zeta < \xi} g_\zeta$; since $\{\mathfrak{D}_\xi : \xi < \mathrm{cf}(\kappa)\}$ is a continuous chain we have $g_\xi : \mathfrak{D}_\xi \to \mathfrak{B}$. Now for $\xi < \mathrm{cf}(\kappa)$ we define $g_{\xi+1}$. We consider the diagram

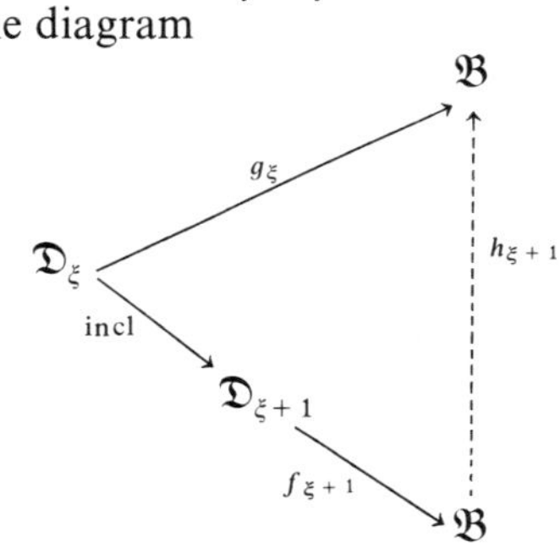

Since $\mathfrak{B}$ is κ-homogeneous, there is an automorphism $h_{\xi+1}$ of $\mathfrak{B}$ such that $h_{\xi+1} \circ f_{\xi+1} \circ \mathrm{incl} = g_\xi$. We set

$$g_{\xi+1} = h_{\xi+1} \circ f_{\xi+1}\,.$$

Clearly $g_\xi \subset g_{\xi+1}$. This completes the recursive definition. We let

$$g : \mathfrak{D} \to \mathfrak{B}$$

be given by $g = \bigcup_{\xi < \mathrm{cf}(\kappa)} g_\xi$. Clearly g is an embedding of $\mathfrak{D}$ into $\mathfrak{B}$ and the proof is complete.

We remark that the proof of Theorem 4.6 shows that if $\omega \le \kappa \le \alpha = \alpha^{\underline{\kappa}}$, $\mathbb{K}$ is a Jónsson class with $(\mathrm{VI})_\kappa$ such that $|\mathbb{I}(\mathbb{K}_\kappa)| \le \alpha$, $\mathfrak{B} \in \mathbb{K}$ and $\mathfrak{B}$ is κ-homogeneous and λ-universal for all $\lambda < \kappa$, then $\mathfrak{B}$ is κ-universal.

4.7. Corollary. *Let $\alpha \geq \omega$ and let $\mathbb{K}$ be a Jónsson class with $(\mathrm{VI})_{\alpha^+}$ such that $|\mathbb{I}(\mathbb{K}_{\alpha^+})| \leq 2^\alpha$. If $\mathfrak{A}$ is a structure in $\mathbb{K}$ and $|\mathfrak{A}| \leq 2^\alpha$, then there is a structure $\mathfrak{B}$ in $\mathbb{K}$ such that*

$$\mathfrak{A} \subset \mathfrak{B}\,,$$
$$|\mathfrak{B}| = 2^\alpha, \quad \textit{and}$$
$$\mathfrak{B} \textit{ is } \alpha^+\textit{-homogeneous-universal}.$$

Proof. We have

$$(2^\alpha)^{\underset{\smile}{\alpha^+}} = (2^\alpha)^\alpha = 2^\alpha$$

so that Theorem 4.6 applies with 2^α and α^+ replacing α and κ respectively.

4.8. Lemma. *Let α be an infinite cardinal and let $\mathbb{K}$ be a Jónsson class with $(\mathrm{VI})_\alpha$. Let $\mathfrak{A} \in \mathbb{K}$ with $|\mathfrak{A}| = \alpha$, and let $\{\mathfrak{A}_\beta : \beta < \alpha\}$ be a specializing chain for $\mathfrak{A}$. Set*

$$\mathbb{C} = \bigcup_{\beta < \alpha} \mathbb{P}^{\mathbb{K}}_{\beta^+}(\mathfrak{A}_\beta)\,.$$

Then $\mathbb{C}$ has the following properties:

(a) $\mathbb{C} \subset \mathbb{P}^{\mathbb{K}}_\alpha(\mathfrak{A})$*;*

(b) $\mathbb{I}(\mathbb{K}_\alpha) \subset \mathbb{I}(\mathbb{C})$*;*

(c) $\mathbb{P}^{\mathbb{K}}(\mathbb{C}) \subset \mathbb{C}$*;*

(d) $\mathbb{C}$ *is* $\mathrm{cf}(\alpha)$*-directed (i.e., given $\mathbb{C}' \subset \mathbb{C}$ with $|\mathbb{C}'| < \mathrm{cf}(\alpha)$, there is $\mathfrak{C} \in \mathbb{C}$ such that $\mathbb{C}' \subset \mathbb{P}^{\mathbb{K}}(\mathfrak{C})$);*

(e) *$\mathfrak{A}$ is the union of a continuous chain of elements of $\mathbb{C}$; and*

(f) *$\mathfrak{A}$ is $\mathbb{C}$-homogeneous (i.e., given $\mathfrak{B}_0 \in \mathbb{K}$ and $\mathfrak{B}_1 \in \mathbb{K}$ with $\mathfrak{B}_0 \subset \mathfrak{B}_1$, and embeddings $f : \mathfrak{B}_0 \to \mathfrak{A}$ and $g : \mathfrak{B}_1 \to \mathfrak{A}$ such that $f[\mathfrak{B}_0] \in \mathbb{C}$ and $g[\mathfrak{B}_1] \in \mathfrak{C}$, there is $\mathfrak{C} \in \mathbb{C}$ such that $f[\mathfrak{B}_0] \subset \mathfrak{C}$ and $g[\mathfrak{B}_1] \subset \mathfrak{C}$, and there is an automorphism h of $\mathfrak{C}$ such that $h \circ f = g \mid \mathfrak{B}_0$).*

Proof. (a) Let $\mathfrak{C} \in \mathbb{C}$. Then $\mathfrak{C} \in \mathbb{P}^{\mathbb{K}}_{\beta^+}(\mathfrak{A}_\beta)$ for some $\beta < \alpha$, so that $|\mathfrak{C}| \leq \beta < \alpha$ and $\mathfrak{C} \in \mathbb{P}^{\mathbb{K}}_\alpha(\mathfrak{A})$.

(b) Let $\mathfrak{C} \in \mathbb{K}_\alpha$; say $|\mathfrak{C}| = \beta < \alpha$. Since $\mathfrak{A}_\beta$ is β^+-universal, we have $\mathbb{I}(\{\mathfrak{C}\}) \subset \mathbb{I}(\mathbb{P}^{\mathbb{K}}_{\beta^+}(\mathfrak{A}_\beta)) \subset \mathbb{I}(\mathbb{C})$.

(c) Let $\mathfrak{B} \in \mathbb{C}$, and let $\mathfrak{C} \subset \mathfrak{B}$ with $\mathfrak{C} \in \mathbb{K}$. Then $\mathfrak{B} \in \mathbb{P}^{\mathbb{K}}_{\beta^+}(\mathfrak{A}_\beta)$ for some $\beta < \alpha$, so that $\mathfrak{C} \in \mathbb{P}^{\mathbb{K}}_{\beta^+}(\mathfrak{A}_\beta) \subset \mathbb{C}$.

(d) Let $\mathbb{C}' \subset \mathbb{C}$ with $|\mathbb{C}'| < \mathrm{cf}(\alpha)$. Then $\mathbb{C}' = \{\mathfrak{C}_\xi : \xi < \lambda\}$ for some cardinal $\lambda < \mathrm{cf}(\alpha)$. For every $\xi < \lambda$ we have $\mathfrak{C}_\xi \in \mathbb{P}^{\mathbb{K}}_{\beta(\xi)^+}(\mathfrak{A}_{\beta(\xi)})$ for some $\beta(\xi) < \alpha$. We set $\beta = \sup_{\xi < \lambda} \beta(\xi)$; since $\lambda < \mathrm{cf}(\alpha)$, it follows that $\beta < \alpha$ and that $\mathfrak{A}_{\beta(\xi)} \subset \mathfrak{A}_\beta$ for all $\xi < \lambda$. Set $\tilde{C} = \bigcup_{\xi < \lambda} C_\xi$; since $|C_\xi| \leq \beta(\xi)$ for $\xi < \lambda$, we have $|\tilde{C}| \leq \beta$.

If $|\mathfrak{A}_\beta| = \alpha$ then by $(\mathrm{VI})_\alpha$ there is $\mathfrak{C} \in \mathbb{P}^{\mathbb{K}}_\alpha(\mathfrak{A}_\beta)$ such that $\tilde{C} \subset C$. If $\beta^+ = \alpha$ then $\mathfrak{C} \in \mathbb{P}^{\mathbb{K}}_{\beta^+}(\mathfrak{A}_\beta) \subset \mathbb{C}$; otherwise, setting $|\mathfrak{C}| = \gamma < \alpha$, it follows that $\mathfrak{C} \in \mathbb{P}^{\mathbb{K}}_{\gamma^+}(\mathfrak{A}_\gamma) \subset \mathbb{C}$.

If $|\mathfrak{A}_\beta| = \gamma < \alpha$ we set $\mathfrak{C} = \mathfrak{A}_\beta$ so that $\mathfrak{C} \in \mathbb{P}^{\mathbb{K}}_{\gamma^+}(\mathfrak{A}_\gamma) \subset \mathbb{C}$.

In any case we have $\mathfrak{C} \in \mathbf{C}$ and $\mathbf{C}' \subset \mathbb{P}^{\mathbb{K}}(\mathfrak{C})$.

(e) We claim first that there is a family $\{C_\xi : \xi < \mathrm{cf}(\alpha)\}$ of subsets of A such that

$$C_\xi \subset A_{\beta(\xi)} \quad \text{for some } \beta(\xi) < \alpha\,,$$
$$|C_\xi| \leq \beta(\xi)\,,$$
$$C_\xi \subset C_\zeta \quad \text{for } \xi < \zeta < \mathrm{cf}(\alpha)\,, \quad \text{and}$$
$$\bigcup_{\xi < \mathrm{cf}(\alpha)} C_\xi = A\,.$$

Indeed, we have that $A = \bigcup_{\beta < \alpha} A_\beta$. For $\eta < \alpha$ set $A_\eta = A_{|\eta|}$, so that $A = \bigcup_{\eta < \alpha} A_\eta$, and let $A = \bigcup_{\xi < \mathrm{cf}(\alpha)} A_{f(\xi)}$ where $f : \mathrm{cf}(\alpha) \to \alpha$ is an order-preserving function with $\sup_{\xi < \mathrm{cf}(\alpha)} f(\xi) = \alpha$. Let $A = \bigcup_{\xi < \mathrm{cf}(\alpha)} B_\xi$, where $B_\xi \subset B_\eta$ for $\xi < \eta < \mathrm{cf}(\alpha)$, and $|B_\xi| \leq |f(\xi)|$ for $\xi < \mathrm{cf}(\alpha)$; and set $C_\xi = B_\xi \cap A_{f(\xi)}$ and $\beta(\xi) = |f(\xi)|$ for $\xi < \mathrm{cf}(\alpha)$. The family $\{C_\xi : \xi < \mathrm{cf}(\alpha)\}$ satisfies the above condition. We assume without loss of generality that $C_\xi = \bigcup_{\zeta < \xi} C_\zeta$ for non-zero limit ordinals $\xi < \mathrm{cf}(\alpha)$.

We define a family $\{\mathfrak{B}_\xi : \xi < \mathrm{cf}(\alpha)\}$ of structures such that

$$\mathfrak{B}_\xi \in \mathbf{C} \quad \text{for } \xi < \mathrm{cf}(\alpha)\,,$$
$$\mathfrak{B}_\zeta \subset \mathfrak{B}_\xi \quad \text{for } \zeta < \xi < \mathrm{cf}(\alpha)\,,$$
$$C_\xi \subset B_\xi \quad \text{for } \xi < \mathrm{cf}(\alpha)\,, \quad \text{and}$$
$$\mathfrak{B}_\xi = \bigcup_{\zeta < \xi} \mathfrak{B}_\zeta \quad \text{for non-zero limit ordinals } \xi < \mathrm{cf}(\alpha)\,.$$

We proceed by transfinite recursion.

If $|\mathfrak{A}_{\beta(0)}| = \alpha$, then by $(\mathrm{VI})_\alpha$ there is $\mathfrak{B}_0 \in \mathbb{P}^{\mathbb{K}}_\alpha(\mathfrak{A}_{\beta(0)})$ such that $C_0 \subset B_0$. If $\beta(0)^+ = \alpha$, then $\mathfrak{B}_0 \in \mathbb{P}^{\mathbb{K}}_{\beta(0)^+}(\mathfrak{A}_{\beta(0)}) \subset \mathbf{C}$; otherwise, setting $\gamma = |\mathfrak{B}_0| < \alpha$, it follows that $\mathfrak{B}_0 \in \mathbb{P}^{\mathbb{K}}_{\gamma^+}(\mathfrak{A}_\gamma) \subset \mathbf{C}$. If $|\mathfrak{A}_{\beta(0)}| = \gamma(0) < \alpha$, we set $\mathfrak{B}_0 = \mathfrak{A}_{\beta(0)}$. Then $\mathfrak{A}_{\beta(0)} \in \mathbb{P}^{\mathbb{K}}_{\gamma(0)^+}(\mathfrak{A}_{\gamma(0)}) \subset \mathbf{C}$.

In any case there is $\mathfrak{B}_0 \in \mathbf{C}$ such that $C_0 \subset B_0$.

If ξ is a non-zero limit ordinal and $\xi < \mathrm{cf}(\alpha)$, we set $\mathfrak{B}_\xi = \bigcup_{\zeta < \xi} \mathfrak{B}_\zeta$. Then $\mathfrak{B}_\xi \in \mathbf{C}$ from parts (c) and (d) above.

We now let $\xi < \mathrm{cf}(\alpha)$ and we define $\mathfrak{B}_{\xi+1}$. Since $\mathfrak{B}_\xi \in \mathbf{C}$ there is $\gamma(\xi) < \alpha$ such that $B_\xi \subset A_{\gamma(\xi)}$. Further $C_{\xi+1} \subset A_{\beta(\xi+1)}$, and setting $\delta(\xi+1) = \beta(\xi-1) + \gamma(\xi)$ we have $\delta(\xi+1) < \alpha$ and $B_\xi \cup C_{\xi+1} \subset A_{\delta(\xi+1)}$. We now repeat the first recursive step to conclude that there is $\mathfrak{B}_{\xi+1} \in \mathbf{C}$ such that

$$B_\xi \cup C_{\xi+1} \subset B_{\xi+1}\,.$$

This completes the recursive definition.

We now have
$$\mathfrak{A} = \bigcup_{\xi < \mathrm{cf}(\alpha)} \mathfrak{B}_\xi$$
where $\{\mathfrak{B}_\xi : \xi < \mathrm{cf}(\alpha)\}$ is a chain in $\mathbf{C}$.

(f) Given $\mathfrak{B}_0 \subset \mathfrak{B}_1$ in $\mathbb{K}$, and embeddings $f:\mathfrak{B}_0 \to \mathfrak{A}$ and $g:\mathfrak{B}_1 \to \mathfrak{A}$ such that $f[\mathfrak{B}_0] \in \mathbf{C}$ and $g[\mathfrak{B}_1] \in \mathbf{C}$, there is, by (a weak form of) (d), $\mathfrak{C} \in \mathbf{C}$ such that $f[\mathfrak{B}_0] \subset \mathfrak{C}$ and $g[\mathfrak{B}_1] \subset \mathfrak{C}$. There is $\beta < \alpha$ such that $\mathfrak{C} \in \mathbb{P}^{\mathbb{K}}_{\beta^+}(\mathfrak{A}_\beta)$. It is easy to find $\mathfrak{B}_2 \in \mathbb{K}$ with $\mathfrak{B}_1 \subset \mathfrak{B}_2$, and isomorphisms $f_1:\mathfrak{B}_2 \xrightarrow{\cong} \mathfrak{C}$ and $g_1:\mathfrak{B}_2 \xrightarrow{\cong} \mathfrak{C}$, such that $f \subset f_1$ and $g \subset g_1$. From the β^+-homogeneity of $\mathfrak{A}_\beta$ there is an automorphism k of $\mathfrak{A}_\beta$ such that $k \circ f_1 = g_1$. Thus $h = k|\mathfrak{C}$ is an automorphism of $\mathfrak{C}$ that satisfies the requirements of $\mathbf{C}$-homogeneity of $\mathfrak{A}$.

4.9. Theorem. *(Uniqueness of special structures). Let α be an infinite cardinal, let $\mathbb{K}$ be a Jónsson class with* (VI)$_\alpha$, *and let $\mathfrak{A}$ and $\mathfrak{B}$ be two special structures in $\mathbb{K}$ of cardinality α, with specializing chains $\{\mathfrak{A}_\beta : \beta < \alpha\}$ and $\{\mathfrak{B}_\beta : \beta < \alpha\}$, respectively. Set $\mathbf{C} = \bigcup_{\beta<\alpha} \mathbb{P}^{\mathbb{K}}_{\beta}(\mathfrak{A}_\beta)$ and $\mathbf{C}' = \bigcup_{\beta<\alpha} \mathbb{P}^{\mathbb{K}}_{\beta}(\mathfrak{B}_\beta)$. Then*

(a) *$\mathfrak{A}$ is α-universal;*

(b) *$\mathfrak{A}$ is* cf(α)*-homogeneous; and*

(c) *given $\mathfrak{C}_0 \in \mathbf{C}$, $\mathfrak{C}'_0 \in \mathbf{C}'$ and an isomorphism $f_0:\mathfrak{C}_0 \xrightarrow{\cong} \mathfrak{C}'_0$, there is an isomorphism $f:\mathfrak{A} \xrightarrow{\cong} \mathfrak{B}$ such that $f_0 \subset f$. In particular, $\mathfrak{A} \cong \mathfrak{B}$.*

Proof. (a) Let $\mathfrak{A}' \in \mathbb{K}$ with $|\mathfrak{A}'| = \alpha$. By Lemma 4.1 there is a chain $\{\mathfrak{A}'_\xi : \xi < \mathrm{cf}(\alpha)\}$ of $\mathbb{K}$-subsystems of $\mathfrak{A}'$ such that

$$|\mathfrak{A}'_\xi| < \alpha \quad \text{for } \xi < \mathrm{cf}(\alpha),$$
$$\mathfrak{A}' = \bigcup_{\xi < \mathrm{cf}(\alpha)} \mathfrak{A}'_\xi, \quad \text{and}$$
$$\mathfrak{A}'_\zeta = \bigcup_{\xi<\zeta} \mathfrak{A}'_\xi \quad \text{for non-zero limit ordinals } \zeta < \mathrm{cf}(\alpha).$$

By Lemma 4.8 (b) there is an embedding $g_0:\mathfrak{A}'_0 \to \mathfrak{A}$ such that $g_0[\mathfrak{A}'_0] \in \mathbf{C}$. We proceed by transfinite recursion. Let $\xi < \mathrm{cf}(\alpha)$ and suppose that for every $\zeta < \xi$ there are defined embeddings $g_\zeta:\mathfrak{A}'_\zeta \to \mathfrak{A}$ such that

$$g_\zeta[\mathfrak{A}'_\zeta] \in \mathbf{C} \quad \text{and}$$
$$g_\eta \subset g_\zeta \quad \text{for } \eta < \zeta < \xi .$$

If ξ is a non-zero limit ordinal, set

$$g_\xi = \bigcup_{\zeta<\xi} g_\zeta .$$

By Lemma 4.8 (d), (e) we have $g_\xi[\mathfrak{A}'_\xi] \in \mathbf{C}$. If $\xi = \zeta + 1$ then there is (by Lemma 4.8 (b)) an embedding $k_\xi:\mathfrak{A}'_\xi \to \mathfrak{A}$ such that $k_\xi[\mathfrak{A}'_\xi] \in \mathbf{C}$. From Lemma 4.8 (f) there is $\mathfrak{D}_\xi \in \mathbf{C}$ such that $g_\zeta[\mathfrak{A}'_\zeta] \subset \mathfrak{D}_\xi$ and $k_\xi[\mathfrak{A}'_\xi] \subset \mathfrak{D}_\xi$, and there is an automorphism h_ξ of $\mathfrak{D}_\xi$ such that $h_\xi \circ (k_\xi | \mathfrak{A}'_\zeta) = g_\zeta$. Set $g_\xi = h_\xi \circ k_\xi$. Then $g_\xi[\mathfrak{A}'_\xi] \in \mathbf{C}$ by Lemma 4.8 (c), and

$$g_\zeta \subset g_\xi .$$

This completes the recursive definition. We set $g=\bigcup_{\xi<\mathrm{cf}(\alpha)} g_\xi$; then g is an embedding of $\mathfrak{A}'$ into $\mathfrak{A}$, proving that $\mathfrak{A}$ is α-universal.

(c) By Lemma 4.8 (e) there are continuous chains $\{\mathfrak{A}'_\xi:\xi<\mathrm{cf}(\alpha)\}$ and $\{\mathfrak{B}'_\xi:\xi<\mathrm{cf}(\alpha)\}$, such that

$$\mathfrak{A}'_\xi\in\mathbf{C}, \quad \mathfrak{B}'_\xi\in\mathbf{C}', \quad \text{and}$$
$$\bigcup_{\xi<\mathrm{cf}(\alpha)}\mathfrak{A}'_\xi=\mathfrak{A}, \quad \bigcup_{\xi<\mathrm{cf}(\alpha)}\mathfrak{B}'_\xi=\mathfrak{B}.$$

Let $\langle d_\xi:\xi<\mathrm{cf}(\alpha)\rangle$ be a well-ordering of the set

$$\{\langle\zeta,i,j\rangle:\zeta<\mathrm{cf}(\alpha),\, i<2,\, j<2\}=\mathrm{cf}(\alpha)\times 2\times 2.$$

We define structures $\mathfrak{C}_\xi\in\mathbf{C}$, $\mathfrak{D}_\xi\in\mathbf{C}'$ and embeddings f_ξ, g_ξ for $\xi<\mathrm{cf}(\alpha)$, such that

(i) $\mathfrak{C}_0$ is as given, $\mathfrak{D}_0=\mathfrak{C}'_0$, f_0 is as given, and $g_0=f_0^{-1}$;

(ii) f_ξ is defined on a sub-structure of $\mathfrak{C}_\xi$ that belongs to $\mathbf{C}$, and g_ξ is defined on a sub-structure of $\mathfrak{D}_\xi$ that belongs to $\mathbf{C}'$;

(iii) $\mathfrak{C}_\xi\subset\mathfrak{C}_\eta$, $\mathfrak{D}_\xi\subset\mathfrak{C}_\eta$, $f_\xi\subset f_\eta$, and $g_\xi\subset g_\eta$ for $\xi<\eta<\mathrm{cf}(\alpha)$;

(iv) $\mathfrak{C}_\xi=\bigcup_{\eta<\xi}\mathfrak{C}_\eta$, $\mathfrak{D}_\xi=\bigcup_{\eta<\xi}\mathfrak{D}_\eta$, $f_\xi=\bigcup_{\eta<\xi}f_\eta$ and $g_\xi=\bigcup_{\eta<\xi}g_\eta$ for non-zero limit ordinals $\xi<\mathrm{cf}(\alpha)$;

(v) if $d_\xi=\langle\zeta,0,0\rangle$, then $\mathfrak{C}_{\xi+1}\supset\mathfrak{A}'_\zeta$;

(vi) if $d_\xi=\langle\zeta,0,1\rangle$, then $\mathfrak{D}_{\xi+1}\supset\mathfrak{B}'_\zeta$;

(vii) if $d_\xi=\langle\zeta,1,0\rangle$, then the domain of $f_{\xi+1}$ is $\mathfrak{C}_{\xi+1}$, and $g_\xi^{-1}\subset f_{\xi+1}$; and

(viii) if $d_\xi=\langle\zeta,1,1\rangle$, then the domain of $g_{\xi+1}$ is $\mathfrak{D}_{\xi+1}$, and $f_\xi^{-1}\subset g_{\xi+1}$.

We proceed by transfinite recursion. The quadruple $\langle\mathfrak{C}_0,\mathfrak{D}_0,f_0,g_0\rangle$ is defined by (i), and if ξ is a non-zero limit ordinal and $\xi<\mathrm{cf}(\alpha)$ then the quadruple $\langle\mathfrak{C}_\xi,\mathfrak{D}_\xi,f_\xi,g_\xi\rangle$ is defined by (iv). By Lemma 4.8 (c), (d), it follows that $\mathfrak{C}_\xi\in\mathbf{C}$ and $\mathfrak{D}_\xi\in\mathbf{C}'$, and that the domain of f_ξ is in $\mathbf{C}$ and the domain of g_ξ is in $\mathbf{C}'$.

Now for $\xi<\mathrm{cf}(\alpha)$ we define the quadruple $\langle\mathfrak{C}_{\xi+1},\mathfrak{D}_{\xi+1},f_{\xi+1},g_{\xi+1}\rangle$. There are four possibilities.

Case 1. $d_\xi=\langle\zeta,0,0\rangle$. By Lemma 4.8 (d) there is a structure $\mathfrak{C}_{\xi+1}\in\mathbf{C}$ such that $\mathfrak{C}_\xi\subset\mathfrak{C}_{\xi+1}$ and $\mathfrak{A}'_\zeta\subset\mathfrak{C}_{\xi+1}$. We also define $\mathfrak{D}_{\xi+1}=\mathfrak{D}_\xi$, $f_{\xi+1}=f_\xi$, and $g_{\xi+1}=g_\xi$.

Case 2. $d_\xi=\langle\zeta,0,1\rangle$. As in *Case* 1 there is by Lemma 4.8(d) a structure $\mathfrak{D}_{\xi+1}\in\mathbf{C}'$ such that $\mathfrak{D}_\xi\subset\mathfrak{D}_{\xi+1}$ and $\mathfrak{B}'_\zeta\subset\mathfrak{D}_{\xi+1}$. We also define $\mathfrak{C}_{\xi+1}=\mathfrak{C}_\xi$, $f_{\xi+1}=f_\xi$, and $g_{\xi+1}=g_\xi$.

Case 3. $d_\xi=\langle\zeta,1,0\rangle$. By Lemma 4.8 (d) there is a structure $\mathfrak{C}_{\xi+1}\in\mathbf{C}$ such that $\mathfrak{C}_\xi\subset\mathfrak{C}_{\xi+1}$ and $g_\xi[\mathfrak{D}_\xi]\subset\mathfrak{C}_{\xi+1}$. By Lemma 4.8 (a), (b), (c), there is an embedding $h_{\xi+1}:\mathfrak{C}_{\xi+1}\to\mathfrak{B}$ such that $h_{\xi+1}[\mathfrak{C}_{\xi+1}]\in\mathbb{C}'$. We consider the diagram

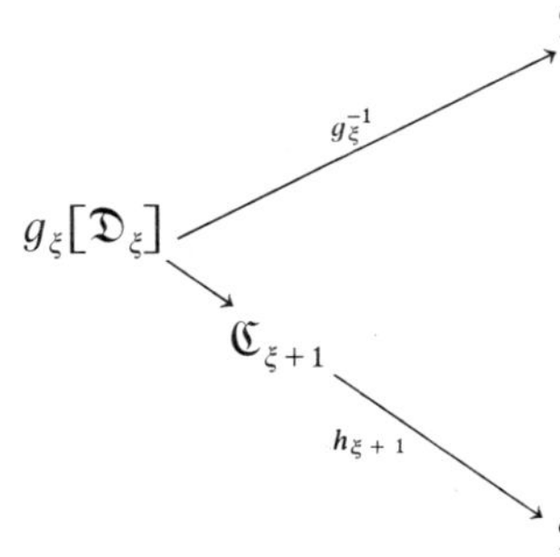

where the unnamed function is inclusion. By Lemma 4.8 (f) there are a structure $\mathfrak{D}_{\xi+1}\in\mathbb{C}'$ such that $\mathfrak{D}_\xi\subset\mathfrak{D}_{\xi+1}$ and $h_{\xi+1}[\mathfrak{C}_{\xi+1}]\subset\mathfrak{D}_{\xi+1}$, and an automorphism $e_{\xi+1}$ of $\mathfrak{D}_{\xi+1}$ such that

$$e_{\xi+1}\circ(h_{\xi+1}\,|\,g_\xi[\mathfrak{D}_\xi])=g_\xi^{-1}.$$

We define $f_{\xi+1}=e_{\xi+1}\circ h_{\xi+1}$. By definition $g_\xi^{-1}\subset f_{\xi+1}$. We finally set $g_{\xi+1}=g_\xi$.

Case 4. $d_\xi=\langle\zeta,1,1\rangle$. This case is similar to *Case* 3, and the details are omitted.

The recursive definition is complete. Then $\mathfrak{A}=\bigcup_{\xi<\mathrm{cf}(\alpha)}\mathfrak{C}_\xi$ by (v), and $\mathfrak{B}=\bigcup_{\xi<\mathrm{cf}(\alpha)}\mathfrak{D}_\xi$ by (vi). We set $f=\bigcup_{\xi<\mathrm{cf}(\alpha)}f_\xi$. By (iii) and (vii) the domain of f is $\mathfrak{A}$; by (iii), (vii), and (viii) the image of f is $\mathfrak{B}$; and by (i) we have $f_0\subset f$.

(b) Let $\mathfrak{A}_0\in\mathbb{K}$ with $|\mathfrak{A}_0|<\mathrm{cf}(\alpha)$, and let $f,g:\mathfrak{A}_0\to\mathfrak{A}$ be two embeddings. It is clear that $f[\mathfrak{A}_0]\in\mathbb{C}$ and $g[\mathfrak{A}_0]\in\mathbb{C}$. By Lemma 4.8 (f) there is $\mathfrak{A}_1\in\mathbb{C}$ such that $f[\mathfrak{A}_0]\subset\mathfrak{A}_1$ and $g[\mathfrak{A}_0]\subset\mathfrak{A}_1$, and there is an automorphism k of $\mathfrak{A}_1$ for which $k\circ g=f$. By (c) above there is an automorphism h of $\mathfrak{A}$ such that $k\subset h$. Thus $h\circ g=f$, and $\mathfrak{A}$ is cf(α)-homogeneous.

4.10. Theorem. (a) *(Existence and uniqueness of special structures). Let α be an infinite cardinal such that $\alpha=2^{\underline{\alpha}}$, let $\mathbb{K}$ be a Jónsson class with $(\mathrm{VI})_{\beta^+}$ for some $\beta<\alpha$, and let $|\mathbb{I}(\mathbb{K}_\alpha)|\le\alpha$. Then there is a unique special system of cardinality α in $\mathbb{K}$.*

(b) *(Existence and uniqueness of homogeneous-universal structures). Let α be an infinite cardinal such that $\alpha=\alpha^{\underline{\alpha}}$, let $\mathbb{K}$ be a Jónsson class with $(\mathrm{VI})_\alpha$, and let $|\mathbb{I}(\mathbb{K}_\alpha)|\le\alpha$. Then there is a unique α-homogeneous-universal system of cardinality α in $\mathbb{K}$.*

Proof. (a) By Corollary 4.7 there is $\mathfrak{B}_\beta \in \mathbb{K}$ such that

$$|\mathfrak{B}_\beta| = 2^\beta \quad \text{and}$$
$$\mathfrak{B}_\beta \text{ is } \beta^+\text{-homogeneous-universal}.$$

Set $\mathfrak{B}_{\beta'} = \mathfrak{B}_\beta$ for $\beta' \le \beta$, and let $\beta < \gamma < \alpha$. Suppose that for all cardinals $\delta < \gamma$ there are defined $\mathfrak{B}_\delta \in \mathbb{K}$ such that

$$|\mathfrak{B}_\delta| \le \alpha,$$
$$\mathfrak{B}_\delta \subset \mathfrak{B}_{\delta'} \quad \text{for } \delta < \delta' < \gamma, \quad \text{and}$$
$$\mathfrak{B}_\delta \text{ is } \delta^+\text{-homogeneous-universal.}$$

Set $\tilde{\mathfrak{B}}_\gamma = \bigcup_{\delta<\gamma} \mathfrak{B}_\delta$. Then $\tilde{\mathfrak{B}}_\gamma \in \mathbb{K}$ and $|\tilde{\mathfrak{B}}_\gamma| \le \sum\{2^\delta : \delta < \gamma\} \le 2^\gamma$. By Corollary 4.7, there is $\mathfrak{B}_\gamma \in \mathbb{K}$ such that

$$\tilde{\mathfrak{B}}_\gamma \subset \mathfrak{B}_\gamma,$$
$$|\mathfrak{B}_\gamma| = 2^\gamma, \quad \text{and}$$
$$\mathfrak{B}_\gamma \text{ is } \gamma^+\text{-homogeneous-universal.}$$

We set $\mathfrak{B} = \bigcup\{\mathfrak{B}_\gamma : \gamma < \alpha\}$. Then $\mathfrak{B} \in \mathbb{K}$, $|\mathfrak{B}| = \alpha$, and $\{\mathfrak{B}_\gamma : \gamma < \alpha\}$ is a specializing chain for $\mathfrak{B}$. This proves the existence of a special system of cardinality α in $\mathbb{K}$. The uniqueness follows by Theorem 4.9.

(b) The existence follows by Theorem 4.6, while the uniqueness follows by Theorem 4.9, since it is clear (as has been remarked earlier) that an α-homogeneous-universal system of cardinality α is special of cardinality α.

The following simple computation gives the most frequently used condition under which Theorem 4.10 is applied.

4.11. Lemma. *Let $\alpha \ge \omega$ and let $\mathbb{K}$ be a class of (similar) relational systems of type $\tau \in \omega^\rho$. Then*

$$|\mathbb{I}(\mathbb{K}_\alpha)| \le 2^{|\rho|} \cdot 2^{\underline{\alpha}}.$$

In particular, if $|\rho| < \alpha$, then $|\mathbb{I}(\mathbb{K}_\alpha)| \le 2^{\underline{\alpha}}$.

Proof. Since every relational system is isomorphic with a system whose universe (underlying set) is a cardinal number, it is sufficient to prove that

$$|\{\mathfrak{A} \in \mathbb{K} : A = \beta \text{ and } \beta < \alpha\}| \le 2^{|\rho|} \cdot 2^{\underline{\alpha}}.$$

For $\mathfrak{A} = \langle \beta, R_i^{\mathfrak{A}} \rangle_{i<\rho} \in \mathbb{K}$ with $\beta < \alpha$ we have $R_i^{\mathfrak{A}} \in \mathscr{P}(\beta^{\tau(i)})$ and $2^{\beta^{\tau(i)}} \le \omega \cdot 2^\beta$. Thus

$$\begin{aligned}|\mathbb{I}(\mathbb{K}_\alpha)| &\le \textstyle\sum_{\beta<\alpha}\left[\prod_{i<\rho} 2^{(\beta^{\tau(i)})}\right] \le \sum_{\beta<\alpha}\left[\prod_{i<\rho} \omega \cdot 2^\beta\right] \\ &= \textstyle\sum_{\beta<\alpha} (\omega \cdot 2^\beta)^{|\rho|} = \sum_{\beta<\alpha} 2^{|\rho|} \cdot \omega \cdot 2^\beta = 2^{|\rho|} \cdot \sum_{\beta<\alpha} \omega \cdot 2^\beta = 2^{|\rho|} \cdot 2^{\underline{\alpha}}.\end{aligned}$$

Remark. It will be noticed that in most of the proofs given above the structure of relational systems we have considered did not play a significant rôle. It is not difficult to isolate the properties that have been used, and to present the material of this section, with virtually no changes in the proofs, from a more general point of view. A slight advantage of this approach is that full expansions need not be introduced when dealing with elementary types (in § 11). We give below only the general definitions on the basis of which one may proceed. This material will not be used in any other part of the book.

Let $\mathbb{K}$ be a class of sets, together with a set of one-to-one functions $\mathrm{Emb}(A,B)$ from A to B (called the *set of embeddings from A to B*) for A,B in $\mathbb{K}$, satisfying the following properties.

(1) The identify function $A \to A$ is in $\mathrm{Emb}(A,A)$ for $A \in \mathbb{K}$;

(2) If A,B and C are in $\mathbb{K}$, $f \in \mathrm{Emb}(A,B)$ and $g \in \mathrm{Emb}(B,C)$, then $g \circ f \in \mathrm{Emb}(A,C)$;

(3) If A and B are in $\mathbb{K}$, $f \in \mathrm{Emb}(A,B)$ and if f is an onto function, then $f^{-1} \in \mathrm{Emb}(B,A)$.

(4) If A, B and C are in $\mathbb{K}$, $f \in \mathrm{Emb}(A,B)$, $f[A] \subset C \subset B$, and the inclusion $C \to B$ is in $\mathrm{Emb}(C,B)$, then the function $\tilde{f}: A \to C$ defined by restricting the range of f is in $\mathrm{Emb}(A,C)$.

If A and B are in $\mathbb{K}$, $f \in \mathrm{Emb}(A,B)$ and f is an onto function, then f is an *isomorphism*; we denote the set of isomorphisms from A onto B by $\mathrm{Iso}(A,B)$.

Let α be an infinite cardinal. A *generalized Jónsson class* (*with* $(\mathrm{VI})_\alpha$) is a class $\mathbb{K}$ of sets together with a set of one-to-one functions $\mathrm{Emb}(A,B)$ from A to B for A,B in $\mathbb{K}$, satisfying the four conditions above and the following.

(I) $\mathbb{K}$ contains sets of arbitrarily large cardinality.

(II) If A,B is in $\mathbb{K}$ and $f \in \mathrm{Emb}(A,B)$ then

(a) $f[A]$ is in $\mathbb{K}$, and

(b) there are C in $\mathbb{K}$ and $g \in \mathrm{Iso}(B,C)$ such that $A \subset C$, the inclusion function $A \to C$ is in $\mathrm{Emb}(A,C)$, and the diagram

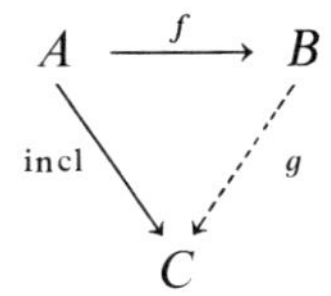

commutes.

(III) If A and B are in $\mathbb{K}$ then there is C in $\mathbb{K}$ such that $\mathrm{Emb}(A,C) \neq \emptyset$ and $\mathrm{Emb}(B,C) \neq \emptyset$.

(IV) If A,B and C are in $\mathbb{K}$, $f \in \mathrm{Emb}(C,A)$ and $g \in \mathrm{Emb}(C,B)$, then there are $D \in \mathbb{K}$, $f_1 \in \mathrm{Emb}(A,D)$ and $g_1 \in \mathrm{Emb}(B,D)$ such that $f_1 \circ f = g_1 \circ g$.

(V) If $\{A_\xi : \xi < \zeta\}$ and $\{B_\xi : \xi < \zeta\}$ are chains in $\mathbb{K}$ and $f_\xi \in \mathrm{Emb}(A_\xi, B_\xi)$ for $\xi < \zeta$ with $f_\xi \subset f_\eta$ for $\xi < \eta < \zeta$, then $\bigcup_{\xi<\zeta} A_\xi \in \mathbb{K}$ and $\bigcup_{\xi<\zeta} B_\xi \in \mathbb{K}$ and

$$\bigcup_{\xi<\zeta} f_\xi \in \mathrm{Emb}(\bigcup_{\xi<\zeta} A_\xi, \bigcup_{\xi<\zeta} B_\xi).$$

$(\mathrm{VI})_\alpha$ If A is in $\mathbb{K}$ and $B \in \mathscr{P}_\alpha(A)$, then there is $C \in \mathbb{K}$ such that

$|C| < \alpha$,
$B \subset C \subset A$,
the inclusion $C \to A$ is in $\mathrm{Emb}(C, A)$.

With this definition the results of this section can be established, with essentially no changes in proofs, for generalized Jónsson classes. (The analogue of) Theorem 4.10 is valid with the following definitions (and with "generalized Jónsson class" replacing "Jónsson class" in its statement).

Let α be an infinite cardinal, let $\mathbb{K}$ be a generalized Jónsson class, and let $A \in \mathbb{K}$.

(a) A is *α-universal in* $\mathbb{K}$ if $\mathrm{Emb}(B, A) \neq \emptyset$ for $B \in \mathbb{K}_{\alpha^+}$.

(b) A is *α-homogeneous in* $\mathbb{K}$ if for $B \in \mathbb{K}_\alpha$ and $f, g \in \mathrm{Emb}(B, A)$ there is $h \in \mathrm{Iso}(A, A)$ such that $h \circ f = g$.

(c) If A is both α-homogeneous and α-universal in $\mathbb{K}$, then A is *α-homogeneous-universal in* $\mathbb{K}$.

(d) A is *special (of cardinality α) in* $\mathbb{K}$ if $|A| = \alpha$ and A is equal to the union of a chain $\{A_\beta : \beta < \alpha\}$ where

$A_\beta \in \mathbb{K}$ and
A_β is β^+-homogeneous-universal

for $\beta < \alpha$. The chain $\{A_\beta : \beta < \alpha\}$ is called a *specializing chain for A*.

Notes for § 4

Cantor [95] (§ 9) created and Hausdorff [08] (§ 23) generalized the method of back-and-forth induction for linearly ordered sets (cf. Notes for § 5). Erdős-Gillman-Henriksen [55] adapted this method for real-closed fields.

The amalgamation property (IV) was first formulated explicitly by Fraïssé [54]. It appears in Jónsson [56], [60] (in somewhat stronger form) and in Morley-Vaught [62]. The property is studied in Jónsson [65]. The name for condition $(\mathrm{VI})_\alpha$ is explained by Theorem 11.4 below.

The general notion of α-universal was defined in Jónsson [56], of α-homogeneous in Jónsson [60], and of a more general form of special (of cardinality α) in Morley-Vaught [62]. The present simpler notion of special is an obvious analogue of Chang-Keisler [66] (6.2). Special structures of cardinality α were introduced because the existence of α-homogeneous-universal structures of cardinality is proved only for cardinals α such that $\alpha=\alpha^{\underline{\alpha}}$, a condition that is not known to be satisfied for any uncountable cardinal without some set-theoretic assumptions such as GCH or the existence of strongly inaccessible cardinals (cf. Theorem 1.27 and its Corollary 1.28).

In our treatment of the theory of Jónsson classes we follow Morley-Vaught [62] (§ 2). Lemmas 4.1 and 4.2 are proved in Jónsson [56] (Lemmas 2.5 and 2.6 respectively). Lemmas 4.3, 4.4 and Theorems 4.5, 4.6 correspond to Morley-Vaught [62] (Lemma 2.9, Theorem 2.10(a), Theorem 2.10(b), and a consequence of Lemma 2.7(c) respectively). Lemma 4.8 is a technical and routine verification of the conditions of Definition 2.2 of Morley-Vaught [62] by the notion of special as defined here. The Cantor-like back-and-forth induction, proving the uniqueness of special structures in Theorem 4.9, corresponds to Morley-Vaught [62] (Theorem 2.3).

Theorem 4.10, the main result of this section, is proved under a slightly more restrictive cardinality condition (*i.e.*, condition $(\mathrm{VI})_{\beta+}$ for some $\beta<\alpha$, rather than $(\mathrm{VI})_\alpha$) than is the corresponding Theorem 2.8 of Morley-Vaught [62]; this reflects our stronger notion of special.

To illustrate the power of the general theory of Jónsson classes, we note that by combining the theory (Theorem 4.10(b)) with the classical theory of real-closed fields (Artin-Schreier [26], Erdős-Gillman-Henriksen [55]) we obtain a simple proof of the theorem that there is a real-closed η_α-field of cardinality α if and only if $\alpha=\alpha^{\underline{\alpha}}$; this result was proved by Alling [61], [62] using complex methods of valuation theory.

The terminology "homogeneous" was first used by Urysohn; he proved in Urysohn [25], [27] the existence of a universal, separable, complete, "ω-homogeneous" metric space. It is curious and interesting that the use of the term "homogeneous" by Jónsson in his general theory is apparently independent of Urysohn's.

The last remark outlines a more general setting for proving the results of this section as described by Fleischer [66] and Sevee [72].

§ 5. The Jónsson Class of Ordered Sets

This is the principal example of a Jónsson class. Arguments such as the back-and-forth induction of Cantor and Hausdorff occur here in their initial and prototypical form (Theorem 5.3). There is also a useful analogy between the simple arguments concerning the η_α-property of ordered sets and the more technically detailed arguments involving the (H_α) and (R_α) properties of Boolean algebras studied in § 6 below; in this sense the material of the present section is preparatory to § 6.

As might be expected, the special situation studied here has several advantages over the general theory of Jónsson classes. The class of ordered sets is one of the few Jónsson classes for which the homogeneous-universal structures are described in concrete form; they are represented as sets of dyadic sequences ordered lexicographically (Theorem 5.12). An interesting byproduct is that an α-universal-homogeneous ordered set of cardinality α exists *if and only if* $\alpha=\alpha^{\underline{\alpha}}$ (Theorem 5.12). This implies a similar result for the Jónsson class of Boolean algebras (Theorem 6.12).

The class of non-empty (linearly) ordered sets is denoted $\mathbb{OS}$. Thus if a structure $\langle A, \leq\rangle$ is in $\mathbb{OS}$ then $\langle A, \leq\rangle$ is a relational system with index 1 and similarity type τ with $\tau(0)=2$; and if $\langle A, \leq\rangle, \langle B, \leq\rangle \in \mathbb{OS}$ and $f:\langle A, \leq\rangle \to \langle B, \leq\rangle$, then f is an embedding of relational systems (as defined in § 4) if and only if f is an embedding of ordered sets (as defined in § 1).

5.1. Theorem. *The class* $\mathbb{OS}$ *is a Jónsson class with* $(\mathrm{VI})_\omega$.

Proof. We refer to the defining conditions of a Jónsson class given in § 4.

(I) Every cardinal number is linearly ordered by $\in$.

(II) is obvious.

(III) If $\langle A, \underset{A}{\leq}\rangle, \langle B, \underset{B}{\leq}\rangle \in \mathbb{OS}$ we set

$$C = A \times \{0\} \cup B \times \{1\}$$

and we define a relation $\leq$ on C as follows:

if $a,b\in A$, then $\langle a,0\rangle \leq \langle b,0\rangle$ if and only if $a \underset{A}{\leq} b$;
if $a,b\in B$, then $\langle a,1\rangle \leq \langle b,1\rangle$ if and only if $a \underset{B}{\leq} b$;
if $a\in A$ and $b\in B$, then $\langle a,0\rangle \leq \langle b,1\rangle$.

Then $\langle C, \leq\rangle \in \mathbb{OS}$, and the functions $a \to \langle a, 0\rangle$ and $b \to \langle b, 1\rangle$ are ordered-set embeddings of A and B respectively into C.

(IV) Let $A, B, C \in \mathbb{OS}$ and let $f: C \to A$ and $g: C \to B$ be embeddings. We are to find $D \in \mathbb{OS}$ and embeddings $f_1: A \to D$ and $g_1: B \to D$ such that $f_1 \circ f = g_1 \circ g$.

We suppose (on account of (II)) without loss of generality that the sets A, B and C are pairwise disjoint. We claim that if $a \in A \setminus f[C]$ and $b \in B \setminus g[C]$, then either

$$f[g^{-1}(P(b))] \subset P(a) \quad \text{or}$$
$$g[f^{-1}(P(a))] \subset P(b)$$

(where $P(a) = \{c \in A : c < a\}$ and $P(b)$ is defined analogously). Otherwise there are $c_0, c_1 \in C$ such that

$$g(c_0) < b, \quad f(c_0) \geq a, \quad f(c_1) < a \quad \text{and} \quad g(c_1) \geq b,$$

and since f and g are embeddings it follows that

$$c_0 < c_1 \quad \text{and} \quad c_1 < c_0,$$

a contradiction. The claim is established.

We set $D = C \cup (A \setminus f[C]) \cup (B \setminus g[C])$ and we define a binary relation $\leq$ on D as follows:

if $a \in A \setminus f[C]$ and $b \in B \setminus g[C]$, then
  $a < b$ if $g[f^{-1}(P(a))] \subset P(b)$ and
  $b < a$ otherwise;
if $a \in A \setminus f[C]$ and $c \in C$, then
  $a < c$ if $a \underset{A}{<} f(c)$ and
  $c < a$ if $f(c) \underset{A}{<} a$;
if $b \in B \setminus f[C]$ and $c \in C$, then
  $b < c$ if $b \underset{B}{<} f(c)$ and
  $c < b$ if $f(c) \underset{B}{<} b$; and
for any two elements of each of the sets $C, A \setminus f[C]$ and $B \setminus f[C]$, the order $<$ on D agrees with the order already given.

Finally we set

$$d_0 \leq d_1 \quad \text{if and only if } d_0 < d_1 \quad \text{or } d_0 = d_1 \quad \text{for } d_0, d_1 \in D.$$

We define a function $f_1: A \to D$ as follows:

$$f_1(f(c)) = c \quad \text{for } c \in C;$$
$$f_1(a) = a \quad \text{for } a \in A \setminus f[C];$$

a function $g_1: B \to D$ is defined similarly. It is easy to see that $\langle D, \leq \rangle \in \mathbb{OS}$, that f_1 and g_1 are well-defined embeddings of B and C respectively into D, and that $f_1 \circ f = g_1 \circ g$.

(V) is obvious.

$(\mathrm{VI})_\omega$ Every subset of an ordered set is an ordered set.

The proof is complete.

It follows from Theorem 4.10(b), Corollary 4.11 and Theorem 5.1 that for every infinite cardinal α such that $\alpha = \alpha^{\underline{\alpha}}$ there is an α-homogeneous-universal element of $\mathbb{OS}$ of cardinality α; further, this is unique up to ordered-set isomorphism. Such an ordered set is denoted $\mathfrak{H}_\alpha$.

Notation. If $\langle A, \leq \rangle \in \mathbb{OS}$ and $B, C \subset A$, we write

$$B < C$$

if $b < c$ for $b \in B$ and $c \in C$. We write $B < c$ instead of $B < \{c\}$, and $b < C$ instead of $\{b\} < C$.

We note that if $B, C \subset A$ and $B = \emptyset$ or $C = \emptyset$, then $B < C$.

Definitions. Let $\alpha \geq \omega$ and $\langle A, \leq \rangle \in \mathbb{OS}$. Then $\langle A, \leq \rangle$ is an *η_α-set* if for $B, C \subset A$ with $B < C$ and $|B| + |C| < \alpha$, there is $a \in A$ such that

$$B < a < C.$$

Let $\langle A, \leq \rangle \in \mathbb{OS}$ and $B \subset A$. Then B is *coinitial in A* if for $a \in A$ there is $b \in B$ such that $b \leq a$; and B is *cofinal in A* if for $a \in A$ there is $b \in B$ such that $a \leq b$.

From the remark preceding these definitions it follows that if $\langle A, \leq \rangle$ is an η_α-set, then for every subset B of A which is cofinal or coinitial we have $|B| \geq \alpha$; in particular, $|A| \geq \alpha$.

5.2. Lemma. *If $\alpha \geq \omega$ then every α-homogeneous-universal ordered set is an η_α-set.*

Proof. Let $\langle A, \leq \rangle$ be an α-homogeneous-universal ordered set and let $B, C \in \mathscr{P}_\alpha(A)$ with $B < C$. We choose a set t such that $t \notin A$, we set $D = B \cup C \cup \{t\}$, and we define an order $\preccurlyeq$ on D as follows:

$$\begin{aligned} &b_0 \preccurlyeq b_1 \quad \text{if } b_0 \leq b_1, \quad \text{for } b_0, b_1 \in B; \\ &c_0 \preccurlyeq c_1 \quad \text{if } c_0 \leq c_1, \quad \text{for } c_0, c_1 \in C; \quad \text{and} \\ &B \preccurlyeq t \preccurlyeq C. \end{aligned}$$

Since $\langle D, \leq \rangle \in \mathbb{OS}$ and $|D| \leq \alpha$ there is an ordered-set embedding $f: D \to A$. Denote by e the natural inclusion of $B \cup C$ into D, so that $f \circ e: B \cup C \to A$ is an embedding. Since A is α-homogeneous there is an ordered-set automorphism h of A such that $h \circ f \circ e$ is equal to the inclusion of $B \cup C$ into A, and since $e[B] \prec t \prec e[C]$ in D we have

$$B < h(f(t)) < C$$

in A.

5.3. Theorem *(Cantor-Hausdorff). If $\alpha \geq \omega$, then any two η_α-sets of cardinality α are isomorphic as ordered sets.*

Proof. Let $\langle A, \leq \rangle$ and $\langle B, \leq \rangle$ be η_α-sets of cardinality α and let $\{\langle a_\xi : \xi < \alpha \rangle\}$ and $\{\langle b_\xi : \xi < \alpha \rangle\}$ be well-orderings of A and B respectively. For $\xi < \alpha$ we define sets A_ξ and B_ξ, and a function $f_\xi : A_\xi \to B_\xi$, such that

(i) $A_0 = \emptyset$, $B_0 = \emptyset$, and $f_0 = \emptyset$;

(ii) $A_\zeta \subset A_\xi$, $B_\zeta \subset B_\xi$, and $f_\zeta \subset f_\xi$ for $\zeta < \xi < \alpha$;

(iii) if ξ is a non-zero limit ordinal and $\xi < \alpha$, then $A_\xi = \bigcup_{\zeta < \xi} A_\zeta$, $B_\xi = \bigcup_{\zeta < \xi} B_\zeta$, and $f_\xi = \bigcup_{\zeta < \xi} f_\zeta$;

(iv) $|A_\xi| \leq |\xi|$, $|B_\xi| \leq |\xi|$ for $\xi < \alpha$;

(v) f_ξ is an ordered-set isomorphism of A_ξ onto B_ξ for $\xi < \alpha$;

(vi) if $\xi = \zeta + 2n + 1$ with ζ a limit ordinal and $n < \omega$, then $a_{\zeta+n} \in A_\xi$; and

(vii) if $\xi = \zeta + 2n + 2$ with ζ a limit ordinal and $n < \omega$, then $b_{\zeta+n} \in B_\xi$.

We proceed by recursion. We define A_ξ, B_ξ and f_ξ by (i) for $\xi = 0$, and by (iii) for ξ a non-zero limit ordinal.

Now let ξ be a non-limit ordinal and $\xi < \alpha$. There are a limit ordinal $\zeta < \alpha$ and $k < \omega$ such that $\xi = \zeta + k$. To define A_ξ, B_ξ and f_ξ we consider two cases.

Case 1. There is $n < \omega$ such that $k = 2n + 1$. If $a_{\zeta+n} \in A_{\zeta+2n}$ we set

$$A_\xi = A_{\zeta+2n}, \quad B_\xi = B_{\zeta+2n}, \quad \text{and} \quad f_\xi = f_{\zeta+2n}.$$

If $a_{\zeta+n} \notin A_{\zeta+2n}$ we set

$$C = \{a \in A_{\zeta+2n} : a < a_{\zeta+n}\} \quad \text{and} \quad D = \{a \in A_{\zeta+2n} : a_{\zeta+n} < a\}.$$

Since $C < D$ in A and $f_{\zeta+2n} : A_{\zeta+2n} \to B_{\zeta+2n}$ is an ordered-set isomorphism we have

$$f_{\zeta+2n}[C] < f_{\zeta+2n}[D]$$

in B; since $|f_{\zeta+2n}[C] \cup f_{\zeta+2n}[D]| < \alpha$ there is $b \in B$ such that

$$f_{\zeta+2n}[C] < b < f_{\zeta+2n}[D].$$

We set

$$A_\xi = A_{\zeta+2n} \cup \{a_{\zeta+n}\}, \quad B_\xi = B_{\zeta+2n} \cup \{b\}, \quad \text{and}$$

$f_\xi = f_{\zeta+n} \cup \{\langle a_{\zeta+n}, b \rangle\}$.

Case 2. There is $n < \omega$ such that $k = 2n + 2$. If $b_{\zeta+n} \in B_{\zeta+2n+1}$ we set

$$A_\xi = A_{\zeta+2n+1}, \quad B_\xi = B_{\zeta+2n+1}, \quad f_\xi = f_{\zeta+2n+1}.$$

If $b_{\zeta+n} \notin B_{\zeta+2n+1}$ we set

$$C = \{b \in B_{\zeta-2n+1} : b < b_{\zeta+n}\} \quad \text{and} \quad D = \{b \in B_{\zeta+2n+1} : b_{\zeta+n} < b\}.$$

Since $C < D$ in B and $f^{-1}_{\zeta+2n+1} : B_{\zeta+2n+1} \to A_{\zeta+2n+1}$ is an ordered-set isomorphism we have

$$f^{-1}_{\zeta+2n+1}[C] < f^{-1}_{\zeta+2n+1}[D]$$

in A; since $|f_{\zeta+2n+1}[C] \cup f_{\zeta+2n+1}[D]| < \alpha$ there is $a \in A$ such that

$$f_{\zeta+2n+1}[C] < a < f_{\zeta+2n+1}[D].$$

We set

$$A_\xi = A_{\zeta+2n+1} \cup \{a\}, \qquad B_\xi = B_{\zeta+2n+1} \cup \{b_{\zeta+n}\}, \quad \text{and}$$

$f_\xi = f_{\zeta+2n+1} \cup \{\langle a, b_{\zeta+n}\rangle\}$.

This completes the definitions by recursion. It is clear that $\bigcup_{\xi<\alpha} f_\xi$ is an ordered-set isomorphism of $\langle A, \leq\rangle$ onto $\langle B, \leq\rangle$.

Definitions. Let η be an ordinal. If $a, b \in \{0,1\}^\eta$ and $a \neq b$ we set

$$\xi(a,b) = \min\{\xi < \eta : a_\xi \neq b_\xi\},$$

and we set $a \leq b$ if

$$a = b \quad \text{or}$$

$$a \neq b \quad \text{and} \quad a_{\xi(a,b)} < b_{\xi(a,b)}.$$

Since η is well-ordered it follows that $\leq$ is a well-defined, linear order on $\{0,1\}^\eta$. An element of $\{0,1\}^\eta$ is a *dyadic sequence of length* η; $\leq$ is the *lexicographic order* on $\{0,1\}^\eta$; and for a cardinal α we set

$$\Lambda(\alpha) = \langle\{0,1\}^\alpha, \leq\rangle.$$

We write $a \in \Lambda(\alpha)$ and $a \in \{0,1\}^\alpha$ interchangeably. The smallest and largest elements of $\Lambda(\alpha)$ are denoted $\mathbb{0}$ and $\mathbb{1}$ respectively.

5.4. Theorem. *If $\alpha \geq \omega$ and A is a well-ordered or an anti-well-ordered subset of $\Lambda(\alpha)$ then $|A| \leq \alpha$.*

Proof. We prove the statement if A is well-ordered; the modifications of the proof for A anti-well-ordered are left to the reader. We assume without loss of generality (by omitting a finite set of elements) that A has no last element. We assume that $|A| > \alpha$, and we claim that there is an ordinal $\xi < \alpha$ for which there are order-embeddings $f, g : \alpha^+ \to A$ such that

(i) $f(\eta) < g(\eta) < f(\eta+1)$ and

(ii) $\xi(f(\eta), g(\eta)) = \xi$

for $\eta < \alpha^+$. We let $s : A \to A$ be defined by

$$s(a) \text{ is the immediate successor of } a \text{ in } A,$$

and we set

$$A_\xi = \{a \in A : \boldsymbol{\xi}(a, s(a)) = \xi\}$$

for $\xi < \alpha$. Since $A = \bigcup_{\xi < \alpha} A_\xi$ there is $\xi < \alpha$ such that $|A_\xi| > \alpha$. By Theorem 1.13 there are a unique ordinal ρ and a unique order-isomorphism $h: \rho \to A_\xi$. We note that $\alpha^+ \leq \rho$ and we set

$$f = h | \alpha^+ ,$$

$$g = s \circ f .$$

It is clear that the triple $\langle \xi, f, g \rangle$ satisfies condition (ii); and that $f(\eta) < s(f(\eta)) = g(\eta) \leq f(\eta + 1)$ for $\eta < \alpha^+$; further since $f(\eta)_\xi = 0 < 1 = g(\eta)_\xi$ for $\eta < \alpha^+$ we have $f(\eta + 1) \neq g(\eta)$. The proof of the claim is complete.

Now let ξ be the least ordinal less than α for which there are order-embeddings f, g satisfying conditions (i) and (ii) above.

We set

$$B_\zeta = \{g(\eta) : \eta < \alpha^+ \text{ and } \boldsymbol{\xi}(g(\eta), f(\eta + 1)) = \zeta\}$$

for $\zeta < \xi$. Since $g(\eta)_\xi = 1$, $f(\eta + 1)_\xi = 0$ and $g(\eta) < f(\eta + 1)$, it follows that $\boldsymbol{\xi}(g(\eta), f(\eta + 1)) < \xi$ for $\eta < \alpha^+$; hence

$$g[\alpha^+] = \bigcup_{\zeta < \xi} B_\zeta .$$

Thus there is $\zeta < \xi$ such that $|B_\zeta| = \alpha^+$. By Theorem 1.13 there are a unique ordinal ρ' and a unique order-isomorphism $h' : \rho' \to B_\zeta$. We note that $\alpha^+ \leq \rho'$, we set

$$f' = h' | \alpha^+ ,$$

and we define

$$g' : \alpha^+ \to A$$

by $g'(\eta) = f(g^{-1}(f'(\eta)) + 1)$. Since $f'[\alpha^+] \subset g[\alpha^+]$ the function g' is well-defined. It is clear that

f' and g' are order-embeddings of α^+ into A,

$f'(\eta) < g'(\eta)$ for $\eta < \alpha^+$, and

$\boldsymbol{\xi}(f'(\eta), g'(\eta)) = \zeta$ for $\eta < \alpha^+$;

furthermore,

$$g'(\eta) = f(g^{-1}(f'(\eta)) + 1) < g(g^{-1}(f'(\eta)) + 1) \leq f'(\eta + 1)$$

since $f'(\eta + 1)$ is the immediate successor of $f'(\eta)$ in $f'[\alpha^+]$, $g(g^{-1}(f'(\eta)) + 1)$ is the immediate successor of $f'(\eta)$ in $g[\alpha^+]$, and $f'[\alpha^+] \subset g[\alpha^+]$.

It follows that the triple $\langle \zeta, f', g' \rangle$ satisfies conditions (i) and (ii) above with $\zeta < \xi$, contradicting the definition of ξ. The proof is complete.

5.5. Lemma. *Let $\alpha \geq \omega$ and $\langle A, \leq \rangle$ an η_{α^+}-set. Then there is an order-embedding of $\Lambda(\alpha)$ into $\langle A, \leq \rangle$.*

Hence $|A| \geq 2^\alpha$.

Proof. Let $\langle b^\xi : \xi < 2^\alpha \rangle$ be a well-ordering of $\Lambda(\alpha)$ and set

$$B_\xi = \{b^\eta \in \Lambda(\alpha) : \eta \leq \xi\}$$

for $\xi < 2^\alpha$. We choose $a^0 \in A$ and for $\xi < 2^\alpha$ we define a function $f_\xi : B_\xi \to A$ such that

(i) $f_0(b^0) = a^0$;

(ii) $f_\eta \subset f_\xi$ for $\eta < \xi < 2^\alpha$; and

(iii) f_ξ is an ordered-set embedding for $\xi < 2^\alpha$.

We proceed by recursion, defining f_ξ by (i) for $\xi = 0$. If $0 < \xi < 2^\alpha$ and f_η is defined for $\eta < \xi$ we set

$$g = \bigcup_{\eta < \xi} f_\eta$$

(so that $g = f_\zeta$ if $\xi = \zeta + 1$), and we choose subsets C and D of $\Lambda(\alpha)$ such that

C is a cofinal subset of $\{b^\eta : \eta < \xi$ and $b^\eta < b^\xi$ in $\Lambda(\alpha)\}$,

D is a coinitial subset of $\{b^\eta : \eta < \xi$ and $b^\xi < b^\eta$ in $\Lambda(\alpha)\}$,

C is well-ordered (as a subset of $\Lambda(\alpha)$), and

D is anti-well-ordered (as a subset of $\Lambda(\alpha)$).

Since $C < b^\xi < D$ in $\Lambda(\alpha)$ and $g : C \cup D \to A$ is an ordered-set embedding we have

$$g[C] < g[D]$$

in A; since $|C| + |D| \leq \alpha$ by Theorem 5.4 and A is an η_{α^+}-set there is $a^\xi \in A$ such that

$$g[C] < a^\xi < g[D].$$

We set

$$f_\xi = g \cup \{\langle b^\xi, a^\xi \rangle\}.$$

We note that, since C is cofinal in $\{b^\eta : \eta < \xi$ and $b^\eta < b^\xi\}$, we have $f_\xi(b^\eta) < f_\xi(b^\xi)$ for $\eta < \xi$ with $b^\eta < b^\xi$. Similarly $f_\xi(b^\xi) < f_\xi(b^\eta)$ for $\eta < \xi$ with $b^\xi < b^\eta$. Hence $f_\xi : B_\xi \to A$ is an ordered-set embedding.

Finally we define

$$f = \bigcup_{\xi < 2^\alpha} f_\xi .$$

It is clear that f is an ordered-set embedding of $\Lambda(\alpha)$ into $\langle A, \leq \rangle$.

5.6. Lemma. *If α is an infinite singular cardinal and $\langle A, \leq \rangle$ is an η_α-set, then $\langle A, \leq \rangle$ is an η_{α^+}-set.*

Proof. We let $B, C \subset A$ such that $B < C$ and $|B| + |C| \leq \alpha$ and we prove that there is $a \in A$ such that $B < a < C$. We note that there is $B_0 \subset B$ such that

$$|B_0| < \alpha, \quad \text{and}$$

$$B_0 \text{ is cofinal in } B.$$

Indeed there is $B_1 \subset B$ such that B_1 is cofinal in B and B_1 is well-ordered by $\leq$. If B_1 has a greatest element we set $B_0 = \{\max B_1\}$. If B_1 does not have a greatest element then there are a limit ordinal ξ and an order-isomorphism f of ξ onto B_1 (Theorem 1.13). We have $\mathrm{cf}(\xi) \leq \xi < \alpha^+$, and since $\mathrm{cf}(\xi)$ is a regular cardinal and α is singular we have $\mathrm{cf}(\xi) < \alpha$. Thus there is $D \subset \xi$ such that $|D| < \alpha$ and D is cofinal in ξ; we set

$$B_0 = f[D].$$

Similarly there is $C_0 \subset C$ such that

$$|C_0| < \alpha, \quad \text{and}$$
$$C_0 \text{ is coinitial in } C.$$

Finally, since $B_0 < C_0$ and $\langle A, \leq \rangle$ is an η_α-set there is $a \in A$ such that $B_0 < a < C_0$. It is clear that

$$B < a < C.$$

Definitions. Let $\alpha \geq \omega$ and $a, b \in \Lambda(\alpha)$. Then $\langle a, b \rangle$ is an *(ordered) pair of consecutive elements of* $\Lambda(\alpha)$ if

$$a < b, \quad \text{and}$$
$$\text{there is no } c \in \Lambda(\alpha) \text{ such that } a < c < b.$$

If $\langle a, b \rangle$ is a pair of consecutive elements of $\Lambda(\alpha)$ we say that b is the *immediate successor* of a and that a is the *immediate predecessor* of b; we write $b = a^+$ and $a = b^-$. We set

$$\mathfrak{L}(\alpha) = \{a \in \Lambda(\alpha) : \text{there is } b \in \Lambda(\alpha) \text{ such that } a = b^-\} \quad \text{and}$$
$$\mathfrak{U}(\alpha) = \{b \in \Lambda(\alpha) : \text{there is } a \in \Lambda(\alpha) \text{ such that } b = a^+\}.$$

The elements of $\mathfrak{L}(\alpha)$ and $\mathfrak{U}(\alpha)$ are called the *lower* and *upper elements* of $\Lambda(\alpha)$, respectively.

5.7. Theorem. *Let* $\alpha \geq \omega$.

(a) $\langle a, b \rangle$ *is a pair of consecutive elements of* $\Lambda(\alpha)$ *if and only if*

(i) $a \neq b$ *and* $a_{\xi(a,b)} < b_{\xi(a,b)}$, *and*

(ii) $a_\xi = 1$, $b_\xi = 0$ *for* $\xi(a,b) < \xi < \alpha$.

(b) $\mathfrak{L}(\alpha) = \{a \in \Lambda(\alpha) :$ *there is* $\xi < \alpha$ *such that* $\alpha_\xi = 0$ *and* $\alpha_\zeta = 1$ *for* $\xi < \zeta < \alpha\}$,

$\mathfrak{U}(\alpha) = \{a \in \Lambda(\alpha) :$ *there is* $\xi < \alpha$ *such that* $\alpha_\xi = 1$ *and* $\alpha_\zeta = 0$ *for* $\xi < \zeta < \alpha\}$.

(c) $|\mathfrak{L}(\alpha)| = |\mathfrak{U}(\alpha)| = 2^{\underline{\alpha}}$.

Proof. (a) Part (i) follows from the definition of the order relation $\leq$. Suppose there is ξ such that

$$\boldsymbol{\xi}(a,b)<\xi<\alpha \quad \text{and}$$
$$a_\xi=0\,,$$

and let c be defined by

$$\begin{aligned} c_\eta&=a_\eta \quad \text{for } \eta\leq\boldsymbol{\xi}(a,b)\\ &=1 \quad \text{for } \boldsymbol{\xi}(a,b)<\eta<\alpha\,. \end{aligned}$$

Then $a<c<b$, a contradiction; thus $a_\xi=1$ for $\boldsymbol{\xi}(a,b)<\xi<\alpha$. A similar argument shows that $b_\xi=0$ for $\boldsymbol{\xi}(a,b)<\xi<\alpha$.

To prove the converse of (a) we assume that (i) and (ii) are satisfied. From part (i) it follows that $a<b$ in $\boldsymbol{\Lambda}(\alpha)$. Suppose there is $c\in\boldsymbol{\Lambda}(\alpha)$ such that

$$a<c<b\,.$$

Since $a_\xi=c_\xi=b_\xi$ for $\xi<\boldsymbol{\xi}(a,b)$ and

$$a_{\boldsymbol{\xi}(a,b)}=0<1=b_{\boldsymbol{\xi}(a,b)}\,,$$

we have either $c_{\boldsymbol{\xi}(a,b)}\neq a_{\boldsymbol{\xi}(a,b)}$ or $c_{\boldsymbol{\xi}(a,b)}\neq b_{\boldsymbol{\xi}(a,b)}$. We suppose without loss of generality that $c_{\boldsymbol{\xi}(a,b)}\neq a_{\boldsymbol{\xi}(a,b)}$. Then $c_{\boldsymbol{\xi}(a,b)}=b_{\boldsymbol{\xi}(a,b)}$ and we have $\boldsymbol{\xi}(b,c)>\boldsymbol{\xi}(a,b)$. Since $c<b$ we have

$$c_{\boldsymbol{\xi}(b,c)}=0<1=b_{\boldsymbol{\xi}(b,c)}\,,$$

but from (ii) we have $b_{\boldsymbol{\xi}(b,c)}=0$, a contradiction.

(b) follows from (a).

(c) We set

$$\mathfrak{L}_\xi(\alpha)=\{a\in\boldsymbol{\Lambda}(\alpha): a_\xi=0, \text{ and } a_\zeta=1 \text{ for } \xi<\zeta<\alpha\}$$

for $\xi<\alpha$. Since

$$\begin{aligned} &|\mathfrak{L}_\xi(\alpha)|=2^{|\xi|} \quad \text{for } \xi<\alpha\,,\\ &\mathfrak{L}_\xi(\alpha)\cap\mathfrak{L}_\eta(\alpha)=\emptyset \quad \text{for } \xi<\eta<\alpha\,, \quad \text{and}\\ &\mathfrak{L}(\alpha)=\textstyle\bigcup_{\xi<\alpha}\mathfrak{L}_\xi(\alpha) \end{aligned}$$

by part (b), we have $|\mathfrak{L}(\alpha)|=\sum_{\xi<\alpha}|\mathfrak{L}_\xi(\alpha)|=\sum_{\xi<\alpha}2^{|\xi|}=2^{\underline{\alpha}}$, by Lemma 1.22(f). The proof that $|\mathfrak{U}(\alpha)|=2^{\underline{\alpha}}$ is similar.

We note from part (b) above that

$$\mathfrak{L}(\alpha)\cap\mathfrak{U}(\alpha)=\emptyset\,.$$

An ordered set $\langle A,\leq\rangle$ is *Dedekind-complete* if every non-empty subset of A has a supremum and an infimum in $\langle A,\leq\rangle$.

It is easy to prove that if $\langle A,\leq\rangle$ has a least element then it is Dedekind-complete if and only if

every non-empty subset of $\langle A,\leq\rangle$ has a supremum in $\langle A,\leq\rangle$.

We note that a non-empty, Dedekind-complete ordered set has a least element and a greatest element; furthermore, these are $\sup\emptyset$ and $\inf\emptyset$ respectively.

If $\langle A, \leq\rangle$ is an ordered set and $C\subset A$, then C is *dense in* A if for $a,b\in A$ each of the sets

$$\{p\in A: a<p<b\},$$
$$\{p\in A: p<b\},$$
$$\{p\in A: a<p\}$$

that is non-empty has non-empty intersection with C.

Thus, C is dense in A if and only if C is a dense subset of the set A with the order topology.

5.8. Theorem. *Let* $\alpha\geq\omega$.

(a) $\Lambda(\alpha)$ *is Dedekind-complete. Indeed if* $\emptyset\neq A\subset\Lambda(\alpha)$ *and* $b=\langle b_\xi:\xi<\alpha\rangle$ *is defined recursively by*

$$b_\xi=\sup\{a_\xi: a\in A \text{ and } a_\zeta=b_\zeta \text{ for } \zeta<\xi\} \text{ for } \xi<\alpha,$$

then $b=\sup A$.

(b) $\mathfrak{U}(\alpha)$ *is dense in* $\Lambda(\alpha)\setminus\mathfrak{L}(\alpha)$; *and*

(c) $\Lambda(\alpha)\setminus\mathfrak{L}(\alpha)$ *is Dedekind-complete.*

Proof. (a) We prove first that b is an upper bound for A. If $a\in A$ and $b<a$ we have

$$b_{\xi(a,b)}=0<1=a_{\xi(a,b)}.$$

If $\xi(a,b)=0$ then $a_0=1$ and hence $b_0=1$; and if $\xi(a,b)>0$ then $b_{\xi(a,b)}=1$ because $a\in A$, $a_{\xi(a,b)}=1$, and $a_\zeta=b_\zeta$ for $\zeta<\xi(a,b)$. From this contradiction it follows that $a\leq b$ for $a\in A$.

If $c\in\Lambda(\alpha)$ and $c<b$ then since

$$c_{\xi(c,b)}=0<1=b_{\xi(c,b)}$$

there is $a\in A$ such that $a_{\xi(c,b)}=1$ and $a_\zeta=b_\zeta$ for $\zeta<\xi(c,b)$. Since $\xi(a,c)=\xi(c,b)$ and

$$c_{\xi(a,c)}=0<1=a_{\xi(a,c)}$$

we have $c<a$. Thus c is not an upper bound for A.

(b) Let $a,b\in\Lambda(\alpha)\setminus\mathfrak{L}(\alpha)$ with $a<b$. Since $a\notin\mathfrak{L}(\alpha)$ the set $\{\xi<\alpha: a_\xi=0\}$ has no largest element. Thus there is $\xi<\alpha$ such that

$$\xi(a,b)<\xi, \text{ and}$$
$$a_\xi=0.$$

We define

$$\begin{aligned} c_\zeta &= a_\zeta \quad \text{for } \zeta<\xi \\ &= 1 \quad \text{for } \zeta=\xi \\ &= 0 \quad \text{for } \xi<\zeta<\alpha . \end{aligned}$$

It is clear that $c\in\mathfrak{U}(\alpha)\subset\Lambda(\alpha)\setminus\mathfrak{L}(\alpha)$ and

$$a<c<b .$$

(c) Let $\emptyset\neq A\subset\Lambda(\alpha)\setminus\mathfrak{L}(\alpha)$ and denote by b the supremum of A in $\Lambda(\alpha)$; such an element b exists by part (a). If $b\in\Lambda(\alpha)\setminus\mathfrak{L}(\alpha)$ then b is the supremum of A in $\Lambda(\alpha)\setminus\mathfrak{L}(\alpha)$, and if $b\in\mathfrak{L}(\alpha)$ then b^+ is the supremum of A in $\Lambda(\alpha)\setminus\mathfrak{L}(\alpha)$.

A formula similar and dual to the one given for $\sup A$ in Theorem 5.8(a) exists for $\inf A$. Indeed

$$(\inf A)_\xi=\inf\{a_\xi : a\in A \text{ and } a_\zeta=(\inf A)_\zeta \text{ for } \zeta<\xi\} \text{ for } \xi<\alpha.$$

5.9. Corollary. *Let* $\alpha\geq\omega$.

(a) *If* $A\subset\mathfrak{U}(\alpha)$ *and* $|A|<\mathrm{cf}(\alpha)$ *then there is* $\xi<\alpha$ *such that*

$$(\sup A)_\zeta=0 \quad \textit{for } \xi<\zeta<\alpha ;$$

(b) *if* $B\subset\mathfrak{L}(\alpha)$ *and* $|B|<\mathrm{cf}(\alpha)$ *then there is* $\xi<\alpha$ *such that*

$$(\inf B)_\zeta=1 \quad \textit{for } \xi<\zeta<\alpha; \quad \textit{and}$$

(c) *if* $A\cup B\subset\mathfrak{U}(\alpha)$, $A<B$ *and* $|A|+|B|<\mathrm{cf}(\alpha)$, *then*

$$\begin{aligned} &\sup A<\inf B, \quad \textit{and} \\ &\sup A,\ \inf B\notin\mathfrak{L}(\alpha) . \end{aligned}$$

Proof. (a) For $a\in A$ there is $\xi_a<\alpha$ such that

$$a_\zeta=0 \quad \text{for } \xi_a<\zeta<\alpha .$$

We set $\xi=\sup\{\xi_a : a\in A\}$; then $\xi<\alpha$ and $(\sup A)_\zeta=0$ for $\xi<\zeta<\alpha$.

(b) is proved similarly and dually to part (a).

(c) We set $a=\sup A$ and $b=\inf B$.

(i) It is clear that $a\leq b$.

If B has a least element then $b=\min B$, and since $c<b$ for $c\in A$ we have $c\leq b^-$ for $c\in A$ and hence $a\leq b^-<b$.

If B has no least element then

$$b=\inf\{c : c\in B\}=\inf\{c^- : c\in B\}$$

and since $c^-\in\mathfrak{L}(\alpha)$ for $c\in B$ there is, by part (b), $\xi<\alpha$ such that $b_\zeta=1$ for $\xi<\zeta<\alpha$. By part (a) there is $\xi'<\alpha$ such that $a_\zeta=0$ for $\xi'<\zeta<\alpha$. Hence $a\neq b$ and we have $a<b$.

(ii) It follows from part (a) and Theorem 5.7(a) that $a \notin \mathfrak{L}(\alpha)$. If $b \in \mathfrak{L}(\alpha)$ then since $B \subset \mathfrak{U}(\alpha)$ and $b \leq c$ for $c \in B$ we have in fact $b < c$ and hence $b^+ \leq c$ for $c \in B$, a contradiction proving that $b \notin \mathfrak{L}(\alpha)$.

5.10. Lemma. *If* $\alpha \geq \omega$, *then* $\mathfrak{U}(\alpha)$ *is an* η_α*-set if and only if* α *is a regular cardinal.*

Proof. Let $\mathfrak{U}(\alpha)$ be an η_α-set. We choose a well-ordered cofinal subset A of $\mathfrak{U}(\alpha)$; by Theorem 5.4, $|A| \leq \alpha$. If α is singular then, by Lemma 5.6, $\mathfrak{U}(\alpha)$ is an η_{α^+}-set and hence $|A| \geq \alpha^+$. Thus α is regular.

Let α be a regular cardinal and let $A, B \subset \mathfrak{U}(\alpha)$ with $A < B$ and $|A| + |B| < \alpha = \mathrm{cf}(\alpha)$. By Corollary 5.9(c), $\sup A < \inf B$ and $\sup A, \inf B \in \Lambda(\alpha) \setminus \mathfrak{L}(\alpha)$; and by Theorem 5.8(b) there is $c \in \mathfrak{U}(\alpha)$ such that $\sup A < c < \inf B$. Thus $A < c < B$.

It follows from Lemma 5.2 and Corollary 4.7 that for $\alpha \geq \omega$ there is an η_{α^+}-set of cardinality 2^α. We now prove that there is a set of dyadic sequences with these properties.

5.11. Corollary. *If* $\alpha \geq \omega$ *then* $\mathfrak{U}(\alpha^+)$ *is an* η_{α^+}*-set of cardinality* 2^α.

Proof. By Lemma 5.11, $\mathfrak{U}(\alpha^+)$ is an η_{α^+}-set; and by Theorem 5.7(c), $|\mathfrak{U}(\alpha^+)| = 2^{\underline{\alpha^+}} = 2^\alpha$.

The following theorem, the central result of the section, describes the homogeneous-universal structures $\mathfrak{H}_\alpha$; it should be compared to the general Theorem 4.10.

5.12. Theorem. *Let* $\alpha \geq \omega$.

(a) *(Existence of* $\mathfrak{H}_\alpha$*). There is an* α*-homogeneous-universal ordered set* $\mathfrak{H}_\alpha$ *of cardinality* α *if and only if* $\alpha = \alpha^{\underline{\alpha}}$;

(b) *(Characterization of* $\mathfrak{H}_\alpha$*). An ordered set is order-isomorphic to* $\mathfrak{H}_\alpha$ *if and only if it is an* η_α*-set of cardinality* α*; and*

(c) *(Representation of* $\mathfrak{H}_\alpha$*). If* $\alpha = \alpha^{\underline{\alpha}}$ *then* $\mathfrak{H}_\alpha$ *is order-isomorphic to* $\mathfrak{U}(\alpha)$.

Proof. We claim first that if there is an η_α-set $\langle A, \leq \rangle$ such that $|A| = \alpha$, then $\alpha = \alpha^{\underline{\alpha}}$. Indeed

$$\alpha \geq 2^\beta \quad \text{for } \beta < \alpha$$

by Lemma 5.5, and if α is singular then $\langle A, \leq \rangle$ is an η_{α^+}-set by Lemma 5.6 and hence $|A| \geq 2^\alpha$ by Lemma 5.5; thus $\alpha = \alpha^{\underline{\alpha}}$ by Theorem 1.27(a).

(a) If $\mathfrak{H}_\alpha$ exists then $\mathfrak{H}_\alpha$ is an η_α-set by Lemma 5.2; hence $\alpha = \alpha^{\underline{\alpha}}$ by the claim proved above. Conversely if $\alpha = \alpha^{\underline{\alpha}}$ then (since by Theorem 5.1 the class $\mathbf{OS}$ is a Jónsson class) there is by Theorem 4.10(b) and Corollary 4.11 an α-homogeneous-universal ordered set of cardinality α.

(b) follows from what has been proved and Theorem 5.3.

(c) The cardinal α is regular by Theorem 1.27(a), and hence $\mathfrak{U}(\alpha)$ is an η_α-set by Lemma 5.10. Since $|\mathfrak{U}(\alpha)|=2^{\underline{\alpha}}=\alpha$ by Theorem 5.7(c) and Theorem 1.27(a), it follows from part (b) that $\mathfrak{U}(\alpha)$ is order-isomorphic to $\mathfrak{H}_\alpha$.

We end this section on ordered sets with a definition and theorem to be used in § 15.

Definition. Let $\alpha \geq \omega$ and $a \in \Lambda(\alpha)$. Then a is an *upper κ-point* if there is a non-zero *limit* ordinal $\xi < \alpha$ such that

(i) $a_\eta = 0$ for $\xi \leq \eta < \alpha$ and

(ii) for $\eta < \xi$ there is ζ such that $\eta < \zeta < \xi$ and $a_\zeta = 1$.

Similarly, a is a *lower κ-point* if there is a non-zero limit ordinal $\xi < \alpha$ such that

(i) $a_\eta = 1$ for $\xi \leq \eta < \alpha$ and

(ii) for $\eta < \xi$ there is ζ such that $\eta < \zeta < \xi$ and $a_\zeta = 0$.

The sets of upper and lower κ-points of $\Lambda(\alpha)$ are denoted $K^+(\alpha)$ and $K^-(\alpha)$ respectively. An element of $K^+(\alpha) \cup K^-(\alpha)$ is a *κ-point*.

We note that $K^+(\alpha) \cap K^-(\alpha) = \emptyset$, and that

$$(K^+(\alpha) \cup K^-(\alpha)) \cap (\mathfrak{U}(\alpha) \cup \mathfrak{L}(\alpha)) = \emptyset .$$

5.13. Lemma. *Let $\alpha \geq \omega$ and $a \in \Lambda(\alpha)$.*

(a) *If $a \in K^+(\alpha)$ there is $A \subset \Lambda(\alpha)$ such that $A < a$, $|A| < \alpha$ and $a = \sup A$;*

(b) *if there is $A \subset \Lambda(\alpha)$ such that $A < a$, $0 < |A| < \mathrm{cf}(\alpha)$ and $a = \sup A$, then $a \in K^+(\alpha)$.*

Proof. (a) Let ξ be the limit ordinal with (i) and (ii) of the definition and for $\zeta < \xi$ let $a^\zeta \in \Lambda(\alpha)$ be defined by

$$\begin{aligned} a_\eta^\zeta &= a_\eta \quad \text{for } \eta \leq \zeta \\ &= 0 \quad \text{for } \zeta < \eta < \alpha . \end{aligned}$$

Then $a^\zeta \leq a^{\zeta'}$ for $\zeta < \zeta' < \xi$, and it is clear by Theorem 5.8(a) that

$$a = \sup \{a^\zeta : \zeta < \xi\} .$$

(b) Let $a = \sup A$ with $A < a$ and $0 < |A| < \mathrm{cf}(\alpha)$, and set

$$\zeta = \sup \{\boldsymbol{\xi}(b,c) : b, c \in A \text{ and } b \neq c\} .$$

Since $|A| < \mathrm{cf}(\alpha)$ and $\boldsymbol{\xi}(b,c) < \alpha$ for $b, c \in A$ and $b \neq c$ we have $\zeta < \alpha$. We claim that $a_\xi = 0$ for $\zeta \leq \xi < \alpha$. If there is ξ such that $\zeta \leq \xi < \alpha$ and $a_\xi = 1$ then by Theorem 5.8(a) there is $b \in A$ such that $b_\eta = a_\eta$ for $\eta < \xi$ and $b_\xi = 1$. There is $c \in A$ such that $b < c < a$, and since $b_\eta = a_\eta$ for $\eta \leq \xi$ we have $b_\eta = c_\eta$ for $\eta \leq \xi$ and hence

$$\boldsymbol{\xi}(b,c) > \xi \geq \zeta ,$$

contradicting the definition of ζ. This proves the claim.

Set $\xi = \min\{\zeta' < \alpha : a_\eta = 0$ for $\zeta' \leq \eta < \alpha\}$. If ξ is not a limit ordinal then $a \in \mathfrak{U}(\alpha)$ and hence $b < a^-$ for $b \in A$; thus ξ is a limit ordinal, so for $\eta < \xi$ there is η' such that $\eta < \eta' < \xi$ and $a_{\eta'} = 1$. The proof is complete.

A result similar and dual to Lemma 5.13 holds for lower κ-points.

5.14. Corollary. *Let α be an infinite, regular cardinal and $a \in \Lambda(\alpha)$.*
(a) *$a \in K^+(\alpha)$ if and only if there is $A \subset \Lambda(\alpha)$ such that $\emptyset \neq A < a$, $|A| < \alpha$, and $a = \sup A$;*
(b) *$a \in K^-(\alpha)$ if and only if there is $A \subset \Lambda(\alpha)$ such that $a < A \neq \emptyset$, $|A| < \alpha$, and $a = \inf A$.*

Proof. Part (a) follows from Lemma 5.13 and (b) is proved similarly.

Remark. We note the following properties of the space $\Lambda(\alpha)$ with the order topology (defined at the end of § 2).
(a) *$\Lambda(\alpha)$ is a compact space* (by Theorem 5.8(a) and an imitation of the standard proof that $[0,1]$ is compact).
(b) *$\Lambda(\alpha)$ is totally disconnected* (by Theorems 5.7(a) and 5.8(b)).
(c) $w(\Lambda(\alpha)) \leq 2^{\underline{\alpha}}$. Indeed the family

$$\{\{p \in \Lambda(\alpha) : a < p < b\} : a, b \in \mathfrak{L}(\alpha) \cup \mathfrak{U}(\alpha)\} \cup \{\{p \in \Lambda(\alpha) : a < p\} : a \in \mathfrak{L}(\alpha) \cup \mathfrak{U}(\alpha)\} \cup \{\{p \in \Lambda(\alpha) : p < b\} : b \in \mathfrak{L}(\alpha) \cup \mathfrak{U}(\alpha)\}$$

is a base for $\Lambda(\alpha)$ of cardinality $2^{\underline{\alpha}}$.
(d) *If α is a regular cardinal, then an element of $\Lambda(\alpha)$ is a P_α-point in $\Lambda(\alpha)$ if and only if it is not a κ-point* (by Corollary 5.14 and the fact that an element of a linearly ordered set is a P_α-point if and only if it is neither the limit of an increasing sequence of length less than α nor the limit of a decreasing sequence of length less than α). *Thus the set of non-P_α-points of $\Lambda(\alpha)$ has cardinality $2^{\underline{\alpha}}$.*
(e) *If α is a regular cardinal, then every element of $\Lambda(\alpha)$ has local weight* α (by part (d) and the fact that $K^+(\alpha) \cap K^-(\alpha) = \emptyset$).

Notes for § 5

The argument used to verify the amalgamation property for $\mathbf{OS}$ in the proof of Theorem 5.1 is adapted from the analogous result on the class of partially ordered sets given by Jónsson [56] (Lemma 3.3).

Cantor [95] (§ 9) proved (in different terminology) that every η_ω-set of cardinality ω is order-isomorphic to the set of rational numbers. The back-and-forth inductive method used by Cantor, in which an isomorphism is defined recursively by extending embeddings "one element at a time" by appeal to the η_ω-property, was generalized by

Hausdorff [14] (Kap. VI, § 8) who introduced η_α-sets and proved Theorem 5.3 and that any η_α-set is α-universal (Hausdorff [08] (§ 23)).

Hausdorff [08] (p. 182) proved that every linearly ordered set is isomorphic to an ordered set of triadic sequences (*i.e.*, to a subset of 3^α, ordered lexicographically, for some α). The fact that $\Lambda(\alpha)$ is α-universal was announced by Sierpiński [40] and proved by Cuesta [47] and Sierpiński [49]. Theorems 5.4 and 5.8 (a) and Corollary 5.11 are proved in Sierpiński [49] (Lemmas II, I, and III, respectively). For the remaining case of limit cardinals α, Lemma 5.10 is proved in Gillman [56] (Theorem 2). The statement that $\mathfrak{U}(\alpha) \cup K^+(\alpha)$ is dense in $\Lambda(\alpha)$ is essentially proved in Sierpiński [22] (cf. Theorem 5.8). The ordered pairs of consecutive elements of $\Lambda(\alpha)$ are described briefly by Sierpiński [65] (page 461) and in detail (for $\alpha = \omega^+$) by Gillman and Jerison [60] (§ 13.16).

Gillman [60] (Theorem 3) also notes that if there is an η_α-set of cardinality α then $\alpha = \alpha^{\underline{\alpha}}$ (cf. Theorem 5.12 (a), (b)). That Hausdorff's η_α-sets are precisely the α-homogeneous-universal sets (Lemma 5.2 and Theorem 5.12 (b)) is remarked in Jónsson [60] (p. 138).

Jónsson [60] (Remark 1, page 141) then establishes the statement, weaker than Gillman's, that if there is an α-homogeneous-universal system of cardinality α in $\mathfrak{OS}$ then α is regular.

Sierpiński [65] has shown that for $\alpha \geq \omega$ there are 2^α pairwise non-isomorphic elements of $\mathfrak{OS}$ of cardinality α, and Gillman [55], [56] has shown that if $\alpha^+ \neq 2^\alpha$ there are two non-isomorphic η_{α^+}-sets of cardinality 2^α.

§ 6. The Jónsson Class of Boolean Algebras

The general results of the theory of Jónsson classes, developed in § 4, are applied here to the Jónsson class of (proper) Boolean algebras. The α-homogeneous-universal Boolean algebra $\mathfrak{C}_\alpha$ of cardinality α and its Stone space S_α, which exist if and only if $\alpha=\alpha^{\underline{\alpha}}$, are studied in considerable detail and characterized by properties given in terms of the partial order of a Boolean algebra; these properties are similar to the η_α-property of ordered sets (studied in § 5). We study the Jónsson class of Boolean algebras not mainly to illustrate the general theory of Jónsson classes, but rather in order to describe the space S_α itself; in later sections some of its properties will be set in analogy or contrast to those of certain spaces of ultrafilters. There are intriguing parallels between some rather refined properties of S_α and corresponding properties of the space $U(\alpha^+)$ (to be defined in § 7) of uniform ultrafilters on α^+. In the particular case $\alpha=\omega^+=2^\omega$ the space S_{ω^+} indeed is homeomorphic to the space (of non-principal ultrafilters on ω) $\boldsymbol{\beta}(\omega)\backslash\omega$ (cf. the introduction and results of § 14 and § 15).

In § 10 we prove the existence of good ultrafilters on certain Boolean algebras associated with S_α by methods that use the results of the present section (mainly Lemma 6.7).

The class of proper Boolean algebras is denoted $\mathbb{BA}$. It is clear that if

$$\mathfrak{B}=\langle B, \wedge, \vee, ' \rangle \in \mathbb{BA}$$

and $\wedge$ and $\vee$ are considered as relations on $B\times B\times B$ (rather than as operations on $B\times B$) and $'$ is considered as a relation on $B\times B$, then $\mathfrak{B}$ is a relational system with index 3 and similarly type τ such that $\tau(\wedge)=\tau(\vee)=3$ and $\tau(')=2$. Further if $\mathfrak{B},\mathfrak{C}\in\mathbb{BA}$ and $f:\mathfrak{B}\to\mathfrak{C}$, then f is an embedding of relational systems (as defined in § 4) if and only if f is a Boolean algebra embedding (as defined in § 2).

6.1. Theorem. *The class* $\mathbb{BA}$ *is a Jónsson class with* $(\mathrm{VI})_\omega$.

Proof. We refer to the defining conditions of a Jónsson class given in § 4.

(I) For every cardinal α the field of sets $\mathscr{P}(\alpha)$ is a Boolean algebra.

(II) This is obvious.

(III) Since the Boolean algebra $\mathbb{2}$ is a Boolean subalgebra of every element of $\mathbb{BA}$, this is a consequence of (IV).

(IV) We let $\mathfrak{A}, \mathfrak{B}, \mathfrak{C} \in \mathbb{BA}$ and $f: \mathfrak{C} \to \mathfrak{A}$ and $g: \mathfrak{C} \to \mathfrak{B}$ be embeddings, and we find $\mathfrak{D} \in \mathbb{BA}$ and embeddings $f_1: \mathfrak{A} \to \mathfrak{D}$ and $g_1: \mathfrak{B} \to \mathfrak{D}$ such that $f_1 \circ f = g_1 \circ g$.

By Theorem 2.25 the functions $\mathsf{S}(f)$ and $\mathsf{S}(g)$ defined by

$$\mathsf{S}(f)(p) = \{c \in \mathfrak{C} : f(c) \in p\} \quad \text{for } p \in \mathsf{S}(\mathfrak{A}) \quad \text{and}$$
$$\mathsf{S}(g)(q) = \{c \in \mathfrak{C} : g(c) \in q\} \quad \text{for } q \in \mathsf{S}(\mathfrak{B}),$$

are continuous functions from $\mathsf{S}(\mathfrak{A})$ and $\mathsf{S}(\mathfrak{B})$ respectively onto $\mathsf{S}(\mathfrak{C})$. We set

$$K = \{\langle p, q\rangle \in \mathsf{S}(\mathfrak{A}) \times \mathsf{S}(\mathfrak{B}) : \mathsf{S}(f)(p) = \mathsf{S}(g)(q)\}, \quad \text{and}$$
$$\phi_0(\langle p, q\rangle) = p \quad \text{and} \quad \phi_1(\langle p, q\rangle) = q \quad \text{for } \langle p, q\rangle \in K.$$

It is clear that

K is a compact, totally disconnected space,

$\phi_0[K] = \mathsf{S}(\mathfrak{A})$ and $\phi_1[K] = \mathsf{S}(\mathfrak{B})$, and

ϕ_0 and ϕ_1 are continuous functions.

We now consider the diagram

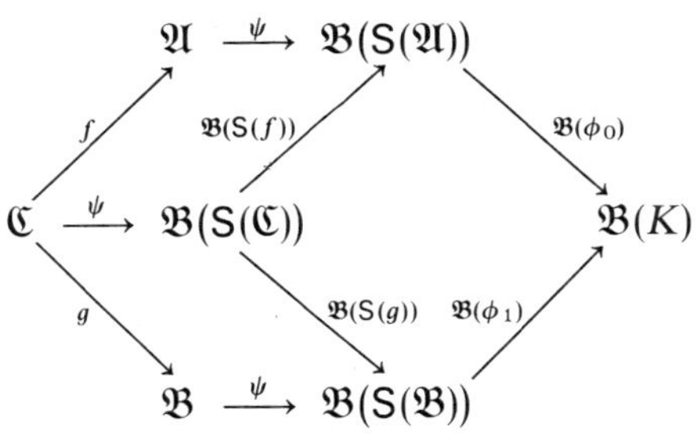

where ψ is the function of Theorem 2.21 and the functions $\mathfrak{B}(\phi_0)$, $\mathfrak{B}(\phi_1)$, $\mathfrak{B}(\mathsf{S}(f))$ and $\mathfrak{B}(\mathsf{S}(g))$ are defined as in the paragraphs preceding Theorem 2.29. The commutativity of the three rectangular diagrams follows from Theorem 2.29 (e), (f). Since ϕ_0 and ϕ_1 are onto functions the homomorphisms $\mathfrak{B}(\phi_0)$ and $\mathfrak{B}(\phi_1)$ are embeddings by Theorem 2.29 (d). We define

$$\mathfrak{D} = \mathfrak{B}(K), \quad f_1 = \mathfrak{B}(\phi_0) \circ \psi \quad \text{and} \quad g_1 = \mathfrak{B}(\phi_1) \circ \psi.$$

Then $\mathfrak{D} \in \mathbb{BA}$, and $f_1 \circ f = g_1 \circ g$.

(V) This is obvious.

$(\mathrm{VI})_\omega$ This follows from Lemma 2.14 (a).

We remark that if $\mathfrak{B} \in \mathbb{BA}$ then there is no Boolean algebra embedding of the improper Boolean algebra into $\mathfrak{B}$. This accounts for the exclusion of the improper Boolean algebra from the class $\mathbb{BA}$.

It follows from Theorem 4.10(b), Corollary 4.11 and Theorem 6.1 that for every infinite cardinal α such that $\alpha = \alpha^{\underline{\alpha}}$ there is an α-homogeneous-universal Boolean algebra $\mathfrak{C}_\alpha$ of cardinality α, unique up to Boolean algebra isomorphism. The main object of this section is the description of $\mathfrak{C}_\alpha$ and of its Stone space (Theorem 6.12).

The Stone Space of the Homogeneous-Universal Boolean Algebras

Definitions. An element a of a Boolean algebra $\mathfrak{B}$ is an *atom* of $\mathfrak{B}$ if $a \neq \mathbb{0}$ and there is no $b \in \mathfrak{B}$ such that $\mathbb{0} < b < a$. If the Boolean algebra $\mathfrak{B}$ has no atoms than $\mathfrak{B}$ is *atomless.*

An element p of a space X is *isolated* if $\{p\}$ is an open subset of X. If the space X has no isolated elements then X is *perfect.*

We note the following simple consequences of these definitions.

(a) *If $\mathfrak{B}$ is a Boolean algebra and $a \in \mathfrak{B}$, then a is an atom if and only if the principal filter $\langle\{a\}\rangle$ generated by a is an ultrafilter of $\mathfrak{B}$;*

(b) *if a is an atom of $\mathfrak{B}$, then $\langle\{a\}\rangle$ is an isolated element of* $\mathsf{S}(\mathfrak{B})$*;*

(c) *if p is an isolated element of* $\mathsf{S}(\mathfrak{B})$, *then $\{p\}$ is an atom of* $\mathfrak{B}(\mathsf{S}(\mathfrak{B}))$*;*

(d) *$\mathfrak{B}$ is an atomless Boolean algebra if and only if* $\mathsf{S}(\mathfrak{B})$ *is a perfect space.*

6.2. Lemma. *If $\alpha \geq \omega$ and $\mathfrak{B}$ is an α-homogeneous-universal Boolean algebra, then $\mathfrak{B}$ is atomless.*

Proof. Let a be a non-zero element of $\mathfrak{B}$ and denote by $\mathfrak{A}$ the Boolean subalgebra of $\mathfrak{B}$ generated by a, *i.e.*, set $\mathfrak{A} = \{\mathbb{0}, \mathbb{1}, a, a'\}$. There are a finite Boolean algebra $\mathfrak{C}$, an embedding $f: \mathfrak{A} \to \mathfrak{C}$, and $c \in \mathfrak{C}$ such that

$$\mathbb{0} < c < f(a),$$

and since $\mathfrak{B}$ is ω-universal there is an embedding $g: \mathfrak{C} \to \mathfrak{B}$. Let e denote the natural inclusion of $\mathfrak{A}$ into $\mathfrak{B}$. Since $\mathfrak{B}$ is α-homogeneous there is an automorphism h of $\mathfrak{B}$ such that $e = h \circ g \circ f$, and since $\mathbb{0} < c < f(a)$ in $\mathfrak{C}$ we have $\mathbb{0} < g(c) < g(f(a))$ and hence

$$\mathbb{0} < h(g(c)) < h(g(f(a))) = e(a) = a$$

in $\mathfrak{B}$. Thus a is not an atom of $\mathfrak{B}$.

A subset A of a Boolean algebra is said to be *directed upwards* if for $F \in \mathscr{P}_\omega^*(A)$ there is $a \in A$ such that $\bigvee F \leq a$; and A is *directed downwards* if for $F \in \mathscr{P}_\omega^*(A)$ there is $a \in A$ such that $a \leq \bigwedge F$.

Definition. Let $\alpha \geq \omega$. A Boolean algebra $\mathfrak{B}$ satisfies *condition* (H_α) if for every $A_0, A_1 \subset \mathfrak{B}$ such that

$$A_0 < A_1,$$
$$|A_0| + |A_1| < \alpha,$$
$$A_0 \subset \mathfrak{B} \setminus \{\mathbb{1}\} \quad \text{and} \quad A_1 \subset \mathfrak{B} \setminus \{\mathbb{0}\}, \quad \text{and}$$
$$A_0 \text{ is directed upwards and } A_1 \text{ is directed downwards,}$$

there is $b \in \mathfrak{B} \setminus \{\mathbb{0}, \mathbb{1}\}$ such that

$$A_0 < b < A_1.$$

6.3. Lemma. *If* $\alpha \geq \omega$ *then every* α*-homogeneous-universal Boolean algebra satisfies condition* (H_α).

Proof. Let $\mathfrak{B}$ be an α-homogeneous-universal Boolean algebra. Using Stone's duality (Theorem 2.21 (d)) we identify $\mathfrak{B}$ with $\mathfrak{B}(\mathsf{S}(\mathfrak{B}))$; this latter is, by Lemma 2.20, a subalgebra of $\mathscr{P}(\mathsf{S}(\mathfrak{B}))$.

Let $A_0 \subset \mathfrak{B} \setminus \{\mathbb{1}\}$ and $A_1 \subset \mathfrak{B} \setminus \{\mathbb{0}\}$ be such that

$$A_0 < A_1,$$
$$|A_0| + |A_1| < \alpha, \quad \text{and}$$
$$A_0 \text{ is directed upwards and } A_1 \text{ is directed downwards.}$$

We assume without loss of generality that $\bigvee F \in A_0$ for $F \in \mathscr{P}^*_\omega(A_0)$ and $\bigwedge F \in A_1$ for $F \in \mathscr{P}^*_\omega(A_1)$. We denote by $\mathfrak{B}_0$ the Boolean subalgebra of $\mathfrak{B}$ generated by $A_0 \cup A_1$, and we define a subset T of $\mathsf{S}(\mathfrak{B})$ as follows.

Case 1. A_0 has a largest element a_0 and A_1 has a smallest element a_1. Since $A_0 < A_1$ we have $a_0 < a_1$, and since $\mathfrak{B}$ is atomless there is $b \in \mathfrak{B}$ such that

$$\mathbb{0} < b < a_1 \setminus a_0.$$

We set $T = a_0 \cup b$. It is clear that

$$A_0 < T < A_1 \quad \text{in } \mathscr{P}(\mathsf{S}(\mathfrak{B})).$$

Case 2. A_0 has no largest element. We set

$$T = \bigcup A_0.$$

Then $a_0 \leq T$ in $\mathscr{P}(\mathsf{S}(\mathfrak{B}))$ for $a_0 \in A_0$, and since A_0 has no largest element we have $a_0 < T$ for $a_0 \in A_0$. We note also that $T \leq a_1$ for $a_1 \in A_1$, since if $p \in T \setminus a_1$ then there is $a_0 \in A_0$ such that $p \in a_0 \setminus a_1$. We claim further that $T < A_1$ in $\mathscr{P}(\mathsf{S}(\mathfrak{B}))$. Indeed if $T \in A_1$ then $T \in \mathfrak{B}(\mathsf{S}(\mathfrak{B}))$, and since A_0 is an open cover of T and $\mathsf{S}(\mathfrak{B})$ is compact there is $F \in \mathscr{P}^*_\omega(A_0)$ such that

$$T = \bigcup F = \bigvee F \in A_0$$

and hence A_0 has a largest element. It follows that

$$A_0 < T < A_1 \text{ in } \mathscr{P}(\mathsf{S}(\mathfrak{B})).$$

Case 3. A_1 has no smallest element. We set

$$T = \bigcap A_1 .$$

An argument analogous to the one in *Case* 2 proves that

$$A_0 < T < A_1 \text{ in } \mathscr{P}(\mathsf{S}(\mathfrak{B})).$$

Now let $\mathfrak{C}$ be the subfield of $\mathscr{P}(\mathsf{S}(\mathfrak{B}))$ generated by $A_0 \cup A_1 \cup \{T\}$, and let e be the natural inclusion of $\mathfrak{B}_0$ into $\mathfrak{C}$, so that $f \circ e: \mathfrak{B}_0 \to \mathfrak{B}$ is an embedding. Since $\mathfrak{B}$ is α-homogeneous there is an automorphism h of $\mathfrak{B}$ such that $h \circ f \circ e$ is equal to the inclusion of $\mathfrak{B}_0$ into $\mathfrak{B}$, and since $e[A_0] < T < e[A_1]$ in $\mathfrak{C}$ we have

$$A_0 = (h \circ f \circ e)[A_0] < h(f(T)) < (h \circ f \circ e)[A_1] = A_1$$

in $\mathfrak{B}$. The proof is complete.

Definition. Let X be a space such that $\mathfrak{B}(X)$ is a base for the topology of X, and let U be an open subspace of X. The *type* of U, denoted $\tau(U)$, is the least cardinal number β such that U is equal to the union of β elements of $\mathfrak{B}(X)$.

6.4. Lemma. *Let X be a compact, totally disconnected space and let U be an open subset of X. If $\tau(U) > 1$, then $\tau(A) \leq \tau(U)$ for every cozero subset A of U.*

Proof. Let $f \in C(U)$ be such that $A = U \setminus f^{-1}(\{0\})$ and set

$$A_n = \{p \in U : |f(p)| \geq 1/n\}$$

for $1 \leq n < \omega$. Since A_n is closed in U and U is a union of $\tau(U)$ compact sets, the set A_n is a union of $\tau(U)$ compact sets; hence there is a family $\mathscr{C}_n$ of subsets of X such that

$$\mathscr{C}_n \subset \mathfrak{B}(X),$$
$$A_n \subset \bigcup \mathscr{C}_n \subset A, \quad \text{and}$$
$$|\mathscr{C}_n| \leq \tau(U)$$

for $1 \leq n < \omega$. Then $A = \bigcup_{1 \leq n < \omega} (\bigcup \mathscr{C}_n)$ and we have

$$\tau(A) \leq \sum_{1 \leq n < \omega} |\mathscr{C}_n| \leq \omega \cdot \tau(U) = \tau(U).$$

We next give topological conditions on $\mathsf{S}(\mathfrak{B})$ equivalent to condition (H_α) for $\mathfrak{B}$. As usual, we identify $\mathfrak{B}$ with $\mathfrak{B}(\mathsf{S}(\mathfrak{B}))$.

For $U \subset \mathsf{S}(\mathfrak{B})$ we denote by $\bar{U}$ the closure of U in $\mathsf{S}(\mathfrak{B})$.

6.5. Lemma. *Let α be an infinite cardinal and $\mathfrak{B}$ a Boolean algebra. The following statements* (a), (b) *and* (c) *are equivalent.*

(a) $\mathfrak{B}$ *satisfies condition* (H_α)*;*

(b) 1. $\mathfrak{B}$ *is atomless,*

2. *if U is an open subset of $\mathsf{S}(\mathfrak{B})$ such that $1 < \tau(U) < \alpha$, then $\bar{U}$ is not an open subset of $\mathsf{S}(\mathfrak{B})$,*

3. *if U and V are open subsets of $\mathsf{S}(\mathfrak{B})$ such that $\tau(U) + \tau(V) < \alpha$ and $U \cap V = \emptyset$, then $\bar{U} \cap \bar{V} = \emptyset$; and*

(c) 1. $\mathfrak{B}$ *is atomless,*

2. *if U is an open subset of $\mathsf{S}(\mathfrak{B})$ such that $1 < \tau(U) < \alpha$, then $\bar{U}$ is not an open subset of $\mathsf{S}(\mathfrak{B})$,*

3. *if U is an open subset of $\mathsf{S}(\mathfrak{B})$ such that $\tau(U) < \alpha$, then U is C^*-embedded in $\mathsf{S}(\mathfrak{B})$.*

Proof. (a)⇒(b). To prove (b)1 let $\mathbb{0} < a$ in $\mathfrak{B}$ and apply condition (H_α) with $A_0 = \{\mathbb{0}\}$ and $A_1 = \{a\}$.

To prove (b)2 and (b)3 let U and V be open subsets of $\mathsf{S}(\mathfrak{B})$ such that

$$U = \bigcup_{i \in I} a_i \quad \text{and} \quad V = \bigcup_{j \in J} b_j,$$

where $\{a_i : i \in I\} \subset \mathfrak{B}(\mathsf{S}(\mathfrak{B}))$, $\{b_j : j \in J\} \subset \mathfrak{B}(\mathsf{S}(\mathfrak{B}))$ and $|I| + |J| < \alpha$. We assume without loss of generality that each of the families $\{a_i : i \in I\}$ and $\{b_j : j \in J\}$ is closed under finite unions. If $\bar{U}$ is open we obtain a contradiction by applying condition (H_α) with

$$A_0 = \{a_i : i \in I\} \quad \text{and} \quad A_1 = \{\bar{U}\};$$

and if $\bar{U} \cap \bar{V} \neq \emptyset$ we obtain a contradiction by applying condition (H_α) with

$$A_0 = \{a_i : i \in I\} \quad \text{and} \quad A_1 = \{\mathsf{S}(\mathfrak{B}) \setminus b_j : j \in J\}.$$

(b)⇒(c). We verify (c)3. If A and B are subsets of U which are completely separated in U then there are disjoint cozero subsets C and D of U such that

$$A \subset C \quad \text{and} \quad B \subset D;$$

since $\tau(C) + \tau(D) < \alpha$ by Lemma 6.4, we have $\bar{C} \cap \bar{D} = \emptyset$ by (b)3; hence $\bar{C}$ and $\bar{D}$ are completely separated in $\mathsf{S}(\mathfrak{B})$ by Tietze's extension theorem. It follows from Urysohn's extension theorem (Corollary 2.8) that U is C^*-embedded in $\mathsf{S}(\mathfrak{B})$.

(c)⇒(b). We verify (b)3. If U and V are disjoint open subsets of $\mathsf{S}(\mathfrak{B})$ such that $\tau(U)+\tau(V)<\alpha$ then $\tau(U\cup V)<\alpha$ and by (c) 3 there is $f\in C^*(\mathsf{S}(\mathfrak{B}))$ such that

$$U\subset f^{-1}(\{0\}) \quad \text{and} \quad V\subset f^{-1}(\{1\}).$$

(b)⇒(a). Let $A_0\subset\mathfrak{B}\setminus\{\mathbb{1}\}$ and $A_1\subset\mathfrak{B}\setminus\{\mathbb{0}\}$ with $A_0<A_1$ and $|A_0|+|A_1|<\alpha$ and with A_0 directed upward and A_1 directed downward. We suppose, adjoining $\mathbb{0}$ to A_0 and $\mathbb{1}$ to A_1 if necessary, that neither A_0 nor A_1 is empty, and we suppose also that A_0 and A_1 are closed under finite unions and finite intersections, respectively. To find $b\in\mathfrak{B}$ such that $A_0<b<A_1$ we distinguish three cases and we appeal in each case to the appropriate condition in (b).

Case 1. A_0 has a largest element a_0 and A_1 has a smallest element a_1. By (b) 1 there is $c\in\mathfrak{B}$ such that

$$\mathbb{0}<c<a_1\setminus a_0,$$

and we set $b=a_0\vee c$. Then

$$A_0<b<A_1.$$

Case 2. Either A_0 has a largest element a_0 and A_1 has no smallest element, or A_0 has no largest element and A_1 has a smallest element a_1. In the latter sub-case we set $U=\bigcup A_0$ and we note that

$$1<\tau(U)\leq|A_0|<\alpha.$$

Since $\bar{U}$ is not open in $\mathsf{S}(\mathfrak{B})$ (by (b)2), we have $\bar{U}\subsetneqq a_1$ and there is $b\in(\mathfrak{B}(\mathsf{S}(\mathfrak{B})))$ such that

$$U\subsetneqq b\subsetneqq a_1.$$

Then $A_0<b<A_1$ in $\mathfrak{B}$. The former sub-case is treated similarly.

Case 3. A_0 has no largest element and A_1 has no least element. We set

$$U=\bigcup\{a:a\in A_0\} \quad \text{and} \quad V=\bigcup\{\mathsf{S}(\mathfrak{B})\setminus a:a\in A_1\},$$

so that $U\cap V=\emptyset$ and $\tau(U)+\tau(V)<\alpha$. Since $\bar{U}\cap\bar{V}=\emptyset$ (by (b)3), there is $b\in\mathfrak{B}(\mathsf{S}(\mathfrak{B}))$ such that

$$\bar{U}\subset b\subset\mathsf{S}(\mathfrak{B})\setminus\bar{V}.$$

By (b)2 we have $\bar{U}\subsetneqq b\subsetneqq\mathsf{S}(\mathfrak{B})\setminus\bar{V}$, and hence

$$A_0<b<A_1.$$

6.6. Lemma. *Let α be an infinite cardinal and $\mathfrak{B}$ a Boolean algebra such that if U is open in $\mathsf{S}(\mathfrak{B})$ and $1<\tau(U)<\alpha$ then $\bar{U}$ is not open in $\mathsf{S}(\mathfrak{B})$. If $\beta<\alpha$ and $\{a_\xi:\xi<\beta\}$ is a family of open-and-closed subsets of $\mathsf{S}(\mathfrak{B})$ with the finite intersection property, then*

$$\operatorname{int}_{\mathsf{S}(\mathfrak{B})}\left(\bigcap\nolimits_{\xi<\beta}a_\xi\right)\neq\emptyset.$$

Proof. If there is $F \in \mathscr{P}_{\omega}^{*}(\beta)$ such that

$$\bigcap_{\xi<\beta} a_{\xi} = \bigcap_{\xi \in F} a_{\xi}$$

then the statement is clear. If there is no such F we set

$$U = \bigcup_{\xi<\beta} (\mathsf{S}(\mathfrak{B}) \setminus a_{\xi})$$

and we note that since

$$1 < \tau(U) \leq \beta < \alpha$$

the set $\bar{U}$ is not open. Hence $\mathsf{S}(\mathfrak{B}) \setminus \bar{U}$ is not closed, so that

$$\mathrm{int}_{\mathsf{S}(\mathfrak{B})}\left(\bigcap_{\xi<\beta} a_{\xi}\right) = \mathsf{S}(\mathfrak{B}) \setminus \bar{U} \neq \emptyset .$$

6.7. Lemma. *Let α be an infinite cardinal and $\mathfrak{B}$ an atomless Boolean algebra such that if U is open in $\mathsf{S}(\mathfrak{B})$ and $1 < \tau(U) < \alpha$ then $\bar{U}$ is not open in $\mathsf{S}(\mathfrak{B})$. If $\beta < \alpha$ and $\{a_{\xi} : \xi < \beta\} \subset \mathfrak{B} \setminus \{\mathbb{0}\}$ then there is $\{b_{\xi} : \xi < \beta\} \subset \mathfrak{B} \setminus \{\mathbb{0}\}$ such that*

$$b_{\xi} \leq a_{\xi} \quad \text{for } \xi < \beta \quad \text{and}$$
$$b_{\xi} \wedge b_{\xi'} = \mathbb{0} \quad \text{for } \xi < \xi' < \beta .$$

Proof. We proceed by recursion. Set $b_0^{(0)} = a_0$, let $\xi < \beta$ and suppose that for $\eta < \xi$ we have defined $\{b_{\eta}^{(\zeta)} : \eta \leq \zeta < \xi\} \subset \mathfrak{B} \setminus \{\mathbb{0}\}$ such that

$$b_{\eta}^{(\eta)} \leq a_{\eta} \quad \text{for } \eta < \xi ,$$
$$b_{\eta}^{(\zeta')} \leq b_{\eta}^{(\zeta)} \quad \text{for } \eta \leq \zeta < \zeta' < \xi , \quad \text{and}$$
$$b_{\eta}^{(\zeta)} \wedge b_{\eta'}^{(\zeta)} = \mathbb{0} \quad \text{for } \eta < \eta' \leq \zeta < \xi .$$

Thus we have the following triangular array of non-zero elements of $\mathfrak{B}$, in which the infimum of any two elements of the same column is $\mathbb{0}$:

$$\begin{array}{lllll}
a_0 = b_0^{(0)} \geq b_0^{(1)} \geq \cdots & \geq b_0^{(\eta)} \geq \cdots & \geq b_0^{(\zeta)} \geq \ldots \\
a_1 \geq \qquad b_1^{(1)} \geq \cdots & \geq b_1^{(\eta)} \geq \cdots & \geq b_1^{(\zeta)} \geq \ldots \\
\ldots\ldots\ldots\ldots\ldots\ldots & & \\
a_{\eta} \geq & \quad b_{\eta}^{(\eta)} \geq \cdots & \geq b_{\eta}^{(\zeta)} \geq \ldots \\
\ldots\ldots\ldots\ldots\ldots\ldots & & .
\end{array}$$

We now define $b_{\eta}^{(\xi)}$ for $\eta \leq \xi$. Since for $\eta < \xi$ the family $\{b_{\eta}^{(\zeta)} : \eta \leq \zeta < \xi\}$, considered as a family of subsets of $\mathsf{S}(\mathfrak{B})$, has the finite intersection property, by Lemma 6.6 there is $c_{\eta}^{(\xi)} \in \mathfrak{B} \setminus \{\mathbb{0}\}$ such that

$$c_{\eta}^{(\xi)} \subset \bigcap_{\eta \leq \zeta < \xi} b_{\eta}^{(\zeta)} .$$

To define $b_{\eta}^{(\xi)}$ for $\eta \leq \xi$ we distinguish two cases.

Case 1. $a_\xi \cap (\bigcup_{\eta<\xi} c_\eta^{(\xi)}) = \emptyset$. We set

$$b_\xi^{(\xi)} = a_\xi, \quad \text{and}$$
$$b_\eta^{(\xi)} = c_\eta^{(\xi)} \quad \text{for } 0 \le \eta < \xi .$$

Case 2. There is $\eta' < \xi$ such that $a_\xi \cap c_{\eta'}^{(\xi)} \neq \emptyset$. Since $\mathfrak{B}$ is atomless there is $c \in \mathfrak{B} \setminus \{\mathbb{0}\}$ such that

$$c \subset a_\xi \cap c_{\eta'}^{(\xi)};$$

we set

$$b_\xi^{(\xi)} = c,$$
$$b_{\eta'}^{(\xi)} = c_{\eta'}^{(\xi)} \setminus c, \quad \text{and}$$
$$b_\eta^{(\xi)} = c_\eta^{(\xi)} \quad \text{for } \eta < \xi \quad \text{with } \eta \neq \eta' .$$

It is then clear that

$$b_\xi^{(\xi)} \le a_\xi,$$
$$b_\eta^{(\zeta')} \le b_\eta^{(\zeta)} \qquad \text{for } \eta \le \zeta \le \zeta' \le \xi, \quad \text{and}$$
$$b_\eta^{(\zeta)} \wedge b_{\eta'}^{(\zeta)} = \mathbb{0} \quad \text{for } \eta < \eta' \le \zeta \le \xi .$$

By Lemma 6.6, for $\eta < \beta$ there is $b_\eta \in \mathfrak{B} \setminus \{\mathbb{0}\}$ such that

$$b_\eta \subset \bigcap_{\eta \le \zeta < \beta} b_\eta^{(\zeta)} .$$

The family $\{b_\eta : \eta < \beta\}$ is as required.

Definition. Let $\alpha \ge \omega$. A Boolean algebra $\mathfrak{B}$ *satisfies condition* (R_α) if the following statement holds: if

$$A_0 \subset \mathfrak{B} \setminus \{\mathbb{1}\}, \quad A_1 \subset \mathfrak{B} \setminus \{\mathbb{0}\}, \quad A_2 \subset \mathfrak{B},$$
$$|A_0| + |A_1| + |A_2| < \alpha,$$

A_0 is directed upwards and A_1 is directed downwards,

$A_0 < A_1$, and

the inequalities $a_2 \le a_0$ and $a_1 \le a_2$ are both false for $a_0 \in A_0$, $a_1 \in A_1$ and $a_2 \in A_2$,

then there is $b \in \mathfrak{B}$ such that

$A_0 < b < A_1$ and

the inequalities $a_2 \le b$ and $b \le a_2$ are both false for $a_2 \in A_2$.

6.8. Lemma. *If* $\alpha \ge \omega$ *and* $\mathfrak{B}$ *is a Boolean algebra satisfying condition* (H_α), *then* $\mathfrak{B}$ *satisfies condition* (R_α).

Proof. Let A_0, A_1 and A_2 be subsets of $\mathfrak{B}$ as in definition of (R_α). We suppose without loss of generality that $\mathbb{0} \in A_0$ and $\mathbb{1} \in A_1$, and we suppose also (since otherwise the statement to be proved is obvious) that $A_2 \neq \emptyset$.

Since A_0 is directed upwards and $a_2 \setminus a_0 \neq \mathbb{0}$ for $a_2 \in A_2$ and $a_0 \in A_0$, by Lemmas 6.5 and 6.6 there is for $a_2 \in A_2$ an element $f(a_2)$ of $\mathfrak{B}(\mathsf{S}(\mathfrak{B}))$ such that

$$\emptyset \neq f(a_2) \subset a_2 \setminus (\bigcup A_0);$$

similarly for $a_2 \in A_2$ there is $g(a_2) \in \mathfrak{B}(\mathsf{S}(\mathfrak{B}))$ such that

$$\emptyset \neq g(a_2) \subset (\bigcap A_1) \setminus a_2;$$

and by condition (H_α) there is $c \in \mathfrak{B}(\mathsf{S}(\mathfrak{B}))$ such that

$$\emptyset \neq c \subset (\bigcap A_1) \setminus (\bigcup A_0).$$

We suppose, using Lemma 6.7 to refine the elements of the family

$$\{f(a_2): a_2 \in A_2\} \cup \{g(a_2): a_2 \in A_2\} \cup \{c\},$$

that

$$f(a_2) \cap g(\tilde{a}_2) = f(a_2) \cap c = g(a_2) \cap c = \emptyset$$

for $a_2, \tilde{a}_2 \in A_2$.

We now define

$$C_0 = \{a_0 \cup (\bigcup g[F]): a_0 \in A_0 \text{ and } F \in \mathscr{P}^*_\omega(A_2)\} \quad \text{and}$$
$$C_1 = \{a_1 \setminus (\bigcup f[F]): a_1 \in A_1 \text{ and } F \in \mathscr{P}^*_\omega(A_2)\}.$$

If $c_0 = a_0 \cup (\bigcup g[F]) \in C_0$ and $c_1 = a_1 \setminus (\bigcup f[G]) \in C_1$ with $a_0 \in A_0$, $a_1 \in A_1$ and $F, G \in \mathscr{P}^*_\omega(A_2)$, then since $a_0 < a_1$ and $(\bigcup g[F]) \cap (\bigcup f[G]) = \emptyset$ we have $c_0 \leq c_1$ in $\mathfrak{B}(\mathsf{S}(\mathfrak{B}))$; and $c_0 \neq c_1$, since $\emptyset \neq c \subset c_1 \setminus c_0$. It follows that $C_0 < C_1$.

It is clear that

$|C_0| + |C_1| < \alpha$,

$C_0 \subset \mathfrak{B} \setminus \{\mathbb{1}\}$ and $C_1 \subset \mathfrak{B} \setminus \{\mathbb{0}\}$ (since $\mathbb{0} \neq c \subset c_1 \setminus c_0$ for $c_0 \in C_0$ and $c_1 \in C_1$), and

C_0 is directed upwards and C_1 is directed downwards.

Hence since $\mathfrak{B}$ satisfies condition (H_α) there is $b \in \mathfrak{B}$ such that

$$C_0 < b < C_1.$$

It is clear that $A_0 < b < A_1$. If $a_2 \in A_2$ then since

$$g(a_2) = \emptyset \cup g(a_2) \in C_0 \quad \text{and} \quad (f(a_2))' = \mathsf{S}(\mathfrak{B}) \setminus f(a_2) \in C_1$$

the inequalities $a_2 \leq b$ and $b \leq a_2$ are both false. The proof is complete.

We recall that if $\mathfrak{B}$ is a Boolean algebra and $C \subset \mathfrak{B}$, then the Boolean subalgebra of $\mathfrak{B}$ generated by C is denoted $\langle\!\langle C \rangle\!\rangle$.

6.9. Lemma. *Let X and Y be sets, $\{u_\eta : \eta < \beta\}$ and $\{v_\eta : \eta < \beta\}$ families of subsets of X and Y respectively such that*

$$u = \bigcup_{\eta<\beta} u_\eta \neq X \quad \text{and} \quad v = \bigcup_{\eta<\beta} v_\eta \neq Y,$$

and f a one-to-one function of u onto v such that

$$v_\eta = f[u_\eta] \quad \text{for } \eta < \beta .$$

Let $\mathscr{R}$ and $\mathscr{S}$ denote the fields of sets on X and Y respectively generated by $\{u_\eta : \eta < \beta\}$ and $\{v_\eta : \eta < \beta\}$.

(a) *There is a (unique) Boolean algebra isomorphism $\phi : \mathscr{R} \to \mathscr{S}$ such that*
($*\mathscr{R}$) *if $d \in \mathscr{R}$ and there is $\eta < \beta$ such that $d \subset u_\eta$, then $\psi(d) = f[d]$.*

(b) *Let $\tilde{\mathscr{R}}$ and $\tilde{\mathscr{S}}$ be fields of sets on X and Y containing $\mathscr{R}$ and $\mathscr{S}$ respectively, $\tilde{\psi} : \tilde{\mathscr{R}} \to \tilde{\mathscr{S}}$ an isomorphism such that*
($*\tilde{\mathscr{R}}$) *if $d \in \tilde{\mathscr{R}}$ and there is $\eta < \beta$ such that $d \subset u_\eta$, then $\tilde{\psi}(d) = f[d]$,*
let $c_\eta \subset u_\eta$ for $\eta < \beta$, and denote by $\overline{\mathscr{R}}$ and $\overline{\mathscr{S}}$ the fields of sets on X and Y respectively generated by $\tilde{\mathscr{R}} \cup \{c_\eta : \eta < \beta\}$ and $\tilde{\mathscr{S}} \cup \{f[c_\eta] : \eta < \beta\}$. Then there is a Boolean algebra isomorphism $\overline{\psi} : \overline{\mathscr{R}} \to \overline{\mathscr{S}}$ such that

$$\tilde{\psi} \subset \overline{\psi} \quad \text{and}$$

($*\overline{\mathscr{R}}$) *if $d \in \overline{\mathscr{R}}$ and there is $\eta < \beta$ such that $d \subset u_\eta$, then $\overline{\psi}(d) = f[d]$.*

Proof. (a) We set $g(u_\eta) = v_\eta$ for $\eta < \beta$ and we note that if $p \in u$ and $\eta < \beta$ then $p \in u_\eta$ if and only if $f(p) \in g(u_\eta)$. This shows

(1) $$p \in \bigcap_{k \le n} \varepsilon_k u_{\eta_k} \quad \text{if and only if } f(p) \in \bigcap_{k \le n} \varepsilon_k g(u_{\eta_k})$$

for $p \in u$, $n < \omega$ and $\eta_k < \beta$, $\varepsilon_k \in \{-1, 1\}$ for $k \le n$.

We claim that if $\bigcap_{k \le n} \varepsilon_k u_{\eta_k} \neq \emptyset$ then $\bigcap_{k \le n} \varepsilon_k g(u_{\eta_k}) \neq \emptyset$. Indeed if $\varepsilon_k = -1$ for $k \le n$ then $\bigcap_{k \le n} \varepsilon_k g(u_{\eta_k}) \neq \emptyset$ since $v \neq Y$, and if there is $\bar{k} \le n$ such that $\varepsilon_{\bar{k}} = 1$ then $\bigcap_{k \le n} \varepsilon_k g(u_{\eta_k}) \neq \emptyset$ by (1). This proves the claim; its converse is proved similarly. It now follows from Lemma 2.16 that there is a (unique) isomorphism ψ of $\mathscr{R}$ onto $\mathscr{S}$. Property ($*\mathscr{R}$) follows from (1) and Lemma 2.14(a).

(b) We set $\bar{g}(r) = \tilde{\psi}(r)$ for $r \in \tilde{\mathscr{R}}$ and $\bar{g}(c_\eta) = f[c_\eta]$ for $\eta < \beta$; then $\bar{g}$ is a well-defined, one-to-one function from $\tilde{\mathscr{R}} \cup \{c_\eta : \eta < \beta\}$ onto $\tilde{\mathscr{S}} \cup \{f[c_\eta] : \eta < \beta\}$ by property ($*\tilde{\mathscr{R}}$). We note that if $p \in u$ and $\eta < \beta$ then $p \in c_\eta$ if and only if $f(p) \in \bar{g}(c_\eta)$. This shows

(2) $$p \in \bigcap_{k \le n} \varepsilon_k c_{\eta_k} \quad \text{if and only if } f(p) \in \bigcap_{k \le n} \varepsilon_k \bar{g}(c_{\eta_k})$$

for $p \in u$, $n < \omega$ and $\eta_k < \beta$, $\varepsilon_k \in \{-1, 1\}$ for $k \le n$.

We claim for $r\in\tilde{\mathscr{R}}$ that if $r\cap(\bigcap_{k<n}\varepsilon_k c_{\eta_k})\neq\emptyset$ then $\bar{g}(r)\cap(\bigcap_{k<n}\varepsilon_k\bar{g}(c_{\eta_k}))\neq\emptyset$. If $(u\cap r)\cap(\bigcap_{k<n}\varepsilon_k c_{\eta_k})\neq\emptyset$ the conclusion follows from (2) and property $(*\tilde{\mathscr{R}})$, and if

$$\emptyset\neq r\cap(\bigcap_{k<n}\varepsilon_k c_{\eta_k})\subset X\setminus u$$

then $\varepsilon_k=-1$ for $k<n$ and from

$$\emptyset\neq r\cap(\bigcap_{k<n}\varepsilon_k c_{\eta_k})=r\cap(\bigcap_{k<n}\varepsilon_k u_{\eta_k})$$

we have

$$\begin{aligned}\emptyset\neq\tilde{\psi}(r\cap(\bigcap_{k<n}\varepsilon_k u_{\eta_k}))&=\tilde{\psi}(r)\cap\bigcap_{k<n}\varepsilon_k\tilde{\psi}(u_{\eta_k})\\&=\bar{g}(r)\cap(\bigcap_{k<n}\varepsilon_k f[u_{\eta_k}])\subset\bar{g}(r)\cap(\bigcap_{k<n}\varepsilon_k f[c_{\eta_k}])\\&=\bar{g}(r)\cap\bigcap_{k<n}\varepsilon_k\bar{g}(c_{\eta_k}).\end{aligned}$$

This proves the claim. Its converse is proved similarly. It now follows from Lemma 2.16 that there is a (unique) isomorphism $\bar{\psi}:\bar{\mathscr{R}}\to\bar{\mathscr{S}}$ such that

$$\tilde{\psi}\subset\bar{g}\subset\bar{\psi}.$$

Property $(*\bar{\mathscr{R}})$ follows from property $(*\tilde{\mathscr{R}})$, (2) and Lemma 2.14(a).

6.10. Lemma. *Let α be an infinite cardinal and let X, Y, β, $\{u_\eta:\eta<\beta\}$, $\{v_\eta:\eta<\beta\}$, $u, v, f, \mathscr{R}, \mathscr{S}, \psi, \tilde{\mathscr{R}}, \tilde{\mathscr{S}}$ and $\tilde{\psi}$ be as in the statement of Lemma 6.9 with $|\tilde{\mathscr{R}}|<\alpha$. Let $\mathscr{A}$ and $\mathscr{B}$ be fields of sets on X and Y respectively such that $\tilde{\mathscr{R}}\subset\mathscr{A}$, $\tilde{\mathscr{S}}\subset\mathscr{B}$ and $\mathscr{B}$ has property (H_α), and let a be an element of $\mathscr{A}$ such that $f[a\cap u_\eta]\in\mathscr{B}$ for $\eta<\beta$.*

Then there are $b\in\mathscr{B}$ and a Boolean algebra isomorphism

$$\phi:\langle\!\langle\tilde{\mathscr{R}}\cup\{a\}\rangle\!\rangle\to\langle\!\langle\tilde{\mathscr{S}}\cup\{b\}\rangle\!\rangle$$

such that

$$\tilde{\psi}\subset\phi,\quad \textit{and}$$

$*(\langle\!\langle\tilde{\mathscr{R}}\cup\{a\}\rangle\!\rangle)$ if $d\in\langle\!\langle\tilde{\mathscr{R}}\cup\{a\}\rangle\!\rangle$ and there is $\eta<\beta$ such that $d\subset u_\eta$, then $\phi(d)=f[d]$.

Proof. Set $c_\eta=a\cap u_\eta$ for $\eta<\beta$ and denote by $\bar{\mathscr{R}}$ and $\bar{\mathscr{S}}$ the Boolean subalgebras of $\mathscr{A}$ and $\mathscr{B}$ generated by $\tilde{\mathscr{R}}\cup\{c_\eta:\eta<\beta\}$ and $\tilde{\mathscr{S}}\cup\{f[c_\eta]:\eta<\beta\}$ respectively. There is by Lemma 6.9(b) an isomorphism $\bar{\psi}:\bar{\mathscr{R}}\to\bar{\mathscr{S}}$ such that $\tilde{\psi}\subset\bar{\psi}$ and property $(*\bar{\mathscr{R}})$ holds.

We note that $\bar{\mathscr{R}}\subset\langle\!\langle\tilde{\mathscr{R}}\cup\{a\}\rangle\!\rangle$, so that $\langle\!\langle\bar{\mathscr{R}}\cup\{a\}\rangle\!\rangle=\langle\!\langle\tilde{\mathscr{R}}\cup\{a\}\rangle\!\rangle$. Hence if $a\in\bar{\mathscr{R}}$, then $a\in\tilde{\mathscr{R}}$ and we complete the proof by setting

$$b=\tilde{\psi}(a)\quad\text{and}\quad\phi=\tilde{\psi}.$$

Thus we assume that $a\notin\bar{\mathscr{R}}$.

We set

$$A_0 = \{r \in \bar{\mathscr{R}} : r \subset a\}$$
$$A_1 = \{r \in \bar{\mathscr{R}} : a \subset r\}, \quad \text{and}$$
$$A_2 = \bar{\mathscr{R}} \setminus (A_0 \cup A_1).$$

Then

$A_0 \subset \bar{\mathscr{R}} \setminus \{\mathbb{0}\}, \quad A_1 \subset \bar{\mathscr{R}} \setminus \{\mathbb{1}\}, \quad A_2 \subset \bar{\mathscr{R}},$

$|A_0| + |A_1| + |A_2| < \alpha,$

A_0 is directed upwards and A_1 is directed downwards (since $a \notin \mathscr{R}$),

$A_0 < A_1$, and

the inequalities $a_2 \leq a_0$ and $a_1 < a_2$ are both false for $a_0 \in A_0$, $a_1 \in A_1$, and $a_2 \in A_2$.

Since $\bar{\psi}$ is a Boolean-algebra embedding of $\bar{\mathscr{R}}$ into $\mathscr{B}$, corresponding statements hold for the sets $\bar{\psi}[A_0]$, $\bar{\psi}[A_1]$ and $\bar{\psi}[A_2]$, so since $\mathscr{B}$ satisfies condition (R_α) (by Lemma 6.8) there is $b \in \mathscr{B}$ such that

$\bar{\psi}[A_0] < b < \bar{\psi}[A_1]$ and

the inequalities $b \leq \bar{\psi}(a_2)$ and $\bar{\psi}(a_2) \leq b$ are both false for $a_2 \in A_2$.

We define a function

$$h : \bar{\mathscr{R}} \cup \{a\} \to \bar{\mathscr{S}} \cup \{b\}$$

by

$$h(r) = \bar{\psi}(r) \quad \text{for } r \in \bar{\mathscr{R}} \quad \text{and}$$
$$h(a) = b,$$

and we claim that if $r \in \bar{\mathscr{R}}$ and $\varepsilon \in \{-1, 1\}$ then

$$r \cap \varepsilon a = \emptyset \quad \text{if and only if } h(r) \cap \varepsilon h(a) = \emptyset.$$

Indeed if $r \cap a = \emptyset$ then $r' \in A_1$ and we have

$$b \subset \bar{\psi}(r') = (\bar{\psi}(r))' = (h(r))'$$

and hence $h(r) \cap h(a) = \emptyset$; and if $r \setminus a = \emptyset$ then $r \in A_0$ and we have

$$\bar{\psi}(r) < b$$

and hence $h(r) \setminus h(a) = \emptyset$. This shows that if $r \cap \varepsilon a = \emptyset$ then $h(r) \cap \varepsilon h(a) = \emptyset$; the reverse implication is proved similarly. It now follows from Lemma 2.16 that there is a (unique) isomorphism

$$\phi : \langle\!\langle \bar{\mathscr{R}} \cup \{a\} \rangle\!\rangle \to \langle\!\langle \bar{\mathscr{S}} \cup \{b\} \rangle\!\rangle$$

such that

$$\overline{\psi} \subset h \subset \phi .$$

We now verify property $*\langle\!\langle \tilde{\mathscr{R}} \cup \{a\} \rangle\!\rangle$. If $d \in \langle\!\langle \tilde{\mathscr{R}} \cup \{a\} \rangle\!\rangle$ and there is $\eta < \beta$ such that $d \subset u_\eta$, then by Lemma 2.14(b) there are $r_0, r_1 \in \tilde{\mathscr{R}}$ such that

$$d = (r_0 \cap a) \cup (r_1 \setminus a)$$

and we have

$$\begin{aligned} d = d \cap u_\eta &= (r_0 \cap a \cap u_\eta) \cup (r_1 \cap (u_\eta \setminus a)) \\ &= (r_0 \cap c_\eta) \cup (r_1 \cap (u_\eta \setminus c_\eta)) \in \langle\!\langle \tilde{\mathscr{R}} \cup \{c_\eta\} \rangle\!\rangle \subset \overline{\mathscr{R}} \end{aligned}$$

and hence $\phi(d) = \overline{\psi}(d) = f[d]$ by property $(*\overline{\mathscr{R}})$.

We note finally that since $\phi[\tilde{\mathscr{R}}] = \tilde{\mathscr{S}}$ and $\phi(a) = b$ we have

$$\langle\!\langle \overline{\mathscr{S}} \cup \{b\} \rangle\!\rangle = \phi \langle\!\langle \overline{\mathscr{R}} \cup \{a\} \rangle\!\rangle = \phi \langle\!\langle \tilde{\mathscr{R}} \cup \{a\} \rangle\!\rangle \subset \langle\!\langle \tilde{\mathscr{S}} \cup \{b\} \rangle\!\rangle ;$$

thus ϕ is the required isomorphism of $\langle\!\langle \tilde{\mathscr{R}} \cup \{a\} \rangle\!\rangle$ onto $\langle\!\langle \tilde{\mathscr{S}} \cup \{b\} \rangle\!\rangle$. The proof is complete.

6.11. Theorem. *Let α be an infinite cardinal and $\mathfrak{A}$ and $\mathfrak{B}$ Boolean algebras such that*

$$|\mathfrak{A}| = |\mathfrak{B}| = \alpha \quad \textit{and}$$
$$\mathfrak{A} \textit{ and } \mathfrak{B} \textit{ satisfy condition } (\mathrm{H}_\alpha).$$

Let u and v be open subsets of $\mathsf{S}(\mathfrak{A})$ and $\mathsf{S}(\mathfrak{B})$ respectively such that

$$u \neq \mathsf{S}(\mathfrak{A}) \quad \textit{and} \quad v \neq \mathsf{S}(\mathfrak{B}) \quad \textit{and}$$
$$\tau(u) < \alpha \quad \textit{and} \quad \tau(v) < \alpha ,$$

and let f be a homeomorphism of u onto v.

Then there is a Boolean algebra isomorphism

$$\phi : \mathfrak{A} \to \mathfrak{B}$$

such that

$$f^{-1} \subset \mathsf{S}(\phi) : \mathsf{S}(\mathfrak{B}) \to \mathsf{S}(\mathfrak{A}) .$$

Proof. We set $X = \mathsf{S}(\mathfrak{A})$ and $Y = \mathsf{S}(\mathfrak{B})$ and we identify $\mathfrak{A}$ and $\mathfrak{B}$ with $\mathfrak{B}(X)$ and $\mathfrak{B}(Y)$ respectively. There are $\beta < \alpha$ and a subset $\{u_\eta : \eta < \beta\}$ of $\mathfrak{A}$ such that

$$u = \bigcup_{\eta < \beta} u_\eta ;$$

we set

$$v_\eta = f[u_\eta] \quad \text{for } \eta < \beta ,$$

so that

$$v = \bigcup_{\eta < \beta} v_\eta .$$

Let $\langle a_\xi : \xi < \alpha \rangle$ and $\langle b_\xi : \xi < \alpha \rangle$ be well-orderings of $\mathfrak{A}$ and $\mathfrak{B}$ respectively. For $\xi < \alpha$ we define Boolean subalgebras $\mathfrak{A}_\xi$ and $\mathfrak{B}_\xi$ of $\mathfrak{A}$ and $\mathfrak{B}$ respectively, and $\phi_\xi : \mathfrak{A}_\xi \to \mathfrak{B}_\xi$, such that

(i) $\mathfrak{A}_0 = \langle\!\langle \{u_\eta : \eta < \beta\} \rangle\!\rangle$ and $\mathfrak{B}_0 = \langle\!\langle \{v_\eta : \eta < \beta\} \rangle\!\rangle$;

(ii) ϕ_ξ is a Boolean algebra isomorphism of $\mathfrak{A}_\xi$ into $\mathfrak{B}_\xi$ for $\xi < \alpha$;

(iii) $\mathfrak{A}_\zeta \subset \mathfrak{A}_\xi$, $\mathfrak{B}_\zeta \subset \mathfrak{B}_\xi$, and $\phi_\zeta \subset \phi_\xi$ for $\zeta < \xi < \alpha$;

(iv) if ξ is a non-zero limit ordinal and $\xi < \alpha$, then $\mathfrak{A}_\xi = \bigcup_{\zeta < \xi} \mathfrak{A}_\zeta$, $\mathfrak{B}_\xi = \bigcup_{\zeta < \xi} \mathfrak{B}_\zeta$, and $\phi_\xi = \bigcup_{\zeta < \xi} \phi_\zeta$;

(v) $\mathfrak{A}_\xi$ is generated (as a Boolean algebra) by a set of cardinality at most $\beta + |\xi|$, for $\xi < \alpha$;

(vi) if $\xi = \zeta + 2n + 1$ with ζ a limit ordinal and $n < \omega$, then $a_{\zeta + n} \in \mathfrak{A}_\xi$;

(vii) if $\xi = \zeta + 2n + 2$ with ζ a limit ordinal and $n < \omega$, then $b_{\zeta + n} \in \mathfrak{B}_\xi$;

and

(viii) if $a \in \mathfrak{A}_\xi$ and there is $\eta < \beta$ such that $a \subset u_\eta$, then $\phi_\xi(a) = f[a]$.

We proceed by recursion. We define $\mathfrak{A}_0$ and $\mathfrak{B}_0$ by (i) and ϕ_0 by Lemma 6.9(a) (with $\mathscr{R}$, $\mathscr{S}$ and ψ replaced by $\mathfrak{A}_0$, $\mathfrak{B}_0$ and ϕ_0 respectively). For non-zero limit ordinals $\xi < \alpha$ we define $\mathfrak{A}_\xi$, $\mathfrak{B}_\xi$ and ϕ_ξ by (iv).

We now define $\mathfrak{A}_{\xi+1}$, $\mathfrak{B}_{\xi+1}$ and $\phi_{\xi+1}$ for $\xi < \alpha$. There are two cases to consider.

Case 1. $\xi = \zeta + 2n + 1$ with ζ a limit ordinal and $n < \omega$. In Lemma 6.10 we replace

$$\mathscr{R}, \mathscr{S}, \psi, \tilde{\mathscr{R}}, \tilde{\mathscr{S}}, \tilde{\psi}, \text{ and } a$$

by

$$\mathfrak{A}_0, \mathfrak{B}_0, \phi_0, \mathfrak{A}_{\zeta+2n}, \mathfrak{B}_{\zeta+2n}, \phi_{\zeta+2n} \text{ and } a_{\zeta+n}$$

respectively and we set

$$\mathfrak{A}_\xi = \langle\!\langle \mathfrak{A}_{\zeta+2n} \cup \{a_{\zeta+n}\} \rangle\!\rangle \quad \text{and } \mathfrak{B}_\xi = \langle\!\langle \mathfrak{B}_{\zeta+2n} \cup \{b\} \rangle\!\rangle$$

with b the element of $\mathfrak{B}$ given by Lemma 6.10; and we define ϕ_ξ to be the isomorphism ϕ given by Lemma 6.10.

Case 2. $\xi = \zeta + 2n + 2$ with ζ a limit ordinal and $n < \omega$. We note from (viii) that if $b \in \mathfrak{B}_{\zeta+2n+1}$ and there is $\eta < \beta$ such that $b \subset v_\eta$, then $\phi_{\zeta+2n+1}^{-1}(b) \subset f^{-1}(b)$. In Lemma 6.10 we interchange X and Y, we replace f by f^{-1}, we replace

$$\mathscr{R}, \mathscr{S}, \psi, \tilde{\mathscr{R}}, \tilde{\mathscr{S}}, \tilde{\psi} \text{ and } a$$

respectively by

$$\mathfrak{B}_0, \mathfrak{A}_0, \phi_0^{-1}, \mathfrak{B}_{\zeta+2n+1}, \mathfrak{A}_{\zeta+2n+1}, \phi_{\zeta+2n+1}^{-1} \text{ and } b_{\zeta+n}$$

and we set

$$\mathfrak{B}_\xi = \langle\!\langle \mathfrak{B}_{\zeta+2n+1} \cup \{b_{\zeta+n}\} \rangle\!\rangle \quad \text{and } \mathfrak{A}_\xi = \langle\!\langle \mathfrak{A}_{\zeta+2n+1} \cup \{a\} \rangle\!\rangle$$

with a the element of $\mathfrak{A}$ given by Lemma 6.10; and with ϕ the isomorphism of $\mathfrak{B}_\xi$ onto $\mathfrak{A}_\xi$ extending $\phi_{\zeta+2n+1}^{-1}$ given by Lemma 6.10, we set $\phi_\xi = \phi^{-1}$.

This completes the definition of $\mathfrak{A}_\xi$, $\mathfrak{B}_\xi$ and ϕ_ξ for $\xi<\alpha$. We set

$$\phi=\bigcup_{\xi<\alpha}\phi_\xi .$$

It is clear that ϕ is a Boolean algebra isomorphism of $\mathfrak{A}$ onto $\mathfrak{B}$.

For $q\in v\subset \mathsf{S}(\mathfrak{B})$ there is $\eta<\beta$ such that $q\in v_\eta$ and we have

$$\begin{aligned} f^{-1}(\{q\}) &= f^{-1}(\bigcap\{b\in\mathfrak{B}: q\in b \text{ and } b\subset v_\eta\}) \\ &= \bigcap\{f^{-1}(b): q\in b\in\mathfrak{B} \text{ and } b\subset v_\eta\} \\ &= \bigcap\{a\in\mathfrak{A}: q\in f[a] \text{ and } a\subset u_\eta\} \\ &= \bigcap\{a\in\mathfrak{A}: q\in \phi(a) \text{ and } a\subset u_\eta\} \\ &= \bigcap\{a\in\mathfrak{A}: q\in \phi(a)\} = \mathsf{S}(\phi)(q), \end{aligned}$$

since $q\in\phi(a)$ in $\mathsf{S}(\mathfrak{B})$ if and only if $\phi(a)\in q$ in $\mathfrak{B}(\mathsf{S}(\mathfrak{B}))=\mathfrak{B}$. The proof is complete.

The following theorem, characterizing $\mathfrak{C}_\alpha$ in terms of condition (H_α) and giving a necessary and sufficient set-theoretic condition for its existence, is the main result of this section.

6.12. Theorem. *Let* $\alpha\geq\omega$.

(a) *There is an* α*-homogeneous-universal Boolean algebra of cardinality* α *if and only if* $\alpha=\alpha^{\underline{\alpha}}$;

(b) *a Boolean algebra of cardinality* α *is isomorphic to* $\mathfrak{C}_\alpha$ *if and only if it satisfies condition* (H_α).

Proof. (a) We claim first that if there is a Boolean algebra $\mathfrak{B}$ satisfying condition (H_α) such that $|\mathfrak{B}|=\alpha$, then $\alpha=\alpha^{\underline{\alpha}}$. By Theorem 5.8 it is sufficient to verify that if C is a maximal linearly ordered subset of $\mathfrak{B}\setminus\{\mathbb{0},\mathbb{1}\}$ then C is an η_α-set of cardinality α. Let $A_0<A_1$ with $A_0\cup A_1\subset C$ and $|A_0|+|A_1|<\alpha$ and suppose that there is no $c\in C$ such that $A_0<c<A_1$. Then since $\mathfrak{B}$ satisfies condition (H_α) there is $b\in\mathfrak{B}\setminus C$ such that $A_0<b<A_1$; we show that $C\cup\{b\}$ is linearly ordered. Indeed if $c\in C$ then either there is $a_0\in A_0$ such that $a_0\nless c$ (and hence $c\leq a_0<b$) or there is $a_1\in A_1$ such that $c\nless a_1$ (and hence $b<a_1\leq c$). It follows that C is an η_α-set of cardinality α.

If (the α-homogeneous-universal Boolean algebra of cardinality α) $\mathfrak{C}_\alpha$ exists then $\mathfrak{C}_\alpha$ satisfies condition (H_α) by Lemma 6.3; hence $\alpha=\alpha^{\underline{\alpha}}$ by the claim just proved and Theorem 5.8. Conversely if $\alpha=\alpha^{\underline{\alpha}}$ then since (by Theorem 6.1) the class $\mathbb{BA}$ is a Jónsson class (with finite index) there is by Theorem 4.10(b) and Corollary 4.11 an α-homogeneous Boolean algebra of cardinality α.

(b) follows from Lemma 6.3 and the fact (given by setting $u=v=f=\emptyset$ in Theorem 6.11) that any two Boolean algebras of cardinality α and satisfying condition (H_α) are isomorphic as Boolean algebras.

Properties of the Space $\mathbb{S}_\alpha$

Definition. Let $\alpha \geq \omega$, let X be a space and let $\mathbb{S}$ be a class of spaces.

(a) X is *α-co-universal for* $\mathbb{S}$ if for every element K of $\mathbb{S}$ such that $w(K) \leq \alpha$ there is a continuous function from X onto K;

(b) X is *α-co-homogeneous for* $\mathbb{S}$ if for every element K of S such that $w(K) < \alpha$ and for every two continuous functions f_0 and f_1 from X onto K there is a homeomorphism h of X such that $f_0 = f_1 \circ h$.

These definitions are dual to (a special case of) those given in § 4.

6.13. Lemma. *If $\alpha \geq \omega$ and X is a compact space which is α-co-homogeneous and α-co-universal for the class of compact, totally disconnected spaces, then X is α-co-homogeneous and α-co-universal for the class of compact spaces.*

Proof. We note first that if Y and Z are compact, totally disconnected spaces such that $w(Y) < \alpha$ and $w(Z) < \alpha$, and if f and g are continuous functions from Y and X respectively onto Z, then there is a continuous function k from X onto Y such that $g = f \circ k$. Indeed since X is α-co-universal for the class of compact, totally disconnected spaces there is a continuous function h from X onto Y, and since X is α-co-homogeneous for the class of compact, totally disconnected spaces there is a homeomorphism e of X such that $g = f \circ h \circ e$; we set $k = h \circ e$.

We prove that X is α-co-homogeneous for the class of compact spaces. Let K be a compact space such that $w(K) < \alpha$ and let f_0 and f_1 be continuous functions from X onto K. By Lemma 2.37 for $i < 2$ there are a compact, totally disconnected space Y_i such that $w(Y_i) = w(K)$ and continuous, onto functions $g_i: X \to Y_i$ and $h_i: Y_i \to K$ such that $f_i = h_i \circ g_i$. We set

$$Y = \{\langle p, q \rangle \in Y_0 \times Y_1 : h_0(p) = h_1(q)\}$$

and as usual we denote by π_i the projection function from $Y_0 \times Y_1$ onto Y_i for $i < 2$. It is easy to verify that $\pi_i[Y] = Y_i$ for $i < 2$; it follows from the first paragraph of this proof (with Y_i, π_i and g_i replacing Z, f and g respectively) that there is a continuous function k_i from X onto Y such that $g_i = \pi_i \circ k_i$ for $i < 2$. Since Y is a compact, totally disconnected space such that $w(Y) < \alpha$ and X is α-co-homogeneous for compact, totally disconnected spaces there is a homeomorphism h of X such that $k_0 = k_1 \circ h$. Hence

$$f_0 = h_0 \circ g_0 = h_0 \circ \pi_0 \circ k_0 = h_0 \circ \pi_0 \circ k_1 \circ h = h_1 \circ \pi_1 \circ k_1 \circ h = h_1 \circ g_1 \circ h = f_1 \circ h.$$

If K is a compact space such that $w(K) \leq \alpha$ then by Corollary 2.38 there are a compact, totally disconnected space Y such that $w(Y) \leq \alpha$

and a continuous function h from Y onto K. Since X is α-co-universal for the class of compact, totally disconnected spaces there is a continuous onto function $f: X \to Y$, and then $h \circ f: X \to K$ is a continuous, onto function. Thus X is α-co-universal for the class of compact spaces.

If $\alpha = \alpha^{\underline{\alpha}} \geq \omega$ then we set $\mathsf{S}_\alpha = \mathsf{S}(\mathfrak{C}_\alpha)$, *i.e.*, S_α is the Stone space of the α-homogeneous-universal Boolean algebra $\mathfrak{C}_\alpha$ of cardinality α.

6.14. Theorem. *Let* $\alpha = \alpha^{\underline{\alpha}} \geq \omega$.

(a) $|\mathsf{S}_\alpha| = 2^\alpha$ *and* $\boldsymbol{w}(\mathsf{S}_\alpha) = \alpha$;

(b) S_α *is α-co-homogeneous and α-co-universal for the class of compact spaces;*

(c) *if* $\alpha > \omega$ *and X is a compact space such that* $\boldsymbol{w}(X) = \alpha$ *and X is α-co-homogeneous and α-co-universal for the class of compact, totally disconnected spaces, then (X is totally disconnected and) X is homeomorphic with* S_α.

Proof. Since $\mathfrak{C}_\alpha$ is an α-homogeneous-universal Boolean algebra it follows from Stone's duality that S_α is α-co-homogeneous and α-co-universal for the class of compact, totally disconnected spaces. Thus statement (b) follows from Lemma 6.13. Since 2^α is compact and $\boldsymbol{w}(2^\alpha) = \alpha$ there is a continuous function from S_α onto 2^α and hence $|\mathsf{S}_\alpha| \geq 2^\alpha$; the reverse inequality is given by the inclusion $\mathsf{S}_\alpha \subset \mathscr{P}(\mathfrak{C}_\alpha)$. The relation $\boldsymbol{w}(\mathsf{S}_\alpha) = \alpha$ follows from Lemma 2.24, so it remains only to prove statement (c).

If X is totally disconnected then $\mathfrak{B}(X)$ is an α-homogeneous-universal Boolean algebra of cardinality α, so that $\mathfrak{B}(X) = \mathfrak{C}_\alpha$ and hence X is homeomorphic with S_α. Thus it is enough to show that X is totally disconnected.

Let $p, q \in X$ with $p \neq q$ and let f be a continuous function from X into $[0,1]$ such that $f(p) \neq f(q)$. Since $f[X]$ is compact there are by Corollary 2.38 a compact, totally disconnected space Y such that $\boldsymbol{w}(Y) = \boldsymbol{w}(f[X])$ and a continuous function h from Y onto $f[X]$, and since

$$\boldsymbol{w}(f[X]) \leq \boldsymbol{w}([0,1]) = \omega < \alpha$$

there is by the first paragraph of the proof of Lemma 6.13 (with $f[X]$, f and h replacing Z, g and f respectively) a continuous function k from X onto Y such that $f = h \circ k$. Since $k(p) \neq k(q)$ and Y is totally disconnected there is $A \in \mathfrak{B}(Y)$ such that $k(p) \in A$ and $k(q) \notin A$; then $k^{-1}(A) \in \mathfrak{B}(X)$, $p \in k^{-1}(A)$ and $q \notin k^{-1}(A)$. It follows that X is totally disconnected.

The Cantor Set. The power set 2^ω is called the *Cantor set*. We make four observations relating this space to Theorem 6.14 and Corollary 6.18 below.

(1) $2^\omega = \mathsf{S}_\omega$. The space 2^ω has no isolated elements; there is no open subset U of 2^ω such that $1 < \tau(U) < \omega$; and if U and V are disjoint open subsets of 2^ω such that $1 < \tau(U) + \tau(V) < \omega$ then U and V are closed and hence $\bar{U} \cap \bar{V} = \emptyset$. It follows from Lemma 6.5 that $\mathfrak{B}(2^\omega)$ satisfies condition (H_ω). Since $|\mathfrak{B}(2^\omega)| = \boldsymbol{w}(2^\omega) = \omega$ by Lemma 2.24, we have $\mathfrak{C}_\omega = \mathfrak{B}(2^\omega)$ (and hence $\mathsf{S}_\omega = 2^\omega$) by Theorem 6.12(b).

(2) For every compact space K such that $\boldsymbol{w}(K) \leq \omega$, *i.e.*, for every compact metric space, there is a continuous function from 2^ω onto K. This follows from (1) and Theorem 6.14(a).

(3) Not every compact space X such that $\boldsymbol{w}(X) = \omega$ and X is ω-co-homogeneous and ω-co-universal for the class of compact spaces is homeomorphic with 2^ω. Thus the condition $\alpha > \omega$ in the statement of Theorem 6.14 (c) is essential. Let Y be any non-empty compact connected metric space and set $X = 2^\omega \times Y$. By (1) and Lemma 6.13 the space X is ω-co-universal for the class of compact spaces. If f is a continuous function from X onto a totally disconnected space then f is constant on every set $\{p\} \times Y$ for $p \in 2^\omega$. Thus X is ω-co-homogeneous for the class of compact, totally disconnected spaces, hence (by Lemma 6.13) for the class of compact spaces.

(4) The Cantor set 2^ω is (clearly) point-homogeneous (a notion defined below, before Corollary 6.18). We show in Corollary 6.18 below that the space S_α is not point-homogeneous for $\alpha = \alpha^{\underline{\alpha}} > \omega$.

Let α be an infinite cardinal. A space X has the *α-Baire category property* if the intersection of any family of at most α dense, open subsets of X is dense in X.

The classical theorem of Baire states that a locally compact space has the ω-Baire category property. The following generalization is useful in a number of cases.

6.15. Theorem. *Let α be an infinite cardinal and let X be a locally compact space such that every non-empty intersection of a family of less than α open subsets of X has non-empty interior. Then X has the α-Baire category property.*

Proof. Let $G = \bigcap_{\xi < \alpha} G_\xi$ where G_ξ is open and dense in X for $\xi < \alpha$ and let K be a compact subset of X with non-empty interior; we prove that $K \cap G \neq \emptyset$. We define a family $\{K_\xi : \xi < \alpha\}$ such that

$K_0 = K$,

K_ξ is a compact subset of X with non-empty interior for $\xi < \alpha$,

$K_{\xi+1} \subset (\operatorname{int} K_\xi) \cap G_\xi$ for $\xi < \alpha$, and

$K_\xi = \bigcap_{\zeta < \xi} K_\zeta$ for non-zero limit ordinals $\xi < \alpha$.

We proceed recursively. We define $K_0=K$; if ξ is a non-zero limit ordinal and $\xi<\alpha$ we set $K_\xi=\bigcap_{\zeta<\xi}K_\zeta$. The recursive assumptions imply that K_ξ is a non-empty compact subset of X and $K_\xi=\bigcap_{\zeta<\xi}\operatorname{int}K_\zeta$. Now for $\xi<\alpha$ we define $K_{\xi+1}$. Since $\operatorname{int}K_\xi\neq\emptyset$ and G_ξ is open and dense in X, it follows that $(\operatorname{int}K_\xi)\cap G_\xi$ is open and non-empty. Since X is locally compact there is a compact set $K_{\xi+1}$ with non-empty interior such that $K_{\xi+1}\subset(\operatorname{int}K_\xi)\cap G_\xi$. This completes the recursive definition. Since K is compact we have

$$\emptyset\neq\bigcap_{\xi<\alpha}K_\xi=\bigcap_{\xi<\alpha}K_{\xi+1}\subset K\cap(\bigcap_{\xi<\alpha}G_\xi)=K\cap G.$$

6.16. Corollary. *Let $\alpha\geq\omega$ and let $\mathfrak{B}$ be a Boolean algebra satisfying condition* (H_α). *Then* $\mathsf{S}(\mathfrak{B})$ *has the α-Baire category property.*

Proof. The space $\mathsf{S}(\mathfrak{B})$ is compact. Let $\beta<\alpha$ and let $\{U_\xi:\xi<\beta\}$ be a family of open subsets of $\mathsf{S}(\mathfrak{B})$ such that $\bigcap_{\xi<\beta}U_\xi\neq\emptyset$. Let $p\in\bigcap_{\xi<\beta}U_\xi$ and let V_ξ be an open-and-closed subset of $\mathsf{S}(\mathfrak{B})$ such that $p\in V_\xi\subset U_\xi$ for $\xi<\beta$. Then $p\in V=\bigcap_{\xi<\beta}V_\xi\subset\bigcap_{\xi<\beta}U_\xi$ and, from Lemma 6.5, the interior of V is non-empty. According to Theorem 6.15 the space $\mathsf{S}(\mathfrak{B})$ has the α-Baire category property.

We recall from § 2 that for $\alpha\geq\omega$ an element p of *a* space X is a P_α-point of X if the intersection of every family of less than α neighborhoods of p is a (not necessarily open) neighborhood of p. We set

$$P_\alpha(X)=\{p\in X: p \text{ is a } P_\alpha\text{-point}\}$$

for X a space and $\alpha\geq\omega$.

6.17. Theorem. *Let $\alpha=\alpha^{\underline{\alpha}}\geq\omega$. Then*

(a) *$P_\alpha(\mathsf{S}_\alpha)$ is a dense subset of S_α, and*

(b) $|P_\alpha(\mathsf{S}_\alpha)|=2^\alpha$.

Proof. (a) It is clear for $p\in\mathsf{S}_\alpha$ that p is a non-P_α-point if and only if there is an open subset U of S_α such that $\tau(U)<\alpha$ and p is in the boundary of U (*i.e.*, $p\in\bar{U}\setminus U$); hence

$$(*)\qquad P_\alpha(\mathsf{S}_\alpha)=\bigcap\{\mathsf{S}_\alpha\setminus(\bar{U}\setminus U): U \text{ is open in } \mathsf{S}_\alpha \text{ and } \tau(U)<\alpha\}.$$

Since $|\mathfrak{B}(\mathsf{S}_\alpha)|=|\mathfrak{C}_\alpha|=\alpha$ we have

$$|\{U: U \text{ is open in } \mathsf{S}_\alpha \text{ and } \tau(U)<\alpha\}|\leq\alpha^{\underline{\alpha}}=\alpha,$$

and since $\mathsf{S}_\alpha\setminus(\bar{U}\setminus U)$ is dense in S_α for U open in S_α it follows from (*) that $P_\alpha(\mathsf{S}_\alpha)$ is equal to the intersection of (at most) α dense, open subsets of S_α. Thus $P_\alpha(\mathsf{S}_\alpha)$ is dense in S_α by Corollary 6.16.

(b) We note from Theorem 6.14(a) that $|P_\alpha(\mathsf{S}_\alpha)|\leq|\mathsf{S}_\alpha|=2^\alpha$. We prove that $|P_\alpha(\mathsf{S}_\alpha)\cap U|\geq 2^\alpha$ for $\emptyset\neq U\in\mathfrak{B}(\mathsf{S}_\alpha)$; from this it follows that $|P_\alpha(\mathsf{S}_\alpha)|=2^\alpha$ and, again, that $P_\alpha(\mathsf{S}_\alpha)$ is dense in S_α.

We set $2^{\underset{\smile}{\alpha}}=\bigcup\{2^{\xi}:\xi<\alpha\}$ and if $s\in 2^{\underset{\smile}{\alpha}}$ we denote by dom(s) the domain of s. If $A\in\mathfrak{B}(\mathsf{S}_\alpha)\setminus\{\emptyset\}$ then by Lemma 6.5 A has no isolated elements and thus there are two non-empty disjoint open-and-closed subsets $A^{(0)}$ and $A^{(1)}$ of A. Let $\langle A_\xi:\xi<\alpha\rangle$ be a well-ordering of $\mathfrak{B}(\mathsf{S}_\alpha)\setminus\{\emptyset\}$. We define a family $\{U_s: s\in 2^{\underset{\smile}{\alpha}}\}$ of elements of $\mathfrak{B}(\mathsf{S}_\alpha)\setminus\{\emptyset\}$ such that

(i) $U_\emptyset=U$;

(ii) $U_s\supset U_t$ for $s,t\in 2^{\underset{\smile}{\alpha}}$ and $s\subset t$;

(iii) $U_{s\cup\{\langle\mathrm{dom}(s),0\rangle\}}\cap U_{s\cup\{\langle\mathrm{dom}(s),1\rangle\}}=\emptyset$ for $s\in 2^{\underset{\smile}{\alpha}}$; and

(iv) if $U_s\cap A_{\mathrm{dom}(s)}\neq\emptyset$ then $U_s\subset A_{\mathrm{dom}(s)}$ for $s\in 2^{\underset{\smile}{\alpha}}$.

We proceed by transfinite recursion (on the domain of s). If dom$(s)=\emptyset$ then $s=\emptyset$ and $U_\emptyset$ is defined by (i). If dom$(s)=\xi+1<\alpha$ then $U_{s|\xi}$ is a non-empty element of $\mathfrak{B}(\mathsf{S}_\alpha)$ and we set

$$\begin{aligned} U_s &= U_{s|\xi}^{(s(\xi))} && \text{if } U_{s|\xi}^{(s(\xi))}\cap A_{\mathrm{dom}(s)}=\emptyset \\ &= U_{s|\xi}^{(s(\xi))}\cap A_{\mathrm{dom}(s)} && \text{if } U_{s|\xi}^{(s(\xi))}\cap A_{\mathrm{dom}(s)}\neq\emptyset\,. \end{aligned}$$

If dom(s) is a non-zero limit ordinal then the compactness of S_α implies that $\bigcap_{\eta<\mathrm{dom}(s)}U_{s|\eta}\neq\emptyset$, and Lemma 6.5 implies that $\mathrm{int}\bigcap_{\eta<\mathrm{dom}(s)}U_{s|\eta}\neq\emptyset$. We let V_s be a non-empty open-and-closed subset of $\mathrm{int}\bigcap_{\eta<\mathrm{dom}(s)}U_{s|\eta}$ and we set

$$\begin{aligned} U_s &= V_s && \text{if } V_s\cap A_{\mathrm{dom}(s)}=\emptyset\,, \\ &= V_s\cap A_{\mathrm{dom}(s)} && \text{if } V_s\cap A_{\mathrm{dom}(s)}\neq\emptyset\,. \end{aligned}$$

This completes the definitions by transfinite recursion. It follows from (ii) that

$$\bigcap_{\xi<\alpha}U_{f|\xi}\neq\emptyset$$

for $f\in 2^\alpha$. Furthermore if $p\in\bigcap_{\xi<\alpha}U_{f|\xi}$ and $p\neq q$ then there is $\xi<\alpha$ such that $p\in A_\xi$ and $q\notin A_\xi$; since $A_\xi\cap U_{f|\xi}\neq\emptyset$ it follows from (iv) that $U_{f|\xi}\subset A_\xi$ and hence $q\notin\bigcap_{\xi<\alpha}U_{f|\xi}$. Thus there is $p_f\in\mathsf{S}_\alpha$ such that

$$\{p_f\}=\bigcap_{\xi<\alpha}U_{f|\xi}\quad\text{for } f\in 2^\alpha\,.$$

It is clear from (iii) that $\{p_f: f\in 2^\alpha\}$ is faithfully indexed.

We finally claim that p_f is a P_α-point of S_α for $f\in 2^\alpha$. Indeed let $f\in 2^\alpha$, $\beta<\alpha$, and $\{V_\zeta:\zeta<\beta\}$ a family of open-and-closed neighborhoods of p_f. There is $\xi(\zeta)<\alpha$ such that $V_\zeta=A_{\xi(\zeta)}$ for $\zeta<\beta$. Since α is regular by Corollary 1.27(a) there is $\xi<\alpha$ such that $\xi(\zeta)<\xi$ for $\zeta<\beta$. Then

$$p_f\in U_{f|\xi}\subset\bigcap_{\eta<\xi}U_{f|\eta}\subset\bigcap_{\zeta<\beta}V_\zeta\,,$$

so that p_f is a P_α-point of S_α.

The proof is complete.

A space X is called *point-homogeneous* if for every $p,q\in X$ there is an auto-homeomorphism h of X such that $h(p)=q$.

6.18. Corollary. *If $\alpha=\alpha^{\underline{\alpha}}>\omega$, then S_α is not point-homogeneous.*

Proof. Since $h[P_\alpha(\mathsf{S}_\alpha)]=P_\alpha(\mathsf{S}_\alpha)$ for every homeomorphism of S_α onto S_α and $P_\alpha(\mathsf{S}_\alpha)\neq\emptyset$ it is enough to show that $P_\alpha(\mathsf{S}_\alpha)\neq\mathsf{S}_\alpha$, *i.e.*, that S_α is not a P_α-space. Since S_α is α-co-universal for the class of compact spaces there is a continuous function f from S_α onto $[0,1]$, and if S_α is a P_α-space (and hence a P_{ω^+}-space) then

$$\{f^{-1}(\{r\}):r\in[0,1]\}$$

is an open cover of S_α with no finite subcover. Thus S_α is not a P_α-space. The proof is complete.

We note that if $\emptyset\neq A\in\mathfrak{B}(\mathsf{S}_\alpha)$ then A is homeomorphic with S_α. Indeed since $\mathfrak{B}(A)$ satisfies condition (H_α) we have $|\mathfrak{B}(A)|\geq\alpha$, so from $\mathfrak{B}(A)\subset\mathfrak{B}(\mathsf{S}_\alpha)$ and $|\mathfrak{B}(\mathsf{S}_\alpha)|=|\mathfrak{C}_\alpha|=\alpha$ it follows that $|\mathfrak{B}(A)|=\alpha$; hence $\mathfrak{B}(A)$ is isomorphic to $\mathfrak{C}_\alpha$ by Theorem 6.12(b).

6.19. Theorem. *Let $\alpha=\alpha^{\underline{\alpha}}\geq\omega$. There is a space Ω_α, unique up to homeomorphism, such that Ω_α is equal to a union $\bigcup_{\xi<\alpha}S_\xi$ such that*

$$S_\xi \text{ is homeomorphic to } \mathsf{S}_\alpha \text{ for } \xi<\alpha,$$

$$S_\xi \text{ is open-and-closed in } \Omega_\alpha \text{ for } \xi<\alpha, \text{ and}$$

$$S_\xi\supsetneqq S_\zeta \quad \text{for } \zeta<\xi<\alpha.$$

Proof. Let $p\in P_\alpha(\mathsf{S}_\alpha)$, A_0 an open-and-closed neighborhood of p in S_α such that $A_0\subsetneqq\mathsf{S}_\alpha$, for $0<\xi<\alpha$ let A_ξ be any element of $\mathfrak{B}(\mathsf{S}_\alpha)$ such that $p\in A_\xi\subsetneqq\bigcap_{\zeta<\xi}A_\zeta$, and set

$$S_\xi=\mathsf{S}_\alpha\setminus A_\xi \quad \text{for } \xi<\alpha \quad \text{and } \Omega_\alpha=\bigcup_{\xi<\alpha}S_\xi.$$

Now let T be equal to a union $\bigcup_{\xi<\alpha}T_\xi$ such that

$$T_\xi \text{ is homeomorphic to } \mathsf{S}_\alpha \text{ for } \xi<\alpha,$$

$$T_\xi \text{ is open-and-closed in } T \text{ for } \xi<\alpha, \quad \text{and}$$

$$T_\xi\supsetneqq T_\zeta \quad \text{for } \zeta<\xi<\alpha.$$

We define a family $\{h_\xi:\xi<\alpha\}$ of functions such that

(i) h_ξ is a homeomorphism of S_ξ onto T_ξ for $\xi<\alpha$, and

(ii) $h_\zeta\subset h_\xi$ for $\zeta<\xi<\alpha$.

We proceed by recursion. Let h_0 be a homeomorphism of S_0 onto T_0. We now define h_ξ for $\xi<\alpha$. We set

$$f_\xi=\bigcup_{\zeta<\xi}h_\zeta, \quad u_\xi=\bigcup_{\zeta<\xi}S_\zeta, \quad \text{and } v_\xi=\bigcup_{\zeta<\xi}T_\zeta,$$

and we note from the recursive hypotheses that f_ξ is a homeomorphism of u_ξ onto v_ξ. Since S_ξ is compact and S_ζ is an open subset of S_ξ for $\zeta<\xi$ and $S_\zeta\subsetneqq S_{\zeta'}\subsetneqq S_\xi$ for $\zeta<\zeta'<\xi$, we have $u_\xi\subsetneqq S_\xi$; similarly $v_\xi\subsetneqq T_\xi$. Since

$\tau(u_\xi) \leq |\xi| < \alpha$ and $\tau(v_\xi) \leq |\xi| < \alpha$ there is by Theorem 6.11 a homeomorphism $g: T_\xi \to S_\xi$ such that $f_\xi^{-1} \subset g$. We set $h_\xi = g^{-1}$.

Finally, we set $h = \bigcup_{\xi<\alpha} h_\xi$. It is clear that h is a homeomorphism of Ω_α onto T. The proof is complete.

6.20. Corollary. *Let* $\alpha = \alpha^{\underline{\alpha}} \geq \omega$, $p \in P_\alpha(\mathsf{S}_\alpha)$, *and* U *an open subset of* S_α *such that* $1 < \tau(U) < \alpha$. *Then*

(a) $\mathsf{S}_\alpha \setminus \{p\}$ *is homeomorphic to* Ω_α, *and*

(b) $\mathsf{S}_\alpha \setminus \bar{U}$ *is homeomorphic to* Ω_α.

Proof. We show that each of the two spaces is equal to a union $\bigcup_{\xi<\alpha} S_\xi$ such that

$$S_\xi \text{ is homeomorphic to } \mathsf{S}_\alpha \text{ for } \xi < \alpha,$$
$$S_\xi \text{ is open-and-closed for } \xi < \alpha, \quad \text{and}$$
$$S_\xi \supsetneqq S_\zeta \quad \text{for } \zeta < \xi < \alpha.$$

(a) Let $\langle A_\xi : \xi < \alpha \rangle$ be a well-ordering of the family of open-and-closed neighborhoods of p in S_α such that $A_0 \neq \mathsf{S}_\alpha$, set $B_0 = A_0$ and recursively for $\xi < \alpha$ let B_ξ be an element of $\mathfrak{B}(\mathsf{S}_\alpha)$ such that

$$p \in B_\xi \subsetneqq A_\xi \cap (\bigcap_{\zeta<\xi} B_\zeta).$$

We set $S_\xi = \mathsf{S}_\alpha \setminus B_\xi$.

(b) We note from Lemma 6.5 that $\bar{U} \neq \mathsf{S}_\alpha$. Let $\langle A_\xi : \xi < \alpha \rangle$ be a well-ordering of the family of non-empty open-and-closed subsets of S_α that are subsets of $\mathsf{S}_\alpha \setminus \bar{U}$, and set $S_0 = A_0$. To define S_ξ for $0 < \xi < \alpha$ we proceed by recursion. We set $V_\xi = A_\xi \cup (\bigcup_{\zeta<\xi} S_\zeta)$ and we note that $\tau(U) + \tau(V_\xi) < \alpha$ and $U \cap V_\xi = \emptyset$. Hence $\bar{U} \cap \bar{V}_\xi = \emptyset$ by Lemma 6.5 and it is clear from the compactness of S_α and the fact that $\mathfrak{B}(\mathsf{S}_\alpha)$ satisfies condition (H_α) that there is $S_\xi \in \mathfrak{B}(\mathsf{S}_\alpha)$ such that

$$\bar{U} \cap S_\xi = \emptyset \quad \text{and } V_\xi \subsetneqq S_\xi.$$

The family $\{S_\xi : \xi < \alpha\}$ is as required.

6.21. Corollary. *Let* $\alpha = \alpha^{\underline{\alpha}} \geq \omega$.

(a) *If* $p, q \in P_\alpha(\mathsf{S}_\alpha)$, *then there is an auto-homeomorphism* h *of* S_α *such that* $h(p) = q$;

(b) *the space* $P_\alpha(\mathsf{S}_\alpha)$ *is point-homogeneous.*

Proof. (a) It follows from Corollary 6.20(a) that there is a homeomorphism f from $\mathsf{S}_\alpha \setminus \{p\}$ onto $\mathsf{S}_\alpha \setminus \{q\}$. We define

$$h = f \cup \{\langle p, q \rangle\}.$$

(b) Since $h[P_\alpha(\mathsf{S}_\alpha)] = P_\alpha(\mathsf{S}_\alpha)$ for every auto-homeomorphism h of S_α, statement (b) follows from statement (a).

It follows from Lemma 6.5(c) and Theorem 6.12(b) that if $\alpha = \alpha^\alpha \geq \omega$ and U is an open subset of S_α such that $\boldsymbol{\tau}(U) < \alpha$, then U is C^*-embedded in S_α, *i.e.*, $\bar{U} = \boldsymbol{\beta}(U)$. We conclude this section with a stronger theorem from which it follows that if U is such a subset of S_α and f is a continuous function from U into a compact space K, then there is a continuous function $g: \mathsf{S}_\alpha \to K$ such that $f \subset g$.

6.22. Theorem. *If* $\alpha = \alpha^\alpha \geq \omega$ *and* U *is an open subset of* S_α *such that* $1 \leq \boldsymbol{\tau}(U) < \alpha$, *then* $\bar{U}$ *is a retract of* S_α.

Proof. If $\boldsymbol{\tau}(U) = 1$ the statement is obvious. We assume that $\boldsymbol{\tau}(U) \geq \omega$. There are $\beta < \alpha$ and $\{U_\eta : \eta < \beta\} \subset \mathfrak{B}(\mathsf{S}_\alpha)$ such that $U = \bigcup_{\eta < \beta} U_\eta$.

Let $\langle A_\xi : \xi < \alpha \rangle$ be a well-ordering of $\mathfrak{B}(\bar{U})$ and let $\mathscr{R}$ and $\mathscr{S}$ denote the Boolean subalgebras of $\mathfrak{B}(\bar{U})$ and $\mathfrak{B}(\mathsf{S}_\alpha)$ respectively generated by $\{U_\eta : \eta < \beta\}$. Since $U \neq \bar{U}$ by Lemma 6.5(c)2, there is by Lemma 6.9(a) a Boolean algebra isomorphism $\psi: \mathscr{R} \to \mathscr{S}$ such that

(*$\mathscr{R}$) if $d \in \mathscr{R}$ and there is $\eta < \beta$ such that $d \subset U_\eta$, then $\psi(d) = d$.

For $\xi < \alpha$ we denote by $\mathscr{R}_\xi$ the Boolean subalgebra of $\mathfrak{B}(\bar{U})$ generated by $\mathscr{R} \cup \{A_\zeta : \zeta < \xi\}$. We define a family $\{\psi_\xi : \xi < \alpha\}$ of functions such that

(i) ψ_ξ is a Boolean algebra embedding of $\mathscr{R}_\xi$ into $\mathfrak{B}(\mathsf{S}_\alpha)$ for $\xi < \alpha$;

(ii) $\psi_0 = \psi$;

(iii) $\psi_\xi = \bigcup_{\zeta < \xi} \psi_\zeta$ if ξ is a non-zero limit ordinal and $\xi < \alpha$;

(iv) $\psi_\zeta \subset \psi_\xi$ for $\zeta < \xi < \alpha$; and

(v) if $d \in \mathscr{R}_\xi$ and there is $\eta < \beta$ such that $d \subset U_\eta$, then $\psi_\xi(d) = d$.

We proceed by recursion. We define ψ_0 by (ii) and we define ψ_ξ by (iii) if ξ is a non-zero limit ordinal and $\xi < \alpha$. Now we define $\psi_{\xi+1}$ for $\xi < \alpha$. By Lemma 6.10, with

$$X, Y, \mathscr{A}, \mathscr{B}, u, v, f, \tilde{\psi}, \tilde{\mathscr{R}}, \tilde{\mathscr{S}} \text{ and } a$$

replaced by

$$\bar{U}, \mathsf{S}_\alpha, \mathfrak{B}(\bar{U}), \mathfrak{B}(\mathsf{S}_\alpha), U, U, \mathrm{id}_U, \psi_\xi, \mathscr{R}_\xi, \psi_\xi[\mathscr{R}_\xi] \text{ and } A_\xi$$

respectively, there is a Boolean algebra embedding

$$\phi: \mathscr{R}_{\xi+1} \to \mathfrak{B}(\mathsf{S}_\alpha)$$

such that $\psi_\xi \subset \phi$ and

(*$\mathscr{R}_{\xi+1}$) if $d \in \mathscr{R}_{\xi+1}$ and there is $\eta < \beta$ such that $d \subset U_\eta$, then $\phi(d) = d$.

We set $\psi_{\xi+1} = \varphi$.

Finally we set $\psi = \bigcup_{\xi < \alpha} \psi_\xi$. It is clear that ψ is a Boolean algebra embedding of $\mathfrak{B}(\bar{U})$ into $\mathfrak{B}(\mathsf{S}_\alpha)$ such that

(*$\mathfrak{B}(\bar{U})$) if $d \in \mathfrak{B}(\bar{U})$ and there is $\eta < \beta$ such that $d \subset U_\eta$, then $\psi(d) = d$.

It follows from Theorem 2.29(b) that $\mathsf{S}(\psi)$ is a continuous function from S_α onto $\bar{U}$. Further for $p \in U$ there is $\eta < \beta$ such that $p \in U_\eta$ and we have

$$\begin{aligned} \mathsf{S}(\psi)(p) = \psi^{-1}(p) &= \psi^{-1}\left(\bigcap\{B \in \mathfrak{B}(\mathsf{S}_\alpha): p \in B\}\right) \\ &= \bigcap\{\psi^{-1}(B): p \in B \in \mathfrak{B}(\mathsf{S}_\alpha) \text{ and } B \subset U_\eta\} \\ &= \bigcap\{B: p \in B \in \mathfrak{B}(\mathsf{S}_\alpha) \text{ and } B \subset U_\eta\} = \{p\}. \end{aligned}$$

Thus $\mathsf{S}(\psi)(p) = p$ for $p \in \bar{U}$ and $\mathsf{S}(\psi)$ is a retraction from S_α onto $\bar{U}$. The proof is complete.

Notes for § 6

The fact that Boolean algebras form a Jónsson class involves only the simple problem of verifying the amalgamation property. The validity of this property has been noted independently by various authors including Eva Kallin (cf. Keisler [66] (last paragraph)), Daigneault [64] (Theorems 2.7 and 6.3), and Dwinger-Yaqub [63].

The results of this section are mostly taken from Negrepontis [69a]. Specifically, Lemmas 6.3, 6.5, 6.6, 6.7 and 6.8 appear as Negrepontis [69a] Lemma 1.1, Corollary 1.5, Negrepontis [71] (Lemma 2), and Negrepontis [69a] (Lemma 1.6), respectively. For $\alpha = \omega^+ = 2^\omega$ and for the Boolean algebra $\mathfrak{B}(\boldsymbol{\beta}(\omega)\setminus\omega)$ the analogue of Lemma 6.8 was first established in the original and significant paper of W. Rudin [56]. The central results of the section, Theorems 6.11 and 6.12, appear in Negrepontis [69a] (Theorem 1.7). For $\alpha = \omega^+ = 2^\omega$ and for the Boolean algebra $\mathfrak{B}(\boldsymbol{\beta}(\omega)\setminus\omega)$ these ideas were developed by W. Rudin [56], Parovičenko [63], Hoffman-Ramsay [65], Keisler [66], M.E. Rudin [66], and Comfort-Negrepontis [68].

Lemma 6.7, the analogue of the disjoint refinement Lemma (7.5) proved below for sets, is used in an essential way for the proof of existence of adequate ultrafilters on special Boolean algebras (10.16—10.19).

Lemma 6.13, Theorem 6.14, and 6.16—6.18 appear in Negrepontis [69a] (2.2—2.4, Theorems 2.7 and 3.5 respectively). For $\alpha = \omega^+ = 2^\omega$ and for the Boolean algebra $\mathfrak{B}(\boldsymbol{\beta}(\omega)\setminus\omega)$ the analogues of 6.16—6.18 were established by W. Rudin [63]. Theorem 6.19 and Corollary 6.20 appear in Comfort-Negrepontis [68] (Theorem 2.1, Corollary 2.2) for $\alpha = \omega^+ = 2^\omega$ and $\mathfrak{B}(\boldsymbol{\beta}(\omega)\setminus\omega)$, and in Negrepontis [69a] (Corollary 3.4, Corollary 3.6) in general. Esenin-Vol'pin [49] proved directly, assuming GCH, the existence of an α-co-universal compact space of weight α for $\alpha \geq \omega$. His result follows from the general results of § 4, § 6. A related result for metric spaces was proved earlier by Sierpiński [45] (cf. Morley-

Vaught [62] (remarks on p. 41) and Hung [72] (Introduction)). Theorem 6.22 is proved in Negrepontis [69a] (Theorem 3.1 and Corollary 3.2).

The results referred to above should be compared with 14.12—14.19 and 15.10—15.11 below.

The normed algebra $C(\mathsf{S}_\alpha)$ of real-valued continuous functions on S_α with the uniform norm has a homogeneous-universal property described in Negrepontis [69b] (§ 3).

§ 7. Basic Facts on Ultrafilters

With this introductory section we begin in earnest our study of ultrafilters. Some basic definitions already mentioned are collected here once more for emphasis. In § 2, z-filters on spaces and filters on Boolean algebras were defined and considered; from now on we deal almost exclusively with filters on cardinals.

Our starting point is this fundamental application of Zorn's lemma: every filter is contained in an ultrafilter (Theorem 7.1). According to this fundamental triviality the objects we wish to study actually exist. An application of the existence of families of large oscillation provides the basic computation of the cardinality of sets of ultrafilters (Theorem 7.3 and its Corollary). Deeper applications of the existence of families of large oscillation form a significant part of this chapter (and appear in §§ 10 and 13). The simple and useful "disjoint refinement lemma", proved by a typical diagonal argument, and some of its consequences, are given in 7.5—7.9. The main application of these results appears in § 9.

The elementary properties of various types of regular ultrafilters (given in 7.10—7.11) are used frequently later, in particular in § 12, for the basic computations of cardinalities of ultraproducts. The topology of the space $U_\kappa(\alpha)$ of κ-uniform ultrafilters on α (described in 7.12—7.15) is studied in detail in §§ 12, 14, and 15, especially in § 14.

The last part of this section deals with some facts concerning filters rather than just ultrafilters; these include the direct image of a filter under a function (7.17—7.19), and products of filters and their relation to the canonical function $\boldsymbol{\phi}: \boldsymbol{\beta}(\alpha\times\beta)\to\boldsymbol{\beta}(\alpha)\times\boldsymbol{\beta}(\beta)$ (7.20—7.24). A part of § 16 is devoted to the study of $\boldsymbol{\phi}$ when $\alpha=\beta=\omega$.

Definitions. Let α be a cardinal and let $\mathscr{F}\subset\mathscr{P}(\alpha)$.

(i) $\mathscr{F}$ has the *finite intersection property* if

$\mathscr{F}\neq\emptyset$ and

$\bigcap_{k\leq n}A_k\neq\emptyset$ whenever $n<\omega$ and $A_k\in\mathscr{F}$ for $k\leq n$.

(ii) $\mathscr{F}$ is a *filter base* if

$\mathscr{F} \neq \emptyset$,

$A \neq \emptyset$ for $A \in \mathscr{F}$, and

if $n < \omega$ and $A_k \in \mathscr{F}$ for $k \leq n$ then there is $A \in \mathscr{F}$ such that

$$A \subset \bigcap_{k \leq n} A_k.$$

(iii) $\mathscr{F}$ is *closed under finite intersections* if

$$A_0 \cap A_1 \in \mathscr{F} \quad \text{whenever } A_0, A_1 \in \mathscr{F}.$$

(iv) $\mathscr{F}$ is a *filter on* α if

$\mathscr{F} \neq \emptyset$,

$A \neq \emptyset$ for $A \in \mathscr{F}$,

$A_0 \cap A_1 \in \mathscr{F}$ for $A_0, A_1 \in \mathscr{F}$, and

if $A \in \mathscr{F}$ and $A \subset B \subset \alpha$ then $B \in \mathscr{F}$.

(v) $\mathscr{F}$ is an *ultrafilter on* α if

$\mathscr{F}$ is a filter on α and

$\mathscr{F}$ is not properly contained in any filter on α.

It is clear that
$\mathscr{F}$ is a filter $\Rightarrow \phi \notin \mathscr{F}$ and $\mathscr{F}$ is closed under finite intersections $\Rightarrow \mathscr{F}$ is a filter-base $\Rightarrow \mathscr{F}$ has the finite intersection property.

Of course, a filter on α is precisely a (proper) filter of $\mathscr{P}(\alpha)$.

We note that if $\mathscr{F} \subset \mathscr{P}(\alpha)$ and $\mathscr{F}$ has the finite intersection property then $\langle \mathscr{F} \rangle$ (as defined in § 2) is a filter on α; this is the *filter on* α *generated by* $\mathscr{F}$.

The fundamental application of Zorn's lemma, given in § 2 for the more general case of filters on Boolean algebras, implies that every filter on α is contained in an ultrafilter on α. More generally we have the following result.

7.1. Theorem. *Let* α *be a cardinal and let* $\mathscr{F}$ *be a family of subsets of* α *with the finite intersection property. Then there is an ultrafilter* p *on* α *such that* $\mathscr{F} \subset p$.

We recall from § 2 that the set of ultrafilters on α is (identified with) the set of elements of the Stone-Čech compactification $\boldsymbol{\beta}(\alpha)$ of α. For $p \in \alpha$, the ultrafilter associated with p is

$$\{A \subset \alpha : p \in A\}.$$

Furthermore if $A \in \mathscr{P}(\alpha)$ and $p \in \boldsymbol{\beta}(\alpha)$ then

$$A \in p \quad \text{if and only if } p \in \mathrm{cl}_{\boldsymbol{\beta}(\alpha)} A .$$

Let p be an ultrafilter on α. The *norm* of p, denoted by $\|p\|$, is defined by

$$\|p\| = \min\{|A| : A \in p\} .$$

Thus $1 \leq \|p\| \leq \alpha$ for all ultrafilters p on α.

Definition. Let κ and α be cardinals and let p be an ultrafilter on α. Then p is *κ-uniform* if $\|p\| \geq \kappa$.

We denote by $U_\kappa(\alpha)$ the set of κ-uniform ultrafilters on α.

An ω-uniform ultrafilter (on α) is said to be *non-principal*; other ultrafilters on α are *principal*. An α-uniform ultrafilter on α is *uniform*. The set $U_\alpha(\alpha)$ of uniform ultrafilters on α is denoted $U(\alpha)$.

Since an ultrafilter on α is principal if and only if it is (identified with) an element of α, we have $\boldsymbol{\beta}(\alpha) \setminus \alpha = U_\omega(\alpha)$ for each cardinal α; this set is empty if and only if $\alpha < \omega$.

Let α and κ be cardinals and let $\mathscr{F} \subset \mathscr{P}(\alpha)$. Then $\mathscr{F}$ has the *κ-uniform finite intersection property* if

$$\mathscr{F} \neq \emptyset \quad \text{and}$$

$$\left|\bigcap\nolimits_{k \leq n} A_k\right| \geq \kappa \quad \text{whenever } n < \omega \quad \text{and } A_k \in \mathscr{F} \quad \text{for } k \leq n.$$

If $\mathscr{F} \subset \mathscr{P}(\alpha)$ and $\mathscr{F}$ has the α-uniform finite intersection property, then $\mathscr{F}$ is said to have the *uniform finite intersection property*.

It is clear that $\mathscr{F}$ has the finite intersection property if and only if $\mathscr{F}$ has the 1-uniform finite intersection property.

We define $\mathscr{P}^\kappa(\alpha) = \{A \in \mathscr{P}(\alpha) : |\alpha \setminus A| < \kappa\}$. Clearly if $\omega \leq \kappa \leq \alpha$ then $\mathscr{P}^\kappa(\alpha)$ is a filter on α. The filter $\mathscr{P}^\alpha(\alpha)$ is called the *(generalized) Fréchet filter on α.*

7.2. Lemma. *Let $\omega \leq \kappa \leq \alpha$. Then*

(a) *an ultrafilter p on α is κ-uniform if and only if $\mathscr{P}^\kappa(\alpha) \subset p$;*

(b) *there is a κ-uniform ultrafilter on α, i.e., $U_\kappa(\alpha) \neq \emptyset$; and*

(c) *each family of subsets of α with the κ-uniform finite intersection property is contained in a κ-uniform ultrafilter on α.*

Proof. (a) Let p be κ-uniform and let $A \in \mathscr{P}^\kappa(\alpha)$. Then $|\alpha \setminus A| < \kappa$ and hence $\alpha \setminus A \notin p$. Since p is an ultrafilter we have $A \in p$.

Conversely let $\mathscr{P}^\kappa(\alpha) \subset p$ and let $A \in p$. If $|A| < \kappa$ then $\alpha \setminus A \in \mathscr{P}^\kappa(\alpha) \subset p$ and so $\emptyset = A \cap (\alpha \setminus A) \in p$; thus $|A| \geq \kappa$.

(b) Since $\omega \leq \kappa \leq \alpha$ the family $\mathscr{P}^\kappa(\alpha)$ is a filter on α. From Theorem 7.1 there is an ultrafilter p on α such that $\mathscr{P}^\kappa(\alpha) \subset p$, and p is κ-uniform by (a).

(c) Let $\mathscr{F}$ be a family of subsets of α with the κ-uniform finite intersection property. We claim that $\mathscr{F} \cup \mathscr{P}^\kappa(\alpha)$ has the finite intersection property. Let $m, n < \omega$ and let $\{A_k : k \leq n\} \subset \mathscr{F}$ and $\{B_i : i \leq m\} \subset \mathscr{P}^\kappa(\alpha)$. Then $|\bigcap_{k \leq n} A_k| \geq \kappa$ and $|\alpha \setminus \bigcap_{i \leq m} B_i| < \kappa$, so

$$\bigcap_{k \leq n} A_k \cap \bigcap_{i \leq m} B_i \neq \emptyset .$$

By Theorem 7.1 there is an ultrafilter p on α such that $\mathscr{F} \cup \mathscr{P}^\kappa(\alpha) \subset p$, and p is κ-uniform by (a) above.

We note the following special cases of this lemma.

(1) If α is an infinite cardinal then there is a non-principal ultrafilter on α;

(2) if α is an infinite cardinal then an ultrafilter p on α is non-principal if and only if $\mathscr{P}^\omega(\alpha) \subset p$;

(3) if α is an infinite cardinal then there is a uniform ultrafilter on α;

(4) if α is an infinite cardinal then an ultrafilter p on α is uniform if and only if $\mathscr{P}^\alpha(\alpha) \subset p$; and

(5) if α is an infinite cardinal and $\mathscr{F}$ is a family of subsets of α with the uniform finite intersection property, then $\mathscr{F}$ is contained in a uniform ultrafilter on α.

Definition. Let α and κ be cardinals. A filter $\mathscr{F}$ on α is *κ-complete* if $\bigcap \mathscr{G} \in \mathscr{F}$ whenever $\mathscr{G} \subset \mathscr{F}$ and $|\mathscr{G}| < \kappa$.

Every filter is ω-complete.

A filter is *countably incomplete* if it is not ω^+-complete.

We use in what follows the fact that if p is an ultrafilter on α and p is not κ-complete, then there are $\beta < \kappa$ and a family $\{A_\xi : \xi < \beta\} \subset p$ such that

$$A_0 = \alpha,$$

$$A_\xi \subset A_\zeta \quad \text{for } \zeta < \xi < \beta, \quad \text{and}$$

$$\bigcap_{\xi < \beta} A_\xi = \emptyset.$$

In fact let β be the smallest cardinal for which there is a family $\{B_\xi : \xi < \beta\} \subset p$ such that $\bigcap_{\xi < \beta} B_\xi \notin p$ and define

$$A_0 = \alpha \quad \text{and}$$

$$A_\xi = (\bigcap_{\eta \leq \xi} B_\eta) \setminus (\bigcap_{\eta < \beta} B_\eta) \text{ for } 0 < \xi < \beta.$$

We now make an application of the results on families of functions of κ-large oscillation, given in § 3, to the exact computation of the number of ultrafilters.

7.3. Theorem. *Let $\omega \leq \kappa \leq \alpha = \alpha^{\underline{\kappa}}$. There is a family Φ of κ-complete filters on α such that*

$$|\Phi| = 2^{2^\alpha}, \quad \textit{and}$$

if $\mathscr{F}_0, \mathscr{F}_1 \in \Phi$ and $\mathscr{F}_0 \neq \mathscr{F}_1$, then there is $A \in \mathscr{P}(\alpha)$ such that $A \in \mathscr{F}_0$ and $\alpha \setminus A \in \mathscr{F}_1$.

Proof. By Theorem 3.16 there is a family $\mathscr{S}$ of κ-large oscillation of functions from α to $\{0,1\}$ such that $|\mathscr{S}| = 2^\alpha$. For $\mathscr{C} \in \mathscr{P}(\mathscr{S})$, we set

$$\mathscr{B}(\mathscr{C}) = \{f^{-1}(\{0\}) : f \in \mathscr{C}\} \cup \{f^{-1}(\{1\}) : f \in \mathscr{S} \setminus \mathscr{C}\}$$

and

$$\mathscr{F}(\mathscr{C}) = \{A \in \mathscr{P}(\alpha) : \text{there is } B \in \mathscr{B}(\mathscr{C}) \text{ such that } B \subset A\}.$$

Since $\mathscr{S}$ is of κ-large oscillation the family $\mathscr{F}(\mathscr{C})$ is, for each $\mathscr{C} \in \mathscr{P}(\mathscr{S})$, a κ-complete filter on α. Now let $\mathscr{C}, \mathscr{C}' \in \mathscr{P}(\mathscr{S})$ with $\mathscr{C} \neq \mathscr{C}'$; there is $f \in (\mathscr{C} \setminus \mathscr{C}') \cup (\mathscr{C}' \setminus \mathscr{C})$, say $f \in \mathscr{C} \setminus \mathscr{C}'$. Then $f^{-1}(\{0\}) \in \mathscr{B}(\mathscr{C})$ and $\alpha \setminus f^{-1}(\{0\}) = f^{-1}(\{1\}) \in \mathscr{B}(\mathscr{C}')$. Thus the family

$$\Phi = \{\mathscr{F}(\mathscr{C}) : \mathscr{C} \in \mathscr{P}(\mathscr{S})\}$$

satisfies the conditions of the theorem.

7.4. Corollary. *If α is an infinite cardinal, then $|\boldsymbol{\beta}(\alpha)| = 2^{2^\alpha}$.*

Proof. We have $|\boldsymbol{\beta}(\alpha)| \leq 2^{2^\alpha}$ because $\boldsymbol{\beta}(\alpha) \subset \mathscr{P}(\mathscr{P}(\alpha))$.

Since $\omega \leq \alpha = \alpha^{\underline{\omega}}$ there is by Theorem 7.3 a family Φ of filters on α such that

$$|\Phi| = 2^{2^\alpha}, \quad \text{and}$$

if $\mathscr{F}_0, \mathscr{F}_1 \in \Phi$ and $\mathscr{F}_0 \neq \mathscr{F}_1$, then there is $A \in \mathscr{P}(\alpha)$ such that $A \in \mathscr{F}_0$ and $\alpha \setminus A \in \mathscr{F}_1$.

For every $\mathscr{F} \in \Phi$ there is an ultrafilter $p_{\mathscr{F}}$ on α such that $\mathscr{F} \subset p_{\mathscr{F}}$. Then $\{p_{\mathscr{F}} : \mathscr{F} \in \Phi\} \subset \boldsymbol{\beta}(\alpha)$, and if $\mathscr{F}_0, \mathscr{F}_1 \in \Phi$ and $\mathscr{F}_0 \neq \mathscr{F}_1$, then $p_{\mathscr{F}_0} \neq p_{\mathscr{F}_1}$. Thus $|\boldsymbol{\beta}(\alpha)| \geq |\Phi| = 2^{2^\alpha}$.

The following simple result, proved by a diagonal argument, is used repeatedly in the sequel.

7.5. Lemma *(Disjoint refinement lemma). Let α be an infinite cardinal and let $\{A_\xi : \xi < \alpha\}$ be a family of sets such that $|A_\xi| = \alpha$ for $\xi < \alpha$. Then there is a family $\{B_\xi : \xi < \alpha\}$ of sets such that*

$$B_\xi \subset A_\xi \quad \textit{for } \xi < \alpha,$$

$$|B_\xi| = \alpha \quad \textit{for } \xi < \alpha, \quad \textit{and}$$

$$B_\xi \cap B_\zeta = \emptyset \quad \textit{for } \xi < \zeta < \alpha.$$

Proof. We proceed by transfinite recursion. Let $p_{0,0}\in A_0$. Let $\zeta<\alpha$ and suppose that for $\xi'\leq\xi<\zeta$ we have defined $p_{\xi',\xi}\in A_{\xi'}$. We choose $p_{\zeta',\zeta}$ for $\zeta'\leq\zeta$ such that

$$p_{\zeta',\zeta}\in A_{\zeta'}\setminus\{p_{\xi',\xi}:\xi'\leq\xi<\zeta\} \quad \text{and}$$
$$p_{\zeta',\zeta}\neq p_{\zeta'',\zeta} \quad \text{for } \zeta'<\zeta''\leq\zeta,$$

and we set

$$B_\xi=\{p_{\xi,\zeta}:\xi\leq\zeta<\alpha\} \quad \text{for } \xi<\alpha.$$

It is easily verified that the family $\{B_\xi:\xi<\alpha\}$ is as required.

7.6. Corollary. *Let $\omega\leq\beta\leq\alpha$ and let $\{A_\xi:\xi<\alpha\}$ be a family of subsets of α such that $|A_\xi|=\alpha$ for $\xi<\alpha$. Then there is a family $\{B_\zeta:\zeta<\beta\}$ of subsets of α such that*

$$B_\zeta\cap B_{\zeta'}=\emptyset \quad \textit{for } \zeta<\zeta'<\beta, \quad \textit{and}$$
$$|A_\xi\cap B_\zeta|=\alpha \quad \textit{for } \xi<\alpha, \quad \zeta<\beta.$$

Proof. By Lemma 7.5 it is clear that there is a family

$$\{C_{\xi,\zeta}:\xi<\alpha,\ \zeta<\beta\}$$

of subsets of α such that

$$C_{\xi,\zeta}\subset A_\xi,$$
$$C_{\xi,\zeta}\cap C_{\xi',\zeta'}=\emptyset \quad \text{for } \langle\xi,\zeta\rangle\neq\langle\xi',\zeta'\rangle, \quad \text{and}$$
$$|C_{\xi,\zeta}|=\alpha$$

for $\langle\xi,\zeta\rangle$, $\langle\xi',\zeta'\rangle\in\alpha\times\beta$. We define

$$B_\zeta=\bigcup_{\xi<\alpha}C_{\xi,\zeta}$$

for $\zeta<\beta$; it is clear that the conditions of the corollary are satisfied.

7.7. Theorem. *Let α be an infinite cardinal and let $\mathscr{F}$ be a family of subsets of α such that*

$$|\mathscr{F}|\leq\alpha, \quad \textit{and}$$
$$\mathscr{F} \textit{ has the uniform finite intersection property.}$$

Then

(a) *there is $E\in\mathscr{P}(\alpha)$ such that each of the families*

$$\mathscr{F}\cup\{E\}, \qquad \mathscr{F}\cup\{\alpha\setminus E\}$$

has the uniform finite intersection property; and

(b) *there are 2^{2^α} distinct uniform ultrafilters on α containing $\mathscr{F}$.*

Proof. We suppose without loss of generality that $\mathscr{F}$ is closed under finite intersections. Let $\langle A_\xi : \xi < \alpha \rangle$ be a well-ordering of $\mathscr{F}$ and let

$$\{B_{\xi,\zeta} : \xi < \alpha,\ \zeta < \alpha\}$$

be a family of subsets of α such that

$$B_{\xi,\zeta} \subset A_\xi,$$
$$B_{\xi,\zeta} \cap B_{\xi',\zeta'} = \emptyset \quad \text{for } \langle \xi,\zeta \rangle \neq \langle \xi',\zeta' \rangle, \quad \text{and}$$
$$|B_{\xi,\zeta}| = \alpha$$

for $\langle \xi,\zeta \rangle, \langle \xi',\zeta' \rangle \in \alpha \times \alpha$.

Now for $p \in \boldsymbol{\beta}(\alpha)$ we define

$$\mathscr{F}_p = \mathscr{F} \cup \{\bigcup \{B_{\xi,\zeta} : \xi < \alpha,\ \zeta \in C\} : C \in p\},$$

and we claim that $\mathscr{F}_p$ has the uniform finite intersection property. To see this let $m, n < \omega$ and let $\{C_k : k \le n\} \subset p$ and $\{A_{\xi(i)} : i \le m\} \subset \mathscr{F}$. There is $\xi' < \alpha$ such that $\bigcap_{i \le m} A_{\xi(i)} = A_{\xi'}$ and with

$$C = \bigcap_{k \le n} C_k \in p$$

there is $\zeta' \in C$. It follows that

$$B_{\xi',\zeta'} \subset A_{\xi'} \cap (\bigcup \{B_{\xi,\zeta} : \xi < \alpha,\ \zeta \in C\}).$$

Now let p and q be distinct elements of $\boldsymbol{\beta}(\alpha)$ and let $C \in \mathscr{P}(\alpha)$ with $C \in p$ and $\alpha \backslash C \in q$. We set

$$E = \bigcup \{B_{\xi,\zeta} : \zeta \in C \text{ and } \xi < \alpha\}$$

and

$$E' = \bigcup \{B_{\xi,\zeta} : \zeta \in \alpha \backslash C \text{ and } \xi < \alpha\}$$

and we have $E \cap E' = \emptyset$, $E \in \mathscr{F}_p$, and $E' \in \mathscr{F}_q$. It follows that $\mathscr{F} \cup \{E\}$ and $\mathscr{F} \cup \{E'\}$, and hence also $\mathscr{F} \cup \{\alpha \backslash E\}$, have the uniform finite intersection property; so (a) is proved.

Because the family $\mathscr{F}_p$ has the uniform finite intersection property for $p \in \boldsymbol{\beta}(\alpha)$, there is (by Lemma 7.2(c)) a uniform ultrafilter $\tilde{p}$ on α such that $\mathscr{F}_p \subset \tilde{p}$. Since $\tilde{p}$ and $\tilde{q}$ contain complementary subsets of α if p and q are distinct elements of $\boldsymbol{\beta}(\alpha)$, the function $p \to \tilde{p}$ from $\boldsymbol{\beta}(\alpha)$ to the set of uniform ultrafilters containing $\mathscr{F}$ is one-to-one. Statement (b) now follows from the fact that $|\boldsymbol{\beta}(\alpha)| = 2^{2^\alpha}$ (Corollary 7.4).

7.8. Corollary. *Let* $\alpha \ge \omega$. *Then*

(a) *the cardinality of each filter base for a uniform ultrafilter on* α *is greater than* α; *and*

(b) $|U(\alpha)| = 2^{2^\alpha}$.

Proof. Both statements follow from Theorem 7.7.

7.9. Corollary. *Let α be an infinite cardinal and let $\mathscr{F}$ be a family of subsets of α such that*

$$|\mathscr{F}| \leq \alpha, \quad \textit{and}$$

$$\mathscr{F} \textit{ has the uniform finite intersection property.}$$

Then there is a sequence $\{A_n : n < \omega\}$ of subsets of α such that

$$\bigcap_{n<\omega} A_n = \emptyset, \quad \textit{and}$$

the family $\mathscr{F} \cup \{A_n : n < \omega\}$ has the uniform finite intersection property.

Proof. With no loss of generality, we assume that $\mathscr{F}$ is closed under finite intersections. Let $\langle F_\xi : \xi < \alpha \rangle$ be a well-ordering of $\mathscr{F}$. By Lemma 7.5 there is a family $\{B_\xi : \xi < \alpha\}$ of subsets of α such that $B_\xi \subset F_\xi$, $|B_\xi| = \alpha$, and $B_\xi \cap B_\zeta = \emptyset$ for $\xi < \zeta < \alpha$. For $\xi < \alpha$, we let $\{A_{\xi,n} : n < \omega\}$ be a sequence of subsets of B_ξ such that

$$|A_{\xi,n}| = \alpha \quad \text{for } n < \omega,$$

$$A_{\xi,n} \supset A_{\xi,n+1} \quad \text{for } n < \omega, \quad \text{and}$$

$$\bigcap_{n<\omega} A_{\xi,n} = \emptyset.$$

We set

$$A_n = \bigcup_{\xi<\alpha} A_{\xi,n}$$

for $n < \omega$. Then $\bigcap_{n<\omega} A_n = \emptyset$, and $\mathscr{F} \cup \{A_n : n < \omega\}$ has the uniform finite intersection property because $A_{\xi,n} \subset A_n \cap F_\xi$ for $n < \omega$ and $\xi < \alpha$.

Definition. Let α, γ and κ be infinite cardinals and let p be an ultrafilter on α. Then p is *(κ, γ)-regular* if there is a family

$$\{A_\xi : \xi < \gamma\} \subset p$$

such that

$$\text{if } I \subset \gamma \text{ and } |I| = \kappa, \text{ then } \bigcap_{\xi \in I} A_\xi = \emptyset.$$

The family $\{A_\xi : \xi < \gamma\}$ is called a *(κ, γ)-regular family for p.*

If p is an (ω, α)-regular ultrafilter (on α) we say simply that p is *regular.*

7.10. Lemma. *Let α, κ, and γ be infinite cardinals and let p be an ultrafilter on α. Then*

(a) *p is (κ^+, κ)-regular;*

(b) *if p is countably incomplete, then p is (ω, ω)-regular;*

(c) *if p is a uniform ultrafilter on α, then p is $(\text{cf}(\alpha), \text{cf}(\alpha))$-regular;*

(d) *if p is $(\text{cf}(\kappa), \text{cf}(\kappa))$-regular, then p is (κ, κ)-regular; and*

(e) *if $\omega \leq \kappa \leq \kappa'$ and $\omega \leq \gamma' \leq \gamma$ and if p is a (κ, γ)-regular ultrafilter, then p is (κ', γ')-regular.*

Proof. (a) and (b) are clear from the definitions.

(c) Let $f:\mathrm{cf}(\alpha)\to\alpha$ be such that

$$f(\eta)\leq f(\eta') \quad \text{for } \eta<\eta'<\mathrm{cf}(\alpha), \quad \text{and}$$
$$\sup_{\eta<\mathrm{cf}(\alpha)} f(\eta)=\alpha .$$

Then the family $\{\alpha\setminus f(\eta):\eta<\mathrm{cf}(\alpha)\}$ is a $(\mathrm{cf}(\alpha),\mathrm{cf}(\alpha))$-regular family for every uniform ultrafilter on α.

(d) Let $\{A_\eta:\eta<\mathrm{cf}(\kappa)\}$ be a $(\mathrm{cf}(\kappa),\mathrm{cf}(\kappa))$-regular family for p, and let $f:\mathrm{cf}(\kappa)\to\kappa$ be such that

$$f(0)=0,$$
$$f(\eta)\leq f(\eta') \quad \text{for } \eta<\eta'<\mathrm{cf}(\kappa), \quad \text{and}$$
$$\sup_{\eta<\mathrm{cf}(\kappa)} f(\eta)=\kappa .$$

Now for $f(\eta)\leq\xi<f(\eta+1)$ with $\eta<\mathrm{cf}(\kappa)$ we set $B_\xi=A_{f(\eta)}$, and we note that if $I\subset\alpha$ and $|I|=\alpha$ then I is cofinal in α, so that $\bigcap_{\xi\in I}B_\xi=\emptyset$. Thus $\{B_\xi:\xi<\alpha\}$ is an (α,α)-regular family for p.

(e) If $\{A_\xi:\xi<\gamma\}$ is a (κ,γ)-regular family for p, then $\{A_\xi:\xi<\gamma'\}$ is a (κ',γ')-regular family for p.

7.11. Lemma. *Let α be an infinite cardinal. Then*

(a) *there is a regular ultrafilter on α, and*

(b) *every regular ultrafilter on α is countably incomplete and uniform.*

Proof. (a) Since α is infinite we have $\alpha=|\mathscr{P}_\omega(\alpha)|$. Let Φ be a function from α onto $\mathscr{P}_\omega(\alpha)$, and for $\xi<\alpha$ set

$$A_\zeta=\{\xi<\alpha:\zeta\in\Phi(\xi)\} .$$

To show that $\{A_\zeta:\zeta<\alpha\}$ has the finite intersection property let $n<\omega$ and $\zeta_0,\dots,\zeta_n<\alpha$, and let $\xi<\alpha$ be such that $\Phi(\xi)=\{\zeta_0,\dots,\zeta_n\}$; then

$$\xi\in\bigcap_{k\leq n}A_{\zeta_k} .$$

Thus there is by Theorem 7.1 an ultrafilter p on α such that

$$\{A_\zeta:\zeta<\alpha\}\subset p ,$$

and it is clear that $\{A_\zeta:\zeta<\alpha\}$ is a regular family for p. Thus p is a regular ultrafilter on α.

(b) It is clear that a regular ultrafilter p is countably incomplete. To prove that p is uniform, let $\{A_\zeta:\zeta<\alpha\}$ be a regular family for p and set $F_\xi=\{\zeta<\alpha:\xi\in A_\zeta\}$ for $\xi<\alpha$; thus $|F_\xi|<\omega$ for $\xi<\alpha$. Now let $A\in\mathscr{P}_\alpha(\alpha)$. Then $|\bigcup_{\xi\in A}F_\xi|<\alpha$, so there is $\zeta\in\alpha\setminus\bigcup_{\xi\in A}F_\xi$. Since $A_\zeta\subset\alpha\setminus A$ and $A_\zeta\in p$ we have $A\notin p$. Thus p is uniform.

Let α and κ be any cardinals and let $A\in\mathscr{P}(\alpha)$. We define

$$\hat{A}=\mathrm{cl}_{\beta(\alpha)}A\cap U_\kappa(\alpha) .$$

We note that the notation does not indicate the dependence of $\hat{A}$ on the cardinal κ; in every case we make clear the value for κ that is being used.

7.12. Lemma. *Let* $\omega \leq \kappa \leq \alpha$. *Then*

(a) $U_\kappa(\alpha)$ *is a non-empty, compact subspace of* $\boldsymbol{\beta}(\alpha)$;

(b) *if* V *is an open-and-closed subset of* $U_\kappa(\alpha)$ *then there is* $A \in \mathscr{P}(\alpha)$ *such that* $V = \hat{A}$;

(c) *if* $A \in \mathscr{P}(\alpha)$ *then* $\hat{A} = \emptyset$ *if and only if* $|A| < \kappa$;

(d) *if* $A, B \in \mathscr{P}(\alpha)$ *then*

(i) $\hat{A} \cap \hat{B} = (A \cap B)\hat{}$,

(ii) $\hat{A} \subset \hat{B}$ *if and only if* $|A \setminus B| < \kappa$, *and*

(iii) $\hat{A} = \hat{B}$ *if and only if* $|(A \setminus B) \cup (B \setminus A)| < \kappa$.

Thus the function

$$A \to \hat{A} = \mathrm{cl}_{\boldsymbol{\beta}(\alpha)} A \cap U_\kappa(\alpha)$$

is a Boolean homomorphism of $\mathscr{P}(\alpha)$ *onto* $\mathfrak{B}(U_\kappa(\alpha))$ *whose associated filter is* $\mathscr{P}^\kappa(\alpha)$; *and* $\mathfrak{B}(U_\kappa(\alpha))$ *is isomorphic to* $\mathscr{P}(\alpha)/\mathscr{P}^\kappa(\alpha)$.

Proof. (a) That $U_\kappa(\alpha) \neq \emptyset$ is given by Lemma 7.2(b).

If $p \in \boldsymbol{\beta}(\alpha) \setminus U_\kappa(\alpha)$ then there is $A \in p$ such that $|A| < \kappa$. Then $\mathrm{cl}_{\boldsymbol{\beta}(\alpha)} A$ is an open-and-closed subset of $\boldsymbol{\beta}(\alpha)$ such that $p \in \mathrm{cl}_{\boldsymbol{\beta}(\alpha)} A$, and $\mathrm{cl}_{\boldsymbol{\beta}(\alpha)} A \cap U_\kappa(\alpha) = \emptyset$. Thus $U_\kappa(\alpha)$ is closed in $\boldsymbol{\beta}(\alpha)$, so $U_\kappa(\alpha)$ is compact.

(b) For every $p \in V$ there is $A_p \in \mathscr{P}(\alpha)$ such that

$$p \in \hat{A}_p = \mathrm{cl}_{\boldsymbol{\beta}(\alpha)} A_p \cap U_\kappa(\alpha) \subset V.$$

Thus $\{A_p : p \in V\}$ is an open cover of V and since V is compact (by part (a)) there are $n < \omega$ and elements $p_k \in V$ for $k \leq n$ such that $\bigcup_{k \leq n} \hat{A}_{p_k} = V$. Then with $A = \bigcup_{k \leq n} A_{p_k}$ we have $\hat{A} = V$.

(c) If $|A| < \kappa$ then $\mathrm{cl}_{\boldsymbol{\beta}(\alpha)} A \cap U_\kappa(\alpha) = \emptyset$ and thus $\hat{A} = \emptyset$. If $|A| \geq \kappa$ then $\hat{A} \neq \emptyset$ by Lemma 7.2(b) (with A replacing α).

(d) (i) follows from the fact that

$$\mathrm{cl}_{\boldsymbol{\beta}(\alpha)}(A \cap B) = \mathrm{cl}_{\boldsymbol{\beta}(\alpha)} A \cap \mathrm{cl}_{\boldsymbol{\beta}(\alpha)} B$$

for any (zero-) sets A and B of α (Theorem 2.7).

(ii) Let $\hat{A} \subset \hat{B}$ and assume that $|A \setminus B| \geq \kappa$; then by part (c) there is $p \in (A \setminus B)\hat{} \subset \hat{A}$; but $p \notin \hat{B}$ because

$$(A \setminus B)\hat{} \cap \hat{B} = [(A \setminus B) \cap B]\hat{} = \hat{\emptyset} = \emptyset$$

by (i).

Conversely if $|A \setminus B| < \kappa$, then

$$\hat{A} = (A \setminus B)\hat{} \cup (B \cap A)\hat{} = \emptyset \cup (B \cap A)\hat{} \subset B\hat{}$$

by (c) and (i).

(iii) follows from (ii).

Definition. The *local weight* of X at p, denoted $\chi(p,X)$ or $\chi(p)$, is the least cardinal equal to the cardinal number of a (filter) base for the neighborhoods of p.

We note that if $p\in Y\subset X$ then $\chi(p,Y)\le\chi(p,X)$. We note also that it follows from Lemma 2.4 (a) that if $\alpha\ge\omega$ and $p\in\boldsymbol{\beta}(\alpha)$, then $\chi(p,\boldsymbol{\beta}(\alpha))\le 2^\alpha$.

It follows from Corollary 7.8 (a) that if $p\in U(\alpha)$ then $\chi(p,U(\alpha))\ge\alpha^+$. Corollary 7.15 below, which implies the existence of uniform ultrafilters p on α such that $\chi(p,U(\alpha))=2^\alpha$, is essentially another application of the existence of families of large oscillation (Theorem 3.16 and Corollary 3.18). We also use the following general topological lemma.

7.13. Lemma. *Let $\alpha\ge\omega$ and let X and Y be compact spaces and f a continuous function from X onto Y. Let $\{V_\xi:\xi<\alpha\}$ be a family of neighborhoods of $p\in Y$ such that*

$$\text{if } I\in\mathscr{P}(\alpha) \text{ and } |I|=\omega \text{ then } \operatorname{int}(\textstyle\bigcap_{\xi\in I}V_\xi)=\emptyset\,.$$

Then there is $q\in f^{-1}(\{p\})$ such that $\chi(q,X)\ge\alpha$.

Proof. We set

$$U_I=\operatorname{int}_X(\textstyle\bigcap_{\xi\in I}f^{-1}(V_\xi)) \quad\text{for } I\in\mathscr{P}(\alpha)\,.$$

We note that if U is open in X and $f^{-1}(\{p\})\subset U$ then $p\notin f[X\backslash U]$ and hence $p\in\operatorname{int}f[U]$. Since

$$\begin{aligned}\operatorname{int}f[U_I]&=\operatorname{int}f[\operatorname{int}_X\textstyle\bigcap_{\xi\in I}f^{-1}(V_\xi)]\subset\operatorname{int}f[\bigcap_{\xi\in I}f^{-1}(V_\xi)]\\&=\operatorname{int}(\textstyle\bigcap_{\xi\in I}V_\xi)=\emptyset\,,\end{aligned}$$

it follows that

$$f^{-1}(\{p\})\not\subset U_I \quad\text{for } I\in\mathscr{P}(\alpha) \text{ and } |I|=\omega\,.$$

Thus the family

$$\{f^{-1}(\{p\})\backslash U_I: I\in\mathscr{P}(\alpha) \text{ and } |I|=\omega\}$$

is a family of non-empty compact subsets of X with the finite intersection property. Hence there is $q\in X$ such that

$$f(q)=p\,,\quad\text{and}$$

$$q\notin U_I \quad\text{for } I\in\mathscr{P}(\alpha) \text{ and } |I|=\omega\,.$$

To prove that $\chi(q,X)\ge\alpha$ we let $\mathscr{N}$ be a base for the neighborhoods of q and we set

$$F_U=\{\xi<\alpha: U\subset f^{-1}(V_\xi)\} \quad\text{for } U\in\mathscr{N}\,.$$

Since $q\notin U_I$ for $|I|=\omega$ it follows that

$$F_U\in\mathscr{P}_\omega(\alpha) \quad\text{for } U\in\mathscr{N}\,;$$

and since $\mathcal{N}$ is a base for the neighborhoods of q it follows that

$$\bigcup\{F_U : U \in \mathcal{N}\} = \alpha .$$

Thus $|\mathcal{N}| \geq \alpha$.

The following simple fact is frequently useful.

7.14. Lemma. *If* $\alpha \geq \omega$ *and* X *is a compact space with* $\boldsymbol{d}(X) \leq \alpha$ *then the Gleason space* $\mathsf{G}(X)$ *of* X *is homeomorphic to a subspace of* $U(\alpha)$.

Proof. We show

(i) $\mathsf{G}(X)$ is homeomorphic to a subspace of $\boldsymbol{\beta}(\alpha)$, and

(ii) $\boldsymbol{\beta}(\alpha)$ is homeomorphic to a subspace of $U(\alpha)$.

(i) Since $\boldsymbol{d}(\mathsf{G}(X)) = \boldsymbol{d}(X) \leq \alpha$ by Theorem 2.41(b) and $\mathsf{G}(X)$ is compact, there is a continuous function f from $\boldsymbol{\beta}(\alpha)$ onto $\mathsf{G}(X)$. If K is a closed subset of $\boldsymbol{\beta}(\alpha)$ such that $f|K$ is irreducible, then by the remark following the proof of Theorem 2.41 the function

$$f|K : K \to \mathsf{G}(X)$$

is a homeomorphism.

(ii) Let $\{A_\xi : \xi < \alpha\}$ be a family of subsets of α such that

$$|A_\xi| = \alpha \quad \text{for } \xi < \alpha , \quad \text{and}$$
$$A_\xi \cap A_\zeta = \emptyset \quad \text{for } \xi < \zeta < \alpha ,$$

and let $p_\xi \in (\mathrm{cl}_{\boldsymbol{\beta}(\alpha)} A_\xi) \cap U(\alpha)$ for $\xi < \alpha$. It is clear that the discrete set $\{p_\xi : \xi < \alpha\}$ is C^*-embedded in $\boldsymbol{\beta}(\alpha)$ (and *a fortiori* in $U(\alpha)$); hence $\mathrm{cl}_{U(\alpha)} \{p_\xi : \xi < \alpha\}$, a subset of $U(\alpha)$, is homeomorphic with $\boldsymbol{\beta}(\alpha)$.

7.15. Corollary. *If* $\alpha \geq \omega$ *then*

$$|\{p \in U(\alpha) : \boldsymbol{\chi}(p, U(\alpha)) = 2^\alpha\}| = 2^{2^\alpha} .$$

Proof. We denote by G the Gleason space of 2^{2^α}; it follows from Corollary 3.18 that $\boldsymbol{d}(2^{2^\alpha}) \leq \alpha$ and we have from Lemma 7.14 that G is (homeomorphic to) a subspace of $U(\alpha)$.

For $p \in 2^{2^\alpha}$ we set $V_\eta = \pi_\eta^{-1}(\{p_\eta\})$ for $\eta < 2^\alpha$ and we note that the family $\{V_\eta : \eta < 2^\alpha\}$ satisfies the conditions of Lemma 7.13 (with Y, α replaced by 2^{2^α}, 2^α, respectively). We denote by $\pi : G \to 2^{2^\alpha}$ the canonical projection. By Lemma 7.13 there is $q_p \in G$ such that

$$\pi(q_p) = p , \quad \text{and}$$
$$\boldsymbol{\chi}(q_p, G) = 2^\alpha ,$$

and *a fortiori* $\boldsymbol{\chi}(q_p, U(\alpha)) = 2^\alpha$. Thus we have

$$2^{2^\alpha} = |\{q_p : p \in 2^{2^\alpha}\}| \leq |\{p \in U(\alpha) : \boldsymbol{\chi}(p, U(\alpha)) = 2^\alpha\}| \leq 2^{2^\alpha} .$$

7.16. Lemma. *If* $\omega \leq \kappa \leq \alpha$ *then the space* $U_\kappa(\alpha)$ *does not have isolated elements.*

Proof. If V is a non-empty, open subset of $U_\kappa(\alpha)$ then there is $A \in \mathscr{P}(\alpha)$ such that

$$\hat{A} \subset V \quad \text{and}$$
$$|A| = \kappa .$$

Let A_0 and A_1 be subsets of A such that $A_0 \cap A_1 = \emptyset$ and $|A_0| = |A_1| = \kappa$. Then $\hat{A}_0 \cap \hat{A}_1 = \emptyset$, $\hat{A}_0 \neq \emptyset$, $\hat{A}_1 \neq \emptyset$ and $\hat{A}_0 \cup \hat{A}_1 \subset \hat{A}$ (using Lemma 7.12) and hence

$$1 < |\hat{A}| \leq |V| .$$

Thus $U_\kappa(\alpha)$ has no isolated elements.

7.17. Lemma. *Let $\mathscr{F}$ and $\mathscr{G}$ be filters on α such that $\mathscr{F} \subsetneqq \mathscr{G}$. Then there is an ultrafilter p on α such that*

$$\mathscr{F} \subset p \quad \text{and} \quad \mathscr{G} \not\subset p .$$

Proof. Let $A \in \mathscr{G} \setminus \mathscr{F}$. We claim that $\mathscr{F} \cup \{\alpha \setminus A\}$ has the finite intersection property. If not then there are $n < \omega$ and $A_k \in \mathscr{F}$ for $k \leq n$ such that

$$A_0 \cap \cdots \cap A_n \cap (\alpha \setminus A) = \emptyset ,$$

so that $A_0 \cap \cdots \cap A_n \subset A$. The contradiction establishes the claim.

There is by Lemma 7.1 an ultrafilter p on α such that $\mathscr{F} \cup \{\alpha \setminus A\} \subset p$. Clearly $\mathscr{F} \subset p$ and $\mathscr{G} \not\subset p$.

If $\mathscr{F}$ is a filter on α we define

$$F_{\mathscr{F}} = \{p \in \boldsymbol{\beta}(\alpha) : \mathscr{F} \subset p\} .$$

Thus if p is an ultrafilter then $F_p = \{p\}$; and according to Lemma 7.17 if $\mathscr{F}$ is not an ultrafilter then $|F_{\mathscr{F}}| \geq 2$.

7.18. Lemma. *Let α be a cardinal and let $\mathscr{F}$ and $\mathscr{G}$ be filters on α. Then*

(a) $F_{\mathscr{F}} = \bigcap \{\mathrm{cl}_{\beta(\alpha)} A : A \in \mathscr{F}\}$;

(b) *if* $\mathscr{F} \subsetneqq \mathscr{G}$ *then* $F_{\mathscr{G}} \subsetneqq F_{\mathscr{F}}$; *and*

(c) *if* $A \in \mathscr{P}(\alpha)$ *then*

$$A \in \mathscr{F} \quad \textit{if and only if} \quad F_{\mathscr{F}} \subset \mathrm{cl}_{\boldsymbol{\beta}(\alpha)} A .$$

Proof. (a) If $p \in F_{\mathscr{F}}$ then $A \in p$ whenever $A \in \mathscr{F}$, so

$$p \in \bigcap \{\mathrm{cl}_{\boldsymbol{\beta}(\alpha)} A : A \in \mathscr{F}\} .$$

Conversely if $p \in \bigcap \{\mathrm{cl}_{\boldsymbol{\beta}(\alpha)} A : A \in \mathscr{F}\}$ and $A \in \mathscr{F}$, then $A \in p$. Thus $\mathscr{F} \subset p$, so $p \in F_{\mathscr{F}}$.

(b) follows from Lemma 7.17.

(c) If $A\in\mathscr{F}$ then $F_{\mathscr{F}}\subset \mathrm{cl}_{\boldsymbol{\beta}(\alpha)}A$ from part (a). Conversely let $F_{\mathscr{F}}\subset \mathrm{cl}_{\boldsymbol{\beta}(\alpha)}A$ and suppose that $A\notin\mathscr{F}$. Then $\mathscr{F}\cup\{A\}$ has the finite intersection property, and with $\mathscr{G}=\langle\mathscr{F}\cup\{A\}\rangle$ we have $\mathscr{F}\subsetneqq\mathscr{G}$ and $F_{\mathscr{F}}=F_{\mathscr{G}}$. This contradicts (b) above.

We remark that if $\mathscr{F}$ is a filter on α and $f\in\beta^\alpha$ then the family

$$\mathscr{G}=\{f[A]:A\in\mathscr{F}\}$$

is a filter base on β. Indeed $\mathscr{G}\neq\emptyset$ since $f[\alpha]\in\mathscr{G}$; $\emptyset\notin\mathscr{G}$ because $f[A]\neq\emptyset$ when $A\in\mathscr{F}$; and if $n<\omega$ and $\{A_k:k\le n\}\subset\mathscr{F}$ then

$$\emptyset\neq f[\textstyle\bigcap_{k\le n}A_k]\subset\bigcap_{k\le n}f[A_k]\,.$$

We define the *direct image* $f[\mathscr{F}]$ of $\mathscr{F}$ under f to be the filter on β generated by the family

$$\{f[A]:A\in\mathscr{F}\}\,.$$

Thus

$$f[\mathscr{F}]=\{B\in\mathscr{P}(\beta):\text{there is }A\in\mathscr{F}\text{ such that }f[A]\subset B\}\,;$$

it is clear that $f[\mathscr{F}]=\{B\in\mathscr{P}(\beta):f^{-1}(B)\in\mathscr{F}\}$.

7.19. Lemma. (a) *If p is an ultrafilter on α and $f\in\beta^\alpha$ then $\bar{f}(p)=f[p]$.*

(b) *If $\mathscr{F}$ is a filter on α and $f\in\beta^\alpha$ then*

$$f[\mathscr{F}]=\{B\in\mathscr{P}(\beta):\bar{f}[F_{\mathscr{F}}]\subset \mathrm{cl}_{\boldsymbol{\beta}(\beta)}B\}\,.$$

Proof. (a) We first verify that $f[p]$ is an ultrafilter on β. Let $B\in\mathscr{P}(\beta)$ and assume that $B\notin f[p]$. Then $f^{-1}(B)\notin p$, so $\alpha\backslash f^{-1}(B)\in p$ because p is an ultrafilter on α. Hence

$$f[\alpha\backslash f^{-1}(B)]\in f[p]\subset\beta\backslash B.$$

We next prove that $f[p]\subset\bar{f}(p)$. If $B\in f[p]$, so that $f^{-1}(B)\in p$, then $p\in \mathrm{cl}_{\boldsymbol{\beta}(\alpha)}(f^{-1}(B))$ and therefore, because the function $\bar{f}$ is continuous, we have

$$\bar{f}(p)\in\bar{f}[\mathrm{cl}_{\boldsymbol{\beta}(\alpha)}(f^{-1}(B))]\subset \mathrm{cl}_{\boldsymbol{\beta}(\beta)}(\bar{f}[f^{-1}(B)])\subset \mathrm{cl}_{\boldsymbol{\beta}(\beta)}B\,;$$

thus $B\in\bar{f}(p)$.

Since $f[p]$ and $\bar{f}(p)$ are ultrafilters on β and $f[p]\subset\bar{f}(p)$, we have $\bar{f}(p)=f[p]$.

(b) If $B\in f[\mathscr{F}]$ then there is $A\in\mathscr{F}$ such that $f[A]\subset B$. From Lemma 7.18(a) we have $F_{\mathscr{F}}\subset \mathrm{cl}_{\boldsymbol{\beta}(\alpha)}A$, and hence from the continuity of $\bar{f}$

$$\bar{f}[F_{\mathscr{F}}]\subset\bar{f}[\mathrm{cl}_{\boldsymbol{\beta}(\alpha)}A]\subset \mathrm{cl}_{\boldsymbol{\beta}(\beta)}f[A]\subset \mathrm{cl}_{\boldsymbol{\beta}(\beta)}B\,.$$

For the converse let $B\in\mathscr{P}(\beta)$ be such that $\bar{f}[F_{\mathscr{F}}]\subset \mathrm{cl}_{\boldsymbol{\beta}(\beta)} B$. We are to show that $f^{-1}(B)\in\mathscr{F}$, and for this by Lemma 7.18(c) it is sufficient to prove that $F_{\mathscr{F}}\subset \mathrm{cl}_{\boldsymbol{\beta}(\beta)} f^{-1}(B)$. If $p\in F_{\mathscr{F}}$, so that $\bar{f}(p)\in\bar{f}[F_{\mathscr{F}}]\subset \mathrm{cl}_{\boldsymbol{\beta}(\beta)} B$ and $B\in\bar{f}(p)$, then $p\in \mathrm{cl}_{\boldsymbol{\beta}(\beta)} f^{-1}(B)$ since otherwise $p\in \mathrm{cl}_{\boldsymbol{\beta}(\beta)} f^{-1}(\beta\backslash B)$ and then

$$\emptyset=(\beta\backslash B)\cap B\in\bar{f}(p),$$

a contradiction. The proof of the lemma is complete.

Definition. Let α and β be cardinals, and let $\mathscr{F}$ and $\mathscr{G}$ be filters on α and β respectively. We define

$$\mathscr{F}\cdot\mathscr{G}=\{A\in\mathscr{P}(\alpha\times\beta):\{\xi<\alpha:\{\zeta<\beta:\langle\xi,\zeta\rangle\in A\}\in\mathscr{G}\}\in\mathscr{F}\}, \quad \text{and}$$
$$\mathscr{F}\times\mathscr{G}=\{A\in\mathscr{P}(\alpha\times\beta):\text{there are } B\in\mathscr{F},\ C\in\mathscr{G} \text{ such that } B\times C\subset A\}.$$

Recall that if X and Y are spaces then the projection functions

$$\pi_0: X\times Y\to X \quad \text{and } \pi_1: X\times Y\to Y$$

are given by $\pi_0(p,q)=p$ and $\pi_1(p,q)=q$.

7.20. Lemma. *Let $\mathscr{F}$ be a filter on α and $\mathscr{G}$ a filter on β. Then*
(a) *$\mathscr{F}\cdot\mathscr{G}$ and $\mathscr{F}\times\mathscr{G}$ are filters on $\alpha\times\beta$;*
(b) *$\mathscr{F}\times\mathscr{G}\subset\mathscr{F}\cdot\mathscr{G}$; and*
(c) *if $\beta\leq\alpha$ and $\mathscr{P}^{\alpha}(\alpha)\subset\mathscr{F}$ then $\mathscr{P}^{\alpha}(\alpha\times\beta)\subset\mathscr{F}\times\mathscr{G}$.*

Proof. (a) The verification that $\mathscr{F}\times\mathscr{G}$ is a filter is routine; we only show that $\mathscr{F}\times\mathscr{G}$ is closed under finite intersections. Let $n<\omega$ and let $\{A_k:k\leq n\}\subset\mathscr{F}\times\mathscr{G}$. Then for $k\leq n$ there are $B_k\in\mathscr{F}$ and $C_k\in\mathscr{G}$ such that $B_k\times C_k\subset A_k$, and with $B=\bigcap_{k\leq n}B_k$ and $C=\bigcap_{k\leq n}C_k$ we have $B\in\mathscr{F}$ and $C\in\mathscr{G}$ and $B\times C\subset\bigcap_{k\leq n}A_k$.

We next prove that $\mathscr{F}\cdot\mathscr{G}$ is a filter on $\alpha\times\beta$. It is clear that $\alpha\times\beta\in\mathscr{F}\cdot\mathscr{G}$ and that $\emptyset\notin\mathscr{F}\cdot\mathscr{G}$. If $A,B\in\mathscr{F}\cdot\mathscr{G}$ then from the equality

$$\{\zeta<\beta:\langle\xi,\zeta\rangle\in A\cap B\}=\{\zeta<\beta:\langle\xi,\zeta\rangle\in A\}\cap\{\zeta<\beta:\langle\xi,\zeta\rangle\in B\},$$

which is satisfied for $\xi<\alpha$, it follows that

$$\begin{aligned}&\{\xi<\alpha:\{\zeta<\beta:\langle\xi,\zeta\rangle\in A\cap B\}\in\mathscr{G}\}\\ &=\{\xi<\alpha:\{\zeta<\beta:\langle\xi,\zeta\rangle\in A\}\in\mathscr{G}\}\cap\{\xi<\alpha:\{\zeta<\beta:\langle\xi,\zeta\rangle\in B\}\in\mathscr{G}\}\in\mathscr{F},\end{aligned}$$

and thus $A\cap B\in\mathscr{F}\cdot\mathscr{G}$.

And if $A\in\mathscr{F}\cdot\mathscr{G}$ and $A\subset B\subset\alpha\times\beta$, then

$$\{\xi<\alpha:\{\zeta<\beta:\langle\xi,\zeta\rangle\in B\}\in\mathscr{G}\}\supset\{\xi<\alpha:\{\zeta<\beta:\langle\xi,\zeta\rangle\in A\}\in\mathscr{G}\}\in\mathscr{F}$$

and thus $B\in\mathscr{F}\cdot\mathscr{G}$.

(b) Let $B \in \mathscr{F}$ and $C \in \mathscr{G}$. Since

$$\begin{aligned}\{\zeta<\beta:\langle\xi,\zeta\rangle\in B\times C\} &= C \quad \text{if } \xi\in B\\ &= \emptyset \quad \text{if } \xi\notin B\end{aligned}$$

we have

$$\{\xi<\alpha:\{\zeta<\beta:\langle\xi,\zeta\rangle\in B\times C\}\in\mathscr{G}\}=B\in\mathscr{F},$$

and therefore $B\times C\in\mathscr{F}\cdot\mathscr{G}$. Statement (b) follows.

(c) Let $A\in\mathscr{P}_\alpha(\alpha\times\beta)$. Then $\pi_0[A]\in\mathscr{P}_\alpha(\alpha)$ and hence $\alpha\setminus\pi_0[A] \in\mathscr{P}^\alpha(\alpha)\subset\mathscr{F}$. Thus we have that

$$(\alpha\times\beta)\setminus A\supset(\alpha\setminus\pi_0[A])\times\beta\in\mathscr{F}\times\mathscr{G}.$$

The filter $\mathscr{F}\cdot\mathscr{G}$ is called the *product* of $\mathscr{F}$ and $\mathscr{G}$.

If α and β are cardinals we define the function

$$t:\beta\times\alpha\to\alpha\times\beta$$

by the rule $t(\zeta,\xi)=\langle\xi,\zeta\rangle$. Since t is a homeomorphism between (discrete) spaces, its unique Stone extension

$$\bar{t}:\boldsymbol{\beta}(\beta\times\alpha)\to\boldsymbol{\beta}(\alpha\times\beta)$$

is a homeomorphism.

Now let $\mathscr{F}$ and $\mathscr{G}$ be filters on α and β respectively. Then $\mathscr{G}\cdot\mathscr{F}$ is a filter on $\beta\times\alpha$ and therefore $t[\mathscr{G}\cdot\mathscr{F}]$ is a filter on $\alpha\times\beta$.

Explicitly we have

$$t[\mathscr{G}\cdot\mathscr{F}]=\{A\in\mathscr{P}(\alpha\times\beta):\{\zeta<\beta:\{\xi<\alpha:\langle\xi,\zeta\rangle\in A\}\in\mathscr{F}\}\in\mathscr{G}\}.$$

It is clear that $t[\mathscr{G}\times\mathscr{F}]=\mathscr{F}\times\mathscr{G}$ and thus from Lemma 7.20(b) we have

$$\mathscr{F}\times\mathscr{G}\subset t[\mathscr{G}\cdot\mathscr{F}].$$

In general $\mathscr{F}\cdot\mathscr{G}$ and $t[\mathscr{G}\cdot\mathscr{F}]$ are distinct filters even if $\mathscr{F}=\mathscr{G}$ and $\mathscr{F}$ is an ultrafilter (cf. Lemma 7.22 below).

If α and β are cardinals then we denote by

$$\boldsymbol{\phi}:\boldsymbol{\beta}(\alpha\times\beta)\to\boldsymbol{\beta}(\alpha)\times\boldsymbol{\beta}(\beta)$$

the Stone extension of the natural embedding (in fact the inclusion) of $\alpha\times\beta$ into $\boldsymbol{\beta}(\alpha)\times\boldsymbol{\beta}(\beta)$.

7.21. Lemma. *Let p be an ultrafilter on α and q an ultrafilter on β. Then*

(a) *$p\cdot q$ and $\bar{t}(q\cdot p)$ are ultrafilters on $\alpha\times\beta$;*

(b) *$\bar{\pi}_0(p\cdot q)=\bar{\pi}_0(\bar{t}(q\cdot p))=p$ and $\bar{\pi}_1(p\cdot q)=\bar{\pi}_1(\bar{t}(q\cdot p))=q$;*

(c) *$\boldsymbol{\phi}^{-1}(\{\langle p,q\rangle\})=\{r\in\boldsymbol{\beta}(\alpha\times\beta):p\times q\subset r\}$*
$=\{r\in\boldsymbol{\beta}(\alpha\times\beta):\bar{\pi}_0(r)=p$ and $\bar{\pi}_1(r)=q\}$; and

(d) *$\{p\cdot q,\bar{t}(q\cdot p)\}\subset\boldsymbol{\phi}^{-1}(\{\langle p,q\rangle\})$.*

Proof. (a) It is sufficient to prove that $p \cdot q$ is an ultrafilter on $\alpha \times \beta$. Let $A \in \mathscr{P}(\alpha \times \beta)$ and suppose that $A \notin p \cdot q$. We set

$$T_\xi = \{\zeta < \beta : \langle \xi, \zeta \rangle \in A\} \quad \text{for } \xi < \alpha \quad \text{and}$$
$$S = \{\xi < \alpha : T_\xi \in q\}.$$

According to the defining property of $p \cdot q$ we have $S \notin p$, so since p is an ultrafilter on α it follows that

$$\alpha \setminus S = \{\xi < \alpha : T_\xi \notin q\} \in p.$$

Because q is an ultrafilter on β we have $T_\xi \notin q$ if and only if $\{\zeta < \beta : \langle \xi, \zeta \rangle \in (\alpha \times \beta) \setminus A\} \in q$ for $\xi < \alpha$. Thus

$$\{\xi < \alpha : \{\zeta < \beta : \langle \xi, \zeta \rangle \in (\alpha \times \beta) \setminus A\} \in q\} \in p,$$

and therefore $(\alpha \times \beta) \setminus A \in p \cdot q$.

(b) According to Lemmas 7.19(a) and 7.20(b) and part (a) we have

$$\bar{\pi}_0(p \cdot q) = \pi_0[p \cdot q] \supset \pi_0[p \times q] = p \quad \text{and}$$
$$\bar{\pi}_1(p \cdot q) = \pi_1[p \cdot q] \supset \pi_1[p \times q] = q.$$

Hence $\bar{\pi}_0(p \cdot q) = p$ and $\bar{\pi}_1(p \cdot q) = q$.

Further since $\pi_0 \circ t = \pi_1 : \beta \times \alpha \to \alpha$ and $\pi_1 \circ t = \pi_0 : \beta \times \alpha \to \beta$ we have

$$\bar{\pi}_0(\bar{t}(q \cdot p)) = \bar{\pi}_1(q \cdot p) = p \quad \text{and } \bar{\pi}_1(\bar{t}(q \cdot p)) = \bar{\pi}_0(q \cdot p) = q.$$

(c) We have

$$\phi(\xi, \zeta) = \langle \xi, \zeta \rangle = \langle \pi_0(\xi, \zeta), \pi_1(\xi, \zeta) \rangle \quad \text{for } \langle \xi, \zeta \rangle \in \alpha \times \beta$$

and therefore

$$\boldsymbol{\phi}(r) = \langle \bar{\pi}_0(r), \bar{\pi}_1(r) \rangle \quad \text{for } r \in \boldsymbol{\beta}(\alpha \times \beta);$$

hence

$$\boldsymbol{\phi}^{-1}(\{\langle p, q \rangle\}) = \{r \in \boldsymbol{\beta}(\alpha \times \beta) : \bar{\pi}_0(r) = p, \bar{\pi}_1(r) = q\}.$$

We now claim that if $r \in \boldsymbol{\phi}^{-1}(\{\langle p, q \rangle\})$, then $p \times q \subset r$. We have that $\boldsymbol{\phi}(r) = \langle p, q \rangle$, and hence $\pi_0 \circ \boldsymbol{\phi}(r) = p$ and $\pi_1 \circ \boldsymbol{\phi}(r) = q$. Let $B \in p$ and $C \in q$. Since $\pi_0 \circ \phi(r) = \bar{\pi}_0(r) = \pi_0[r] = p$ and $\pi_1 \circ \phi(r) = \bar{\pi}_1(r) = \pi_1[r] = q$ (from Lemma 7.19(a)) it follows that there are $A_0, A_1 \in r$ such that

$$\pi_0[A_0] \subset B \quad \text{and } \pi_1[A_1] \subset C.$$

We set $A = A_0 \cap A_1$. Then $A \in r$ and $A \subset B \times C$, so $B \times C \in r$. It follows that $p \times q \subset r$.

If $p\times q\subset r\in\boldsymbol{\beta}(\alpha\times\beta)$ then from Lemma 7.19(a) we have

$$\bar{\pi}_0(r)=\pi_0[r]\supset\pi_0[p\times q]=p \quad\text{and}\quad \bar{\pi}_1(r)=\pi_1[r]\supset\pi_1[p\times q]=q\,;$$

hence

$$\{r\in\boldsymbol{\beta}(\alpha\times\beta):p\times q\subset r\}\subset\{r\in\boldsymbol{\beta}(\alpha\times\beta):\bar{\pi}_0(r)=p \quad\text{and}\quad \bar{\pi}_1(r)=q\}\,.$$

(d) follows from (b) and (c).

Let α be a cardinal number. We define

$$U_\alpha=\{\langle\xi,\zeta\rangle\in\alpha\times\alpha:\xi<\zeta\}\,,$$
$$L_\alpha=\{\langle\xi,\zeta\rangle\in\alpha\times\alpha:\zeta<\xi\}\,,\quad\text{and}$$
$$\Delta_\alpha=\{\langle\xi,\xi\rangle\in\alpha\times\alpha:\xi<\alpha\}\,.$$

We also define the *diagonal* function

$$d:\alpha\to\alpha\times\alpha$$

by $d(\xi)=\langle\xi,\xi\rangle$.

7.22. Lemma. *Let α be an infinite cardinal and let p and q be uniform ultrafilters on α. Then*

(a) $U_\alpha\in p\cdot q$ *and* $L_\alpha\in\bar{t}(q\cdot p)$; *thus* $p\times q$ *is not an ultrafilter and* $|\boldsymbol{\phi}^{-1}(\{\langle p,q\rangle\})|\geq 2$.

(b) $U_\alpha\in p\cdot p$, $L_\alpha\in\bar{t}(p\cdot p)$, *and* $\Delta_\alpha\in\bar{d}(p)$; *thus* $p\times p$ *is not an ultrafilter and* $|\boldsymbol{\phi}^{-1}(\{\langle p,p\rangle\})|\geq 3$.

Proof. (a) To prove that $U_\alpha\in p\cdot q$ we note that

$$|\alpha\backslash\{\zeta<\alpha:\langle\xi,\zeta\rangle\in U_\alpha\}|=|\xi+1|\in\mathscr{P}_\alpha(\alpha)\,;$$

thus since q is uniform we have

$$\{\zeta<\alpha:\langle\xi,\zeta\rangle\in U_\alpha\}\in q \quad\text{for } \xi<\alpha\,,$$

and hence

$$\{\xi<\alpha:\{\zeta<\alpha:\langle\xi,\zeta\rangle\in U_\alpha\}\in q\}=\alpha\in p\,.$$

In an analogous way, using the fact that p is uniform, it is proved that $L_\alpha\in\bar{t}(q\cdot p)$.

Because $L_\alpha\cap U_\alpha=\emptyset$ it follows that $p\cdot q\neq\bar{t}(q\cdot p)$. From Lemma 7.20(b) we have that $p\times q\subset p\cdot q$; furthermore $q\times p\subset q\cdot p$ and hence $p\times q=t[q\times p]\subset t[q\cdot p]=\bar{t}(q\cdot p)$ (using Lemmas 7.19(a) and 7.20(b)). Since $p\cdot q\neq\bar{t}(q\cdot p)$ and $p\times q\subset p\cdot q$ and $p\times q\subset\bar{t}(q\cdot p)$ it follows that $p\times q$ is not an ultrafilter. From Lemma 7.21(c) we have $|\boldsymbol{\phi}^{-1}(\{\langle p,q\rangle\})|\geq 2$.

(b) From part (a) we have $U_\alpha\in p\cdot p$ and $L_\alpha\in\bar{t}(p\cdot p)$, and (hence) that $p\times p$ is not an ultrafilter; furthermore $\Delta_\alpha=d[\alpha]\in d[p]=\bar{d}(p)$.

Since $L_\alpha \cap U_\alpha = L_\alpha \cap \Delta_\alpha = U_\alpha \cap \Delta_\alpha = \emptyset$ it follows that the three ultrafilters $p \cdot p$, $\bar{t}(p \cdot p)$, and $\bar{d}(p)$ are distinct. We note that $p \times p \subset \bar{d}(p)$, for if $B, C \in p$ then $B \cap C \in p$ and then, with $A = d[B \cap C]$, we have $A \in d[p] = \bar{d}(p)$ and $A \subset B \times C$. Thus from Lemma 7.21 (c) we have $|\phi^{-1}(\{\langle p,p \rangle\})| \geq 3$.

We remark that the inequalities given in the above lemma are in general the best possible. We prove in Theorem 16.4 below that if $\alpha = \beta = \omega$ and if $\omega^+ = 2^\omega$, then there are ultrafilters p and q on ω such that

$$|\phi^{-1}(\{\langle p,q \rangle\})| = 2 \quad \text{and}$$
$$|\phi^{-1}(\{\langle p,p \rangle\})| = |\phi^{-1}(\{\langle q,q \rangle\})| = 3 .$$

7.23. Theorem. *Let p be an ultrafilter on α and let q be an ultrafilter on β. Then the following are equivalent.*

(a) $|\phi^{-1}(\{\langle p,q \rangle\})| = 1$;

(b) *$p \times q$ is an ultrafilter on $\alpha \times \beta$;*

(c) *$\alpha \times \beta$ is C^*-embedded in $(\alpha \times \beta) \cup \{\langle p,q \rangle\}$; and*

(d) *for every family $\{C_\xi : \xi < \alpha\} \subset q$ there is $B \in p$ such that $\bigcap_{\xi \in B} C_\xi \in q$.*

Proof. (a)⇒(b) follows from Lemmas 7.17 and 7.21 (c).

(b)⇒(c) follows from the fact that if $p \times q$ is an ultrafilter on $\alpha \times \beta$ then $\phi[p \times q] = \phi(p \times q) = \langle p,q \rangle$.

(c)⇒(a) follows from the (characteristic) C^*-embedding property of the Stone-Čech compactification.

(b)⇒(d). Let $\{C_\xi : \xi < \alpha\} \subset q$. Then with

$$A = \{\langle \xi, \zeta \rangle \in \alpha \times \beta : \zeta \in C_\xi\}$$

we claim that $A \in p \times q$. If $D \in p$ and $E \in q$ then $(D \times (E \cap C_\xi)) \cap A \neq \emptyset$ and thus $(p \times q) \cup \{A\}$ has the finite intersection property; since $p \times q$ is an ultrafilter it follows that $A \in p \times q$, and hence there are $B \in p$ and $C \in q$ such that $B \times C \subset A$. Statement (d) follows from the inclusion $C \subset \bigcap_{\xi \in B} C_\xi$.

(d)⇒(b). Let $A \in \mathscr{P}(\alpha \times \beta)$. Since $p \cdot q$ is an ultrafilter on $\alpha \times \beta$ (Lemma 7.21 (a)) either $A \in p \cdot q$ or $(\alpha \times \beta) \setminus A \in p \cdot q$. With no loss of generality we assume that $A \in p \cdot q$ (and prove that $A \in p \times q$). We define

$$B_0 = \{\xi < \alpha : \pi_1[\pi_0^{-1}(\{\xi\}) \cap A] \in q\} ,$$

and we have $B_0 \in p$ by the definition of $p \cdot q$. Set

$$\begin{aligned} C_\xi &= \pi_1[\pi_0^{-1}(\{\xi\}) \cap A] && \text{if } \xi \in B_0 , \\ &= \beta && \text{if } \xi \in \alpha \setminus B_0 . \end{aligned}$$

Then $\{C_\xi : \xi < \alpha\} \subset q$ and by (d) there is $B_1 \in p$ such that

$$\bigcap_{\xi \in B_1} C_\xi \in q .$$

We set

$$B = B_0 \cap B_1, \quad \text{and}$$
$$C = \bigcap_{\xi \in B} C_\xi .$$

Then $B \in p$ and $C \supset \bigcap_{\xi \in B_1} C_\xi \in q$. Then

$$B \times C \in p \times q \quad \text{and}$$
$$B \times C \subset A,$$

so that $A \in p \times q$. This completes the proof.

7.24. Corollary. *Let p and q be ultrafilters on α and β, respectively. Then*

(a) *if q is α^--complete (in particular, if q is principal) then $p \times q$ is an ultrafilter; and*

(b) *if p and q are countably incomplete then $p \times q$ is not an ultrafilter.*

Proof. (a) It is sufficient to verify condition (d) of Theorem 7.23 above. For each family $\{C_\xi : \xi < \alpha\} \subset q$ we set $B = \alpha \in p$, and then

$$\bigcap_{\xi \in B} C_\xi \in q$$

because q is α^+-complete.

(b) Let p and q be countably incomplete ultrafilters, so there are families $\{A_n : n < \omega\}$ and $\{B_n : n < \omega\}$ such that

$$\alpha = A_0 \supset A_1 \supset \cdots \supset A_n \supset \cdots,$$
$$\beta = B_0 \supset B_1 \supset \cdots \supset B_n \supset \cdots,$$
$$A_n \in p \quad \text{and} \quad B_n \in q \quad \text{for } n < \omega, \quad \text{and}$$
$$\bigcap_{n < \omega} A_n = \bigcap_{n < \omega} B_n = \emptyset .$$

For $\xi < \alpha$ let $n(\xi)$ be the least $n < \omega$ such that $\xi \notin A_n$, so that $\{\xi < \alpha : n(\xi) = n\} = A_{n-1} \setminus A_n \notin p$ for $n \geq 1$. We set

$$C_\xi = \bigcap_{n < n(\xi)} B_n .$$

Then $C_\xi \in q$ and for each $B \in p$ we have

$$\bigcap_{\xi \in B} C_\xi = \bigcap_{n < \omega} B_n = \emptyset \notin q .$$

Thus condition (d) of Theorem 7.23 fails and we conclude that $p \times q$ is not an ultrafilter on $\alpha \times \beta$.

Notes for § 7

Theorem 7.1 is due to Tarski [30a] (Lemma 4).

The existence of a non-principal ultrafilter on ω implies [without using the axiom of choice] the existence of a Lebesgue non-measurable subset of ℝ (Sierpiński [38b], Semadeni [64]). [Therefore in the model of set theory in Solovay [70] in which every subset of ℝ is Lebesgue measurable, there is no non-principal ultrafilter on ω.]

The existence of a non-principal ultrafilter on ω has been used in measure theory, by Dieudonné [48], [51] to prove the Radon-Nikodým theorem and by Ionescu Tulcea and Ionescu Tulcea [69] (Chapter IV) in the theory of lifting.

Theorem 7.3 follows from Theorem 3.16; thus for $\kappa=\omega$ it can be considered a Corollary to the fundamental paper of Hausdorff [36]. Tarski [39] (Satz 3.19ff.) proves Theorem 7.3 and various related results. Corollary 7.4 is proved by Pospíšil [37], who shows the existence of a space X such that $|X|=2^{2^\alpha}$ and $\boldsymbol{d}(X)=\alpha$; for $\alpha=\omega$ a shorter similar argument is in Mrówka [59].

The disjoint refinement lemma (7.5) has been noticed independently by various authors, including Kuratowski [47] (Lemme 1), Erdős, Gillman and Henriksen [55] (Lemma 4.6), and Keisler [61] (Lemma 2.1). Corollary 7.6 is from Keisler [65b] (Lemma 7). Theorem 7.7(b) and Corollary 7.9 are from Keisler [64b] (Theorem 5.2 and Lemma 5B). Corollary 7.8(a) appears to have been generally known; see for example Gillman and Jerison [60] (4G. 2).

It follows from Theorem 7.8(a) that if α is an infinite cardinal and $\alpha^+=2^\alpha$ then $\chi(p,U(\alpha))=2^\alpha$ for every $p\in U(\alpha)$. For $\kappa=\omega$ the same conclusion follows from Martin's axion, according to Martin-Solovay [70] (page 154), and Ellentuck-Rucker [72] (Corollary 5). According to Kunen [72], however, it is consistent with ZFC that $\omega^+<2^\omega$ and that there is $p\in U(\omega)$ such that $\chi(p,U(\omega))=\omega^+$.

Regular ultrafilters arise naturally in the course of the proof of the compactness theorem using ultraproducts. See in this connection the abstract Morel, Scott and Tarski [55], Kochen [55] (Theorem 2.7) and Frayne, Morel and Scott [62] (Theorem 2.10). The proof we give of Theorem 11.7 makes use of the existence of a regular ultrafilter (Lemma 7.11(a)). An early explicit proof of the existence of regular ultrafilters is given by Gillman and Jerison [60] (Theorem 12.7), whose use is similar to that in Lemma 13.1 below.

Keisler [65a] (Question 3B) and Gillman (see Problem 82 of Erdős and Hajnal [71]) have asked whether every uniform ultrafilter on an (uncountable) α is regular. This has been answered in the affirmative for $\alpha\neq\omega^+$ assuming Gödel's axiom of constructibility $V=L$ by Prikry [70] (according to whom the statement that every uniform ultrafilter on ω^+ is regular is also consistent with $ZFC+\neg(V=L)$). Jensen (unpublished) extended the result of Prikry for $\alpha=\omega_n$ (the nth successor of ω). In the negative direction it has been noted by Silver (see Keisler [67a] (footnote to page 28)) and proved by Adler [71] that if α is measurable and uncountable then there is a countably incomplete uniform ultrafilter on α that is not regular.

The concept of a (κ,γ)-regular ultrafilter and Lemma 7.10 were given by Keisler [64a] and used to determine the cardinality of certain ultraproducts (cf. Theorem 12.15 below and its corollaries).

The case $\alpha=\kappa=\omega$ of Lemma 7.12 was proved by W. Rudin [56]. Lemma 7.14 follows the main idea of Negrepontis [67a]; it is used for the proof of Corollary 12.4 below. Lemma 7.13, Corollary 7.15 and their proofs are from Pospíšil [39] (Theorem II). Juhász [67], [69] gives two proofs of Corollary 7.15 and in addition establishes the following:

$$|\{p\in\boldsymbol{\beta}(\alpha): 2^{\chi(p,\boldsymbol{\beta}(\alpha))}<2^{2^\alpha}\}|<2^{2^\alpha}; \quad \text{and}$$

$$\text{assuming GCH,} \quad |\{p\in\boldsymbol{\beta}(\alpha): \chi(p,\boldsymbol{\beta}(\alpha))=2^\beta\}|=\alpha^\beta\cdot 2^{2^\alpha}$$

for $\omega\leq\beta\leq\alpha$.

Kunen [72] (Theorem 2.8) also proves the existence of a uniform ultrafilter on α with local character 2^α.

The filter multiplication $\mathscr{F}\cdot\mathscr{G}$ was introduced (with different notation) to prove Lemma 15.20 below by Kochen [61] (Theorem 6.7) and Frayne, Morel and Scott [62] (Theorem 1.10). Our notation follows Blass [70], who is responsible for most of the results in 7.21—7.23. Specifically Lemmas 7.21, 7.22, Theorem 7.23 (b)$\Leftrightarrow$(d) appear in Blass [70] (Corollary 20.1, Corollary 20.12, Theorem 3.6, respectively).

§ 8. Large Cardinals

We study in this section some elementary aspects of the theory of large cardinals, with emphasis on properties of the weakly compact and measurable cardinals. Both these classes of cardinals are defined by conditions concerning the existence of certain ultrafilters. Some elementary knowledge of such cardinals is needed later on (in §§ 14, 15). In view of the extensive recent developments in this area, our study of these cardinals for their own interest is far from complete and in fact rather modest; on the other hand it is more than sufficient for our later uses.

Much of this section is devoted to characterizations of weakly compact cardinals. (Among those needed later on are: α is ramifiable; $(2^\alpha)_\alpha$ is α-compact.) These characterizations are organized in two groups, each centering around the concept of a strongly inaccessible ramifiable cardinal. The *combinatorial* equivalences (8.1—8.10) include the Ramsey property and the Hausdorff property. Here we develop the rudiments of the Erdős-Rado partition calculus and we prove the "arrow relation"

$$\beth_n(\alpha)^+ \to (\alpha^+)_\alpha^{n+1}$$

(Corollary 8.10), which for $n=1$ has already been applied in § 3 in connection with Souslin numbers. The *Boolean-algebraic* and *topological* equivalences include analogues of Alexander's subbase property, Tychonoff's product property, the α-compactness of $(2^\alpha)_\alpha$, the extendability of α-complete filters in α-generated fields of sets of cardinality α, and others. The results are summarized in Theorem 8.23. Another topological characterization is given in Corollary 15.7 below. The long list of equivalences in Theorem 8.23 does not include, however, some of the most interesting properties, such as the compactness theorem for certain infinitary languages.

Concerning measurable cardinals α we here anticipate the ultraproduct construction (given in § 11) and for $\alpha > \omega$ we establish the fundamental fact on the existence of normal ultrafilters on α (Scott's Theorem 8.26). Normal ultrafilters are Ramsey (Theorem 8.27); this fact is used to establish the combinatorial property $\alpha \to (\alpha)_2^{<\omega}$ of uncountable measurable cardinals α (Erdős-Hajnal; Corollary 8.28), and the immense size of measurable cardinals (Hanf-Scott, Rowbottom; Theorem 8.30). Ramsey ultrafilters receive a close look in the next section, in connection with the Rudin-Keisler partial order of ultrafilters (Theorem 9.6).

The final part of this section deals with the Chang-Kunen-Prikry result on descendingly incomplete ultrafilters (Theorem 8.35), a result evocative of large cardinals.

Weakly Compact Cardinals: Combinatorial Equivalences

The arrow notation. Let α, κ, and λ be cardinals and $n<\omega$. The *arrow* notation

$$\alpha \to (\kappa)^n_\lambda$$

denotes the following partition relation: if

$$[\alpha]^n = \bigcup_{i<\lambda} P_i$$

then there are $A \in \mathscr{P}(\alpha)$ and $i<\lambda$ such that

$$|A| = \kappa, \quad \text{and}$$
$$[A]^n \subset P_i\,.$$

The set A is *homogeneous* for the family $\{P_i : i<\lambda\}$. Clearly there is no loss of generality in assuming that

$$P_i \cap P_j = \emptyset \quad \text{for } i<j<\lambda\,.$$

The following fact is obvious: if $\alpha' \geq \alpha$, $\kappa' \leq \kappa$, $\lambda' \leq \lambda$, and $\alpha \to (\kappa)^n_\lambda$ then $\alpha' \to (\kappa')^n_{\lambda'}$.

We sometimes indicate that $\alpha \to (\kappa)^n_\lambda$ fails by writing $\alpha \nrightarrow (\kappa)^n_\lambda$.

Definition. An infinite cardinal α is a *Ramsey* cardinal if $\alpha \to (\alpha)^2_2$.

Definition. An infinite cardinal α is a *Hausdorff* cardinal if for every linear order $\preccurlyeq$ on α there is a subset of α of cardinality α which is either well-ordered or anti-well-ordered by $\preccurlyeq$.

We note that ω is a Hausdorff cardinal. Indeed an infinite countable linearly ordered set is either well-ordered (in which case by Theorem 1.13 it is isomorphic as an ordered set to an infinite countable ordinal) or it is not well-ordered (in which case by Theorem 1.16 it contains an infinite decreasing sequence of elements).

8.1. Theorem. *A Ramsey cardinal is a Hausdorff cardinal.*

Proof. Let α be a Ramsey cardinal and let $\preccurlyeq$ be a linear order on α.

We set

$$P_0=\{\{\xi,\zeta\}: \text{either } \xi<\zeta<\alpha \text{ and } \xi\preccurlyeq\zeta, \text{ or } \zeta<\xi<\alpha \text{ and } \zeta\preccurlyeq\xi\}, \text{ and}$$

$$P_1=\{\{\xi,\zeta\}: \text{either } \xi<\zeta<\alpha \text{ and } \zeta\preccurlyeq\xi, \text{ or } \zeta<\xi<\alpha \text{ and } \xi\preccurlyeq\zeta\}.$$

Thus P_0 is the set of unordered pairs of distinct elements of α on which the well-ordering of α agrees with the linear order $\preccurlyeq$, and P_1 is the set of such pairs on which the two orderings disagree. We have

$$[\alpha]^2=P_0\cup P_1,$$

and since α is a Ramsey cardinal there is $A\in[\alpha]^\alpha$ such that either $[A]^2\subset P_0$ or $[A]^2\subset P_1$. If $[A]^2\subset P_0$ then A is well-ordered by $\preccurlyeq$, and if $[A]^2\subset P_1$ then A is anti-well-ordered by $\preccurlyeq$.

8.2. Theorem. *A Hausdorff cardinal is strongly inaccessible.*

Proof. We let α be a Hausdorff cardinal and we prove the following two statements.

(i) α is a regular cardinal.

Suppose that $\mathrm{cf}(\alpha)<\alpha$ and let $\{A_\xi:\xi<\mathrm{cf}(\alpha)\}$ be a family of subsets of α such that

$$|A_\xi|<\alpha \quad \text{for } \xi<\mathrm{cf}(\alpha),$$
$$A_\xi\cap A_\zeta=\emptyset \quad \text{for } \xi<\zeta<\mathrm{cf}(\alpha), \quad \text{and}$$
$$\alpha=\textstyle\bigcup_{\xi<\mathrm{cf}(\alpha)}A_\xi.$$

We define a (binary) relation $\preccurlyeq$ on α by the following rule:

$\eta\prec\eta'$ if and only if either there are $\xi<\xi'<\mathrm{cf}(\alpha)$ such that $\eta\in A_\xi$ and $\eta'\in A_{\xi'}$ or there is $\xi<\mathrm{cf}(\alpha)$ such that $\eta,\eta'\in A_\xi$ and $\eta'<\eta$.

It is clear that $\preccurlyeq$ is a linear order on α. If A is a subset of α well-ordered by $\preccurlyeq$ then, because $\preccurlyeq$ is an anti-well-ordering on A_ξ for $\xi<\mathrm{cf}(\alpha)$, we have $|A\cap A_\xi|<\omega$ for $\xi<\mathrm{cf}(\alpha)$ and hence

$$|A|=\textstyle\sum_{\xi<\mathrm{cf}(\alpha)}|A\cap A_\xi|\le\omega\cdot\mathrm{cf}(\alpha)<\alpha;$$

if A is anti-well-ordered, then there is $\zeta<\mathrm{cf}(\alpha)$ such that the greatest element of A (in the order $\preccurlyeq$) is an element of A_ζ, and we have $A\subset\bigcup_{\xi\le\zeta}A_\xi$ and hence

$$|A|\le\textstyle\sum_{\xi\le\zeta}|A_\xi|<\alpha.$$

Thus α is not a Hausdorff cardinal, a contradiction.

(ii) α is a strong limit cardinal.

Let β be a cardinal such that $\alpha \leq 2^\beta$. Then there is a subset A of $\Lambda(\beta)$ such that $|A| = \alpha$ (where, as in § 5, $\Lambda(\beta)$ denotes the set 2^β ordered lexicographically). Since A is linearly ordered (in the order induced by $\Lambda(\beta)$) and α is a Hausdorff cardinal there is a subset B of A such that

$$|B| = \alpha, \quad \text{and}$$
$$B \text{ is either well-ordered or anti-well-ordered.}$$

According to Theorem 5.4 a well-ordered or anti-well-ordered subset of $\Lambda(\beta)$ is of cardinality at most β. Thus $\alpha = |B| \leq \beta$.

The proof of the theorem is complete.

Definitions. A *ramification system* is a partially ordered set $\mathfrak{A} = \langle A, \preccurlyeq \rangle$ such that

$$\text{if } A \neq \emptyset \text{ then } A \text{ has a least element, and}$$
$$P(a) \text{ is well-ordered for } a \in A$$

(where as in § 1

$$P(a) = \{b \in A : b \preccurlyeq a \text{ and } b \neq a\} \quad \text{for } a \in A).$$

The *order* of a is the order-type of $P(a)$ for $a \in A$; the *order of the ramification system* $\mathfrak{A} = \langle A, \preccurlyeq \rangle$ is the least ordinal ξ such that

$$\text{the order of } P(a) \text{ is less than } \xi \text{ for } a \in A.$$

We set

$$\mathscr{A}_\xi = \{a \in A : \text{order of } P(a) \text{ is } \xi\} \quad \text{for an ordinal } \xi.$$

An infinite cardinal α is a *ramifiable* cardinal if for every ramification system $\mathfrak{A} = \langle A, \preccurlyeq \rangle$ such that

$$\text{the order of } \mathfrak{A} \text{ is } \alpha \text{ and}$$
$$|\mathscr{A}_\xi| < \alpha \quad \text{for } \xi < \alpha,$$

there is a subset B of A such that

$$|B| = \alpha \quad \text{and}$$
$$B \text{ is well-ordered by } \preccurlyeq.$$

Definition. Let $\alpha \geq \omega$ and let $\mathfrak{A} = \langle \alpha, \preccurlyeq \rangle$ be a ramification system (of order κ).

For $\eta \leq \xi < \kappa$ and $a \in \mathscr{A}_\xi$ we denote by $a|\eta$ the unique element of $\mathfrak{A}$ such that

$$a|\eta \preccurlyeq a, \quad \text{and}$$
$$a|\eta \in \mathscr{A}_\eta .$$

We note that if $\zeta,\eta<\kappa$, $a\in\mathscr{A}_\xi$, $b\in\mathscr{A}_\eta$ and the relations $a\preccurlyeq b$ and $b\preccurlyeq a$ both fail, then there is a unique ordinal $\xi(a,b)$ such that

$$a|\xi(a,b)\neq b|\xi(a,b) \quad \text{and}$$
$$\text{if } \zeta<\xi(a,b) \quad \text{then } a|\zeta=b|\zeta\,;$$

it is clear that $\xi(a,b)\leq\xi$ and $\xi(a,b)\leq\eta$.

The *Brouwer-Kleene lexicographic order* $\lll$ on $\mathfrak{A}$ is the binary relation defined as follows:

$$a\lll b \quad \text{if and only if}$$

(i) either $a\preccurlyeq b$,

(ii) or $a\not\preccurlyeq b$, $b\not\preccurlyeq a$, and $a|\xi(a,b)<b|\xi(a,b)$ (where $<$ denotes the natural well-ordering of α).

We set

$$R(a)=\{b\in\mathfrak{A}: a\preccurlyeq b\} \quad \text{for } a\in A.$$

8.3. Lemma. *Let $\alpha\geq\omega$ and let $\mathfrak{A}=\langle\alpha,\preccurlyeq\rangle$ be a ramification system. The Brouwer-Kleene lexicographic order $\lll$ on $\mathfrak{A}$ is a linear order such that*

$$R(a) \textit{ is an interval of } \langle\mathfrak{A},\lll\rangle \textit{ for } a\in\alpha.$$

Proof. We first prove that $\lll$ is a partial order.

The reflexive property $a\lll a$ is clear since $a\preccurlyeq a$ for $a\in A$.

For the anti-symmetric property let $a,b\in\mathfrak{A}$ and $a\lll b$ and $b\lll a$. There are two possibilities to consider.

Case 1. $a\preccurlyeq b$. Since $b\lll a$ we have either

$$b\preccurlyeq a, \quad \text{or}$$
$$b\not\preccurlyeq a, \quad a\not\preccurlyeq b, \quad \text{and } b|\xi(a,b)<a|\xi(a,b).$$

Thus in fact $b\preccurlyeq a$ and we have $a=b$ from the anti-symmetry property of $\preccurlyeq$.

Case 2. $a\not\preccurlyeq b$. Then we must have $b\not\preccurlyeq a$ and $a|\xi(a,b)<b|\xi(a,b)$. Since $b\not\preccurlyeq a$ and $b\lll a$ we have also $b|\xi(a,b)<a|\xi(a,b)$, a contradiction.

To prove that $\lll$ is a transitive relation let $a,b,c\in\mathfrak{A}$ with $a\lll b$ and $b\lll c$. There are four possibilities to consider.

Case 1. $a\preccurlyeq b$ and $b\preccurlyeq c$. Then $a\preccurlyeq c$ and hence $a\lll c$.

Case 2. $a\preccurlyeq b$ and $b\not\preccurlyeq c$. If $a\preccurlyeq c$ we have $a\lll c$, so we assume $a\not\preccurlyeq c$. If $c\preccurlyeq a$ then from $a\preccurlyeq c$ we have $a=c$, a contradiction; hence $c\not\preccurlyeq a$. And if $c\preccurlyeq b$ then from $c\lll b\lll c$ we have $b=c$, contradicting the assumption $b\not\preccurlyeq c$; thus $c\not\preccurlyeq b$. Then $\xi(a,c)$ and $\xi(b,c)$ are defined, $\xi(a,c)=\xi(b,c)$, and we have

$$a|\xi(a,c)=b|\xi(a,c)=b|\xi(b,c)<c|\xi(b,c)=c|\xi(a,c),$$

so that $a\lll c$.

Case 3. $a \not\preccurlyeq b$ and $b \preccurlyeq c$. If $a \preccurlyeq c$ we have $a \ll c$, so we assume $a \not\preccurlyeq c$. If $b \preccurlyeq a$ then from $b \ll a \ll b$ we have $a = b$, contradicting the assumption $a \not\preccurlyeq b$; thus $b \not\preccurlyeq a$. And if $c \preccurlyeq a$ then from $b \preccurlyeq c$ we have $b \preccurlyeq a$, a contradiction; thus $c \not\preccurlyeq a$. Then $\xi(a,c)$ and $\xi(b,c)$ are defined, $\xi(a,c) = \xi(b,c)$, and we have

$$a|\xi(a,c) = a|\xi(a,b) < b|\xi(a,b) = c|\xi(a,b) = c|\xi(a,c),$$

so that $a \ll c$.

Case 4. $a \not\preccurlyeq b$ and $b \not\preccurlyeq c$. If $a \preccurlyeq c$ we have $a \ll c$, so we assume $a \not\preccurlyeq c$. We note (as in *Cases* 1 and 2) that if $b \preccurlyeq a$ then $a = b$; if $c \preccurlyeq b$ then $b = c$; and if $c \preccurlyeq a$ then $a = b$. Thus $b \not\preccurlyeq a$, $c \not\preccurlyeq b$ and $c \not\preccurlyeq a$, and $\xi(a,b)$, $\xi(b,c)$ and $\xi(a,c)$ are all defined. We consider three cases.

(i) $\xi(a,b) = \xi(b,c)$. Then $\xi(a,b) = \xi(a,c)$ and we have

$$a|\xi(a,c) = a|\xi(a,b) < b|\xi(a,b) = b|\xi(b,c) < c|\xi(b,c) = c|\xi(a,c),$$

so that $a \ll c$.

(ii) $\xi(a,b) < \xi(b,c)$. Then $\xi(a,b) = \xi(a,c)$ and we have

$$a|\xi(a,c) = a|\xi(a,b) < b|\xi(a,b) = c|\xi(a,b) = c|\xi(a,c),$$

so that $a \ll c$.

(iii) $\xi(b,c) < \xi(a,b)$. Then $\xi(a,c) = \xi(b,c)$ and we have (as in (ii))

$$a|\xi(a,c) = a|\xi(b,c) = b|\xi(b,c) < c|\xi(b,c) = c|\xi(a,c),$$

so that $a \ll c$.

We next prove that $\ll$ is a linear order on α. Let $a, b \in \alpha$. If $a \preccurlyeq b$ or $b \preccurlyeq a$ then $a \ll b$ or $b \ll a$, and if $a \not\preccurlyeq b$ and $b \not\preccurlyeq a$ then $\xi(a,b)$ is defined and we have

$$\begin{aligned} a \ll b \quad &\text{if } a|\xi(a,b) < b|\xi(a,b), \quad \text{and} \\ b \ll a \quad &\text{if } b|\xi(a,b) < a|\xi(a,b). \end{aligned}$$

We finally prove that

$$R(a) \text{ is an interval of } \langle \mathfrak{A}, \ll \rangle.$$

We let $b, c, d \in \mathfrak{A}$ be such that

$$\begin{aligned} &b \ll c \ll d, \quad \text{and} \\ &a \preccurlyeq b, \quad a \preccurlyeq d \end{aligned}$$

and we prove that $a \preccurlyeq c$. Suppose that $a \not\preccurlyeq c$. We claim that

$$c \text{ is not comparable in the order } \preccurlyeq \text{ with } a, b, \text{ or } d.$$

Indeed

$a \not\preccurlyeq c$ by assumption;
$c \not\preccurlyeq a$ since if $c \preccurlyeq a$ then $c \leqslant a \leqslant c$ and hence $c = a$;
$b \not\preccurlyeq c$ since if $b \preccurlyeq c$ then $a \preccurlyeq c$;
$c \not\preccurlyeq b$ since if $c \preccurlyeq b$ then a and c are comparable in the order $\preccurlyeq$;
$d \not\preccurlyeq c$ since if $d \preccurlyeq c$ then from $d \leqslant c \leqslant d$ we have $d = c$ and hence $a \preccurlyeq c$; and
$c \not\preccurlyeq d$ since if $c \preccurlyeq d$ then a and c are comparable in the order $\preccurlyeq$.

It follows that $\xi(a,c)$, $\xi(b,c)$ and $\xi(d,c)$ are defined, $\xi(a,c) = \xi(b,c) = \xi(d,c)$, and (since $a \leqslant c \leqslant d$) that

$$a|\xi(a,c) < c|\xi(a,c) = c|\xi(d,c) < d|\xi(d,c);$$

but since $a \preccurlyeq d$ we have

$$a|\xi(a,c) = d|\xi(a,c) = d|\xi(d,c),$$

a contradiction. The proof is complete.

8.4. Theorem. *A Hausdorff cardinal is ramifiable.*

Proof. Let α be a Hausdorff cardinal and let $\mathfrak{A} = \langle A, \preccurlyeq \rangle$ be a ramification system of order α such that

$$|\mathscr{A}_\xi| < \alpha \quad \text{for } \xi < \alpha.$$

We assume without loss of generality that $A = \alpha$. Let $\leqslant$ be the Brouwer-Kleene linear order of $\mathfrak{A}$. Since α is a Hausdorff cardinal there is a subset B of α such that

$|B| = \alpha$, and
B is either well-ordered or anti-well-ordered by $\leqslant$.

We assume without loss of generality that

either $\langle B, \leqslant \rangle$ is order-isomorphic to α or $\langle B, \geqslant \rangle$ is order-isomorphic to α.

We claim that

$$\left|\{a \in \mathscr{A}_\xi : |R(a) \cap B| = \alpha\}\right| = 1 \quad \text{for } \xi < \alpha.$$

We first prove that $\{a \in \mathscr{A}_\xi : |R(a) \cap B| = \alpha\} \neq \emptyset$ for $\xi < \alpha$. Suppose on the contrary that there is $\xi < \alpha$ such that $|R(a) \cap B| < \alpha$ for $a \in \mathscr{A}_\xi$. Then from

$$B = \bigcup_{\zeta \le \xi} (B \cap \mathscr{A}_\zeta) \cup \bigcup \{R(a) \cap B : a \in \mathscr{A}_\xi\},$$

we have

$$|B| \le \sum_{\zeta \le \xi} |\mathscr{A}_\zeta| + \sum_{a \in \mathscr{A}_\xi} |R(a) \cap B| < \alpha,$$

since α is regular and $|\mathscr{A}_\zeta| < \alpha$ for $\zeta < \alpha$.

Next we claim that $|\{a \in \mathscr{A}_\xi : |R(a) \cap B| = \alpha\}| \leq 1$. Let $\xi < \alpha$ and let $a, a' \in \mathscr{A}_\xi$, and

$$|R(a) \cap B| = |R(a') \cap B| = \alpha .$$

Since $\ll$ is a linear order, we assume without loss of generality that $a \ll a'$.

If $\langle B, \ll \rangle$ is order-isomorphic to α then $R(a) \cap B$ is cofinal in $\langle B, \ll \rangle$ since $|R(a) \cap B| = \alpha$, and thus there is $b \in R(a) \cap B$ such that $a' \ll b$; thus $a \ll a' \ll b$, and it follows from Lemma 8.3 that $a' \in R(a)$, *i.e.*, $a' \preccurlyeq a$. Since $a, a' \in \mathscr{A}_\xi$ we have $a = a'$. The case that $\langle B, \gg \rangle$ is order-isomorphic to α is analogous (with the rôles of a and a' reversed). The proof of the claim is complete.

Let $a_\xi \in \mathscr{A}_\xi$ be such that $|R(a_\xi) \cap B| = \alpha$ for $\xi < \alpha$ and define

$$C = \{a_\xi : \xi < \alpha\} .$$

If $\xi < \zeta < \alpha$ then $R(a_\zeta) \cap B \subset R(a_\zeta | \xi) \cap B$, so that $|R(a_\zeta | \xi) \cap B| = \alpha$ and from the claim we have $a_\zeta | \xi = a_\xi$, *i.e.*, $a_\xi \prec a_\zeta$.

Thus α is ramifiable and the proof is complete.

8.5. Corollary *(D. König). ω is a ramifiable cardinal.*

Here is a direct proof of the classical theorem of D. König. Let $\langle A, \preccurlyeq \rangle$ be a ramification system of order ω and such that $|\mathscr{A}_n| < \omega$ for $n < \omega$. We proceed recursively: if $n < \omega$ and $a_0, \ldots, a_{n-1}$ have been defined we let $a_n \in \mathscr{A}_n$ be such that

$$a_{n-1} \prec a_n, \quad \text{and}$$
$$|\{a \in A : a_n \prec a\}| = \omega .$$

Then $\{a_n : n < \omega\}$ is an infinite subset of A which is well-ordered by $\preccurlyeq$.

The Aronszajn and Specker theorem (Corollary 14.32) states that if $\alpha = \alpha^{\underline{\alpha}} \geq \omega$ then α^+ is not a ramifiable cardinal. Thus in particular ω^+ is not ramifiable.

The following theorem is at the heart of the partition calculus. On the one hand it relates partitions to ramification systems (and thus it is of help in the proof that every strongly inaccessible ramifiable cardinal is a Ramsey cardinal (Corollary 8.7)). On the other hand, starting with a partition of $[\alpha]^{n+1}$ it produces a subset A of α such that the partition class of each element of $[A]^{n+1}$ is determined by its first n elements. Thus the partition relations derived from this process ultimately depend on some form of classical pigeon-hole principle of Dedekind (occurring for $n = 1$).

If $\mathfrak{A} = \langle A, \preccurlyeq \rangle$ is a ramification system, $\xi < \alpha$, and $a \in \mathscr{A}_\xi$ we set

$$\mathscr{S}(a) = \{b \in \mathscr{A}_{\xi+1} : a \preccurlyeq b\} ,$$

the set of *immediate successors* of a.

8.6. Theorem. *Let* $1 \leq \lambda \leq \lambda + \omega \leq \kappa \leq \alpha$, $n < \omega$ *and assume that every ramification system* $\mathfrak{A}$ *(of subsets of* α *under reverse set inclusion) of cardinality* α *such that*

$$\text{if } a \in \mathscr{A}_\xi \text{ then } |\mathscr{S}(a)| \leq \lambda^{|\xi+n|^n} \text{ for } \xi < \alpha$$

contains a chain of cardinality κ. *Let*

$$[\alpha]^{n+1} = \bigcup_{\eta < \lambda} P_\eta \quad \text{and}$$
$$P_\eta \cap P_{\eta'} = \emptyset \quad \text{for } \eta < \eta' < \lambda .$$

Then there is $A \in \mathscr{P}(\alpha)$ *such that*

$|A| = \kappa$, *and*

if $F, G \in [A]^{n+1}$ *and* $F \setminus \{\max F\} = G \setminus \{\max G\}$ *then there is (a unique)* $\eta < \lambda$ *such that* $F, G \in P_\eta$.

Proof. By transfinite recursion we define (using notation whose correctness will become evident) for $\xi < \alpha$ a family $\mathscr{A}_\xi$ of subsets of α, and for $A \in \mathscr{A}_\xi$ a subset M_A of α and an equivalence relation $\underset{A}{\equiv}$ on $A \setminus M_A$, such that

(i) $\mathscr{A}_0 = \{\alpha\}$,

(ii) $\mathscr{A}_\xi = \{\bigcap_{\zeta < \xi} A_\zeta : A_\zeta \in \mathscr{A}_\zeta$ for $\zeta < \xi$ and $\bigcap_{\zeta < \xi} A_\zeta \neq \emptyset\}$ for non-zero limit ordinals $\xi < \alpha$,

(iii) if $A \in \mathscr{A}_\xi$ then

$$M_A = n \cup \{\min B : A \subset B, B \in \bigcup_{\zeta \leq \xi} \mathscr{A}_\zeta\} \quad \text{for } \xi < \alpha ,$$

(iv) if $A \in \mathscr{A}_\xi$ and $\zeta, \zeta' \in A \setminus M_A$ then $\zeta \underset{A}{\equiv} \zeta'$ if

$$F \cup \{\zeta\} \in P_\eta \quad \text{if and only if } F \cup \{\zeta'\} \in P_\eta$$

for $F \in [M_A]^n$ and $\eta < \lambda$, and

(v) $\mathscr{A}_{\xi+1} = \{B : B$ is an $\underset{A}{\equiv}$-equivalence class for some $A \in \mathscr{A}_\xi\}$ for $\xi < \alpha$.

It is trivial to prove by transfinite induction that

$\mathscr{A}_\xi$ is a family of pairwise disjoint non-empty subsets of α for $\xi < \alpha$, and

if $A \in \mathscr{A}_\zeta$ and $\xi < \zeta < \alpha$ then there is $B \in \mathscr{A}_\xi$ such that $A \subset B$.

For $\xi < \alpha$ and $A \in \mathscr{A}_\xi$ we set

$$\mathscr{S}(A) = \{B \in \mathscr{A}_{\xi+1} : B \subset A\} ,$$

and we prove by transfinite induction that

$$\bigcup \mathscr{S}(A) = A \setminus (n \cup \{\min A\}) .$$

If $\xi<\alpha$ then there is a unique $A_\zeta \in \mathscr{A}_\zeta$ such that $A \subset A_\zeta$ for $\zeta<\xi$; by the inductive assumption

$$\bigcup \mathscr{S}(A_\zeta) = A_\zeta \setminus (n \cup \{\min A_\zeta\}) \quad \text{for } \zeta<\xi\,,$$

and $M_A = (\bigcup_{\zeta<\xi} M_{A_\zeta}) \cup \{\min A\}$ by (iii). Thus

$$\bigcup \mathscr{S}(A) = A \setminus M_A = A \setminus (n \cup \{\min A\})\,.$$

It follows that

(*) $\qquad \bigcup \mathscr{S}(A) = A \setminus \{\min A\} \quad \text{for } A \in \mathscr{A}_\xi \quad \text{and } 0<\xi<\alpha\,.$

We define

$$\mathfrak{A} = \bigcup_{\xi<\alpha} \mathscr{A}_\xi\,.$$

It is clear that $\mathfrak{A}$ is a ramification system (of subsets of α under reverse set inclusion). We further define $\phi: \mathfrak{A} \to \alpha$ by $\phi(A) = \min A$. It follows from (*) that ϕ is a one-to-one function. Furthermore if $\xi<\zeta<\alpha$, $A \in \mathscr{A}_\xi$, $B \in \mathscr{A}_\zeta$ and $B \subset A$, then $A \neq B$, since $\min A \in A$ and $\min A \notin B$; thus $\min A < \min B$ and ϕ is order-preserving.

It follows that $\mathscr{A}_\xi$ is the set of elements of $\mathfrak{A}$ of order ξ for $\xi<\alpha$, and in fact if $A \in \mathscr{A}_\xi$ then M_A is a subset of α order-isomorphic to the ordinal sum $(n-1)+(\xi+1)$. We next note that an $\underset{A}{\equiv}$-equivalence class is determined by a function from $[M_A]^n$ to λ. In fact if $\zeta \in A \setminus \{\min A\}$ define $\eta_{A,\zeta}: [M_A]^n \to \lambda$ by the condition $F \cup \{\zeta\} \in P_{\eta_{A,\zeta}(F)}$ for $F \in [M_A]^n$; then $\zeta \underset{A}{\equiv} \zeta'$ if and only if $\eta_{A,\zeta} = \eta_{A,\zeta'}$ for $\zeta, \zeta' \in A \setminus \{\min A\}$. From these remarks it follows that

$$|\mathscr{S}(A)| \leq \lambda^{|[M_A]^n|} \leq \lambda^{|\xi+n|^n}\,.$$

We note finally that

$$|\mathfrak{A}| = \alpha\,.$$

Indeed if $\mathscr{A}_\xi \neq \emptyset$ for $\xi<\alpha$ this statement is obvious, and if $\mathscr{A}_\xi = \emptyset$ for some $\xi<\alpha$ then $\phi[\mathfrak{A}] = \alpha \setminus \{1, \ldots, n-1\}$.

By the assumption of the theorem there is a family $\{A_\xi : \xi<\kappa\}$ such that $A_\xi \in \mathscr{A}_\xi$ for $\xi<\kappa$, and $A_{\xi'} \subset A_\xi$ for $\xi<\xi'<\kappa$. We set $a_\xi = \min A_\xi$ for $\xi<\kappa$, we define

$$A = \{a_\xi : \xi<\kappa\}\,,$$

and we note that $|A| = \kappa$ since the function ϕ is one-to-one. We claim that if $F, G \in [A]^{n+1}$ and $F \setminus \{\max F\} = G \setminus \{\max G\}$ then there is (a unique) $\eta<\lambda$ such that $F, G \in P_\eta$. Let

$$a_{\xi_0} < \cdots < a_{\xi_{n-1}} < a_\xi \quad \text{and}$$
$$a_{\xi_0} < \cdots < a_{\xi_{n-1}} < a_{\xi'}$$

be the natural order of the elements of F and G respectively (as subsets of α), and assume without loss of generality that $\xi \leq \xi'$. Then $A_{\xi'} \subset A_{\xi_{n-1}}$ and hence $a_\xi, a_{\xi'} \in A_\xi$, so that $a_\xi \overline{\overline{\ \ }}_{\xi \bar{A}_{\xi_{n-1}}} a_{\xi'}$ (by (v)); the claim now follows from (iv). The proof is complete.

The power of the theorem just proved is revealed by the following corollaries.

8.7. Corollary. *If α is a strong limit and ramifiable cardinal then $\alpha \to (\alpha)^n_\beta$ for $n < \omega$ and $\beta < \mathrm{cf}(\alpha)$.*

Proof. $\alpha \to (\alpha)^0_\beta$ is vacuously true and $\alpha \to (\alpha)^1_\beta$ holds for $\beta < \mathrm{cf}(\alpha)$.

We let $\beta < \mathrm{cf}(\alpha)$, $n < \omega$, we assume that $\alpha \to (\alpha)^n_\beta$ holds, and we prove that $\alpha \to (\alpha)^{n+1}_\beta$ holds. It is easy to see (from the fact that α is a strong limit cardinal) that if $\mathfrak{A}$ is a ramification system of cardinality α such that

$$\text{if } a \in \mathscr{A}_\xi \quad \text{then } |\mathscr{S}(a)| \leq \beta^{|\xi+n|^n} \quad \text{for } \xi < \alpha,$$

then the order of $\mathfrak{A}$ is at least α. Hence the conditions of Theorem 8.6 are satisfied with α, α, and β replacing α, κ, and λ, respectively. Let $[\alpha]^{n+1} = \bigcup_{\eta < \beta} P_\eta$, where $P_\eta \cap P_{\eta'} = \emptyset$ for $\eta < \eta' < \beta$. Then there is $A \subset \alpha$ such that

$|A| = \alpha$, and

(*) if $F, G \in [A]^{n+1}$ and $F \setminus \{\max F\} = G \setminus \{\max G\}$ then there is (a unique) $\eta < \beta$ such that $F, G \in P_\eta$.

We define

$$Q_\eta = \{H \in [A]^n : H \cup \{\xi\} \in P_\eta \text{ for } \max H < \xi < \alpha\}$$

for $\eta < \beta$. Then $[A]^n = \bigcup_{\eta < \beta} Q_\eta$ by (*). By the recursive assumption there are $B \subset A$ and $\bar{\eta} < \beta$ such that

$$|B| = \alpha \quad \text{and}$$

$$[B]^n \subset Q_{\bar{\eta}}.$$

Then $[B]^{n+1} \subset P_{\bar{\eta}}$.

8.8. Corollary *(Ramsey). $\omega \to (\omega)^n_m$ for $n, m < \omega$.*

Proof. From D. König's Theorem (8.5) and Corollary 8.7.

8.9. Corollary *(The stepping-up lemma of Erdős-Rado). Let $1 \leq \lambda \leq \omega + \lambda \leq \kappa \leq \alpha \leq \beta$ and $n < \omega$, and assume that*

$$\alpha \to (\kappa)^n_\lambda \quad \textit{and}$$

$$\textstyle\sum_{\xi < \alpha} \lambda^{|\xi|} < \beta.$$

Then

$$\beta \to (\kappa)^{n+1}_\lambda.$$

Proof. We first note that $\beta > \alpha$. If $\mathfrak{A}$ is a ramification system of cardinality β such that

$$\text{if } a \in \mathscr{A}_\xi \quad \text{then } |\mathscr{S}(a)| \leq \lambda^{|\xi+n|^n} \quad \text{for } \xi < \beta ,$$

then it is easy to prove inductively that

$$|\mathscr{A}_\xi| \leq (\lambda + \omega)^{|\xi|} \quad \text{for } \xi < \alpha .$$

Hence every such system $\mathfrak{A}$ contains a chain of cardinality α; indeed if $\mathscr{A}_\alpha \neq \emptyset$ there is nothing to prove and if $\mathscr{A}_\xi = \emptyset$ for $\alpha \leq \xi$ then

$$|\mathfrak{A}| = |\textstyle\bigcup_{\xi<\alpha} \mathscr{A}_\xi| \leq \sum_{\xi<\alpha} (\lambda+\omega)^{|\xi|} < \beta ,$$

a contradiction.

Now let $[\beta]^{n+1} = \bigcup_{\eta<\lambda} P_\eta$ with $P_\eta \cap P_{\eta'} = \emptyset$ for $\eta < \eta' < \lambda$. From Theorem 8.6, with β, α, and λ replacing α, κ, and λ, respectively, there is $A \in \mathscr{P}(\beta)$ such that

$$|A| = \alpha, \quad \text{and}$$

$$(*) \qquad \text{if } F, G \in [A]^{n+1} \quad \text{and } F \setminus \{\max F\} = G \setminus \{\max G\}$$

then there is (a unique) $\eta < \lambda$ such that $F, G \in P_\eta$.

We define

$$Q_\eta = \{H \in [A]^n : H \cup \{\xi\} \in P_\eta \text{ for } \max H < \xi < \alpha\}$$

for $\eta < \lambda$. Then $[A]^n = \bigcup_{\eta<\lambda} Q_\eta$ by (*). From the assumption $\alpha \to (\kappa)^n_\lambda$ it follows that there are $B \subset A$ and $\bar{\eta} < \lambda$ such that

$$|B| = \kappa, \quad \text{and}$$

$$[B]^n \subset Q_{\bar{\eta}} .$$

It is clear that $[B]^{n+1} \subset P_{\bar{\eta}}$ and the proof is complete.

Definition. Let α be a cardinal. We define

$$\beth_0(\alpha) = \alpha, \quad \text{and}$$

$$\beth_\xi(\alpha) = \sum_{\zeta<\xi} 2^{\beth_\zeta(\alpha)} \quad \text{for non-zero ordinals } \xi .$$

Thus in particular $\beth_{\xi+1}(\alpha) = 2^{\beth_\xi(\alpha)}$ for ordinals ξ. We note that $\beth_\xi = \beth_\xi(\omega)$ for every ordinal ξ.

8.10. Corollary *(Erdős-Rado). Let α be an infinite cardinal and $n < \omega$. Then*

$$\beth_n(\alpha)^+ \to (\alpha^+)^{n+1}_\alpha ;$$

in particular $(2^\alpha)^+ \to (\alpha^+)^2_\alpha$.

Proof. We proceed by recursion on n. The statement $\alpha^+ \to (\alpha^+)^1_\alpha$ holds because α^+ is a regular cardinal. We assume now that $0 < n < \omega$ and

$\beth_{n-1}(\alpha)^+ \to (\alpha^+)^n_\alpha$, and we prove that $\beth_n(\alpha)^+ \to (\alpha^+)^{n+1}_\alpha$. We apply Corollary 8.9 with $\beth_n(\alpha)^+$, $\beth_{n-1}(\alpha)^+$, α^+, and α replacing β, α, κ, and λ, respectively. It is sufficient to prove that

$$\sum_{\xi < \beth_{n-1}(\alpha)^+} \alpha^{|\xi|} < \beth_n(\alpha)^+ .$$

Indeed we have

$$\begin{aligned}\sum_{\xi < \beth_{n-1}(\alpha)^+} \alpha^{|\xi|} &\le \beth_{n-1}(\alpha)^+ \cdot \alpha^{\beth_{n-1}(\alpha)} = \beth_{n-1}(\alpha)^+ \cdot 2^{\beth_{n-1}(\alpha)} \\ &= 2^{\beth_{n-1}(\alpha)} = \beth_n(\alpha) < \beth_n(\alpha)^+ .\end{aligned}$$

The partition relation $(2^\alpha)^+ \to (\alpha^+)^2_\alpha$ has been used in § 3 to obtain an upper bound for the Souslin number of product spaces.

Weakly Compact Cardinals: Boolean-Algebraic and Topological Equivalences

Definitions. A filter $\mathscr{F}$ of a Boolean algebra $\mathfrak{B}$ is *α-complete* if for every subset C of $\mathscr{F}$ of cardinality less than α, $\bigwedge C$ exists and is an element of $\mathscr{F}$. In particular, $\mathfrak{B}$ is *α-complete* if for every subset C of $\mathfrak{B}$ of cardinality less than α, $\bigwedge C$ exists and is an element of $\mathfrak{B}$.

Let $\phi: \mathfrak{A} \to \mathfrak{B}$ be a homomorphism of α-complete Boolean algebras, and let $\beta < \alpha$ and $a_\xi \in \mathfrak{A}$ for $\xi < \beta$. Then it is clear that

$$\phi(\textstyle\bigwedge^{\mathfrak{A}}_{\xi<\beta} a_\xi) \le \bigwedge^{\mathfrak{B}}_{\xi<\beta} \phi(a_\xi) .$$

If for every such subset $\{\alpha_\xi : \xi < \beta\}$ of A we have

$$\phi(\textstyle\bigwedge^{\mathfrak{A}}_{\xi<\beta} a_\xi) = \bigwedge^{\mathfrak{B}}_{\xi<\beta} \phi(a_\xi)$$

then ϕ is an *α-homomorphism* (*from* $\mathfrak{A}$ *to* $\mathfrak{B}$). An *α-embedding* is an embedding which is an α-homomorphism.

8.11. Lemma. *Let $\alpha \ge \omega$, let $\mathfrak{B}$ be an α-complete Boolean algebra, and let T be a non-empty set of α-complete ultrafilters on $\mathfrak{B}$. The function*

$$\phi : \mathfrak{B} \to \mathscr{P}(T)$$

given by

$$\phi(a) = \{p \in T : a \in p\}$$

is an α-homomorphism.

Proof. It is simple to verify directly that h is a homomorphism (cf. Theorem 2.17(a)). Let $\beta < \alpha$, $\{a_\xi : \xi < \beta\} \subset \mathfrak{B}$, and $a = \bigwedge_{\xi<\beta} a_\xi$. It is sufficient to prove that $\phi(a) = \bigcap_{\xi<\beta} \phi(a_\xi)$. The inclusion $\phi(a) \subset \bigcap_{\xi<\beta} \phi(a_\xi)$ is true for any Boolean homomorphism ϕ. If $p \in \bigcap_{\xi<\beta} \phi(a_\xi)$ then $a_\xi \in p$

for $\xi<\beta$ and, since p is α-complete, it follows that $a=\bigwedge_{\xi<\beta}a_\xi\in p$. Hence $p\in\phi(a)$, and the reverse inclusion is proved.

A field of sets $\mathscr{S}$ is an *α-complete field of sets* if $\mathscr{S}$ is closed under the operation of intersection of a subfamily of $\mathscr{S}$ of cardinality less than α; *i.e.*, if $\bigcap_{\xi<\beta}A_\xi\in\mathscr{S}$ whenever $\beta<\alpha$ and $A_\xi\in\mathscr{S}$ for $\xi<\beta$. An α-complete Boolean algebra is *α-representable* if it is isomorphic to an α-complete field of sets.

8.12. Lemma. *Let $\alpha\geq\omega$ and let $\mathfrak{B}$ be an α-complete Boolean algebra. Then $\mathfrak{B}$ is α-representable if and only if every non-zero element of $\mathfrak{B}$ is an element of an α-complete ultrafilter of $\mathfrak{B}$.*

Proof. If $\mathfrak{B}$ is α-representable and a is a non-zero element of $\mathfrak{B}$ then there is $p\in a$; the ultrafilter of $\mathfrak{B}$ determined by p, *i.e.*, the family $\{b\in\mathfrak{B}:p\in b\}$, is α-complete and contains a.

Conversely let P_α denote the set of all α-complete ultrafilters on $\mathfrak{B}$ and let $\phi:\mathfrak{B}\to\mathscr{P}(P_\alpha)$ be the α-homomorphism given by Lemma 8.11. Then ϕ is an embedding, for if $a\in\mathfrak{B}$ and $a\neq\mathbb{0}$ then by the assumption there is $p\in P_\alpha$ such that $a\in p$, so that $p\in\phi(a)\neq\mathbb{0}$. Thus ϕ is an α-embedding of $\mathfrak{B}$ into the α-complete field of sets $\mathscr{P}(P_\alpha)$.

We recall that if $\mathscr{F}$ is a filter of a Boolean algebra $\mathfrak{B}$ then the canonical homomorphism $\pi:\mathfrak{B}\to\mathfrak{B}/\mathscr{F}$ associated with $\mathscr{F}$ is given by $\pi(a)=\{b\in\mathfrak{B}:a\underset{\mathscr{F}}{=}b\}$.

8.13. Lemma. *Let $\alpha\geq\omega$ and let $\mathscr{F}$ be a filter of an α-complete Boolean algebra $\mathfrak{B}$. Then $\mathscr{F}$ is an α-complete filter if and only if the canonical homomorphism associated with $\mathscr{F}$ is an α-homomorphism.*

Proof. If π is an α-homomorphism, $\beta<\alpha$, and $a_\xi\in\mathscr{F}$ for $\xi<\beta$, then we set $a=\bigwedge^{\mathfrak{B}}_{\xi<\beta}a_\xi$ and we have

$$\pi(a)=\bigwedge_{\xi<\beta}\pi(a_\xi)=\mathbb{1}\,,$$

so that $a\in\mathscr{F}$. Thus $\mathscr{F}$ is an α-complete filter.

If $\mathscr{F}$ is an α-complete filter, $\beta<\alpha$, and $a_\xi\in\mathfrak{B}$ for $\xi<\beta$, then we set $a=\bigwedge_{\xi<\beta}a_\xi$ and we clearly have $\pi(a)\leq\pi(a_\xi)$ for $\xi<\beta$. We prove that $\pi(a)$ is the infimum in $\mathfrak{B}/\mathscr{F}$ of the set $\{\pi(a_\xi):\xi<\beta\}$. Suppose then that $\pi(b)\leq\pi(a_\xi)$ so that

$$\pi(b\wedge a_\xi)=\pi(b)\wedge\pi(a_\xi)=\pi(b)\quad\text{for }\xi<\beta\,.$$

Then $b\wedge a_\xi\underset{\mathscr{F}}{=}b$, so there is $c_\xi\in\mathscr{F}$ such that

$$(*)\qquad b\wedge a_\xi\wedge c_\xi=b\wedge c_\xi\quad\text{for }\xi<\beta\,.$$

We set $c=\bigwedge_{\xi<\beta}c_\xi$ and we have $c\in\mathscr{F}$, and

$$b\wedge a\wedge c=b\wedge c$$

from (*). Thus $b \wedge a \underset{\mathscr{F}}{=} b$, so that

$$\pi(b) \wedge \pi(a) = \pi(b \wedge a) = \pi(b)$$

and $\pi(b) \leq \pi(a)$. It follows that $\pi(a) = \bigwedge \{\pi(a_\xi) : \xi < \beta\}$.

Definition. Let α be an infinite cardinal and let $\mathfrak{B}$ be an α-complete Boolean algebra. Then $\mathfrak{B}$ is *α-distributive* if the following condition is satisfied: If

$$|I| < \alpha,$$
$$B_i \subset \mathfrak{B} \text{ for } i \in I, \text{ and}$$
$$|B_i| < \alpha \quad \text{for } i \in I,$$

then $\bigvee \{\bigwedge_{i \in I} b_i : b_i \in B_i \text{ for } i \in I\}$ exists and

$$\bigvee \{\bigwedge_{i \in I} b_i : b_i \in B_i \text{ for } i \in I\} = \bigwedge_{i \in I} (\bigvee B_i).$$

It is clear that an α-complete Boolean algebra $\mathfrak{B}$ is α-distributive if and only if it satisfies the following condition: if $|I| < \alpha$, $|J_i| < \alpha$ for $i \in I$, $b_{i,j} \in \mathfrak{B}$ for $i \in I$ and $j \in J_i$, and we set $K = \prod_{i \in I} J_i$, then

$$(\bigvee_{f \in K} \bigwedge_{i \in I} b_{i,f(i)} \text{ exists and})$$
$$\bigvee_{f \in K} \bigwedge_{i \in I} b_{i,f(i)} = \bigwedge_{i \in I} \bigvee_{j \in J_i} b_{i,j}.$$

We remark that if $\mathfrak{B}$ is α-distributive, $\mathbb{0} \neq a \in \mathfrak{B}$, $|I| < \alpha$, $B_i \subset \mathfrak{B}$, $|B_i| < \alpha$, and $\bigvee B_i = a$ for $i \in I$, then

$$\bigvee \{\bigwedge_{i \in I} b_i : b_i \in B_i \text{ for } i \in I \text{ and } \bigwedge_{i \in I} b_i \neq \mathbb{0}\} = a.$$

It is clear that if a Boolean algebra is α-representable then it is α-distributive.

8.14. Lemma. *Let $\alpha \geq \omega$ and let $\mathscr{F}$ be a $(2^\alpha)^+$-complete filter on a $(2^\alpha)^+$-complete, α^+-distributive Boolean algebra $\mathfrak{B}$. Then $\mathfrak{B}/\mathscr{F}$ is α^+-distributive.*

Proof. Let $\pi : \mathfrak{B} \to \mathfrak{B}/\mathscr{F}$ be the canonical homomorphism associated with $\mathscr{F}$. We let $|I| \leq \alpha$, $|J_i| \leq \alpha$ for $i \in I$, $b_{i,j} \in \mathfrak{B}/\mathscr{F}$ for $i \in I$ and $j \in J_i$, and $K = \prod_{i \in I} J_i$ and we prove that

$$\bigvee_{f \in K} \bigwedge_{i \in I} b_{i,f(i)} = \bigwedge_{i \in I} \bigvee_{j \in J_i} b_{i,j}.$$

Let $a_{i,j} \in \mathfrak{B}$ be such that $\pi(a_{i,j}) = b_{i,j}$ for $i \in I$, $j \in J_i$. Since $\mathfrak{B}$ is α^+-distributive

$$\bigvee_{f \in K} \bigwedge_{i \in I} a_{i,f(i)} = \bigwedge_{i \in I} \bigvee_{j \in J_i} a_{i,j}.$$

We note that $|K| \leq \alpha^\alpha = 2^\alpha$. Since π is a $(2^\alpha)^+$-homomorphism from Lemma 8.13, the required equality follows.

An α-complete Boolean algebra $\mathfrak{B}$ is *weakly α-representable* if $\mathfrak{B}$ is isomorphic to the quotient of an α-complete field of sets by an α-complete filter, *i.e.*, if there is an α-complete field of sets $\mathscr{S}$ and an α-homomorphism ϕ of $\mathscr{S}$ onto $\mathfrak{B}$.

8.15. Corollary. *If α is a strongly inaccessible cardinal, then every weakly α-representable Boolean algebra is α-distributive.*

We recall that the Stone space of $\mathscr{P}(2^\alpha)/\mathscr{P}^{\alpha^+}(2^\alpha)$ is the space $U_{\alpha^+}(2^\alpha)$ of α^+-uniform ultrafilters on 2^α. Concerning the distributivity properties of these Boolean algebras we have the following result.

8.16. Theorem. *Let α be an infinite cardinal.*

(a) *The Boolean algebra $\mathscr{P}(2^\alpha)/\mathscr{P}^{\alpha^+}(2^\alpha)$ is not α^+-distributive.*

(b) *If α is strongly inaccessible then the Boolean algebra $\mathscr{P}(\alpha)/\mathscr{P}^\alpha(\alpha)$ is α-distributive.*

(c) *Assume the generalized continuum hypothesis and let α be regular. Then α is (strongly) inaccessible if and only if the Boolean algebra $\mathscr{P}(\alpha)/\mathscr{P}^\alpha(\alpha)$ is α-distributive.*

Proof. (a) We replace 2^α by the set α^α of all functions from α to α, we set $\mathfrak{B}=\mathscr{P}(\alpha^\alpha)/\mathscr{P}^{\alpha^+}(\alpha^\alpha)$ and we denote by π the canonical homomorphism associated with $\mathscr{P}^{\alpha^+}(\alpha^\alpha)$. For $\xi,\zeta<\alpha$ we set

$$A_{\xi,\zeta}=\{f\in\alpha^\alpha : f(\xi)=\zeta\}$$

so that $\bigcup_{\zeta<\alpha}A_{\xi,\zeta}=\alpha^\alpha$ for $\xi<\alpha$ and hence $\bigcap_{\xi<\alpha}\bigcup_{\zeta<\alpha}A_{\xi,\zeta}=\alpha^\alpha$. Since, by Lemma 8.13, π is an α^+-homomorphism, we have

$$(*) \qquad \bigwedge_{\xi<\alpha}\Big(\bigvee_{\zeta<\alpha}\pi(A_{\xi,\zeta})\Big)=\mathbb{1}\,.$$

Furthermore for $f\in\alpha^\alpha$ we have

$$\bigcap_{\xi<\alpha}A_{\xi,f(\xi)}=\{g\in\alpha^\alpha : g(\xi)=f(\xi) \text{ for each } \xi<\alpha\}=\{f\}\,,$$

so that $\bigcap_{\xi<\alpha}A_{\xi,f(\xi)}\in\mathscr{P}_{\alpha^+}(\alpha^\alpha)$ and hence

$$\bigwedge_{\xi<\alpha}\pi(A_{\xi,f(\xi)})=\pi\Big(\bigcap_{\xi<\alpha}A_{\xi,f(\xi)}\Big)=\mathbb{0}\,.$$

Thus we have

$$(**) \qquad \bigvee_{f\in\alpha^\alpha}\bigwedge_{\xi<\alpha}\pi(A_{\xi,f(\xi)})=\mathbb{0}\,.$$

It follows from (*) and (**) that $\mathfrak{B}$ is not α^+-distributive.

(b) follows from Corollary 8.15.

(c) follows from (a) and (b).

Definitions. Let $\mathfrak{B}$ be a Boolean algebra.

(i) If $\mathscr{A}\subset\mathfrak{B}$ and $c\in\mathfrak{B}$ then $\mathscr{A}$ is a *partition* of c if

$$\bigvee\mathscr{A}=c, \quad\text{and}$$

$$a_0\wedge a_1=\mathbb{0} \quad\text{for } a_0,a_1\in\mathscr{A} \quad\text{and } a_0\neq a_1\,.$$

A *partition* of $\mathfrak{B}$ is a partition of $\mathbb{1}$.

(ii) If $\mathscr{A}_0, \mathscr{A}_1$ are partitions of $\mathfrak{B}$ then $\mathscr{A}_1$ is a *refinement* of $\mathscr{A}_0$ (denoted $\mathscr{A}_0 \preccurlyeq \mathscr{A}_1$) if for $a_1 \in \mathscr{A}_1$ there is (a unique) $a_0 \in \mathscr{A}_0$ such that $a_1 \leq a_0$.

(iii) If $\{\mathscr{A}_i : i \in I\}$ is a family of partitions of $\mathfrak{B}$ then

$$\bigwedge_{i\in I} \mathscr{A}_i = \{\bigwedge_{i\in I} a_i : a_i \in \mathscr{A}_i \text{ for } i \in I \text{ and } \bigwedge_{i\in I} a_i \text{ exists}\}.$$

8.17. Lemma. *Let α be a strongly inaccessible cardinal, let $\mathfrak{B}$ be an α-distributive Boolean algebra of cardinality α, let $\langle b_\xi : \xi < \alpha \rangle$ be a well-ordering of $\mathfrak{B}$, let $c \in \mathfrak{B}$ and $c \neq \mathbb{0}$ and suppose that there is no atom $a \in \mathfrak{B}$ such that $a \leq c$. Then there is a family $\{\mathscr{A}_\xi : \xi < \alpha\}$ of subsets $\mathfrak{B}$ such that*

(i) $\mathscr{A}_0 = \{c\}$,

(ii) $\mathscr{A}_\xi = \bigwedge_{\zeta<\xi} \mathscr{A}_\zeta$ *for non-zero limit ordinals* $\xi < \alpha$,

(iii) $\mathscr{A}_\zeta \prec \mathscr{A}_\xi$ *for* $\zeta < \xi < \alpha$,

(iv) $\{c \wedge b_\xi, c \setminus b_\xi\} \prec \mathscr{A}_{\xi+1}$ *for* $\xi < \alpha$,

(v) $\mathscr{A}_\xi$ *is a partition of c consisting of non-zero elements,*

(vi) $2 \leq |\{b \in \mathscr{A}_{\xi+1} : b \leq a\}| \leq 4$ *for* $a \in \mathscr{A}_\xi$, *and*

(vii) $|\mathscr{A}_\xi| < \alpha$ *for* $\xi < \alpha$.

Proof. Using the fact that there is no atom $a \leq c$, we define a function

$$\phi : \{b \in \mathfrak{B} : b \leq c\} \to \mathfrak{B}$$

such that

$$\phi(\mathbb{0}) = \mathbb{0}, \quad \text{and}$$
$$\text{if } b \leq c \quad \text{and } b \neq \mathbb{0} \quad \text{then } \mathbb{0} \neq \phi(b) < b.$$

We proceed by transfinite recursion. We define $\mathscr{A}_0 = \{c\}$; and if ξ is a non-zero limit ordinal and $\xi < \alpha$ we define $\mathscr{A}_\xi$ by condition (ii). Because of (vii) and the fact that α is strongly inaccessible we have $|\mathscr{A}_\xi| < \alpha$. Since $\mathfrak{B}$ is α-distributive and by (v), $\mathscr{A}_\xi$ is a partition of c consisting of non-zero elements. Finally, (iii) is obvious.

Now let $\xi < \alpha$; for $a \in \mathscr{A}_\xi$ we set

$$\mathscr{S}(a) = \{\phi(a \wedge b_\xi), a \wedge b_\xi \setminus \phi(a \wedge b_\xi), \phi(a \setminus b_\xi), a \setminus b_\xi \setminus \phi(a \setminus b_\xi)\} \setminus \{\mathbb{0}\}.$$

It is clear that

$$2 \leq |\mathscr{S}(a)| \leq 4, \quad \text{and}$$
$$\bigvee \mathscr{S}(a) = a.$$

We define

$$\mathscr{A}_{\xi+1} = \bigcup \{\mathscr{S}(a) : a \in A_\xi\}.$$

It is clear that the recursive assumptions are satisfied.

Definition. An infinite cardinal α is *weakly compact* if α is strongly inaccessible and every α-complete, α-distributive Boolean algebra of cardinality at most α is α-representable.

We note that it follows from (the easy part) of Stone's duality that ω is weakly compact.

8.18. Corollary. *If α is a strongly inaccessible, ramifiable cardinal then α is weakly compact.*

Proof. According to Lemma 8.12 it is sufficient to prove that if $\mathfrak{B}$ is an α-distributive Boolean algebra of cardinality α and c is a non-zero element of $\mathfrak{B}$ then c is an element of an α-complete ultrafilter of $\mathfrak{B}$. We assume with no loss of generality that there is no atom a of $\mathfrak{B}$ such that $a \leq c$. Let $\langle b_\xi : \xi < \alpha \rangle$ be a well-ordering of $\mathfrak{B}$ and apply Lemma 8.17 to obtain a family $\{\mathscr{A}_\xi : \xi < \alpha\}$ of subsets of $\mathfrak{B}$ satisfying conditions (i) to (vii) (of that lemma). Set $A = \bigcup_{\xi<\alpha} \mathscr{A}_\xi$ and consider the partial order $\preccurlyeq$ on A given by the reverse Boolean inclusion: $a \preccurlyeq b$ if $a \geq b$ for $a, b \in A$. It is clear that $\langle A, \preccurlyeq \rangle$ is a ramification system such that

$$\mathscr{A}_\xi \text{ is the set of elements of } \langle A, \preccurlyeq \rangle \text{ of order } \xi \text{ for } \xi < \alpha.$$

Since α is a ramifiable cardinal there is a subset D of A such that

$$|D| = \alpha, \quad \text{and}$$
$$D \text{ is well-ordered under } \preccurlyeq .$$

We define

$$p = \{b \in \mathfrak{B} : d \leq b \text{ for some } d \in D\} .$$

It is easy to verify that

$$p \text{ is an } \alpha\text{-complete ultrafilter of } \mathfrak{B}, \text{ and}$$
$$c \in p .$$

8.19. Lemma. *If α is weakly compact, then every α-complete filter of an α-complete field of sets of cardinality at most α is contained in an α-complete ultrafilter.*

Proof. Let $\mathscr{S}$ be an α-complete field of sets such that $|\mathscr{S}| \leq \alpha$ and let $\mathscr{F}$ be an α-complete filter of $\mathscr{S}$. We denote by $\pi_{\mathscr{F}} : \mathscr{S} \to \mathscr{S}/\mathscr{F}$ the canonical homomorphism associated with $\mathscr{F}$. Since α is strongly inaccessible $\mathscr{S}/\mathscr{F}$ is α-distributive (and of cardinality at most α) (from Corollary 8.15), and since α is weakly compact it follows that $\mathscr{S}/\mathscr{F}$ is α-representable. By Lemma 8.12 there is an α-complete ultrafilter p of $\mathscr{S}/\mathscr{F}$. Then it is clear that $\pi_{\mathscr{F}}^{-1}(p)$ is an α-complete ultrafilter of $\mathscr{S}$ containing $\mathscr{F}$.

Definitions. A space X is *α-compact* if every open cover of X has a subcover of cardinality less than α.

An infinite cardinal α has *Alexander's property* if it satisfies the following condition: if X is a space with an α-subbase $\mathscr{A}$ such that

$|\mathscr{A}| \leq \alpha$, and

every cover of X by elements of $\mathscr{A}$ has a subcover of cardinality less than α,

then X is α-compact.

8.20. Theorem. *Let $\alpha = \alpha^{\underline{\alpha}} \geq \omega$ and assume that every (proper) α-complete filter in an α-complete field of sets of cardinality at most α is contained in an α-complete ultrafilter. Then α has Alexander's property.*

Proof. We let X be a space with an α-subbase $\mathscr{A}$ such that

$|\mathscr{A}| \leq \alpha$, and

every cover of X by elements of $\mathscr{A}$ has a subcover of cardinality less than α.

and we prove that X is α-compact. Let $\mathscr{B}$ be the base for X determined by $\mathscr{A}$, *i.e.*,

$$\mathscr{B} = \{\textstyle\bigcap_{i \in I} A_i : A_i \in \mathscr{A} \text{ for } i \in I \text{ and } |I| < \alpha\}.$$

It is sufficient to prove that if $\mathscr{C} \subset \mathscr{B}$ and $\mathscr{C}$ is a cover of X then there is a subcover of $\mathscr{C}$ of cardinality less than α. Suppose on the contrary that there is $\mathscr{C} \subset \mathscr{B}$ such that

$\mathscr{C}$ is a cover of X, and

every subcover of $\mathscr{C}$ has cardinality α.

Let $\mathscr{S}$ be the least α-complete field of subsets of X containing $\mathscr{A}$. Since $|\mathscr{A}| \leq \alpha$ and $\alpha = \alpha^{\underline{\alpha}}$ it follows that $|\mathscr{S}| \leq \alpha$. We define

$$\mathscr{F} = \{X \backslash C : C \in \mathscr{C}\}$$

and we note that

$\mathscr{F} \subset \mathscr{S}$, and

the α-complete filter of $\mathscr{S}$ *α-generated* by $\mathscr{F}$ (*i.e.*, the set of elements of $\mathscr{S}$ which contain an intersection of fewer than α elements of $\mathscr{F}$) is proper.

By assumption there is an α-complete ultrafilter p of $\mathscr{S}$ such that $\mathscr{F} \subset p$. We define

$$\mathscr{D} = \{A \in \mathscr{A} : X \backslash A \in p\}$$

and we claim that $\mathscr{D}$ is not a cover of X. Indeed if $\mathscr{D}$ is a cover of X then since $\mathscr{D} \subset \mathscr{A}$ the assumptions on $\mathscr{A}$ imply that there is a subcover of $\mathscr{D}$ of cardinality less than α, a contradiction to the α-completeness of the ultrafilter p. Let $q \in X \backslash \bigcup \mathscr{D}$; since $\mathscr{C}$ is a cover of X there is $C \in \mathscr{C}$ such that $q \in C$, and since $\mathscr{C} \subset \mathscr{B}$ there is a family $\{A_i : i \in I\} \subset \mathscr{A}$

with $|I|<\alpha$ such that

$$C=\bigcap_{i\in I}A_i.$$

Thus

$$\bigcup\{X\setminus A_i : i\in I\}=X\setminus C\in\mathscr{F}\subset p,$$

and since p is α-complete and $|I|<\alpha$, there is $i\in I$ such that $X\setminus A_i\in p$. We then have

$$q\in A_i \quad\text{and}\quad A_i\in\mathscr{D},$$

contradicting the fact that $q\in X\setminus\bigcup\mathscr{D}$. The proof is complete.

8.21. Lemma. *Let α be a cardinal satisfying Alexander's property. Then*

(a) *if X is a space such that*

$$w(X)\le\alpha, \quad \textit{and}$$
$$X \textit{ is } \alpha\textit{-compact},$$

then X_α is α-compact; and

(b) *if $\{X_i : i\in I\}$ is a family of spaces such that*

$$X_i \textit{ is } \alpha\textit{-compact for } i\in I,$$
$$w(X_i)\le\alpha \quad \textit{for } i\in I, \quad \textit{and}$$
$$|I|\le\alpha,$$

then $(\prod_{i\in I}X_i)_\alpha$ is α-compact.

Proof. (a) Let $\mathscr{A}$ be a base for X such that $|\mathscr{A}|=w(X)$. Then $\mathscr{A}$ is an α-subbase for X_α, and every cover of X_α by elements of $\mathscr{A}$ has a subcover of cardinality less than α. Alexander's property implies that X_α is α-compact

(b) Let $\mathscr{B}_i$ be a base for X_i such that $|\mathscr{B}_i|=w(X_i)$ for $i\in I$. We set $X=(\prod_{i\in I}(X_i)_\alpha)_\alpha$ and we define

$$\mathscr{A}=\{\pi_i^{-1}(B) : B\in\mathscr{B}_i \text{ and } i\in I\}.$$

Then $\mathscr{A}$ is an α-subbase for X and $|\mathscr{A}|\le\alpha$.

Let $\mathscr{C}$ be a cover of X such that $\mathscr{C}\subset\mathscr{A}$. We set

$$\mathscr{C}_i=\{B\in\mathscr{B}_i : \pi_i^{-1}(B)\in\mathscr{C}\}$$

for $i\in I$ and we note that $\mathscr{C}=\bigcup_{i\in I}\mathscr{C}_i$. We claim that there is $i\in I$ such that

$$\mathscr{C}_i \text{ is a cover of } (X_i)_\alpha.$$

If the claim fails then there is $p_i\in(X_i)_\alpha\setminus\bigcup\mathscr{C}_i$ for $i\in I$. We set $p=\langle p_i : i\in I\rangle\in X$ and we note that $p\in X\setminus\bigcup\mathscr{C}$, contradicting the fact that $\mathscr{C}$ is a cover of X. The claim is proved.

It follows from part (a) that $(X_i)_\alpha$ is α-compact, so there is $\mathscr{D}_i\subset\mathscr{C}_i$ such that

$$|\mathscr{D}_i|<\alpha, \quad\text{and}$$
$$\mathscr{D}_i \text{ is a cover of } (X_i)_\alpha.$$

Then $\{\pi_i^{-1}(B): B \in \mathscr{D}_i\}$ is a subcover of $\mathscr{C}$ of cardinality less than α.

These properties of the α-subbase $\mathscr{A}$ imply (since α has Alexander's property) that X is α-compact. Finally the identity function

$$X \to (\prod_{i\in I} X_i)_\alpha$$

is clearly continuous and hence $(\prod_{i\in I} X_i)_\alpha$ is α-compact.

8.22. Theorem. *If α is an infinite cardinal such that $(2^\alpha)_\alpha$ is α-compact then α is a strongly inaccessible ramifiable cardinal.*

Proof. If α is singular then $(2^\alpha)_\alpha = (2^\alpha)_{\alpha^+}$ and hence $(2^\alpha)_\alpha$, a discrete space of cardinality 2^α, is not even 2^α-compact. Thus α is regular. To prove that α is a strong limit cardinal let $\beta < \alpha$ and define

$$C_q = \{p \in (2^\alpha)_\alpha : p_\xi = q_\xi \text{ for } \xi < \beta\}$$

for $q \in 2^\beta$. We note that $C_q = \bigcap_{\xi<\beta}\{p \in (2^\alpha)_\alpha : p_\xi = q_\xi\}$ and thus C_q is a non-empty open-and-closed subset of $(2^\alpha)_\alpha$ for $q \in 2^\beta$; thus the family $\{C_q : q \in 2^\beta\}$ is an open cover of $(2^\alpha)_\alpha$ and a family of pairwise disjoint, non-empty sets. Since $(2^\alpha)_\alpha$ is α-compact it follows that $2^\beta < \alpha$. Thus α is strongly inaccessible.

Now we let $\mathfrak{A} = \langle A, \preccurlyeq \rangle$ be a ramification system of order α such that $|\mathscr{A}_\xi| < \alpha$ for $\xi < \alpha$, where as usual we denote by $\mathscr{A}_\xi$ the set of elements of $\mathfrak{A}$ of order ξ; we prove that A contains a subset of cardinality α that is well-ordered under $\preccurlyeq$. Since $|A| = \alpha$, we assume without loss of generality that $A = \alpha$. For $\xi < \alpha$ and $a \in \mathscr{A}_\xi$ we define

$$F(a) = \{p \in (2^\alpha)_\alpha : p(b) = 1 \text{ if } b \prec a;\text{ and } p(b) = 0 \text{ if } b \in \bigcup_{\eta<\xi}\mathscr{A}_\eta \text{ and } b \prec a \text{ fails}\}.$$

Now $|\bigcup_{\eta<\xi}\mathscr{A}_\eta| < \alpha$ since α is regular, and therefore $F(a)$ is an open-and-closed subset of $(2^\alpha)_\alpha$ for $a \in \mathscr{A}_\xi$ and $\xi < \alpha$. It is clear that $F(a) \neq \emptyset$ for $a \in \mathscr{A}_\xi$ and $\xi < \alpha$. We define

$$G_\xi = \bigcup\{F(a) : a \in \mathscr{A}_\xi\} \quad \text{for } \xi < \alpha .$$

Since $|\mathscr{A}_\xi| < \alpha$ and $(2^\alpha)_\alpha$ is a P_α-space it follows that G_ξ is a non-empty open-and-closed set for $\xi < \alpha$. It is clear that

$$G_\xi \subset G_\eta \quad \text{for } \eta < \xi < \alpha ;$$

furthermore, if $I \in \mathscr{P}_\alpha(\alpha)$, then the regularity of α implies that there is $\eta < \alpha$ such that $\xi < \eta$ for $\xi \in I$, so that $\phi \neq G_\eta \subset \bigcap_{\xi\in I} G_\xi$. Because $(2^\alpha)_\alpha$ is α-compact we have

$$\bigcap_{\xi<\alpha} G_\xi \neq \emptyset .$$

If $p \in \bigcap_{\xi<\alpha} G_\xi$ then for $\xi < \alpha$

$$\{b \in \alpha : b \in \bigcup_{\eta<\xi}\mathscr{A}_\eta \text{ and } p(b) = 1\}$$

is order-isomorphic to ξ under $\preccurlyeq$ and hence

$$\{b \in \alpha : p(b) = 1\} \text{ is order-isomorphic to } \alpha \text{ under } \preccurlyeq .$$

This completes the proof.

8.23. Theorem. *For an infinite cardinal α the following statements are equivalent.*

(a) *α is a Ramsey cardinal (i.e., $\alpha \to (\alpha)_2^2$);*

(b) *$\alpha \to (\alpha)_\beta^n$ for $n < \omega$ and $\beta < \alpha$;*

(c) *α is a Hausdorff cardinal;*

(d) *α is a strongly inaccessible ramifiable cardinal;*

(e) *α is weakly compact;*

(f) *$\alpha = \alpha^{\underline{\alpha}}$ and every α-complete filter of an α-complete field of sets of cardinality at most α is contained in an α-complete ultrafilter;*

(g) *α has Alexander's property;*

(h) *if X is a space such that $\mathbf{w}(X) \le \alpha$ and X is α-compact, then X_α is α-compact;*

(i) *if $\{X_i : i \in I\}$ is a family of spaces such that X_i is α-compact and $\mathbf{w}(X_i) \le \alpha$ for $i \in I$ and $|I| \le \alpha$, then $(\prod_{i \in I} X_i)_\alpha$ is α-compact;*

(j) *$(2^\alpha)_\alpha$ is α-compact; and*

(k) *α is regular and $(2^\alpha)_{(\alpha)}$ is α-compact.*

Proof. We consider the diagram of implications below, where the number next to an arrow of implication is the number of the theorem in this section establishing the implication.

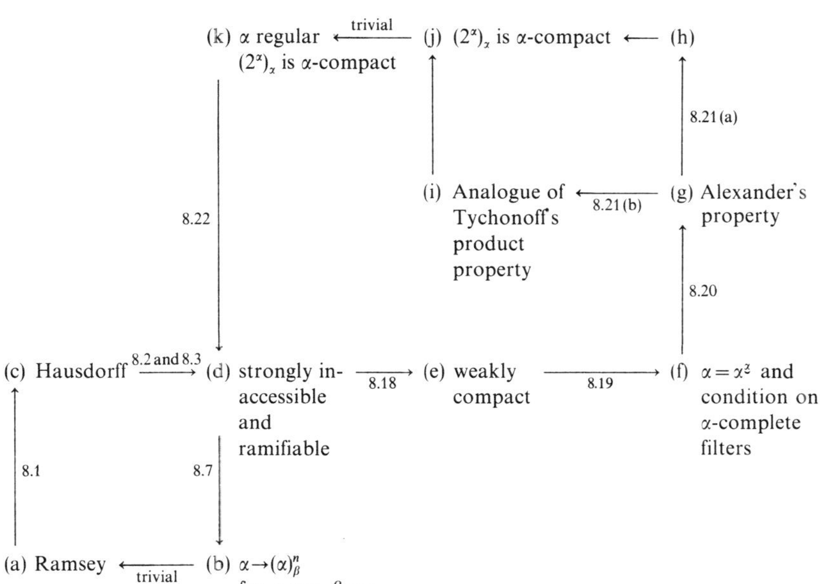

The following are the only implications that need comment.

(b)⇒(a) is trivial.

(h)⇒(j). Clearly $w(2^\alpha)=\alpha$. From the classical Tychonoff product theorem (Corollary 2.7) the space 2^α is compact and hence α-compact.

(i)⇒(j). The space 2 is compact, hence α-compact, and of weight $2<\alpha$.

(j)⇒(k) follows from the fact that, according to the first part of Theorem 8.22, α is regular.

(k)⇒(d) from Theorem 8.22.

The proof of the theorem is complete.

Measurable Cardinals

Definition. A cardinal α is a *measurable* cardinal if there is an α-complete, non-principal ultrafilter on α.

Clearly ω is a measurable cardinal.

We will make use of the definition of (a simple case of) the ultraproduct, given in §11 below. Since what we need is so little, we make the presentation self-contained.

Let α be a cardinal and p an ultrafilter on α; we define the binary relation $\underset{p}{\equiv}$ on α^α by

$$f \underset{p}{\equiv} g \quad \text{if } \{\xi<\alpha: f(\xi)=g(\xi)\}\in p\,.$$

It is easy to see that $\underset{p}{\equiv}$ is an equivalence relation on α^α. If $f\in\alpha^\alpha$ we denote by f/p the $\underset{p}{\equiv}$-equivalence class of f, and we let α^α/p be the set of $\underset{p}{\equiv}$-equivalence classes. We define a binary relation $\leq_p$ on α^α/p by

$$f/p\leq_p g/p \quad \text{if } \{\xi<\alpha: f(\xi)\leq g(\xi)\}\in p$$

for $f,g\in\alpha^\alpha$. It is easy to see that $\leq_p$ is a well-defined binary relation on α^α/p (*i.e.*, if $f/p=f'/p$ and $g/p=g'/p$ and $f/p\leq_p g/p$ then $f'/p\leq_p g'/p$ for $f,f',g,g'\in\alpha^\alpha$).

We claim that

$$\leq_p \text{ is a linear order on } \alpha^\alpha/p.$$

That $\leq_p$ is a partial order is easy to see. Now let $f,g\in\alpha^\alpha$; then

$$\{\xi<\alpha: f(\xi)\leq g(\xi)\}\cup\{\xi<\alpha: g(\xi)\leq f(\xi)\}=\alpha\,,$$

and since p is an ultrafilter one of the two sets $\{\xi<\alpha: f(\xi)\leq g(\xi)\}$, $\{\xi<\alpha: g(\xi)\leq f(\xi)\}$ is an element of p. Thus either $f/p\leq_p g/p$ or $g/p\leq_p f/p$.

(This is the only place where we have used the fact that p is an ultrafilter, rather than simply a filter.)

8.24. Lemma. *If* $\alpha > \omega$ *and* p *is an* ω^+*-complete ultrafilter on* α *then* $\leq_p$ *is a well-ordering of* α^α/p.

Proof. Let $\{f_n : n<\omega\}$ be a sequence of elements of α^α such that

$$f_{n+1}/p <_p f_n/p \quad \text{for } n<\omega .$$

Then

$$A_n = \{\xi < \alpha : f_{n+1}(\xi) < f_n(\xi)\} \in p$$

for $n<\omega$. Since p is ω^+-complete, we have $\bigcap_{n<\omega} A_n \in p$ and there is $\eta \in \bigcap_{n<\omega} A_n$. Then

$$f_{n+1}(\eta) < f_n(\eta) \quad \text{for } n<\omega ,$$

contradicting the well-ordering of α.

If α and p are as in Lemma 8.24 then $\langle \alpha^\alpha/p, \leq_p\rangle$ is well-ordered and hence there is a unique ordinal ξ_p and a unique order-isomorphism $\phi_p : \langle \alpha^\alpha/p, \leq_p\rangle \to \xi_p$.

Let $e : \alpha \to \alpha^\alpha$ be the diagonal function defined by

$$e(\xi) = \langle \xi, \dots \xi, \dots\rangle \quad \text{for } \xi < \alpha .$$

It is clear that the function $\xi \to e(\xi)/p$ is a one-to-one, order-preserving function from α into $\langle \alpha^\alpha/p, \leq_p\rangle$.

We denote as usual by id the identity function from α to α.

8.25. Lemma. *Let* $\alpha > \omega$ *and let* p *be an* α*-complete ultrafilter on* α. *Then*

(a) *the function* $\xi \to \phi_p(e(\xi)/p)$ *is the natural inclusion of* α *into* ξ_p; *and*

(b) *if (in addition)* p *is non-principal, then* $\phi_p(\mathrm{id}/p) \geq \alpha$ *and hence* $\alpha < \xi_p$.

Proof. (a) By Lemma 8.24 α^α/p is well-ordered. Let ζ be the least ordinal such that $\zeta < \alpha$ and $\phi_p(e(\zeta)/p) \neq \zeta$. Since e and ϕ are one-to-one, we have $\zeta < \phi_p(e(\zeta)/p)$. Let $g \in \alpha^\alpha$ such that $\phi_p(g/p) = \zeta$; then $g <_p e(\zeta)$ and hence $\{\xi < \alpha : g(\xi) < \zeta\} \in p$. Since p is α-complete and $|\zeta| < \alpha$, there is $\eta < \zeta$ such that

$$\{\xi < \alpha : g(\xi) = \eta\} \in p .$$

Thus $g/p = e(\eta)/p$. The inductive assumption implies that

$$\eta = \phi_p(e(\eta)/p) = \phi_p(g/p) = \zeta ,$$

a contradiction. Thus

$$\phi_p(e(\zeta)/p) = \zeta \quad \text{for } \zeta < \alpha .$$

(b) We note that

$$\mathrm{id}/p \neq e(\zeta)/p \quad \text{for } \zeta < \alpha,$$

since $\{\xi<\alpha: \xi=\zeta\}=\{\zeta\}\notin p$. By part (a) we have $\phi_p(\mathrm{id}/p)\geq\alpha$.

Definition. If $\alpha>\omega$ and p is a (non-principal) α-complete ultrafilter on α, then p is *normal* if

$$\phi_p(\mathrm{id}/p)=\alpha.$$

The following is a simple but fundamental result about uncountable measurable cardinals.

8.26. Theorem (*Scott*). *If $\alpha>\omega$ and p is a non-principal α-complete ultrafilter on α and $f\in\alpha^\alpha$ is such that*

$$\phi_p(f/p)=\alpha,$$

then $\bar{f}(p)$ is a normal ultrafilter on α. In particular, if α is an uncountable measurable cardinal then there is a normal ultrafilter on α.

Proof. We prove the following statements.

(i) $\bar{f}(p)$ is α-complete.

From Lemma 7.16(a) we have

$$\bar{f}(p)=\{A\in\mathscr{P}(\alpha): f^{-1}(A)\in p\}.$$

It follows that if $\beta<\alpha$ and $\{A_\xi:\xi<\beta\}\subset\bar{f}(p)$ then $f^{-1}(A_\xi)\in p$ for $\xi<\beta$ and

$$f^{-1}\left(\bigcap_{\xi<\beta} A_\xi\right)=\bigcap_{\xi<\beta} f^{-1}(A_\xi)\in p,$$

so $\bigcap_{\xi<\beta} A_\xi\in\bar{f}(p)$.

(ii) $\bar{f}(p)$ is non-principal.

If there is $\xi<\alpha$ such that $\{\xi\}\in\bar{f}(p)$ then

$$\{\eta<\alpha: f(\eta)=\xi\}=f^{-1}(\{\xi\})\in p$$

so that $f/p=e(\xi)/p$. Thus $\bar{f}(p)$ is not the principal ultrafilter $\{A\subset\alpha:\xi\in A\}$.

(iii) if $g\in\alpha^\alpha$ and $g/\bar{f}(p)<_{\bar{f}(p)}\mathrm{id}/\bar{f}(p)$, then $g/\bar{f}(p)=e(\xi)/\bar{f}(p)$ for some $\xi<\alpha$.

Let $g\in\alpha^\alpha$ with $g/\bar{f}(p)<_{\bar{f}(p)}\mathrm{id}/\bar{f}(p)$ and let A be an element of $\bar{f}(p)$ such that

$$g(\xi)<_p\mathrm{id}(\xi)=\xi$$

for $\xi\in A$. Now $f^{-1}(A)\in p$, and for $\eta\in f^{-1}(A)$ we have $f(\eta)\in A$ and then

$$(g\circ f)(\eta)=g(f(\eta))<f(\eta).$$

Hence

$$g\circ f/p = e(\xi)/p$$

for some $\xi<\alpha$, so that

$$f^{-1}(g^{-1}(\{\xi\})) = (g\circ f)^{-1}(\{\xi\})\in p$$

and therefore

$$g^{-1}(\{\xi\})\in \bar{f}(p);$$

thus $g/\bar{f}(p) = e(\xi)/\bar{f}(p)$.

Definition. If α is a cardinal and p a uniform ultrafilter on α, then p is a *Ramsey* ultrafilter if it satisfies the following condition: if

$$[\alpha]^n = P_0\cup P_1 \quad \text{and}$$
$$P_0\cap P_1=\emptyset,$$

then there are $A\in\mathscr{P}(\alpha)$ and $i<2$ such that

$$A\in p, \quad \text{and}$$
$$[A]^n\subset P_i$$

for $n<\omega$.

8.27. Theorem *(Rowbottom). If p is a normal ultrafilter on α then p is a Ramsey ultrafilter.*

Proof. We proceed by induction. There is nothing to prove if $n=0$. Let $n<\omega$, assume that the Ramsey condition is true for n and let

$$[a]^{n+1} = P_0\cup P_1 \quad \text{and}$$
$$P_0\cap P_1=\emptyset.$$

We define for $\xi<\alpha$

$$P_0(\xi) = \{F\in[\alpha]^n : \text{either } \xi\in F, \text{ or } \xi\notin F \text{ and } F\cup\{\xi\}\in P_0\} \quad \text{and}$$
$$P_1(\xi) = \{F\in[\alpha]^n : \xi\notin F \text{ and } F\cup\{\xi\}\in P_1\}.$$

By the inductive assumption there is $A_\xi\in p$ and $\varepsilon(\xi)<2$ such that

$$[A_\xi]^n\subset P_{\varepsilon(\xi)}(\xi)$$

for $\xi<\alpha$. Thus $\varepsilon\in 2^\alpha$ and there is $i<2$ such that $\varepsilon^{-1}(\{i\}) = B\in p$. We define

$$A = \{\zeta\in B : \zeta\in A_\xi \text{ for } \xi<\zeta, \xi\in B\}$$

and we prove the following two statements.

(i) $A\in p$.

We define $f:\alpha\to\alpha$ as follows: if $\zeta\in A$ we set $f(\zeta)=\zeta$; if $\zeta\in B\setminus A$, we set

$$f(\zeta) = \min\{\xi\in B : \zeta\notin A_\xi\};$$

and if $\zeta\in\alpha\setminus B$, we set $f(\zeta)=0$.

We note that if $B\setminus A\in p$ then $\{\zeta<\alpha: f(\zeta)<\zeta\}\in p$, and since p is normal there is $\xi\in B$ such that $f^{-1}(\{\xi\})\in p$. Thus $\{\zeta<\alpha:\zeta\notin A_\xi\}=\alpha\setminus A_\xi\in p$, a contradiction.

(ii) $[A]^{n+1}\subset P_i$.

Let $F\in[A]^{n+1}$ and set $\xi=\min F$ and $G=F\setminus\{\xi\}$. Then $G\in[A]^n$, $\xi\in B$, and $\xi<\zeta$ for $\zeta\in G$; hence if $\zeta\in G$ then $\zeta\in A_\xi$, so

$$G\in[A_\xi]^n\subset P_i(\xi)$$

and $F\in P_i$.

The proof of the theorem is complete.

Definition. An infinite cardinal α satisfies the arrow relation

$$\alpha\to(\kappa)^{<\omega}_\lambda$$

if the following condition is satisfied: if

$$\mathscr{P}_\omega(\alpha)=\bigcup_{i<\lambda}P_i \quad \text{and}$$
$$P_i\cap P_j=\emptyset \quad \text{for } i<j<\lambda,$$

then there are $A\in\mathscr{P}(\alpha)$ and $\varepsilon\in\lambda^\omega$ such that

$$|A|=\kappa \quad \text{and}$$
$$[A]^n\subset P_{\varepsilon(n)} \quad \text{for } n<\omega.$$

An infinite cardinal α is a *strongly Ramsey* cardinal if $\alpha\to(\alpha)^{<\omega}_2$.

It is clear that a strongly Ramsey cardinal is Ramsey and hence strongly inaccessible.

8.28. Corollary *(Erdős-Hajnal). If α is a uncountable measurable cardinal then α is a strongly Ramsey cardinal.*

Proof. Let

$$\mathscr{P}_\omega(\alpha)=P_0\cup P_1 \quad \text{and}$$
$$P_0\cap P_1=\emptyset.$$

We set $P_{i,n}=P_i\cap[\alpha]^n$ for $n<\omega$, $i<2$. From Theorem 8.26 there is a normal ultrafilter p on α, which is Ramsey by Theorem 8.27 above. Thus there are $A_n\in p$ and $\varepsilon(n)<2$ such that

$$[A_n]^n\subset P_{\varepsilon(n),n} \quad \text{for } n<\omega.$$

We set $A=\bigcap_{n<\omega}A_n$; since p is α-complete and $\alpha>\omega$ we have $A\in p$. Clearly $[A]^n\subset P_{\varepsilon(n),n}\subset P_{\varepsilon(n)}$ for $n<\omega$.

8.29. Theorem *(Rowbottom). If $\kappa\geq\omega$ and $\alpha\to(\kappa)^{<\omega}_\lambda$* then *$\alpha\to(\kappa)^{<\omega}_{\lambda^\omega}$. In particular $\omega\not\to(\omega)^{<\omega}_2$.*

Proof. Let $\phi:\omega\times\omega\to\omega$ be such that

$$\phi \text{ is one-to-one and onto, and}$$
$$\phi(k,n)\geq n \quad \text{for } k,n<\omega.$$

Let

$$\mathscr{P}_\omega(\alpha)=\bigcup_{i<\lambda^\omega}P_i \quad \text{and}$$
$$P_i\cap P_j=\emptyset \quad \text{for } i<j<\lambda^\omega.$$

We define

$$Q_\eta=\{\{\xi_0<\cdots<\xi_{m-1}\}\in\mathscr{P}_\omega(\alpha):m\geq 1 \text{ and } \{\xi_0<\cdots<\xi_{\pi_1((\phi^{-1})(\{m\}))}\}\in P_i \text{ and } i(\pi_0(\phi^{-1}(\{m\})))=\eta\} \quad \text{for } \eta<\lambda.$$

We note that Q_η is well-defined for $\eta<\lambda$ since $\pi_1(\phi^{-1}(\{m\}))\leq m$ for $m<\omega$. Then

$$\mathscr{P}_\omega(\alpha)=\bigcup_{\eta<\lambda}Q_\eta$$

and thus by assumption there are $A\in\mathscr{P}(\alpha)$ and $\varepsilon\in\lambda^\omega$ such that

$$|A|=\kappa \quad \text{and}$$
$$[A]^m\subset Q_{\varepsilon(m)} \quad \text{for } m<\omega.$$

With no loss of generality we assume that A contains no largest element. We define

$$\sigma:\omega\to\lambda^\omega$$

by $\sigma(n)=\{\langle k,\varepsilon(\phi(k,n))\rangle:k<\omega\}$ and we claim that

$$[A]^n\subset P_{\sigma(n)} \quad \text{for } n<\omega.$$

Let $\{\xi_0<\cdots<\xi_{n-1}\}\in[A]^n$ and let $i<\lambda^\omega$ be such that $\{\xi_0<\cdots<\xi_{n-1}\}\in P_i$. For $k<\omega$ we have $\phi(k,n)\geq n$ and since A has no largest element there are $\xi_n,\ldots,\xi_{\phi(k,n)-1}\in A$ such that $\xi_{n-1}<\xi_n<\cdots<\xi_{\phi(k,n)-1}$. Then

$$\{\xi_0<\cdots<\xi_{\phi(k,n)-1}\}\in Q_{\varepsilon(\phi(k,n))}$$

and therefore $i(k)=\varepsilon(\phi(k,n))$. It follows that $i=\sigma(n)$ and $[A]^n\subset P_{\sigma(n)}$ for $n<\omega$.

The special case follows from the obvious fact that $\omega\nrightarrow(\omega)^{<\omega}_\omega$. The proof of the theorem is complete.

8.30. Theorem *(Hanf-Scott, Rowbottom). If p is a normal ultrafilter on α then*

$$\{\beta<\alpha:\beta\to(\beta)^{<\omega}_2\}\in p.$$

Proof. We first show that

$$\{\xi<\alpha:\xi=|\xi|\}\in p.$$

For suppose not; then since $|\xi|\leq\xi$ for $\xi<\alpha$ we have $\{\xi<\alpha:|\xi|<\xi\}\in p$. Let $f:\alpha\to\alpha$ be given by $f(\xi)=|\xi|$ for $\xi<\alpha$; then $f/p<_p\mathrm{id}/p$, and since

p is a normal ultrafilter there is $\beta<\alpha$ such that $\beta^+\setminus\beta=\{\xi<\alpha:|\xi|=\beta\}\in p$, a contradiction since p is an α-complete ultrafilter and thus uniform on α.

Assume now that the conclusion of the theorem fails, *i.e.*, that $\{\beta<\alpha:\beta\to(\beta)_2^{<\omega}\}\notin p$. Then, with the fact just shown, we have

$$I=\{\beta<\alpha:\beta\nrightarrow(\beta)_2^{<\omega}\}\in p\,.$$

For $\beta\in I$ there is a partition $\{P_{\beta,0},P_{\beta,1}\}$ of $\mathscr{P}_\omega(\beta)$ such that

if $A\in\mathscr{P}(\beta)$ and $|A|=\beta$ then there is $n(\beta,A)<\omega$ such that

$$[A]^{n(\beta,A)}\cap P_{\beta,i}\neq\emptyset \quad \text{for } i<2\,.$$

We define

$$\begin{aligned}(1)\qquad Q_{n,0}&=\{F\in[\alpha]^n:\{\beta<\alpha:F\in[\beta]^n\cap P_{\beta,0}\}\in p\},\quad\text{and}\\ Q_{n,1}&=[\alpha]^n\setminus Q_{n,0}\end{aligned}$$

for $n<\omega$. Since $\alpha\to(\alpha)_2^{<\omega}$ (Corollary 8.28) there are $A\in\mathscr{P}(\alpha)$ and $\varepsilon\in 2^\omega$ such that

$$\begin{aligned}&|A|=\alpha\quad\text{and}\\ (2)\qquad &[A]^n\subset Q_{n,\varepsilon(n)}\quad\text{for } n<\omega\,.\end{aligned}$$

We define $g:\alpha\to\alpha$ by

$$\begin{aligned}g(\beta)&=|A\cap\beta|\quad\text{if } \beta\in I,\quad\text{and}\\ g(\xi)&=0\qquad\quad\ \text{if } \xi\in\alpha\setminus I\,.\end{aligned}$$

Then $g(\beta)\le\beta$ for $\beta\in I$; we claim that

$$I_1=\{\beta\in I:g(\beta)=\beta\}\in p\,.$$

Indeed suppose not, so that $\{\beta\in I:g(\beta)<\beta\}\in p$. Then since p is normal, there is $\gamma<\alpha$ such that

$$\{\beta\in I:|A\cap\beta|=\gamma\}\in p\,,$$

a contradiction to the fact that p is uniform.

The defining property of the partition $\{P_{\beta,0},P_{\beta,1}\}$ implies that for $\beta\in I_1$ there is (a least) $n(\beta)<\omega$ such that

$$[A\cap\beta]^{n(\beta)}\cap P_{\beta,i}\neq\emptyset\quad\text{for } i<2\,.$$

We set

$$I_{1,n}=\{\beta\in I_1:n(\beta)=n\}\quad\text{for } n<\omega\,.$$

Thus we have $I_1=\bigcup_{n<\omega}I_{1,n}\in p$ and since p is α-complete (and hence ω^+-complete) there is $n<\omega$ such that

$$I_2=I_{1,n}\in p\,.$$

Thus

$$[A\cap\beta]^n\cap P_{\beta,i}\neq\emptyset$$

and thus there is

(3) $$F_{\beta,i}\in[A\cap\beta]^n\cap P_{\beta,i} \quad \text{for } \beta\in I_2 \quad \text{and } i<2.$$

Let $\xi_{\beta,i,1}<\cdots<\xi_{\beta,i,n}$ be the elements of $F_{\beta,i}$ in their natural ordering (as elements of α) for $\beta\in I_2$ and $i<2$. We define functions

$$f_{i,k}:\alpha\to\alpha \quad \text{for } 1\leq k\leq n \quad \text{and } i<2$$

by the rule

$$f_{i,k}(\xi)=0 \quad \text{if } \xi\notin I_2, \quad \text{and}$$
$$f_{i,k}(\beta)=\xi_{\beta,i,k} \quad \text{for } \beta\in I_2.$$

Then we have

$$f_{i,k}(\beta)<\beta \quad \text{for } \beta\in I_2, \quad 1\leq k\leq n.$$

The normality of p implies that there are $I_3\in\mathscr{P}(\alpha)$ and $\xi_{i,k}<\alpha$ such that

$$I_3\subset I_2, \quad I_3\in p, \quad \text{and}$$
$$f_{i,k}(\beta)=\xi_{i,k} \quad \text{for } 1\leq k\leq n \quad \text{and } i<2.$$

Let $F_i=\{\xi_{i,k}:1\leq k\leq n\}$ for $i<2$; then

(4) $$F_{\beta,i}=F_i \quad \text{for } \beta\in I_3,\ i<2.$$

From (3) and (4) we have

$$I_3\subset\{\beta<\alpha:F_i\in[\beta]^n\cap P_{\beta,i}\}\in p \quad \text{for } i<2.$$

and hence from (1)

$$F_i\in Q_{n,i} \quad \text{for } i<2.$$

This contradicts the fact, implied by (2) and (3), that

$$F_0,F_1\in[A]^n\subset Q_{n,\varepsilon(n)}.$$

The proof of the theorem is complete.

Remark. (a) It follows from Corollary 8.28 and Theorem 8.23 that a measurable cardinal is strongly inaccessible. We give below the direct classical proof of this fact.

Theorem. *A measurable cardinal is strongly inaccessible.*

Proof. Let α be a measurable cardinal and let p be a non-principal α-complete ultrafilter on α.

We first prove that α is regular. Suppose that $\mathrm{cf}(\alpha)<\alpha$ and let $\{\alpha_\xi:\xi<\mathrm{cf}(\alpha)\}$ be a family of cardinals such that

$$\alpha_\xi < \alpha \quad \text{for } \xi < \text{cf}(\alpha),$$
$$\alpha_\xi < \alpha_\zeta \quad \text{for } \xi < \zeta < \text{cf}(\alpha), \quad \text{and}$$
$$\textstyle\sum_{\xi < \text{cf}(\alpha)} \alpha_\xi = \alpha .$$

If $\alpha \setminus \alpha_\xi \in p$ for $\xi < \text{cf}(\alpha)$, then

$$\textstyle\bigcap_{\xi < \text{cf}(\alpha)} \alpha \setminus \alpha_\xi = \emptyset \in p$$

since $\text{cf}(\alpha) < \alpha$ and p is α-complete. Thus $\alpha_\xi \in p$ for some $\xi < \text{cf}(\alpha)$. Since p is non-principal, we have $\alpha_\xi \setminus \{\eta\} \in p$ for $\eta < \alpha_\xi$. Thus

$$\textstyle\bigcap_{\eta < \alpha_\xi} \alpha_\xi \setminus \{\eta\} = \emptyset \in p$$

since $\alpha_\xi < \alpha$ and p is α-complete. This contradiction proves that α is regular.

We next prove that if $\beta < \alpha$ then $2^\beta < \alpha$. Suppose that $\beta < \alpha$ and $\alpha \le 2^\beta$. We set $q = \{B \in \mathscr{P}(2^\beta) : B \cap \alpha \in p\}$. Then q is an α-complete (and hence β^+-complete) non-principal ultrafilter on 2^β. For $\xi < \beta$ we set

$$\mathscr{S}_\xi = \{A \subset \beta : \xi \in A\} \quad \text{and} \quad \mathscr{S}'_\xi = \{A \subset \beta : \xi \notin A\},$$

and we set

$$S = \{\xi < \beta : \mathscr{S}_\xi \in q\} .$$

We define

$$\mathscr{S} = \textstyle\bigcap_{\xi \in S} \mathscr{S}_\xi \cap \bigcap_{\xi \notin S} \mathscr{S}'_\xi .$$

Then $\mathscr{S} \in q$ because the ultrafilter q is β^+-complete. But $\mathscr{S} = \{S\}$, so q is principal. This contradiction completes the proof of the theorem.

(b) It follows from Theorems 8.23 and 8.30 that if α is an uncountable measurable cardinal and p is a normal ultrafilter on α then

$$\{\beta < \alpha : \beta \text{ is strongly inaccessible}\} \in p .$$

We give a direct proof of this fact as well, making use only of part (a) of this remark.

Theorem. *If p is a normal ultrafilter on α then*

$$\{\beta < \alpha : \beta \textit{ is strongly inaccessible}\} \in p .$$

Proof. As we have seen in the first paragraph of the proof of Theorem 8.30, the fact that p is normal implies that

$$I_0 = \{\xi < \alpha : \xi = |\xi|\} \in p .$$

We define

$$I_1 = \{\beta < \alpha : \beta \text{ is a regular cardinal}\}, \quad \text{and}$$
$$I_2 = \{\beta < \alpha : \beta \text{ is a strong limit cardinal}\}$$

and we prove the following statements.

(i) $I_1 \in p$.

If $I_1 \notin p$ then $I_0 \setminus I_1 \in p$; we define

$$f: \alpha \to \alpha$$

by the rule

$$\begin{aligned} f(\beta) &= \mathrm{cf}(\beta) \quad \text{if } \beta \in I_0 \setminus I_1, \\ f(\xi) &= 0 \qquad\quad \text{if } \xi \in \alpha \setminus (I_0 \setminus I_1). \end{aligned}$$

Then $f(\beta) < \beta$ for $\beta \in I_0 \setminus I_1$ and $I_0 \setminus I_1 \in p$, so $f/p <_p \mathrm{id}/p$. Since p is a normal ultrafilter there is $\gamma < \alpha$ such that

$$J = \{\beta < \alpha : \mathrm{cf}(\beta) = \gamma\} \in p.$$

Then for $\beta \in J$ there is a family $\{\kappa_{\beta,i} : i < \gamma\}$ of cardinals such that

$$\begin{aligned} &\kappa_{\beta,i} < \beta \quad \text{for } i < \gamma \quad \text{and} \\ &\textstyle\sum_{i<\gamma} \kappa_{\beta,i} = \beta. \end{aligned}$$

We define for $i < \gamma$

$$f_i: \alpha \to \alpha$$

by the rule:

$$\begin{aligned} f_i(\beta) &= \kappa_{\beta,i} \quad \text{if } \beta \in J, \quad \text{and} \\ f_i(\xi) &= 0 \qquad \text{if } \xi \in \alpha \setminus J. \end{aligned}$$

Then $f_i(\beta) < \beta$ for $i < \gamma$ and $\beta \in J$, so $f_i/p <_p \mathrm{id}/p$ for $i < \gamma$. Since p is a normal ultrafilter there is $\kappa_i < \alpha$ such that

$$A_i = \{\beta < \alpha : \kappa_{\beta,i} = \kappa_i\} \in p \quad \text{for } i < \gamma.$$

We set

$$\begin{aligned} \kappa &= \textstyle\sum_{i<\gamma} \kappa_i \quad \text{and} \\ A &= J \cap \textstyle\bigcap_{i<\gamma} A_i \end{aligned}$$

and we note that $\kappa < \alpha$ since α is regular, and that $A \in p$ since p is α-complete. Finally $A \setminus \kappa \in p$ since p is α-complete and for $\beta \in A \setminus \kappa$ we have

$$\beta = \textstyle\sum_{i<\gamma} \kappa_{\beta,i} = \sum_{i<\gamma} \kappa_i = \kappa,$$

a contradiction. Thus $I_1 \in p$.

(ii) $I_2 \in p$.

If $I_2 \notin p$ then $I_0 \setminus I_2 \in p$. For $\beta \in I_0 \setminus I_2$ there is a cardinal κ_β such that

$$\kappa_\beta < \beta \leq 2^{\kappa_\beta}.$$

We let $f: \alpha \to \alpha$ be given by the rule:

$$\begin{aligned} f(\beta) &= \kappa_\beta \quad \text{if } \beta \in I_0 \setminus I_2, \quad \text{and} \\ f(\xi) &= 0 \quad\;\; \text{if } \xi \in \alpha \setminus (I_0 \setminus I_2). \end{aligned}$$

Then $f(\beta)<\beta$ for $\beta\in I_0\setminus I_2$ and $I_0\setminus I_2\in p$, so we have $f/p<_p\mathrm{id}/p$. Since p is a normal ultrafilter there is $\kappa<\alpha$ such that

$$\{\beta<\alpha:\kappa_\beta=\kappa\}\in p\,.$$

We note however that $\{\beta<\alpha:\kappa_\beta=\kappa\}\subset 2^\kappa\setminus\kappa\subset 2^\kappa$ and that $2^\kappa<\alpha$ since α is a strong limit cardinal, in contradiction to the fact that p is α-complete.

The proof is complete.

We examine briefly some concepts related to measurability.

Definition. A cardinal α is called *Ulam-measurable* if there is an ω^+-complete, non-principal ultrafilter on α.

We note that ω is not an Ulam-measurable cardinal. The connection between measurable and Ulam-measurable cardinals is described in the following simple result.

8.31. Theorem. *A cardinal is Ulam-measurable if and only if it is greater than or equal to the first uncountable measurable cardinal.*

Proof. It is clear that if a cardinal is Ulam-measurable then any greater cardinal is also Ulam-measurable. Thus we only need prove that the first Ulam-measurable cardinal α is equal to the first uncountable measurable cardinal. There is an ω^+-complete, non-principal ultrafilter p on α. Let β be the first cardinal such that p is not β-complete. Then $\omega^+<\beta\leq\alpha$; we suppose that $\beta<\alpha$. There is a family $\{A_\xi:\xi<\beta\}$ of elements of p such that $\bigcap_{\xi<\beta}A_\xi\notin p$. We assume without loss of generality

$$A_0=\alpha\,;$$
$$\textstyle\bigcap_{\xi<\beta}A_\xi=\emptyset \text{ (since } p \text{ is an ultrafilter); and}$$
$$\textstyle A_\xi\subsetneqq\bigcap_{\xi'<\xi}A_{\xi'} \quad \text{for } \xi<\beta \text{ (by the choice of } \beta).$$

We set $B_\xi=\bigcap_{\xi'<\xi}A_{\xi'}\setminus A_\xi$ for $\xi<\beta$, so that

$$\textstyle\bigcup_{\xi<\beta}B_\xi=\alpha\,;$$
$$B_\xi\cap B_{\xi'}=\emptyset \quad \text{for } \xi<\xi'<\beta\,; \text{ and}$$
$$B_\xi\notin p \quad \text{for } \xi<\beta\,.$$

We set $q=\{C\in\mathscr{P}(\beta):\bigcup_{\xi\in C}B_\xi\in p\}$. It is easy to verify that q is an ω^+-complete, non-principal ultrafilter on β, contradicting the choice of α.

An infinite cardinal α is *strongly compact* if for every infinite cardinal β, every α-complete filter on β can be extended to an α-complete ultrafilter on β.

An infinite cardinal α is *strongly measurable* if α is regular and every α-complete filter on α can be extended to an α-complete ultrafilter on α.

We note that a strongly compact cardinal is strongly measurable.

We also note that ω is a strongly compact cardinal.

For $\omega \leq \kappa \leq \alpha$ let $\Omega_\kappa(\alpha)$ denote the set of κ-complete ultrafilters on α.

8.32. Theorem. (a) *If α is measurable then*

$$2^\alpha \leq |\Omega_\alpha(\alpha)| \leq 2^{2^\alpha}.$$

(b) *If $\omega \leq \kappa \leq \alpha = \alpha^{\underline{\kappa}}$ and every κ-complete filter on α is contained in a κ-complete ultrafilter on α, then*

$$|\Omega_\kappa(\alpha)| = 2^{2^\alpha}.$$

(c) *If $\omega \leq \kappa \leq \alpha = \alpha^{\underline{\kappa}}$ and κ is strongly compact, then*

$$|\Omega_\kappa(\alpha)| = 2^{2^\alpha}.$$

(d) *If α is strongly measurable, then*

$$|\Omega_\alpha(\alpha)| = 2^{2^\alpha}.$$

Proof. (a) follows from Corollary 12.3 (b) below (since $\alpha = \alpha^{\underline{\alpha}}$) and the fact that $|\Omega_\alpha(\alpha)| \leq |\boldsymbol{\beta}(\alpha)| \leq 2^{2^\alpha}$.

(b) We have $|\Omega_\kappa(\alpha)| \leq 2^{2^\alpha}$.

Since $\alpha = \alpha^{\underline{\kappa}}$ there is by Theorem 7.3 a family Φ of κ-complete filters on α such that

$|\Phi| = 2^{2^\alpha}$, and

if $\mathscr{F}_0, \mathscr{F}_1 \in \Phi$ and $\mathscr{F}_0 \neq \mathscr{F}_1$ then there is $A \in \mathscr{P}(\alpha)$ such that $A \in \mathscr{F}_0$ and $\alpha \backslash A \in \mathscr{F}_1$.

For every $\mathscr{F} \in \Phi$ there is a κ-complete ultrafilter $p_{\mathscr{F}}$ on α such that $\mathscr{F} \subset p_{\mathscr{F}}$. Then $\{p_{\mathscr{F}} : \mathscr{F} \in \Phi\} \subset \Omega_\kappa(\alpha)$ and if $\mathscr{F}_0, \mathscr{F}_1 \in \Phi$ and $\mathscr{F}_0 \neq \mathscr{F}_1$ then $p_{\mathscr{F}_0} \neq p_{\mathscr{F}_1}$. Thus $|\Omega_\kappa(\alpha)| \geq |\Phi| = 2^{2^\alpha}$.

(c) follows from (b).

(d) follows from (b).

Descendingly Incomplete Ultrafilters

8.33. Lemma *(Ulam). If $\alpha \geq \omega$ then there is a family*

$$\{A_{\xi,\eta} : \xi < \alpha, \eta < \alpha^+\}$$

of subsets of α^+ such that

$A_{\xi,\eta} \cap A_{\xi,\eta'} = \emptyset$ *for* $\xi < \alpha$ *and* $\eta < \eta' < \alpha^+$, *and*

$|\alpha^+ \backslash \bigcup_{\xi<\alpha} A_{\xi,\eta}| \leq \alpha$ *for* $\eta < \alpha^+$.

Proof. If $\zeta<\alpha^+$ then $|\zeta|\leq\alpha$ and thus there is a one-to-one function

$$\phi_\zeta:\zeta\to\alpha.$$

We define

$$A_{\xi,\eta}=\{\zeta<\alpha^+:\eta<\zeta \text{ and } \phi_\zeta(\eta)=\xi\}$$

for $\xi<\alpha$, $\eta<\alpha^+$ and we verify the following statements.

(i) $A_{\xi,\eta}\cap A_{\xi,\eta'}=\emptyset$ for $\xi<\alpha$ and $\eta<\eta'<\alpha^+$.

If $\xi<\alpha$, $\eta<\eta'<\alpha^+$ and $\zeta\in A_{\xi,\eta}\cap A_{\xi,\eta'}$ then $\phi_\zeta(\eta)=\zeta=\phi_\zeta(\eta')$, contradicting the fact that ϕ_ζ is a one-to-one function.

(ii) $|\alpha^+\setminus\bigcup_{\xi<\alpha}A_{\xi,\eta}|\leq\alpha$ for $\eta<\alpha^+$.

If $\eta<\alpha^+$ and $\zeta\in\alpha^+\setminus\bigcup_{\xi<\alpha}A_{\xi,\eta}$ then $\zeta\leq\eta$ and hence

$$|\alpha^+\setminus\bigcup_{\xi<\alpha}A_{\xi,\eta}|\leq|\eta|\leq\alpha.$$

8.34. Lemma. *Let $\alpha\geq\omega$, let $\{A_{\xi,\eta}:\xi<\alpha,\eta<\alpha^+\}$ be a family of subsets of α^+ satisfying the conditions of Lemma* 8.33, *and set*

$$B_{\xi,\eta}=\bigcup_{\zeta<\xi}A_{\zeta,\eta} \quad \textit{for } \xi<\alpha,\eta<\alpha^+.$$

If $A\in\mathscr{P}(\alpha^+)$, $\xi<\alpha$, and $|A|>|\xi|$, then $\bigcap_{\eta\in A}B_{\xi,\eta}=\emptyset$.

Proof. Suppose that there are $\xi<\alpha$, $A\in\mathscr{P}(\alpha^+)$, and $i<\alpha^+$ such that

$$|A|>\xi \quad\text{and}$$
$$i\in\bigcap_{\eta\in A}B_{\xi,\eta}.$$

For $\eta\in A$ there is $\zeta_\eta<\xi$ such that $i\in A_{\zeta_\eta,\eta}$. Since $|A|>|\xi|$ there are $\eta,\eta'\in A$ such that

$$\eta\neq\eta' \quad\text{and}\quad \zeta_\eta=\zeta_{\eta'}.$$

Thus $i\in A_{\zeta_\eta,\eta}\cap A_{\zeta_\eta,\eta'}$, contrary to Lemma 8.33.

Definition. Let α and β be infinite cardinals. An ultrafilter p on α is *β-descendingly incomplete* if there is a family $\{A_\xi:\xi<\beta\}$ of elements of p such that

$$A_\xi\subset A_\zeta \quad\text{for } \zeta<\xi<\beta, \text{ and}$$
$$\bigcap_{\xi<\beta}A_\xi=\emptyset.$$

We note that p is β-descendingly incomplete if and only if p is $\mathrm{cf}(\beta)$-descendingly incomplete. For let $\phi:\mathrm{cf}(\beta)\to\beta$ be an order-preserving function such that $\phi(0)=0$ and $\sup\{\phi(\zeta):\zeta<\mathrm{cf}(\beta)\}=\beta$. If the family $\{A_\xi:\xi<\beta\}$ makes p β-descendingly incomplete then $\{A_{\phi(\zeta)}:\zeta<\mathrm{cf}(\beta)\}$ makes p $\mathrm{cf}(\beta)$-descendingly incomplete. Conversely if $\{A_\zeta:\zeta<\mathrm{cf}(\beta)\}$ makes p $\mathrm{cf}(\beta)$-descendingly incomplete and we define for $\xi<\beta$

$$B_\xi = A_\zeta \quad \text{for } \phi(\zeta) \le \xi < \phi(\zeta+1),$$

then the family $\{B_\xi : \xi<\beta\}$ makes p β-descendingly incomplete.

8.35. Theorem. *Let $\alpha \ge \omega$ and let p be a uniform ultrafilter on α^+. Then, either*

(a) *p is α-descendingly incomplete, or*

(b) *there is a regular cardinal $\beta \le \alpha$ such that p is (β, α^+)-regular.*

Proof. Let $\{A_{\xi,\eta} : \xi<\alpha, \eta<\alpha^+\}$ be subsets of α^+ satisfying the conditions of Lemma 8.33 and let $B_{\xi,\eta}$ be defined for $\xi<\alpha$, $\eta<\alpha^+$ as in Lemma 8.34. There are two possibilities to consider.

Case 1. There is $\eta<\alpha^+$ such that $B_{\xi,\eta} \notin p$ for $\xi<\alpha$. We set

$$B = \alpha^+ \setminus \bigcup_{\xi<\alpha} A_{\xi,\eta},$$

and we note that $|B| \le \alpha$ by Lemma 8.33. We define

$$B_\xi = \alpha^+ \setminus (B \cup B_{\xi,\eta}) \quad \text{for } \xi<\alpha.$$

Because $|B| \le \alpha$, we have $B \cup B_{\xi,\eta} \notin p$ for $\xi<\alpha$, so that $B_{\xi'} \subset B_\xi$ for $\xi<\xi'<\alpha$, and we have

$$\begin{aligned}\textstyle\bigcap_{\xi<\alpha} B_\xi &= (\alpha^+ \setminus B) \cap \textstyle\bigcap_{\xi<\alpha} (\alpha^+ \setminus B_{\xi,\eta}) \\ &= (\alpha^+ \setminus B) \cap (\alpha^+ \setminus \textstyle\bigcup_{\xi<\alpha} B_{\xi,\eta}) \\ &= (\alpha^+ \setminus B) \cap B = \emptyset .\end{aligned}$$

Thus p is α-descendingly incomplete.

Case 2. For $\eta<\alpha^+$ there is $\xi<\alpha$ such that $B_{\xi,\eta} \in p$. Let $\xi_\eta<\alpha$ be such that $B_{\xi_\eta,\eta} \in p$ for $\eta<\alpha^+$. Then there are a subset B of α^+ and $\bar\xi<\alpha$ such that

$$|B| = \alpha^+, \quad \text{and}$$
$$\xi_\eta = \bar\xi \quad \text{for } \eta \in B.$$

By Lemma 8.34

$$\text{if } I \in \mathscr{P}(\alpha^+) \text{ and } |I| > |\bar\xi| \text{ then } \textstyle\bigcap_{\eta\in I} B_{\xi,\eta} = \emptyset .$$

We set $\beta = |\bar\xi|^+$ and we note that the family $\{B_{\xi,\eta} : \eta \in B\}$ makes p (β, α^+)-regular.

8.36. Corollary. *Let $\alpha \ge \omega$ and let p be a uniform ultrafilter on α^+.*

(a) *If α is a regular cardinal then p is α-descendingly incomplete; and*

(b) *if α is a singular cardinal then*

either p is α-descendingly incomplete,

or there is $\beta<\alpha$ such that p is γ-descendingly incomplete for all regular cardinals γ such that $\beta<\gamma<\alpha$.

Proof. We first remark that if β is an infinite regular cardinal and p is a (β,β)-regular ultrafilter, then p is a β-descendingly incomplete ultrafilter. Indeed let $\{A_\xi : \xi<\beta\}$ be a family of elements of p such that

$$\text{if } I\subset\beta \text{ and } |I|=\beta \text{ then } \bigcap_{\xi\in I} A_\xi=\emptyset,$$

and set

$$B_\xi=\bigcup_{\xi<\zeta<\beta} A_\zeta \quad \text{for } \xi<\beta;$$

it is clear that $\{B_\xi : \xi<\beta\}\subset p$ is such that

$$B_\zeta\subset B_\xi \quad \text{for } \xi<\zeta<\beta \quad \text{and}$$
$$\bigcap_{\xi<\beta} B_\xi=\emptyset.$$

If statement (a) is false then by Theorem 8.35 there is a regular cardinal $\beta\leq\alpha$ such that p is (β,α^+)-regular. Then p is (α,α)-regular by Lemma 7.10(e), so p is α-descendingly incomplete by the initial remark of this proof. The proof of (a) is complete.

To prove (b) we note from Theorem 8.35 that if p is not α-descendingly incomplete then there is a regular cardinal $\beta\leq\alpha$ such that p is (β,α^+)-regular. Since α is singular we have $\beta<\alpha$. If γ is a regular cardinal and $\beta<\gamma<\alpha$ then p is (γ,γ)-regular by Lemma 7.10(e) and hence p is γ-descendingly incomplete by the initial remark of this proof. The proof of (b) is complete.

Notes for § 8

Theorem 8.1 is proved in Erdős-Tarski [61] (Theorem 1.2). Theorem 8.2 is proved in Erdős-Tarski [61] (Theorem 1.1), where it is attributed in implicit form to Hausdorff [08]. Some of the results proved in Erdős-Tarski [61] were announced without proof in Erdős-Tarski [43] (problems 1—6). Theorem 8.4 is due to Hanf [64]; a simplified proof, employing Lemma 8.3 and using a generalization of the Brouwer-Kleene order, was given by Hajnal [64].

The Brouwer-Kleene lexicographic order was considered for countable ramification systems by Brouwer [24] and Kleene [55] (p. 203).

The original reference for Corollary 8.5 is D. König [27].

The negative result of Aronszajn (Kurepa [36]) and Specker [51], that if $\alpha=\alpha^{\underline{\alpha}}\geq\omega$ then α^+ is not ramifiable, is proved in Corollary 14.32 below.

The partition calculus, which is considered as a generalization of the Dedekind-Dirichlet pigeon-hole principle and whose first main result is Ramsey's theorem (8.8) (Ramsey [30] (Theorem A)), was introduced and studied extensively by Erdős, Rado, Hajnal, and others. A comprehensive treatment is found in Erdős-Rado [56] and Erdős-Hajnal-Rado [65]. Our treatment of the results on partition calculus essentially follows the outline given in Morley [68a] (§ 4), and has benefited from a letter (May, 1972) by K. Prikry to the authors.

Theorem 8.6 is closely related to the ramification lemma of Erdős-Hajnal-Rado [65] (Lemma 6.1). Its precise statement is given by Morley [68] (Theorem 4.8). Corollary 8.9 is proved (under somewhat different conditions) in Erdős-Rado [56] (Theorem 39(i)) and Erdős-Hajnal-Rado [65] (Lemma 8.1). Corollary 8.10 appears (under somewhat different conditions) in Erdős-Rado [56] (Theorem 39(ii)) and Kurepa [59]. A different proof, using the saturated structures of model theory (§ 11), has been announced by Simpson [70] and appears in Keisler [71] (Section 14). We note that the special case $(2^{\alpha})^{+} \to (\alpha^{+})^{2}_{\alpha}$ of Corollary 8.10 appears in Erdős [42] (Theorem I). Related results on partition calculus were also proved independently by Kurepa [39], [59a], [59b] (cf. the review by Erdős [62]).

The standard information on α-complete, α-distributive, etc. Boolean algebras contained in 8.11—8.15 mostly follows the book by Sikorski [64] (§§ 19—24). We note however that Sikorski's definitions of α-complete, α-distributive, and α-representable correspond to our definitions of α^{+}-complete, α^{+}-distributive, and weakly α^{+}-representable, respectively (in agreement with Keisler-Tarski [64], for example). Lemma 8.14 is due to Pierce [57] (Corollary 6.5) and Smith and Tarski [57] (Theorem 2.10). Corollary 8.15 appears explicitly in Keisler-Tarski [64] (Theorem 0.10). Theorem 8.16(a) is essentially proved in Pierce [58b] (§ 5) (cf. Sikorski [64] (§ 20, Example J)); a stronger result, assuming GCH, is proved by Banach and Kuratowski [29] (cf. Sikorski [64] (§ 31, Example C)). Corollary 8.18 is proved in Erdős-Tarski [61] (Theorem 4.3). Lemma 8.19 is implied by results in Keisler-Tarski [64] (§ 4).

Theorem 8.20, Lemma 8.21, and Theorem 8.22 are essentially due to Parovičenko [57], [59], [67], [68], who also proved the equivalences of Theorem 8.23 (except those concerning (a), (b), (e) and (f)). Unfortunately the early papers [57] and [59] of Parovičenko were published in a Russian journal not widely circulated, were not translated, and still appear not to be generally known. Since we are not familiar ourselves with these two papers we do not know in particular the method that is used to establish the interesting implication contained in Theorem 8.4. The implications (f)$\Rightarrow$(j)$\Rightarrow$(d) of Theorem 8.23 are also proved by Monk and Scott [64].

Various other significant equivalences for weakly compact cardinals are not given here. Principal among them is the compactness of infinitary languages studied by Hanf [63], [64a].

The thesis of Silver [66] (Theorem 1.13) contains various characterizations of weakly compact cardinals. In particular he outlines a proof that condition (f) of our Theorem 8.23 implies the compactness statement for infinitary languages using a method of ultraproducts (cf. the proof of the usual compactness theorem, 11.7 below). Dickmann [70] (Chapter 2, § 3C) also has an extensive treatment of such equivalences.

It is not known if there is an accessible ramifiable cardinal. In this direction Silver [66] (Theorem 1.16) proved that if α is a real-measurable cardinal then α is ramifiable. See Jech [71a] (§ 1) for more details on the problem of the existence of an accessible ramifiable cardinal.

Lemmas 8.24 and 8.25 appear in Keisler-Tarski [64] (Lemmas 1.3, 1.5, and 1.6). Theorem 8.26, due to Scott, is proved in Keisler-Tarski [64] (Theorem 1.8). For consistency results concerning the possible cardinalities of the set of normal ultrafilters, see Kunen [70a] (especially p. 201). Theorem 8.27 is essentially proved in the thesis of Rowbottom [64] (Theorem 4.8); the proof given here is a simplified version due to Silver [60] (Lemma 1.4.3). See also Martin-Solovay [69] (2.4, Lemma 1). Corollary 8.28 was first proved by a direct argument in Erdős-Hajnal [58] (Theorem 9a). Theorem 8.29 is attributed by Dickmann [70] (Lemma 5.18) to Rowbottom.

The first (explicit) result on the relative size of large cardinals, established by Hanf [63] (thesis) and [64a], implies that the first uncountable weakly compact cardinal (if it exists) is greater than the first uncountable strongly inaccessible cardinal. The method of proof involves infinitary languages and is not given here.

Theorem 8.30 is a special case of a result announced by Hanf and Scott [61]. The proof given here is from the thesis of Rowbottom [64] (Theorem 4.11). The proof of Rowbottom follows the ultrapower method of Keisler [62] and Tarski [62] (given in detail in Keisler-Tarski [64]). In particular the proof of part (b) of the long remark following Theorem 8.30 is found in Keisler [62] (Theorem 1). Erdős-Hajnal [62], using the combinatorial methods of Erdős-Hajnal [58], also give a direct proof of the fact that the first uncountable measurable cardinal is greater than the first uncountable strongly inaccessible cardinal.

Solovay [71], using methods similar to those of Keisler-Tarski [64], proves similar results for real-measurable cardinals.

The content of part (a) of the remark following Theorem 8.30 (also a consequence of 8.28 and 8.23) is the classical theorem that a measurable cardinal is strongly inaccessible. This has been proved by Ulam [30] and Tarski [38b] (p. 153).

In connection with 8.24—8.30 see also the historical remarks of Keisler [65a] (§ 2).

Model-theoretic characterizations of measurable cardinals are given in Keisler [62] (Theorem 2) and Dickmann [70] (Chapter 2, Theorem 3.9). For an exposition of a number of combinatorial results on measurable cardinals not given here the reader is referred to Kleinberg [73].

We do not attempt to enter here into a discussion of the consistency statements that are known for the existence and relative size of large cardinals. The most intensively studied consistency results concern consistency relative to the axiom of constructibility V=L. Somewhere between the notions of measurable and weakly compact cardinals the work of various people has discovered a breaking point. Thus Rowbottom [64], refining the result of Scott [61] on the inconsistency of V=L with the existence of a measurable cardinal, proved the inconsistency of the existence of a cardinal α such that $\alpha\rightarrow(\omega^+)_2^{<\omega}$ with V=L. Silver [70], on the other hand, has proved that if there is ("in the real world") a cardinal α such that $\alpha\rightarrow(\omega)_2^{<\omega}$, then α satisfies the same arrow relation in V=L.

Theorem 8.32 appears in Kunen [70a] (Lemma 7.2 and subsequent remarks) and Comfort-Negrepontis [72a].

Lemma 8.33 is proved by Ulam [30] (§ 1); the family of sets $\{A_{\xi,\eta}: \xi<\alpha, \eta<\alpha^-\}$ with the properties of 8.33 is referred to as an *Ulam matrix*. Ulam used such matrices to prove that a real-measurable cardinal is weakly inaccessible. We remark in this connection that analogues of Ulam matrices have been used by Prikry [68], [70], Erdős-Ulam [68], and Hajnal [69].

Corollary 8.36 was first proved by Chang [67], using a form of Ulam matrices and assuming the generalized continuum hypothesis. The proofs of 8.34—8.36, due to Prikry, are from Kunen and Prikry [71]; the same paper contains another proof, due to Kunen and using a combinatorial lemma of Lusin [47] on orthogonal (*i.e.*, almost disjoint) sets, of Theorem 8.35 and Corollary 8.36 for regular cardinals α. Corollary 8.36 has also been obtained independently by Čudnovskiĭ and Čudnovskiĭ [71]. See also Prikry [73] on descendingly complete ultrafilters.

We thank Neil Hindman for a careful, constructive reading of this section.

§ 9. The Rudin-Keisler Order on Types of Ultrafilters

In this and the next few sections we will introduce and study concepts that help clarify the structure of ultrafilters. The Rudin-Keisler partial order on (isomorphism types of) ultrafilters on a set is one such fundamental concept; through it we can compare different ultrafilters, and we can learn something about the intrinsic structure of an ultrafilter by examining its relative position in the Rudin-Keisler partial order.

The type $\tau(0)$ of the principal ultrafilters is the least element in the Rudin-Keisler order, and a next logical step of inquiry is to look for the immediate successors of $\tau(0)$, *i.e.*, the minimal types in $\tau[\boldsymbol{\beta}(\alpha)\setminus\alpha]$. But our reason for looking into these minimal types is based on a more fundamental consideration. Temporarily restricting ourselves to $\alpha=\omega$, it is historically true that the (topological) study of ultrafilters closely related to those of minimal type, the P-points of $\boldsymbol{\beta}(\omega)\setminus\omega$, has preceded and motivated the study of the Rudin-Keisler partial order; thus, these minimal ultrafilters are "natural" objects, mathematically interesting independent of the Rudin-Keisler order. Furthermore, the proof of their existence, assuming the continuum hypothesis, proceeds by building such ultrafilters by a transfinite recursive argument that has wide applicability in other cases; and, at every step in the recursion, a diagonal argument is used. Topologically the whole process can be described (and it will be studied from this point of view in § 15) as a proof of existence by a Baire category type of argument, and in this sense the minimal types, although they possess an extremal property and are therefore exceptional, in fact form a "residual" set; they are "typical" elements, and in some extended sense "almost" every ultrafilter belongs to a minimal type.

These considerations extend to all infinite cardinals α, provided we assume $\alpha^+=2^\alpha$ and restrict our attention to the minimal types of $\tau[U(\alpha)]$ corresponding to uniform ultrafilters on α.

Essentially the same recursive argument is used in Theorem 9.13 to prove the existence of α^+-good ultrafilters on α; their model-theoretic properties, examined in § 13, are of fundamental importance. Thus we can say, again from a topological point of view, that "almost" every uniform ultrafilter on α is α^+-good and belongs to a minimal type. (In § 10, however, the existence of good ultrafilters will be proved without any use of the generalized continuum hypothesis. Not so for the minimal types.) There is no doubt that the good ultrafilters and the minimal ultrafilters are the two most significant types of ultrafilters from a structural, topological, or model-theoretic point of view, and though their defining properties are evidently unrelated they are similar as to method of construction.

These remarks help to indicate the central reason for our interest in minimal ultrafilters: the proof of their existence is, up to a point, methodologically similar to the proof of the existence of good ultrafilters; and through the Rudin-Keisler order (whose definition they motivate) they may help clarify the structure of the set of all ultrafilters. Whether they can be used for a complete classification of types, however, is open to doubt.

We note that on ω every ultrafilter is good, and that there is a proof of the existence of good ultrafilters that does not make use of any segment of the generalized continuum hypothesis. Thus it appears to be "easier" for an ultrafilter to be good than minimal; this has suggested the possibility that every minimal ultrafilter is good. Most of the work in disproving this statement (for cardinals whose cofinality is uncountable) will be done in this section (Theorem 9.19); the actual theorem will be proved in § 10.

We have been using the term "minimal type" rather loosely in this introduction, but the remarks up to here are intended for types minimal in $\tau[U(\alpha)]$. If we ask for types of uniform ultrafilters minimal in $\tau[\boldsymbol{\beta}(\alpha)\backslash\alpha]$ we get nothing new in the case $\alpha=\omega$, but when $\alpha>\omega$ then these types exist if α is a measurable cardinal, their existence can be proved without any use of the generalized continuum hypothesis, and they coincide with the types of normal ultrafilters. A long list of equivalent conditions (including the condition that the filter be a Ramsey ultrafilter) is given in Theorem 9.6.

The Rudin-Keisler Order

Elements p and q of a space X are said to be *type-equivalent*, or of the same type, relative to the space X if there is an auto-homeomorphism h of X such that $h(p)=q$. When p and q are type-equivalent (relative to X) we write $p\approx q$.

Type-equivalence is evidently an equivalence relation on X; an equivalence class is called a *type* of X, and the set of types of X is denoted by $\mathbb{T}(X)$. The quotient equivalence function is denoted by

$$\tau: X \to \mathbb{T}(X)$$

and $\tau(p)$ is called the *type* of p, for $p \in X$.

These general definitions indicate the context in which the problems that we consider make sense; however, the results of this section deal only with a special case of these definitions, as follows.

Let p and q be ultrafilters on (a set of cardinality) α, so p and q are elements of the space $\boldsymbol{\beta}(\alpha)$ and we may apply the definitions and symbols introduced for any space to the case $X = \boldsymbol{\beta}(\alpha)$. Thus p is type-equivalent to q if there is an auto-homeomorphism h of $\boldsymbol{\beta}(\alpha)$ such that $h(p) = q$; this latter condition is clearly equivalent to the statement that there is a one-to-one function π from α onto α (*i.e.*, a *permutation* of α) such that the continuous Stone extension $\bar{\pi}: \boldsymbol{\beta}(\alpha) \to \boldsymbol{\beta}(\alpha)$ has the property that $\bar{\pi}(p) = q$. (We recall here the following notation that we use systematically: a bar above a function always denotes the continuous Stone extension of the function.)

If p is an ultrafilter on β and $\beta \leq \alpha$, then since the inclusion $\beta \subset \alpha$ induces the inclusion (by the natural identification) $\boldsymbol{\beta}(\beta) \subset \boldsymbol{\beta}(\alpha)$ we may regard p as an ultrafilter on α; thus if q is an ultrafilter on α we may compare the types of p and q.

Definition. Let α be a cardinal. The *Rudin-Keisler pre-order* $\preccurlyeq$ *on* $\boldsymbol{\beta}(\alpha)$ is the binary relation on $\boldsymbol{\beta}(\alpha)$ given by $p \preccurlyeq q$ if there is $f \in \alpha^\alpha$ such that $\bar{f}(q) = p$. It is clear that $\preccurlyeq$ is reflexive and transitive.

The *Rudin-Keisler partial order* (also denoted by $\preccurlyeq$) *on* $\mathbb{T}(\boldsymbol{\beta}(\alpha))$ is the binary relation on $\mathbb{T}(\boldsymbol{\beta}(\alpha))$ defined by

$$\sigma \preccurlyeq \tau \text{ if there are } p, q \in \boldsymbol{\beta}(\alpha) \text{ such that } \tau(p) = \sigma, \ \tau(q) = \tau, \text{ and } p \preccurlyeq q.$$

It is clear that $\preccurlyeq$ is a well-defined reflexive and transitive relation on $\mathbb{T}(\boldsymbol{\beta}(\alpha))$; we will prove below that $\preccurlyeq$ is also *anti-symmetric*, *i.e.*, that if σ and τ are types in $\mathbb{T}(\boldsymbol{\beta}(\alpha))$ such that $\sigma \preccurlyeq \tau$ and $\tau \preccurlyeq \sigma$, then $\sigma = \tau$ (Corollary 9.2(b)).

Sometimes we use the notation $p \prec q$ to mean that $p \preccurlyeq q$ and p is not type-equivalent to q; as usual for partial order relations, the notation $\sigma \prec \tau$ means that $\sigma \preccurlyeq \tau$ and $\sigma \neq \tau$.

We note the following facts for $\alpha \geq \omega$.

(a) $|\tau(p)| \leq 2^\alpha$ for $p \in \boldsymbol{\beta}(\alpha)$;

(b) $|\{\tau(q) \in \mathbb{T}(\boldsymbol{\beta}(\alpha)) : \tau(q) \preccurlyeq \tau(p)\}| \leq 2^\alpha$ for $p \in \boldsymbol{\beta}(\alpha)$;

(c) $|\mathbb{T}(\boldsymbol{\beta}(\alpha))| = 2^{2^\alpha}$; and

(d) $|\{\tau(q) \in \mathbb{T}(\boldsymbol{\beta}(\alpha)) : \tau(p) \preccurlyeq \tau(q)\}| = 2^{2^\alpha}$ for $p \in \boldsymbol{\beta}(\alpha)$.

Indeed (a) and (b) follow from the fact that there are 2^α functions from α to α; and (a) together with the fact that $|\boldsymbol{\beta}(\alpha)|=2^{2^\alpha}$ implies (c). For $p,q\in\boldsymbol{\beta}(\alpha)$ we have

$$\overline{\pi}_0(p\cdot q)=p\,,\qquad \overline{\pi}_1(p\cdot q)=q\;;$$

hence, noting that $p\cdot q$ is essentially an ultrafilter on α, we have $p\leqslant p\cdot q$ and $p\cdot q\neq p\cdot q'$ for $p,q,q'\in\boldsymbol{\beta}(\alpha)$ and $q\neq q'$. This together with (c) proves (d).

It follows from (d) that, for $\alpha\geq\omega$, $\mathbb{T}(\boldsymbol{\beta}(\alpha))$ has no largest element; in contrast, $\tau(0)=\alpha$ is the smallest element of $\mathbb{T}(\boldsymbol{\beta}(\alpha))$.

9.1. Lemma. *Let α be a cardinal and let $f\in\alpha^\alpha$ be such that $f(\xi)\neq\xi$ for $\xi<\alpha$. Then there are subsets A_0, A_1, and A_2 of α such that*

$$\alpha=A_0\cup A_1\cup A_2\,,$$
$$A_i\cap A_j=\emptyset\quad \textit{for } i,j\leq 2\quad \textit{and } i\neq j\,,\quad \textit{and}$$
$$A_i\cap f[A_i]=\emptyset\quad \textit{for } i\leq 2\,.$$

Proof. For $\xi<\alpha$ and $n<\omega$ we define $f^n(\xi)$ recursively by $f^0(\xi)=\xi$, $f^{n+1}(\xi)=f(f^n(\xi))$. We define a binary relation $\sim$ on α by

$$\xi\sim\zeta\quad\text{if and only if } f^n(\xi)=f^m(\zeta)\quad\text{for some } m,n<\omega\,,$$

and we show that $\sim$ is an equivalence relation on α. It is clear that $\sim$ is reflexive and symmetric; to verify transitivity we assume that $\xi\sim\zeta$ and $\zeta\sim\eta$, so that there are $n,m,k,l<\omega$ such that $f^n(\xi)=f^m(\zeta)$ and $f^k(\zeta)=f^l(\eta)$. If $m=k$ then $f^n(\xi)=f^m(\zeta)=f^k(\zeta)=f^l(\eta)$; if $m<k$ then $f^{n+k-m}(\xi)=f^{m+k-m}(\zeta)=f^l(\eta)$; and if $m>k$ then $f^{l+m-k}(\eta)=f^{k+m-k}(\zeta)$ $=f^n(\xi)$. Thus $\xi\sim\eta$.

Now let $\mathscr{R}$ be the set of $\sim$-equivalence classes of α. It is clear that if $r\in\mathscr{R}$ then $f[r]\subset r$. For every $r\in\mathscr{R}$ we consider two possibilities as follows.

Case 1. There are $\xi_r\in r$ and $n>0$ such that $f^n(\xi_r)=\xi_r$. We define

$$n_r=\min\{n<\omega: n>0 \text{ and } f^n(\xi_r)=\xi_r\}\quad\text{and}$$
$$C_r=\{f^k(\xi_r): k<n_r\}\,.$$

If $\xi\in r\backslash C_r$ then there are $k<n_r$ and $m<\omega$ such that $f^m(\xi)=f^k(\xi_r)$. We define for $\xi\in r\backslash C_r$

$$k(\xi)=\min\{k<n_r: \text{there is } m<\omega\quad\text{such that } f^m(\xi)=f^k(\xi_r)\}\,,$$

and

$$m(\xi)=\min\{m<\omega: f^m(\xi)=f^{k(\xi)}(\xi_r)\}\,.$$

We claim that

(*) if $\xi\in r\setminus C_r$ and $m(\xi)>1$ then $k(f(\xi))=k(\xi)$ and $m(f(\xi))=m(\xi)-1$.

It is clear that $k(\xi)\leq k(f(\xi))$, and the equation $f^{k(\xi)}(\xi_r)=f^{m(\xi)-1}(f(\xi))$ implies that $k(f(\xi))\leq k(\xi)$; it is clear also that $m(\xi)\leq m(f(\xi))+1$, and the equation $f^{k(f(\xi))}(\xi_r)=f^{k(\xi)}(\xi_r)=f^{m(\xi)-1}(f(\xi))$ implies that $m(f(\xi))\leq m(\xi)-1$.

We define

$$\begin{aligned}A_{r,0}&=\{f^k(\xi_r): k \text{ is even and } k<n_r-1\}\\&\quad\cup\{\xi\in r\setminus C_r: k(\xi)+m(\xi)\text{ is even}\},\\A_{r,1}&=\{f^k(\xi_r): k \text{ is odd and } k<n_r-1\}\\&\quad\cup\{\xi\in r\setminus C_r: k(\xi)+m(\xi)\text{ is odd}\},\quad\text{and}\\A_{r,2}&=\{f^{n_r-1}(\xi_r)\}.\end{aligned}$$

It is clear that

$$\begin{aligned}&r=A_{r,0}\cup A_{r,1}\cup A_{r,2}\quad\text{and}\\&A_{r,i}\cap A_{r,j}=\emptyset\quad\text{for } i,j\leq 2\quad\text{and } i\neq j,\end{aligned}$$

and also that if $\xi\in A_{r,i}$ and either $\xi\in C_r$ or $\xi\in r\setminus C_r$ with $m(\xi)=1$, then $f(\xi)\notin A_{r,i}$ for $i\leq 2$. The fact that $f(\xi)\notin A_{r,i}$ if $\xi\in A_{r,i}\setminus C_r$ and $m(\xi)>1$ follows from (*) for $i\leq 2$.

Case 2. *Case* 1 fails. Let $\xi_r\in r$; then for every $\xi\in r$ there are $k,m<\omega$ such that $f^m(\xi)=f^k(\xi_r)$. We define for $\xi\in r$

$$\begin{aligned}&k(\xi)=\min\{k<\omega:\text{there is } m<\omega \text{ such that } f^m(\xi)=f^k(\xi_r)\},\quad\text{and}\\&m(\xi)=\min\{m<\omega: f^m(\xi)=f^{k(\xi)}(\xi_r)\}.\end{aligned}$$

Then we note that

$$\begin{aligned}&\text{if } m(\xi)=0\quad\text{then } m(f(\xi))=0\quad\text{and } k(f(\xi))=k(\xi)+1,\quad\text{and}\\&\text{if } m(\xi)>0\quad\text{then } m(f(\xi))=m(\xi)-1\quad\text{and } k(f(\xi))=k(\xi),\end{aligned}$$

so that

$$k(\xi)+m(\xi)\quad\text{is even if and only if } k(f(\xi))+m(f(\xi))\quad\text{is odd}.$$

We define

$$\begin{aligned}A_{r,0}&=\{\xi\in r: k(\xi)+m(\xi)\text{ is even}\},\\A_{r,1}&=\{\xi\in r: k(\xi)+m(\xi)\text{ is odd}\},\\A_{r,2}&=\emptyset.\end{aligned}$$

It is clear that

$$r = A_{r,0} \cup A_{r,1} \cup A_{r,2},$$
$$A_{r,i} \cap A_{r,j} = \emptyset \quad \text{for } i,j \leq 2 \quad \text{and } i \neq j, \quad \text{and}$$
$$A_{r,i} \cap f[A_{r,i}] = \emptyset \quad \text{for } i \leq 2.$$

Finally we define

$$A_i = \bigcup \{A_{r,i} : r \in \mathscr{R}\} \quad \text{for } i \leq 2;$$

it is clear that these sets satisfy the requirements of the lemma.

9.2. Theorem. *Let α be a cardinal, let $f \in \alpha^\alpha$, and let $p \in \boldsymbol{\beta}(\alpha)$. Then*

(a) *$\bar{f}(p) = p$ if and only if $\{\xi < \alpha : f(\xi) = \xi\} \in p$; and*

(b) *$\bar{f}(p) \approx p$ if and only if there is $A \in p$ such that $f|A$ is one-to-one.*

Proof. (a) We set $A = \{\xi < \alpha : f(\xi) = \xi\}$ and let id denote the identity function on $\boldsymbol{\beta}(\alpha)$.

Both $\bar{f}$ and id are continuous functions on $\boldsymbol{\beta}(\alpha)$ which are equal on A and hence on $\mathrm{cl}_{\boldsymbol{\beta}(\alpha)} A$. If $A \in p$ then $p \in \mathrm{cl}_{\boldsymbol{\beta}(\alpha)} A$, so that $\bar{f}(p) = \mathrm{id}(p) = p$.

Suppose, conversely, that $A \notin p$. Then $\alpha \backslash A \in p$ and $f(\xi) \neq \xi$ for $\xi \in \alpha \backslash A$; we assume without loss of generality that $A = \emptyset$ (and hence $f(\xi) \neq \xi$ for $\xi < \alpha$). According to Lemma 9.1 there are subsets A_0, A_1, and A_2 of α such that

$$\alpha = A_0 \cup A_1 \cup A_2,$$
$$A_i \cap A_j = \emptyset \quad \text{for } i,j \leq 2 \quad \text{and } i \neq j, \quad \text{and}$$
$$A_i \cap f[A_i] = \emptyset \quad \text{for } i \leq 2.$$

Since $\alpha \in p$ and p is an ultrafilter it follows that exactly one of the sets A_0, A_1, and A_2 is an element of p. If $i \leq 2$ and $A_i \in p$ then $p \in \mathrm{cl}_{\boldsymbol{\beta}(\alpha)} A_i$ and it follows from the continuity of f that

$$\bar{f}(p) \in \mathrm{cl}_{\boldsymbol{\beta}(\alpha)}(f[A_i]) \subset \mathrm{cl}_{\boldsymbol{\beta}(\alpha)}(\bigcup \{A_j : j \leq 2 \text{ and } j \neq i\}),$$

and hence that $\bar{f}(p) \neq p$.

(b) If $\bar{f}(p) \approx p$ then there is a permutation π of α such that $\bar{\pi}(\bar{f}(p)) = p$. Then $\overline{\pi \circ f}(p) = p$ and hence

$$A = \{\xi < \alpha : \pi \circ f(\xi) = \xi\} \in p.$$

It is clear that $f|A$ is one-to-one.

Suppose conversely that there is $A \in p$ such that $f|A$ is one-to-one. If $p \in \alpha$ then $\bar{f}(p) = f(p) \in \alpha$ and thus $p \approx \bar{f}(p)$. If $p \in \boldsymbol{\beta}(\alpha) \backslash \alpha$ then A is infinite and thus there are subsets A_0 and A_1 of A such that

$$A = A_0 \cup A_1,$$
$$A_0 \cap A_1 = \emptyset,$$
$$|A| = |A_0| = |A_1|, \quad \text{and}$$
$$A_0 \in p.$$

Then $|\alpha\setminus A_0|=|\alpha\setminus f[A_0]|$, and if π is a permutation of α such that

$$\pi[\alpha\setminus A_0]=\alpha\setminus f[A_0] \quad \text{and}$$
$$\pi|A_0=f|A_0\,,$$

then $\overline{\pi}(p)=\overline{f}(p)$ and thus $\overline{f}(p)\approx p$.

9.3. Corollary. *Let p and q be ultrafilters on α. Then $p\approx q$ if and only if $p\leqslant q$ and $q\leqslant p$. Hence the Rudin-Keisler partial order on $\mathbb{T}(\boldsymbol{\beta}(\alpha))$ satisfies the anti-symmetric property.*

Proof. If $p\approx q$ then $p\leqslant q$ and $q\leqslant p$.

If $p\leqslant q$ and $q\leqslant p$ then there are functions $f,g\in\alpha^\alpha$ such that $\overline{f}(p)=q$ and $\overline{g}(q)=p$. Thus $\overline{g\circ f}(p)=\overline{g}(\overline{f}(p))=\overline{g}(q)=p$, and thus according to Theorem 9.2(a) we have

$$A=\{\xi<\alpha: g\circ f(\xi)=\xi\}\in p\,.$$

Then $f|A$ is one-to-one, and $p\approx q$ by Theorem 9.2(b).

Rudin-Keisler Minimal Types in $\tau[\boldsymbol{\beta}(\alpha)\setminus\alpha]$

In the next theorem (9.6 below) we obtain various characterizations of those ultrafilters on α whose types are minimal in the Rudin-Keisler order on $\tau[\boldsymbol{\beta}(\alpha)\setminus\alpha]$. In doing so we restrict attention without any essential loss to uniform ultrafilters. For if M_κ is the set of types $\tau\in\mathbb{T}(\boldsymbol{\beta}(\alpha))$ for which there are $A\in\mathscr{P}(\alpha)$ and $p\in\boldsymbol{\beta}(\alpha)$ such that

$A\in p$ (*i.e.*, $p\in \mathrm{cl}_{\boldsymbol{\beta}(\alpha)} A=\boldsymbol{\beta}(A)$),

$\|p\|=|A|=\kappa$,

$\tau(p)=\tau$, and

$\tau(p)$ is minimal in the Rudin-Keisler order of $\tau[\boldsymbol{\beta}(A)\setminus A]$,

then the set of minimal types of $\tau[\boldsymbol{\beta}(\alpha)\setminus\alpha]$ is equal to $\bigcup\{M_\kappa:\omega\le\kappa\le\alpha\}$.

It turns out that if there are uniform ultrafilters on α whose types are minimal in $\tau[\boldsymbol{\beta}(\alpha)\setminus\alpha]$ then α is a measurable cardinal (and if $\alpha>\omega$ then every such ultrafilter is type-equivalent to a normal ultrafilter).

Definitions. An ultrafilter p on α is *κ-uniformly selective* if for every $f\in\alpha^\alpha$ there is $A\in p$ such that either $f|A$ is one-to-one or $|f[A]|<\kappa$.

If p is ω-uniformly selective then p is simply *selective;* and if p is α-uniformly selective then p is *uniformly selective.*

It is clear that a uniform ultrafilter p on α is selective if and only if for every $f\in\alpha^\alpha$ there is $A\in p$ such that $f|A$ is either one-to-one or constant.

A *partition* of α is a family

$$d=\{d_\xi : \xi<\alpha\}$$

of subsets of α such that

$$\bigcup_{\xi<\alpha} d_\xi=\alpha, \quad \text{and}$$
$$d_\xi \cap d_\zeta=\emptyset \quad \text{for } \xi<\zeta<\alpha.$$

9.4. Lemma. *Let $\omega\le\kappa\le\alpha$ and let p be a uniform ultrafilter on α. The following are equivalent.*

(a) *$\tau(p)$ is minimal in the Rudin-Keisler order of $\tau[U_\kappa(\alpha)]$;*

(b) *p is κ-uniformly selective;*

(c) *for every partition $d=\{d_\xi:\xi<\alpha\}$ of α either there is $S\in\mathscr{P}_\kappa(\alpha)$ such that $\bigcup_{\xi\in S} d_\xi\in p$ or there is $A\in p$ such that $|A\cap d_\xi|\le 1$ for $\xi<\alpha$;*

(d) *for every partition $d=\{d_\xi:\xi<\alpha\}$ of α there is $A\in p$ such that $|\{\xi<\alpha:|A\cap d_\xi|>1\}|<\kappa$.*

Proof. (a)$\Rightarrow$(b). Let $f:\alpha\to\alpha$ be a function. Since $\tau(p)$ is minimal in the Rudin-Keisler order of $\tau[U_\kappa(\alpha)]$ and $\bar f(p)\preccurlyeq p$, either $\bar f(p)\approx p$ or $\bar f(p)\notin U_\kappa(\alpha)$. If $\bar f(p)\approx p$, then by Theorem 9.2(a) there is $A\in p$ such that $f|A$ is one-to-one; and if $\bar f(p)\notin U_\kappa(\alpha)$ then there is $B\in\mathscr{P}_\kappa(\alpha)$ such that $B\in\bar f(p)$, so that $A=f^{-1}(B)\in p$ and $|f[A]|=|B|<\kappa$. Thus p is κ-uniformly selective.

(b)$\Rightarrow$(c). Let $d=\{d_\xi:\xi<\alpha\}$ be a partition of α, and define $f:\alpha\to\alpha$ by the condition $f[d_\xi]=\{\xi\}$ (for $d_\xi\neq\emptyset$ and $\xi<\alpha$). Since p is κ-uniformly selective there is $A\in p$ such that either $f|A$ is one-to-one or $|f[A]|<\kappa$. If $f|A$ is one-to-one then $|A\cap d_\xi|\le 1$ for $\xi<\alpha$; and if $|f[A]|<\kappa$, then $A\subset f^{-1}(f[A])=\bigcup_{\xi\in f[A]} d_\xi\in p$. Thus (c) is satisfied.

(c)$\Rightarrow$(d). Let $d=\{d_\xi:\xi<\alpha\}$ be a partition of α. If there is $S\in\mathscr{P}_\kappa(\alpha)$ such that $\bigcup_{\xi\in S} d_\xi\in p$, then we set $A=\bigcup_{\xi\in S} d_\xi$ and note that $|\{\xi<\alpha:|d_\xi\cap A|>1\}|\le|S|<\kappa$; and if there is $A\in p$ such that $|A\cap d_\xi|\le 1$ for $\xi<\alpha$, then $|\{\xi<\alpha:|d_\xi\cap A|>1\}|=0$.

(d)$\Rightarrow$(a). We let $f\in\alpha^\alpha$ be such that $\bar f(p)\in U_\kappa(\alpha)$ and we prove that $\bar f(p)\approx p$. We set $d_\xi=f^{-1}(\{\xi\})$ for $\xi<\alpha$ and $d=\{d_\xi:\xi<\alpha\}$. Since d is a partition of α there is $A\in p$ such that $|\{\xi<\alpha:|A\cap d_\xi|>1\}|<\kappa$. We set

$$S=\{\xi<\alpha:|A\cap d_\xi|>1\}$$

and we have $S\notin\bar f(p)$ since $\bar f(p)\in U_\kappa(\alpha)$; hence

$$\alpha\setminus S=\{\xi<\alpha:|A\cap d_\xi|\le 1\}\in\bar f(p).$$

Thus $A\cap f^{-1}(\alpha\setminus S)\in p$ and $f|(A\cap f^{-1}(\alpha\setminus S))$ is one-to-one; it follows from Theorem 9.2(b) that $\bar f(p)\approx p$.

9.5. Lemma. *Let $\omega \leq \kappa \leq \alpha$ and let p be a κ-uniformly selective, κ-complete uniform ultrafilter on α. Then*

(a) *p is α-complete (and thus α is a measurable cardinal); and*

(b) *if $\alpha > \omega$ then p is type-equivalent to a normal ultrafilter on α.*

Proof. (a) If the statement fails then (by the remarks preceding Theorem 7.3) there are $\beta < \alpha$ and a family $\{A_\eta : \eta < \beta\}$ such that

$$A_\eta \in p \quad \text{for } \eta < \beta,$$
$$\alpha = A_0, \quad A_{\eta'} \subset A_\eta \quad \text{for } \eta < \eta' < \beta, \quad \text{and}$$
$$\bigcap_{\eta < \beta} A_\eta = \emptyset.$$

We set $d_\eta = A_\eta \setminus A_{\eta+1}$ for $\eta < \beta$. It is clear that

$$d_\eta \notin p \quad \text{for } \eta < \beta,$$
$$d_\eta \cap d_{\eta'} = \emptyset \quad \text{for } \eta < \eta' < \beta, \quad \text{and}$$
$$\bigcup_{\eta < \beta} d_\eta = \alpha.$$

We define $f: \alpha \to \alpha$ by the condition $f[d_\eta] = \{\eta\}$ (for $d_\eta \neq \emptyset$ and $\eta < \beta$). Since p is κ-uniformly selective there is $A \in p$ such that either $f|A$ is one-to-one or $|f[A]| < \kappa$. Since $f^{-1}(\{\eta\}) = d_\eta \notin p$ for $\eta < \beta$ and p is κ-complete the latter possibility cannot occur. Thus there is $A \in p$ such that $f|A$ is one-to-one. Then $|A \cap d_\eta| \leq 1$ for $\eta < \beta$, and $|A| = \alpha$ because p is uniform. It follows that $\beta = \alpha$, a contradiction.

(b) Since $\alpha > \omega$ and p is α-complete, the ultrapower α^α/p is well-ordered (Lemma 8.24). Let $f \in \alpha^\alpha$ be such that f/p is the least non-constant element of α^α/p; then $\bar{f}(p)$ is a normal ultrafilter on α (Theorem 8.26), and in particular $\bar{f}(p) \in U_\kappa(\alpha)$. Since a κ-uniformly selective ultrafilter on α has minimal type in the Rudin-Keisler order of $\tau[U_\kappa(\alpha)]$ (Lemma 9.4) it follows that $\bar{f}(p) \approx p$, *i.e.*, p is type-equivalent to a normal ultrafilter on α.

Definitions. Let $\omega \leq \kappa \leq \alpha$ and let p be a uniform ultrafilter on α. Then,

p is a *Ramsey* ultrafilter if for every $n < \omega$ and every partition $\{P_0, P_1\}$ of $[\alpha]^n$, there is $A \in p$ such that either $[A]^n \subset P_0$ or $[A]^n \subset P_1$. (The set A is *homogeneous* for the partition $\{P_0, P_1\}$.)

p is a *weakly Ramsey* ultrafilter if for every partition $\{P_0, P_1\}$ of $[\alpha]^2$ there is $A \in p$ such that either $[A]^2 \subset P_0$ or $[A]^2 \subset P_1$; and,

p is a *quasi-normal* ultrafilter if for every family $\{A_\xi : \xi < \alpha\} \subset p$ there is $A \in p$ such that if $\xi, \zeta \in A$ and $\xi < \zeta$ then $\zeta \in A_\xi$.

9.6. Theorem. *Let α be an infinite cardinal and let p be a uniform ultrafilter on α. The following statements are equivalent.*

(a) *$\tau(p)$ is minimal in the Rudin-Keisler order of $\tau[\boldsymbol{\beta}(\alpha)\setminus\alpha]$;*

(b) *p is a selective ultrafilter;*

(c) *there is a cardinal κ, with $\omega\leq\kappa\leq\alpha$, such that p is κ-complete and $\tau(p)$ is minimal in the Rudin-Keisler order of $\tau[U_\kappa(\alpha)]$;*

(d) *for every partition $d=\{d_\xi:\xi<\alpha\}$ of α there is $A\in p$ such that*

$$|\{\xi<\alpha:|d_\xi\cap A|>1\}|\leq 1\,;$$

(e) *p is a Ramsey ultrafilter;*

(f) *p is a weakly Ramsey ultrafilter;*

(g) *for every ramification system $\langle\alpha,\preccurlyeq\rangle$ there is $A\in p$ such that A is either a chain or an anti-chain in $\langle\alpha,\preccurlyeq\rangle$;*

(h) $|\boldsymbol{\phi}^{-1}(\{\langle p,p\rangle\})|=3$ *(where $\boldsymbol{\phi}$ denotes the natural function $\boldsymbol{\phi}:\boldsymbol{\beta}(\alpha\times\alpha)\to\boldsymbol{\beta}(\alpha)\times\boldsymbol{\beta}(\alpha)$); and*

(i) *p is a quasi-normal ultrafilter.*

If in addition $\alpha>\omega$ then the following statement is also equivalent.

(j) *p is type-equivalent to a normal ultrafilter.*

Proof. The proof of the theorem proceeds according to the following diagram of implications.

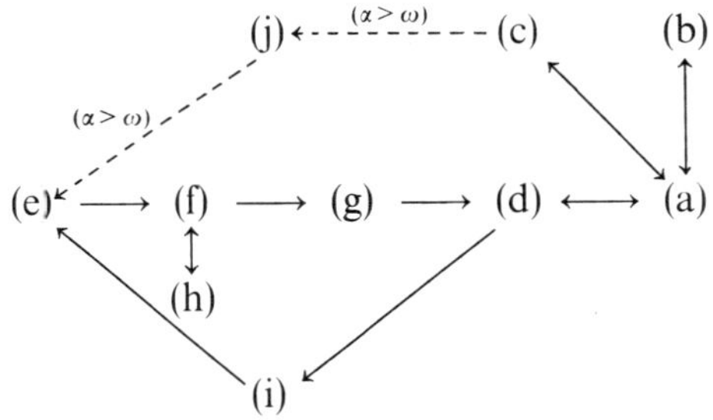

The equivalences (a)$\Leftrightarrow$(b)$\Leftrightarrow$(d) follow from Lemma 9.4.

(e)$\Rightarrow$(f) is trivial.

(f)$\Rightarrow$(g). Let $\langle\alpha,\preccurlyeq\rangle$ be a ramification system. We set

$$P_0=\{\{\xi,\zeta\}\in[\alpha]^2:\text{either }\xi\preccurlyeq\zeta\text{ or }\zeta\preccurlyeq\xi\},\quad\text{and}$$
$$P_1=[\alpha]^2\setminus P_0\,.$$

Since p is a weakly Ramsey ultrafilter there is $A\in p$ such that either $[A]^2\subset P_0$, in which case A is a chain of $\langle\alpha,\preccurlyeq\rangle$, or $[A]^2\subset P_1$, in which case A is an anti-chain of $\langle\alpha,\preccurlyeq\rangle$.

(g)$\Rightarrow$(d). Let $\{d_\xi:\xi<\alpha\}$ be a partition of α. We define the binary relation $\preccurlyeq$ on α by

$$\zeta\preccurlyeq\eta\quad\text{if }\zeta\leq\eta\quad\text{and there is }\xi<\alpha\quad\text{such that }\zeta,\eta\in d_\xi\,.$$

It is clear that $\langle\alpha, \preccurlyeq\rangle$ is a ramification system and thus there is $A \in p$ such that either A is a chain or A is an anti-chain of $\langle\alpha, \preccurlyeq\rangle$. If A is a chain we have

$$|\{\xi<\alpha : |d_\xi \cap A|>1\}|=1\,,$$

and if A is an anti-chain we have

$$\{\xi<\alpha : |d_\xi \cap A|>1\}=\emptyset\,.$$

(d)⇒(i). Let $\{A_\xi : \xi<\alpha\} \subset p$. If $\bigcap_{\xi<\alpha} A_\xi \in p$, we set $A=\bigcap_{\xi<\alpha} A_\xi$ and it is clear that A satisfies the defining property of quasi-normality.

We assume that $\bigcap_{\xi<\alpha} A_\xi \notin p$. We assume with no loss of generality (replacing A_ξ by $A_\xi \setminus \bigcap_{\eta<\alpha} A_\eta$ for $\xi<\alpha$) that $\bigcap_{\xi<\alpha} A_\xi=\emptyset$.

We define

$$f : \alpha \to \alpha$$

by the rule $f(\zeta)=\min\{\xi<\alpha : \zeta \notin A_\xi\}$. We note that $f^{-1}(\{\xi\}) \cap A_\xi=\emptyset$ for $\xi<\alpha$ and hence f is not a constant function on any element of p. Since p is selective there is $B \in p$ such that $f|B$ is one-to-one. Thus we have $|\{\eta \in B : f(\eta) \leq \zeta\}| \leq |\zeta|<\alpha$, and since α is regular by Lemma 9.5(a), we have $\sup\{\eta \in B : f(\eta) \leq \zeta\}<\alpha$ for $\zeta<\alpha$. We define

$$g : \alpha \to \alpha$$

by the rule $g(\zeta)=\max\{\zeta+1, \sup\{\eta \in B : f(\eta) \leq \zeta\}\}$. We note that

$$\begin{aligned} &\zeta+1 \leq g(\zeta) \quad \text{for } \zeta<\alpha\,, \quad \text{and} \\ (*) \qquad &g(\zeta) \leq g(\eta) \quad \text{for } \zeta<\eta<\alpha\,. \end{aligned}$$

We define

$$d : \alpha \to \alpha$$

recursively by

$$\begin{aligned} &d(0)=0\,, \\ &d(\zeta+1)=g(d(\zeta)) \quad \text{for } \zeta<\alpha\,, \quad \text{and} \\ &d(\zeta)=\sup\{d(\eta) : \eta<\zeta\} \quad \text{for non-zero limit ordinals } \zeta<\alpha. \end{aligned}$$

The fact that α is regular implies that d is well-defined. It is clear that

$$\sup\{d(\zeta) : \zeta<\alpha\}=\alpha\,.$$

We define

$$h : \alpha \to \alpha$$

by $h(\zeta)=\min\{\eta<\alpha : \zeta \leq d(\eta)\}$. We note that

$$\begin{aligned} &\zeta \leq d(h(\zeta)) \quad \text{for } \zeta<\alpha\,, \\ &h(\zeta) \leq h(\eta) \quad \text{for } \zeta<\eta<\alpha\,, \quad \text{and} \\ &|h^{-1}(\{\xi\})| \leq |d(\xi)|<\alpha \quad \text{for } \xi<\alpha\,. \end{aligned}$$

Since p is uniform, h is not a constant function on any element of p; and since p is selective there is $C \in p$ such that $h|C$ is one-to-one.

We note that $B \cap C \in p$; since p is uniform, we have $|B \cap C| = \alpha$, and there is a unique order-isomorphism $\phi: B \cap C \to \alpha$. We set

$$D_0 = \{\zeta \in B \cap C : \phi(\zeta) \text{ is an even ordinal}\}, \quad \text{and}$$
$$D_1 = B \cap C \setminus D_0 .$$

We denote by A the unique set among D_0, D_1 which is an element of p. If $\xi, \zeta \in A$ and $\xi < \zeta$, then from the definitions of ϕ and A it follows that there is $\eta \in B \cap C$ such that $\xi < \eta < \zeta$, and thus since $h|C$ is one-to-one we have $h(\xi) < h(\eta) < h(\zeta)$ and hence

$$h(\xi) + 1 < h(\zeta) \quad \text{and} \quad d(h(\xi) + 1) < \zeta ;$$

since also $\xi \le d(h(\xi))$, from (*) we have

$$g(\xi) \le g(d(h(\xi))) = d(h(\xi) + 1) < \zeta ,$$

and hence $\xi < f(\zeta)$, *i.e.*, $\zeta \in A_\xi$.

(i)⇒(e). We proceed by induction. There is nothing to prove if $n = 0$. Let $n < \omega$, assume that the Ramsey condition is true for n, and let

$$[\alpha]^{n+1} = P_0 \cup P_1 \quad \text{and}$$
$$P_0 \cap P_1 = \emptyset .$$

We define for $\xi < \alpha$

$$P_0(\xi) = \{F \in [\alpha]^n : \text{either } \xi \in F, \text{ or } \xi \notin F \text{ and } F \cup \{\xi\} \in P_0\}$$

and

$$P_1(\xi) = \{F \in [\alpha]^n : \xi \notin F \text{ and } F \cup \{\xi\} \in P_1\} .$$

By the inductive assumption there is $A_\xi \in p$ and $\varepsilon(\xi) < 2$ such that

$$[A_\xi]^n \subset P_{\varepsilon(\xi)}(\xi)$$

for $\xi < \alpha$. Then $\varepsilon \in 2^\alpha$ and there is $i < 2$ such that $\varepsilon^{-1}(\{i\}) = B \in p$. Then

$$A = \{\zeta \in B : \zeta \in A_\xi \text{ for } \xi < \zeta\} \in p$$

since p is quasi-normal. We prove that $[A]^{n+1} \subset P_i$. Indeed let $F \in [A]^{n+1}$, set $\xi = \min F$ and $G = F \setminus \{\xi\}$. Thus $G \in [A]^n$, $\xi \in B$, and $\xi < \zeta$ for $\zeta \in G$; hence if $\zeta \in G$ then $\zeta \in A_\xi$, so

$$G \in [A_\xi]^n \subset P_i(\xi)$$

and $F \in P_i$.

(f)⇒(h). We make use of the notation of Lemma 7.20(b). We have

$$\begin{aligned}&\phi^{-1}(\{\langle p,p\rangle\})\supset\{p\cdot p, \bar{t}(p\cdot p), \bar{d}(p)\},\\ &U_\alpha\in p\cdot p, \quad L_\alpha\in\bar{t}(p\cdot p), \quad \Delta_\alpha\in\bar{d}(p), \quad \text{and}\\ &\{U_\alpha, L_\alpha, \Delta_\alpha\} \text{ is a pairwise disjoint family of sets.}\end{aligned}$$

It is enough to verify that

$$\begin{aligned}p\cdot p&=\langle p\times p\cup\{U_\alpha\}\rangle,\\ \bar{t}(p\cdot p)&=\langle p\times p\cup\{L_\alpha\}\rangle, \quad \text{and}\\ \bar{d}(p)&=\langle p\times p\cup\{\Delta_\alpha\}\rangle.\end{aligned}$$

The third of these three statements is clear and the verification of the second statement is similar to the first.

Since p is a uniform ultrafilter, $p\times p\cup\{U_\alpha\}$ is a filter base on $\alpha\times\alpha$. It is thus enough to prove that for $A\in\mathscr{P}(\alpha\times\alpha)$ there is $B\in p$ such that

$$\text{either } (B\times B)\cap U_\alpha\subset A \quad \text{or } (B\times B)\cap U_\alpha\subset(\alpha\times\alpha)\setminus A.$$

Let $u: U_\alpha\to[\alpha]^2$ be the function given by $u(\xi,\zeta)=\{\xi,\zeta\}$; it is clear that u is one-to-one and onto. For $A\in\mathscr{P}(\alpha\times\alpha)$ we set

$$\begin{aligned}P_0&=u[U_\alpha\cap A], \quad \text{and}\\ P_1&=u[U_\alpha\setminus A],\end{aligned}$$

so that $\{P_0,P_1\}$ is a partition of $[\alpha]^2$. Since p is a weakly Ramsey ultrafilter there is $B\in p$ such that

$$\text{either } [B]^2\subset P_0 \quad \text{or } [B]^2\subset P_1.$$

If $[B]^2\subset P_0$ then $(B\times B)\cap U_\alpha\subset A$ and if $[B]^2\subset P_1$ then $(B\times B)\cap U_\alpha \subset(\alpha\times\alpha)\setminus A$.

(h)⇒(f). We make use of the notation of Lemma 7.20(b). Let $u: U_\alpha\to[\alpha]^2$ be the function given by $u(\xi,\zeta)=\{\xi,\zeta\}$, and let

$$\begin{aligned}[\alpha]^2&=P_0\cup P_1 \quad \text{and}\\ P_0\cap P_1&=\emptyset.\end{aligned}$$

Then $u^{-1}(P_0)\in\mathscr{P}(\alpha\times\alpha)$, and since $|\phi^{-1}(\{\langle p,p\rangle\})|=3$ it follows that $(p\times p)\cup\{U_\alpha\}$ is contained in a unique ultrafilter. Hence there is $B\in p$ such that

$$\text{either } (B\times B)\cap U_\alpha\subset u^{-1}(P_0) \quad \text{or } (B\times B)\cap U_\alpha\subset(\alpha\times\alpha)\setminus u^{-1}(P_0)=u^{-1}(P_1),$$

i.e., either $[B]^2\subset P_0$ or $[B]^2\subset P_1$.

We have proved the equivalence of the statements (a), (b), (d), (e), (f), (g), (h), (i). The equivalence (a)$\Leftrightarrow$(c) is trivial if $\alpha=\omega$. We assume now that $\alpha>\omega$.

(a)$\Rightarrow$(c) is trivial.

(c)$\Rightarrow$(j) follows from Lemma 9.5.

(j)$\Rightarrow$(e) follows from Theorem 8.27.

Good Ultrafilters

Definitions. Let α and β be cardinals.

A function $\phi:\mathscr{P}_\omega(\beta)\to\mathscr{P}(\alpha)$ is *monotone* if

$$\phi(F)\subset\phi(G) \quad \text{for } F,G\in\mathscr{P}_\omega(\beta) \quad \text{and } G\subset F\,.$$

A function $\phi:\mathscr{P}_\omega(\beta)\to\mathscr{P}(\alpha)$ is *multiplicative* if

$$\phi(F\cup G)=\phi(F)\cap\phi(G) \quad \text{for } F,G\in\mathscr{P}_\omega(\beta)\,.$$

For two functions $\phi,\psi:\mathscr{P}_\omega(\beta)\to\mathscr{P}(\alpha)$ we define $\psi\le\phi$ if $\psi(F)\subset\phi(F)$ for $F\in\mathscr{P}_\omega(\beta)$.

An ultrafilter p on α is *β-good* if for every $\gamma<\beta$ and every monotone function $\phi:\mathscr{P}_\omega(\gamma)\to p$ there is a multiplicative function $\psi:\mathscr{P}_\omega(\gamma)\to p$ such that $\psi\le\phi$. If there is a cardinal β such that p is not β-good then the *degree of goodness* $\boldsymbol{G}(p)$ is given by

$$\boldsymbol{G}(p)=\min\{\beta: p \text{ is not } \beta^+\text{-good}\}\,.$$

9.7. Lemma. *An ultrafilter p is α^+-good if and only if for every (monotone) function $\phi:\mathscr{P}_\omega(\alpha)\to p$ there is a multiplicative function $\psi:\mathscr{P}_\omega(\alpha)\to p$ such that $\psi\le\phi$.*

Proof. Let $\beta\le\alpha$ and let ϕ be a monotone function from $\mathscr{P}_\omega(\beta)$ to p. We define $\phi':\mathscr{P}_\omega(\alpha)\to p$ by $\phi'(F)=\phi(F\cap\beta)$. Clearly ϕ' is monotone and thus by hypothesis there is a multiplicative function $\psi':\mathscr{P}_\omega(\alpha)\to p$ such that $\psi'\le\phi'$. Let $e:\mathscr{P}_\omega(\beta)\to\mathscr{P}_\omega(\alpha)$ be the natural inclusion induced by the inclusion $\beta\subset\alpha$, and set $\psi=\psi'\circ e$. Then ψ is multiplicative and $\psi\le\phi$.

9.8. Lemma. *Every ultrafilter is ω^+-good.*

Proof. Let p be an ultrafilter (on some cardinal). By Lemma 9.7, it is enough to consider monotone functions $\phi:\mathscr{P}_\omega(\omega)\to p$. For $F\in\mathscr{P}_\omega(\omega)$ we set

$$n(F)=\min\{n<\omega: F\subset n\}\in\mathscr{P}_\omega(\omega)\,,$$

and we define

$$\psi:\mathscr{P}_\omega(\omega)\to p$$

by $\psi(F)=\phi(n(F))$. Since $F\subset n(F)$ and ϕ is monotone we have $\psi(F)=\phi(n(F))\subset\phi(F)$ for $F\in\mathscr{P}_\omega(\omega)$, *i.e.*, $\psi\le\phi$.

To prove that ψ is multiplicative let $F,G\in\mathscr{P}_\omega(\omega)$; then

$$n(F\cup G)=n(F)\cup n(G)=\max\{n(F),n(G)\},$$

and hence

$$\begin{aligned}\psi(F\cup G)&=\phi(n(F\cup G))=\phi(\max\{n(F),n(G)\})\\&=\phi(n(F))\cap\phi(n(G))=\psi(F)\cap\psi(G)\end{aligned}$$

since ϕ is monotone.

9.9. Theorem. *If p is a countably incomplete, β^+-good ultrafilter on α, then p is (ω,β)-regular.*

Proof. Since p is countably incomplete there is a family $\{A_n:n<\omega\}$ of elements of p such that $A_0=\alpha$, $A_{n+1}\subset A_n$ for $n<\omega$, and $\bigcap_{n<\omega}A_n=\emptyset$. We define $\phi:\mathscr{P}_\omega(\beta)\to p$ by

$$\phi(F)=A_{|F|}.$$

Then ϕ is monotone, and since p is β^+-good there is a multiplicative function $\psi:\mathscr{P}_\omega(\beta)\to p$ such that $\psi\le\phi$. For $\zeta<\alpha$ we set

$$F(\zeta)=\{\xi<\beta:\zeta\in\psi(\{\xi\})\}.$$

We set $n(\zeta)=\min\{n<\omega:\zeta\notin A_n\}$, so that $n(\zeta)<\omega$ for $\zeta<\alpha$, and we claim that

$$|F(\zeta)|<n(\zeta)\quad\text{for } \zeta<\alpha.$$

Indeed suppose that for some $\zeta<\alpha$ we have $|F(\zeta)|\ge n(\zeta)$; then there is $G\subset F(\zeta)$ such that $|G|=n(\zeta)$, and since ψ is multiplicative we have

$$\zeta\in\bigcap_{\xi\in G}\psi(\{\xi\})=\psi(G)\subset\phi(G)=A_{n(\zeta)},$$

a contradiction. Thus $F(\zeta)\in\mathscr{P}_\omega(\beta)$ for $\zeta<\alpha$. It follows that every intersection of infinitely many elements of the family $\{\psi(\{\xi\}):\xi<\beta\}$ of elements of p is empty; therefore p is (ω,β)-regular.

9.10. Corollary. *Let p be a countably incomplete ultrafilter on α. Then*

(a) $\omega^+\le\boldsymbol{G}(p)\le\alpha^+$;

(b) *if* $\boldsymbol{G}(p)=\alpha^+$, *then p is regular (and therefore uniform).*

Proof. (a) We have $\omega^+\le\boldsymbol{G}(p)$ from Lemma 9.8. If $\boldsymbol{G}(p)>\alpha^+$ then p is α^{++}-good and is therefore by Theorem 9.9 an (ω,α^+)-regular ultrafilter on α; it is clear that such an ultrafilter does not exist.

(b) is the special case $\beta=\alpha$ of Theorem 9.9.

9.11. Lemma. *Let α be an infinite cardinal, let $\mathscr{F}$ be a family of subsets of α with the uniform finite intersection property such that $|\mathscr{F}|\le\alpha$, and* *let* $d=\{d_\zeta:\zeta<\alpha\}$ *be a partition of* α. *Then there is a subset* F *of* α *such that*

$$\mathscr{F}\cup\{F\} \text{ has the uniform finite intersection property},$$

and

$$|\{\zeta<\alpha:|F\cap d_\zeta|>1\}|<\alpha\,.$$

Proof. With no loss of generality we assume that $\mathscr{F}$ is closed under finite intersections. Let $\langle E_\xi:\xi<\alpha\rangle$ be a well-ordering of $\mathscr{F}$. There are two cases to consider.

Case 1. There is $\xi<\alpha$ such that $|\{\zeta<\alpha:|E_\xi\cap d_\zeta|>1\}|<\alpha$. In this case we set $F=E_\xi$.

Case 2. $|\{\zeta<\alpha:|E_\xi\cap d_\zeta|>1\}|=\alpha$ for every $\xi<\alpha$. We set

$$A_\xi=\{\zeta<\alpha:|E_\xi\cap d_\zeta|>1\}\,,$$

so that $|A_\xi|=\alpha$ for $\xi<\alpha$. By Lemma 7.5, there is a family $\{B_\xi:\xi<\alpha\}$ of subsets of α such that $B_\xi\subset A_\xi$, $|B_\xi|=\alpha$, and $B_\xi\cap B_\eta=\emptyset$ for $\xi<\eta<\alpha$. We choose $p_{\xi,\zeta}\in E_\xi\cap d_\zeta$ for $\xi<\alpha$ and $\zeta\in B_\xi$, and we set

$$F=\{p_{\xi,\zeta}:\xi<\alpha \text{ and } \zeta\in B_\xi\}\,.$$

We note that if $\xi<\alpha$ and $\zeta,\zeta'\in B_\xi$ and $\zeta\neq\zeta'$, then $p_{\xi,\zeta}\neq p_{\xi,\zeta'}$. Since

$$\{p_{\xi,\zeta}:\zeta\in B_\xi\}\subset E_\xi\cap F$$

it follows that $|E_\xi\cap F|=\alpha$ for $\xi<\alpha$, and since $\mathscr{F}$ is closed under finite intersections the family $\mathscr{F}\cup\{F\}$ has the uniform finite intersection property. Furthermore if $\zeta<\alpha$ and $\zeta\notin\bigcup_{\xi<\alpha}B_\xi$ then $F\cap d_\zeta=\emptyset$; and if $\zeta<\alpha$ and $\zeta\in B_\xi$ for some (necessarily unique) $\xi<\alpha$, then $F\cap d_\zeta=\{p_{\xi,\zeta}\}$. Thus, in this case, $|F\cap d_\zeta|\leq 1$ for $\zeta<\alpha$.

9.12. Lemma. *Let* α *be an infinite cardinal, let* $\mathscr{F}$ *be a family of subsets of* α *with* $|\mathscr{F}|\leq\alpha$, *and let* ϕ *be a monotone function from* $\mathscr{P}_\omega(\alpha)$ *into* $\mathscr{P}(\alpha)$ *such that*

$$\mathscr{F}\cup\phi[\mathscr{P}_\omega(\alpha)] \text{ has the uniform finite intersection property.}$$

Then there is a function $\psi:\mathscr{P}_\omega(\alpha)\to\mathscr{P}(\alpha)$ *such that*

$\psi\leq\phi$,

ψ *is multiplicative, and*

$\mathscr{F}\cup\psi[\mathscr{P}_\omega(\alpha)]$ *has the uniform finite intersection property.*

Proof. With no loss of generality we assume that $\mathscr{F}$ is closed under finite intersections. Let

$$h:\mathscr{P}_\omega(\alpha)\times\mathscr{F}\to\mathscr{P}(\alpha)$$

be given by $h(F,E)=\phi(F)\cap E$. Since

$$|\mathscr{P}_\omega(\alpha)\times\mathscr{F}|=\alpha \quad \text{and}$$
$$|h(F,E)|=\alpha \quad \text{for } \langle F,E\rangle\in\mathscr{P}_\omega(\alpha)\times\mathscr{F},$$

there is by Lemma 7.5 a family $\{d(F,E):\langle F,E\rangle\in\mathscr{P}_\omega(\alpha)\times\mathscr{F}\}$ of subsets of α such that

$$|d(F,E)|=\alpha,$$
$$d(F,E)\subset h(F,E), \quad \text{and}$$
$$d(F,E)\cap d(F',E')=\emptyset$$

for $\langle F,E\rangle$, $\langle F',E'\rangle\in\mathscr{P}_\omega(\alpha)\times\mathscr{F}$ and $\langle F,E\rangle\neq\langle F',E'\rangle$. We define

$$\psi:\mathscr{P}_\omega(\alpha)\rightarrow\mathscr{P}(\alpha)$$

by $\psi(F)=\bigcup\{d(G,E):\langle G,E\rangle\in\mathscr{P}_\omega(\alpha)\times\mathscr{F},\ F\subset G\}$, and we verify the following statements.

(i) $\psi\leq\phi$.

For $F\in\mathscr{P}_\omega(\alpha)$ and $\langle G,E\rangle\in\mathscr{P}_\omega(\alpha)\times\mathscr{F}$ with $F\subset G$ we have

$$d(G,E)\subset h(G,E)=\phi(G)\cap E\subset\phi(G)\subset\phi(F),$$

and thus $\psi(F)\subset\phi(F)$.

(ii) ψ is multiplicative.

It is clear from the definition of ψ that $\psi(F\cup F')\subset\psi(F)\cap\psi(F')$ for $F,F'\in\mathscr{P}_\omega(\alpha)$. For the reverse inclusion let $\xi\in\psi(F)\cap\psi(F')$; then there are $\langle G,E\rangle$, $\langle G',E'\rangle\in\mathscr{P}_\omega(\alpha)\times\mathscr{F}$ such that

$$F\subset G, \quad F'\subset G', \quad \text{and}$$
$$\xi\in d(G,E)\cap d(G',E').$$

Since $d(G,E)\cap d(G',E')\neq\emptyset$ it follows that $\langle G,E\rangle=\langle G',E'\rangle$, hence that $F\cup F'\subset G$ and $\xi\in\psi(F\cup F')$.

(iii) $\mathscr{F}\cup\psi[\mathscr{P}_\omega(\alpha)]$ has the uniform finite intersection property. Indeed $d(F,E)\subset\psi(F)\cap E$ for $\langle F,E\rangle\in\mathscr{P}_\omega(\alpha)\times\mathscr{F}$. If $E\in\mathscr{F}$, $n<\omega$, and $F_k\in\mathscr{P}_\omega(\alpha)$ for $k<n$, we set $F=\bigcup_{k<n}F_k$ and we have

$$\bigcap_{k<n}\psi(F_k)\cap E=\psi(F)\cap E\supset d(F,E),$$

so that $|\bigcap_{k<n}\psi(F_k)\cap E|=\alpha$.

9.13. Theorem. *Let α be an infinite cardinal and $\alpha^+=2^\alpha$. Any family of subsets of α with the uniform finite intersection property, and of cardinality at most α, is contained in 2^{2^α} (distinct) ultrafilters on α each of which is countably incomplete, α^+-good, and of minimal type in the Rudin-Keisler order of $\tau[U(\alpha)]$.*

Proof. There are (at most) $(2^\alpha)^\alpha = 2^\alpha = \alpha^+$ monotone functions on $\mathscr{P}_\omega(\alpha)$ into $\mathscr{P}(\alpha)$ and there are $\alpha^\alpha = 2^\alpha = \alpha^+$ partitions of α. Let $\langle d_\xi : \xi < \alpha^+ \rangle$ be a well-ordering of the set consisting of all such functions and all such partitions.

Let $\mathscr{F}$ be a family of subsets of α with the uniform finite intersection property such that $|\mathscr{F}| \leq \alpha$. According to Corollary 7.9 we assume without loss of generality that there is a sequence $\{A_n : n < \omega\}$ of elements of $\mathscr{F}$ such that $\bigcap_{n<\omega} A_n = \emptyset$.

For a family $\mathscr{G}$ of subsets of α with the uniform finite intersection property and such that $|\mathscr{G}| \leq \alpha$ there are, according to Theorem 7.7(a), families $\mathscr{G}^{(i)}$ for $i < 2$ of subsets of α such that

$$|\mathscr{G}^{(i)}| \leq \alpha ,$$

$$\mathscr{G}^{(i)} \text{ has the uniform finite intersection property, and}$$

$$\mathscr{G} \subset \mathscr{G}^{(i)}$$

for $i < 2$, and

$$\text{there is } A \in \mathscr{P}(\alpha) \quad \text{such that } A \in \mathscr{G}^{(0)} \quad \text{and } \alpha \setminus A \in \mathscr{G}^{(1)} .$$

We define families $\{\mathscr{F}_\xi(h) : \xi < \alpha^+\}$ for $h \in 2^{(\alpha^+)}$ such that

(i) $\mathscr{F}_0(h) = \mathscr{F}$;

(ii) $\mathscr{F}_\xi(h)$ is a family of subsets of α with the uniform finite intersection property, and $|\mathscr{F}_\xi(h)| \leq \alpha$ for $\xi < \alpha^+$;

(iii) $\mathscr{F}_\xi(h) \subset \mathscr{F}_\zeta(h)$ for $\xi < \zeta < \alpha^+$;

(iv) $\mathscr{F}_\xi(h) = \bigcup_{\zeta < \xi} \mathscr{F}_\zeta(h)$ for non-zero limit ordinals $\xi < \alpha^+$;

(v) if d_ξ is a monotone function on $\mathscr{P}_\omega(\alpha)$ into $\mathscr{P}(\alpha)$ and $d_\xi[\mathscr{P}_\omega(\alpha)] \cup (\mathscr{F}_\xi(h))^{(h(\xi))}$ does not have the uniform finite intersection property, then

$$\mathscr{F}_{\xi+1}(h) = (\mathscr{F}_\xi(h))^{(h(\xi))};$$

(vi) if d_ξ is a monotone function on $\mathscr{P}_\omega(\alpha)$ into $\mathscr{P}(\alpha)$ and $d_\xi[\mathscr{P}_\omega(\alpha)] \cup (\mathscr{F}_\xi(h))^{(h(\xi))}$ has the uniform finite intersection property, then there is a multiplicative function

$$\psi_{\xi+1} : \mathscr{P}_\omega(\alpha) \to \mathscr{P}(\alpha) \quad \text{such that}$$

$$\psi_{\xi+1} \leq d_\xi , \quad \text{and}$$

$$\mathscr{F}_{\xi+1}(h) = (\mathscr{F}_\xi(h))^{(h(\xi))} \cup \psi_{\xi+1}[\mathscr{P}_\omega(\alpha)];$$

(vii) if d_ξ is a partition of α into (at most) α elements, say $d_\xi = \{d_{\xi,\zeta} : \zeta < \alpha\}$, then there is $F_{\xi+1} \in \mathscr{P}(\alpha)$ such that

$$|\{\zeta < \alpha : |F_{\xi+1} \cap d_{\xi,\zeta}| > 1\}| < \alpha , \quad \text{and}$$

$$\mathscr{F}_{\xi+1}(h) = (\mathscr{F}_\xi(h) \cup \{F_{\xi+1}\})^{(h(\xi))} .$$

We proceed by transfinite recursion. We set

$$\mathscr{F}_0(h) = \mathscr{F} .$$

For a non-zero limit ordinal $\xi<\alpha^+$, we define $\mathscr{F}_\xi(h)$ by condition (iv). We now let $\xi<\alpha^+$ and we define $\mathscr{F}_{\xi+1}(h)$; there are three possibilities.

Case 1. d_ξ is a monotone function of $\mathscr{P}_\omega(\alpha)$ into $\mathscr{P}(\alpha)$ and $d_\xi[\mathscr{P}_\omega(\alpha)] \cup (\mathscr{F}_\xi(h))^{(h(\xi))}$ does not have the uniform finite intersection property. We define $\mathscr{F}_{\xi+1}(h)$ by condition (v).

Case 2. d_ξ is a monotone function of $\mathscr{P}_\omega(\alpha)$ into $\mathscr{P}(\alpha)$ and $d_\xi[\mathscr{P}_\omega(\alpha)] \cup (\mathscr{F}_\xi(h))^{(h(\xi))}$ has the uniform finite intersection property. We use Lemma 9.12 to define $\mathscr{F}_{\xi+1}(h)$ as in conditions (ii) and (vi).

Case 3. d_ξ is a partition of α into (at most) α elements. We use Lemma 9.11 to define $\mathscr{F}_{\xi+1}(h)$ as in conditions (ii) and (vii).

The definition of the families $\{\mathscr{F}_\xi(h): \xi<\alpha^+\}$ for $h\in 2^{(\alpha^+)}$ is now complete. For $h\in 2^{(\alpha^+)}$ we set

$$\mathscr{F}(h)=\bigcup_{\xi<\alpha^+}\mathscr{F}_\xi(h).$$

It is clear from conditions (ii) and (iii) that $\mathscr{F}(h)$ is a family of subsets of α with the uniform finite intersection property. Hence there is a uniform ultrafilter $p(h)$ on α such that $\mathscr{F}(h)\subset p(h)$.

The ultrafilter $p(h)$ is countably incomplete since $\{A_n: n<\omega\} \subset\mathscr{F}\subset p(h)$; it is α^+-good by conditions (v) and (vi) and Lemma 9.7; and it is uniformly selective, and hence of minimal type in the Rudin-Keisler order of $\tau[U(\alpha)]$, by condition (vii).

Furthermore if $h,k\in 2^{(\alpha^+)}$ and $h\neq k$ we set $\xi=\min\{\zeta<\alpha^+: h(\zeta)\neq k(\zeta)\}$ and we have

$\mathscr{F}_\xi(h)=\mathscr{F}_\xi(k)$, and

there is $A\in\mathscr{P}(\alpha)$ such that $A\in(\mathscr{F}_\xi(h))^{h(\xi)}\subset\mathscr{F}_{\xi+1}(h)\subset p(h)$ and $\alpha\backslash A\in(\mathscr{F}_\xi(k))^{k(\xi)}\subset\mathscr{F}_{\xi+1}(k)\subset p(k)$;

thus $p(h)\neq p(k)$.

The proof of the theorem is complete.

We remark that Theorems 9.6 and 9.13 provide all the information we have on the existence of uniform ultrafilters on α whose type is minimal in the Rudin-Keisler order of $\tau[U_\kappa(\alpha)]$ for some κ with $\omega\le\kappa\le\alpha$.

We now draw from Lemma 9.11 some topological consequences concerning the space of uniform ultrafilters. We need the following definitions.

Definitions. Let α be an infinite cardinal.

An *α-partition* $d=\{d_\xi: \xi<\alpha\}$ of α is a family of subsets of α such that

$$\bigcup_{\xi<\alpha} d_\xi=\alpha,$$

$$d_\xi\cap d_\zeta=\emptyset \quad \text{for } \xi<\zeta<\alpha, \quad \text{and}$$

$$|d_\xi|=\alpha \quad \text{for } \xi<\alpha.$$

An ultrafilter p on α is *uniform over an α-partition* $d=\{d_\xi:\xi<\alpha\}$ of α if for every $A\in p$ we have

$$|\{\xi<\alpha:|A\cap d_\xi|=\alpha\}|=\alpha.$$

The set of all uniform ultrafilters on α which are not uniform over any α-partition on α is denoted by $P(\alpha)$.

A family $\{V_\xi:\xi<\alpha\}$ of α distinct, non-empty open-and-closed subsets of $U(\alpha)$ is *strongly disjoint* if there is an α-partition $d=\{d_\xi:\xi<\alpha\}$ of α such that

$$V_\xi=\hat{d}_\xi(=\mathrm{cl}_{\boldsymbol{\beta}(\alpha)}d_\xi\cap U(\alpha)) \quad \text{for } \xi<\alpha.$$

If X is a space and $\{V_\xi:\xi<\alpha\}$ is a family of non-empty subsets of X and $p\in X$, then p is in the *uniform closure* of $\{V_\xi:\xi<\alpha\}$ if for every neighborhood V of p we have

$$|\{\xi<\alpha:V\cap V_\xi\neq\emptyset\}|=\alpha.$$

The space X has *property Φ_α* if the uniform closure of any family of α pairwise disjoint non-empty open-and-closed subsets of X contains no non-empty intersection of at most α open-and-closed subsets of X.

We remark that if p is a uniform ultrafilter on α whose type is minimal in the Rudin-Keisler order of $\tau[U(\alpha)]$, then $p\in P(\alpha)$ (Lemma 9.4). From Lemma 9.14(c) below it then follows that if $p\in\boldsymbol{\beta}(\omega)\backslash\omega$ is of minimal type in the Rudin-Keisler order of $\tau[\boldsymbol{\beta}(\omega)\backslash\omega]$, then p is a P-point of $\boldsymbol{\beta}(\omega)\backslash\omega$.

9.14. Lemma. (a) *If $\alpha\geq\omega$, p is an ultrafilter on α and $d=\{d_\xi:\xi<\alpha\}$ is an α-partition of α, then p is uniform over d if and only if p is in the uniform closure of $\{\hat{d}_\xi:\xi<\alpha\}$.*

(b) *If α is an infinite, regular cardinal then every family of α pairwise disjoint, non-empty, open-and-closed subsets of $U(\alpha)$ is strongly disjoint.*

(c) $P(\omega)=P_{\omega^+}(\boldsymbol{\beta}(\omega)\backslash\omega)$.

Proof. (a) follows from the equality

$$\{\xi<\alpha:|A\cap d_\xi|=\alpha\}=\{\xi<\alpha:\hat{A}\cap\hat{d}_\xi\neq\emptyset\}$$

for all $A\in\mathscr{P}(\alpha)$.

(b) Let $\{V_\xi:\xi<\alpha\}$ be a family of pairwise disjoint non-empty, open-and-closed subsets of $U(\alpha)$. We choose $A_\xi\in\mathscr{P}(\alpha)$ such that $\hat{A}_\xi=V_\xi$ for $\xi<\alpha$. Set $\beta=|\alpha\backslash\bigcup_{\xi<\alpha}A_\xi|$ and let $\langle s_\xi:\xi<\beta\rangle$ be a well-ordering of $\alpha\backslash\bigcup_{\xi<\alpha}A_\xi$. We set

$$\begin{aligned} B_\xi&=A_\xi\cup\{s_\xi\} \quad \text{for } \xi<\beta\\ &=A_\xi \quad \text{for} \quad \text{for } \beta\leq\xi<\alpha, \end{aligned}$$

so that $\bigcup_{\xi<\alpha} B_\xi = \alpha$ and $\hat{B}_\xi = \hat{A}_\xi$ for $\xi < \alpha$. Now for $\xi < \alpha$ set

$$d_\xi = B_\xi \setminus \bigcup_{\zeta<\xi} B_\zeta .$$

Since the family $\{V_\xi : \xi < \alpha\}$ is pairwise disjoint we have $|B_\xi \cap B_\zeta| < \alpha$ for $\zeta < \xi < \alpha$, and the regularity of α implies that $|B_\xi \setminus d_\xi| < \alpha$ for $\xi < \alpha$. It is then clear that $\hat{d}_\xi = \hat{B}_\xi = \hat{A}_\xi = V_\xi$ for $\xi < \alpha$.

(c) Let p be a P_{ω^+}-point of $\boldsymbol{\beta}(\omega)\setminus\omega$ and let $d = \{d_n : n < \omega\}$ be an ω-partition of ω. We prove that there is $A \in p$ such that

$$|\{n < \omega : |A \cap d_n| = \omega\}| < \omega .$$

If $d_n \in p$ for some $n < \omega$ then with $A = d_n$ we have

$$|\{k < \omega : |A \cap d_k| = \omega\}| = 1 .$$

If $d_n \notin p$ for $n < \omega$ then $d_0 \cup \cdots \cup d_n \notin p$ for $n < \omega$ and hence, since p is an ultrafilter, we have

$$\bigcup_{n<k<\omega} d_k \in p \quad \text{for } n < \omega .$$

We set $V_n = (\bigcup_{n<k<\omega} d_k)\hat{}$, so that $p \in V_n$ for $n < \omega$. Since p is a P-point there is $A \in p$ such that $\hat{A} \subset \bigcap_{n<\omega} V_n$, and then $|A \cap d_n| < \omega$ for $n < \omega$ and we have

$$\{n < \omega : |A \cap d_n| = \omega\} = \emptyset .$$

Conversely let $\{V_n : n < \omega\}$ be a sequence of open-and-closed neighborhoods of p in $\boldsymbol{\beta}(\omega)\setminus\omega$. Without any loss of generality we assume that

$$\boldsymbol{\beta}(\omega)\setminus\omega = V_0 \supsetneqq \cdots \supsetneqq V_n \supsetneqq V_{n+1} \supsetneqq \cdots .$$

Let $A_n \in \mathscr{P}(\omega)$ be such that $\hat{A}_n = V_n$ for $n < \omega$. Without loss of generality we assume that

$$\omega = A_0 \supset \cdots \supset A_n \supset A_{n+1} \supset \cdots .$$

We define $d_n = A_n \setminus A_{n+1}$ for $n < \omega$ and we set $d = \{d_n : n < \omega\}$. Clearly d is an ω-partition of ω such that $d_n \notin p$ for $n < \omega$. Hence there is $A \in p$ such that

$$|\{n < \omega : |A \cap d_n| = \omega\}| < \omega .$$

Let $F = \{n < \omega : |A \cap d_n| = \omega\}$; then $\bigcup_{n\in F} d_n \notin p$ since $|F| < \omega$ and $d_n \notin p$ for $n < \omega$. Then with

$$B = A \setminus \bigcup_{n\in F} d_n$$

we have $B \in p$ and $\hat{B} \subset V_n$ for $n < \omega$, as required.

The following result describes a Baire category type of property for $U(\alpha)$; it is closely related to the α^+-Baire category property of $(U(\alpha))_{\alpha^+}$ which follows from Theorems 15.8 and 15.14.

9.15. Corollary. *Let α be an infinite cardinal.*

(a) *A non-empty intersection of at most α open-and-closed subsets of $U(\alpha)$ is not contained in the uniform closure of any α-partition of α.*

(b) *If α is a regular cardinal then $U(\alpha)$ has property Φ_α.*

(c) *If $G=\bigcap_{\xi<\alpha^+} G_\xi$ where G_ξ is open in $U(\alpha)$ and $U(\alpha)\setminus G_\xi$ is contained in the uniform closure of an α-partition of α for $\xi<\alpha^+$, then G is dense in $U(\alpha)$.*

(d) *If $\alpha^+=2^\alpha$ then $P(\alpha)$ is dense in $U(\alpha)$.*

Proof. (a) Let $\{F_\xi:\xi<\alpha\}$ be a family of open-and-closed subsets of $U(\alpha)$ such that $\bigcap_{\xi<\alpha} F_\xi \neq \emptyset$ and let $d=\{d_\zeta:\zeta<\alpha\}$ be an α-partition of α. We let $A_\xi\in\mathscr{P}(\alpha)$ be such that $\hat{A}_\xi=F_\xi$ for $\xi<\alpha$. Since the family $\mathscr{F}=\{A_\xi:\xi<\alpha\}$ has the uniform finite intersection property, from Lemma 9.11 there is $A\in\mathscr{P}(\alpha)$ such that $\mathscr{F}\cup\{A\}$ has the uniform finite intersection property and

$$|\{\zeta<\alpha:|A\cap d_\zeta|>1\}|<\alpha\,.$$

Then

$$\bigcap_{\xi<\alpha} F_\xi\cap\hat{A}\neq\emptyset\,,$$

and it follows from Lemma 9.14(a) that

$\hat{A}$ is disjoint from the uniform closure of $\{\hat{d}_\zeta:\zeta<\alpha\}$.

(b) This follows from part (a) and Lemma 9.14(b).

(c) Let $A\in\mathscr{P}(\alpha)$ be of cardinality α. We prove that $\hat{A}\cap G\neq\emptyset$. We define a family $\{V_\xi:\xi<\alpha^+\}$ such that

$V_0=\hat{A}$,

V_ξ is the non-empty intersection of at most α open-and-closed subsets of $U(\alpha)$ for $\xi<\alpha^+$,

$V_{\xi+1}\subset V_\xi\cap G_\xi$ for $\xi<\alpha^+$, and

$V_\xi=\bigcap_{\zeta<\xi} V_\zeta$ for non-zero limit ordinals $\xi<\alpha^+$.

We proceed recursively. We define $V_0=\hat{A}$. For non-zero limit ordinals $\xi<\alpha^+$ we set $V_\xi=\bigcap_{\zeta<\xi} V_\zeta$. The compactness of $U(\alpha)$ and the recursive assumptions imply that $V_\xi\neq\emptyset$. Now let $\xi<\alpha^+$. We define $V_{\xi+1}$. By part (a) we have that $V_\xi\cap G_\xi\neq\emptyset$. Let $p\in V_\xi\cap G_\xi$ and let $A_{\xi+1}\in\mathscr{P}(\alpha)$ be such that $p\in\hat{A}_{\xi+1}\subset G_\xi$. Then $|A_{\xi+1}|=\alpha$, $\hat{A}_{\xi+1}\subset G_\xi$, and $\hat{A}_{\xi+1}\cap V_\xi\neq\emptyset$, and we define

$$V_{\xi+1}=V_\xi\cap\hat{A}_{\xi+1}\,.$$

This completes the recursive definitions. We define $V=\bigcap_{\xi<\alpha^+} V_\xi$. The compactness of $U(\alpha)$ implies that $V\neq\emptyset$ and clearly $V\subset\hat{A}\cap G$.

(d) It is clear that

$$P(\alpha)=\bigcap\{U(\alpha)\setminus C_d : d \text{ is an } \alpha\text{-partition of } \alpha\},$$

where for any α-partition d of α we have set

$$C_d=\{p\in U(\alpha): p \text{ is uniform over } d\}.$$

The cardinality of the set of all α-partitions of α is 2^α; thus if $\alpha^+=2^\alpha$ then part (c) implies that $P(\alpha)$ is a dense subset of $U(\alpha)$.

9.16. Lemma. *Let α and β be cardinals, $e:\boldsymbol{\beta}(\beta)\to\boldsymbol{\beta}(\alpha)$ a continuous function, $q\in\boldsymbol{\beta}(\beta)$ and $p=e(q)$.*

(a) *If $B\in q$ and $A_\zeta\in e(\zeta)$ for $\zeta<\beta$ then $\bigcup_{\zeta\in B}A_\zeta\in p$.*

(b) *Let $r:\alpha\to\beta$ be such that $\bar{r}\circ e$ is the identity on $\boldsymbol{\beta}(\beta)$ and let $A\in\mathscr{P}(\alpha)$; then*

$$A\in p \quad \textit{if and only if} \quad \{\zeta<\beta: A\cap r^{-1}(\{\zeta\})\in e(\zeta)\}\in q.$$

Proof. (a) It is enough to prove that if $A\in p$ then $A\cap(\bigcup_{\zeta\in B}A_\zeta)\neq\emptyset$ (by Lemma 2.3(a)). If $A\in p$ then, since e is a continuous function and $\mathrm{cl}_{\boldsymbol{\beta}(\alpha)}A$ is a neighborhood of p, it follows that $e^{-1}(\mathrm{cl}_{\boldsymbol{\beta}(\alpha)}A)$ is a neighborhood of q. There is $\bar{\zeta}\in B\cap e^{-1}(\mathrm{cl}_{\boldsymbol{\beta}(\alpha)}A)$; then $A\in e(\bar{\zeta})$ and thus

$$\emptyset\neq A\cap A_{\bar{\zeta}}\subset A\cap(\bigcup_{\zeta\in B}A_\zeta).$$

(b) If $A\in p$, then $p\in\mathrm{cl}_{\boldsymbol{\beta}(\alpha)}A$ and hence $q\in e^{-1}(\mathrm{cl}_{\boldsymbol{\beta}(\alpha)}A)$; since $e^{-1}(\mathrm{cl}_{\boldsymbol{\beta}(\alpha)}A)$ is a non-empty open-and-closed subset of $\boldsymbol{\beta}(\beta)$, it follows that $e^{-1}(\mathrm{cl}_{\boldsymbol{\beta}(\alpha)}A)\cap\beta\in q$. We show that $e^{-1}(\mathrm{cl}_{\boldsymbol{\beta}(\alpha)}A)\cap\beta\subset\{\zeta<\beta: A\cap r^{-1}(\{\zeta\})\in e(\zeta)\}$; indeed if $\zeta\in e^{-1}(\mathrm{cl}_{\beta(\alpha)}A)\cap\beta$ then $e(\zeta)\in\mathrm{cl}_{\beta(\alpha)}A$, *i.e.*, $A\in e(\zeta)$, and since $r^{-1}(\{\zeta\})\in e(\zeta)$ it follows that $r^{-1}(\{\zeta\})\cap A\in e(\zeta)$. The reverse implication follows immediately from part (a).

We now examine a topological consequence of Theorem 9.2.

Definition. Let α be a cardinal and $D\subset\boldsymbol{\beta}(\alpha)$. The set D is *strongly discrete* if there is a family $\{A_d: d\in D\}$ of subsets of α such that

$$A_d\cap A_{d'}=\emptyset \quad \text{for } d,d'\in D, \quad d\neq d', \quad \text{and}$$
$$A_d\in d \quad \text{for } d\in D.$$

9.17. Corollary. *Let α be a cardinal, $e:\boldsymbol{\beta}(\alpha)\to\boldsymbol{\beta}(\alpha)$ a topological embedding, $p\in\boldsymbol{\beta}(\alpha)$ and $e(p)=p$.*

(a) *If there is a function $r:\alpha\to\alpha$ such that $\bar{r}\circ e$ is the identity on $\boldsymbol{\beta}(\alpha)$ then*

$$\{\xi<\alpha: e(\xi)=\xi\}\in p.$$

(b) *If $e[\alpha]$ is a strongly discrete subset of $\boldsymbol{\beta}(\alpha)$ then*

$$\{\xi<\alpha: e(\xi)=\xi\}\in p.$$

Proof. (a) We set

$$A=\{\xi<\alpha: r(\xi)=\xi\}$$

and we note that $A\in p$ from Theorem 9.2(a). We claim that

$$\text{if } \xi\in A \quad \text{then } A\cap r^{-1}(\{\xi\})=\{\xi\}.$$

Indeed $\xi\in A\cap r^{-1}(\{\xi\})$, and if $\zeta\in A\cap r^{-1}(\{\xi\})$ then $r(\zeta)=\xi$ and $r(\zeta)=\zeta$, so that $\zeta=\xi$.

Since $A\in p$, it follows from Lemma 9.16(b) that

$$\{\xi<\alpha: A\cap r^{-1}(\{\xi\})\in e(\xi)\}\in p.$$

It is now clear that

$$A\cap\{\xi<\alpha: A\cap r^{-1}(\{\xi\})\in e(\xi)\}\subset\{\xi<\alpha: e(\xi)=\xi\}\in p.$$

(b) If $p\in\alpha$ there is nothing to prove, so let $\alpha\geq\omega$ and $p\in\boldsymbol{\beta}(\alpha)\setminus\alpha$. Let $\{A_\xi:\xi<\alpha\}$ be a family of subsets of α such that

$$A_\xi\cap A_\zeta=\emptyset \quad \text{for } \xi<\zeta<\alpha, \quad \text{and}$$
$$A_\xi\in e(\xi) \quad \text{for } \xi<\alpha.$$

We define $r:\alpha\to\alpha$ by the rule

$$r[A_\xi]=\{\xi\} \quad \text{for } \xi<\alpha,$$
$$r[\alpha\setminus\textstyle\bigcup_{\xi<\alpha}A_\xi]=\{0\}.$$

It is clear that

$$\bar{r}\circ e=\text{identity on } \boldsymbol{\beta}(\alpha),$$

so the result follows from part (a).

9.18. Lemma. *Let α be an infinite cardinal and $\beta\leq\alpha$, let $\mathscr{F}(\zeta)$ be a family of subsets of α such that*

$|\mathscr{F}(\zeta)|\leq\alpha$ *and*

$\mathscr{F}(\zeta)$ *has the uniform finite intersection property for* $\zeta<\beta$,

and let $f_\zeta\in\alpha^\alpha$ be such that

$$|f_\zeta[A]|=\alpha \quad \textit{for } A\in\mathscr{F}(\zeta), \quad \zeta<\beta.$$

Then there is a family $\{F_\zeta:\zeta<\beta\}$ of subsets of α such that

$F_\zeta\cap F_\eta=\emptyset$ *for* $\zeta<\eta<\beta$, *and*

$\mathscr{F}(\zeta)\cup\{f_\zeta^{-1}(F_\zeta)\}$ *has the uniform finite intersection property for* $\zeta<\beta$.

Proof. With no loss of generality, we assume that each $\mathscr{F}(\zeta)$ is closed under finite intersections. Let $\langle A_{\zeta,\xi}:\xi<\alpha\rangle$ be a well-ordering of $\mathscr{F}(\zeta)$, so that $|f_\zeta[A_{\zeta,\xi}]|=\alpha$ for $\zeta<\beta$ and $\xi<\alpha$. Since $|\beta\times\alpha|=\alpha$ there is by Lemma 7.5 a family $\{B_{\zeta,\xi}:\langle\zeta,\xi\rangle\in\beta\times\alpha\}$ of subsets of α such that

$$|B_{\zeta,\xi}|=\alpha\,,$$
$$B_{\zeta,\xi}\cap B_{\zeta',\xi'}=\emptyset \quad \text{for } \langle\zeta,\xi\rangle\neq\langle\zeta',\xi'\rangle, \quad \text{and}$$
$$B_{\zeta,\xi}\subset f_\zeta[A_{\zeta,\xi}]$$

for $\langle\zeta,\xi\rangle$, $\langle\zeta',\xi'\rangle\in\beta\times\alpha$. We set

$$F_\zeta=\bigcup\nolimits_{\xi<\alpha}B_{\zeta,\xi}$$

for $\zeta<\beta$ and we note that

$$F_\zeta\cap F_\eta=\emptyset \quad \text{for } \zeta<\eta<\beta, \quad \text{and}$$
$$B_{\zeta,\xi}\subset F_\zeta\cap f_\zeta[A_{\zeta,\xi}] \quad \text{for } \langle\zeta,\xi\rangle\in\beta\times\alpha\,.$$

Thus

$$\alpha=|B_{\zeta,\xi}|\leq|F_\zeta\cap f_\zeta[A_{\zeta,\xi}]|\leq|f_\zeta^{-1}(F_\zeta)\cap A_{\zeta,\xi}|$$

for $\langle\zeta,\xi\rangle\in\beta\times\alpha$, and it follows that $\mathscr{F}(\zeta)\cup\{f_\zeta^{-1}(F_\zeta)\}$ has the uniform finite intersection property for $\zeta<\beta$.

9.19. Theorem *(Blass). Let α be an infinite cardinal, $\alpha^+=2^\alpha$, and $\beta\leq\alpha$. There is a family $\{p_\zeta:\zeta<\beta\}$ of ultrafilters on α such that*

(a) *p_ζ is countably incomplete, α^+-good, and of minimal type in the Rudin-Keisler order of $\tau[U(\alpha)]$ for $\zeta<\beta$; and*

(b) *for any family $\{f_\zeta:\zeta<\beta\}\subset\alpha^\alpha$, if there is a family $\{A_\zeta:\zeta<\beta\}$ of subsets of α such that*

$$A_\zeta\in p_\zeta \quad \textit{for } \zeta<\beta, \quad \textit{and}$$
$$f_\zeta|A_\zeta \textit{ is one-to-one for } \zeta<\beta\,,$$

then the set $\{\overline{f}_\zeta(p_\zeta):\zeta<\beta\}$ is faithfully indexed and strongly discrete in $\boldsymbol{\beta}(\alpha)$.

Proof. There are $2^\alpha=\alpha^+$ subsets of α, there are (at most) $(2^\alpha)^\alpha=2^\alpha=\alpha^+$ monotone functions on $\mathscr{P}_\omega(\alpha)$, there are $\alpha^\alpha=2^\alpha=\alpha^+$ partitions of α into (at most) α elements, and there are $(\alpha^\alpha)^\beta=\alpha^\alpha=\alpha^+$ functions from β into α^α. Let $\langle d_\xi:\xi<\alpha^+\rangle$ be a well-ordering of the set consisting of all such subsets, monotone functions, partitions, and functions as described.

Let $\mathscr{F}_0$ be a uniform filter base on α of cardinality α such that

$\mathscr{P}^\alpha(\alpha)\subset\mathscr{F}_0$, and

there is a sequence $\{B_n:n<\omega\}\subset\mathscr{F}_0$ such that $\bigcap_{n<\omega}B_n=\emptyset$.

We define families $\{\mathscr{F}_\xi(\zeta): \xi<\alpha^+, \zeta<\beta\}$ such that

(i) $\mathscr{F}_0(\zeta)=\mathscr{F}_0$ for $\zeta<\beta$;

(ii) $\mathscr{F}_\xi(\zeta)$ is a family of subsets of α with the uniform finite intersection property, and $|\mathscr{F}_\xi(\zeta)|\leq\alpha$ for $\xi<\alpha^+$, $\zeta<\beta$;

(iii) $\mathscr{F}_\xi(\zeta)\subset\mathscr{F}_\eta(\zeta)$ for $\xi<\eta<\alpha^+$, $\zeta<\beta$;

(iv) $\mathscr{F}_\xi(\zeta)=\bigcup_{\eta<\xi}\mathscr{F}_\eta(\zeta)$ for $\zeta<\beta$ and for non-zero limit ordinals $\xi<\alpha^+$;

(v) if $d_\xi\in\mathscr{P}(\alpha)$, then either $d_\xi\in\mathscr{F}_{\xi+1}(\zeta)$ or $\alpha\backslash d_\xi\in\mathscr{F}_{\xi+1}(\zeta)$ for $\zeta<\beta$;

(vi) if d_ξ is a monotone function on $\mathscr{P}_\omega(\alpha)$ into $\mathscr{P}(\alpha)$ and $d_\xi[\mathscr{P}_\omega(\alpha)]\cup\mathscr{F}_\xi(\zeta)$ does not have the uniform finite intersection property, then $\mathscr{F}_{\xi+1}(\zeta)=\mathscr{F}_\xi(\zeta)$;

(vii) if d_ξ is a monotone function on $\mathscr{P}_\omega(\alpha)$ into $\mathscr{P}(\alpha)$ and $d_\xi[\mathscr{P}_\omega(\alpha)]\cup\mathscr{F}_\xi(\zeta)$ has the uniform finite intersection property, then there is a multiplicative function

$$\psi_{\xi+1}(\zeta):\mathscr{P}_\omega(\alpha)\to\mathscr{P}(\alpha)$$

such that

$$\psi_{\xi+1}(\zeta)\leq d_\xi, \quad \text{and}$$
$$\mathscr{F}_{\xi+1}(\zeta)=\psi_{\xi+1}(\zeta)[\mathscr{P}_\omega(\alpha)]\cup\mathscr{F}_\xi(\zeta);$$

(viii) if d_ξ is a partition of α into (at most) α elements, say $d_\xi=\{d_{\xi,\eta}:\eta<\alpha\}$, then there is $F_{\xi+1}(\zeta)\in\mathscr{P}(\alpha)$ such that

$$|\{\eta<\alpha:|F_{\xi+1}(\zeta)\cap d_{\xi,\eta}|>1\}|<\alpha, \quad \text{and}$$
$$\mathscr{F}_{\xi-1}(\zeta)=\mathscr{F}_\xi(\zeta)\cup\{F_{\xi+1}(\zeta);$$

(ix) if $d_\xi:\beta\to\alpha^\alpha$ and there are $\zeta'<\beta$ and $A\in\mathscr{F}_\xi(\zeta')$ such that $|d_\xi(\zeta')[A]|<\alpha$, then $\mathscr{F}_{\xi+1}(\zeta)=\mathscr{F}_\xi(\zeta)$ for $\zeta<\beta$; and

(x) if $d_\xi:\beta\to\alpha^\alpha$ and $|d_\xi(\zeta)[A]|=\alpha$ for all $A\in\mathscr{F}_\xi(\zeta)$ and $\zeta<\beta$, then there is a family $\{F_{\xi+1}(\zeta):\zeta<\beta\}$ of subsets of α such that

$$F_{\xi+1}(\zeta)\cap F_{\xi+1}(\zeta')=\emptyset \quad \text{for } \zeta<\zeta'<\beta, \quad \text{and}$$
$$\mathscr{F}_{\xi-1}(\zeta)=\mathscr{F}_\xi(\zeta)\cup\{(d_\xi(\zeta))^{-1}(F_{\xi+1}(\zeta))\}.$$

We proceed by transfinite recursion. We set

$$\mathscr{F}_0(\zeta)=\mathscr{F}_0 \quad \text{for } \zeta<\beta.$$

If ξ is a non-zero limit ordinal and $\xi<\alpha^+$, we define $\{\mathscr{F}_\xi(\zeta):\zeta<\beta\}$ by condition (iv). We now let $\xi<\alpha^+$ and we define $\mathscr{F}_{\xi+1}(\zeta)$ for $\zeta<\beta$. There are six possibilities.

Case 1. $d_\xi\in\mathscr{P}(\alpha)$. For $\zeta<\beta$, either $\mathscr{F}_\xi(\zeta)\cup\{d_\xi\}$ has the finite intersection property, in which case we set $\mathscr{F}_{\xi+1}(\zeta)=\mathscr{F}_\xi(\zeta)\cup\{d_\xi\}$; or $\mathscr{F}_\xi(\zeta)$

$\cup\{\alpha\backslash d_\xi\}$ has the finite intersection property, in which case we set $\mathscr{F}_{\xi+1}(\zeta)=\mathscr{F}_\xi(\zeta)\cup\{\alpha\backslash d_\xi\}$.

Case 2. d_ξ is a monotone function on $\mathscr{P}_\omega(\alpha)$ to $\mathscr{P}(\alpha)$ and $d_\xi[\mathscr{P}_\omega(\alpha)]\cup\mathscr{F}_\xi(\zeta)$ does not have the uniform finite intersection property. We set $\mathscr{F}_{\xi+1}(\zeta)=\mathscr{F}_\xi(\zeta)$.

Case 3. d_ξ is a monotone function on $\mathscr{P}_\omega(\alpha)$ to $\mathscr{P}(\alpha)$ and $d_\xi[\mathscr{P}_\omega(\alpha)]\cup\mathscr{F}_\xi(\zeta)$ has the uniform finite intersection property. We use Lemma 9.12 to define $\mathscr{F}_{\xi+1}(\zeta)$ satisfying conditions (ii) and (vii).

Case 4. d_ξ is a partition of α into at most α elements. We use Lemma 9.11 to define $\mathscr{F}_{\xi+1}(\zeta)$ satisfying conditions (ii) and (viii).

Case 5. $d_\xi\in(\alpha^\alpha)^\beta$ and there are $\zeta'<\beta$ and $A\in\mathscr{F}_\xi(\zeta')$ such that $|d_\xi(\zeta')[A]|<\alpha$. We set $\mathscr{F}_{\xi+1}(\zeta)=\mathscr{F}_\xi(\zeta)$ for $\zeta<\beta$.

Case 6. $d_\xi\in(\alpha^\alpha)^\beta$ and $|d_\xi(\zeta)[A]|=\alpha$ for $A\in\mathscr{F}_\xi(\zeta)$ and $\zeta<\beta$. We use Lemma 9.18 to define $\mathscr{F}_{\xi+1}(\zeta)$ satisfying conditions (ii) and (x).

The definition of the families $\{\mathscr{F}_\xi(\zeta):\xi<\alpha^+,\zeta<\beta\}$ is now complete.

We set

$$p_\zeta=\bigcup_{\xi<\alpha^+}\mathscr{F}_\xi(\zeta)$$

for $\zeta<\beta$. We note that p_ζ is an ultrafilter on α by conditions (ii), (iii), and (v); that p_ζ is α^+-good by condition (vii) and Lemma 9.7; and that p_ζ is uniformly selective by condition (vii), and hence, by Theorem 9.4, of minimal type in the Rudin-Keisler order of $\tau[U(\alpha)]$.

We finally verify property (b) of the theorem. Let $\{f_\zeta:\zeta<\beta\}\subset\alpha^\alpha$ and suppose that there is a family $\{A_\zeta:\zeta<\beta\}$ of subsets of α such that $A_\zeta\in p_\zeta$ and $f_\zeta|A_\zeta$ is one-to-one for $\zeta<\beta$. It follows that $|f_\zeta[A]|=\alpha$ for $A\in p_\zeta$ and $\zeta<\beta$. There is $\xi<\alpha^+$ such that $\{f_\zeta:\zeta<\beta\}=d_\xi$ and since $\mathscr{F}_\xi(\zeta)\subset p_\zeta$ we have $|f_\zeta[A]|=\alpha$ for $A\in\mathscr{F}_\xi(\zeta)$ and $\zeta<\beta$. Hence *Case* 6 was used to define $\mathscr{F}_{\xi+1}(\zeta)$ satisfying condition (x) and we have

$$\mathscr{F}_{\xi+1}(\zeta)=\mathscr{F}_\xi(\zeta)\cup\{(d_\xi(\zeta))^{-1}(F_{\xi+1}(\zeta))\}\subset p_\zeta$$

for $\zeta<\beta$ where

$$(*)\qquad F_{\xi+1}(\zeta)\cap F_{\xi+1}(\eta)=\emptyset \quad\text{for } \zeta<\eta<\beta\,.$$

Since

$$f_\zeta^{-1}(F_{\xi+1}(\zeta))=(d_\xi(\zeta))^{-1}(F_{\xi+1}(\zeta))\in p_\zeta$$

we have $F_{\xi+1}(\zeta)\in\bar{f}_\zeta(p_\zeta)$, which with (*) implies that the set $\{\bar{f}_\zeta(p_\zeta):\zeta<\beta\}$ is faithfully indexed and strongly discrete.

The proof of the theorem is complete.

Notes for § 9

Types of ultrafilters (on ω) were first defined and considered by W. Rudin [56] (Theorem 1.5). Frolík [67a], [67b] (§ 1) uses them in connection with the non-homogeneity of $\boldsymbol{\beta}(\omega)\backslash\omega$ (cf. § 16 below). The fundamental properties of the Rudin-Keisler order were studied by M.E. Rudin [66], [71] and by Keisler [67b] (who used it to provide some counterexamples to certain statements in model theory; cf. Booth [70] (introductory remarks to § 3)).

Lemma 9.1 is proved by Katětov [67]. A related result is given in Frolík [68b]. Theorem 9.2 and Corollary 9.3 (as well as Corollary 9.17) have been found independently by various workers in the subject, including Frolík [68a] (Theorem A), M.E. Rudin [71] (III. A), Keisler, Blass [70] (Theorem 2.5 and corollaries).

Most of Theorem 9.6 is due to Kunen, in particular the implication (d)$\Rightarrow$(i) (cf. Blass [70] (Proposition 10.6) for a complete proof, and Booth [69] ((5)$\Rightarrow$(1) of Theorem 1.9 and remarks before 1.9)). The implication (i)$\Rightarrow$(e) is essentially due to Rowbottom [74] (Theorem 4.8) (cf. Theorem 8.27 above and Notes for § 8). The equivalence of (h) is proved by Blass [70] (Proposition 10.10). A direct proof of the equivalence (i)$\Leftrightarrow$(j) is given in Keisler-Tarski [64] (Lemma 1.16).

It was Choquet in his papers [68a] and especially in [68b] (Proposition 12 and Théorème 19) (using a different terminology and motivated from a problem in probability theory; cf. also Mokobodzki [69]) who obtained, assuming the continuum hypothesis, the proof of existence of a selective ultrafilter on ω.

The concept of good ultrafilter is due to Keisler [64b], where the results 9.7—9.10 also appear. Lemma 9.11 is a slight improvement of Negrepontis [69c] (Lemma 3.2). A similar statement appears in the proof of Blass [70] (Theorem 8.3). Lemma 9.12 is due to Keisler [64b] (Lemma 4B).

The part of Theorem 9.13 concerning the existence of α^+-good ultrafilters is due to Keisler [64b] (Theorem 4.6). A statement somewhat weaker than Theorem 9.13, with $P(\alpha)$-points instead of ultrafilters whose type is minimal in the Rudin-Keisler order, is proved by Negrepontis [69c] (Theorem 3.1). Theorem 9.13 (with inessential differences) is stated in Blass [70] (Corollary 8.8). In this connection cf. Theorem 15.14(b).

Theorem 9.13 includes a considerable amount of information and is optimal in various senses; its proof however depends on a segment of the generalized continuum hypothesis. In § 10 we examine what parts of 9.13 can be proved without any special set-theoretic assumptions.

The following statement, announced by Kunen [72] (remark following Theorem 2.2), sets a limitation on any attempt to prove the existence of a uniformly selective ultrafilter: *There is a model of set theory (specifically, a model obtained by adjoining ω^{++} random reals to a model of set theory in which the continuum hypothesis holds) in which $\tau[\boldsymbol{\beta}(\omega)\backslash\omega]$ has no minimal elements in the Rudin-Keisler order.*

For regular cardinals α such that $\alpha^+ = 2^\alpha$ the following result of Balcar and Vopěnka has been announced (without proof) in Hájek [71] (pp. 75—76): There are 2^{2^α} fine ultrafilters on α (where an ultrafilter p on α is *fine* if for every $f \in \alpha^\alpha$ there is $A \in p$ such that $f|A$ is either constant or strictly increasing).

Lemma 9.14 and Corollary 9.15 are given in Negrepontis [69c] (Proposition 4.1 and Theorem 4.2).

Lemma 9.18 and Theorem 9.19 are due to Blass [70] (Lemma 17.4). They are used below to prove Corollary 10.8 (Blass [70] (Theorem 17.3)).

Ketonen [73] examines properties of the Rudin-Keisler order for α-complete non-principal ultrafilters on a measurable cardinal α.

§ 10. Good Ultrafilters

The questions studied in this section are roughly the same as those of § 9, and our aim is to prove as many of the results in § 9 as possible without making use of any segment of the generalized continuum hypothesis or, in fact, of any special axioms of set theory. The central result of this section is Kunen's Theorem 10.4. In the absence of GCH this is the best available replacement of Theorem 9.13; Theorem 10.6 (and its Corollary 10.7) provide additional information. The fundamental technique that enables us to avoid GCH is the existence of families of large oscillation of high cardinality, proved in Theorem 3.16. For an infinite cardinal α, such a family of large oscillation of cardinality 2^α forms the basis for a guided transfinite process of 2^α steps, at each of which we give up a negligible number of the functions of the initial family in order to obtain in return the satisfaction of a bit or a quantum of the conditions that we wish to achieve on the filter under definition. The existence of good ultrafilters, and various more refined and detailed statements, follow from these arguments, but the existence of minimal ultrafilters in the Rudin-Keisler order of $\tau[U(\alpha)]$ does not (and according to the Notes for § 9 cannot) follow. The best that can be proved is the existence of (at least) 2^α ultrafilters on α with pairwise non-comparable types; this proves that the "Souslin number" of $\tau[U(\alpha)]$ is at least $(2^\alpha)^+$.

The technique of families of large oscillation is also used to provide information concerning the directedness of $\tau[U(\alpha)]$ (in a sense "dual" or "orthogonal" to the part of Theorem 10.4 dealing with the Souslin number of $\tau[U(\alpha)]$). These results are combined with a consequence of a general combinatorial theorem of Hajnal (10.14) to yield a best possible result on the limits of the directness properties of $\tau[U(\alpha)]$ (Corollary 10.15).

The applications of the existence of families of large oscillation are not exhausted in this section; a further fundamental application is Shelah's characterization of elementary equivalence in terms of ultrapowers, proved below (Theorem 13.19).

An interesting consequence of Theorem 9.19 and of the results of this section (including Theorem 10.5) is a result of Blass (Corollary 10.8) implying, for cardinals α such that $\mathrm{cf}(\alpha)>\omega$ and $\alpha^+=2^\alpha$, the existence of a regular ultrafilter on α whose type is minimal in the Rudin-Keisler order on $\tau[U(\alpha)]$ but which is not α^+-good.

The final result of this section (Theorem 10.19) uses the translation of the techniques of § 9 and the characteristic topological properties of the Stone space of α-homogeneous-universal Boolean algebras, established in § 6, to prove, for cardinals $\alpha=\alpha^{\underline{\alpha}}>\omega$, the existence of α^+-good ultrafilters on certain powers of the α-homogeneous-universal Boolean algebra of cardinality α, and more generally the existence of adequate ultrafilters (defined in the statement of Theorem 10.19) for powers of special Boolean algebras of cardinality $\alpha=2^{\underline{\alpha}}>\omega$. The analogue of Lemma 7.5 is Lemma 6.6 (already proved), and the analogue of Lemma 9.19 is Lemma 10.16.

Families of Large Oscillation Modulo Filters; the Fundamental Existence Theorem of Good Ultrafilters

The following definition generalizes the definition given in § 3 of a family of large oscillation.

Definition. Let α, β and κ be cardinals, $\mathcal{S}\subset\beta^\alpha$, and $\mathcal{F}$ a filter on α. Then $\mathcal{S}$ is *(a family of functions) of κ-large oscillation* mod $\mathcal{F}$ if for $\lambda<\kappa$, a family $\{h_i : i<\lambda\}$ of distinct elements of $\mathcal{S}$, a subset $\{\zeta_i : i<\lambda\}$ of β, and $A\in\mathcal{F}$, we have

$$\{\xi<\alpha : h_i(\xi)=\zeta_i \text{ for } i<\lambda\}\cap A\neq\emptyset .$$

If $\mathcal{S}$ is of ω-large oscillation mod $\mathcal{F}$, then $\mathcal{S}$ is *of large oscillation* mod $\mathcal{F}$.

We note that $\mathcal{S}$ is of κ-large oscillation mod $\{\alpha\}$ if and only if $\mathcal{S}$ is of κ-large oscillation.

Further, if $\mathcal{S}$ is a *non-empty* family of κ-large oscillation mod $\mathcal{F}$ with $\kappa\geq\omega$, then $\mathcal{F}$ is a (proper) filter and $\mathcal{F}$ is not an ultrafilter.

We use the following notation given in § 2 and § 7: if $\mathcal{F}\subset\mathcal{P}(\alpha)$ then we denote by $\langle\mathcal{F}\rangle$ the (possibly improper) filter of $\mathcal{P}(\alpha)$ generated by $\mathcal{F}$. As already noted, $\langle\mathcal{F}\rangle$ is a proper filter if and only if $\mathcal{F}$ has the finite intersection property.

The following three lemmas are used in proving (in Theorem 10.4, without assuming any segment of the generalized continuum hypothesis) the existence of (at least) 2^α countably incomplete α^+-good ultrafilters on α whose types are pairwise non-comparable in the Rudin-Keisler order of $\tau[U(\alpha)]$. Of them, Lemma 10.1 is proved in a more general situation than presently needed, in anticipation of its use in the proof of Theorem 13.19 below.

10.1. Lemma. *Let α, β and κ be cardinals with $\kappa \geq \omega$, let $\mathscr{S} \subset \beta^{\alpha}$, let $\mathscr{F}$ be a filter on α such that*

$$\mathscr{S} \textit{ is of } \kappa\textit{-large oscillation} \bmod \mathscr{F},$$

and let $A \in \mathscr{P}(\alpha)$. Then there are a family $\mathscr{S}' \subset \mathscr{S}$ and a filter $\mathscr{F}' \supset \mathscr{F}$ such that

$$|\mathscr{S} \setminus \mathscr{S}'| < \kappa,$$
$$\mathscr{S}' \textit{ is of } \kappa\textit{-large oscillation} \bmod \mathscr{F}', \textit{ and}$$
$$\textit{either } A \in \mathscr{F}' \quad \textit{or } \alpha \setminus A \in \mathscr{F}'.$$

Proof. *Case* 1. $\mathscr{S}$ is of κ-large oscillation $\bmod \langle \mathscr{F} \cup \{A\} \rangle$. We set $\mathscr{S}' = \mathscr{S}$ and $\mathscr{F}' = \langle \mathscr{F} \cup \{A\} \rangle$.

Case 2. $\mathscr{S}$ is not of κ-large oscillation $\bmod \langle \mathscr{F} \cup \{A\} \rangle$. Then there are $\lambda < \kappa$, a family $\{h_i : i < \lambda\}$ of distinct elements of $\mathscr{S}$, a subset $\{\zeta_i : i < \lambda\}$ of β, and $F \in \mathscr{F}$ such that

$$\{\xi < \alpha : h_i(\xi) = \zeta_i \text{ for } i < \lambda\} \cap A \cap F = \emptyset,$$

so that

$$(*) \qquad \{\xi < \alpha : h_i(\xi) = \zeta_i \text{ for } i < \lambda\} \cap F \subset \alpha \setminus A.$$

We define

$$\mathscr{S}' = \mathscr{S} \setminus \{h_i : i < \lambda\} \quad \text{and}$$
$$\mathscr{F}' = \langle \mathscr{F} \cup \{\alpha \setminus A\} \rangle.$$

To prove that $\mathscr{S}'$ is of κ-large oscillation $\bmod \mathscr{F}'$ let $\mu < \kappa$, $\{f_j : j < \mu\}$ a family of distinct elements of $\mathscr{F}'$, $\{\eta_j : j < \mu\}$ a subset of β and $F \in \mathscr{F}$; we prove that

$$\{\xi < \alpha : f_j(\xi) = \eta_j \text{ for } j < \lambda\} \cap (\alpha \setminus A) \cap F \neq \emptyset.$$

Since $\lambda + \mu < \kappa$,

$$\{f_j : j < \mu\} \cap \{h_i : i < \lambda\} = \emptyset,$$

and $\mathscr{S}$ is of κ-large oscillation $\bmod \mathscr{F}$, we have

$$\{\xi < \alpha : h_i(\xi) = \zeta_i \text{ for } i < \lambda \text{ and } f_j(\xi) = \eta_j \text{ for } j < \mu\} \cap F \neq \emptyset.$$

The claim now follows from (*).

We remark, for later use, that the proof of Lemma 10.1 gives the following information: if $\mathscr{F}$ is generated (as a filter) by at most γ elements, then $\mathscr{F}'$ is generated (as a filter) by at most $\gamma + 1$ elements.

10.2. Lemma. *Let $\alpha \geq 2$, $\mathscr{S} \subset \alpha^\alpha$, $f \in \alpha^\alpha$ and let $\mathscr{F}$ and $\mathscr{G}$ be filters on α such that*

$\mathscr{S}$ is a non-empty family of large oscillation mod $\mathscr{F}$ *and* mod $\mathscr{G}$.

Then there are a family $\mathscr{S}' \subset \mathscr{S}$, filters $\mathscr{F}' \supset \mathscr{F}$ and $\mathscr{G}' \supset \mathscr{G}$, and $B \in \mathscr{P}(\alpha)$ such that

$$|\mathscr{S} \setminus \mathscr{S}'| < \omega,$$
$\mathscr{S}'$ is of large oscillation mod $\mathscr{F}'$ *and* mod $\mathscr{G}'$,
$$B \in \mathscr{F}', \quad \text{and}$$
$$\alpha \setminus f^{-1}(B) \in \mathscr{G}'.$$

Proof. We choose a function ϕ of α onto 2 and an element $h \in \mathscr{S}$, and we set $A = (\phi \circ h)^{-1}(\{0\})$. We distinguish two possibilities.

Case 1. $\mathscr{S} \setminus \{h\}$ is of large oscillation $\mathrm{mod}\langle \mathscr{G} \cup \{\alpha \setminus f^{-1}(A)\}\rangle$.
In this case we set

$$\mathscr{S}' = \mathscr{S} \setminus \{h\},$$
$$\mathscr{F}' = \langle \mathscr{F} \cup \{A\}\rangle,$$
$$\mathscr{G}' = \langle \mathscr{G} \cup \{\alpha \setminus f^{-1}(A)\}\rangle, \quad \text{and}$$
$$B = A,$$

and we verify that $\mathscr{S}'$ is of large oscillation mod $\mathscr{F}'$.

We let $n < \omega$, $\{h_i : i < n\}$ a family of distinct elements $\mathscr{S}'$, $\{\zeta_i : i < n\}$ a subset of α, and $F \in \mathscr{F}$, and prove that

$$\{\xi < \alpha : h_i(\xi) = \zeta_i \text{ for } i < n\} \cap A \cap F \neq \emptyset.$$

Indeed, let $\zeta < \alpha$ be such that $\phi(\zeta) = 0$. Since $n+1 < \omega$ and $\{h_i : i < n\} \cup \{h\}$ is a family of distinct elements of $\mathscr{S}$ and $\mathscr{S}$ is of large oscillation mod $\mathscr{F}$, we have

$$\{\xi < \alpha : h_i(\xi) = \zeta_i \text{ for } i < n \text{ and } h(\xi) = \zeta\} \cap F \neq \emptyset.$$

Since $\xi \in A$ for $h(\xi) = \zeta$, the result follows.

Case 2. $\mathscr{S} \setminus \{h\}$ is not of large oscillation $\mathrm{mod}\langle \mathscr{G} \cup \{\alpha \setminus f^{-1}(A)\}\rangle$.
In this case there are $n < \omega$, a family $\{h_i : i < n\}$ of distinct elements of $\mathscr{S} \setminus \{h\}$, a subset $\{\zeta_i : i < n\}$ of α and $G \in \mathscr{G}$ such that

$$\{\xi < \alpha : h_i(\xi) = \zeta_i \text{ for } i < n\} \cap G \cap (\alpha \setminus f^{-1}(A)) = \emptyset,$$

so that

$$(*) \qquad \{\xi < \alpha : h_i(\xi) = \zeta_i \text{ for } i < n\} \cap G \subset f^{-1}(A).$$

We set

$$\begin{aligned}&\mathcal{S}'=\mathcal{S}\setminus(\{h_i:i<n\}\cup\{h\}),\\&B=\alpha\setminus A,\\&\mathcal{F}'=\langle\mathcal{F}\cup\{B\}\rangle,\quad\text{and}\\&\mathcal{G}'=\langle\mathcal{G}\cup\{\alpha\setminus f^{-1}(B)\}\rangle,\end{aligned}$$

and we verify that $\mathcal{S}'$ is of large oscillation mod $\mathcal{F}'$ and mod $\mathcal{G}'$. We thus let $m<\omega$, $\{f_j:j<m\}$ a family of distinct elements of $\mathcal{S}'$, $\{\eta_j:j<m\}$ a subset of α, $F\in\mathcal{F}$, and $G\in\mathcal{G}$, and we claim that

$$\begin{aligned}&\{\xi<\alpha:f_j(\xi)=\eta_j\text{ for }j<m\}\cap B\cap F\neq\emptyset\quad\text{and}\\&\{\xi<\alpha:f_j(\xi)=\eta_j\text{ for }j<m\}\cap(\alpha\setminus f^{-1}(B))\cap G\neq\emptyset.\end{aligned}$$

Indeed, let $\zeta<\alpha$ be such that $\phi(\zeta)=1$. Since $n+m+1<\omega$,

$$\{h_i:i<n\}\cup\{f_j:j<m\}\cup\{h\}$$

is a family of distinct elements of $\mathcal{S}$, and $\mathcal{S}$ is of large oscillation mod $\mathcal{F}$ and mod $\mathcal{G}$, we have

(1) $$\{\xi<\alpha:f_j(\xi)=\eta_j\text{ for }j<m\text{ and }h(\xi)=\zeta\}\cap F\neq\emptyset$$

and

(2) $$\{\xi<\alpha:h_i(\xi)=\zeta_i\text{ for }i<n\text{ and }f_j(\xi)=\eta_j\text{ for }j<m\}\cap G\neq\emptyset.$$

The first claim follows from (1) and the fact that $\xi\in B$ for $h(\xi)=\zeta$, and the second claim follows from (2).

It is clear that $B\in\mathcal{F}'$ and that $\alpha\setminus f^{-1}(B)\in\mathcal{G}'$. The proof is complete.

The following lemma differs from the two previous ones in that its proof makes apparently essential use of the assumption that $\mathcal{S}\subset\alpha^\alpha$, *i.e.*, that $\beta=\alpha$. It is the appropriate replacement of Lemma 9.11 in the attempt to avoid all use of the generalized continuum hypothesis in proving the existence of good ultrafilters.

10.3. Lemma. *Let α be an infinite cardinal, let $\mathcal{S}\subset\alpha^\alpha$, let $\mathcal{F}$ be a filter on α such that*

$$\mathcal{S}\text{ is a non-empty family of large oscillation mod }\mathcal{F},$$

and let $\phi:\mathcal{P}_\omega(\alpha)\to\mathcal{F}$ be a monotone function. Then there are a family $\mathcal{S}'\subset\mathcal{S}$, a filter $\mathcal{F}'\supset\mathcal{F}$, and a function $\psi:\mathcal{P}_\omega(\alpha)\to\mathcal{F}'$ such that

$$\begin{aligned}&|\mathcal{S}\setminus\mathcal{S}'|=1,\\&\mathcal{S}'\text{ is of large oscillation mod }\mathcal{F}',\\&\psi\leq\phi,\quad\text{and}\\&\psi\text{ is multiplicative}.\end{aligned}$$

Proof. We choose $h \in \mathcal{S}$ and set

$$\mathcal{S}' = \mathcal{S} \setminus \{h\}.$$

Let $\langle F_\xi : \xi < \alpha \rangle$ be a well-ordering of $\mathcal{P}_\omega(\alpha)$, and define

$$\psi : \mathcal{P}_\omega(\alpha) \to \mathcal{P}(\alpha)$$

by the rule

$$\psi(F) = \bigcup \{\phi(F_\zeta) \cap h^{-1}(\{\zeta\}) : \zeta < \alpha,\ F \subset F_\zeta\}.$$

We set

$$\mathcal{F}' = \langle \mathcal{F} \cup \{\psi(F) : F \in \mathcal{P}_\omega(\alpha)\} \rangle$$

and we prove the following three statements.

(i) ψ is multiplicative.

If $F, G \in \mathcal{P}_\omega(\alpha)$ and $\eta \in \psi(F \cup G)$ then there is $\zeta < \alpha$ such that $\eta \in \phi(F_\zeta) \cap h^{-1}(\{\zeta\})$ and $F \cup G \subset F_\zeta$. Then $F \subset F_\zeta$ and $G \subset F_\zeta$, so that $\eta \in \psi(F) \cap \psi(G)$. Thus

$$\psi(F \cup G) \subset \psi(F) \cap \psi(G).$$

If, conversely, $\eta \in \psi(F) \cap \psi(G)$, there are $\zeta, \zeta' < \alpha$ such that

$$\eta \in \phi(F_\zeta) \cap h^{-1}(\{\zeta\}) \cap \phi(F_{\zeta'}) \cap h^{-1}(\{\zeta'\})$$

and $F \subset F_\zeta$ and $G \subset F_{\zeta'}$. Clearly $\zeta = \zeta'$, so

$$F \cup G \subset F_\zeta = F_{\zeta'}$$

and $\eta \in \psi(F \cup G)$. It follows that

$$\psi(F \cup G) = \psi(F) \cap \psi(G)$$

for $F, G \in \mathcal{P}_\omega(\alpha)$, *i.e.*, that ψ is multiplicative.

(ii) $\mathcal{S}'$ is of large oscillation mod $\mathcal{F}'$.

Since $\psi(\bigcup_{k<n} F^{(k)}) = \bigcap_{k<n} \psi(F^{(k)})$ for $\{F^{(k)} : k < n\} \subset \mathcal{P}_\omega(\alpha)$ from part (i), it is enough to prove that if $m < \omega$, $\{h_i : i < m\}$ a family of distinct elements of $\mathcal{S}'$, $\{\zeta_i : i < m\}$ a subset of α, $F \in \mathcal{F}$ and $F_\zeta \in \mathcal{P}_\omega(\alpha)$, then

$$\{\xi < \alpha : h_i(\xi) = \zeta_i \text{ for } i < m\} \cap F \cap \psi(F_\zeta) \neq \emptyset.$$

Since $m + 1 < \omega$, $\{h_i : i < m\} \cup \{h\}$ is a family of distinct elements of $\mathcal{S}$, $F \cap \phi(F_\zeta) \in \mathcal{F}$, and $\mathcal{S}$ is of large oscillation mod $\mathcal{F}$, we have

$$\{\xi < \alpha : h_i(\xi) = \zeta_i \text{ for } i < m \text{ and } h(\xi) = \zeta\} \cap F \cap \phi(F_\zeta) \neq \emptyset.$$

The result now follows from the fact that

$$\phi(F_\zeta) \cap h^{-1}(\{\zeta\}) \subset \psi(F_\zeta).$$

(iii) $\psi \leq \phi$.

If $F \subset F_\zeta \in \mathscr{P}_\omega(\alpha)$ then $\phi(F_\zeta) \subset \phi(F)$, and thus

$$\begin{aligned}\psi(F) &= \bigcup \{\phi(F_\zeta) \cap h^{-1}(\{\zeta\}) : \zeta < \alpha,\ F \subset F_\zeta\} \\ &\subset \bigcup \{\phi(F_\zeta) : \zeta < \alpha, F \subset F_\zeta\} \subset \phi(F)\end{aligned}$$

for $F \in \mathscr{P}_\omega(\alpha)$.

The proof is complete.

10.4. Theorem *(Kunen). Let α be an infinite cardinal. There exist at least 2^α countably incomplete, α^+-good ultrafilters on α whose types are pairwise non-comparable in the Rudin-Keisler order of $\tau[U(\alpha)]$.*

Proof. We let $\mathscr{D}$ be the family which is equal to the union of the following three families:

$$\{\langle A, \xi \rangle : A \in \mathscr{P}(\alpha),\ \xi < 2^\alpha\},$$

$$\{\langle \phi, \xi \rangle : \phi \text{ is monotone, } \phi : \mathscr{P}_\omega(\alpha) \to \mathscr{P}(\alpha),\ \xi < 2^\alpha\}, \quad \text{and}$$

$$\{\langle f, \xi, \zeta \rangle : f \in \alpha^\alpha,\ \xi < 2^\alpha,\ \zeta < 2^\alpha \text{ and } \xi \neq \zeta\}.$$

It is clear that $|\mathscr{D}| = 2^\alpha$. Let $\langle d_\eta : \eta < 2^\alpha \rangle$ be a well-ordering of $\mathscr{D}$ such that each element of the form $\langle \phi, \xi \rangle$, where ϕ is a monotone function from $\mathscr{P}_\omega(\alpha)$ into $\mathscr{P}(\alpha)$ and $\xi < 2^\alpha$, appears 2^α times.

By Corollary 3.17 there is a family $\mathscr{S} \subset \alpha^\alpha$ such that

$$|\mathscr{S}| = 2^\alpha, \quad \text{and}$$

$$\mathscr{S} \text{ is of large oscillation.}$$

We choose $h \in \mathscr{S}$ and we set

$$A_n = \{\xi < \alpha : n < h(\xi) < \omega\}$$

for $n < \omega$. Clearly $\{A_n : n < \omega\}$ has the finite intersection property, and $\bigcap_{n<\omega} A_n = \emptyset$.

We define families $\{\mathscr{S}_\eta : \eta < 2^\alpha\}$ and filters $\{\mathscr{F}_{\xi,\eta} : \xi < 2^\alpha, \eta < 2^\alpha\}$, such that

(i) $\mathscr{S}_0 = \mathscr{S} \setminus \{h\}$, $\mathscr{F}_{\xi,0} = \langle A_n : n < \omega \}\rangle$ for $\xi < 2^\alpha$;

(ii) $\mathscr{S}_\eta \supset \mathscr{S}_\zeta$ and $\mathscr{F}_{\xi,\eta} \subset \mathscr{F}_{\xi,\zeta}$ for $\eta < \zeta < 2^\alpha$, $\xi < 2^\alpha$;

(iii) $|\mathscr{S} \setminus \mathscr{S}_\eta| \leq \omega + |\eta|$ for $\eta < 2^\alpha$;

(iv) $\mathscr{S}_\eta$ is of large oscillation mod $\mathscr{F}_{\xi,\eta}$ for $\xi < 2^\alpha$, $\eta < 2^\alpha$;

(v) $\mathscr{S}_\eta = \bigcap_{\zeta<\eta} \mathscr{S}_\zeta$ and $\mathscr{F}_{\xi,\eta} = \bigcup_{\zeta<\eta} \mathscr{F}_{\xi,\zeta}$ for $\xi < 2^\alpha$ and non-zero limit ordinals $\eta < 2^\alpha$;

(vi) if $d_\eta = \langle A, \xi \rangle$, where $A \in \mathscr{P}(\alpha)$ and $\xi < 2^\alpha$, then either $A \in \mathscr{F}_{\xi,\eta+1}$ or $\alpha \setminus A \in \mathscr{F}_{\xi,\eta+1}$;

(vii) if $d_\eta = \langle \phi, \xi \rangle$, where ϕ is a monotone function from $\mathscr{P}_\omega(\alpha)$ into $\mathscr{P}(\alpha)$ and $\xi < 2^\alpha$ and $\phi[\mathscr{P}_\omega(\alpha)] \subset \mathscr{F}_{\xi,\eta}$, then there is a multiplicative function $\psi_{\eta+1} : \mathscr{P}_\omega(\alpha) \to \mathscr{F}_{\xi,\eta+1}$, such that $\psi_{\eta+1} \leq \phi$; and,

(viii) if $d_\eta = \langle f, \xi, \zeta \rangle$, where $f \in \alpha^\alpha$, $\xi < 2^\alpha$, $\zeta < 2^\alpha$ and $\xi \neq \zeta$, then there is $B_{\eta+1} \in \mathscr{P}(\alpha)$ such that

$$B_{\eta+1} \in \mathscr{F}_{\xi,\eta+1}, \quad \text{and}$$
$$\alpha \setminus f^{-1}(B_{\eta+1}) \in \mathscr{F}_{\zeta,\eta+1}.$$

We proceed by transfinite recursion. We define $\mathscr{S}_0$ and $\{\mathscr{F}_{\xi,0} : \xi < 2^\alpha\}$ by condition (i); it is easily verified that $\mathscr{S}_0$ is of large oscillation mod $\langle \{A_n : n < \omega\} \rangle$. For non-zero limit ordinals $\eta < 2^\alpha$, we define $\mathscr{S}_\eta$ and $\{\mathscr{F}_{\xi,\eta} : \xi < 2^\alpha\}$ by condition (v). We now let $\eta < 2^\alpha$, and we define $\mathscr{S}_{\eta+1}$ and $\mathscr{F}_{\xi,\eta+1}$ for $\xi < 2^\alpha$. There are three possibilities.

Case 1. $d_\eta = \langle A, \xi \rangle$, where $A \in \mathscr{P}(\alpha)$ and $\xi < 2^\alpha$. We set $\mathscr{F}_{\zeta,\eta+1} = \mathscr{F}_{\zeta,\eta}$ for $\zeta < 2^\alpha$ and $\zeta \neq \xi$. We define $\mathscr{S}_{\eta+1}$ and $\mathscr{F}_{\xi,\eta+1}$ using Lemma 10.1 in such a way that

$$|\mathscr{S}_\eta \setminus \mathscr{S}_{\eta+1}| < \omega,$$
$$\mathscr{S}_\eta \text{ is of large oscillation mod } \mathscr{F}_{\xi,\eta+1}, \text{ and}$$
$$\text{either } A \in \mathscr{F}_{\xi,\eta+1} \text{ or } \alpha \setminus A \in \mathscr{F}_{\xi,\eta+1}.$$

Case 2. $d_\eta = \langle \phi, \xi \rangle$, where $\phi : \mathscr{P}_\omega(\alpha) \to \mathscr{P}(\alpha)$ is a monotone function and $\xi < 2^\alpha$. If $\zeta < 2^\alpha$ and $\zeta \neq \xi$ we set $\mathscr{F}_{\zeta,\eta+1} = \mathscr{F}_{\zeta,\eta}$. If $\phi[\mathscr{P}_\omega(\alpha)] \subset \mathscr{F}_{\xi,\eta}$ is false, we set $\mathscr{S}_{\eta+1} = \mathscr{S}_\eta$ and $\mathscr{F}_{\xi,\eta+1} = \mathscr{F}_{\xi,\eta}$. And if $\phi[\mathscr{P}_\omega(\alpha)] \subset \mathscr{F}_{\xi,\eta}$, then we define $\mathscr{S}_{\eta+1}$, $\mathscr{F}_{\xi,\eta+1}$ and $\psi_{\eta+1}$, using Lemma 10.3, in such a way that

$$|\mathscr{S}_\eta \setminus \mathscr{S}_{\eta+1}| = 1,$$
$$\mathscr{S}_{\eta+1} \text{ is of large oscillation mod } \mathscr{F}_{\xi,\eta+1},$$
$$\psi_{\eta+1} : \mathscr{P}_\omega(\alpha) \to \mathscr{F}_{\xi,\eta+1},$$
$$\psi_{\eta+1} \leq \phi, \quad \text{and}$$
$$\psi_{\eta+1} \text{ is multiplicative.}$$

Case 3. $d_\eta = \langle f, \xi, \zeta \rangle$, where $f \in \alpha^\alpha$, $\xi < 2^\alpha$, and $\xi \neq \zeta$. If $i < 2^\alpha$ with $i \neq \xi$ and $i \neq \zeta$, we define $\mathscr{F}_{i,\eta+1} = \mathscr{F}_{i,\eta}$. We define $\mathscr{S}_{\eta+1}$, $\mathscr{F}_{\xi,\eta+1}$, $\mathscr{F}_{\zeta,\eta+1}$ (and $B_{\eta+1} \in \mathscr{P}(\alpha)$) using Lemma 10.2 in such a way that

$$|\mathscr{S}_\eta \setminus \mathscr{S}_{\eta+1}| < \omega,$$
$$\mathscr{S}_{\eta+1} \text{ is of large oscillation mod } \mathscr{F}_{\xi,\eta+1} \text{ and mod } \mathscr{F}_{\zeta,\eta+1},$$
$$B_{\eta+1} \in \mathscr{F}_{\xi,\eta+1}, \quad \text{and}$$
$$\alpha \setminus f^{-1}(B_{\eta+1}) \in \mathscr{F}_{\zeta,\eta+1}.$$

The recursive definitions are now complete. We define

$$p_\xi = \bigcup_{\eta < 2^\alpha} \mathscr{F}_{\xi,\eta} \quad \text{for } \xi < 2^\alpha.$$

By conditions (i), (ii), (iv) and (vi), p_ξ is a countably incomplete ultrafilter on α for $\xi<2^\alpha$. From the fact that if $\phi:\mathscr{P}_\omega(\alpha)\to p_\xi$ is monotone then $\langle\phi,\xi\rangle$ appears 2^α times in the well-ordering of $\mathscr{D}$ (and $\mathrm{cf}(2^\alpha)>\alpha$) it follows that p_ξ is α^+-good for $\xi<2^\alpha$. Finally if $\xi,\zeta<2^\alpha$ with $\xi\neq\zeta$ and $f\in\alpha^\alpha$, by condition (viii) there is $\eta<2^\alpha$ such that $d_\eta=\langle f,\xi,\zeta\rangle$ and hence

$$B_\eta\in\mathscr{F}_{\xi,\eta+1}\subset p_\xi \quad\text{and}\quad \alpha\backslash f^{-1}(B_\eta)\in\mathscr{F}_{\zeta,\eta+1}\subset p_\zeta;$$

thus $\bar{f}(p_\zeta)\neq p_\xi$ by Lemma 7.19 and it follows that the elements of the set $\{p_\xi:\xi<2^\alpha\}$ are of pairwise non-comparable type in the Rudin-Keisler order of $\tau[U(\alpha)]$.

The proof of the theorem is complete.

Additional Existence Results

10.5. Theorem. *Let $\omega\leq\beta\leq\alpha$, let r be a function from α onto β, and let e be a continuous function from $\boldsymbol{\beta}(\beta)$ into $\boldsymbol{\beta}(\alpha)$ such that $\bar{r}\circ e$ is the identity function on $\boldsymbol{\beta}(\beta)$. If $q\in\boldsymbol{\beta}(\beta)$ and $p=e(q)\in\boldsymbol{\beta}(\alpha)$ and q is countably incomplete, then p and q have the same degree of goodness, i.e., $\boldsymbol{G}(p)=\boldsymbol{G}(q)$.*

Proof. We first prove that $\boldsymbol{G}(p)\leq\boldsymbol{G}(q)$. Let $\gamma\geq\omega$, suppose that p is γ^+-good, and let $\phi:\mathscr{P}_\omega(\gamma)\to q$ be a monotone function. We define $\phi':\mathscr{P}_\omega(\gamma)\to\mathscr{P}(\alpha)$ by $\phi'(F)=r^{-1}(\phi(F))$. Then in fact

$$\phi':\mathscr{P}_\omega(\gamma)\to p$$

by Lemma 9.16(b), and since ϕ' is monotone there is a multiplicative function

$$\psi':\mathscr{P}_\omega(\gamma)\to p$$

such that $\psi'\leq\phi'$. We define $\psi:\mathscr{P}_\omega(\gamma)\to\mathscr{P}(\beta)$ by

$$\psi(F)=e^{-1}(\mathrm{cl}_{\boldsymbol{\beta}(\alpha)}\psi'(F))\cap\beta\,.$$

Since e is a topological embedding of $\boldsymbol{\beta}(\beta)$ into $\boldsymbol{\beta}(\alpha)$ and $\mathrm{cl}_{\boldsymbol{\beta}(\alpha)}\psi'(F)$ is a neighborhood in $\boldsymbol{\beta}(\alpha)$ of p for $F\in\mathscr{P}_\omega(\gamma)$ we have

$$\psi:\mathscr{P}_\omega(\gamma)\to q\,.$$

Furthermore if $F,G\in\mathscr{P}_\omega(\gamma)$ then from Theorem 2.6 we have

$$\begin{aligned}\psi(F\cup G)&=e^{-1}(\mathrm{cl}_{\boldsymbol{\beta}(\alpha)}\psi'(F\cup G))\cap\beta\\&=e^{-1}(\mathrm{cl}_{\boldsymbol{\beta}(\alpha)}(\psi'(F)\cap\psi'(G)))\cap\beta\\&=e^{-1}(\mathrm{cl}_{\boldsymbol{\beta}(\alpha)}\psi'(F))\cap e^{-1}(\mathrm{cl}_{\boldsymbol{\beta}(\alpha)}\psi'(G))\cap\beta\\&=\psi(F)\cap\psi(G)\,,\end{aligned}$$

so that ψ is multiplicative.

Finally for $F \in \mathscr{P}_\omega(\gamma)$ we have $\psi'(F) \subset \phi'(F)$ and then

$$e^{-1}(\mathrm{cl}_{\boldsymbol{\beta}(\alpha)} \psi'(F)) \subset e^{-1}(\mathrm{cl}_{\boldsymbol{\beta}(\alpha)} \phi'(F)) = e^{-1}(\mathrm{cl}_{\boldsymbol{\beta}(\alpha)} r^{-1}(\phi(F)))$$
$$\subset e^{-1}(\overline{r}^{-1} \mathrm{cl}_{\boldsymbol{\beta}(\beta)} \phi(F)) = \mathrm{cl}_{\boldsymbol{\beta}(\beta)} \phi(F),$$

so that

$$\psi(F) = e^{-1}(\mathrm{cl}_{\boldsymbol{\beta}(\alpha)} \psi'(F)) \cap \beta \subset \mathrm{cl}_{\boldsymbol{\beta}(\beta)} \phi(F) \cap \beta = \phi(F)$$

and $\psi \leq \phi$. It follows that q is γ^+-good. Thus $\boldsymbol{G}(p) \leq \boldsymbol{G}(q)$.

We prove next that $\boldsymbol{G}(q) \leq \boldsymbol{G}(p)$. Let $\gamma \geq \omega$, suppose that q is γ^+-good, and let $\phi: \mathscr{P}_\omega(\gamma) \to p$ be a monotone function. Let $\phi': \mathscr{P}_\omega(\gamma) \to \mathscr{P}_\omega(\beta)$ be given by $\phi'(F) = \{\zeta < \beta : \phi(F) \in e(\zeta)\}$. We note that if $F \in \mathscr{P}_\omega(\gamma)$ and $\phi'(F) \notin q$ then

$$B = \{\zeta < \beta : \alpha \backslash \phi(F) \in e(\zeta)\} \in q.$$

It follows from Lemma 9.16(b) (with $B_\zeta = \alpha \backslash \phi(F)$ for $\zeta \in B$) that

$$\alpha \backslash \phi(F) = \textstyle\bigcup_{\zeta \in B} B_\zeta \in p;$$

thus we have in fact

$$\phi': \mathscr{P}_\omega(\gamma) \to q.$$

It is clear that ϕ' is a monotone function.

Since q is countably incomplete there is a sequence $\{A_n : n < \omega\} \subset q$ such that

$$\beta = A_0 \supset A_1 \supset \cdots \supset A_n \supset \cdots, \quad \text{and}$$
$$\textstyle\bigcap_{n<\omega} A_n = \emptyset.$$

We define

$$\phi'': \mathscr{P}_\omega(\gamma) \to q$$

by the condition $\phi''(F) = \phi'(F) \cap A_{|F|}$; clearly ϕ'' is a monotone function. Since q is γ^+-good, there is a multiplicative function

$$\psi'': \mathscr{P}_\omega(\gamma) \to q$$

such that $\psi'' \leq \phi''$. For $\zeta < \beta$ we set

$$F(\zeta) = \{\eta < \gamma : \zeta \in \psi''(\{\eta\})\}.$$

We set $n(\zeta) = \min\{n < \omega : \zeta \notin A_n\}$, so that $n(\zeta) < \omega$ for $\zeta < \beta$, and we claim that

$$|F(\zeta)| < n(\zeta) \quad \text{for } \zeta < \beta.$$

Indeed suppose that for some $\zeta < \beta$ we have $|F(\zeta)| \geq n(\zeta)$; then there is $G \subset F(\zeta)$ such that $|G| = n(\zeta)$, and since ψ'' is multiplicative we have

$$\zeta \in \textstyle\bigcap_{\eta \in G} \psi''(\{\eta\}) = \psi''(G) \subset \phi''(G) \subset A_{n(\zeta)},$$

a contradiction. Thus $F(\zeta) \in \mathscr{P}_\omega(\gamma)$ for $\zeta < \beta$.

We define $\psi:\mathscr{P}_\omega(\gamma)\to\mathscr{P}(\alpha)$ by the rule

$$\psi(F)=\bigcup\{\phi(F(\zeta))\cap r^{-1}(\{\zeta\}):\zeta\in\psi''(F)\},$$

and we verify the following statements.

(i) $\psi:\mathscr{P}_\omega(\gamma)\to p$.

We have

$$\zeta\in\psi''(F(\zeta))\subset\phi''(F(\zeta))\subset\phi'(F(\zeta))$$

and thus

$$\phi(F(\zeta))\cap r^{-1}(\{\zeta\})\in e(\zeta)$$

for $\zeta<\beta$. It follows from Lemma 9.16(b) that if $F\in\mathscr{P}_\omega(\gamma)$ then

$$\psi(F)=\bigcup_{\zeta\in\psi''(F)}\phi(F(\zeta))\cap r^{-1}(\{\zeta\})\in p.$$

(ii) $\psi\le\phi$.

Let $F\in\mathscr{P}_\omega(\gamma)$ and $\xi\in\psi(F)$, so that $\xi\in\phi(F(\zeta))\cap r^{-1}(\{\zeta\})$ for some $\zeta\in\psi''(F)$. It is easy to see, using the facts that ψ'' is multiplicative and $\psi''(\zeta)\neq\emptyset$, that $F\subset F(\zeta)$; hence $\xi\in\phi(F(\zeta))\subset\phi(F)$.

(iii) ψ is multiplicative.

Let $F,G\in\mathscr{P}_\omega(\gamma)$, and $\xi<\alpha$. If $\xi\in\psi(F\cup G)$, then $\xi\in\phi(F(\zeta))\cap r^{-1}(\{\zeta\})$ for some $\zeta\in\psi''(F\cup G)=\psi''(F)\cap\psi''(G)$, and thus $\xi\in\psi(F)\cap\psi(G)$. And if $\xi\in\psi(F)\cap\psi(G)$ then $\xi\in\phi(F(\zeta))\cap r^{-1}(\{\zeta\})$ for some $\zeta\in\psi''(F)$ and $\xi\in\phi(F(\zeta'))\cap r^{-1}(\{\zeta'\})$ for some $\zeta'\in\psi''(G)$. It is clear that $\zeta=\zeta'$, so that $\zeta\in\psi''(F)\cap\psi''(G)=\psi''(F\cup G)$ and hence $\xi\in\psi(F\cup G)$.

This completes the proof of the theorem.

10.6. Theorem. *Let $\omega\le\beta\le\alpha$. Any family of subsets of α with the uniform finite intersection property and of cardinality at most α is contained in 2^{2^α} distinct uniform ultrafilters each of which is countably incomplete and has degree of goodness equal to β^+.*

Proof. We let $\mathscr{F}$ be the given family of subsets and we assume without loss of generality that $\mathscr{F}$ is a uniform filter base on α. Let $\langle A_\xi:\xi<\alpha\rangle$ be a well-ordering of the given family. By Corollary 7.6 there is a family $\{B_\zeta:\zeta<\beta\}$ of subsets of α such that

$$B_\zeta\cap B_{\zeta'}=\emptyset \quad \text{for } \zeta<\zeta'<\beta, \quad \text{and}$$

$$|A_\xi\cap B_\zeta|=\alpha \quad \text{for } \xi<\alpha \quad \text{and } \zeta<\beta.$$

We suppose without loss of generality that $\bigcup_{\zeta<\beta}B_\zeta=\alpha$. We set $\mathscr{B}_\zeta=\{A_\xi\cap B_\zeta:\xi<\alpha\}$ for $\zeta<\beta$. Clearly $\mathscr{B}_\zeta$ is a uniform filter base on α and of cardinality at most α. Thus by Theorem 7.7(b) $\mathscr{B}_\zeta$ is contained in 2^{2^α} distinct uniform ultrafilters on α; we denote this (faithfully indexed) family of ultrafilters by

$$\{p_{\zeta,\eta}:\eta<2^{2^\alpha}\} \quad \text{for } \zeta<\beta.$$

By Theorem 10.4 there is a countably incomplete β^+-good ultrafilter q on β. We let $r:\alpha\to\beta$ be defined by the condition $r[B_\zeta]=\{\zeta\}$ for $\zeta<\beta$; and we let

$$e_\eta:\boldsymbol{\beta}(\beta)\to\boldsymbol{\beta}(\alpha)$$

be defined by the conditions

$$e_\eta(\zeta)=p_{\zeta,\eta} \quad \text{for } \zeta<\beta\,, \quad \text{and}$$

$$e_\eta \text{ is continuous}$$

for $\eta<2^{2^\alpha}$. Since $e_\eta(\zeta)=p_{\zeta,\eta}\in \mathrm{cl}_{\boldsymbol{\beta}(\alpha)}B_\zeta$ we have that $\bar{r}\circ e_\eta$ is the identity on $\boldsymbol{\beta}(\beta)$. We set $p_\eta=e_\eta(q)$ and we note that $\boldsymbol{G}(p_\eta)=\boldsymbol{G}(q)=\beta^+$ for $\eta<2^{2^\alpha}$ from Theorem 10.5; furthermore p_η is countably incomplete since $\bar{r}(p_\eta)=q$ and q is countably incomplete.

To prove that $\{A_\xi:\xi<\alpha\}\subset p_\eta$ let $\xi<\alpha$ and $A\in p_\eta$; then there is $\zeta<\beta$ such that $A\in p_{\zeta,\eta}$ and $A_\xi\cap B_\zeta\in\mathscr{B}_\zeta\subset p_{\zeta,\eta}$, so that $A\cap A_\xi\cap B_\zeta\in p_{\zeta,\eta}$ and hence $A\cap A_\xi\neq\emptyset$. Finally it is easy to see that if $\eta<\eta'<2^{2^\alpha}$ then

$$\mathrm{cl}_{\boldsymbol{\beta}(\alpha)}\{p_{\zeta,\eta}:\zeta<\beta\}\cap\mathrm{cl}_{\boldsymbol{\beta}(\alpha)}\{p_{\zeta,\eta'}:\zeta<\beta\}=\emptyset\,;$$

since $p_\eta\in\mathrm{cl}_{\boldsymbol{\beta}(\alpha)}\{p_{\zeta,\eta}:\zeta<\beta\}$ and $p_{\eta'}\in\mathrm{cl}_{\boldsymbol{\beta}(\alpha)}\{p_{\zeta,\eta'}:\zeta<\beta\}$ we have $p_\eta\neq p_{\eta'}$. The proof is complete.

Remark. We note that the method of proof of Theorem 10.6 implies that

$$\text{if } \alpha=\beta \quad \text{then } p_\eta \text{ is not uniformly selective for } \eta<2^{2^\alpha}\,.$$

Indeed, if $\eta<2^{2^\alpha}$ and $A\in p_\eta$ then $|\{\zeta<\alpha:A\in p_{\zeta,\eta}\}|=\alpha$ and hence, since $B_\zeta\in p_{\zeta,\eta}$ and $p_{\zeta,\eta}$ is uniform, we have

$$|\{\zeta<\alpha:|A\cap B_\zeta|=\alpha\}|=\alpha\,.$$

We thus have the following consequence of (the method of proof of) Theorem 10.6.

10.7. Corollary. *Let $\alpha\geq\omega$. Any family of subsets of α with the uniform finite intersection property and of cardinality at most α is contained in 2^{2^α} distinct countably incomplete α^+-good ultrafilters which are not of minimal type in the Rudin-Keisler order of $\tau[U(\alpha)]$.*

The following result is a corollary to Theorems 9.19, 10.4, and 10.5, and it implies, assuming that $\mathrm{cf}(\alpha)>\omega$ and $\alpha^+=2^\alpha$, the existence of ultrafilters on α which are Rudin-Keisler minimal but not α^+-good.

10.8. Corollary *(Blass). Let α and β be cardinals such that $\omega\leq\beta<\mathrm{cf}(\alpha)$, and assume $\alpha^+=2^\alpha$. Then there is an ultrafilter p on α such that*

p is regular,

$\boldsymbol{G}(p)=\beta^+$, and

p is uniformly selective.

Hence, assuming $\alpha^+=2^\alpha$ *and* $\mathrm{cf}(\alpha)>\omega$, *there is a uniformly selective, regular ultrafilter on* α *which is not* α^+*-good.*

Proof. By Theorem 9.19 there is a family $\{p_\zeta:\zeta<\beta\}$ of ultrafilters on α such that

p_ζ is countably incomplete and α^+-good,

p_ζ is uniformly selective, and

(*) for any family $\{f_\zeta:\zeta<\beta\}\subset\alpha^\alpha$, if there is a family $\{B_\zeta:\zeta<\beta\}$ of subsets of α such that

$$B_\zeta\in p_\zeta \quad \text{for } \zeta<\beta \quad \text{and}$$

$$f_\zeta|B_\zeta \text{ is one-to-one for } \zeta<\beta\,,$$

then the set $\{\overline{f}_\zeta(p_\zeta):\zeta<\beta\}$ is faithfully indexed and strongly discrete in $\boldsymbol{\beta}(\alpha)$.

We note from (*), by taking $f_\zeta=$identity on α for $\zeta<\beta$, that $\{p_\zeta:\zeta<\beta\}$ is a faithfully indexed and strongly discrete subset of $\boldsymbol{\beta}(\alpha)$. Thus there is a partition $d=\{d_\zeta:\zeta<\beta\}$ of α such that

$$d_\zeta\in p_\zeta \quad \text{for } \zeta<\beta\,.$$

We define

$$e:\boldsymbol{\beta}(\beta)\rightarrow\boldsymbol{\beta}(\alpha)$$

by the conditions $e(\zeta)=p_\zeta$ for $\zeta<\beta$ and e continuous on $\boldsymbol{\beta}(\beta)$; and we define

$$r:\alpha\rightarrow\beta$$

by the rule $r[d_\zeta]=\{\zeta\}$ for $\zeta<\beta$. It is clear that $\overline{r}\circ e$ is the identity function on $\boldsymbol{\beta}(\beta)$.

According to Theorem 10.4 there is a countably incomplete β^+-good ultrafilter q on β. We set

$$p=e(q)\,,$$

we note that $\boldsymbol{G}(p)=\beta^+$ by Theorem 10.5, and we prove the following two statements.

(i) p is regular.

Since p_ζ is countably incomplete and α^+-good, it is regular by Corollary 9.10(b). Thus there is a family $\{A_{\zeta,\xi}:\xi<\alpha\}$ of elements of p_ζ such that

$$\text{if } I\subset\alpha \quad \text{and } |I|\geq\omega \quad \text{then } \textstyle\bigcap_{\xi\in I}A_{\zeta,\xi}=\emptyset$$

for $\zeta<\beta$. Since $d_\zeta\in p_\zeta$ we assume with no loss of generality that

$$A_{\zeta,\xi}\subset d_\zeta \quad \text{for } \xi<\alpha \quad \text{and } \zeta<\beta\,.$$

We set

$$A_\xi = \bigcup_{\zeta<\beta} A_{\zeta,\xi} \quad \text{for } \xi<\alpha,$$

and we note from Lemma 9.16(b) that $A_\xi \in p$ for $\xi<\alpha$. Furthermore, if $I \subset \alpha$ and $|I| \geq \omega$ then

$$\bigcap_{\xi\in I} A_\xi = \bigcap_{\xi\in I} \bigcup_{\zeta<\beta} A_{\zeta,\xi} = \bigcup_{\zeta<\beta} \bigcap_{\xi\in I} A_{\zeta,\xi} = \emptyset,$$

since $A_{\zeta,\xi} \subset d_\zeta$ for $\zeta<\beta$, $\xi \in I$ and $\{d_\zeta : \zeta<\beta\}$ is a partition of α.

(ii) p is uniformly selective.

Let $f \in \alpha^\alpha$ and for $\zeta<\beta$ choose $B_\zeta \in p_\zeta$ such that either $f|B_\zeta$ is one-to-one or $|f[B_\zeta]|<\alpha$; with no loss of generality we assume that $B_\zeta \subset d_\zeta$ for $\zeta<\beta$. We set $C=\{\zeta<\beta : |f[B_\zeta]|<\alpha\}$, and we consider two cases.

Case 1. $C \in q$.

Then $\bigcup_{\zeta\in C} B_\zeta \in p$ by Lemma 9.16(b) and since $\beta<\mathrm{cf}(\alpha)$ we have

$$|f[\bigcup_{\zeta\in C} B_\zeta]| \leq \sum_{\zeta\in C} |f[B_\zeta]| < \alpha.$$

Case 2. $C \notin q$.

We define

$$g : \alpha \to \alpha$$

by

$$\begin{aligned} g(\xi) &= f(\xi) && \text{for } \xi \in \textstyle\bigcup_{\zeta\in\beta\setminus C} d_\zeta, \\ &= \xi && \text{for } \xi \in \textstyle\bigcup_{\zeta\in C} d_\zeta. \end{aligned}$$

Then $g|B_\zeta$ is one-to-one for $\zeta<\beta$, and hence by (*) the set $\{\overline{g}(p_\zeta) : \zeta<\beta\}$ is faithfully indexed and strongly discrete in $\boldsymbol{\beta}(\alpha)$. Thus there is a family $\{C_\zeta : \zeta<\beta\}$ of subsets of α such that

$$\begin{aligned} &C_\zeta \in \overline{g}(p_\zeta) && \text{for } \zeta<\beta, \text{ and} \\ &C_\zeta \cap C_{\zeta'} = \emptyset && \text{for } \zeta<\zeta'<\beta; \end{aligned}$$

with no loss of generality we assume that

$$C_\zeta \subset g[B_\zeta] \quad \text{for } \zeta<\beta.$$

Then $g^{-1}(C_\zeta) \in p_\zeta$ and hence $g^{-1}(C_\zeta) \cap B_\zeta \in p_\zeta$ for $\zeta<\beta$; since $\beta\setminus C \in q$ we have

$$B = \bigcup_{\zeta\in\beta\setminus C} (f^{-1}(C_\zeta) \cap B_\zeta) \in p$$

from Lemma 9.16(b). It is clear that $f|B$ is one-to-one.

It follows that p is uniformly selective and the proof of the theorem is complete.

Directedness Properties of the Rudin-Keisler Order

10.9. Theorem. *Let κ be a regular cardinal such that $\omega \leq \kappa \leq \alpha = \alpha^{\underline{\kappa}}$, and suppose that every κ-complete filter on α is contained in a κ-complete ultrafilter. Then $\tau[\Omega_\kappa(\alpha)]$ is $(2^\alpha)^+$-directed.*

Proof. We first prove that there is continuous function

$$g: \Omega_\kappa(\alpha) \rightarrow (\Omega_\kappa(\alpha))^{(2^\alpha)}$$

such that $g[\alpha] \subset \alpha^{(2^\alpha)}$.

From Corollary 3.18 there is α function

$$f: \alpha \rightarrow \alpha^{(2^\alpha)}$$

such that $f[\alpha]$ is dense in $(\alpha^{(2^\alpha)})_{(\kappa)}$. We denote by $\bar{f}$ the (continuous) Stone extension

$$\bar{f}: \boldsymbol{\beta}(\alpha) \rightarrow (\boldsymbol{\beta}(\alpha))^{(2^\alpha)}$$

of f, we set $g = \bar{f} | \Omega_\kappa(\alpha)$, and we verify that

$$g[\Omega_\kappa(\alpha)] \subset (\Omega_\kappa(\alpha))^{(2^\alpha)}.$$

Indeed let $p \in \Omega_\kappa(\alpha)$ and set $q_\eta = (\pi_\eta \circ g)(p)$ for $\eta < 2^\alpha$. We claim that q_η is κ-complete. If $\lambda < \kappa$ and $A_i \in q_\eta$ for $i < \lambda$ then $(\pi_\eta \circ g)^{-1}(A_i) \in p$ for $i < \lambda$, and since p is κ-complete we have

$$B = \bigcap_{i<\lambda} (\pi_\eta \circ g)^{-1}(A_i) \in p$$

and hence $(\pi_\eta \circ g)[B] \subset \bigcap_{i<\lambda} A_i$; further since $p \in \mathrm{cl}_{\boldsymbol{\beta}(\alpha)} B$ we have

$$q_\eta = (\pi_\eta \circ g)(p) \in \mathrm{cl}_{\boldsymbol{\beta}(\alpha)} (\pi_\eta \circ g)[B]$$

and hence $(\pi_\eta \circ g)[B] \in q_\eta$. It follows that q_η is κ-complete, *i.e.*, $q_\eta \in \Omega_\kappa(\alpha)$ for $\eta < 2^\alpha$.

Finally we prove that g is a function onto $(\Omega_\kappa(\alpha))^{(2^\alpha)}$. Let $q = \langle q_\eta : \eta < 2^\alpha \rangle$ and denote by $\mathscr{F}$ the family of open-and-closed subsets of $((\Omega_\kappa(\alpha))^{(2^\alpha)})_{(\kappa)}$ containing q. Since κ is regular and $\Omega_\kappa(\alpha)$ is a P_κ-space (according to the simple Lemma 14.34 below) it follows that $((\Omega_\kappa(\alpha))^{(2^\alpha)})_{(\kappa)}$ is a P_κ-space; since $g[\alpha]$ is dense in $((\Omega_\kappa(\alpha))^{(2^\alpha)})_{(\kappa)}$ every element of $\mathscr{F}$ intersects $g[\alpha]$. Thus

$$\{g^{-1}(A) \cap \alpha : A \in \mathscr{F}\}$$

is a κ-complete filter on α, so by assumption it is contained in an element p of $\Omega_\kappa(\alpha)$. We claim that $g(p) = q$; indeed if A is a neighborhood of q in $(\Omega_\kappa(\alpha))^{(2^\alpha)}$ then $g^{-1}(A) \cap \alpha \in p$, hence $p \in \mathrm{cl}_{\boldsymbol{\beta}(\alpha)}(g^{-1}(A) \cap \alpha)$, and therefore

$$g(p) \in \mathrm{cl}_{\boldsymbol{\beta}(\alpha)}(A \cap \alpha^{(2^\alpha)}),$$

establishing the claim.

We now complete the proof of the theorem. Let $q_\eta \in \Omega_\kappa(\alpha)$ for $\eta < 2^\alpha$ and set $q = \langle q_\eta : \eta < 2^\alpha \rangle \in (\Omega_\kappa(\alpha))^{(2^\alpha)}$. By what has been proved there is $p \in \Omega_\kappa(\alpha)$ such that $g(p) = q$, and since $(\pi_\eta \circ g)(p) = q_\eta$ we have $q_\eta \lessdot p$ for $\eta < 2^\alpha$.

As special cases of Theorem 10.9 we have the following statements.

10.10. Corollary. *Let* $\alpha \geq \omega$.

(a) *The set* $\mathbb{T}(\boldsymbol{\beta}(\alpha))$ *is* $(2^\alpha)^+$*-directed;*

(b) *if* κ *is a strongly compact cardinal and* $\alpha = \alpha^{\underline{\kappa}}$, *then* $\tau[\Omega_\kappa(\alpha)]$ *is* $(2^\alpha)^+$*-directed;*

(c) *if* α *is a strongly measurable cardinal, then* $\tau[\Omega_\alpha(\alpha)]$ *is* $(2^\alpha)^+$*-directed.*

We have seen already that $|\tau[\Omega_\kappa(\alpha)]| \leq 2^{2^\alpha}$ and that every element of $\tau[\Omega_\kappa(\alpha)]$ has at most 2^α predecessors in the Rudin-Keisler order on $\mathbb{T}(\boldsymbol{\beta}(\alpha))$. Since a linearly ordered set has a well-ordered cofinal subset, we have the following consequence of Theorem 10.9.

10.11. Corollary. *Let* κ *be a regular cardinal such that* $\omega \leq \kappa \leq \alpha = \alpha^{\underline{\kappa}}$, *and suppose that every* κ*-complete filter on* α *is contained in a* κ*-complete ultrafilter. If* $(2^\alpha)^+ = 2^{2^\alpha}$, *then* $\tau[\Omega_\kappa(\alpha)]$ *has a well-ordered cofinal subset with* 2^{2^α} *elements.*

10.12. Lemma. *Let* $\alpha \geq \omega$.

(a) $\tau[U(\alpha)]$ *is cofinal in* $\mathbb{T}(\boldsymbol{\beta}(\alpha))$;

(b) *the set of types of countably incomplete* α^+*-good ultrafilters on* α *is cofinal in the set of types of countably incomplete ultrafilters on* α.

Proof. Let $p \in \boldsymbol{\beta}(\alpha)$. There is a family $\{D_\xi : \xi < \alpha\}$ of subsets of α such that

$$D_\xi \cap D_\zeta = \emptyset \quad \text{for } \xi < \zeta < \alpha,$$
$$|D_\xi| = \alpha \quad \text{for } \xi < \alpha, \quad \text{and}$$
$$\textstyle\bigcup_{\xi < \alpha} D_\xi = \alpha.$$

Let $f_\xi : D_\xi \to \alpha$ be a one-to-one and onto function and let $p_\xi \in \mathrm{cl}_{\boldsymbol{\beta}(\alpha)} D_\xi = \boldsymbol{\beta}(D_\xi)$ be the unique element such that $\overline{f}_\xi(p_\xi) = p$ for $\xi < \alpha$. We define

$$e : \boldsymbol{\beta}(\alpha) \to \boldsymbol{\beta}(\alpha)$$

by the conditions $e(\xi) = p_\xi$ for $\xi < \alpha$ and e continuous on $\boldsymbol{\beta}(\alpha)$; and we define

$$r : \alpha \to \alpha$$

by the rule $r[D_\xi] = \{\xi\}$ for $\xi < \alpha$. It is clear that $\overline{r} \circ e$ is the identity function on $\boldsymbol{\beta}(\alpha)$.

(a) Let $q \in U(\alpha)$ and set $s = e(q)$. Since $\overline{r}(s) = q$ we have $s \in U(\alpha)$, and since $\overline{f}(p_\xi) = p$ for $\xi < \alpha$ and $s \in \mathrm{cl}_{\boldsymbol{\beta}(\alpha)}\{p_\xi : \xi < \alpha\}$ we have $\overline{f}(s) = p$ and hence $p \preccurlyeq s$.

(b) Let q be an α^+-good countably incomplete ultrafilter on α and set $s = e(q)$. It follows from Theorem 10.5 that $\boldsymbol{G}(s) = \boldsymbol{G}(q) = \alpha^+$, and s is countably incomplete since $\overline{r}(s) = q$; finally since $\overline{f}(p_\xi) = p$ for $\xi < \alpha$ and $s \in \mathrm{cl}_{\boldsymbol{\beta}(\alpha)}\{p_\xi : \xi < \alpha\}$, we have $\overline{f}(s) = p$ and hence $p \preccurlyeq s$.

10.13. Corollary. *Let $\alpha \geq \omega$.*

(a) *The set $\tau[U(\alpha)]$, the set of types of countably incomplete ultrafilters on α, and the set of types of α^+-good countably incomplete ultrafilters on α are each $(2^\alpha)^+$-directed;*

(b) *if α is smaller than the first uncountable measurable cardinal and $(2^\alpha)^+ = 2^{2^\alpha}$, then $\mathbb{T}(\boldsymbol{\beta}(\alpha))$ contains a well-ordered cofinal subset of 2^{2^α} types of α^+-good ultrafilters.*

Proof. (a) $\tau[U(\alpha)]$ is $(2^\alpha)^+$-directed by Corollary 10.10(a) and Lemma 10.12(a); the set of types of countably incomplete ultrafilters on α is $(2^\alpha)^+$-directed by Lemma 10.12(a); the set of types of α^+-good countably incomplete ultrafilters on α is $(2^\alpha)^+$-directed by Lemma 10.12(b).

(b) By Theorem 8.31 every non-principal ultrafilter on α is countably incomplete; thus the set of types of countably incomplete ultrafilters on α is $\mathbb{T}(\boldsymbol{\beta}(\alpha)) \setminus \{\tau(0)\}$. Part (b) now follows from Lemma 10.12(a) and part (a).

10.14. Theorem *(Hajnal). Let $\kappa < \alpha$ and*

$$f : \alpha \to \mathscr{P}_\kappa(\alpha)$$

a function such that $\xi \notin f(\xi)$ for $\xi < \alpha$. Then there is $A \in \mathscr{P}(\alpha)$ such that

$$|A| = \alpha, \quad \text{and}$$

$$\xi \notin f(\zeta) \quad \text{for } \xi, \zeta \in A \quad \text{and } \xi \neq \zeta \quad (\text{i.e., } A \text{ is free for } f).$$

Proof. Case 1. α is a regular cardinal.
Suppose that there is no set A satisfying the conclusion of the theorem. We define a family $\{B_i : i < \kappa\} \subset \mathscr{P}_\alpha(\alpha)$ recursively as follows. Let $i < \kappa$ and assume that B_j has been defined for $j < i$. We choose $A_i \subset \alpha \setminus \bigcup_{j<i} B_j$ such that A_i is maximal with respect to the property of being free for f. Such a set A_i exists by Zorn's lemma and we have $|A_i| < \alpha$. We define

$$B_i = A_i \cup \left(\bigcup\{f(\xi) : \xi \in A_i\} \setminus \bigcup_{j<i} B_j\right).$$

This completes the recursive definition of the sets B_i, $i < \kappa$.

Since α is regular we have

$$|B_i| \leq |A_i| + \textstyle\sum_{\xi \in A_i} |f(\xi)| < \alpha$$

and hence $|\bigcup_{i<\kappa} B_i| < \alpha$; thus there is $\xi \in \alpha \setminus \bigcup_{i<\kappa} B_i$. We claim that $f(\xi) \cap B_i \neq \emptyset$ for $i<\kappa$. Indeed if $f(\xi) \cap B_i = \emptyset$ then $A_i \cup \{\xi\}$ is easily seen to be free for f, contradicting the maximality of A_i. Since $B_i \cap B_j = \emptyset$ for $j<i<\kappa$ we have $|f(\xi)| \geq \kappa$, a contradiction.

Case 2. α is a singular cardinal.
Let $\{\alpha_\eta : \eta < \mathrm{cf}(\alpha)\}$ be a family of cardinals such that

$$\alpha_\eta < \alpha \quad \text{for } \eta < \mathrm{cf}(\alpha),$$
$$\alpha_\eta \text{ is regular for } \eta < \mathrm{cf}(\alpha),$$
$$\alpha = \sum_{\eta < \mathrm{cf}(\alpha)} \alpha_\eta, \quad \text{and}$$
$$(\kappa + \mathrm{cf}(\alpha))^+ < \alpha_0 < \alpha_1 < \cdots < \alpha_\eta < \cdots.$$

We set $\beta = (\kappa + \mathrm{cf}(\alpha))^+$. Let $\{A_\eta : \eta < \mathrm{cf}(\alpha)\}$ be a family of subsets of α such that

$$A_\eta \cap A_{\eta'} = \emptyset \quad \text{for } \eta' < \eta < \mathrm{cf}(\alpha),$$
$$|A_\eta| = \alpha_\eta \quad \text{for } \eta < \mathrm{cf}(\alpha), \quad \text{and}$$
$$\bigcup_{\eta < \mathrm{cf}(\alpha)} A_\eta = \alpha.$$

We define functions

$$f_\eta : A_\eta \to \mathscr{P}_\kappa(A_\eta)$$

by $f_\eta(\xi) = f(\xi) \cap A_\eta$ for $\eta < \mathrm{cf}(\alpha)$. By *Case* 1 there is $B_\eta \subset A_\eta$ such that

$$|B_\eta| = \alpha_\eta, \quad \text{and}$$

(1) B_η is free for f_η (and hence for f) for $\eta < \mathrm{cf}(\alpha)$.
We define

$$C_\eta = B_\eta \setminus \bigcup \{f(\xi) : \xi \in A_{\eta'} \text{ and } \eta' < \eta\}$$

and we note, since $|f(\xi)| < \kappa$ for $\xi < \alpha$ and $|A_{\eta'}| = \alpha_{\eta'} < \alpha_\eta$, that $|C_\eta| = \alpha_\eta$ for $\eta < \mathrm{cf}(\alpha)$. Furthermore we have

(2) if $\eta' < \eta < \mathrm{cf}(\alpha)$ and $\xi' \in C_{\eta'}$ and $\xi \in C_\eta$, then $\xi \notin f(\xi')$.
We define families $\{D_\eta : \eta < \mathrm{cf}(\alpha)\}$ and $\{D_{\eta,i} : \eta < \mathrm{cf}(\alpha), i < \beta\}$ of subsets of α and a family $\{i(\eta) : \eta < \mathrm{cf}(\alpha)\}$ of cardinals such that

$$D_0 = C_0,$$
$$D_{\eta,i} \cap D_{\eta,j} = \emptyset \quad \text{for } \eta < \mathrm{cf}(\alpha) \quad \text{and } i < j < \beta,$$
$$|D_{\eta,i}| = \alpha_\eta,$$
$$D_\eta = \bigcup_{i<\beta} D_{\eta,i},$$
$$i(\eta) < \beta, \quad \text{and}$$
$$f(\xi) \cap (\bigcup_{\eta' < \eta} D_{\eta',i}) = \emptyset \quad \text{for } i(\eta) \leq i < \beta \quad \text{and } \xi \in D_\eta$$

for $\eta < \mathrm{cf}(\alpha)$.

We proceed by transfinite recursion. We set $D_0 = C_0$ and we let $\{D_{0,i}: i < \beta\}$ be a family of subsets of D_0 such that $|D_{0,i}| = \alpha_0$ for $i < \beta$, $\bigcup_{i<\beta} D_{0,i} = D_0$, and $D_{0,i} \cap D_{0,j} = \emptyset$ for $i < j < \beta$. Such a family of subsets exists since $|D_0| = |C_0| = \alpha_0 > \beta$. We set $i(0) = 0$.

We now let $\eta < \mathrm{cf}(\alpha)$ and we set

$$S_\eta = \bigcup_{\eta' < \eta} D_{\eta'}, \quad \text{and}$$
$$S_{\eta,i} = \bigcup_{\eta' < \eta} D_{\eta',i} \quad \text{for } i < \beta.$$

It is clear that $S_\eta = \bigcup_{i<\beta} S_{\eta,i}$ and that $S_{\eta,i} \cap S_{\eta,j} = \emptyset$ for $i < j < \beta$; thus

$$f(\xi) \cap S_\eta = \bigcup_{i<\beta} (f(\xi) \cap S_{\eta,i}) \quad \text{for } \xi \in C_\eta.$$

Furthermore we note that

if $\xi \in C_\eta$ then there is $i(\xi,\eta) < \beta$ such that $f(\xi) \cap S_{\eta,i} = \emptyset$ for $i(\xi,\eta) \le i < \beta$.

Indeed, this last implication follows from the facts that $|f(\xi) \cap S_\eta| < \kappa < \beta$, that $\{f(\xi) \cap S_{\eta,i}: i < \beta\}$ is a pairwise disjoint family of sets, and that β is regular. We set

$$T_{\eta,i} = \{\xi \in C_\eta : i(\xi,\eta) \le i\} \quad \text{for } i < \beta,$$

and we have

$$C_\eta = \bigcup_{i<\beta} T_{\eta,i}, \quad \text{and}$$
$$T_{\eta,i} \subset T_{\eta,j} \quad \text{for } i < j < \beta.$$

Since $|C_\eta| = \alpha_\eta > \beta$ and α_η is regular there is $i(\eta) < \beta$ such that $|T_{\eta,i(\eta)}| = \alpha_\eta$; we define

$$D_\eta = T_{\eta,i(\eta)}.$$

We let $\{D_{\eta,i}: i < \beta\}$ be a family of subsets of D_η such that $|D_{\eta,i}| = \alpha_\eta$ for $i < \beta$, $\bigcup_{i<\beta} D_{\eta,i} = D_\eta$, and $D_{\eta,i} \cap D_{\eta,j} = \emptyset$ for $i < j < \beta$. Such a family of subsets exists since $|D_\eta| = \alpha_\eta > \beta$. This completes the recursive definitions.

We set

$$I_\eta = \{i(\eta'): \eta < \eta' < \mathrm{cf}(\alpha)\},$$

so that $|I_\eta| \le \mathrm{cf}(\alpha) < \beta$ for $\eta < \mathrm{cf}(\alpha)$. Let $i^*(\eta)$ be such that

$$\sup I_\eta < i^*(\eta) < \beta$$

for $\eta < \mathrm{cf}(\alpha)$. We define

$$E_\eta = \bigcup_{i^*(\eta) \le i < \beta} D_{\eta,i} \quad \text{for } i < \beta, \quad \text{and}$$
$$A = \bigcup_{\eta < \mathrm{cf}(\alpha)} E_\eta.$$

It is clear that $|E_\eta|=\alpha_\eta$ for $\eta<\mathrm{cf}(\alpha)$. Since $E_\eta\subset D_\eta\subset C_\eta\subset B_\eta\subset A_\eta$ it follows from the defining properties of $\{A_\eta:\eta<\mathrm{cf}(\alpha)\}$ that

$$|A|=\alpha\,.$$

We claim that

(3) if $\eta<\eta'<\mathrm{cf}(\alpha)$ and $\xi\in C_\eta$, $\xi'\in C_{\eta'}$, then $\xi\notin f(\xi')$.
Indeed $E_{\eta'}\subset D_{\eta'}$, so $f(\xi')\cap S_{\eta',i}=\emptyset$ for $i(\eta')\leq i$ and *a fortiori* for $i^*(\eta')\leq i$. Thus

$$f(\xi')\cap E_\eta=\bigcup_{i^*(\eta)\leq i<\beta}(f(\xi')\cap D_{\eta,i})=\emptyset \quad \text{for } \eta<\eta'.$$

From (1), (2) and (3), it is now easy to see that A is a free set for f. The proof of the theorem is complete.

10.15. Corollary. *Let α be an infinite cardinal. The following statements are equivalent.*
(a) $(2^\alpha)^+<2^{2^\alpha}$;
(b) *any set of 2^{2^α} ultrafilters on α contains a subset of 2^{2^α} ultrafilters whose types are pairwise non-comparable in the Rudin-Keisler order.*

Proof. (a)⇒(b). Let $S\subset\boldsymbol{\beta}(\alpha)$ with $|S|=2^{2^\alpha}$. We define $f(p)=\{q\in S:p\prec q\}$. Clearly $f(p)\in\mathscr{P}_{(2^\alpha)^+}(S)$ and $p\notin f(p)$. By Theorem 10.14 and our assumption $(2^\alpha)^+<2^{2^\alpha}$, there is $A\subset S$ such that

$$|A|=2^{2^\alpha}, \quad \text{and}$$
$$\text{if } p,q\in A \text{ and } p\neq q, \text{ then } p\not\prec q \text{ and } q\not\prec p\,.$$

The implication follows from statement (a) preceding Lemma 9.1.
(b)⇒(a). Use (the second statement of) Corollary 10.11 with $\kappa=\omega$.

Adequate Ultrafilters on Special Boolean Algebras

Notation and Definitions. Let $\mathscr{S}$ be a field of sets and set $\mathscr{S}^*=\mathscr{S}\setminus\{\emptyset\}$. Recall that a subset $\mathscr{F}$ of $\mathscr{S}^*$ is *closed under finite intersections* if the intersection of any non-empty finite subset of $\mathscr{F}$ is an element of $\mathscr{F}$, and an ultrafilter p on $\mathscr{S}$ is *countably incomplete* if there is a sequence $\{A_n:n<\omega\}$ of elements of p such that $\bigcap_{n<\omega}A_n=\emptyset$.

We introduce the notion of a good ultrafilter on a field of sets in analogy with the definition given in § 9, as follows:

A function $\phi:\mathscr{P}_\omega(\beta)\to\mathscr{S}$ is *monotone* if $\phi(F)\subset\phi(F')$ for $F,F'\in\mathscr{P}_\omega(\beta)$ with $F\supset F'$; and ϕ is *multiplicative* if $\phi(F\cup F')=\phi(F)\cap\phi(F')$ for $F,F'\in\mathscr{P}_\omega(\beta)$. An ultrafilter p on $\mathscr{S}$ is *α-good* if for every $\beta<\alpha$ and every monotone function $\phi:\mathscr{P}_\omega(\beta)\to p$, there is a multiplicative function $\psi:\mathscr{P}_\omega(\beta)\to p$ such that $\psi\leq\phi$, *i.e.* such that $\psi(F)\subset\phi(F)$ for $F\in\mathscr{P}_\omega(\beta)$.

If $\mathscr{F} \subset \mathscr{S}^*$ and α is an infinite cardinal, we denote by $\operatorname{Mon}(\alpha, \mathscr{F})$ the set of all monotone functions $\phi: \mathscr{P}_\omega(\beta) \to \mathscr{F}$ for $\beta < \alpha$, and by $\operatorname{Mult}(\alpha, \mathscr{F})$ the set of all multiplicative functions $\phi: \mathscr{P}_\omega(\beta) \to \mathscr{F}$ for $\beta < \alpha$. If $h: \mathscr{S} \to \mathscr{T}$ is an embedding of fields of sets, we denote by

$$\operatorname{Mon}(\alpha, h): \operatorname{Mon}(\alpha, \mathscr{F}) \to \operatorname{Mon}(\alpha, h[\mathscr{F}])$$

and

$$\operatorname{Mult}(\alpha, h): \operatorname{Mult}(\alpha, \mathscr{F}) \to \operatorname{Mult}(\alpha, h[\mathscr{F}])$$

the natural induced embeddings. For a fixed embedding h, an ultrafilter p on $\mathscr{T}$ is called $(\alpha, \mathscr{F})$-*good* if $h[\mathscr{F}] \subset p$ and for every $\phi \in \operatorname{Mon}(\alpha, \mathscr{F})$, there is $\psi \in \operatorname{Mult}(\alpha, p)$ such that $\psi \leq \operatorname{Mon}(\alpha, h)\phi$.

Let $\mathfrak{A}$ be a Boolean algebra and let δ be an infinite cardinal. We identify, *via* Stone's duality, the power $\mathfrak{A}^\delta$ with the field of all open-and-closed subsets of $\mathsf{S}(\mathfrak{A}) \times \delta$, the set δ having (as usual) the discrete topology. Thus $\mathfrak{A}^\delta$ and $\mathfrak{B}(\mathsf{S}(\mathfrak{A}) \times \delta)$ are isomorphic (and usually are identified in the sequel). An ultrafilter p on $\mathfrak{A}^\delta$ will be called δ-*uniform* if for every $A \in p$ the set

$$\{i < \delta : (\mathsf{S}(\mathfrak{A}) \times \{i\}) \cap A \neq \emptyset\}$$

has cardinality δ. Let $h: \mathfrak{A} \to \mathfrak{A}_1$ be a Boolean algebra embedding. The induced continuous function from $\mathsf{S}(\mathfrak{A}_1)$ onto $\mathsf{S}(\mathfrak{A})$ is denoted $\mathsf{S}(h)$, and we define $\mathsf{S}(h) \times \mathrm{id}_\delta$: $\mathsf{S}(\mathfrak{A}_1) \times \delta \to \mathsf{S}(\mathfrak{A}) \times \delta$ by the rule

$$(\mathsf{S}(h) \times \mathrm{id}_\delta)(s, i) = \langle \mathsf{S}(h)s, i \rangle .$$

10.16. Lemma. *Let* $\beta < \alpha$, $\mathfrak{A}$ *an* α*-homogeneous-universal Boolean algebra, and* $\mathscr{F}$ *a subset of* $\mathfrak{B}^*(\mathsf{S}(\mathfrak{A}) \times \delta)$ *such that* $\mathscr{F}$ *is closed under finite intersections and of cardinality less than* α. *For every monotone function*

$$\phi: \mathscr{P}_\omega(\beta) \to \mathfrak{B}^*(\mathsf{S}(\mathfrak{A}) \times \delta)$$

such that

$$\phi(F) \cap A \neq \emptyset \quad \textit{for } F \in \mathscr{P}_\omega(\beta) \quad \textit{and } A \in \mathscr{F},$$

there is a multiplicative function

$$\psi: \mathscr{P}_\omega(\beta) \to \mathfrak{B}^*(\mathsf{S}(\mathfrak{A}) \times \delta)$$

such that

$$\psi(F) \cap A \neq \emptyset \quad \textit{for } F \in \mathscr{P}_\omega(\beta) \quad \textit{and } A \in \mathscr{F}, \quad \textit{and}$$

$$\psi \leq \phi .$$

Proof. The first part of the proof follows the method of the proof of Lemma 9.12. We define

$$h: \mathscr{P}_\omega(\beta) \times \mathscr{F} \to \mathfrak{B}^*(\mathsf{S}(\mathfrak{A}) \times \delta)$$

by $h(F,A)=\phi(F)\cap A$, and we set

$$P_i=\{\langle F,A\rangle\in\mathscr{P}_\omega(\beta)\times\mathscr{F}:\phi(F)\cap A\cap(\mathsf{S}(\mathfrak{A})\times\{i\})\neq\emptyset\}\quad\text{for } i<\delta\,.$$

Clearly $|\mathscr{P}_\omega(\beta)\times\mathscr{F}|<\alpha$, and thus $|P_i|<\alpha$ for $i<\delta$. By Lemma 6.7, there is a family

$$\{d_i(F,A):\langle F,A\rangle\in P_i\}$$

of non-empty, open-and-closed subsets of $\mathsf{S}(\mathfrak{A})\times\{i\}$ such that

$$d_i(F,A)\cap d_i(F',A')=\emptyset\quad\text{for } \langle F,A\rangle\neq\langle F',A'\rangle\quad\text{and}$$
$$d_i(F,A)\subset\phi(F)\cap A=h(F,A)\,,$$

for $\langle F,A\rangle$, $\langle F',A'\rangle\in P_i$. We define a function

$$e:\mathscr{P}_\omega(\beta)\times\mathscr{F}\to\mathfrak{B}(\mathsf{S}(\mathfrak{A})\times\delta)$$

by

$$e(F,A)=\bigcup\{d_i(F,A):\langle F,A\rangle\in P_i\text{ and } i<\delta\}\,,$$

and for $F\in\mathscr{P}_\omega(\beta)$ we set

$$\Phi(F)=\bigcup\{e(G,A):G\in\mathscr{P}_\omega(\beta),\ G\supset F,\ A\in\mathscr{F}\}\,.$$

It follows, as in Lemma 9.12, that Φ satisfies the following conditions:

$$\Phi(F)\subset\phi(F)\,,$$
$$\Phi(F\cup F')=\Phi(F)\cap\Phi(F')\,,\quad\text{and}$$
$$\Phi(F)\cap A\neq\emptyset$$

for $F,F'\in\mathscr{P}_\omega(\beta)$ and $A\in\mathscr{F}$.

The proof continues by noting the following property of Φ:

(*) The set $\Phi(F)\setminus\Phi(F')$ is open and of type less than α for

$$F,F'\in\mathscr{P}_\omega(\beta)\,.$$

This follows from the definition of Φ and the fact that every set $e(G,A)$ is open-and-closed in $\mathsf{S}(\mathfrak{A})\times\delta$.

Now recursively we define functions

$$\psi_n:\beta\to\mathfrak{B}^*(\mathsf{S}(\mathfrak{A})\times\delta)$$

for $n<\omega$ as follows. Let $\psi_0(\xi)=\phi(\{\xi\})$; let $n>0$, and suppose that we have defined functions

$$\psi_m:\beta\to\mathfrak{B}^*(\mathsf{S}(\mathfrak{A})\times\delta)$$

for $m<n$ such that

(i) $\Phi(\{\xi\})\subset\psi_{n-1}(\xi)\subset\psi_{n-2}(\xi)\subset\cdots\subset\psi_0(\xi)=\phi(\{\xi\})$

and

(ii) if $|F| \leq n$, say $F = \{\xi_0, \ldots, \xi_{n-1}\}$, then

$$\Phi(F) \subset \psi_{n-1}(\xi_0) \cap \cdots \cap \psi_{n-1}(\xi_{n-1}) \subset \phi(F).$$

We are to define $\psi_n: \beta \to \mathfrak{B}^*(\mathsf{S}(\mathfrak{A}) \times \delta)$. We set

$$\mathscr{P}_{n+2}(\beta, \xi) = \{F \in \mathscr{P}_\omega(\beta): |F| \leq n+1,\ \xi \in F\}$$

and first for $F \in \mathscr{P}_{n+2}(\beta, \xi)$ we define

$$\psi_n^F(\xi) \in \mathfrak{B}^*(\mathsf{S}(\mathfrak{A}) \times \delta).$$

If $F \in \mathscr{P}_{n+2}(\beta, \xi)$ and $|F| < n+1$ we set

$$\psi_n^F(\xi) = \psi_{n-1}(\xi).$$

If $F \in \mathscr{P}_{n+2}(\beta, \xi)$ and $|F| = n+1$, say $F = \{\xi, \xi_0, \ldots, \xi_{n-1}\}$, we set

$$R(F) = \psi_{n-1}(\xi) \cap \psi_{n-1}(\xi_0) \cap \cdots \cap \psi_{n-1}(\xi_{n-1}) \setminus \phi(F),$$

so that $R(F) \in \mathfrak{B}(\mathsf{S}(\mathfrak{A}) \times \delta)$. Set (as usual) $[F]^m = \{G \in \mathscr{P}(F): |G| = m\}$ for $m \leq n$, and note that $R(F) \cap \Phi(F) = \emptyset$ since $\Phi(F) = \phi(F)$. Set $\sum(F) = \emptyset$, and $\sum(G) = R(F) \cap \Phi(G)$ for $G \in [F]^n$. Then $\{\sum(G): G \in [F]^n\}$ is a finite family of pairwise disjoint sets. For $G \in [F]^{n-1}$ set $\sum(G) = (R(F) \cap \Phi(G)) \setminus \bigcup\{\Phi(H): H \in [F]^n\}$, and recursively if $0 < m < n$ and $\sum(H)$ has been defined for $H \in [F]^{m+1}$ then for $G \in [F]^m$ set

$$\textstyle\sum(G) = (R(F) \cap \Phi(G)) \setminus \bigcup\{\Phi(H): H \in [F]^{m+1}\}.$$

Thus we have defined sets $\sum(G)$ for $G \in \mathscr{P}(F)$ such that

$$\textstyle\sum(G) \cap \sum(H) = \emptyset \quad \text{if } G \neq H,$$

$$\textstyle\sum(G) \subset R(F),$$

$\sum(G)$ is an open subspace of $\mathsf{S}(\mathfrak{A}) \times \delta$ such that $\sum(G) \cap (\mathsf{S}(\mathfrak{A}) \times \{i\})$ is of type less than α for $i < \delta$, and

$$\textstyle R(F) \cap \Phi(\{\xi\}) = \bigcup\{\sum(G): \xi \in G\}.$$

(The last property follows from (*) and the fact that $R(F)$ is open-and-closed.) We use Lemma 6.6 to find open-and-closed sets $V(G)$, for $G \in \mathscr{P}(F)$, such that

$$V(G) \cap V(H) = \emptyset \quad \text{if } G \neq H,$$

$$\textstyle\sum(G) \subset V(G) \subset R(F), \quad \text{and}$$

$$\textstyle V(G) = \emptyset \quad \text{if } \sum(G) = \emptyset.$$

Thus in particular $V(F) = \emptyset$. Now we set

$$R_\xi^F = \bigcup\{V(G): G \in \mathscr{P}(F),\ \xi \in F\},$$

and finally

$$\psi_n^F(\xi) = R_\xi^F \cup [\psi_{n-1}(\xi) \setminus (\psi_{n-1}(\xi_0) \cap \psi_{n-1}(\xi_1) \cap \cdots \cap \psi_{n-1}(\xi_{n-1}))] \\ \cup [\psi_{n-1}(\xi) \cap \psi_{n-1}(\xi_0) \cap \cdots \cap \psi_{n-1}(\xi_{n-1}) \cap \phi(F)] .$$

It is clear that $\psi_n^F(\xi) \in \mathfrak{B}(\mathsf{S}(\mathfrak{A}) \times \delta)$, and if $\xi < \beta$ and $F \in \mathscr{P}_{n+2}(\beta, \xi)$ then

$$\Phi(\{\xi\}) \subset \psi_n^F(\xi) .$$

Since $|\mathscr{P}_{n+2}(\beta, \xi)| = \beta < \alpha$ there is, again by Lemma 6.6, an open-and-closed subset $\psi_n(\xi)$ of $\mathsf{S}(\mathfrak{A}) \times \delta$ such that

$$\Phi(\{\xi\}) \subset \psi_n(\xi) \subset \bigcap \{\psi_n^F(\xi) : F \in \mathscr{P}_{n+2}(\beta, \xi)\} .$$

This completes the definition of the function ψ_n.

We now verify that ψ_n satisfies the recursive conditions (i) and (ii). Condition (i) is obvious. For (ii), let $|F| \le n+1$. If $|F| < n+1$, say $F = \{\xi_0, \ldots, \xi_{n-1}\}$ (with possible repetitions), then

$$\Phi(F) \subset \psi_n(\xi_0) \cap \cdots \cap \psi_n(\xi_{n-1}) \subset \psi_{n-1}(\xi_0) \cap \cdots \cap \psi_{n-1}(\xi_{n-1}) \subset \phi(F) ,$$

using the facts that Φ is multiplicative and that $\psi_n \le \psi_{n-1}$. Now let $|F| = n+1$, say $F = \{\xi_0, \ldots, \xi_n\}$. Then

$$\psi_n(\xi_0) \cap \cdots \cap \psi_n(\xi_n) \subset \psi_n^F(\xi_0) \cap \cdots \cap \psi_n^F(\xi_n) \\ \subset (\psi_{n-1}(\xi_0) \cap \cdots \cap \psi_{n-1}(\xi_n) \cap \phi(F)) \cup (R_{\xi_0}^F \cap \cdots \cap R_{\xi_n}^F).$$

If $p \in R_{\xi_0}^F \cap \cdots \cap R_{\xi_n}^F$ then $p \in V(G)$ for some (unique) $G \in \mathscr{P}(F)$ and we have $\xi_0 \in G, \ldots, \xi_n \in G$, *i.e.*, $G = F$. But $\sum(F) = V(F) = \emptyset$. We conclude that $R_{\xi_0}^F \cap \cdots \cap R_{\xi_n}^F = \emptyset$, so

$$\psi_n(\xi_0) \cap \cdots \cap \psi_n(\xi_n) \subset \phi(F)$$

and (ii) is proved.

To complete the proof of the lemma, we use again Lemma 6.6 to find an open-and-closed set $\psi(\{\xi\})$ in $\mathsf{S}(\mathfrak{A}) \times \delta$ such that

$$\Phi(\{\xi\}) \subset \psi(\{\xi\}) \subset \bigcap \{\psi_n(\xi) : n < \omega\} .$$

We define

$$\psi : \mathscr{P}_\omega(\beta) \to \mathfrak{B}^*(\mathsf{S}(\mathfrak{A}) \times \delta)$$

by $\psi(F) = \bigcap_{\xi \in F} \psi(\{\xi\})$. It is clear that ψ satisfies the conclusions of the lemma.

10.17. Lemma. *Let β and δ be infinite cardinals such that $\delta \le \beta$. Let $\mathfrak{A}_0$ be a Boolean algebra such that*

$$|\mathfrak{A}_0| \le 2^\beta ,$$

and $\mathscr{F}_0$ *a family of non-empty, open-and-closed subsets of* $\mathsf{S}(\mathfrak{A}_0)\times\delta$ *such that*

$$\mathscr{F}_0 \text{ is a filter, and}$$
$$\mathsf{S}(\mathfrak{A}_0)\times F\notin\mathscr{F}_0 \quad \text{for } F\in\mathscr{P}_\delta(\delta).$$

Then there is a Boolean algebra embedding

$$h:\mathfrak{A}_0\to\mathfrak{A}_1,$$

and there is an ultrafilter p_1 *of open-and-closed subsets of* $\mathsf{S}(\mathfrak{A}_1)\times\delta$, *such that*

$$p_1 \text{ is } \delta\text{-uniform and countably incomplete},$$
$$(\mathsf{S}(h)\times\mathrm{id}_\delta)^{-1}[\mathscr{F}_0]\subset p_1, \quad \text{and}$$
$$p_1 \text{ is } (\beta^+,\mathscr{F}_0)\text{-good}.$$

Proof. There is by Corollary 4.7 and Theorem 6.1 an embedding

$$h_0:\mathfrak{A}_0\to\mathfrak{A}$$

where $\mathfrak{A}$ is a $(2^\beta)^+$-homogeneous-universal Boolean algebra of cardinality 2^{2^β}. We transfer all considerations to $\mathfrak{A}$ and set

$$\mathscr{F}_0^* = (\mathsf{S}(h_0)\times\mathrm{id}_\delta)^{-1}[\mathscr{F}_0] \quad \text{and}$$
$$\phi^* = \mathrm{Mon}(\beta^+,h_0)\phi \quad \text{for } \phi\in\mathrm{Mon}(\beta^+,\mathscr{F}_0).$$

Since $|\mathscr{F}_0|\le(2^\beta)^\delta=2^\beta$ we have $|\mathrm{Mon}(\beta^+,\mathscr{F}_0)|\le(2^\beta)^\beta=2^\beta$; let $\langle\phi_\eta:\eta<2^\beta\rangle$ be a well-ordering of $\mathrm{Mon}(\beta^+,\mathscr{F}_0)$ and $\langle\phi_\eta^*:\eta<2^\beta\rangle$ the corresponding well-ordering in $\mathfrak{A}$. We proceed by transfinite recursion on 2^β, using Lemma 10.14 (with $\alpha=(2^\beta)^+$) at each stage.

Set $\mathscr{G}_0=\mathscr{F}_0^*\cup\{\mathsf{S}(\mathfrak{A})\times(\delta\setminus F): F\in\mathscr{P}_\delta(\delta)\}$.

For non-zero limit ordinals $\zeta<2^\beta$, set $\mathscr{G}_\zeta=\bigcup_{\eta<\zeta}\mathscr{G}_\eta$.

We define $\mathscr{G}_{\eta+1}$ for $\eta<2^\beta$. Let $\phi_\eta^*:\mathscr{P}_\omega(\gamma)\to\mathscr{F}_0^*$ for some cardinal γ (depending on η) with $\gamma\le\beta$. We apply Lemma 10.14 to find a multiplicative function $\psi_\eta^*:\mathscr{P}_\omega(\gamma)\to\mathfrak{B}^*(\mathsf{S}(\mathfrak{A})\times\delta)$ such that

$$\psi_\eta^*\le\phi_\eta^*, \quad \text{and}$$
$$\psi_\eta^*(F)\cap A\ne\emptyset \quad \text{for } F\in\mathscr{P}_\omega(\gamma), \quad A\in\mathscr{G}_\eta.$$

Let $\mathscr{G}_{\eta+1}$ be the least subfamily of $\mathfrak{B}^*(\mathsf{S}(\mathfrak{A})\times\delta)$ containing

$$\mathscr{G}_\eta\cup\{\psi_\eta^*(F)\cap A: F\in\mathscr{P}_\omega(\gamma),\ A\in\mathscr{G}_\eta\}$$

and closed under finite intersection.

It is easy to prove inductively that $|\mathscr{G}_\eta|\le 2^\beta$ for $\eta<2^\beta$, and that $\mathscr{G}_\eta$ consists of non-empty (open-and-closed) subsets of $\mathsf{S}(\mathfrak{A})\times\delta$. We set $\mathscr{G}_1^*=\bigcup_{\eta<2^\beta}\mathscr{G}_\eta$; thus $|\mathscr{G}_1^*|\le 2^\beta$. Let $\mathfrak{A}_1$ be a Boolean subalgebra of $\mathfrak{A}$ having cardinality 2^β and containing $\mathfrak{A}_0$ and all elements of $\mathfrak{A}$ of the form

$$A \cap (\mathsf{S}(\mathfrak{A}) \times \{i\})$$

for $A \in \mathscr{G}_1^*$, $i < \delta$, and let p_1 be an ultrafilter of open-and-closed subsets of $\mathsf{S}(\mathfrak{A}_1) \times \delta$ containing $\mathscr{G}_1^*$. Then the conclusions of the lemma are clearly satisfied.

By repeated use of Lemma 10.17 we obtain the following result.

10.18. Lemma. *Let β and δ be infinite cardinals such that $\delta \leq \beta$. Let $\mathfrak{A}_0$ be a Boolean algebra such that*

$$|\mathfrak{A}_0| \leq 2^\beta,$$

and let $\mathscr{F}_0$ be a family of non-empty, open-and-closed subsets of $\mathsf{S}(\mathfrak{A}_0) \times \delta$ such that

$$\mathscr{F}_0 \text{ is a filter, and}$$
$$\mathsf{S}(\mathfrak{A}_0) \times D \notin \mathscr{F}_0 \quad \text{for } D \in \mathscr{P}_\delta(\delta).$$

Then there is a Boolean algebra $\mathfrak{A}$ such that

$$|\mathfrak{A}| = 2^\beta \quad \text{and}$$
$$\mathfrak{A} \text{ is } \beta^+\text{-homogeneous-universal},$$

there is a Boolean algebra embedding

$$h: \mathfrak{A}_0 \to \mathfrak{A},$$

and there is an ultrafilter p of open-and-closed subsets of $\mathsf{S}(\mathfrak{A}) \times \delta$ such that

$$p \text{ is countably incomplete and } \delta\text{-uniform},$$
$$(\mathsf{S}(h) \times \mathrm{id}_\delta)^{-1}[\mathscr{F}_0] \subset p, \quad \text{and}$$
$$p \text{ is } \beta^+\text{-good}.$$

Proof. Using Corollary 4.7 and Theorem 6.1, we embed $\mathfrak{A}_0$ into a $(2^\beta)^+$-homogeneous-universal algebra $\mathfrak{D}$ of cardinality 2^{2^β}. We assume that $\mathfrak{A}_0 \subset \mathfrak{D}$, and we assume also without loss of generality that

$$\{\mathsf{S}(\mathfrak{A}) \times (\delta \setminus D) : D \in \mathscr{P}_\delta(\delta)\} \subset \mathscr{F}_0.$$

$\mathfrak{A}_0$ and $\mathscr{F}_0$ are already defined; we proceed by recursion. Let $\xi < \beta^+$ and suppose that for $\zeta < \xi$ we have defined Boolean subalgebras $\mathfrak{A}_\zeta$ of $\mathfrak{D}$ such that $|\mathfrak{A}_\zeta| = 2^\beta$ and ultrafilters p_ζ of open-and-closed subsets of $\mathsf{S}(\mathfrak{A}_\zeta) \times \delta$, and that for $\zeta < \eta < \xi$ we have defined Boolean algebra embeddings $h_{\zeta,\eta}: \mathfrak{A}_\zeta \to \mathfrak{A}_\eta$ in such a way that $(\mathsf{S}(h_{\zeta,\eta}) \times \mathrm{id}_\delta)^{-1}[p_\zeta] \subset p_\eta$. Here and below we abuse notation by considering the embeddings $h_{\zeta,\eta}$ to be inclusions; thus $\{\mathfrak{A}_\zeta : \zeta < \xi\}$ is a chain.

If ξ is a non-zero limit ordinal define

$$\mathfrak{A}_\xi = \bigcup_{\zeta<\xi} \mathfrak{A}_\zeta$$

and for $\zeta<\xi$ let $h_{\zeta,\xi}: \mathfrak{A}_\zeta \to \mathfrak{A}_\xi$ be the induced embeddings. Also in this case let p_ξ be the union of the corresponding chain of the set of ultrafilters $\{p_\zeta : \zeta<\xi\}$ with embeddings given by $(\mathsf{S}(h_{\zeta,\eta}) \times \mathrm{id}_\delta)^{-1}: p_\zeta \to p_\eta$ for $\zeta<\eta<\xi$. It is clear that p_ξ is an ultrafilter of open-and-closed subsets of $\mathsf{S}(\mathfrak{A}_\xi) \times \delta$.

Let $\xi=\zeta+1$. We use Lemma 10.17, with $\mathfrak{A}_\zeta$ and p_ζ replacing $\mathfrak{A}_0$ and $\mathscr{F}_0$ respectively, to obtain: a Boolean algebra $\mathfrak{A}_{\xi,0}$ such that

$$|\mathfrak{A}_{\xi,0}| = 2^\beta,$$

a Boolean algebra embedding

$$h_{\zeta,\xi,0}: \mathfrak{A}_\zeta \to \mathfrak{A}_{\xi,0},$$

and an ultrafilter $p_{\xi,0}$ satisfying the conditions of Lemma 10.17 in the context of our present notation. Because $\mathfrak{D}$ is $(2^\beta)^+$-homogeneous-universal we assume, with no loss of generality, that

$$\mathfrak{A}_\zeta \subset \mathfrak{A}_{\xi,0} \subset \mathfrak{D}$$

and that $h_{\zeta,\xi,0}$ is the indicated inclusion. We now use Corollary 4.7 and Theorem 6.1 to embed $\mathfrak{A}_{\xi,0}$ in a β^+-homogeneous-universal Boolean algebra $\mathfrak{A}_\xi$ of cardinality 2^β. Again we assume, with no loss of generality, that

$$\mathfrak{A}_{\xi,0} \subset \mathfrak{A}_\xi \subset \mathfrak{D}.$$

Let $h_{\xi,0}$ be the inclusion: $\mathfrak{A}_{\xi,0} \to \mathfrak{A}_\xi$, set $h_{\zeta,\xi} = h_{\zeta,\xi,0}$, and let p_ξ be any ultrafilter of open-and-closed subsets of $\mathsf{S}(\mathfrak{A}_\xi) \times \delta$ containing $(\mathsf{S}(h_{\xi,0}) \times \mathrm{id}_\delta)^{-1}[p_{\xi,0}]$. This completes the recursive definition.

We define $\mathfrak{A}$ to be the union of the chain $\{\mathfrak{A}_\xi : \xi<\beta^+\}$, with connecting embeddings $h_{\zeta,\xi}$ for $\zeta<\xi<\beta^+$, and p to be the union of the corresponding chain $\{p_\xi : \xi<\beta^+\}$ of ultrafilters with connecting embeddings $(\mathsf{S}(h_{\zeta,\xi}) \times \mathrm{id}_\delta)^{-1}$, $\zeta<\xi<\beta^+$. Clearly p is an ultrafilter of open-and-closed subsets of $\mathsf{S}(\mathfrak{A}) \times \delta$. Since $|\mathfrak{A}_\xi| = 2^\beta$ for $\xi<\beta^+$, we have $|\mathfrak{A}| = 2^\beta$. It follows from the regularity of the cardinal β^+ that $\mathfrak{A}$ is β^+-homogeneous-universal and that the ultrafilter p is β^+-good, countably incomplete and δ-uniform.

10.19. Theorem. *Let $\omega \le \delta < \alpha = 2^{\underline{\alpha}}$ and let $\mathfrak{C}_\alpha$ denote the unique special Boolean algebra of cardinality α (in the Jónsson class of Boolean algebras). Then $\mathfrak{C}_\alpha$ has a specializing chain*

$$\{\mathfrak{A}_\beta : \beta<\alpha\}$$

such that

$$|\mathfrak{A}_\beta| = 2^\beta \quad \textit{for } \beta < \alpha \quad \textit{and}$$
$$\mathfrak{A}_\beta \textit{ is } \beta^+\textit{-homogeneous-universal for } \beta < \alpha\,;$$

and there is an ultrafilter p of open-and-closed subsets of $\mathsf{S}(\mathfrak{C}_\alpha) \times \delta$ *which is adequate, i.e., one that satisfies the following conditions:*

$$p \textit{ is countably incomplete and } \delta\textit{-uniform, and}$$
$$p_\beta = p \cap \mathfrak{B}^*(\mathsf{S}(\mathfrak{A}_\beta) \times \delta) \textit{ is } \beta^+\textit{-good for all } \beta \textit{ with } \delta \le \beta < \alpha.$$

Proof. According to Corollary 4.7 and Theorem 6.1 and the fact that a β^+-homogeneous-universal Boolean algebra is of cardinality at least 2^β (Corollary 5.8 and proof of Theorem 6.12(a)) there is a specializing chain $\{\mathfrak{D}_\beta : \beta < \alpha\}$ for $\mathfrak{C}_\alpha$ such that $|\mathfrak{D}_\beta| = 2^\beta$ for $\beta < \alpha$.

Set $\mathfrak{A}_\beta = \mathfrak{D}_\beta$ for $\beta < \delta$. Let $\delta \le \beta < \alpha$ and assume recursively that for every cardinal γ such that $\omega \le \gamma < \beta$ we have defined Boolean algebras $\mathfrak{A}_\gamma$ such that

$$|\mathfrak{A}_\gamma| = 2^\gamma \quad \text{for } \gamma < \beta\,,$$
$$\mathfrak{A}_\gamma \text{ is } \gamma^+\text{-homogeneous-universal for } \gamma < \beta\,,$$
$$\mathfrak{A}_\gamma \subset \mathfrak{A}_{\gamma'} \quad \text{for } \gamma < \gamma' < \beta\,,$$

and we have defined ultrafilters p_γ of open-and-closed subsets of $\mathsf{S}(\mathfrak{A}_\gamma) \times \delta$ such that

$$p_\gamma = p_{\gamma'} \cap \mathfrak{B}^*(\mathsf{S}(\mathfrak{A}_\gamma) \times \delta) \quad \text{for } \gamma < \gamma' < \beta\,,$$
$$p_\gamma \text{ is countably incomplete and } \delta\text{-uniform for } \gamma < \beta, \quad \text{and}$$
$$p_\gamma \text{ is } \gamma^+\text{-good for } \delta \le \gamma < \beta.$$

We now define $\mathfrak{A}_\beta$ and p_β. Set $\tilde{\mathfrak{A}}_\beta = \bigcup_{\gamma<\beta} \mathfrak{A}_\gamma$ and $\tilde{p}_\beta = \bigcup_{\gamma<\beta} p_\gamma$. Clearly $|\tilde{\mathfrak{A}}_\beta| \le 2^\beta$, and $\tilde{p}_\beta$ is a filter of open-and-closed subsets of $\mathsf{S}(\tilde{\mathfrak{A}}_\beta) \times \delta$ which is countably incomplete and δ-uniform. Thus Lemma 10.18 applies. Let $\mathfrak{A}_\beta$ and p_β be the Boolean algebra and the β^+-good ultrafilter, respectively, that satisfy the conclusion of Lemma 10.18 in the context of our present notation. This completes the recursive definition of the (specializing) chain

$$\{\mathfrak{A}_\beta : \beta < \alpha\}\,.$$

Then $\bigcup_{\beta<\alpha} \mathfrak{A}_\beta$ is isomorphic to $\mathfrak{C}_\alpha$, and with $p = \bigcup_{\beta<\alpha} p_\beta$ it is clear that p is an adequate, countably incomplete, δ-uniform ultrafilter of open-and-closed sets of $\mathsf{S}(\mathfrak{C}_\alpha) \times \delta$. This completes the proof of the theorem.

It is easy to see that if in addition α is regular, *i.e.*, if $\alpha = \alpha^{\mathfrak{z}} > \omega$, then the adequate ultrafilter p of Theorem 10.19 is in fact α-good.

For an ultrafilter p of open-and-closed subsets of a space, we denote (as in the definition of degree of goodness in § 9) by $\boldsymbol{G}(p)$ the least cardinal β such that p is not β-good (if such a cardinal exists). Thus the preceding paragraph implies that if $\alpha=\alpha^{\underline{\alpha}}>\omega$ then there is an ultrafilter p of open-and-closed subsets of $\mathsf{S}(\mathfrak{C}_\alpha)\times\delta$ such that $\boldsymbol{G}(p)\geq\alpha$. It is not known if $\boldsymbol{G}(p)=\alpha$.

Notes for § 10

The definition of a family of large oscillation modulo a filter is due to Kunen [72]. He proved Lemmas 10.2, 10.3 and (the case $\kappa=\omega$, $\beta=2$ of) 10.1, and for $\alpha\geq\omega$ he proved these two facts (which Theorem 10.4 unifies): There is a countably incomplete α^+-good ultrafilter on α, and there are 2^α ultrafilters whose types are pairwise incomparable in the Rudin-Keisler order of $\tau[U(\alpha)]$.

Theorem 10.5 is from Keisler [65b] (Theorem 6). The case $\beta=\omega$ of Theorem 10.6 and the case for general β (assuming $\beta^+=2^\beta$) are from Keisler [64b] (Theorem 5.3) and Keisler [65b] (Theorem 8) respectively. Corollary 10.8, a solution to a question of Keisler, is from Blass [70] (Theorem 17.3); a similar question has been posed by Negrepontis [69c] (§ 5, Problem 1).

M.E. Rudin [71](III. B.3) and Booth [70] (Theorem 3.5) remark that if $p,q\in\boldsymbol{\beta}(\omega)$ then there is $r\in\boldsymbol{\beta}(\omega)$ such that $p,q\leqslant r$. That $\mathbb{T}(\boldsymbol{\beta}(\alpha))$ is α^+-directed for $\alpha\geq\omega$ is given by Blass [70] (Proposition 5.10). The results 10.9—10.13 on the $(2^\alpha)^+$-directedness of $\tau[U(\alpha)]$ and certain of its subsets are 4.3—4.7 respectively of Comfort and Negrepontis [72a]. Corollary 10.10 was proved independently by Kunen (see Ketonen [72] (Theorem 2.3(2))).

Theorem 10.14 was proved for regular cardinals α by Lazar [36] and S. Piccard [37] and for arbitrary α, assuming the generalized continuum hypothesis, by Erdős [50]. In its full generality, establishing a conjecture of Ruziewicz, Theorem 10.14 is due to Hajnal [61]. Corollary 10.15 was communicated to us by S. Shelah in a letter of December, 1971.

The results of 10.16—10.19 are from Negrepontis [72a] (Part I: Products).

§ 11. Elementary Types

The material of this section is standard (with the exception of Theorem 11.19). We develop the rudiments of model-theoretic notions needed for the purpose of studying the fundamental properties of an ultraproduct; these properties are developed in this section, in § 12 and in § 13 below. The ultraproduct construction is given and the Theorem of Łoś (11.5), together with its Corollaries 11.6 and 11.7 (Compactness theorem), is proved. It is verified that an elementary type is an elementary Jónsson class (Theorem 11.11) whose elementarily α-homogeneous-universal and (more generally) special structures are characterized in terms of saturation (Theorem 11.18).

Theorem 11.19 and Example 11.20 aim to clarify the relation of the Jónsson class of ordered sets (studied in § 5) to its subclass of densely ordered, ordered sets without a first or last element, and of the Jónsson class of Boolean algebras (studied in § 6) to its subclass of atomless Boolean algebras.

Definition. Let ρ be an ordinal and $\tau \in \omega^\rho$ a similarity type. The *(first-order) language (with equality)* $L(\tau)$ consists of the symbols, terms, atomic formulas and formulas defined in the following way.

The *symbols* of $L(\tau)$ are

(a) the *variables* v_n for $n<\omega$;

(b) the 2-ary *equality symbol* $=$, and for every $i<\rho$ the $\tau(i)$-ary *predicate symbol* P_i; and

(c) the symbols of *negation* $\neg$, *conjunction* $\wedge$, and *existential quantification* $\exists$.

Of these (a), (c), and the equality symbol are the *logical* symbols; the rest are the *non-logical* symbols. A 0-ary predicate symbol is a *constant* symbol.

The *terms* of $L(\tau)$ are

(a) the variables v_n for $n<\omega$; and

(b) the constant symbols.

The *atomic formulas* of $L(\tau)$ are

(a) $t=s$, where t and s are terms of $L(\tau)$; and

(b) $P_i t_1 \ldots t_{\tau(i)}$, where $i<\rho$, $\tau(i)>0$ and t_k is a term of $L(\tau)$ for $1 \leq k \leq \tau(i)$.

Finally, the *formulas* of $L(\tau)$ are defined recursively as follows:

(a) an atomic formula is a formula;

(b) if ϕ is a formula, then $\neg\phi$ is a formula;

(c) if ϕ and ψ are formulas, then $\wedge\,\phi\psi$ is a formula; and

(d) if ϕ is a formula, then $\exists v_n \phi$ is a formula for every $n<\omega$.

This completes the definition of the language $L(\tau)$.

The recursive definition of a formula determines a recursive procedure of definition (and an inductive method of proof) of a statement for all formulas, by

(i) defining [proving] the statement for all atomic formulas, and

(ii) defining [proving] the statement for formulas $\neg\phi$, $\wedge\,\phi\psi$, and $\exists v_n \phi$ for $n<\omega$, if it is assumed that the statement has been defined [proved] for formulas ϕ and ψ.

This procedure is referred to as *definition [proof] by recursion [induction] on (the length of) formulas.*

In writing down formulas we add parentheses and commas if this will increase their readability; for example, we write $P_i(t_1, \ldots, t_{\tau(i)})$ for $P_i t_1 \ldots t_{\tau(i)}$, etc. We write (as usual) $\phi \wedge \psi$ for $\wedge\,\phi\psi$, and if $F=\{\phi_0, \ldots, \phi_n\}$ is a finite non-empty set of formulas, we write $\bigwedge_{\phi\in F}\phi$ for $\phi_0 \wedge \cdots \wedge \phi_n$.

Definitions. The set of *subformulas* $\mathrm{sub}(\phi)$ of a formula ϕ is defined recursively as follows:

(a) if ϕ is an atomic formula, then $\mathrm{sub}(\phi)=\{\phi\}$;

(b) $\mathrm{sub}(\neg\phi)=\mathrm{sub}(\phi)\cup\{\neg\phi\}$;

(c) $\mathrm{sub}(\phi\wedge\psi)=\mathrm{sub}(\phi)\cup\mathrm{sub}(\psi)\cup\{\phi\wedge\psi\}$; and

(d) $\mathrm{sub}(\exists v_n\phi)=\mathrm{sub}(\phi)\cup\{\exists v_n\phi\}$.

An occurrence of a variable v_n is *bound* in the formula ϕ if there is $\psi\in\mathrm{sub}(\phi)$ such that $\exists v_n\psi\in\mathrm{sub}(\phi)$; otherwise it is a *free* occurrence. (Thus a variable v_n can have both a bound and a free occurence in a formula ϕ.) A formula in which every occurrence of each variable is bound is called a *sentence.*

If all the free (occurrences of) variables in a formula ϕ are among the (first $n+1$) variables $v_0,\ldots,v_n$, we sometimes write $\phi(v_0,\ldots,v_n)$ for ϕ. The set of all formulas of the language $L(\tau)$ is denoted by $\Phi(L(\tau))$; the set of all formulas of $L(\tau)$ all of whose free (occurrences of) variables are among $v_0, \ldots, v_{n-1}$ is denoted by $\Phi_n(L(\tau))$. Thus $\Phi_0(L(\tau))$ denotes the set of all sentences of $L(\tau)$. The *cardinality* $|L(\tau)|$ *of the language* $L(\tau)$ is defined to be the cardinal $|\rho|+\omega$; clearly $|L(\tau)|=|\Phi(L(\tau))|$.

If there is no danger of confusion we sometimes denote $L(\tau)$ simply by L. If L is a language of type τ, a *structure for* L is a relational system of type τ.

In practice, especially when adherence to the notation as introduced here produces awkward description of a conceptually simple situation, we will deviate from and abuse the notation. Thus if $\mathfrak{A} = \langle A, R_i^{\mathfrak{A}} \rangle_{i<\rho}$ is a system of type τ, and we wish to add relations and/or constants to $\mathfrak{A}$, we do not attempt to define carefully the resulting type, and the ordinal that is equal to its domain; instead we simply adjoin the relations, denoting the resulting system

$$\langle A, R_i^{\mathfrak{A}}, R_j \rangle_{i<\rho, j\in J} \quad \text{or} \quad \langle \mathfrak{A}, R_j \rangle_{j\in J}.$$

When we do this, we always assume that the added indices $j \in J$ are distinct from the elements of ρ and from each other. (The actual relations and constants, however, need not be distinct.) The most common occurence of such an abuse of notation is in connection with the addition of constant symbols for some elements of a structure for the existing language, as follows.

Let L be a language and let $\mathfrak{A}$ be a structure for L. Given a subset C of A, we let L_C be the language whose symbols are the symbols of L together with a distinct constant symbol $\bar{c}$ for every $c \in C$. A structure for the (expanded) language L_C is a relational system of the form

$$\mathfrak{B}' = \langle \mathfrak{B}, b_c \rangle_{c\in C} = \langle B, R_i^{\mathfrak{B}}, b_c \rangle_{i<\rho, c\in C},$$

where $\mathfrak{B}$ is a structure for L and $b_c \in B$ for $c \in C$. The structure $\langle \mathfrak{A}, c \rangle_{c\in C}$ is sometimes denoted by $\mathfrak{A}_C$.

The basic model-theoretic concept of a model for a sentence will now be defined indirectly, by defining recursively what is meant by the satisfaction of a formula in a structure by a sequence of its elements.

Definition (Tarski). Let ρ be an ordinal, $\tau \in \omega^\rho$ a similarity type, ϕ a formula of $L(\tau)$, and let $\mathfrak{A} = \langle A, R_i^{\mathfrak{A}} \rangle_{i<\rho}$ be a relational system of type τ and $a = \langle a_0, \ldots, a_n, \ldots \rangle \in A^\omega$. We define the statement *a satisfies ϕ in $\mathfrak{A}$* (denoted by $\mathfrak{A} \models_{a} \phi$) recursively as follows.

(a) Let ϕ be an atomic formula.

If ϕ is $v_m = v_n$ for some $m, n < \omega$, then $\mathfrak{A} \models_{a} v_m = v_n$ if and only if $a_n = a_m$;

if ϕ is $v_m = P_i$ for some $m < \omega$, $i < \rho$ and $\tau(i) = 0$, then $\mathfrak{A} \models_{a} v_m = P_i$ if and only if $a_m = R_i^{\mathfrak{A}}$;

if ϕ is $P_i = P_j$ for some $i, j < \rho$ and $\tau(i) = \tau(j) = 0$, then $\mathfrak{A} \models_{a} P_i = P_j$ if and only if $R_i^{\mathfrak{A}} = R_j^{\mathfrak{A}}$; and

if ϕ is $P_i(t_1, \ldots, t_{\tau(i)})$ for some $i < \rho$ and $\tau(i) > 0$, and if t_k is a term of $L(\tau)$ for $1 \le k \le \tau(i)$, then $\mathfrak{A} \models_{a} P_i(t_1, \ldots, t_{\tau(i)})$ if and only if $\langle b_1, \ldots, b_{\tau(i)} \rangle \in R_i^{\mathfrak{A}}$, where $b_k = a_{n_k}$ if t_k is equal to the variable v_{n_k} and $b_k = R_j^{\mathfrak{A}}$ if t_k is the constant symbol P_j;

(b) $\mathfrak{A} \models_{\bar{a}} \neg \phi$ if and only if a does not satisfy ϕ in $\mathfrak{A}$;

(c) $\mathfrak{A} \models_{\bar{a}} \phi \wedge \psi$ if and only if $\mathfrak{A} \models_{\bar{a}} \phi$ and $\mathfrak{A} \models_{\bar{a}} \psi$; and

(d) $\mathfrak{A} \models_{\bar{a}} \exists v_n \phi$ if and only if for some $b \in A$, $\langle a_0, \ldots, a_{n-1}, b, a_{n+1}, \ldots \rangle$ satisfies ϕ in $\mathfrak{A}$.

The proof of the following simple fact uses induction on formulas.

11.1. Lemma. *Let ϕ be a formula with all its free variables among $v_0, \ldots, v_n$, let $\mathfrak{A}$ be a structure, and let a and b be elements of $\mathfrak{A}^\omega$ such that $a_k = b_k$ for $k \leq n$. Then*

$$\mathfrak{A} \models_{\bar{a}} \phi \quad \textit{if and only if} \quad \mathfrak{A} \models_{\bar{b}} \phi .$$

As a result of Lemma 11.1 we often write

$$\mathfrak{A} \models \phi[a_0, \ldots, a_n]$$

in place of $\mathfrak{A} \models_{\bar{a}} \phi$, when all the free variables of ϕ are among $v_0, \ldots, v_n$. In particular, if ϕ is a sentence, we simply write

$$\mathfrak{A} \models \phi$$

and we express this fact by saying that ϕ *is true* (or *holds*) in $\mathfrak{A}$, or that $\mathfrak{A}$ is a *model* of ϕ. If Φ is a set of sentences of $L(\tau)$, we say that $\mathfrak{A}$ is a *model* of Φ (denoted by $\mathfrak{A} \models \Phi$) if $\mathfrak{A}$ is a model of ϕ for all $\phi \in \Phi$. We remark that the type of a model of Φ, with $\Phi \subset \Phi_0(L(\tau))$, is always τ; thus this type depends only on the language $L(\tau)$, and not on Φ.

Definitions. Let $\mathfrak{A}$ and $\mathfrak{B}$ be two structures of type τ. We say that $\mathfrak{A}$ is *elementarily equivalent* to $\mathfrak{B}$ (denoted by $\mathfrak{A} \equiv \mathfrak{B}$) if

$$\mathfrak{A} \models \phi \quad \text{if and only if } \mathfrak{B} \models \phi$$

for every sentence ϕ of $L(\tau)$.

It is clear that $\equiv$ is an equivalence relation on the class of structures of type τ; and if $\mathfrak{A}$ and $\mathfrak{B}$ are isomorphic structures then $\mathfrak{A} \equiv \mathfrak{B}$.

We define $Th(\mathfrak{A}) = \{\phi \in \Phi_0(L) : \mathfrak{A} \models \phi\}$; $Th(\mathfrak{A})$ is called the *theory* of $\mathfrak{A}$. It is clear that $\mathfrak{A} \equiv \mathfrak{B}$ if and only if $Th(\mathfrak{A}) = Th(\mathfrak{B})$, and that if $Th(\mathfrak{A}) \subset Th(\mathfrak{B})$ then $Th(\mathfrak{A}) = Th(\mathfrak{B})$.

The *elementary type* of an infinite structure $\mathfrak{A}$ is the class of all structures $\mathfrak{B}$ such that $\mathfrak{A} \equiv \mathfrak{B}$; this class is denoted by $\mathbb{M}(Th(\mathfrak{A}))$.

Definition. Let $\mathfrak{A}$ and $\mathfrak{B}$ be structures of type τ and let $\mathfrak{A} \subset \mathfrak{B}$. We say that $\mathfrak{A}$ is an *elementary substructure* of $\mathfrak{B}$ (and that $\mathfrak{B}$ is an *elementary extension* of $\mathfrak{A}$; in symbols $\mathfrak{A} \prec \mathfrak{B}$) if for every formula ϕ of $L(\tau)$, and every $a \in A^\omega$,

$$\mathfrak{A} \models_{\bar{a}} \phi \quad \text{if and only if } \mathfrak{B} \models_{\bar{a}} \phi .$$

If $f:\mathfrak{A}\to\mathfrak{B}$ is an embedding (of relational systems), we say that f is an *elementary embedding* if $f[\mathfrak{A}]\prec\mathfrak{B}$.

We note that

$$\text{if } \mathfrak{A}\prec\mathfrak{B} \quad \text{then } \mathfrak{A}\equiv\mathfrak{B}.$$

Let $\phi\in\Phi(L(\tau))$, let $\mathfrak{A}$ be a structure of type τ, let $a=\langle a_0,\dots,a_n,\dots\rangle\in A^\omega$, and let $n_1,\dots,n_k<\omega$. We denote by

$$\phi_{v_{n_1}\cdots v_{n_k}}[\bar{a}_1,\dots,\bar{a}_k]$$

the expression obtained by replacing in ϕ every *free* occurrence of v_{n_i} by $\bar{a}_i$ for $1\le i\le k$.

It is easily verified that

$$\phi_{v_{n_1}\dots v_{n_k}}[\bar{a}_1,\dots,\bar{a}_k]\in\Phi(L(\tau)_A), \quad \text{and that}$$
$$\mathfrak{A}\models_a\phi \quad \text{if and only if } \mathfrak{A}_A\models_a\phi_{v_{n_1}\dots v_{n_k}}[\bar{a}_1,\dots,\bar{a}_k].$$

Using these facts, we obtain the following descriptions of elementary embeddings, used frequently in the sequel. The details of the proof will be omitted.

11.2. Lemma. *Let $\mathfrak{A}$ and $\mathfrak{B}$ be structures of type τ such that $\mathfrak{A}\subset\mathfrak{B}$. The following statements are equivalent.*

(a) $\mathfrak{A}\prec\mathfrak{B}$;

(b) $\mathfrak{A}_A\equiv\mathfrak{B}_A$;

(c) $\mathfrak{A}_A\prec\mathfrak{B}_A$;

(d) $\mathfrak{A}_F\equiv\mathfrak{B}_F$ *whenever* $F\in\mathscr{P}_\omega(A)$;

(e) *for all* $\phi\in\Phi(L(\tau))$ *and* $a\in A^\omega$ *such that* $\mathfrak{B}\models_a\exists v_n\phi$, *there is* $b\in A^\omega$ *such that*

$$\mathfrak{B}\models_b\phi \quad \textit{and}$$
$$b_k=a_k \quad \textit{for } n\neq k<\omega; \quad \textit{and}$$

(f) *if* $\phi\in\Phi_1(L(\tau)_A)$ *and* $\mathfrak{B}_A\models\exists v_0\phi$, *then there is* $a\in A$ *such that* $\mathfrak{B}_A\models\phi[a]$.

Definition. Let η be a non-zero ordinal and let

$$\langle\mathfrak{A}_\xi:\xi<\eta\rangle$$

be a chain of systems (of similarity type τ) whose union is $\mathfrak{A}$. The chain is an *elementary* chain if

$$\mathfrak{A}_\xi\prec\mathfrak{A}_\zeta \quad \text{for } \xi<\zeta<\eta.$$

11.3. Theorem *(Tarski-Vaught). The union of an elementary chain is an elementary extension of each element of the chain.*

Proof. Let $\langle \mathfrak{A}_\xi : \xi < \eta \rangle$ be an elementary chain whose union is $\mathfrak{A}$. We must prove that if ϕ is a formula of L then

$$(*) \qquad \mathfrak{A}_\xi \underset{a}{\models} \phi \quad \text{if and only if} \quad \mathfrak{A} \underset{a}{\models} \phi \quad \text{for } \xi < \eta \quad \text{and } a \in A_\xi^\omega .$$

We prove (*) by induction on the formulas of L.

If ϕ is atomic, then the statement (*) holds simply because $\mathfrak{A}_\xi \subset \mathfrak{A}$. Also, (*) holds for $\neg\phi$ and for $\phi \wedge \psi$ if it holds for ϕ and ψ. We now assume that (*) holds for ϕ, and prove it for $\exists v_n \phi$ (with $n < \omega$).

If $\mathfrak{A}_\xi \underset{a}{\models} \exists v_n \phi$ then $\mathfrak{A}_\xi \underset{b}{\models} \phi$ for some $b \in A_\xi^\omega$ such that $b_k = a_k$ for all $k < \omega$ with $k \neq n$; then $\mathfrak{A} \underset{b}{\models} \phi$ by the inductive assumption, and hence $\mathfrak{A} \underset{a}{\models} \exists v_n \phi$. For the converse let $\mathfrak{A} \underset{a}{\models} \exists v_n \phi$, so that $\mathfrak{A} \underset{b}{\models} \phi$ for some $b \in A^\omega$ such that $b_k = a_k$ for $k < \omega$ and $k \neq n$. Since $A = \bigcup_{\zeta < \eta} A_\zeta$, we have $b_n \in A_\zeta$ for some $\zeta < \eta$. We choose $i < \eta$ such that $\xi \leq i$ and $\zeta \leq i$ and we have $b \in A_i^\omega$, and thus, by the inductive assumption, $\mathfrak{A}_i \underset{b}{\models} \phi$. Thus $\mathfrak{A}_i \underset{a}{\models} \exists v_n \phi$. Since $\mathfrak{A}_\xi \prec \mathfrak{A}_i$, it follows that $\mathfrak{A}_\xi \underset{a}{\models} \exists v_n \phi$.

11.4. Theorem *(Downward Löwenheim-Skolem theorem). Let $\alpha \geq (|\rho| + \omega)^+$, let $\mathfrak{A}$ be a structure of type $\tau \in \omega^\rho$, and let $C \subset A$ with $|C| < \alpha$. Then there is a structure $\mathfrak{B}$ such that*

$$C \subset B \subset A,$$
$$|B| < \alpha, \quad \textit{and}$$
$$\mathfrak{B} \prec \mathfrak{A} .$$

Proof. If $|A| < \alpha$, we simply take $\mathfrak{B} = \mathfrak{A}$. Thus we assume $|A| \geq \alpha$. We define a sequence $\{B_n : n < \omega\}$ of subsets of A such that

(i) $B_0 = C$;

(ii) $|B_n| \leq \omega + |\rho| + |C|$ for $n < \omega$; and

(iii) if $\phi \in \Phi_1(L(\tau)_{B_n})$ and $\mathfrak{A}_{B_n} \models \exists v_0 \phi$, then there is $a \in B_{n+1}$ such that $\mathfrak{A}_{B_n} \models \phi[a]$.

We proceed recursively. We set $B_0 = C$. Let $n < \omega$; we define B_{n+1}. Since $|B_n| \leq \omega + |\rho| + |C|$, it follows that $|\Phi_1(L(\tau)_{B_n})| \leq \omega + |\rho| + |C|$; for every $\phi \in \Phi_1(L(\tau)_{B_n})$ such that $\mathfrak{A}_{B_n} \models \exists v_0 \phi$, we choose $a_\phi \in A$ such that $\mathfrak{A}_{B_n} \models \phi[a_\phi]$, and we set

$$B_{n+1} = \{a_\phi : \phi \in \Phi_1(L(\tau)_{B_n}) \ \text{ and } \ \mathfrak{A}_{B_n} \models \exists v_0 \phi\} .$$

Clearly $|B_{n+1}| \leq \omega + |\rho| + |C|$.

This completes the recursive definition. We set

$$B = \bigcup_{n < \omega} B_n ,$$
$$R_i^{\mathfrak{B}} = R_i^{\mathfrak{A}} \cap B^{\tau(i)} \quad \text{if } i < \rho \quad \text{and } \tau(i) > 0,$$
$$R_i^{\mathfrak{B}} = R_i^{\mathfrak{A}} \quad \text{if } i < \rho \quad \text{and } \tau(i) = 0; \quad \text{and}$$
$$\mathfrak{B} = \langle B, R_i^{\mathfrak{B}} \rangle_{i < \rho} .$$

Clearly, $|B| \leq \omega + |\rho| + |C| < \alpha$. We first verify that $\mathfrak{B} \subset \mathfrak{A}$. Clearly $B \neq \emptyset$, since $\mathfrak{A} \models \exists v_0(v_0 = v_0)$; and if $i < \rho$ and $\tau(i) = 0$, then $\mathfrak{A} \models \exists v_0(v_0 = P_i)$, so $\mathfrak{B}$ contains all the constant relations of $\mathfrak{A}$.

To prove that $\mathfrak{B} \prec \mathfrak{A}$, we use Lemma 11.2(f). Let $\phi \in \Phi_1(L(\phi)_B)$ and suppose that $\mathfrak{A}_B \models \exists v_0 \phi$. Since ϕ contains a finite number of symbols, hence a finite number of constants from B, there is $n < \omega$ such that $\mathfrak{A}_{B_n} \models \exists v_0 \phi$. Hence there is $a \in B_{n+1} \subset B$ such that $\mathfrak{A}_{B_n} \models \phi[a]$. Thus, $\mathfrak{A}_B \models \phi[a]$; and it follows that $\mathfrak{B} \prec \mathfrak{A}$.

In the next few pages we define reduced products and ultraproducts, and we prove the basic theorem of Łoś. We then examine some elementary consequences of the theorem of Łoś, as it applies to ultraproducts modulo regular ultrafilters; as a result we obtain the compactness theorem, and the embedding and amalgamation properties for elementary types. The deeper properties of ultraproducts modulo good ultrafilters are postponed until § 13.

Definitions. Let $\{\mathfrak{A}_\xi : \xi < \alpha\}$ be a family of similar relational structures of type $\tau \in \omega^\rho$ and let $\mathscr{F}$ be a filter on α. We define the reduced product of the family $\{\mathfrak{A}_\xi : \xi < \alpha\}$ modulo the filter $\mathscr{F}$ as follows.

For elements f and g of $\prod_{\xi < \alpha} A_\xi$, we write

$$f \underset{\mathscr{F}}{\equiv} g \quad \text{if and only if } \{\xi < \alpha : f(\xi) = g(\xi)\} \in \mathscr{F}.$$

It is not difficult to verify that $\underset{\mathscr{F}}{\equiv}$ is an equivalence relation on $\prod_{\xi < \alpha} A_\xi$. Indeed, if f, g and h are elements of $\prod_{\xi < \alpha} A_\xi$, then:

(i) $f \underset{\mathscr{F}}{\equiv} f$, since $\{\xi < \alpha : f(\xi) = f(\xi)\} = \alpha \in \mathscr{F}$;

(ii) $f \underset{\mathscr{F}}{\equiv} g$ implies $g \underset{\mathscr{F}}{\equiv} f$; and

(iii) if $f \underset{\mathscr{F}}{\equiv} g$ and $g \underset{\mathscr{F}}{\equiv} h$, then

$$\{\xi < \alpha : f(\xi) = h(\xi)\} \supset \{\xi < \alpha : f(\xi) = g(\xi)\} \cap \{\xi < \alpha : g(\xi) = h(\xi)\} \in \mathscr{F},$$

so that $\{\xi < \alpha : f(\xi) = h(\xi)\} \in \mathscr{F}$ and therefore $f \underset{\mathscr{F}}{\equiv} h$.

We let $A = \prod_{\xi < \alpha} A_\xi / \mathscr{F}$ be the set of $\underset{\mathscr{F}}{\equiv}$-equivalence classes of $\prod_{\xi < \alpha} A_\xi$, and for $f \in \prod_{\xi < \alpha} A_\xi$ we denote by $f/\mathscr{F}$ the $\underset{\mathscr{F}}{\equiv}$-equivalence class of f. For $i < \rho$ and $\tau(i) > 0$ the $\tau(i)$-ary relation $R_i^{\mathfrak{A}}$ on A is defined by the rule

$$\langle f_1/\mathscr{F}, \ldots, f_{\tau(i)}/\mathscr{F} \rangle \in R_i^{\mathfrak{A}}$$

if and only if $\{\xi < \alpha : \langle f_1(\xi), \ldots, f_{\tau(i)}(\xi) \rangle \in R_i^{\mathfrak{A}_\xi}\} \in \mathscr{F}$; and for $i < \rho$ and $\tau(i) = 0$, we define (recalling that $R_i^{\mathfrak{A}_\xi} \in A_\xi$ for $\xi < \alpha$)

$$R_i^{\mathfrak{A}} = \langle R_i^{\mathfrak{A}_\xi} : \xi < \alpha \rangle / \mathscr{F}.$$

We now verify that, for $i<\rho$ and $\tau(i)>0$, $R_i^{\mathfrak{A}}$ is well-defined. For $1\leq n\leq\tau(i)$ let f_n and g_n be elements of A such that $f_n/\mathscr{F}=g_n/\mathscr{F}$, and suppose that

$$\{\xi<\alpha:\langle f_1(\xi),\dots,f_{\tau(i)}(\xi)\rangle\in R_i^{\mathfrak{A}_\xi}\}\in\mathscr{F}.$$

Since $\{\xi<\alpha:\langle g_1(\xi),\dots,g_{\tau(i)}(\xi)\rangle\in R_i^{\mathfrak{A}_\xi}\}$ contains the set

$$\bigcap_{1\leq n\leq\tau(i)}\{\xi<\alpha:f_n(\xi)=g_n(\xi)\}\cap\{\xi<\alpha:\langle f_1(\xi),\dots,f_{\tau(i)}(\xi)\rangle\in R_i^{\mathfrak{A}_\xi}\}$$

and this latter belongs to $\mathscr{F}$, we conclude that $\langle g_1/\mathscr{F},\dots,g_{\tau(i)}/\mathscr{F}\rangle\in R_i^{\mathfrak{A}}$ whenever $\langle f_1/\mathscr{F},\dots,f_{\tau(i)}/\mathscr{F}\rangle\in R_i^{\mathfrak{A}}$. By a symmetric argument, the reverse implication also holds. Thus $R_i^{\mathfrak{A}}$ is well-defined for $i<\rho$.

The structure

$$\langle\prod_{\xi<\alpha}A_\xi/\mathscr{F},R_i^{\mathfrak{A}}\rangle_{i<\rho},$$

a structure of type τ, is called the *reduced product* of $\{\mathfrak{A}_\xi:\xi<\alpha\}$ *modulo* $\mathscr{F}$, and is denoted $\prod_{\xi<\alpha}\mathfrak{A}_\xi/\mathscr{F}$. If p is an ultrafilter on α, the reduced product $\prod_{\xi<\alpha}\mathfrak{A}_\xi/p$ is called the *ultraproduct* of $\{\mathfrak{A}_\xi:\xi<\alpha\}$ *modulo* p. If $\mathfrak{A}_\xi=\mathfrak{A}$ for all $\xi<\alpha$, then instead of $\prod_{\xi<\alpha}\mathfrak{A}_\xi/\mathscr{F}$ we write $\mathfrak{A}^\alpha/\mathscr{F}$, and this we call the *reduced power* of $\mathfrak{A}$ *modulo* $\mathscr{F}$; and finally, if p is an ultrafilter, $\mathfrak{A}^\alpha/p$ is called the *ultrapower* of $\mathfrak{A}$ *modulo* p.

We note that the concept of a reduced product modulo a filter is "purely mathematical" and does not involve any "metamathematical" considerations about the language and its formulas. One of the fundamental applications of ultraproducts is the mathematical description of the metamathematical notion of elementary equivalence: Two (similar) structures are elementarily equivalent if and only if they have isomorphic ultrapowers (Corollary 13.21).

The simpler direction of this theorem (Corollary 11.6) will follow from the next result, which expresses the basic model-theoretic property of ultraproducts.

11.5. Theorem *(Łoś). Let α be a cardinal number and p an ultrafilter on α, let $\tau\in\omega^\rho$ be a similarity type and ϕ a formula of $L(\tau)$, and for $\xi<\alpha$ let $\mathfrak{A}_\xi$ be a structure of type τ. For $f=\langle f_0,\dots,f_n,\dots\rangle\in(\prod_{\xi<\alpha}A_\xi)^\omega$ let*

$$f/p=\langle f_0/p,\dots,f_n/p,\dots\rangle \quad \textit{and} \quad f(\xi)=\langle f_0(\xi),\dots,f_n(\xi),\dots\rangle$$

for $\xi<\alpha$. Then

$$\prod_{\xi<\alpha}\mathfrak{A}_\xi\models_{f/p}\phi \quad \textit{if and only if} \quad \{\xi<\alpha:\mathfrak{A}_\xi\models_{f(\xi)}\phi\}\in p.$$

Proof. The proof employs induction on (the length of) ϕ.

(a) Let ϕ be an atomic formula.

If ϕ is $v_m=v_n$, then

$$\prod_{\xi<\alpha}\mathfrak{A}_\xi/p\models_{f/p}v_m=v_n \quad \text{if and only if } f_m/p=f_n/p,$$

i.e., if and only if

$$\{\xi<\alpha: f_m(\xi)=f_n(\xi)\} = \{\xi<\alpha: \mathfrak{A}_\xi \models_{f(\xi)} v_m = v_n\} \in p\,.$$

If ϕ is $v_m = P_i$ for some $m<\omega$ with $i<\rho$ and $\tau(i)=0$, then $\prod_{\xi<\alpha}\mathfrak{A}_\xi/p \models_{f/p} v_m = P_i$ if and only if $f_m/p = \langle R_i^{\mathfrak{A}_\xi}: \xi<\alpha\rangle/p$, *i.e.*, if and only if

$$\{\xi<\alpha: f_m(\xi)=R_i^{\mathfrak{A}_\xi}\} = \{\xi<\alpha: \mathfrak{A}_\xi \models_{f(\xi)} v_m = P_i\} \in p\,.$$

If ϕ is $P_i = P_j$ for some $i,j<\rho$ with $\tau(i)=\tau(j)=0$, then $\prod_{\xi<\alpha}\mathfrak{A}_\xi/p \models_{f/p} P_i = P_j$ if and only if $\langle R_i^{\mathfrak{A}_\xi}: \xi<\alpha\rangle/p = \langle R_j^{\mathfrak{A}_\xi}: \xi<\alpha\rangle/p$, *i.e.*, if and only if

$$\{\xi<\alpha: R_i^{\mathfrak{A}_\xi} = R_j^{\mathfrak{A}_\xi}\} = \{\xi<\alpha: \mathfrak{A}_\xi \models_{f(\xi)} P_i = P_j\} \in p\,.$$

Finally, if ϕ is $P_i(t_1,\ldots,t_{\tau(i)})$ for some $i<\rho$ and $\tau(i)>0$, with t_k a term of $L(\tau)$ for $1\le k\le\tau(i)$, then

$$\prod_{\xi<\alpha}\mathfrak{A}_\xi/p \models_{f/p} P_i(t_1,\ldots,t_{\tau(i)})$$

if and only if $\langle g_1/p,\ldots,g_{\tau(i)}/p\rangle \in R_i$; here R_i stands for $R_i^{\prod_{\xi<\alpha}\mathfrak{A}_\xi/p}$, and $g_k/p = f_{n_k}/p$ if $t_k = v_{n_k}$ and

$$g_k/p = \langle R_j^{\mathfrak{A}_\xi}: \xi<\alpha\rangle/p$$

if $t_k = P_j$. But

$$\langle g_1/p,\ldots,g_{\tau(i)}/p\rangle \in R_i$$

if and only if

$$\{\xi<\alpha: \langle g_1(\xi),\ldots,g_{\tau(i)}(\xi)\rangle \in R_i^{\mathfrak{A}_\xi}\} = \{\xi<\alpha: \mathfrak{A}_\xi \models_{f(\xi)} P_i(t_1,\ldots,t_{\tau(i)})\} \in p\,.$$

(b) Suppose the conclusion of the theorem holds for ϕ. We prove it for $\neg\phi$. By the definition of satisfaction, $\prod_{\xi<\alpha}\mathfrak{A}_\xi/p \models_{f/p} \neg\phi$ if and only if $\prod_{\xi<\alpha}\mathfrak{A}_\xi/p \models_{f/p} \phi$ is false, *i.e.*, if and only if (by the inductive assumption) $\{\xi<\alpha: \mathfrak{A}_\xi \models_{f(\xi)} \phi\} \notin p$. Since p is an *ultrafilter*, this last statement occurs if and only if

$$\{\xi<\alpha: \mathfrak{A}_\xi \models_{f(\xi)} \phi \text{ fails}\} = \{\xi<\alpha: \mathfrak{A}_\xi \models_{f(\xi)} \neg\phi\} \in p\,.$$

(c) Suppose that the conclusion of the theorem holds for ϕ, and let $n<\omega$. We verify the conclusion of the theorem for $\exists v_n\phi$. Assume first that $\prod_{\xi<\alpha}\mathfrak{A}_\xi/p \models_{f/p} \exists v_n\phi$; then, by the definition of satisfaction, $\prod_{\xi<\alpha}\mathfrak{A}_\xi/p \models_{g/p} \phi$ for some $g\in(\prod_{\xi<\alpha}A_\xi)^\omega$ such that $g_k/p = f_k/p$ whenever $k<\omega$ and $k\neq n$. Clearly, we may assume that $g_k = f_k$ for $k<\omega$ and $k\neq n$. Thus, by the inductive assumption,

$$\{\xi<\alpha: \mathfrak{A}_\xi \models_{g(\xi)} \phi\} \subset \{\xi<\alpha: \mathfrak{A}_\xi \models_{f(\xi)} \exists v_n\phi\} \in p\,.$$

For the converse implication assume that the set

$$B=\{\xi<\alpha:\mathfrak{A}_\xi \models_{\overline{f(\xi)}} \exists v_n \phi\} \in p .$$

For $\xi \in B$, there is $a \in A_\xi^\omega$ such that

$$a_{\xi,k}=f(\xi)_k \quad \text{whenever } k<\omega \quad \text{and } k \neq n, \quad \text{and}$$
$$\mathfrak{A}_\xi \models_{\overline{a_\xi}} \phi .$$

We define $g \in (\prod_{\xi<\alpha} A_\xi)^\omega$ by the rule:

$$g(\xi)_k=f(\xi)_k \quad \text{if } k<\omega, \quad k \neq n, \quad \text{and } \xi<\alpha ;$$
$$g(\xi)_n=a_{\xi,n} \quad \text{if } \xi \in B ,$$
$$g(\xi)_n \quad \text{is an arbitrary element of } A_\xi \text{ if } \xi \in \alpha \setminus B.$$

Then since $\{\xi<\alpha:\mathfrak{A}_\xi \models_{\overline{g(\xi)}} \phi\} \supset B \in p$ it follows from the inductive assumption that $\prod_{\xi<\alpha}\mathfrak{A}_\xi/p \models_{\overline{g/p}} \phi$, so that $\prod_{\xi<\alpha}\mathfrak{A}_\xi/p \models_{\overline{f/p}} \exists v_n \phi$ as required.

This completes the proof of the Theorem of Łoś.

Two of the most useful applications of the theorem of Łoś are given below, as Corollary 11.6 and Theorem 11.7.

Let $\mathfrak{A}$ be a structure, α a cardinal and p an ultrafilter on α. For $a \in A$, let $\underline{a} \in A^\alpha$ be given by $\underline{a}_\xi=a$ for $\xi<\alpha$.

The *canonical embedding*

$$e:\mathfrak{A} \to \mathfrak{A}^\alpha/p$$

is given by the diagonal function $e(a)=\underline{a}/p$. Then e is a one-to-one function (since $\{\xi<\alpha:\underline{a}_\xi \neq \underline{b}_\xi\}=\alpha \in p$ whenever a and b are distinct elements of A).

11.6. Corollary. *The canonical embedding $e:\mathfrak{A} \to \mathfrak{A}^\alpha/p$ is an elementary embedding. Hence $\mathfrak{A} \equiv \mathfrak{A}^\alpha/p$.*

Proof. Let ϕ be a formula and let $a \in A^\omega$. If $\mathfrak{A} \models_{\overline{a}} \phi$ then $\mathfrak{A}=\mathfrak{A}_\xi \models_{\overline{e(a)_\xi}} \phi$ for all $\xi<\alpha$, so since $\alpha \in p$ it follows from the theorem of Łoś that $\mathfrak{A}^\alpha/p \models_{\overline{e(a)}} \phi$. Conversely, if $\mathfrak{A}^\alpha/p \models_{\overline{e(a)}} \phi$, then

$$\emptyset \neq \{\xi<\alpha:\mathfrak{A}=\mathfrak{A}_\xi \models_{\overline{a}} \phi\} \in p$$

(again by the theorem of Łoś), and therefore $\mathfrak{A} \models_{\overline{a}} \phi$.

For simplicity, and because the canonical embedding is a naturally defined function, we usually consider $\mathfrak{A}$ as an elementary subsystem of $\mathfrak{A}^\alpha/p$ (by identifying $a \in A$ with $e(a)$).

11.7. Theorem *(Compactness theorem). A set Φ of sentences (of a given language L) has a model if and only if every finite subset of Φ has a model.*

Proof. If Φ has a model, then every (finite) subset of Φ has a model. For the converse we assume, with no loss of generality, that Φ is an infinite set. Let $\alpha=|\Phi|$, let $\langle\phi_\eta:\eta<\alpha\rangle$ be a well-ordering of Φ, and let p be a regular ultrafilter on α. Such an ultrafilter exists by Lemma 7.11 (a). There is a family $\{C_\eta:\eta<\alpha\}$ of elements of p such that for every $\xi<\alpha$ the set

$$F(\xi)=\{\eta<\alpha:\xi\in C_\eta\}$$

is finite. By assumption, there is a structure $\mathfrak{A}_\xi$ such that

$$\mathfrak{A}_\xi\models\{\phi_\eta:\eta\in F(\xi)\}\quad\text{for }\xi<\alpha\,.$$

Now we claim, with

$$\mathfrak{A}=\prod_{\xi<\alpha}\mathfrak{A}_\xi/p\,,$$

that $\mathfrak{A}\models\Phi$. Indeed if $\phi\in\Phi$, so that $\phi=\phi_\eta$ for some $\eta<\alpha$, then we have

$$\{\xi<\alpha:\mathfrak{A}_\xi\models\phi_\eta\}\supset\{\xi<\alpha:\eta\in F(\xi)\}=C_\eta\in p\,;$$

thus $\prod_{\xi<\alpha}\mathfrak{A}_\xi/p\models\phi$ by the theorem of Łoś.

Let $\mathfrak{A}$ be a structure of type $\tau\in\omega^\rho$ and let $a\in A^\omega$. Let $I\subset\rho$, and set $\tau'=\tau|I$. (Strictly speaking τ' is not a similarity type but, since I is isomorphic to a unique ordinal, τ' is easily made into one *via* the (unique) isomorphism to this ordinal.) The *reduct of* $\mathfrak{A}$ *to* τ', denoted $\mathfrak{A}\restriction\tau'$, is the structure

$$\langle A,R_i^{\mathfrak{A}}\rangle_{i\in I}$$

of type τ'.

The following fact will be used in the sequel without explicit mention: if $\phi\in\Phi(L_{\tau'})$ and $a\in A^\omega$, then $\mathfrak{A}\models_a\phi$ if and only if $\mathfrak{A}\restriction\tau'\models_a\phi$. (The simple proof by induction on formulas is omitted.)

The next result is closely related to, and is used below in proving, the embedding property and the amalgamation property for elementary types. It is a special case of the Robinson-Craig Joint Consistency Theorem.

11.8. Theorem. *Let* $\mathfrak{A}$ *and* $\mathfrak{B}$ *be elementarily equivalent structures of type* $\tau\in\omega^\rho$. *Let* I *and* J *be two sets of indices such that* $\rho\cap I=\rho\cap J=I\cap J=\emptyset$, *let* $\{a_i:i\in I\}\subset A$, *and for* $j\in J$ *let* S_j *be either a finitary relation over* B *or an element of* B. *Then there exist a structure* $\mathfrak{C}$ *of type* τ *and a subset* $\{c_i:i\in I\}$ *of* C, *and for* $j\in J$ *there exists* T_j *which is either a finitary relation over* C *or an element of* C, *such that*

(a) $\langle\mathfrak{A},a_i\rangle_{i\in I}\equiv\langle\mathfrak{C},c_i\rangle_{i\in I}$, *and*
(b) $\langle\mathfrak{B},S_j\rangle_{j\in J}\equiv\langle\mathfrak{C},T_j\rangle_{j\in J}$.

Proof. We claim that if $\phi\in Th(\langle\mathfrak{A},a_i\rangle_{i\in I})$, then there is a model of ϕ of the form $\langle\mathfrak{B},b_i\rangle_{i\in I}$.

Let $\phi \in Th(\langle \mathfrak{A}, a_i \rangle_{i \in I})$; and let $i_0, \ldots, i_n \in I$ be such that $\{\bar{a}_{i_0}, \ldots, \bar{a}_{i_n}\}$ is the set of distinct constants of ϕ. Let $N < \omega$ be such that if v_m has a (necessarily bound) occurrence in ϕ, then $m < N$. We denote by $\bar{\phi}$ the sentence

$$\exists v_N \cdots \exists v_{N+n} \phi_{\bar{a}_{i_0} \ldots \bar{a}_{i_n}}(v_N, \ldots, v_{N+n}),$$

where $\phi_{\bar{a}_{i_0} \ldots \bar{a}_{i_n}}(v_N, \ldots, v_{N+n})$ is the formula obtained by replacing in ϕ every occurrence of the constant symbol $\bar{a}_{i_k}$ by the variable v_{N+k} for $k \le n$. Since $\bar{\phi} \in Th(\mathfrak{A}) = Th(\mathfrak{B})$ we have $\mathfrak{B} \models \bar{\phi}$, so that

$$\mathfrak{B} \models_{b} \phi_{\bar{a}_{i_0} \ldots \bar{a}_{i_n}}(v_N, \ldots, v_{N+n})$$

for some $b \in B^{\omega}$. We set

$$b_{i_k, \phi} = b_{N+k} \quad \text{for } k \le n \quad \text{and}$$
$$b_{i, \phi} \in B \quad \text{for } i \in I \setminus \{i_0, \ldots, i_n\}.$$

Clearly $\langle \mathfrak{B}, b_{i,\phi} \rangle_{i \in I} \models \phi$, and the claim is established.

It now follows that $\Phi = Th(\langle \mathfrak{A}, a_i \rangle_{i \in I}) \cup Th(\langle \mathfrak{B}, S_j \rangle_{j \in J})$ has a model. For $Th(\langle \mathfrak{A}, a_i \rangle_{i \in I})$ is closed under finite conjunctions, and if $\phi \in Th(\langle \mathfrak{A}, a_i \rangle_{i \in I})$ then

$$\langle \mathfrak{B}, b_{i,\phi}, S_j \rangle_{i \in I, j \in J}$$

is a model of $\{\phi\} \cup Th(\langle \mathfrak{B}, S_j \rangle_{j \in J})$; thus every finite subset of Φ has a model, so Φ has a model by the compactness theorem (11.7).

Let $\langle \mathfrak{C}, c_i, T_j \rangle_{i \in I, j \in J} \models \Phi$. Then $\langle \mathfrak{C}, c_i, T_j \rangle_{i \in I} \models Th(\langle \mathfrak{A}, a_i \rangle_{i \in I})$, and hence $\langle \mathfrak{C}, c_i \rangle_{i \in I} \models Th(\langle \mathfrak{A}, a_i \rangle_{i \in I})$, *i.e.*, $Th(\langle \mathfrak{C}, c_i \rangle_{i \in I}) \supset Th(\langle \mathfrak{A}, a_i \rangle_{i \in I})$, whence (a) follows. Statement (b) is proved similarly.

11.9. Corollary *(The embedding property). Let $\mathfrak{A}$ and $\mathfrak{B}$ be elementarily equivalent structures. Then there are a structure $\mathfrak{C}$, and elementary embeddings $f: \mathfrak{A} \to \mathfrak{C}$ and $g: \mathfrak{B} \to \mathfrak{C}$.*

Proof. We apply Theorem 11.8 with $\{a : a \in A\}$ and $\{b : b \in B\}$ replacing $\{a_i : i \in I\}$ and $\{S_j : j \in J\}$ respectively. (We assume without loss of generality that the disjointness requirements of Theorem 11.8 are satisfied.) We obtain a structure $\langle \mathfrak{C}, c_a, d_b \rangle_{a \in A, b \in B}$ for which

$$\langle \mathfrak{A}, a \rangle_{a \in A} \equiv \langle \mathfrak{C}, c_a \rangle_{a \in A} \quad \text{and}$$
$$\langle \mathfrak{B}, b \rangle_{b \in B} \equiv \langle \mathfrak{C}, d_b \rangle_{b \in B}.$$

Let $f: \mathfrak{A} \to \mathfrak{C}$ and $g: \mathfrak{B} \to \mathfrak{C}$ be defined by $f(a) = c_a$ and $g(b) = d_b$. To show that f is an embedding let a and a' be distinct elements of A, so that

$$(\bar{a} \neq \bar{a}') \in Th(\langle \mathfrak{A}, a \rangle_{a \in A}) = Th(\langle \mathfrak{C}, c_a \rangle_{a \in A}).$$

Then $f(a) = c_a \neq c_{a'} = f(a')$, so f is one-to-one.

To prove that f is an elementary embedding, let ϕ be a formula all of whose free variables are among $v_0, \ldots, v_n$, and let $a_0, \ldots, a_n \in A$. Each of the following four statements is clearly equivalent to its successor.

$$\mathfrak{A} \models \phi[a_0, \ldots, a_n];$$
$$\langle \mathfrak{A}, a \rangle_{a \in A} \models \phi[\bar{a}_0, \ldots, \bar{a}_n];$$
$$\langle \mathfrak{C}, c_a \rangle_{a \in A} \models \phi[\bar{a}_0, \ldots, \bar{a}_n]$$

(since $\langle \mathfrak{A}, a \rangle_{a \in A} \equiv \langle \mathfrak{C}, c_a \rangle_{a \in A}$);

$$\mathfrak{C} \models \phi[c_{a_0}, \ldots, c_{a_n}];$$
$$\mathfrak{C} \models \phi[f(a_0), \ldots, f(a_n)].$$

Thus f, and similarly g, is an elementary embedding.

11.10. Corollary *(The amalgamation property). Let $\mathfrak{A}, \mathfrak{B}_0$ and $\mathfrak{B}_1$ be structures of the same type and let $f_0: \mathfrak{A} \to \mathfrak{B}_0$ and $f_1: \mathfrak{A} \to \mathfrak{B}_1$ be elementary embeddings. Then there are a structure $\mathfrak{C}$, and elementary embeddings $g_0: \mathfrak{B}_0 \to \mathfrak{C}$ and $g_1: \mathfrak{B}_1 \to \mathfrak{C}$, such that $g_0 \circ f_0 = g_1 \circ f_1$.*

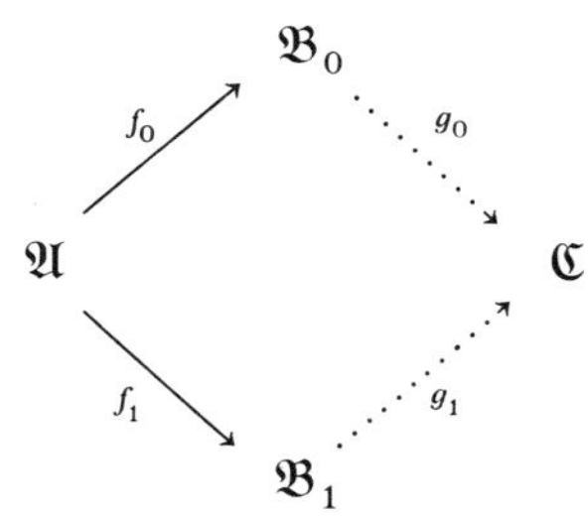

Proof. Let

$$\mathfrak{B}'_0 = \langle \mathfrak{B}_0, f_0(a) \rangle_{a \in A} \quad \text{and}$$
$$\mathfrak{B}'_1 = \langle \mathfrak{B}_1, f_1(a) \rangle_{a \in A},$$

so that $\mathfrak{B}'_0 \equiv \langle \mathfrak{A}, a \rangle_{a \in A} \equiv \mathfrak{B}'_1$, by Lemma 11.2. By the embedding property (Corollary 11.9) there is a structure

$$\mathfrak{C}' = \langle \mathfrak{C}, c_a \rangle_{a \in A},$$

and elementary embeddings

$$g'_0: \mathfrak{B}'_0 \to \mathfrak{C}' \quad \text{and}$$
$$g'_1: \mathfrak{B}'_1 \to \mathfrak{C}'.$$

Let

$$g_0: \mathfrak{B}_0 \to \mathfrak{C} \quad \text{and}$$
$$g_1: \mathfrak{B}_1 \to \mathfrak{C}$$

be the embeddings g_0' and g_1' considered on the reducts $\mathfrak{B}_0$ and $\mathfrak{B}_1$ of $\mathfrak{B}_0'$ and $\mathfrak{B}_1'$ respectively. Clearly g_0 and g_1 are elementary embeddings for which $g_0 \circ f_0 = g_1 \circ f_1$.

A class $\mathbb{K}$ of similar relational systems is called an *elementary Jónsson class* (*with* $(\mathrm{VI})_\kappa$) if it satisfies conditions (I) through $(\mathrm{VI})_\kappa$ of the definition of a Jónsson class (given in § 6), where in conditions (V) and $(\mathrm{VI})_\kappa$ "subsystem" is replaced by "elementary subsystem", and in conditions (III) and (IV) "embedding" is replaced by "elementary embedding".

11.11. Theorem (*Morley-Vaught*). *Let* $\mathbb{K}$ *be an elementary type (of structures of type* $\tau \in \omega^\rho$*). Then* $\mathbb{K}$ *is an elementary Jónsson class with* $(\mathrm{VI})_{(|\rho|+\omega)^+}$.

Proof. (I) follows from Theorem 12.22 below and Corollary 11.6. (It is because of this condition that an elementary type is defined to be the class of all structures elementarily equivalent to a given *infinite* structure; this is needed in order to apply Theorem 12.22;) (II) is trivial; (III) is Corollary 11.9; (IV) is Corollary 11.10; (V) is the Tarski-Vaught Theorem (11.3); and $(\mathrm{VI})_{(|\rho|+\omega)^+}$ is the downward Löwenheim-Skolem Theorem (11.4).

For a formula ϕ of $L(\tau)$, let $n(\phi)$ be the least $n<\omega$ such that all the free variables of ϕ are among $v_0, \ldots, v_{n-1}$, let $\mathfrak{A}$ be a structure of type τ, and for $n(\phi) \geq 1$ set

$$\phi^{\mathfrak{A}} = \{\langle a_0, \ldots, a_{n(\phi)-1}\rangle \in A^{n(\phi)} : \mathfrak{A} \models \phi[a_0, \ldots, a_{n(\phi)-1}]\}$$

We set $\mathfrak{A}^* = \langle \mathfrak{A}, \phi^{\mathfrak{A}} \rangle_{\phi \in \Phi(L(\tau)),\, n(\phi) \geq 1}$; $\mathfrak{A}^*$ is called the *full expansion* of $\mathfrak{A}$.

11.12. Lemma. *Let* $\mathfrak{A}$ *and* $\mathfrak{B}$ *be structures of type* τ. *The following are equivalent:*

(a) $\mathfrak{A} \prec \mathfrak{B}$;

(b) $\mathfrak{A}^* \subset \mathfrak{B}^*$.

Proof. Each of the following three statements is clearly equivalent to its successor.

$\mathfrak{A}^* \subset \mathfrak{B}^*$;

$\phi^{\mathfrak{A}} = \phi^{\mathfrak{B}} \cap A^{n(\phi)}$ for all $\phi \in \Phi(L(\tau))$ with $n(\phi) \geq 1$;

for every $\phi \in \Phi(L(\tau))$ with $n(\phi) \geq 1$, and all $a_0, \ldots, a_{n(\phi)-1} \in A$, we have

$\mathfrak{A} \models \phi[a_0, \ldots, a_{n(\phi)-1}]$ if and only if $\mathfrak{B} \models \phi[a_0, \ldots, a_{n(\phi)-1}]$; and,

$\mathfrak{A} \prec \mathfrak{B}$.

Let $\mathbb{K}$ be an elementary Jónsson class with $(\mathrm{VI})_\kappa$, and let $\mathfrak{A} \in \mathbb{K}$. The structure $\mathfrak{A}$ is called *elementarily* α*-homogeneous*, or *elementarily* *α-universal*, or *elementarily special (of cardinality α)*, if it has the corresponding properties defined in § 6 with "subsystem" replaced by "elementary subsystem" and with "embedding" replaced by "elementary embedding".

11.13. Corollary. *Let $\mathbb{K}$ be an elementary type (of structures with similarity type $\tau \in \omega^\rho$).*

(a) *If $\alpha = 2^{\underline{\alpha}} > |\rho| + \omega$, there is a unique elementarily special structure of cardinality α in $\mathbb{K}$;*

(b) *if $\alpha = \alpha^{\underline{\alpha}} > |\rho| + \omega$, there is a unique elementarily α-homogeneous-universal structure of cardinality α in $\mathbb{K}$.*

Proof. By Theorem 11.11 and Lemma 11.12, the class $\mathbb{K}^* = \{\mathfrak{A}^* : \mathfrak{A} \in \mathbb{K}\}$ is a Jónsson class (with $(\mathrm{VI})_{(|\rho|+\omega)^+}$). The corollary now follows from Theorem 4.10.

[Using Theorem 11.11 and the more general approach to the theory of Jónsson classes outlined at the end of § 4, it is possible to give an alternative proof of Corollary 11.13 which avoids full expansions.]

We give next a characterization of the elementarily α-homogeneous-universal structures of elementary types that corresponds to the characterization, obtained earlier, of the α-homogeneous-universal structures of the class of ordered sets as η_α-sets (Lemma 5.2 and Theorem 5.12(b)). The generalization of the notion of an η_α-set to elementary types is Keisler's notion of an α-saturated structure.

We first prove a simple lemma.

11.14. Lemma. *Let $|\mathfrak{A}| = \alpha > \omega + |\rho|$, and let $\mathfrak{A}$ be elementarily α-universal. Then $\mathfrak{A}$ is elementarily α-homogeneous in the elementary type of $\mathfrak{A}$ if and only if for all $X \subset A$ with $|X| < \alpha$, and for all $f: X \to A$ such that*

$$\langle \mathfrak{A}, x \rangle_{x \in X} \equiv \langle \mathfrak{A}, f(x) \rangle_{x \in X},$$

there is an automorphism h of $\mathfrak{A}$ such that

$$h(x) = f(x) \quad \textit{for } x \in X.$$

Proof. One implication (from right to left) is trivial. For the converse, let $\mathfrak{A}$ be elementarily α-homogeneous. Since $\alpha \geq (\omega + |\rho|)^+$, the elementary type $\mathbb{M}(Th(\mathfrak{A}))$ satisfies $(\mathrm{VI})_\alpha$; hence there are structures $\mathfrak{B}$ and $\mathfrak{C}$ such that

$$\begin{aligned} &\mathfrak{B} \prec \mathfrak{A}, \quad \mathfrak{C} \prec \mathfrak{A}, \\ &|B| < \alpha, \quad |C| < \alpha, \\ &X \subset B, \quad \text{and } f[X] \subset C. \end{aligned}$$

By (III) and $(\mathrm{VI})_\alpha$ there exist a structure $\mathfrak{D} \in \mathbb{M}(Th(\mathfrak{A}))$ such that $|D| < \alpha$, and elementary embeddings g_0 and h_0 of $\mathfrak{B}$ and $\mathfrak{C}$ respectively into $\mathfrak{D}$. It is clear from the elementary α-universality-homogeneity of $\mathfrak{A}$ that there are elementary embeddings g_1 and h_1 of $\mathfrak{D}$ into $\mathfrak{A}$ such that

$$\begin{aligned} g_1 \circ g_0 \quad &\text{is the inclusion of } \mathfrak{B} \text{ into } \mathfrak{A} \text{ and} \\ h_1 \circ h_0 \quad &\text{is the inclusion of } \mathfrak{C} \text{ into } \mathfrak{A}. \end{aligned}$$

By the elementary α-homogeneity of $\mathfrak{A}$ there is an automorphism h of $\mathfrak{A}$ such that

$$h \circ g_1 = h_1 .$$

We thus have the following situation:

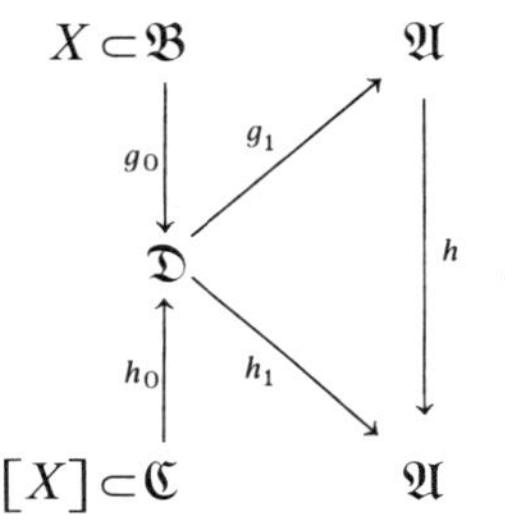

It is clear that $h(x) = f(x)$ for $x \in X$, and the proof of the lemma is complete.

We recall that $\Phi_1(L)$ denotes the set of formulas in the language L whose only free variable is v_0 or which have no free variables.

Definitions. Let $\Phi \subset \Phi_1(L)$ and let $\mathfrak{A}$ be a structure for L. Then Φ is *satisfiable in* $\mathfrak{A}$ if there is $a \in A$ such that

$$\mathfrak{A} \models \phi[a]$$

for $\phi \in \Phi$; and Φ is *finitely satisfiable in* $\mathfrak{A}$ if every finite subset of Φ is satisfiable in $\mathfrak{A}$.

A structure $\mathfrak{A}$ is *α-saturated* if it satisfies the following condition: For $C \in \mathscr{P}_\alpha(A)$ and Φ a subset of $\Phi_1(L)$, if Φ is finitely satisfiable in $\langle \mathfrak{A}, c \rangle_{c \in C}$ then Φ is satisfiable in $\langle \mathfrak{A}, c \rangle_{c \in C}$.

11.15. Theorem. *Let $\mathfrak{A}$ be an elementarily α-homogeneous-universal structure in an elementary type with $\tau \in \omega^\rho$, where $\alpha > |\rho| + \omega$. Then $\mathfrak{A}$ is α-saturated.*

Proof. Let $C \in \mathscr{P}_\alpha(A)$ and let $\Phi \subset \Phi_1(L)$ be finitely satisfiable in $\langle \mathfrak{A}, c \rangle_{c \in C}$. We set

$$\Psi = Th(\langle \mathfrak{A}, c \rangle_{c \in C}) .$$

Let L'_C be the language L_C with a new constant s, and set $\Phi' = \{\phi_{v_0}(s) : \phi \in \Phi\}$ (where $\phi_{v_0}(s)$ is the sentence of L'_C that is obtained by replacing in ϕ each free occurrence of v_0 by s). Every finite subset of $\Phi' \cup \Psi$ has a model, since Φ is finitely satisfiable in $\langle\mathfrak{A}, c\rangle_{c\in C}$, so by the compactness theorem the set $\Phi' \cup \Psi$ has a model. Since $|\rho|+\omega<\alpha$, it follows from the downward Löwenheim-Skolem property $(\mathrm{VI})_\alpha$ that there is a model

$$\langle\mathfrak{B}, b_c, d\rangle_{c\in C}$$

for $\Phi' \cup \Psi$, of cardinality at most α, where b_c is the interpretation of the constant $\bar{c}$ and d is the interpretation of the constant s. Since $\langle\mathfrak{B}, b_c\rangle_{c\in C} \models \Psi$, we have $\mathfrak{A}\equiv\mathfrak{B}$, *i.e.*, $\mathfrak{B}\in\mathbb{M}(Th(\mathfrak{A}))$. Since $\mathfrak{A}$ is elementarily α-universal, there is an elementary embedding $f:\mathfrak{B}\to\mathfrak{A}$. Thus $\langle\mathfrak{B}, b_c\rangle_{c\in C}\equiv\langle\mathfrak{A}, f(b_c)\rangle_{c\in C}$, and since $\langle\mathfrak{A}, c\rangle_{c\in C}\models\Psi$ we have $\langle\mathfrak{B}, b_c\rangle_{c\in C}\equiv\langle\mathfrak{A}, c\rangle_{c\in C}$ and hence $\langle\mathfrak{A}, c\rangle_{c\in C}\equiv\langle\mathfrak{A}, f(b_c)\rangle_{c\in C}$. Since $\mathfrak{A}$ is α-homogeneous, there is by Lemma 11.14 an automorphism h of $\mathfrak{A}$ such that

$$h(c)=f(b_c) \quad \text{for } c\in C.$$

Then with $a=h^{-1}(\{f(d)\})$ we have

$$\langle\mathfrak{A}, c\rangle_{c\in C}\models\phi[a]$$

for all $\phi\in\Phi$. This completes the proof that $\mathfrak{A}$ is α-saturated.

11.16. Corollary. *Let $\mathfrak{A}$ be an elementarily special structure of cardinality α in an elementary type with $\tau\in\omega^\rho$, where $\alpha>|\rho|+\omega$. Then $\mathfrak{A}$ is equal to the union of an elementary chain $\{\mathfrak{A}_\beta:\beta<\alpha\}$, where $\mathfrak{A}_\beta$ is β^+-saturated for $\beta<\alpha$.*

11.17. Theorem. *Let $\mathfrak{A}$ and $\mathfrak{B}$ be elementarily equivalent structures of cardinality α, and suppose that $\mathfrak{A}$ and $\mathfrak{B}$ are equal to the union of elementary chains $\{\mathfrak{A}_\beta:\beta<\alpha\}$ and $\{\mathfrak{B}_\beta:\beta<\alpha\}$, respectively, where $\mathfrak{A}_\beta$ and $\mathfrak{B}_\beta$ are β^+-saturated for $\beta<\alpha$. Then $\mathfrak{A}$ and $\mathfrak{B}$ are isomorphic structures.*

Proof. We first claim that there is a well-ordering $\langle a_\xi:\xi<\alpha\rangle$ of $\mathfrak{A}$ such that $a_\xi\in A_{|\xi|}$ for $\xi<\alpha$. If $\alpha<\omega$, then since $\mathfrak{A}\equiv\mathfrak{A}_\beta$ for $\beta<\alpha$ we have $|\mathfrak{A}|=|\mathfrak{A}_\beta|$ for $\beta<\alpha$, so that $\mathfrak{A}_\beta=\mathfrak{A}$ for $\beta<\alpha$; thus in this case *any* enumeration of $\mathfrak{A}$ satisfies the condition of the claim. If $\alpha\geq\omega$ then, because $\mathfrak{A}\equiv\mathfrak{A}_\beta$ for $\beta<\alpha$, it follows that $|\mathfrak{A}_\beta|\geq\omega$ for $\beta<\alpha$. Let $\langle a_{\beta\xi}:\xi<\alpha\rangle$ be a well-ordering of $\mathfrak{A}_\beta$ for $\beta<\alpha$ and set

$$S_\beta=\{a_{\gamma\zeta}:\gamma\leq\beta \text{ and } \zeta<\beta^+\}.$$

Then clearly

$$|S_\beta|<\omega \quad \text{if } \beta<\omega, \text{ and}$$
$$|S_\beta|\leq\beta^+ \quad \text{if } \omega\leq\beta<\alpha.$$

For $\beta<\omega$ we choose a well-ordering

$$\langle a_\xi:\textstyle\sum_{\gamma<\beta}|S_\gamma|\leq\xi<\sum_{\gamma\leq\beta}|S_\gamma|\}$$

of S_β, and for $\omega \leq \beta < \alpha$ we choose a well-ordering

$$\langle a_\xi : \beta \leq \xi < \beta^+ \rangle$$

of S_β. Then $\langle a_\xi : \xi < \alpha \rangle$ is a well-ordering of $\mathfrak{A}$ which easily seen to satisfy the condition $a_\xi \in A_{|\xi|}$ for $\xi < \alpha$.

We choose also a well-ordering $\langle b_\xi : \xi < \alpha \rangle$ of $\mathfrak{B}$ such that $b_\xi \in B_{|\xi|}$ for $\xi < \alpha$.

We define a'_ξ and b'_ξ for $\xi < \alpha$, such that

(i) $a'_\xi \in A_{|\xi|}$;

(ii) $b'_\xi \in B_{|\xi|}$;

(iii) $\langle \mathfrak{A}, a'_\zeta \rangle_{\zeta<\xi} \equiv \langle \mathfrak{B}, b'_\zeta \rangle_{\zeta<\xi}$;

(iv) if $\xi = \eta - 2n$, where η is a limit ordinal and $n < \omega$, then $a'_\xi = a_{\eta+n}$; and

(v) if $\xi = \eta + 2n + 1$, where η is a limit ordinal and $n < \omega$, then $b'_\xi = b_{\eta+n}$.

We proceed recursively, and consider two possibilities.

Case 1. $\xi = \eta + 2n$ where η is a limit ordinal and $n < \omega$. Then $\langle \mathfrak{A}, a'_\zeta \rangle_{\zeta<\xi} \equiv \langle \mathfrak{B}, b'_\zeta \rangle_{\zeta<\xi}$ by (iii), and $\{a'_\zeta : \zeta < \xi\} \subset A_{|\xi|}$ and $\{b'_\zeta : \zeta < \xi\} \subset B_{|\xi|}$ by (i) and (ii). Since $\mathfrak{A}_{|\xi|} \equiv \mathfrak{A}$ and $\mathfrak{B}_{|\xi|} \equiv \mathfrak{B}$ we have

$$\langle \mathfrak{A}_{|\xi|}, a'_\zeta \rangle_{\zeta<\xi} \equiv \langle \mathfrak{A}, a'_\zeta \rangle_{\zeta<\xi} \quad \text{and} \quad \langle \mathfrak{B}_{|\xi|}, b'_\zeta \rangle_{\zeta<\xi} \equiv \langle \mathfrak{B}, b'_\zeta \rangle_{\zeta<\xi}$$

(cf. Lemma 11.2), so that

$$\langle \mathfrak{A}_{|\xi|}, a'_\zeta \rangle_{\zeta<\xi} \equiv \langle \mathfrak{B}_{|\xi|}, b'_\zeta \rangle_{\zeta<\xi}.$$

We set $a'_\xi = a_{\eta+n}$ and $\Phi = \{\phi_{\overline{a'_\xi}}(v_0) : \phi \in Th(\langle \mathfrak{A}_{|\xi|}, a'_\zeta \rangle_{\zeta<\xi})\}$, where $\phi_{\overline{a'_\xi}}(v_0)$ is the formula that is obtained by replacing every occurrence of the constant symbol $\overline{a'_\xi}$, in the language $L_{\{a'_\xi\}}$, in ϕ by the variable v_0 and where we assume, without any loss of generality, that ϕ does not contain any occurrence of v_0.

We claim next that Φ is finitely satisfiable in $\langle \mathfrak{B}_{|\xi|}, b'_\zeta \rangle_{\zeta<\xi}$. To show this let $F \in \mathscr{P}_\omega(\Phi)$, so that

$$\langle \mathfrak{A}_{|\xi|}, a'_\zeta \rangle_{\zeta<\xi} \models \phi[a'_0] \quad \text{for } \phi \in F$$

and therefore

$$\langle \mathfrak{A}_{|\xi|}, a'_\zeta \rangle_{\zeta<\xi} \models \exists v_0 \bigwedge\nolimits_{\phi \in F} \phi.$$

Since $\langle \mathfrak{A}_{|\xi|}, a'_\zeta \rangle_{\zeta<\xi} \equiv \langle \mathfrak{B}_{|\xi|}, b'_\zeta \rangle_{\zeta<\xi}$ we have

$$\langle \mathfrak{B}_{|\xi|}, b'_\zeta \rangle_{\zeta<\xi} \models \exists v_0 \bigwedge\nolimits_{\phi \in F} \phi,$$

so there is $b \in \mathfrak{B}_0$ such that

$$\langle \mathfrak{B}_{|\xi|}, b'_\zeta \rangle_{\zeta < \xi} \models \phi[b] \quad \text{for } \phi \in F,$$

and the claim is verified.

Because $\mathfrak{B}_{|\xi|}$ is $|\xi|^+$-saturated, Φ is satisfiable in $\langle \mathfrak{B}_{|\xi|}, b'_\zeta \rangle_{\zeta < \xi}$, *i.e.*, there is $b'_\xi \in \mathfrak{B}_{|\xi|}$ such that

$$\langle \mathfrak{B}_{|\xi|}, b'_\zeta \rangle_{\zeta < \xi} \models \phi[b'_\xi] \quad \text{for } \phi \in \Phi .$$

It is clear that

$$\langle \mathfrak{A}_{|\xi|}, a'_\zeta \rangle_{\zeta < \xi + 1} \equiv \langle \mathfrak{B}_{|\xi|}, b'_\zeta \rangle_{\zeta < \xi + 1} .$$

Case 2. $\xi = \eta + 2n + 1$, where η is a limit ordinal and $n < \omega$. We proceed as in *Case* 1 with the rôles of $\mathfrak{A}$ and $\mathfrak{B}$ reversed. The details will be omitted.

This completes the recursive definition of the well-orderings $\langle a'_\xi : \xi < \alpha \rangle$, $\langle b'_\xi : \xi < \alpha \rangle$ of A and B, respectively. It is clear that

$$\langle \mathfrak{A}, a'_\xi \rangle_{\xi < \alpha} \equiv \langle \mathfrak{B}, b'_\xi \rangle_{\xi < \alpha} .$$

Let $h: \mathfrak{A} \to \mathfrak{B}$ be given by $h(a'_\xi) = b'_\xi$ for $\xi < \alpha$. Then h is an elementary embedding (cf. the proof of Corollary 11.9) and h is an isomorphism because it is an onto function. This completes the proof of the theorem.

It is easy to prove directly that each finite structure is α-saturated for all α. Hence if $\mathfrak{A}$ and $\mathfrak{B}$ are elementarily equivalent structures, and $\mathfrak{A}$ is finite, then $\mathfrak{A} \cong \mathfrak{B}$ by Theorem 11.17.

11.18. Theorem. *Let* $\mathbb{K}$ *be an elementary type with type* $\tau \in \omega^\rho$ *where* $|\rho| + \omega < \alpha$, *and let* $\mathfrak{A} \in \mathbb{K}$ *with* $|\mathfrak{A}| = \alpha$. *Then*

(a) $\mathfrak{A}$ *is elementarily special for the type* $\mathbb{K}$ *if and only if* $\mathfrak{A}$ *is equal to the union of an elementary chain* $\{\mathfrak{A}_\beta : \beta < \alpha\}$, *where* $\mathfrak{A}_\beta$ *is* β^+*-saturated for* $\beta < \alpha$*; and*

(b) $\mathfrak{A}$ *is elementarily* α*-homogeneous-universal for the type* $\mathbb{K}$ *if and only if* $\mathfrak{A}$ *is* α*-saturated.*

Proof. The statements follow from Corollary 11.16 and Theorem 11.17.

The next result and the examples that follow it indicate that in a sense the Jónsson classes of ordered sets (§ 5) and of Boolean algebras (§ 6) are instances of the theory of elementary types.

For each cardinal κ we define the cardinal κ^- as follows:

$$\begin{aligned} \kappa^- &= \kappa \quad \text{if } \kappa \text{ is a limit cardinal} \\ &= \lambda \quad \text{if } \kappa = \lambda^+ . \end{aligned}$$

11.19. Theorem. *Let* $\mathbb{K}$ *be a Jónsson class with* $(\mathrm{VI})_\kappa$ *and let* $\mathbb{L}$ *be a class of similar structures such that*

(a) $\mathbb{L}$ satisfies (II), (V) and $(\mathrm{VI})_\kappa$;

(b) $\mathbb{K} \subset \mathbb{L}$;

(c) *for every* $\mathfrak{A} \in \mathbb{L}$ *there are* $\hat{\mathfrak{A}} \in \mathbb{K}$ *and an embedding* $i_\mathfrak{A}: \mathfrak{A} \to \hat{\mathfrak{A}}$, *and for every embedding* $f: \mathfrak{A} \to \mathfrak{B}$ *of structures* $\mathfrak{A}$ *and* $\mathfrak{B}$ *of* $\mathbb{L}$ *there is an embedding* $\hat{f}: \hat{\mathfrak{A}} \to \hat{\mathfrak{B}}$, *such that*

(c1) $|\hat{\mathfrak{A}}| \leq |\mathfrak{A}| + \kappa^-$ *and*

(c2) *the diagram*

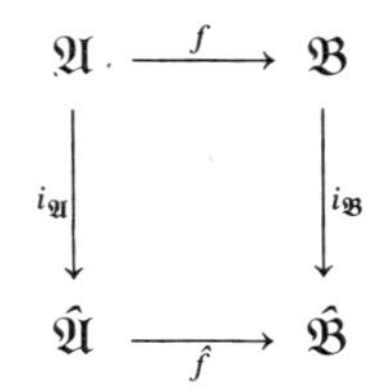

commutes.

Then $\mathbb{L}$ *is a Jónsson class with* $(\mathrm{VI})_\kappa$, *and if* $\alpha = \alpha^{\underline{\alpha}} > \kappa^-$ *then the* α*-homogeneous-universal structures of cardinality* α *for* $\mathbb{K}$ *and* $\mathbb{L}$ *coincide.*

Proof. To prove that $\mathbb{L}$ is a Jónsson class we must verify (I), (III), and (IV). (I) follows from (b). For (III) let $\mathfrak{A}$ and $\mathfrak{B}$ be elements of $\mathbb{L}$, and (using (III) for $\mathbb{K}$) let $\mathfrak{C}$ be an element of $\mathbb{K}$ for which there are embeddings f and g of $\hat{\mathfrak{A}}$ and $\hat{\mathfrak{B}}$ respectively into $\mathfrak{C}$. Then $f \circ i_\mathfrak{A}$ and $g \circ i_\mathfrak{B}$ are embeddings of $\mathfrak{A}$ and $\mathfrak{B}$ respectively into $\mathfrak{C}$.

The amalgation property (IV) for $\mathbb{L}$ is verified by consideration of the diagram

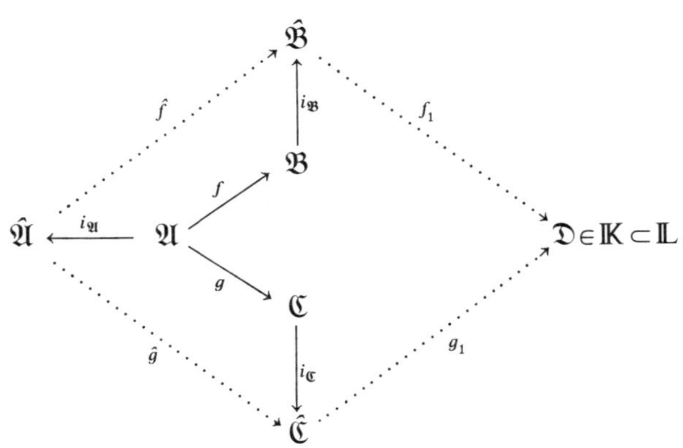

where $f: \mathfrak{A} \to \mathfrak{B}$ and $g: \mathfrak{A} \to \mathfrak{C}$ are embeddings of structures in $\mathbb{L}$, the structures $\hat{\mathfrak{A}}, \hat{\mathfrak{B}}$ and $\hat{\mathfrak{C}}$ and the embeddings into them are given by (c2) above, and the existence of $\mathfrak{D}$ and the embeddings $f_1: \hat{\mathfrak{B}} \to \mathfrak{D}$ and $g_1: \hat{\mathfrak{C}} \to \mathfrak{D}$ are given by (IV) in $\mathbb{K}$.

Let $\alpha=\alpha^{\underline{\alpha}}>\kappa^-$ and let $\mathfrak{S}_\alpha$ be the unique α-homogeneous-universal structure of cardinality α in $\mathbb{L}$. We will prove that $\mathfrak{S}_\alpha\in\mathbb{K}$. Let $\langle p_\xi:\xi<\alpha\rangle$ be a well-ordering of $\mathfrak{S}_\alpha$. Since $\mathbb{L}$ is a Jónsson class with $(\mathrm{VI})_\alpha$ there is $\mathfrak{S}_0\in\mathbb{L}$ such that $p_0\in\mathfrak{S}_0\subset\mathfrak{S}_\alpha$ and $|\mathfrak{S}_0|<\alpha$. By (c1), $|\hat{\mathfrak{S}}_0|<\alpha$. We consider the diagram

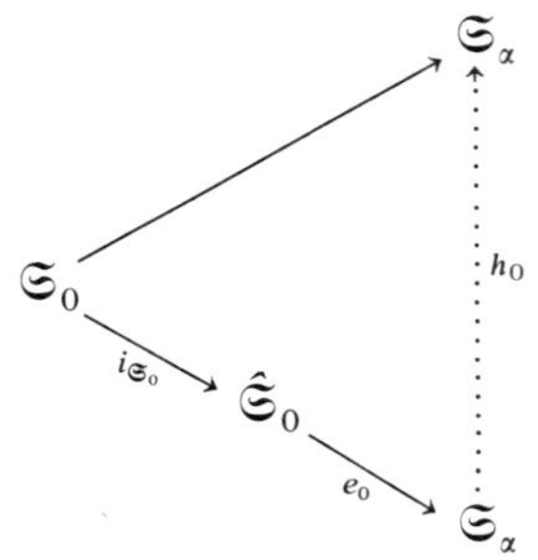

where the unnamed mapping is the natural inclusion and e_0 is an embedding of $\hat{\mathfrak{S}}_0$ into $\mathfrak{S}_\alpha$. (Such an embedding exists because $\hat{\mathfrak{S}}_0\in\mathbb{K}\subset\mathbb{L}$, $|\hat{\mathfrak{S}}_0|<\alpha$, and $\mathfrak{S}_\alpha$ is α-universal in $\mathbb{L}$.) Because $\mathfrak{S}_\alpha$ is α-homogeneous in $\mathbb{L}$ there is an automorphism h_0 of $\mathfrak{S}_\alpha$ such that

$$h_0\circ e_0\circ i_{\mathfrak{S}_0} \quad \text{is the inclusion of } \mathfrak{S}_0 \text{ into } \mathfrak{S}_\alpha.$$

Set $\mathfrak{T}_0=h_0\circ e_0[\hat{\mathfrak{S}}_0]$; then $\mathfrak{T}_0\in\mathbb{K}$, $|\mathfrak{T}_0|<\alpha$, and

$$\mathfrak{S}_0\subset\mathfrak{T}_0\subset\mathfrak{S}_\alpha.$$

We proceed recursively. Let $0<\xi<\alpha$ and assume that for all $\zeta<\xi$ we have defined $\mathfrak{T}_\zeta\in\mathbb{K}$ such that

$$\begin{aligned}&\mathfrak{T}_\zeta\subset\mathfrak{T}_{\zeta'}\subset\mathfrak{S}_\alpha \quad \text{for } \zeta<\zeta'<\xi,\\ &|\mathfrak{T}_\zeta|<\alpha \quad \text{for } \zeta<\xi, \text{ and}\\ &p_\zeta\in\mathfrak{T}_\zeta \quad \text{for } \zeta<\xi.\end{aligned}$$

Set $\tilde{\mathfrak{T}}_\xi=\bigcup_{\zeta<\xi}\mathfrak{T}_\zeta$; then $\tilde{\mathfrak{T}}_\xi\subset\mathfrak{S}_\alpha$, $|\tilde{\mathfrak{T}}_\xi|<\alpha$ and $\tilde{\mathfrak{T}}_\xi\in\mathbb{K}$. We now repeat the argument used in the first step of the recursion; since $\mathbb{L}$ is a Jónsson class there is $\mathfrak{S}_\xi\in\mathbb{L}$ such that $\tilde{\mathfrak{T}}_\xi\subset\mathfrak{S}_\xi$, $|\mathfrak{S}_\xi|<\alpha$, and $p_\xi\in\mathfrak{S}_\xi$; and $|\hat{\mathfrak{S}}_\xi|<\alpha$ by (c1). Proceeding as above with $\mathfrak{T}_0$ we define $\mathfrak{T}_\xi\in\mathbb{K}$ for $\xi<\alpha$. We note that

$$\mathfrak{S}_\alpha=\bigcup_{\xi<\alpha}\mathfrak{T}_\xi\in\mathbb{K}$$

since $\mathbb{K}$ is a Jónsson class. It is now clear that $\mathfrak{S}_\alpha$ is the α-homogeneous-universal structure of cardinality α in $\mathbb{K}$.

The argument given thus far and the uniqueness (for $\alpha=\alpha^{\underline{\alpha}}>\kappa^-$) of the α-homogeneous-universal structure in $\mathbb{K}$ show that the α-homogeneous-universal structure in $\mathbb{K}$ is also α-homogeneous-universal in $\mathbb{L}$.

If $\mathbb{K}$ and $\mathbb{L}$ satisfy the conditions of Theorem 11.19 we say that

$\mathbb{L}$ is a *conservative enlargement* of $\mathbb{K}$.

We will now show informally that the Jónsson classes that we have examined in some detail in the previous chapter, namely the Jónsson classes of ordered sets and of Boolean algebras, are conservative enlargements of Jónsson classes of elementary types; in this sense the study of these classes $\mathbb{OS}$ and $\mathbb{BA}$ is reduced to the case of elementary types. The description of these examples is not self-contained, in that we use without proof certain results from model theory; these are either simple or standard results which are established with the techniques of ultraproducts developed in the present section.

11.20. Examples. (1) Let $\mathbb{Q}$ be the ordered set of rationals. It is known from model theory that

(a) the elementary type $\mathbb{M}(Th(\mathbb{Q}))$ of $\mathbb{Q}$ consists of the class of densely ordered sets without first or last element; and

(b) every embedding of structures in $\mathbb{M}(Th(\mathbb{Q}))$ is an elementary embedding, so that $\mathbb{M}(Th(\mathbb{Q}))$ is, in fact, a Jónsson class. We set

$$\mathbb{K} = \mathbb{M}(Th(\mathbb{Q})), \quad \text{and}$$
$$\mathbb{L} = \mathbb{OS}, \quad \text{the class of ordered sets.}$$

For $\mathfrak{A} \in \mathbb{OS}$ we set $\hat{\mathfrak{A}} = \mathfrak{A} \times \mathbb{Q}$, ordered lexicographically. It is clear from (a) above that $\hat{\mathfrak{A}} \in \mathbb{M}(Th(\mathbb{Q}))$.

We let $i_{\mathfrak{A}}: \mathfrak{A} \to \hat{\mathfrak{A}}$ be given by $i_{\mathfrak{A}}(a) = \langle a, 0 \rangle$ and if $f: \mathfrak{A} \to \mathfrak{B}$ is an embedding of structures in $\mathbb{OS}$ we let $\hat{f}: \hat{\mathfrak{A}} \to \hat{\mathfrak{B}}$ be given by $\hat{f}(a, q) = \langle f(a), q \rangle$. It is easily verified that with these definitions

$$\mathbb{OS} \text{ is a conservative enlargement of } \mathbb{M}(Th(\mathbb{Q})).$$

(2) As in §2 let $\mathfrak{B}(2^{\omega})$ denote the Boolean algebra of open-and-closed subsets of the Cantor set 2^{ω}. It is known from model theory that

(a) the elementary type $\mathbb{M}(Th(\mathfrak{B}(2^{\omega})))$ of $\mathfrak{B}(2^{\omega})$ consists of the class of atomless Boolean algebras; and

(b) every embedding of structures in $\mathbb{M}(Th(\mathfrak{B}(2^{\omega})))$ is an elementary embedding, so that $\mathbb{M}(Th(\mathfrak{B}(2^{\omega})))$ is, in fact, a Jónsson class. We set

$$\mathbb{K} = \mathbb{M}(Th(\mathfrak{B}(2^{\omega}))), \quad \text{and}$$
$$\mathbb{L} = \mathbb{BA}, \quad \text{the class of Boolean algebras.}$$

Even though it is not necessary, we use Stone's duality and topological language to define the required correspondence. Let $\mathfrak{B} \in \mathbb{BA}$, and let $\mathsf{S}(\mathfrak{B})$ denote (as usual) the Stone space of $\mathfrak{B}$. Since the infinite countable power $(\mathsf{S}(\mathfrak{B}))^{\omega}$, with the product topology, does not have isolated elements, the Boolean algebra $\mathfrak{B}((\mathsf{S}(\mathfrak{B}))^\omega)$ is atomless and therefore by (a) above it is an element of $\mathbb{M}(Th(\mathfrak{B}(2^\omega)))$. We set $\hat{\mathfrak{B}}=\mathfrak{B}((\mathsf{S}(\mathfrak{B}))^\omega)$, we let

$$\pi_0:(\mathsf{S}(\mathfrak{B}))^\omega \to \mathsf{S}(\mathfrak{B})$$

be the projection function $\pi_0(p_0,\ldots,p_n,\ldots)=p_0$, and we let $i_{\mathfrak{B}}$ be the composition given by

$$\mathfrak{B} \cong \mathfrak{B}(\mathsf{S}(\mathfrak{B})) \xrightarrow{\mathfrak{B}(\pi_0)} \mathfrak{B}(\mathsf{S}((\mathfrak{B})^\omega))=\hat{\mathfrak{B}}\,.$$

Now if $f:\mathfrak{A}\to\mathfrak{B}$ is a Boolean algebra embedding we let $\hat{f}:\hat{\mathfrak{A}}\to\hat{\mathfrak{B}}$ be given by $\hat{f}=\mathfrak{B}(\mathsf{S}(f)^\omega)$, where

$$\mathsf{S}(f)^\omega\langle p_0,\ldots,p_n,\ldots\rangle=\langle \mathsf{S}(f)(p_0),\ldots,\mathsf{S}(f)(p_n),\ldots\rangle\,.$$

Notes for § 11

The recursive definition of formulas, and the resulting procedures of proof by induction and definition by recursion and the notion of satisfaction, are due to Tarski [36], who laid the foundations of model theory.

Lemma 11.2 collects statements of Tarski-Vaught [57] (Theorems 1.10, 1.11). Theorem 11.3 is given in Tarski-Vaught [57] (Theorem 1.9). The Löwenheim-Skolem theorem (11.4) in its present form is given in Tarski-Vaught [57] (Theorem 2.1), where references are given to earlier results.

The first paper that can be considered to contain the concept of ultraproduct is by Skolem [34], who used a variant of the present notion to construct a countable non-standard model of arithmetic. (In this connection see also Skolem [55].) Hewitt [48], in his study of the ring of continuous real-valued functions on a space, considered what is in effect a generalized (non-discrete) notion of the ultraproduct of real-closed fields (cf. the thesis of Kochen [58] (Chapter 2)). A forerunner of the present notion of an ultraproduct was essentially given by Łoś [55] (p. 104), where the fundamental property of ultraproducts expressed by Theorem 11.5 is also stated. A special case of reduced products was used by Chang and Morel [58].

The proof of the theorem of Łoś (11.5) as given here appears in Kochen [61] (Theorem 5.1). A similar proof appears in Frayne-Morel-Scott [62] (Theorem 2.2). Corollary 11.6 is given in Kochen [61] (Corollary 5.4). The compactness theorem (11.7) is proved here by making use of regular ultrafilters. (Cf. the proofs given in Kochen [58] (Theorem 2.7) and Frayne-Morel-Scott [62] (Theorem 2.10).) For various applications of ultraproducts not considered here, see the surveys of Keisler [65a] and Chang [67b].

Theorem 11.8 is a special case of Robinson [56a] (Theorem 2.9), [65] (Theorem 5.1.6) (which in turn is closely related to Craig's interpolation lemma (Craig [57] (Theorem 5))). Corollary 11.9 is given in Robinson [56b] (Theorem 4.2.2) and Fraïssé [54].

The proof of Corollaries 11.9 and 11.10 is contained in Morley and Vaught [62] (Theorem 3.2).

We note that the embedding and amalagamation property for elementary types (Corollaries 11.9 and 11.10) can be proved using Frayne's lemma (given in Frayne-Morel-Scott [62] (Theorem 2.12) and Kochen [61] (Lemma 9.1)); for such proofs see Bell-Slomson [69] (§10, Theorem 4.2).

Theorem 11.11 is essentially the same as Morley-Vaught [62] (Theorem 3.2). Full expansions are introduced in Morley-Vaught [62] and their Lemma 3.1 is the same as our Lemma 11.12; full expansions can be avoided by considering generalized Jónsson classes (cf. Notes for § 3, Fleischer [66] and Sevee [72]).

The concept of an α-saturated structure is given in Keisler [64c]. The results 11.14—11.17 leading to Theorem 11.18 are proved in Morley-Vaught [62] (Theorems 3.3 and 3.4).

The applications of ultraproducts to model theory and to questions of large cardinals are treated in the detailed and authoritative work of Chang and Keisler [73].

For the relations of ultraproducts to non-standard analysis the reader is referred to Robinson [66], Luxemburg [64], [69], Machover-Hirschfeld [69], and Luxemburg-Robinson [72].

§ 12. Families of Almost Disjoint Sets

In this section we collect a number of results which are loosely related by the use of families of almost disjoint sets.

The first part of the section develops the classical results of Sierpiński and Tarski on the cardinalities of families of almost disjoint sets (Theorem 12.2 and Corollary 12.3).

The Balcar-Vopěnka theorem (12.13) makes systematic use of the properties of maximal α-almost disjoint families on α and of the notion of regressive set to achieve, for uncountable regular cardinals α with $\alpha^+ = 2^\alpha$, the surprising identification, up to homeomorphism, of the Gleason space of the space of uniform ultrafilters on α with the Gleason space of the Stone-Čech compactification of the metric space $(\alpha^+)^\omega$.

It follows from the result of Solovay (Theorem 12.14), for α an uncountable regular cardinal such that $\alpha^+ = 2^\alpha$, that the Boolean algebra of the Gleason space of the space of uniform ultrafilters on α is completely generated by a countable set of elements.

The last part of the section deals with results on cardinalities of ultraproducts. Many of these follow from Keisler's general theorem (12.15), including the most useful one concerning ultraproducts modulo regular ultrafilters (Theorem 12.22). A related theorem is the Keisler-Prikry result on the Eklof property of good ultrafilters (Theorem 12.26).

Cardinalities of Families of Almost Disjoint Sets

Definition. Let $\omega \leq \kappa \leq \alpha$. A family $\mathscr{C}$ of subsets of α is a *κ-almost disjoint family on α* if

$$|C| \geq \kappa \quad \text{for } C \in \mathscr{C}, \quad \text{and}$$
$$|C_0 \cap C_1| < \kappa \quad \text{for } C_0, C_1 \in C \quad \text{and } C_0 \neq C_1.$$

An ω-almost disjoint family (on α) is simply called an *almost disjoint family* (on α).

We denote by $S(\alpha,\kappa)$ the least cardinal β such that there is no family $\mathscr{C}$ such that

$$|\mathscr{C}|=\beta, \quad \text{and}$$
$$\mathscr{C} \text{ is a } \kappa\text{-almost disjoint family on } \alpha.$$

A κ-almost disjoint family $\mathscr{C}$ on α is called *maximal* if for every subset D of α with $|D|\geq\kappa$, there is $C\in\mathscr{C}$ such that $|C\cap D|\geq\kappa$. Equivalently, a κ-almost disjoint family $\mathscr{C}$ on α is maximal if there is no κ-almost disjoint family $\mathscr{D}$ on α such that $\mathscr{C}\subsetneqq\mathscr{D}$.

We note that it follows from Zorn's lemma that every κ-almost disjoint family on α is contained in a maximal such family on α.

It is clear that

$$\alpha^+\leq S(\alpha,\kappa)\leq(2^\alpha)^+ .$$

12.1. Lemma. *Let $\omega\leq\kappa\leq\alpha$; then $S(\alpha,\kappa)=\boldsymbol{S}(U_\kappa(\alpha))$ (i.e., the Souslin number of the space $U_\kappa(\alpha)$) and hence $S(\alpha,\kappa)$ is a regular uncountable cardinal.*

Proof. The equality $S(\alpha,\kappa)=\boldsymbol{S}(U_\kappa(\alpha))$ follows from the observation that for $\mathscr{C}\subset\mathscr{P}(\alpha)$, $\mathscr{C}$ is a κ-almost disjoint family on α if and only if $\{\hat{C}:C\in\mathscr{C}\}$ is a family of pairwise disjoint non-empty open-and-closed subsets of $U_\kappa(\alpha)$, where $\hat{C}=(\mathrm{cl}_{\boldsymbol{\beta}(\alpha)}C)\cap U_\kappa(\alpha)$. That $S(\alpha,\kappa)$ is a regular uncountable cardinal follows from the Erdős-Tarski Theorem 3.5.

The following result contains most of the information known about these numbers.

12.2. Theorem. *Let κ be an infinite cardinal and $\alpha\geq 2$. Then $S(\alpha^{\underline{\kappa}},\kappa)=\boldsymbol{S}(U_\kappa(\alpha^{\underline{\kappa}}))=(\alpha^\kappa)^+$.*

Proof. (a) We first that $S(\alpha^{\underline{\kappa}},\kappa)\geq(\alpha^\kappa)^+$, *i.e.*, that there is a κ-almost disjoint family on $\alpha^{\underline{\kappa}}$ of cardinality α^κ. We denote by $\alpha^{\underline{\underline{\kappa}}}$ the set of all functions $f:\xi\to\alpha$, where $\xi<\kappa$. From Lemma 1.22(f), we have $|\alpha^{\underline{\underline{\kappa}}}|=\alpha^{\underline{\kappa}}$. We set

$$C_s=\{s|\xi:\xi<\kappa\} \quad \text{for } s\in\alpha^\kappa, \quad \text{and}$$
$$\mathscr{C}=\{C_s:s\in\alpha^\kappa\} .$$

We note that if $s\in\alpha^\kappa$ and $\xi<\kappa$, then $s|\xi\in\alpha^\xi\subset\alpha^{\underline{\underline{\kappa}}}$; thus $C_s\in\mathscr{P}(\alpha^{\underline{\underline{\kappa}}})$ and $\mathscr{C}\subset\mathscr{P}(\alpha^{\underline{\underline{\kappa}}})$.

We prove that $|\mathscr{C}|=\alpha^\kappa$. Indeed if $s,t\in\alpha^\kappa$ and $s\neq t$ then $s(\xi)\neq t(\xi)$ for some $\xi<\kappa$; then, since $\kappa\geq\omega$, we have $\xi+1<\kappa$, and $s|\xi+1\neq t|\xi+1$; thus $C_s\neq C_t$.

Finally we prove that $\mathscr{C}$ is a κ-almost disjoint family on $\alpha^{\underline{\kappa}}$. If $s,t\in\alpha^\kappa$ and $s\neq t$, then $s(\xi)\neq t(\xi)$ for some $\xi<\kappa$; then $C_s\cap C_t\subset\{s|\eta:\eta\leq\xi\}$ and we have $|C_s\cap C_t|<\kappa$.

(b) Let $\mathscr{C}$ be a κ-almost disjoint family on $\alpha^{\underline{\kappa}}$. For $A\in\mathscr{C}$, let $A'\subset A$ be such that $|A'|=\kappa$. Clearly if A and B are distinct elements of $\mathscr{C}$, then $A'\neq B'$, since otherwise we would have $A'=B'\subset A\cap B$, in contradiction to the fact that $|A\cap B|<\kappa$. Thus from Lemma 12.1 we have $|\mathscr{C}|\leq\alpha^{\kappa}$.

12.3. Corollary. (a) *If* $\alpha\geq\omega$, *then* $S(\alpha,\omega)=\boldsymbol{S}(\boldsymbol{\beta}(\alpha)\setminus\alpha)=(\alpha^{\omega})^{+}$.

(b) *If* $\alpha\geq\omega$, *then* $S(2^{\underline{\alpha}},\alpha)=\boldsymbol{S}(U_{\alpha}(2^{\underline{\alpha}}))=(2^{\alpha})^{+}$.

(c) *If* $\alpha=2^{\underline{\alpha}}\geq\omega$, *then* $S(\alpha,\alpha)=\boldsymbol{S}(U(\alpha))=(2^{\alpha})^{+}$.

(d) *If* $\alpha\geq\omega$, *then* $S(2^{\alpha},\alpha^{+})=\boldsymbol{S}(U_{\alpha^{+}}(2^{\alpha}))=(2^{(\alpha^{+})})^{+}$.

Proof. (a) follows from Theorem 12.2, taking $\kappa=\omega$ and noting that $\alpha^{\underline{\omega}}=\alpha$; (b) follows from Theorem 12.2, replacing κ, α by α, 2, respectively, and (c) is a special case of (b). Finally, (d) follows from (b) by replacing α by α^{+} and noting that $2^{\underline{\alpha^{+}}}=2^{\alpha}$.

For a result related to those of Corollary 12.3 but proved in a different way, see Corollary 12.20, below.

Let X be a compact space and let $\mathsf{G}(X)$ and $\pi:\mathsf{G}(X)\to X$ denote respectively the Gleason space of X and the associated canonical projection (defined as in § 2). We note that $\boldsymbol{S}(X)=\boldsymbol{S}(\mathsf{G}(X))$. Indeed if $\mathscr{U}$ is a cellular family in X then $\{\pi^{-1}(U):U\in\mathscr{U}\}$ is a cellular family in $\mathsf{G}(X)$, and hence $\boldsymbol{S}(\mathsf{G}(X))\geq\boldsymbol{S}(X)$; and if $\mathscr{V}$ is a cellular family in $\mathsf{G}(X)$ then $\{\mathrm{int}_X\pi(V):V\in\mathscr{V}\}$ is a cellular family in X and hence $\boldsymbol{S}(X)\geq\boldsymbol{S}(\mathsf{G}(X))$.

Now by way of contrast with Corollary 12.3 we have the following fact.

12.4. Corollary. *If* $\alpha\geq\omega$ *then there is* $G\subset U(\alpha)$ *such that*

$|G|=2^{2^{\alpha}}$,

G *is compact extremally disconnected, and*

$\boldsymbol{S}(G)=\omega^{+}$.

Proof. From Lemma 7.14 there is a subset G of $U(\alpha)$ homeomorphic to $\mathsf{G}(2^{2^{\alpha}})$. Since $|U(\alpha)|=2^{2^{\alpha}}$ (by Corollary 7.8(b)) and $\pi[\mathsf{G}(2^{2^{\alpha}})]=2^{2^{\alpha}}$ we have $|G|=2^{2^{\alpha}}$; the space G is compact and extremally disconnected by Theorem 2.33; and $\boldsymbol{S}(G)=\omega^{+}$ by Theorem 3.16 and the remark above.

The Balcar-Vopěnka Theorem

Notations. *For* $\mathscr{D}\subset\mathscr{P}(\alpha)$ *and* $A\in\mathscr{P}(\alpha)$, *we set*

$$\mathscr{D}\restriction A=\{A\cap D:D\in\mathscr{D},|A\cap D|=\alpha\}.$$

If $\mathscr{C}$ and $\mathscr{D}$ are (α-almost disjoint) families of subsets of α, then $\mathscr{C}$ is a *refinement* of $\mathscr{D}$ (denoted $\mathscr{D}\prec\mathscr{C}$) if for every $C\in\mathscr{C}$ there is $D\in\mathscr{D}$ such that $C\subset D$.

We recall that $[\alpha]^\alpha$ denotes the set $\{A \in \mathscr{P}(\alpha) : |A| = \alpha\}$.

12.5. Lemma. *Let $\alpha \geq \omega$, let $\mathscr{C}$ be a maximal α-almost disjoint family on α, let $\mathscr{D}$ be an α-almost disjoint family on α such that*

$$\mathscr{C} \prec \mathscr{D}, \quad \textit{and}$$

let $A \in [\alpha]^\alpha$. Then there is an α-almost disjoint family $\mathscr{D}'$ on α such that

$$\mathscr{D} \subset \mathscr{D}',$$
$$|\mathscr{D}' \setminus \mathscr{D}| \leq 1,$$
$$\mathscr{C} \prec \mathscr{D}', \quad \textit{and}$$

either $\mathscr{D} \upharpoonright A$ is a maximal α-almost disjoint family on A or there is $B \in \mathscr{D}'$ such that $B \subset A$.

Proof. If $\mathscr{D} \upharpoonright A$ is a maximal α-almost disjoint family on A, we set $\mathscr{D}' = \mathscr{D}$.

If $\mathscr{D} \upharpoonright A$ is not a maximal α-almost disjoint family on A, then there is $B \subset A$ such that

$|B| = \alpha$, and

$\mathscr{D} \cup \{B\}$ is an α-almost disjoint family on α.

We note that either $B \in \mathscr{C}$ or $\mathscr{C} \cup \{B\}$ is not an α-almost disjoint family on α (since $\mathscr{C}$ is maximal α-almost disjoint on α). Thus, in either case, there is $C \in \mathscr{C}$ such that $|B \cap C| = \alpha$. We set $\mathscr{D}' = \mathscr{D} \cup \{B \cap C\}$.

The following result is proved by a typical diagonal argument.

12.6. Lemma. *Let $\alpha \geq \omega$ and let $\mathscr{C}$ be a maximal α-almost disjoint family on α. Then $|\mathscr{C}| \neq \mathrm{cf}(\alpha)$.*

Proof. Suppose instead that there is a well-ordering $\{C_\xi : \xi < \mathrm{cf}(\alpha)\}$ of $\mathscr{C}$ such that

$$C_\xi \neq C_\zeta \quad \text{for } \xi < \zeta < \mathrm{cf}(\alpha),$$

and let $f : \mathrm{cf}(\alpha) \to \alpha$ be an unbounded, order-preserving function. Since $|C_\zeta| = \alpha$ and $|C_\zeta \cap C_\xi| < \alpha$ for $\xi < \zeta < \mathrm{cf}(\alpha)$ there is $A_\zeta \subset C_\zeta \setminus \bigcup_{\xi < \zeta} C_\xi$ such that $|A_\zeta| = |f(\zeta)|$ for $\zeta < \mathrm{cf}(\alpha)$. We set $A = \bigcup_{\zeta < \mathrm{cf}(\alpha)} A_\zeta$. Then $\mathscr{C} \cup \{A\}$ is an α-almost disjoint family on α and $A \notin \mathscr{C}$, a contradiction that proves the statement.

12.7. Theorem. *Let α be an infinite regular cardinal such that $\alpha^+ = 2^\alpha$, and let $\mathscr{C}$ be a maximal α-almost disjoint family on α. Then there is a maximal α-almost disjoint family $\mathscr{D}$ on α such that*

$$\mathscr{C} \prec \mathscr{D}, \quad \textit{and}$$

if $A \in [\alpha]^\alpha$ then either there is $D \in \mathscr{D}$ such that $D \subset A$, or $|\{D \in \mathscr{D} : |A \cap D| = \alpha\}| < \alpha$. (Such a family $\mathscr{D}$ is called uniformly selective.)

Proof. Let $\langle A_\eta : \eta < \alpha^+ \rangle$ be a well-ordering of the set $[\alpha]^\alpha$. We define families $\mathscr{D}_\eta$ for $\eta < \alpha^+$, such that

(i) $\mathscr{D}_0 = \emptyset$;

(ii) $\mathscr{D}_\eta = \bigcup_{\zeta < \eta} \mathscr{D}_\zeta$ for non-zero limit ordinals $\eta < \alpha^+$;

(iii) $\mathscr{D}_\eta \subset \mathscr{D}_{\eta+1}$;

(iv) $\mathscr{D}_\eta$ is an α-almost disjoint family on α;

(v) $|\mathscr{D}_\eta| \leq |\eta|$;

(vi) $\mathscr{C} \prec \mathscr{D}_\eta$; and

(vii) either $\mathscr{D}_\eta \upharpoonright A_\eta$ is a maximal α-almost disjoint family on A_η or there is $B_{\eta+1} \in \mathscr{D}_{\eta+1}$ such that $B_{\eta+1} \subset A_\eta$.

We proceed by transfinite recursion. We set $\mathscr{D}_0 = \emptyset$; for non-zero limit ordinals $\eta < \alpha^+$ we define $\mathscr{D}_\eta$ by (ii); and if $\eta < \alpha^+$, we define $\mathscr{D}_{\eta+1}$ using Lemma 12.5.

We set $\mathscr{D} = \bigcup_{\eta < \alpha^+} \mathscr{D}_\eta$; it is clear that $\mathscr{D}$ is an α-almost disjoint family on α such that $\mathscr{C} \prec \mathscr{D}$. We verify the following two statements.

(a) $\mathscr{D}$ is a maximal α-almost disjoint family on α. Indeed let $A \in [\alpha]^\alpha$. Then $A = A_\eta$ for some $\eta < \alpha^+$. By condition (vii) above, either $\mathscr{D}_\eta \upharpoonright A_\eta$ is maximal α-almost disjoint on A_η or there is $B_{\eta+1} \in \mathscr{D}_{\eta+1}$ such that $B_{\eta+1} \subset A_\eta$. In either case, clearly $\mathscr{D}_{\eta+1} \upharpoonright A_\eta \neq \emptyset$ and hence $\mathscr{D} \upharpoonright A_\eta \neq \emptyset$; thus $\mathscr{D}$ is maximal.

(b) $\mathscr{D}$ is uniformly selective on α. Indeed let $A \in [\alpha]^\alpha$ be such that

$$(*) \qquad |\{D \in \mathscr{D} : |A \cap D| = \alpha\}| \geq \alpha .$$

Then $A = A_\eta$ for some $\eta < \alpha^+$. We claim that $\mathscr{D}_\eta \upharpoonright A_\eta$ is not a maximal α-almost family on A_η. Indeed by condition (v) we have $|\mathscr{D}_\eta \upharpoonright A_\eta| \leq \alpha$, and from Lemma 12.6 and the regularity of α it follows that $|\mathscr{D}_\eta \upharpoonright A_\eta| < \alpha$; since the maximality of $\mathscr{D}_\eta \upharpoonright A_\eta$ clearly implies that $\mathscr{D}_\eta \upharpoonright A_\eta = \mathscr{D} \upharpoonright A_\eta$, this contradicts (*).

Since $\mathscr{D}_\eta \upharpoonright A_\eta$ is not maximal, by condition (vii) there is $B_{\eta+1} \in \mathscr{D}_{\eta+1} \subset \mathscr{D}$ such that $B_{\eta+1} \subset A_\eta$. This proves (b) and completes the proof of the theorem.

Definition. Let α be an infinite cardinal, and let $\mathscr{B} \subset [\alpha]^\alpha$. Then $\mathscr{B}$ is a *uniform pseudobase on* α if for every $A \in [\alpha]^\alpha$ there is $B \in \mathscr{B}$ such that $|B \setminus A| < \alpha$.

For $A \in \mathscr{P}(\alpha)$, let $\hat{A}$ denote the set $\mathrm{cl}_{\beta(\alpha)} A \cap U(\alpha)$. It is clear that $\mathscr{B}$ is a uniform pseudobase on α if and only if

$$\{\hat{B} : B \in \mathscr{B}\}$$

is a pseudobase for $U(\alpha)$.

Definitions. Let α be an infinite cardinal and let $A \in [\alpha]^\alpha$. Regarded as an ordered set under the order induced by the natural order on α,

A is order-isomorphic to α by a unique isomorphism (Theorem 1.13) (from A to α) denoted by e_A.

Now let $B \subset A \subset \alpha$ with $|B| = \alpha$. Then B is *regressive in* A if $e_B(\xi) < e_A(\xi)$ for $\xi \in B$.

We set

$$\mathrm{Regr}(A) = \{B \in \mathscr{P}(A) : |B| = \alpha \text{ and } B \text{ is regressive in } A\}$$

for $A \in \mathscr{P}(\alpha)$; and

$$\mathrm{Regr}(\mathscr{D}) = \bigcup \{\mathrm{Regr}(A) : A \in \mathscr{D}\}$$

for $\mathscr{D} \subset [\alpha]^\alpha$.

We note that if

$$\begin{aligned} &\alpha \supset A_0 \supset A_1 \supset \cdots \supset A_n \supset \cdots, \\ &|A_n| = \alpha \quad \text{for } n < \omega, \quad \text{and} \\ &A_{n+1} \quad \text{is regressive in } A_n \text{ for } n < \omega, \end{aligned}$$

then $\bigcap_{n<\omega} A_n = \emptyset$. For if $\xi \in \bigcap_{n<\omega} A_n$, then

$$e_{A_0}(\xi) > e_{A_1}(\xi) > \cdots > e_{A_n}(\xi) > \cdots$$

would be a strictly decreasing infinite sequence of ordinals.

12.8. Lemma. *Let α be an infinite cardinal and let $\mathscr{C}$ be a maximal α-almost disjoint family on α. Then* $\mathrm{Regr}(\mathscr{C})$ *is a uniform pseudobase on α.*

Proof. (a) We set $R = \{\xi < \alpha : \xi \text{ is a non-limit ordinal}\}$, and we prove that $R \in \mathrm{Regr}(\alpha)$. Clearly $|R| = \alpha$. If R is not regressive, then there is a least ordinal $\xi \in R$ such that $e_R(\xi) = \xi$. Since $\xi \in R$ we have $\xi = \zeta + 1$ for some $\zeta < \alpha$, and since e_R is an order-isomorphism there is a unique $\eta < \xi$ such that $\eta \in R$ and $e_R(\eta) = \zeta$; it follows from the choice of ξ that $\zeta = e_R(\eta) < \eta$, and hence $\xi = \zeta + 1 \leq \eta$, contradicting the fact that $\eta < \xi$.

(b) We prove next that if $B \subset A \subset \alpha$ and $|B| = \alpha$, then there is $C \subset B$ such that $|C| = \alpha$ and C is regressive in B (and hence in A). Indeed set $C = e_B^{-1}(R)$, where R is defined as in part (a); it is clear that C is regressive in B.

(c) Let $\mathscr{C}$ be a maximal α-almost disjoint family on α, and let $A \in [\alpha]^\alpha$. There is $C \in \mathscr{C}$ such that $|A \cap C| = \alpha$, and by part (b) there is $B \subset A \cap C$ with $|B| = \alpha$ such that B is regressive in $A \cap C$ and hence in C. Thus $B \in \mathrm{Regr}(\mathscr{C})$ and $B \subset A$.

12.9. Theorem. *For every infinite cardinal α there is a sequence $\{\mathscr{C}_n : n < \omega\}$ such that*

(i) *$\mathscr{C}_n$ is a maximal α-almost disjoint family on α;*
(ii) *$\mathscr{C}_n \prec \mathscr{C}_{n+1}$ for $n < \omega$; and*
(iii) *if $A_n \in \mathscr{C}_n$ for $n < \omega$ and $A_0 \supset A_1 \supset \cdots \supset A_n \supset \cdots$, then $\bigcap_{n<\omega} A_n = \emptyset$.*

(A sequence $\{\mathscr{C}_n : n<\omega\}$ *of maximal* α*-almost disjoint families on* α *satisfying* (ii) *and* (iii) *is called descendingly incomplete.)*

Proof. Set $\mathscr{C}_0=\{\alpha\}$. Let $n<\omega$ and suppose that $\mathscr{C}_0,\dots,\mathscr{C}_n$ have been defined in such a way that (i) and (ii) are satisfied for $k\le n$. Since $\mathscr{C}_n$ is maximal α-almost disjoint on α, $\mathrm{Regr}(\mathscr{C}_n)$ is a uniform pseudobase on α by Lemma 12.8 and hence there is $\mathscr{C}_{n+1}$ such that

$$\mathscr{C}_{n+1} \text{ is maximal } \alpha\text{-almost disjoint on } \alpha, \text{ and}$$
$$\mathscr{C}_{n+1}\subset \mathrm{Regr}(\mathscr{C}_n).$$

This completes the recursive definition of the sequence $\{\mathscr{C}_n : n<\omega\}$. Conditions (i) and (ii) are clearly satisfied. For (iii) let $A_n\in\mathscr{C}_n$ for $n<\omega$, and assume that $A_0\supset A_1\supset\cdots\supset A_n\supset\cdots$. Since $A_{n+1}\in\mathscr{C}_{n+1}\subset\mathrm{Regr}(\mathscr{C}_n)$ we have $A_{n+1}\subset B$ for some $B\in\mathscr{C}_n$, and since $|A_{n+1}|=\alpha$ and $\mathscr{C}_n$ is maximal α-almost disjoint on α, clearly the element B is unique, *i.e.*, $B=A_n$. Hence A_{n+1} is regressive in A_n for $n<\omega$, and it follows from the remark immediately after the definition of regressive sets that $\bigcap_{n<\omega}A_n=\emptyset$.

12.10. Lemma. *Let* α *be a regular uncountable cardinal, let* $\{\mathscr{C}_n : n<\omega\}$ *be a descendingly incomplete sequence of maximal* α*-almost disjoint families on* α*, and let* $A\in[\alpha]^\alpha$*. Then*

$$|\{C\in\mathscr{C}_n : |A\cap C|=\alpha\}|>\alpha$$

for some $n<\omega$.

Proof. We note first that if $\mathscr{C}$ is maximal α-almost disjoint on α, then $\mathscr{C}\restriction A$ is maximal α-almost disjoint on A. For $\mathscr{C}\restriction A$ is α-almost disjoint, and if $B\in[A]^\alpha$ then there is $C\in\mathscr{C}$ such that

$$|B\cap(A\cap C)|=|B\cap C|=\alpha\,.$$

Thus $\mathscr{C}_n\restriction A$ is maximal α-almost disjoint on A, and since α is regular we have

$$|\{C\in\mathscr{C}_n : |A\cap C|=\alpha\}|\neq\alpha$$

for $n<\omega$ by Lemma 12.6. We suppose that

$$|\{C\in\mathscr{C}_n : |A\cap C|=\alpha\}|<\alpha$$

for $n<\omega$ and derive a contradiction.

Since $\mathscr{C}_n\restriction A$ is maximal α-almost disjoint on A we have

$$|A\setminus\bigcup\{A\cap C : C\in\mathscr{C}_n, |A\cap C|=\alpha\}|<\alpha$$

for $n<\omega$; we set

$$B=\bigcap_{n<\omega}(\bigcup\{A\cap C : C\in\mathscr{C}_n, |A\cap C|=\alpha\})$$

and we have $|A \setminus B| < \alpha$ (since $\alpha = \mathrm{cf}(\alpha) > \omega$) and therefore $|B| = \alpha$. We set

$$B_n = \bigcup \{A \cap C \cap D : C, D \in \mathscr{C}_n, C \neq D, |A \cap C| = |A \cap D| = \alpha\}$$

and we have $|B_n| < \alpha$ for $n < \omega$ since α is regular; hence there is

$$\xi \in B \setminus \bigcup_{n<\omega} B_n .$$

For $n < \omega$ there is $C_n \in \mathscr{C}_n$ such that $\xi \in A \cap C_n$ and $|A \cap C_n| = \alpha$. We claim that

$$C_{n+1} \subset C_n \quad \text{for } n < \omega .$$

Indeed since $C_{n+1} \in \mathscr{C}_{n+1}$ and $\mathscr{C}_n \prec \mathscr{C}_{n+1}$ there is $D \in \mathscr{C}_n$ such that $C_{n+1} \subset D$. Clearly

$$\alpha = |A \cap C_{n+1}| \leq |A \cap D| ;$$

hence if $C_n \neq D$ then

$$\xi \in A \cap C_n \cap D \subset B_n ,$$

a contradiction proving the claim.

We now have $\xi \in \bigcap_{n<\omega} C_n$, a contradiction to the fact that $\{\mathscr{C}_n : n < \omega\}$ is descendingly incomplete.

12.11. Theorem. *Assume that $\alpha^+ = 2^\alpha$ and that α is a regular uncountable cardinal. Then there is a descendingly incomplete sequence $\{\mathscr{B}_n : n < \omega\}$ of maximal α-almost disjoint families on α whose union $\mathscr{B} = \bigcup_{n<\omega} \mathscr{B}_n$ is a uniform pseudobase on α.*

Proof. For maximal α-almost disjoint families $\mathscr{D}_0, \ldots, \mathscr{D}_n$ on α (with $n < \omega$), we set

$$\mathscr{D}_0 \wedge \cdots \wedge \mathscr{D}_n = \{D_0 \cap \cdots \cap D_n : D_k \in \mathscr{D}_k \text{ for } k \leq n \text{ and } |D_0 \cap \cdots \cap D_n| = \alpha\} .$$

It is clear that

$$\mathscr{D}_0 \wedge \cdots \wedge \mathscr{D}_n \quad \text{is maximal } \alpha\text{-almost disjoint on } \alpha \text{, and}$$
$$\mathscr{D}_k \prec \mathscr{D}_0 \wedge \cdots \wedge \mathscr{D}_n \quad \text{for } k \leq n .$$

By Theorem 12.9 there is a descendingly incomplete sequence $\{\mathscr{C}_n : n < \omega\}$ of maximal α-almost disjoint families on α. The sequence $\{\mathscr{B}_n : n < \omega\}$ is defined recursively.

By Theorem 12.7 there is a maximal α-almost disjoint family $\mathscr{B}_0$ on α such that $\mathscr{C}_0 \prec \mathscr{B}_0$ and $\mathscr{B}_0$ is uniformly selective. Let $0 < n < \omega$ and suppose that $\mathscr{B}_0, \ldots, \mathscr{B}_n$ are defined. By Theorem 12.7 there is a maximal α-almost disjoint family $\mathscr{B}_{n+1}$ on α such that

$$\mathscr{B}_0 \wedge \cdots \wedge \mathscr{B}_n \wedge \mathscr{C}_{n+1} \prec \mathscr{B}_{n+1}, \quad \text{and}$$
$$\mathscr{B}_{n+1} \quad \text{is uniformly selective.}$$

It is clear that $\{\mathscr{B}_n : n<\omega\}$ is a descendingly incomplete sequence of maximal α-almost disjoint families on α.

To verify that $\mathscr{B}=\bigcup_{n<\omega}\mathscr{B}_n$ is a uniform pseudobase on α let $A\in[\alpha]^\alpha$. By Lemma 12.10, $|\{B\in\mathscr{B}_n : |A\cap B|=\alpha\}|>\alpha$ for some $n<\omega$. Since $\mathscr{B}_n$ is uniformly selective there is $B\in\mathscr{B}_n$ such that $B\subset A$. Thus $\mathscr{B}$ is a uniform pseudobase on α.

12.12. Lemma. *Let $\mathfrak{A}$ and $\mathfrak{B}$ be Boolean algebras generated respectively by non-empty dense sets C and D of non-zero elements. Let f be a function from C onto D such that*

$$a_0\leq a_1 \quad \text{if and only if} \quad f(a_0)\leq f(a_1)$$

for $a_0,a_1\in C$. Then there is a unique isomorphism

$$\phi:\mathfrak{A}\to\mathfrak{B}$$

such that $f\subset\phi$.

Proof. We note first that

$$a_0\wedge a_1=\mathbb{0} \quad \text{if and only if} \quad f(a_0)\wedge f(a_1)=\mathbb{0}$$

for $a_0,a_1\in C$. Indeed if $a_0\wedge a_1\neq\mathbb{0}$ then there is $a\in C$ such that

$$\mathbb{0}<a\leq a_0\wedge a_1$$

and thus

$$f(a)\leq f(a_0)\wedge f(a_1).$$

Since $f(a)\in D\subset\mathfrak{B}\setminus\{\mathbb{0}\}$, it follows that $f(a_0)\wedge f(a_1)\neq\mathbb{0}$. The reverse implication is proved similarly.

According to Lemma 2.16 it is enough to prove that

$$\bigwedge_{k\leq n}\varepsilon_\kappa c_\kappa=\mathbb{0} \quad \text{if and only if} \quad \bigwedge_{k\leq n}\varepsilon_\kappa f(c_\kappa)=\mathbb{0}$$

for $n<\omega$ and $c_\kappa\in C$, $\varepsilon_\kappa\in\{-1,1\}$ for $k\leq n$. To verify this criterion assume that

$$\bigwedge_{k\leq n}\varepsilon_\kappa c_\kappa\neq\mathbb{0}.$$

Then there is $a\in C$ such that

$$\mathbb{0}<a\leq\bigwedge_{k\leq n}\varepsilon_\kappa c_\kappa.$$

We set

$$I=\{k\leq n:\varepsilon_\kappa=1\},$$
$$J=\{k\leq n:\varepsilon_\kappa=-1\}$$

and we note that

if $k\in I$ then $a\leq c_\kappa$ and $f(a)\leq f(c_\kappa)=\varepsilon_\kappa f(c_\kappa)$; and

if $k\in J$ then $a\wedge c_\kappa=\mathbb{0}$ and hence $f(a)\wedge f(c_\kappa)=\mathbb{0}$ and

$f(a) \leq \varepsilon_\kappa f(c_\kappa)$. Thus

$$\mathbb{0} < f(a) \leq \bigwedge_{k \leq n} \varepsilon_\kappa f(c_\kappa).$$

The reverse implication is proved similarly.

12.13. Theorem *(Balcar-Vopěnka). Assume that $\alpha^+ = 2^\alpha$ and that α is an uncountable regular cardinal. The Gleason space of the space $U(\alpha)$ of uniform ultrafilters on α is homeomorphic to the Gleason space of the Stone-Čech compactification of the (metric) space $(\alpha^+)^\omega$.*

Proof. Using Lemmas 12.10 and 12.11 it is easy to prove that the space $U(\alpha)$, and hence by Theorem 2.42(a) its Gleason space $\mathsf{G}(U(\alpha))$, has a pseudobase of the form $\mathscr{B} = \bigcup_{n<\omega} \mathscr{B}_n$, where

(i) $\mathscr{B}_n \prec \mathscr{B}_{n+1}$,

(ii) $\mathscr{B}_n$ consists of non-empty pairwise disjoint open-and-closed sets, and

(iii) $|\mathscr{B}_0| = \alpha^+$, and each $B \in \mathscr{B}_n$ contains α^+ elements of $\mathscr{B}_{n+1}$ for $n < \omega$.

To verify that the space $\mathsf{G}(\boldsymbol{\beta}((\alpha^+)^\omega))$ has a pseudobase $\mathscr{C} = \bigcup_{n<\omega} \mathscr{C}_n$ satisfying conditions (analogous to) (i), (ii) and (iii) above it is enough to show that the space $(\alpha^+)^\omega$ has such a base. We set

$$E(s) = \{t \in (\alpha^+)^\omega : t|n = s\} \quad \text{for } s \in (\alpha^+)^n \text{ and } n < \omega,$$
$$\mathscr{E}_n = \{E(s) : s \in (\alpha^+)^{n+1}\} \quad \text{for } n < \omega, \text{ and}$$
$$\mathscr{E} = \bigcup_{n<\omega} \mathscr{E}_n;$$

the base $\mathscr{E}$ for $(\alpha^+)^\omega$ is as required.

By finite recursion we define a function $f: \mathscr{B} \to \mathscr{C}$ satisfying the condition of Lemma 12.12; thus by Lemma 12.12 there is an isomorphism between the Boolean algebras generated by the families $\mathscr{B}$ and $\mathscr{C}$ (where we consider that $\mathscr{B} \subset \mathfrak{B}(\mathsf{G}(U(\alpha)))$ and $\mathscr{C} \subset \mathfrak{B}(\mathsf{G}(\boldsymbol{\beta}((\alpha^+)^\omega)))$). By Theorem 2.36(b), this isomorphism extends to an isomorphism of $\mathfrak{B}(\mathsf{G}(U(\alpha)))$ and $\mathfrak{B}(\mathsf{G}(\boldsymbol{\beta}((\alpha^+)^\omega)))$; finally by Stone's duality the spaces $\mathsf{G}(U(\alpha))$ and $\mathsf{G}(\boldsymbol{\beta}((\alpha^+)^\omega))$ are homeomorphic.

Definition. A complete Boolean algebra $\mathfrak{B}$ is *completely generated* by a subset A of $\mathfrak{B}$ if $\mathfrak{B}$ is the only complete Boolean subalgebra of $\mathfrak{B}$ that contains A.

12.14. Theorem *(Gaifman, Hales, Solovay). If $\alpha \geq \omega$ then $\mathfrak{R}(\alpha^\omega)$ is completely generated by a countable set.*

Proof. We denote by $\pi_n: \alpha^\omega \to \alpha$ the usual projection function for $n < \omega$, and we set

$$A_{n,\xi} = \pi_n^{-1}(\{\xi\}) \quad \text{for } n < \omega, \quad \xi < \alpha.$$

We note that $\mathfrak{R}(\alpha^\omega)$ is completely generated by the set

$$\mathscr{A}=\{A_{n,\xi}: n<\omega, \xi<\alpha\}.$$

Indeed $\mathscr{A}\subset\mathfrak{B}(\alpha^\omega)\subset\mathfrak{R}(\alpha^\omega)$, and if $A\in\mathfrak{B}(\alpha^\omega)$ then there are a set I and for $i\in I$ a set F_i with $|F_i|<\omega$ such that

$$A=\bigcup_{i\in I}\bigcap_{j\in F_i}\pi_{n_j}^{-1}(\{\xi_j\})=\bigvee_{i\in I}\bigwedge_{j\in F_i}A_{n_j,\xi_j};$$

since $\mathfrak{B}(\alpha^\omega)$ is a dense subalgebra of $\mathfrak{R}(\alpha^\omega)$ it follows that $\mathfrak{R}(\alpha^\omega)$ is completely generated by $\mathscr{A}$. We now set

$$B_{m,n}=\{x\in\alpha^\omega: x_n\le x_m\} \quad \text{for } m,n<\omega$$

and we note that $B_{m,n}\in\mathfrak{B}(\alpha^\omega)$. We define

$$\mathscr{B}=\{B_{m,n}: m,n<\omega\}$$

and we have that $\mathscr{B}$ is a countable subset of $\mathfrak{B}(\alpha^\omega)$. It is easy to verify that

$$\bigvee_{\eta<\xi}A_{n,\eta}=\bigcup_{\eta<\xi}A_{n,\eta}=\pi_n^{-1}(\xi)\in\mathfrak{B}(\alpha^\omega)$$

for $n<\omega$, $\xi<\alpha$ and hence that

$$\bigvee_{\eta<\xi}A_{n,\eta}\vee B_{m,n}=\bigcup_{\eta<\xi}A_{n,\eta}\cup B_{m,n}\in\mathfrak{B}(\alpha^\omega)$$

for $m,n<\omega$ and $\xi<\alpha$. We claim that

$$(*)\qquad \bigwedge_{m<\omega}\Big(\bigvee_{\eta<\xi}A_{n,\eta}\vee B_{m,n}\Big)=\pi_n^{-1}(\xi+1) \quad \text{for } n<\omega, \quad \xi<\alpha.$$

We set $C_{m,n,\xi}=\bigvee_{\eta<\xi}A_{n,\eta}\vee B_{m,n}$ for $m,n<\omega$ and $\xi<\alpha$. To prove (*) let $x\in\alpha^\omega$ with $x_n\le\xi$ and let $y\in\pi_n^{-1}(\{x_n\})$. Since for $m<\omega$ either $y_n\le y_m$ or $y_m<y_n=x_n\le\xi$ we have $y\in\bigcap_{m<\omega}C_{m,n,\xi}$; hence

$$x\in\pi_n^{-1}(\{x_n\})\subset\bigcap_{m<\omega}C_{m,n,\xi}$$

and it follows that $x\in\operatorname{int}\operatorname{cl}(\bigcap_{m<\omega}C_{m,n,\xi})=\bigwedge_{m<\omega}C_{m,n,\xi}$. For the reverse inclusion let $x\in\operatorname{int}(\bigcap_{m<\omega}C_{m,n,\xi})$ and suppose that $x_n>\xi$. There is $N<\omega$ such that

$$\bigcap_{k\le N}\pi_k^{-1}(\{x_k\})\subset\bigcap_{m<\omega}C_{m,n,\xi};$$

without loss of generality we assume $N\ge n$. We define $y\in\alpha^\omega$ by

$$\begin{aligned} y_m&=x_m \quad \text{for } m\le N\\ &=\xi \quad\ \ \text{for } N<m \end{aligned}$$

and we note that since $y\in\bigcap_{k\le N}\pi_k^{-1}(\{x_k\})$ we have $y\in C_{N+1,n,\xi}$. But $y\notin\bigcup_{\eta<\xi}A_{n,\eta}$, and $y\notin B_{N+1,n}$ since

$$y_{N+1}=\xi<x_n=y_n;$$

hence $y\notin C_{N+1,n,\xi}$. This contradiction completes the proof of (*).

It follows from (*), since

$$A_{n,\xi}=\{x\in\alpha^\omega: x_n\le\xi\}\setminus\{x\in\alpha^\omega: x_n<\xi\},$$

that

(**) $$A_{n,\xi}=\bigwedge_{m<\omega}\left(\bigvee_{\eta<\xi}A_{n,\eta}\vee B_{m,n}\right)\setminus\bigvee_{\eta<\xi}A_{n,\eta}$$

for $n<\omega$, $\xi<\alpha$. Now let $\mathfrak{C}$ denote the complete Boolean subalgebra of $\mathfrak{R}(\alpha^\omega)$ completely generated by $\mathscr{B}$. It is easy to prove by transfinite induction, using (**), that

$$A_{n,\xi}\in\mathfrak{C}\quad\text{for } n<\omega,\quad \xi<\alpha,$$

i.e., that $\mathscr{A}\subset\mathfrak{C}$. It follows that $\mathfrak{C}=\mathfrak{R}(\alpha^\omega)$, and therefore that $\mathfrak{R}(\alpha^\omega)$ is completely generated by the countable set $\mathscr{B}$.

Cardinalities of Ultraproducts

For an infinite cardinal α, an ultrafilter p on α, and a family $\{\beta_\xi:\xi<\alpha\}$ of cardinals, $\prod_{\xi<\alpha}\beta_\xi/p$ denotes the ultraproduct of the family $\{\langle\beta_\xi,\varepsilon\rangle:\xi<\alpha\}$ modulo p. If $\{\mathfrak{B}_\xi:\xi<\alpha\}$ is a family of similar relational systems and $|\mathfrak{B}_\xi|=\beta_\xi$ for $\xi<\alpha$, it is clear that

$$|\prod_{\xi<\alpha}\mathfrak{B}_\xi/p|=|\prod_{\xi<\alpha}\beta_\xi/p|.$$

The following obvious fact will be used in the sequel without explicit mention:

$$\min\{\beta_\xi:\xi<\alpha\}\le|\prod_{\xi<\alpha}\beta_\xi/p|\le\prod_{\xi<\alpha}\beta_\xi;$$

in particular, for any cardinal β,

$$\beta\le|\beta^\alpha/p|\le\beta^\alpha.$$

Most computations of the cardinality of ultraproducts (with the exception of 12.21) are consequences of the following general statement.

12.15. Theorem *(Keisler). Let α and γ be non-zero cardinals, p an ultrafilter on α, $\{F_\zeta:\zeta<\gamma\}$ a family of elements of p, and $\{\beta_\xi:\xi<\alpha\}$ a family of cardinals. Define $\lambda_\xi=|\{\zeta<\gamma:\xi\in F_\zeta\}|$ for $\xi<\alpha$. Then*

$$|\prod_{\xi<\alpha}\beta_\xi/p|^\gamma\le|\prod_{\xi<\alpha}(\beta_\xi^{\lambda_\xi})/p|.$$

Proof. The natural function

$$P:\left(\prod_{\xi<\alpha}\beta_\xi\right)^\gamma\to\left(\prod_{\xi<\alpha}\beta_\xi/p\right)^\gamma$$

given by $P(\langle f(\zeta):\zeta<\gamma\rangle)=\langle f(\zeta)/p:\zeta<\gamma\rangle$ is clearly onto. Since $\{\zeta<\gamma:\xi\in F_\zeta\}$ is a subset of γ of cardinality λ_ξ, there is a function

$$e_\xi:\lambda_\xi\to\gamma$$

such that $e_\xi[\lambda_\xi]=\{\zeta<\gamma:\xi\in F_\zeta\}$ for $\xi<\alpha$. We denote by π_ξ the usual projection from $\prod_{\xi<\alpha}\beta_\xi$ onto β_ξ for $\xi<\alpha$. Thus if $f\in(\prod_{\xi<\alpha}\beta_\xi)^\gamma$ we may form the composition

$$\lambda_\xi \xrightarrow{e_\xi} \gamma \xrightarrow{f} \prod_{\xi<\alpha}\beta_\xi \xrightarrow{\pi_\xi} \beta_\xi,$$

so that $\pi_\xi\circ f\circ e_\xi\in\beta_\xi^{\lambda_\xi}$ for $\xi<\alpha$. We define the function

$$Q:(\prod_{\xi<\alpha}\beta_\xi)^\gamma\to\prod_{\xi<\alpha}(\beta_\xi^{\lambda_\xi})/p$$

by the rule

$$Q(f)=\langle\pi_\xi\circ f\circ e_\xi:\xi<\alpha\rangle/p\,,$$

and we claim that if $f,g\in(\prod_{\xi<\alpha}\beta_\xi)^\gamma$ are such that $P(f)\neq P(g)$, then $Q(f)\neq Q(g)$. Indeed, $P(f)=\langle f(\zeta)/p:\zeta<\gamma\rangle$ and $P(g)=\langle g(\zeta)/p:\zeta<\gamma\rangle$, and if $P(f)\neq P(g)$ then there is $\zeta<\gamma$ such that

$$f(\zeta)/p\neq g(\zeta)/p\,,$$

i.e., for which the set

$$A=\{\xi<\alpha:\pi_\xi\circ f(\zeta)\neq\pi_\xi\circ g(\zeta)\}\in p\,.$$

Thus $A\cap F_\zeta\in p$; if $\xi\in A\cap F_\zeta$ there is $\eta<\lambda_\xi$ such that $\zeta=e_\xi(\eta)$ for some $\eta<\lambda_\xi$, and hence $\pi_\xi\circ f\circ e_\xi\neq\pi_\xi\circ g\circ e_\xi$. Thus

$$A\cap F_\zeta\subset\{\xi<\alpha:\pi_\xi\circ f\circ e_\xi\neq\pi_\xi\circ g\circ e_\xi\}\in p\,,$$

so that $Q(f)\neq Q(g)$ and the claim is proved. Finally if $s,t\in(\prod_{\xi<\alpha}\beta_\xi/p)^\gamma$ and $s\neq t$, then $Q[P^{-1}(\{s\})]\cap Q[P^{-1}(\{t\})]=\emptyset$ by the claim, and

$$\emptyset\neq Q[P^{-1}(\{s\})]\subset\prod_{\xi<\alpha}(\beta_\xi^{\lambda_\xi})/p$$

since P is an onto function. The proof is complete.

12.16. Corollary. *Let α and γ be infinite cardinals, p an ultrafilter on α, and $\{\beta_\xi:\xi<\alpha\}$ a family of cardinals. Then*

$$|\prod_{\xi<\alpha}\beta_\xi/p|^\gamma\leq|\prod_{\xi<\alpha}(\beta_\xi^\gamma)/p|\,.$$

In particular, for any cardinal β,

$$|\beta^\alpha/p|^\gamma\leq|(\beta^\gamma)^\alpha/p|\,.$$

Proof. By Lemma 7.10(a) there is a (γ^+,γ)-regular family $\{F_\zeta:\zeta<\gamma\}$ for p. Since $|\{\zeta<\gamma:\xi\in F_\zeta\}|\leq\gamma$ for $\xi<\alpha$, the inequality follows from Theorem 12.15.

12.17. Corollary. *Let* $\alpha\geq\omega$, *p a countably incomplete ultrafilter on* α, *and* $\{\beta_\xi:\xi<\alpha\}$ *a family of infinite cardinals. Then*

$$|\prod_{\xi<\alpha}\beta_\xi/p|^\omega=|\prod_{\xi<\alpha}\beta_\xi/p|.$$

In particular, if β *is any infinite cardinal, then*

$$|\beta^\alpha/p|^\omega=|\beta^\alpha/p|.$$

Proof. By Lemma 7.10(b) there is an (ω,ω)-regular family $\{F_n:n<\omega\}$ for p. Since $\lambda_\xi<\omega$ for $\xi<\alpha$ we have $\beta_\xi^{\lambda_\xi}=\beta_\xi$ and hence

$$|\prod_{\xi<\alpha}\beta_\xi/p|^\omega\leq|\prod_{\xi<\alpha}\beta_\xi/p|$$

from Theorem 12.15 with $\gamma=\omega$. The reverse inequality is obvious.

12.18. Theorem. *Let* α *and* γ *be infinite cardinals, p a uniform ultrafilter on* α, *and* $\{\beta_\xi:\xi<\alpha\}$ *a family of cardinals. Then*

(a) *if* $\mathrm{cf}(\alpha)=\mathrm{cf}(\gamma)$, *then*

$$|\prod_{\xi<\alpha}\beta_\xi/p|^\gamma\leq|\prod_{\xi<\alpha}(\beta_\xi^{\underline{\gamma}})/p|;\quad \textit{and}$$

(b) *if* $\{\zeta<\alpha:\beta_\zeta\geq\beta_\xi\}\in p$ *for* $\xi<\alpha$, *then*

$$|\prod_{\xi<\alpha}(\beta_\xi^{\underline{\alpha}})/p|=|\prod_{\xi<\alpha}\beta_\xi/p|^\alpha=(\prod_{\xi<\alpha}\beta_\xi)^\alpha.$$

In particular, if β *is any cardinal, then*

$$|(\beta^{\underline{\alpha}})^\alpha/p|=|\beta^\alpha/p|^\alpha=\beta^\alpha.$$

Proof. (a) By Lemma 7.10(c) the ultrafilter p is $(\mathrm{cf}(\alpha),\mathrm{cf}(\alpha))$-regular, *i.e.*, $(\mathrm{cf}(\gamma),\mathrm{cf}(\gamma))$-regular, and so by Lemma 7.10(d) there is a (γ,γ)-regular family $\{F_\zeta:\zeta<\gamma\}$ for p. Since $\lambda_\xi<\gamma$ for $\xi<\alpha$ we have $\beta_\xi^{\lambda_\xi}\leq\beta_\xi^{\underline{\gamma}}$ and the inequality follows from Theorem 12.15.

(b) Using part (a) above and setting $\gamma=\alpha$, we get that

$$|\prod_{\xi<\alpha}\beta_\xi/p|^\alpha\leq|\prod_{\xi<\alpha}(\beta_\xi^{\underline{\alpha}})/p|.$$

Since $\beta_\xi\leq|\prod_{\zeta<\alpha}\beta_\zeta/p|$ for $\xi<\alpha$, we have

$$\begin{aligned}|\prod_{\xi<\alpha}(\beta_\xi^{\underline{\alpha}})/p|&\leq\prod_{\xi<\alpha}(\beta_\xi^\alpha)=(\prod_{\xi<\alpha}\beta_\xi)^\alpha\leq(\prod_{\xi<\alpha}(|\prod_{\zeta<\alpha}\beta_\zeta/p|))^\alpha\\&=((|\prod_{\zeta<\alpha}\beta_\zeta/p|)^\alpha)^\alpha=|\prod_{\xi<\alpha}\beta_\xi/p|^\alpha,\end{aligned}$$

and the result follows.

12.19. Lemma. *Let α be an infinite cardinal, and $\mathscr{S} \subset \alpha^\alpha$ such that*

(i) $|\{\xi<\alpha : f(\xi)=g(\xi)\}|<\alpha$ *for $f, g \in \mathscr{S}$ and $f \neq g$; and*

(ii) $\mathscr{S}$ *is maximal with respect to* (i).

Then $|\mathscr{S}|>\alpha$.

In particular, there is a family $\mathscr{S} \subset \alpha^\alpha$, with $|\mathscr{S}|>\alpha$, satisfying (i).

Proof. The last statement follows from the first statement and the fact (proved using Zorn's lemma) that every family $\mathscr{S}$ satisfying (i) is included in a maximal family satisfying (i).

Let $\mathscr{S} \subset \alpha^\alpha$, and suppose that $\mathscr{S}$ satisfies (i) and (ii) and that $|\mathscr{S}| \leq \alpha$. Let $\langle f_\xi : \xi<\alpha\rangle$ be a well-ordering of $\mathscr{S}$.

We define $f \in \alpha^\alpha$, such that

$$f(\xi) \in \alpha \backslash \{f_\zeta(\xi) : \zeta<\xi\}$$

for $\xi<\alpha$. This is always possible since

$$|\{f_\zeta(\xi) : \zeta<\xi\}| \leq |\xi|<\alpha \quad \text{for } \xi<\alpha .$$

It is clear that $f \notin \mathscr{S}$, and that in fact

$$\{\xi<\alpha : f(\xi)=f_\zeta(\xi)\} \subset \zeta+1$$

for $\zeta<\alpha$. Thus $\mathscr{S} \cup \{f\} \neq \mathscr{S}$ and $\mathscr{S} \cup \{f\}$ satisfies (i). This contradiction completes the proof.

We note the following immediate corollary to Lemma 12.19, which should be compared with Corollary 12.3(c).

12.20. Corollary. *If $\alpha \geq \omega$, then $S(\alpha, \alpha)=\boldsymbol{S}(U(\alpha)) \geq \alpha^{++}$.*

12.21. Theorem. *If $\alpha \geq \omega$ and p is a uniform ultrafilter on α, then*

$$|\alpha^\alpha / p|>\alpha .$$

Proof. By Lemma 12.19 there is a family $\mathscr{S} \subset \alpha^\alpha$ such that

$|\{\xi<\alpha : f(\xi)=g(\xi)\}|<\alpha$ for $f, g \in \mathscr{S}$ and $f \neq g$, and $|\mathscr{S}|>\alpha$.

Since p is a uniform ultrafilter on α, it follows that if $f, g \in \mathscr{S}$ and $f \neq g$, then $f/p \neq g/p$. Thus $|\alpha^\alpha/p| \geq |\mathscr{S}|>\alpha$.

The following special case of Theorem 12.15 is one of the most useful results on the cardinalities of ultraproducts. It has already been used to prove that an elementary type satisfies property (I) of an (elementary) Jónsson class (Theorem 11.11).

Notation. If $\{\beta_\xi : \xi<\alpha\}$ is a family of cardinals and p is an ultrafilter on α, the *essential supremum of $\{\beta_\xi : \xi<\alpha\}$ with respect to p* is given by

$$\text{ess.}\sup_p \{\beta_\xi : \xi<\alpha\}=\min \{\sup \{\beta_\xi : \xi \in A\} : A \in p\} .$$

12.22. Theorem. *Let* $\alpha \geq \omega$, p *a regular ultrafilter on* α, $\{\beta_\xi : \xi < \alpha\}$ *a family of non-zero cardinals such that*

$$\{\xi < \alpha : \beta_\xi \geq \omega\} \in p,$$

and define

$$\kappa = \text{ess. sup}_p \{\beta_\xi : \xi < \alpha\}.$$

Then

$$|\textstyle\prod_{\xi<\alpha} \beta_\xi / p| = \kappa^\alpha.$$

In particular, if β *is an infinite cardinal then*

$$|\beta^\alpha / p| = \beta^\alpha.$$

Proof. Since $\{\beta_\xi : \xi < \alpha\}$ is a family of non-zero cardinals such that $\{\xi < \alpha : \beta_\xi \geq \omega\} \in p$, we assume with no loss of generality that $\beta_\xi \geq \omega$ for $\xi < \alpha$.

It follows from Theorem 12.15, by taking $\gamma = \alpha$ and $\{F_\zeta : \zeta < \alpha\}$ an (ω, α)-regular family for p, that

$$(*) \qquad |\textstyle\prod_{\xi<\alpha} \beta_\xi / p| = |\prod_{\xi<\alpha} \beta_\xi / p|^\alpha.$$

We now prove the following statements.

(i) $\kappa \leq |\prod_{\xi<\alpha} \beta_\xi / p|$.

There are two possibilities to consider.

Case 1. $\kappa = \lambda^+$ for some λ.

Then $B = \{\xi < \alpha : \beta_\xi = \kappa\} \in p$, and so

$$|\textstyle\prod_{\xi<\alpha} \beta_\xi / p| = |\prod_{\xi\in B} \beta_\xi / p| = |\kappa^{|B|} / p| = \kappa^{|B|} = \kappa^\alpha,$$

using the particular case of (*) and the fact that a regular ultrafilter is uniform.

Case 2. κ is a limit cardinal.

We note that

$$B_\lambda = \{\xi < \alpha : \lambda < \beta_\xi\} \in p \quad \text{for } \lambda < \kappa.$$

Hence

$$|\textstyle\prod_{\xi<\alpha} \beta_\xi / p| = |\prod_{\xi\in B_\lambda} \beta_\xi / p| \geq |\lambda^{|B_\lambda|} / p| = \lambda^\alpha \geq \lambda$$

for $\lambda < \kappa$, using the particular case of (*) and the fact that a regular ultrafilter is uniform. Thus

$$\kappa \leq |\textstyle\prod_{\xi<\alpha} \beta_\xi / p|.$$

(ii) $|\prod_{\xi<\alpha} \beta_\xi / p| \leq \kappa^\alpha$.

Let $B \in p$ be such that

$$\kappa = \sup\{\beta_\xi : \xi \in B\}.$$

Then

$$|\prod_{\xi<\alpha}\beta_\xi/p| = |\prod_{\xi\in B}\beta_\xi/p| \le |\prod_{\xi\in B}\kappa/p| = \kappa^{|B|} = \kappa^\alpha,$$

using the particular case of (*) and the fact that a regular ultrafilter is uniform.

From (i), (ii) and (*), we have

$$\kappa^\alpha \le |\prod_{\xi<\alpha}\beta_\xi/p|^\alpha = |\prod_{\xi<\alpha}\beta_\xi/p| \le \kappa^\alpha,$$

and hence $|\prod_{\xi<\alpha}\beta_\xi/p| = \kappa^\alpha$.

12.23. Theorem. *If α and β are infinite cardinals, p is a β^+-good ultrafilter on α, $1 \le n_\xi < \omega$ for $\xi<\alpha$, and* ess. $\sup_p\{n_\xi : \xi<\alpha\} = \omega$ *then*

$$2^\beta \le |\prod_{\xi<\alpha} n_\xi/p|.$$

Proof. We define $\phi : \mathscr{P}_\omega(\beta) \to \mathscr{P}(\alpha)$ by

$$\phi(F) = \{\xi<\alpha : n_\xi \ge 2^{|F|}\}.$$

Since ess. $\sup_p\{n_\xi : \xi<\alpha\} = \omega$ it follows that in fact

$$\phi : \mathscr{P}_\omega(\beta) \to p.$$

It is clear that ϕ is a monotone function, and since p is β^+-good there is a multiplicative function

$$\psi : \mathscr{P}_\omega(\beta) \to p$$

such that $\psi \le \phi$. We set

$$F(\xi) = \{\eta<\beta : \xi \in \psi(\{\eta\})\}$$

and we prove that $F(\xi) \in \mathscr{P}_\omega(\beta)$ for $\xi<\alpha$. We note that $\bigcap_{n<\omega}\phi(n) = \emptyset$, so for $\xi<\alpha$ there is a least $n<\omega$, denoted $n(\xi)$, such that $\xi \notin \phi(n)$; we prove that $|F(\xi)| < n(\xi)$. Indeed otherwise there would be a subset $F' \subset F(\xi)$ with $|F'| = n(\xi)$, and hence since ψ is multiplicative

$$\xi \in \bigcap_{\eta\in F'}\psi(\{\eta\}) = \psi(F') \subset \phi(F') = \phi(|F'|) = \phi(n(\xi)),$$

contradicting the fact that $\xi \notin \phi(n(\xi))$.

In fact we prove (the stronger statement) that

$$2^{|F(\xi)|} \le n_\xi \quad \text{for } \xi<\alpha.$$

If $F(\xi) = \emptyset$ then this is true by assumption, and if $F(\xi) = \{\eta_i : i<n\}$ then we have, since ψ is multiplicative, that

$$\xi \in \bigcap_{i<n}\psi(\{\eta_i\}) = \psi(F(\xi)) \subset \phi(F(\xi))$$

for $\xi<\alpha$.

It follows that

$$|\prod_{\xi<\alpha} 2^{F(\xi)}| \le |\prod_{\xi<\alpha} n_\xi|,$$

and therefore to complete the proof it is enough to show that

$$2^{\beta} \leq |\prod_{\xi<\alpha} 2^{F(\xi)}/p|.$$

We define

$$Q: 2^{\beta} \to \prod_{\xi<\alpha} 2^{F(\xi)}/p$$

by $Q(f) = \langle f|F(\xi): \xi<\alpha\rangle/p$, and we prove that Q is a one-to-one function. Let $f, g \in 2^{\beta}$ and $f \neq g$. Then there is $\eta<\beta$ such that $f(\eta) \neq g(\eta)$ and we have

$$\{\xi<\alpha: f|F(\xi) \neq g|F(\xi)\} \supset \psi(\{\eta\}) \in p,$$

hence that $Q(f) \neq Q(g)$. The proof is now complete.

12.24. Corollary. *If* $\alpha \geq \omega$, p *is an* α^{+}*-good ultrafilter on* α, $1 \leq n_{\xi} < \omega$ *for* $\xi<\alpha$, *and* $\operatorname{ess.sup}_p\{n_{\xi}: \xi<\alpha\} = \omega$, *then*

$$|\prod_{\xi<\alpha} n_{\xi}/p| = 2^{\alpha}.$$

Proof. It is clear that

$$|\prod_{\xi<\alpha} n_{\xi}/p| \leq |\prod_{\xi<\alpha} n_{\xi}| \leq \omega^{\alpha} = 2^{\alpha}.$$

The reverse inequality follows from Theorem 12.23.

12.25. Corollary. *If* $\alpha \geq \omega$, p *is a countably incomplete ultrafilter on* α, *and* $\{\beta_{\xi}: \xi<\alpha\}$ *is a family of cardinals, then*

$$\textit{either } |\prod_{\xi<\alpha} \beta_{\xi}/p| < \omega \quad \textit{or } 2^{\omega} \leq |\prod_{\xi<\alpha} \beta_{\xi}/p|.$$

Proof. If $\beta_{\xi}=0$ for some $\xi<\alpha$ then $|\prod_{\xi<\alpha} \beta_{\xi}/p| = 0$.

If $\operatorname{ess.sup}_p\{\beta_{\xi}: \xi<\alpha\} < \omega$ then it is clear from the theorem of Łoś that $|\prod_{\xi<\alpha} \beta_{\xi}/p| < \omega$.

We assume now that $1 \leq \beta_{\xi}$ for $\xi<\alpha$, and $\operatorname{ess.sup}_p\{\beta_{\xi}: \xi<\alpha\} \geq \omega$. We set

$$A = \{\xi<\alpha: \omega \leq \beta_{\xi}\}$$

and we consider two possibilities.

Case 1. $A \in p$.
Then the result follows from Corollary 12.17.

Case 2. $A \notin p$.
Since p is countably incomplete there is a sequence $\{A_n: n<\omega\}$ of subsets of α such that

$$A_n \notin p \quad \text{for } n<\omega,$$
$$\bigcup_{n<\omega} A_n = \alpha \setminus A, \quad \text{and}$$
$$A_n \cap A_m = \emptyset \quad \text{for } n<m<\omega.$$

We set

$$\begin{aligned} n_\xi &= n \quad \text{if } \xi \in A_n \quad \text{for } n<\omega, \\ &= 1 \quad \text{if } \xi \in A. \end{aligned}$$

The result now follows from Theorem 12.23 and Lemma 9.8.

Definition. Let $\alpha \geq \omega$ and let p be an ultrafilter on α. The ultrafilter p has the *Eklof property* if for every family $\{\alpha_{\xi,n}: \xi<\alpha, n<\omega\}$ of cardinals such that

$$\alpha_{\xi,m} \leq \alpha_{\xi,n} \quad \text{for } \xi<\alpha \quad \text{and } n<m<\omega,$$

there is $h \in \omega^\alpha$ such that

$$\begin{aligned} &h \text{ is not constant } \operatorname{mod} p \ (i.e.,\ h^{-1}(\{n\}) \notin p \text{ for } n<\omega), \quad \text{and} \\ &|\textstyle\prod_{\xi<\alpha} \alpha_{\xi,h(\xi)}/p| = \min\{|\textstyle\prod_{\xi<\alpha} \alpha_{\xi,n}/p| : n<\omega\}. \end{aligned}$$

12.26. Theorem. *If $\alpha \geq \omega$ and p is a countably incomplete α^+-good ultrafilter on α then p has the Eklof property.*

Proof. Let $\{\alpha_{\xi,n}: \xi<\alpha, n<\omega\}$ be a family of cardinals such that

$$\alpha_{\xi,m} \geq \alpha_{\xi,n} \quad \text{for } \xi<\alpha \quad \text{and } n<m<\omega.$$

We set

$$\gamma_n = |\textstyle\prod_{\xi<\alpha} \alpha_{\xi,n}/p| \quad \text{for } n<\omega.$$

It is clear that

$$\gamma_0 \geq \gamma_1 \geq \cdots \geq \gamma_n \geq \cdots.$$

We set

$$\gamma = \min\{\gamma_n : n<\omega\}.$$

We must prove that there is $h \in \omega^\alpha$ such that

$$\begin{aligned} &h^{-1}(\{n\}) \notin p \quad \text{for } n<\omega, \quad \text{and} \\ &|\textstyle\prod_{\xi<\alpha} \alpha_{\xi,h(\xi)}/p| = \gamma. \end{aligned}$$

We define

$$\beta_n = \text{ess.}\sup_p\{\alpha_{\xi,n}: \xi<\alpha\} \quad \text{for } n<\omega,$$

and we note that

$$\beta_0 \geq \beta_1 \geq \cdots \geq \beta_n \geq \cdots.$$

We set

$$\beta = \min\{\beta_n : n<\omega\}.$$

We assume without loss of generality that

$$\begin{aligned} \beta_n &= \beta, \quad \text{and} \\ \gamma_n &= \gamma \end{aligned}$$

for $n<\omega$. There are two possibilities to consider.

Case 1. There are $N \in \mathscr{P}(\omega)$ and cardinals $\{\alpha_n : n \in N\}$ such that

$$|N| = \omega, \quad \text{and}$$
$$\{\xi < \alpha : \alpha_{\xi,n} = \alpha_n\} \in p \quad \text{for } n \in N.$$

It is then clear that there is $m < \omega$ such that

$$\{\xi < \alpha : \alpha_{\xi,n} = \gamma\} \in p \quad \text{for } m \leq n < \omega.$$

We assume without loss of generality that

$$A_n = \{\xi < \alpha : \alpha_{\xi,n} = \gamma\} \in p \quad \text{for } n < \omega.$$

We set $A = \bigcap_{n<\omega} A_n$ and we distinguish two sub-cases.

(i) $A \in p$.

Let $\{B_n : n < \omega\}$ be a sequence of pairwise disjoint subsets of A such that

$$B_n \notin p \quad \text{for } n < \omega, \quad \text{and}$$
$$\bigcup_{n<\omega} B_n = A.$$

We set $B = A$.

(ii) $A \notin p$.

We set

$$B_n = A_0 \cap \cdots \cap A_n \setminus A_{n+1} \quad \text{for } n < \omega$$

and we note that

$$B_n \notin p \quad \text{for } n < \omega.$$

We set $B = \bigcup_{n<\omega} B_n$.

In both (i) and (ii) we define $h \in \omega^\alpha$ by the conditions

$$h[(\alpha \setminus B) \cup B_0] \subset \{0\},$$
$$h[B_n] \subset \{n\} \quad \text{for } 1 \leq n < \omega.$$

Then $h^{-1}(\{n\}) \notin p$ for $n < \omega$, and $\{\xi < \alpha : \alpha_{\xi,h(\xi)} = \gamma\} \in p$. The required conclusion follows easily in this case.

Case 2. Not *Case* 1.

We assume without loss of generality that

$$\{\xi < \alpha : \alpha_{\xi,n} = \delta\} \notin p \quad \text{for } n < \omega \quad \text{and for every cardinal } \delta.$$

We claim that β is a non-zero limit cardinal. Suppose on the contrary that $\beta = \kappa^+$ for some κ. Then there is $A \in p$ such that

$$\kappa^+ = \sup\{\alpha_{\xi,0} : \xi \in A\}.$$

We set

$$A_0 = \{\xi \in A : \alpha_{\xi,0} \leq \kappa\},$$
$$A_1 = \{\xi \in A : \alpha_{\xi,0} = \kappa^+\},$$

and we note that $A_0 \cup A_1 = A \in p$. If $A_0 \in p$ then $\sup\{\alpha_{\xi,0} : \xi \in A_0\} \leq \kappa$, hence $\beta \leq \kappa$, a contradiction; and if $A_1 = \{\xi \in A : \alpha_{\xi,0} = \kappa^+\} \in p$, then we have a contradiction to the assumption (of *Case* 2).

Furthermore if $\beta = 0$ then there is $A \in p$ such that

$$0 = \sup\{\alpha_{\xi,0} : \xi \in A\},$$

i.e., $\alpha_{\xi,0} = 0$ for $\xi \in A$, contrary to the assumption (of *Case* 2). The proof of the claim is complete.

Now let $\{\delta_\eta : \eta < \mathrm{cf}(\beta)\}$ be a family of cardinals such that

$$\begin{aligned}&\delta_0 \leq \delta_1 \leq \cdots \leq \delta_\eta \leq \cdots,\\ &\delta_\eta < \beta, \quad \text{and}\\ &\beta = \sup\{\delta_\eta : \eta < \mathrm{cf}(\beta)\}.\end{aligned}$$

We note that $\mathrm{cf}(\beta) \leq \alpha$; indeed, there is $A \in p$ such that

$$\beta = \sup\{\alpha_{\xi,0} : \xi \in A\},$$

and since $|A| \leq \alpha$ it follows that $\mathrm{cf}(\beta) \leq |A| \leq \alpha$.

Since p is countably incomplete there is a sequence $\{A_n : n < \omega\}$ of elements of p such that

$$\begin{aligned}&A_0 \supset A_1 \supset \cdots \supset A_n \supset \cdots, \quad \text{and}\\ &\textstyle\bigcap_{n<\omega} A_n = \emptyset.\end{aligned}$$

We define

$$\phi : \mathscr{P}_\omega(\mathrm{cf}(\beta)) \to \mathscr{P}(\alpha)$$

by

$$\phi(F) = A_{|F|} \cap \{\xi < \alpha : \alpha_{\xi,|F|} \geq \max\{\delta_\eta : \eta \in F\}\}.$$

We note that

$$\phi(\emptyset) = A_0 \in p;$$

and if $F \neq \emptyset$ and $\{\xi < \alpha : \alpha_{\xi,|F|} \geq \max\{\delta_\eta : \eta \in F\}\} \notin p$, then $A = \{\xi < \alpha : \alpha_{\xi,|F|} < \max\{\delta_\eta : \eta \in F\}\} \in p$, and hence

$$\sup\{\alpha_{\xi,|F|} : \xi \in A\} \leq \delta_F < \beta,$$

contradicting the definition of β. Thus, in fact,

$$\phi : \mathscr{P}_\omega(\mathrm{cf}(\beta)) \to p.$$

Since p is α^+-good and $\mathrm{cf}(\beta) \leq \alpha$ there is a multiplicative function

$$\psi : \mathscr{P}_\omega(\mathrm{cf}(\beta)) \to p$$

such that $\psi \leq \phi$. We set

$$F(\xi) = \{\eta < \mathrm{cf}(\beta) : \xi \in \psi(\{\eta\})\}$$

and we claim that

$$F(\xi)\in\mathscr{P}_\omega(\mathrm{cf}(\beta)) \quad \text{for } \xi<\alpha .$$

Indeed let $\xi<\alpha$ and let $n(\xi)$ be the least $n<\omega$ such that $\xi\notin A_n$; we prove that $|F(\xi)|<n(\xi)$. Otherwise there would be a subset $F'\subset F(\xi)$ with $|F'|=n(\xi)$ and hence, since ψ is multiplicative,

$$\xi\in\bigcap_{\eta\in F'}\psi(\{\eta\})=\psi(F')\subset\phi(F')\subset A_{n(\xi)},$$

contradicting the fact that $\xi\notin A_{n(\xi)}$.

We define $h\in\omega^\alpha$ by

$$h(\xi)=|F(\xi)|,$$

and we prove the following two statements.

(i) h is not constant $\mathrm{mod}\, p$.

We note that if $m<\omega$ and $\xi\in\psi(m+1)$ then $\xi\in\psi(\{\eta\})$ for $\eta<m+1$ and hence $h(\xi)\geq m+1$, *i.e.*, $\xi\notin h^{-1}(\{m\})$. Thus

$$\psi(m+1)\cap h^{-1}(\{m\})=\emptyset \quad \text{for } m<\omega$$

and since $\psi(m+1)\in p$ we have $h^{-1}(\{m\})\notin p$ for $m<\omega$.

(ii) $\mathrm{ess.\,sup}_p\{\alpha_{\xi,h(\xi)}:\xi<\alpha\}\geq\beta$.

It is enough to prove that

$$\psi(\{\eta\})\subset\{\xi<\alpha:\alpha_{\xi,h(\xi)}\geq\delta_\eta\} \quad \text{for } \eta<\mathrm{cf}(\beta).$$

Indeed let $\eta<\mathrm{cf}(\beta)$ and $\xi\in\psi(\{\eta\})$. If $F(\xi)=\{\eta_i:i<n\}$ then, since ψ is multiplicative,

$$\xi\in\bigcap_{i<n}\psi(\{\eta_i\})=\psi(F(\xi))\subset\phi(F(\xi)),$$

and hence from the definition of ϕ we have

$$\delta_\eta\leq\alpha_{\xi,|F(\xi)|}=\alpha_{\xi,h(\xi)}.$$

It remains to compute the cardinalities of the ultraproducts involved.

(iii) $|\prod_{\xi<\alpha}\alpha_{\xi,n}/p|=\beta^\alpha$ for $n<\omega$.

There are two possibilities to consider.

Case 1. $\{\xi<\alpha:\alpha_{\xi,n}\geq\omega\}\in p$.

From Theorem 12.22 we have

$$|\prod_{\xi<\alpha}\alpha_{\xi,n}/p|=\beta^\alpha \quad \text{for } n<\omega .$$

Case 2. $\{\xi<\alpha:\alpha_{\xi,n}<\omega\}\in p$.

In this case, we assume with no loss of generality that $\alpha_{\xi,n}<\omega$ for $\xi<\alpha$. We note that $\beta=\omega$, and it follows from Corollary 12.24 that

$$|\prod_{\xi<\alpha}\alpha_{\xi,n}/p|=2^\alpha=\omega^\alpha=\beta^\alpha .$$

(iv) $|\prod_{\xi<\alpha}\alpha_{\xi,h(\xi)}/p|=\beta^\alpha$.

There are two possibilities to consider.

Case 1. $\{\xi<\alpha:\alpha_{\xi,h(\xi)}\geq\omega\}\in p$.
It follows from Theorem 12.22 and part (ii) above that

$$|\prod_{\xi<\alpha}\alpha_{\xi,h(\xi)}/p|=(\text{ess. sup}_p\{\alpha_{\xi,h(\xi)}:\xi<\alpha\})^\alpha\geq\beta^\alpha.$$

On the other hand

$$|\prod_{\xi<\alpha}\alpha_{\xi,h(\xi)}/p|\leq|\prod_{\xi<\alpha}\alpha_{\xi,0}/p|=\beta^\alpha$$

from part (iii) above.

Case 2. $\{\xi<\alpha:\alpha_{\xi,h(\xi)}<\omega\}\in p$.
In this case, we assume with no loss of generality that $\alpha_{\xi,h(\xi)}<\omega$ for $\xi<\alpha$. We note that $\beta=\omega$, and it follows from Corollary 12.24 that

$$|\prod_{\xi<\alpha}\alpha_{\xi,h(\xi)}/p|=2^\alpha=\omega^\alpha=\beta^\alpha.$$

The required equality now follows from (iii) and (iv).

Notes for § 12

The fact that no maximal almost disjoint family on ω is countable goes back to Cantor (cf. Lusin [47]). The fact that there is an α-almost disjoint family on α of cardinality greater than α (essentially Lemma 12.19) was known to Erdős in 1934 (see Erdős, Gillman and Henriksen [55] (Lemma 4.1 and footnote)) and proved by Sierpiński [28] for $\alpha=\alpha^{\underline{\alpha}}$ and Sierpiński [37], [38a] for general $\alpha\geq\omega$. The case $\alpha=\kappa=\omega$ of Theorem 12.2 is from Sierpiński [28]; cf. also Nakamura and Kakutani [43]. Theorem 12.2 is from Tarski [28] (Théorème 7).

In connection with almost disjoint families we note two questions. The first is the exact computation of $\boldsymbol{S}(U(\alpha))$. Results here are incomplete and prove by the diagonal argument of Lemma 12.19 that $\boldsymbol{S}(U(\alpha))\geq\alpha^{++}$ for general α, and by Corollary 12.3(b) (a non-diagonal argument) that $\boldsymbol{S}(U(\alpha))=(2^\alpha)^+$ for $\alpha=2^{\underline{\alpha}}$. Furthermore, Baumgartner [70] has defined a model of set theory in which $\boldsymbol{S}(U(\omega^+))\leq2^{(\omega^+)}$. The second question concerns the possible cardinalities of infinite maximal α-almost disjoint families on α. For $\alpha=\omega$ Martin's axiom implies that the only possible cardinality is 2^ω (Martin and Solovay [70]); and there is a model of set theory such that ω^+ is such a possible cardinality and $\omega^+<2^\omega$ (Hechler [72]). It follows from Lemma 12.19 that for every infinite cardinal α there is an α-almost disjoint family on α of cardinality greater than α; in fact if $\mathscr{S}$ satisfies the conditions of Lemma 12.19, then $\mathscr{S}$ is an α-almost disjoint family on $\alpha\times\alpha$. It has been shown by Erdős and Hechler [*] (Theorem 2.4) that if α is singular and $\beta^{\text{cf}(\alpha)}<\alpha$ for

$\beta < \alpha$ (in particular, if α is a singular strong limit cardinal) then there is a maximal α-almost disjoint family on α of cardinality α.

A weak form of Corollary 12.4 has been given assuming GCH in Hajnal and Juhász [72]; our proof is from Negrepontis [**].

Lemma 12.12 is from Sikorski [63] (Theorem xviii). With this exception the results 12.5—12.13, proving the surprising Theorem 12.13, are from Balcar and Vopěnka [72]. Earlier Vopěnka [71] had obtained Theorem 12.13 assuming the axiom of constructibility. The following corollary of Theorem 12.13 and of the results in Štěpánek and Vopěnka [67] is given in Balcar and Vopěnka [72] (Theorem 6(2)): *If* $\omega < \alpha$, *α is regular,* $\alpha^+ = 2^\alpha$, $\omega < \kappa \leq \alpha^+$ *and κ is regular, then there is a family* $\{N_\eta : \eta < \kappa\}$ *of nowhere dense subsets of* $U(\alpha)$ *such that*

$$N_\eta \subset N_\zeta \quad \textit{for } \eta < \zeta < \kappa, \quad \textit{and}$$

$$\bigcup_{\eta < \kappa} N_\eta = U(\alpha).$$

Gaifman [64] and Hales [64] proved that for arbitrarily large cardinals α there is a complete Boolean algebra of cardinality α completely generated by a countable set. Theorem 12.14 is from Solovay [66]. Kripke [67] has shown that every Boolean algebra can be embedded in a complete Boolean algebra of arbitrarily large cardinality, completely generated by a countable set. This result has been extended to other classes of structures by Stavi [73].

The results 12.15—12.18 are from Keisler [64a]. Our 12.15, 12.16, 12.17 and 12.18(a),(b) correspond respectively to his 2.1, 2.2(ii), 2.2(i) and 2.2(iii), (iv).

A result related to Corollary 12.17, not proved here, is the following statement due to Shelah [70], answering in the affirmative a conjecture of Keisler [67c] (end of § 1): *If p is an ultrafilter on α,* $n_\xi < \omega$ *for* $\xi < \alpha$ *and* $\omega \leq |\prod_{\xi < \alpha} n_\xi / p|$, *then* $|\prod_{\xi < \alpha} n_\xi / p| = |\prod_{\xi < \alpha} n_\xi / p|^\omega$.

Theorem 12.21 is proved in Frayne-Morel-Scott [62] (Theorem 1.23).

Theorem 12.22 is from Keisler-Prikry [74] (Lemma 2); the particular case $|\beta^\alpha / p| = \beta^\alpha$ had been noted earlier by Chang-Keisler [62] (Lemma 2.3), Frayne-Morel-Scott [62] (Lemma 1.26), and Keisler [63] (Theorem 4.1).

Theorem 12.23 and Corollary 12.24 were first proved by Keisler, assuming $\beta^+ = 2^\beta$, in Lemma 2.1 of Keisler [67c]. The present proof was communicated to the authors by Keisler in a letter of June, 1973.

Corollary 12.25 was proved in Frayne-Morel-Scott [62] (Theorem 1.28 ff.) and by Keisler [64a] (as a consequence of Theorem 2.3).

The Eklof property for ultrafilters was introduced by Eklof [73], who proved that every non-principal ultrafilter on ω has the Eklof property. Theorem 12.26 is due to Keisler-Prikry [74] (Theorem 1); in the same paper a number of open questions are given, of which the following is the main one: Does every countably incomplete ultrafilter on an infinite cardinal have the Eklof property?

§ 13. Saturation of Ultraproducts

This section studies the two questions, which are related as it turns out, of the saturation properties of ultraproducts and of the characterization of elementary equivalence in terms of ultrapowers.

The results in their present sweeping generality hide, as is often the case, a long history of earlier partial discoveries and specific, more modest applications. As far as we know the idea (loosely stated) that a reduced product modulo a countably incomplete filter is ω^+-saturated first occurred to and was developed by du Bois-Reymond (in quite different terms, of course). He essentially established Hausdorff's η_1-property (studied in § 5) of the "reduced product" of the set of continuous real-valued functions in a real variable modulo the "Fréchet filter" of the family of complements of compact subsets of the reals. This led to his theory of orders (scales) of infinity, an active branch of the theory of real functions around the turn of the century, to which Hadamard, Pincherle, Borel, and Hardy also contributed. More recently in their study of the maximal ideals of the ring $C(X)$ of real-valued continuous functions on a space X, Hewitt implicitly and Erdős-Gillman-Henriksen explicitly have established the η_1-property of the "ultrapower" $C(X)/M$ of $C(X)$ modulo a maximal ideal (dual ultrafilter) M. Thus this idea, though quite old, has remained until relatively recently in the domain of real analysis and general topology and outside model theory.

The model-theoretic forerunners of the ultraproduct construction are Skolem, with his proof of the existence of a *countable* non-standard model of arithmetic, and Łoś, who expressed in a special case the characteristic property of an ultraproduct (Theorem of Łoś 11.5). The concept was clarified in its proper generality in the papers of Frayne-Morel-Scott and Kochen. The decisive contributions to our knowledge of its saturation properties have been made by Keisler, who created the notion and proved the existence of good ultrafilters for the purpose (Theorems 9.13, 13.9).

Somewhat earlier and independently Jónsson studied conditions on general classes of structures that ensure the existence and uniqueness in power of homogeneous-universal structures (§ 4); this theory was connected with model theory by Morley-Vaught, who proved that an elementary type is essentially a Jónsson class in which the elementarily homogeneous-universal structures are identified with the saturated structures in the sense of Keisler (§11). We note here the remarkable equivalence for elementary types of two far-reaching independent generalizations of Hausdorff's notion of a linearly ordered η_α-set.

The very general problem, due to Tarski, of finding an "interesting" mathematical characterization of the metamathematical notion of elementary equivalence was given a "concrete" and unexpected solution by Keisler, exploiting the saturation products of ultraproducts. His provisional solution was the following: assuming GCH, two similar structures are elementarily equivalent (if and) only if some ultrapower of one is isomorphic to some ultrapower of the other. This result did not give any information on the nature or relation of the two ultrafilters involved. Subsequently the uniqueness in power of the saturated structures, together with the existence and saturation properties of ultraproducts modulo good ultrafilters, led Keisler to the following much stronger statement, assuming GCH: Every ultrapower modulo the *same* (good) ultrafilter is saturated; hence the ultrapowers modulo the same good ultrafilter of any two elementarily equivalent structures of "sufficiently small" cardinality and index are isomorphic (Corollary 13.12).

The only defect of Keisler's results is their dependence on the generalized continuum hypothesis. The avoidance of GCH from these fundamental results was a natural problem attempted by various people; their methods consisted mostly in replacing the notion of ultraproduct by a more general notion (such as limit ultrapower with respect to a family of regular ultrafilters (Corollary 13.4), Kochen's ultralimit, Boolean ultrapower, etc.) with respect to which analogous results could be proved without GCH. Motivated by the success of Kunen in proving the existence without GCH of a good ultrafilter by techniques involving families of large oscillation, Shelah obtained by related methods, in his definitive work, the characterization of elementary equivalence without GCH (Theorem 13.19).

Ultraproducts Modulo Regular Ultrafilters

13.1. Lemma. *Let* $\alpha \geq \omega$, *p a regular ultrafilter on* α, *and* $\mathfrak{A}$ *a structure of type* $\tau \in \omega^\rho$ *with* $|\rho| \leq \alpha$. *We identify* $\mathfrak{A}$ *with an elementary substructure of* $\mathfrak{A}^\alpha/p$ *via the canonical embedding defined in* §11. *If* $B \subset A$ *is such that* $|B| \leq \alpha$ *and if* $\Phi \subset \Phi_1(L(\tau)_B)$ *is such that*

$$\Phi \text{ is finitely satisfiable in } \langle \mathfrak{A}^\alpha/p, b\rangle_{b \in B},$$

then

$$\Phi \text{ is satisfiable in } \langle \mathfrak{A}^\alpha/p, b\rangle_{b\in B}.$$

Proof. Since p is regular, there is a family $\{A_\zeta : \zeta < \alpha\}$ of elements of p such that

$$\text{if } I \subset \alpha \quad \text{and } |I| \geq \omega \quad \text{then } \bigcap_{\eta\in I} A_\eta = \emptyset .$$

For $\xi < \alpha$ we set

$$F(\xi) = \{\zeta < \alpha : \xi \in A_\zeta\} ;$$

it is clear that $F(\xi) \in \mathscr{P}_\omega(\alpha)$ for $\xi < \alpha$. We note that $|\Phi| \leq \alpha$ and we choose a function $h: \alpha \to \Phi$ from α onto Φ. For $\xi < \alpha$ we choose $c_\xi \in A$ such that

$$\langle \mathfrak{A}, b\rangle_{b\in B} \models \phi[c_\xi] \quad \text{for } \phi \in h[F(\xi)],$$

we set

$$c = \langle c_\xi : \xi < \alpha\rangle/p \in \mathfrak{A}^\alpha/p ,$$

and we prove that

$$\langle \mathfrak{A}^\alpha/p, b\rangle_{b\in B} \models \phi[c] \quad \text{for } \phi \in \Phi .$$

Indeed let $\phi \in \Phi$ and choose $\zeta < \alpha$ such that $h(\zeta) = \phi$. If $\xi \in A_\zeta$ then $\zeta \in F(\xi)$ and thus $\langle \mathfrak{A}, b\rangle_{b\in B} \models \phi[c_\xi]$ by the definition of c_ξ. Hence

$$A_\zeta \subset \{\xi < \alpha : \langle \mathfrak{A}, b\rangle_{b\in B} \models \phi[c_\xi]\}$$

and the theorem follows from the theorem of Łoś.

Definition. A *limit ultrapower* of a structure $\mathfrak{A}$ is a structure which is equal to the union of a chain $\{\mathfrak{A}_\xi : \xi < \eta\}$ (where η is a non-zero ordinal) such that

$\mathfrak{A}_0 = \mathfrak{A}$,

$\mathfrak{A}_\xi = \bigcup_{\zeta<\xi} \mathfrak{A}_\zeta$ for non-zero limit ordinals $\xi < \eta$,

$\mathfrak{A}_{\xi+1}$ is an ultrapower of $\mathfrak{A}_\xi$ for $\xi + 1 < \eta$, and

the inclusion $\mathfrak{A}_\xi \to \mathfrak{A}_{\xi+1}$ is the canonical embedding (by which $\mathfrak{A}_\xi$ is identified with an elementary subsystem of $\mathfrak{A}_{\xi+1}$) for $\xi < \eta$.

We note that

an ultrapower of $\mathfrak{A}$ is the special case $\eta = 2$ of a limit ultrapower of $\mathfrak{A}$; and

a limit ultrapower of $\mathfrak{A}$ is an elementary extension of $\mathfrak{A}$.

(The latter statement follows from Corollary 11.6 (to the theorem of Łoś) and the Tarski-Vaught theorem (11.3).)

The fundamental property of limit ultrapowers (with respect to regular ultrafilters) is expressed in the following statement.

13.2. Theorem. *Let $\alpha=2^{\underline{\alpha}}$ and $|\rho|+\omega<\alpha$ and let $\mathfrak{A}$ be a structure of type $\tau\in\omega^\rho$ such that $\omega\leq|\mathfrak{A}|<\alpha$. Then there is a structure $\mathfrak{B}$ such that*

> *$\mathfrak{B}$ is special of cardinality α for the elementary type of $\mathfrak{A}$, and*
>
> *$\mathfrak{B}$ is a limit ultrapower of $\mathfrak{A}$.*

Proof. We define recursively $\mathfrak{A}_\xi$ for $\xi<\alpha$ as follows.

$\mathfrak{A}_0=\mathfrak{A}$;

$\mathfrak{A}_{\xi+1}=\mathfrak{A}_\xi^{\omega+|\rho|+|\xi|}/p_\xi$ where p_ξ is a regular ultrafilter on $\omega+|\rho|+|\xi|$;

and

$\mathfrak{A}_\xi=\bigcup_{\zeta<\xi}\mathfrak{A}_\zeta$ for non-zero limit ordinals $\xi<\alpha$.

(Here as usual we identify $\mathfrak{A}_\xi$ with an elementary subsystem of $\mathfrak{A}_{\xi+1}$ *via* the canonical embedding.) We set

$$\mathfrak{B}=\bigcup_{\xi<\alpha}\mathfrak{A}_\xi .$$

It is clear that $\mathfrak{B}$ is a limit ultrapower of $\mathfrak{A}$. It is easy to verify by transfinite induction that

$$|\mathfrak{A}_\xi|\leq 2^{\omega+|\rho|+|\xi|} \quad \text{for } \xi<\alpha ,$$

and using Lemma 13.1 that

$$\mathfrak{A}_{\beta^+} \quad \text{is } \beta^+\text{-saturated for } \beta<\alpha .$$

Thus

$$\{\mathfrak{A}_{\beta^+}:\beta<\alpha\}$$

is a specializing chain for $\mathfrak{B}$, and hence $\mathfrak{B}$ is special of cardinality (at most) $\alpha=2^{\underline{\alpha}}$.

The following corollary is immediate.

13.3. Corollary. *Let $\mathbb{K}$ be an elementary type with $\tau\in\omega^\rho$ and let $|\rho|+\omega<\alpha=2^{\underline{\alpha}}$. The unique elementarily special structure of cardinality α in $\mathbb{K}$ is (representable as) a limit ultrapower (of a structure of $\mathbb{K}$).*

13.4. Corollary. *Two structures $\mathfrak{A}$ and $\mathfrak{B}$ are elementarily equivalent if and only if some limit ultrapower of $\mathfrak{A}$ is isomorphic to some limit ultrapower of $\mathfrak{B}$.*

Proof. If some limit ultrapower of $\mathfrak{A}$ is isomorphic to some limit ultrapower of $\mathfrak{B}$ then clearly (as remarked earlier) $\mathfrak{A}\equiv\mathfrak{B}$.

Conversely, assume that $\mathfrak{A}\equiv\mathfrak{B}$. If $|\mathfrak{A}|<\omega$ then $\mathfrak{A}\cong\mathfrak{B}$ by the remark following Theorem 11.17, so we assume that $\mathfrak{A}$ (and hence $\mathfrak{B}$) are infinite structures. By Theorem 1.27 there is a cardinal α such that

$$|\rho|+\omega+|\mathfrak{A}|+|\mathfrak{B}|<\alpha=2^{\underline{\alpha}} ,$$

where $\tau \in \omega^\rho$ is the type of $\mathfrak{A}$. By Corollary 13.3 there are structures $\mathfrak{A}_0, \mathfrak{B}_0$ such that

$\mathfrak{A}_0$ is a limit ultrapower of $\mathfrak{A}$,

$\mathfrak{B}_0$ is a limit ultrapower of $\mathfrak{B}$, and

$\mathfrak{A}_0, \mathfrak{B}_0$ are both elementarily special of cardinality α (for the type $\mathbb{M}(Th(\mathfrak{A}))$).

Then $\mathfrak{A}_0 \cong \mathfrak{B}_0$ by Theorem 11.17.

The non-trivial part of this corollary is greatly strengthened in Corollary 13.21 below, with the replacement of "limit ultrapower" by "ultrapower".

13.5. Lemma. *Let $\mathbb{K}$ be an elementary type with $\tau \in \omega^\rho$ where $|\rho| + \omega \leq \alpha$ and let $\mathfrak{A} \in \mathbb{K}$. Then the following two statements are equivalent.*

(a) *$\mathfrak{A}$ is elementarily α-universal;*

(b) *for every $C \subset A$ with $|C| \leq \alpha$ and $\Phi \subset \Phi_0(L_C)$ such that $Th(\mathfrak{A}) \cup \Phi$ has a model, there are $a_c \in A$ for $c \in C$ such that*

$$\langle \mathfrak{A}, a_c \rangle_{c \in C} \models \Phi .$$

Proof. (a)$\Rightarrow$(b). Let $C \subset A$ with $|C| \leq \alpha$, and assume that $\Phi \subset \Phi_0(L_C)$ and $Th(\mathfrak{A}) \cup \Phi$ has a model. There is a structure $\langle \mathfrak{B}, b_c \rangle_{c \in C}$ such that

$$\langle \mathfrak{B}, b_c \rangle_{c \in C} \models Th(\mathfrak{A}) \cup \Phi .$$

By the Löwenheim-Skolem theorem (11.4) we assume without loss of generality (since $|\rho| + \omega \leq \alpha$) that $|\mathfrak{B}| \leq \alpha$. Then $\mathfrak{B} \models Th(\mathfrak{A})$, *i.e.* $\mathfrak{A} \equiv \mathfrak{B}$, and thus there is an elementary embedding $\phi : \mathfrak{B} \to \mathfrak{A}$. It then follows that

$$\langle \mathfrak{A}, \phi(b_c) \rangle_{c \in C} \models \Phi .$$

(b)$\Rightarrow$(a). Let $\mathfrak{B} \equiv \mathfrak{A}$ with $|\mathfrak{B}| \leq \alpha$. We set $\Phi = Th(\langle \mathfrak{B}, b \rangle_{b \in B})$, so that $Th(\mathfrak{A}) \cup \Phi = \Phi$ has a model. Thus there are $a_b \in A$ for $b \in B$ such that

$$\langle \mathfrak{A}, a_b \rangle_{b \in B} \models \Phi .$$

Let $\phi : \mathfrak{B} \to \mathfrak{A}$ be given by $\phi(b) = a_b$. It is easy to verify that ϕ is a (well-defined) elementary embedding.

13.6. Theorem. *Let $\beta \geq \omega$ and p an ultrafilter on α. The following two statements are equivalent.*

(a) *p is (ω, β)-regular;*

(b) *for every structure $\mathfrak{A}$ with $|L_\mathfrak{A}| \leq \beta$, the ultrapower $\mathfrak{A}^\alpha/p$ is elementarily β-universal.*

Proof. (a)$\Rightarrow$(b). We use the equivalences of Lemma 13.5. Let $\mathfrak{B} = \mathfrak{A}^\alpha/p$, $C \subset A$ with $|C| \leq \beta$, and let $\Phi \subset \Phi_0(L_C)$ be such that $Th(\mathfrak{B}) \cup \Phi$ has a model. Since $|L_{\mathfrak{A}}| \leq \beta$ we have $|\Phi| \leq \beta$ and there is a function $h: \beta \to \Phi$ onto. Since p is (ω, β)-regular there is a family $\{A_\eta : \eta < \beta\} \subset p$ such that

$$\text{if } I \subset \beta \quad \text{and } |I| \geq \omega \quad \text{then } \bigcap_{\eta \in I} A_\eta = \emptyset .$$

For $\xi < \alpha$ we set $F(\xi) = \{\eta < \beta : \xi \in A_\eta\}$. Clearly $F(\xi) \in \mathscr{P}_\omega(\beta)$ for $\xi < \alpha$. We note further that $h[F(\xi)] \subset \Phi$ and thus, by assumption, $Th(\mathfrak{A}) \cup h[F(\xi)]$ has a model for $\xi < \alpha$. It is easy to see then that there are $a_{c,\xi} \in A$ for $c \in C$ such that

$$\langle \mathfrak{A}, a_{c,\xi} \rangle_{c \in C} \models h[F(\xi)] \quad \text{for } \xi < \alpha .$$

We set

$$a_c = \langle a_{c,\xi} : \xi < \alpha \rangle / p \in \mathfrak{B}$$

and we claim that

$$\langle \mathfrak{B}, a_c \rangle_{c \in C} \models \Phi .$$

Indeed let $\phi \in \Phi$, so that there is $\eta < \beta$ with $h(\eta) = \phi$. If $\xi \in A_\eta$ then $\eta \in F(\xi)$ and thus

$$\langle \mathfrak{A}, a_{c,\xi} \rangle_{c \in C} \models \phi .$$

Hence $A_\eta \subset \{\xi < \alpha : \langle \mathfrak{A}, a_{c,\xi} \rangle_{c \in C} \models \phi\} \in p$ for $\eta < \beta$, and the claim follows from the theorem of Łoś.

(b)⇒(a). We set

$$R_i = \{F \in \mathscr{P}_\omega(\beta) : i \in F\}$$

for $i < \beta$, and we set

$$\mathfrak{A} = \langle \mathscr{P}_\omega(\beta), R_i^{\mathfrak{A}} \rangle_{i < \beta} .$$

Furthermore, we set $\Phi = \{R_i(v_0) : i < \beta\}$. Then

$|L_{\mathfrak{A}}| = \beta, \quad \Phi \subset \Phi_1(L_{\mathfrak{A}})$,
Φ is finitely satisfiable in $\mathfrak{A}$, and
no infinite subset of Φ is satisfiable in $\mathfrak{A}$.

We set $L = L_{\mathfrak{A}}$, we let c be a new constant and we consider the language L_c. We set

$$\Psi = \{\phi_{v_0}[c] : \phi \in \Phi\} \subset \Phi_0(L_c) .$$

Since $\mathfrak{A}^\alpha/p$ is elementarily β-universal there is $a \in \mathfrak{A}^\alpha/p$ such that

$$\langle \mathfrak{A}^\alpha/p, a \rangle \models \Psi ;$$

hence there is $f \in \mathfrak{A}^\alpha$ such that $\mathfrak{A}^\alpha/p \models_{f/p} \Phi$. We set

$$A_i = \{\xi < \alpha : \mathfrak{A} \models_{f(\xi)} R_i(v_0)\}$$

and we note, using the theorem of Łoś, that

$$A_i \in p \quad \text{for } i<\beta .$$

Furthermore

$$\text{if } I \subset \beta \quad \text{and } |I| \geq \omega \quad \text{then } \bigcap_{i\in I} A_i = \emptyset ,$$

so that p is (ω, β)-regular.

For the purposes of the next theorem we introduce the following definitions of notions that are variations of the notions of α-saturated and elementarily α-universal, respectively.

A structure $\mathfrak{A}$ is α-**saturated* if for every $B \subset A$ and $\Phi \subset \Phi_1(L_B)$ with $|\Phi| < \alpha$, if Φ is finitely satisfiable in $\langle \mathfrak{A}, b\rangle_{b\in B}$ then Φ is satisfiable in $\langle \mathfrak{A}, b\rangle_{b\in B}$.

A structure $\mathfrak{A}$ is α-**universal* if for every $B \subset A$ and $\Phi \subset \Phi_0(L_B)$ with $|\Phi| < \alpha$, if $Th(\mathfrak{A}) \cup \Phi$ has a model then there are $a_b \in A$ for $b \in B$ such that $\langle \mathfrak{A}, a_b\rangle_{b\in B} \models \Phi$.

The following statements are easily verified.

Every structure is ω-*saturated (and ω-*universal);

if $|L_{\mathfrak{A}}| = \omega + |\rho| < \alpha$, then $\mathfrak{A}$ is α-saturated if and only if $\mathfrak{A}$ is α-*saturated; and

if $|L_{\mathfrak{A}}| = \omega + |\rho| < \alpha$, then $\mathfrak{A}$ is elementarily β-universal for all $\beta < \alpha$ if and only if $\mathfrak{A}$ is α-*universal.

13.7. Theorem *(Keisler). Let α, β and γ be infinite cardinals and let p be a (β,γ)-regular ultrafilter on α.*

Let $\mathfrak{A}_\xi$ and $\mathfrak{B}_\xi$ be structures of type $\tau \in \omega^\rho$ such that

$$\omega + |\rho| \leq \gamma,$$
$$\mathfrak{A}_\xi \equiv \mathfrak{B}_\xi,$$
$$\mathfrak{A}_\xi \text{ is } \beta\text{-*saturated, and}$$
$$\mathfrak{B}_\xi \text{ is } \beta\text{-*universal}$$

for $\xi < \alpha$. Define

$$\mathfrak{A} = \prod_{\xi<\alpha} \mathfrak{A}_\xi / p, \qquad \mathfrak{B} = \prod_{\xi<\alpha} \mathfrak{B}_\xi / p .$$

If $\mathfrak{B}$ is γ^+-saturated then $\mathfrak{A}$ is γ^+-saturated.

Proof. Let $C \subset A$ with $|C| \leq \gamma$, and let $\Phi \subset \Phi_1(L_C)$ be finitely satisfiable in $\langle \mathfrak{A}, c\rangle_{c\in C}$.

We note that $|\Phi| \leq |\Phi_1(L_C)| \leq \omega + |\rho| + |C| \leq \gamma$, and thus there is a function

$$h: \gamma \to \Phi \quad \text{onto}.$$

Since p is (β, γ)-regular there is a family $\{A_\eta : \eta < \gamma\} \subset p$ such that

$$\text{if } I \subset \gamma \quad \text{and } |I| \geq \beta \quad \text{then } \bigcap_{\eta\in I} A_\eta = \emptyset .$$

We set

$$F(\xi)=\{\eta<\gamma:\xi\in A_\eta\}$$

and we note, since p is (β,γ)-regular, that $|F(\xi)|<\beta$, and hence that

$$|h[F(\xi)]|<\beta \quad \text{for } \xi<\alpha.$$

We set

$$\Psi(\xi)=\{(\exists v_0)(\phi_0\wedge\cdots\wedge\phi_n):n<\omega \text{ and } \phi_0,\ldots,\phi_n\in h[F(\xi)]\}$$

and we note that

$$\Psi(\xi)\subset\Phi_0(L_C) \quad \text{and}$$
$$|\Psi(\xi)|<\beta$$

for $\xi<\alpha$.

There is $a_c\in A^\alpha$ such that $c=a_c/p$ for $c\in C$. We set

$$\Omega(\xi)=(\Psi(\xi)\cap Th(\langle\mathfrak{A}_\xi,a_c(\xi)\rangle_{c\in C}))\cup\{\neg\psi:\psi\in\Psi(\xi)\setminus Th(\langle\mathfrak{A}_\xi,a_c(\xi)\rangle_{c\in C})\}$$

and we note that

$$|\Omega(\xi)|<\beta, \quad \text{and}$$
$$\langle\mathfrak{A}_\xi,a_c(\xi)\rangle_{c\in C}\models Th(\mathfrak{A}_\xi)\cup\Omega(\xi)$$

for $\xi<\alpha$. Since $\mathfrak{A}_\xi\equiv\mathfrak{B}_\xi$ we have $Th(\mathfrak{A}_\xi)=Th(\mathfrak{B}_\xi)$ and thus $Th(\mathfrak{B}_\xi)\cup\Omega(\xi)$ has a model for $\xi<\alpha$. Since $\mathfrak{B}_\xi$ is β-*universal it follows that there are $b_c(\xi)\in B_\xi$ such that

$$\langle\mathfrak{B}_\xi,b_c(\xi)\rangle_{c\in C}\models\Omega(\xi) \quad \text{for } \xi<\alpha.$$

We note that

(1) $$\Psi(\xi)\cap Th(\langle\mathfrak{A}_\xi,a_c(\xi)\rangle_{c\in C})=\Psi(\xi)\cap Th(\langle\mathfrak{B}_\xi,b_c(\xi)\rangle_{c\in C})$$

for $\xi<\alpha$. We claim that

$$\Phi \text{ is finitely satisfiable in } \langle\mathfrak{B},b_c/p\rangle_{c\in C}.$$

Indeed let $n<\omega$ and $\phi_0,\ldots,\phi_n\in\Phi$. We denote by ψ the sentence $(\exists v_0)(\phi_0\wedge\cdots\wedge\phi_n)$. Since h is onto there are $\eta_0,\ldots,\eta_n<\gamma$ such that $\phi_0=h(\eta_0),\ldots,\phi_n=h(\eta_n)$. Thus if $\xi\in A_{\eta_0}\cap\cdots\cap A_{\eta_n}$ then $\eta_0,\ldots,\eta_n\in F(\xi)$, hence $\phi_0,\ldots,\phi_n\in h[F(\xi)]$; it follows that $\psi\in\Psi(\xi)$, so that

(2) $$\{\xi<\alpha:\psi\in\Psi(\xi)\}\in p.$$

By assumption Φ is finitely satisfiable in $\langle\mathfrak{A},c\rangle_{c\in C}$; hence $\langle\mathfrak{A},c\rangle_{c\in C}\models\psi$ and we have from the theorem of Łoś that

(3) $$\{\xi<\alpha:\langle\mathfrak{A}_\xi,a_c(\xi)\rangle_{c\in C}\models\psi\}\in p.$$

We have from (1), (2), (3) that

$$\{\xi<\alpha:\langle\mathfrak{B}_\xi,b_c(\xi)\rangle_{c\in C}\models\psi\}\in p,$$

and from the theorem of Łoś that

$$\langle \mathfrak{B}, b_c/p \rangle_{c\in C} \models \psi .$$

The claim is proved.

Since $\mathfrak{B}$ is γ^+-saturated and $|C| \leq \gamma$ and Φ is finitely satisfiable in $\langle \mathfrak{B}, b_c/p \rangle_{c\in C}$, it follows that there is $b \in B^\alpha$ such that

$$\Phi \text{ is satisfiable in } \langle \mathfrak{B}, b_c/p \rangle_{c\in C} \text{ by } b/p .$$

We set

$$\Gamma(\xi) = \{\phi \in h[F(\xi)] : \langle \mathfrak{B}_\xi, b_c(\xi) \rangle_{c\in C} \models_{\overline{b(\xi)}} \phi\}$$

and we note that

$$|\Gamma(\xi)| < \beta, \quad \text{and}$$
$$\Gamma(\xi) \text{ is finitely satisfiable in } \langle \mathfrak{A}_\xi, a_c(\xi) \rangle_{c\in C}$$

for $\xi < \alpha$. To prove the latter statement let $n < \omega$ and $\phi_0, \ldots, \phi_n \in \Gamma(\xi)$; thus $\phi_0, \ldots, \phi_n \in h[F(\xi)]$ and $\langle \mathfrak{B}_\xi, b_c(\xi) \rangle_{c\in C} \models_{\overline{b(\xi)}} \phi_0 \wedge \cdots \wedge \phi_n$, so that

$$\exists v_0(\phi_0 \wedge \cdots \wedge \phi_n) \in \Psi(\xi) \cap Th(\langle \mathfrak{B}_\xi, b_c(\xi) \rangle_{c\in C}) \subset Th(\langle \mathfrak{A}_\xi, a_c(\xi) \rangle_{c\in C}),$$

using (1); hence $\phi_0, \ldots, \phi_n$ is satisfiable in $\langle \mathfrak{A}_\xi, a_c(\xi) \rangle_{c\in C}$. Since $\mathfrak{A}_\xi$ is β-*saturated there is $a(\xi) \in \mathfrak{A}_\xi$ such that

$$\langle \mathfrak{A}_\xi, a_c(\xi) \rangle_{c\in C} \models_{\overline{a(\xi)}} \Gamma(\xi) \quad \text{for } \xi < \alpha .$$

We claim that

$$\langle \mathfrak{A}, c \rangle_{c\in C} \models_{\overline{a/p}} \Phi .$$

Indeed let $\phi \in \Phi$. Since h is onto there is $\eta \prec \gamma$ such that $\phi = h(\eta)$, and then

$$A_\eta \subset \{\xi < \alpha : \phi \in h[F(\xi)]\} \in p .$$

From the theorem of Łoś we have

$$\{\xi < \alpha : \langle \mathfrak{B}_\xi, b_c(\xi) \rangle_{c\in C} \models_{\overline{b(\xi)}} \phi\} \in p$$

and it now follows that

$$S_\phi = \{\xi < \alpha : \phi \in \Gamma(\xi)\} \in p .$$

We also have that

$$S_\phi \subset \{\xi < \alpha : \langle \mathfrak{A}_\xi, a_c(\xi) \rangle \models_{\overline{a(\xi)}} \phi\} ,$$

and it follows from the theorem of Łoś that

$$\langle \mathfrak{A}, c \rangle_{c\in C} \models_{\overline{a/p}} \phi ,$$

establishing the claim and completing the proof of the theorem.

13.8. Corollary. *Let α and β be infinite cardinals and let p be an (ω, β)-regular ultrafilter on α.*

Let $\mathfrak{A}_\xi$ and $\mathfrak{B}_\xi$ be structures of type $\tau \in \omega^\rho$ such that

$$\omega + |\rho| \leq \beta, \quad \textit{and}$$
$$\mathfrak{A}_\xi \equiv \mathfrak{B}_\xi \quad \textit{for } \xi < \alpha.$$

Define $\mathfrak{A} = \prod_{\xi<\alpha} \mathfrak{A}_\xi/p$ and $\mathfrak{B} = \prod_{\xi<\alpha} \mathfrak{B}_\xi/p$. Then

$$\mathfrak{A} \textit{ is } \gamma^+\textit{-saturated if and only if } \mathfrak{B} \textit{ is } \gamma^+\textit{-saturated.}$$

Proof. We use Theorem 13.7, taking into account the fact (noted above) that every structure is ω-*saturated (and hence ω-*universal).

Ultraproducts Modulo Good Ultrafilters

13.9. Theorem. *Let α and β be infinite cardinals with $\beta > \omega$, and let p be a countably incomplete β-good ultrafilter on α. Let $\{\mathfrak{A}_\xi : \xi < \alpha\}$ be a family of structures of type $\tau \in \omega^\rho$, with $\rho < \beta$. Then*

$$\prod_{\xi<\alpha} \mathfrak{A}_\xi/p \textit{ is } \beta\textit{-saturated.}$$

Proof. Let $\mathfrak{A} = \prod_{\xi<\alpha} \mathfrak{A}_\xi/p$ and $C \in \mathscr{P}_\beta(A)$, and let $\Phi \subset \Phi_1(L_C)$ be such that Φ is finitely satisfiable in $\langle \mathfrak{A}, c \rangle_{c \in C}$ (where L is the language $L(\tau)$). We are to prove that Φ is satisfiable in $\langle \mathfrak{A}, c \rangle_{c \in C}$. Let $f_c \in \prod_{\xi<\alpha} A_\xi$ be such that $f_c/p = c$ for $c \in C$. It is clear that

$$\langle \mathfrak{A}, c \rangle_{c \in C} = \left(\prod_{\xi<\alpha} \langle \mathfrak{A}_\xi, f_c(\xi) \rangle_{c \in C}\right)/p\,.$$

We note also, since $|\rho| + \omega + |C| < \beta$, that $|\Phi| < \beta$. We define

$$g : \mathscr{P}_\omega(\Phi) \to \mathscr{P}(\alpha)$$

by $g(F) = \{\xi < \alpha : F \text{ is satisfiable in } \langle \mathfrak{A}, f_c(\xi) \rangle_{c \in C}\}$. Clearly g is monotone. For $F \in \mathscr{P}_\omega(\Phi)$ and $F \neq \emptyset$ we let ϕ_F be the sentence

$$\exists v_0 \bigwedge_{\phi \in F} \phi(v_0)\,.$$

Thus $\langle \mathfrak{A}, c \rangle_{c \in C} \models \phi_F$, so $g(F) \in p$ for $F \in \mathscr{P}_\omega(\Phi)$ by the theorem of Łoś.

Since p is countably incomplete, there is a sequence $\{A_n : n < \omega\}$ of elements of p such that

$$\alpha = A_0 \supset A_1 \supset \cdots \supset A_n \supset \cdots, \quad \text{and}$$
$$\bigcap_{n<\omega} A_n = \emptyset\,.$$

We define

$$g' : \mathscr{P}_\omega(\Phi) \to p$$

by $g'(F)=g(F)\cap A_{|F|}$; then g' is monotone, and $g'\leq g$. Since p is β-good, there is a multiplicative function

$$h:\mathscr{P}_{\omega}(\Phi)\to p$$

such that $h\leq g'$.

We set $\Phi(\xi)=\{\phi\in\Phi:\xi\in h(\{\phi\})\}$ and we claim that $|\Phi(\xi)|<\omega$ for $\xi<\alpha$. Indeed, we denote by $n(\xi)$ the least $n<\omega$ such that $\xi\notin A_n$, and we prove that $|\Phi(\xi)|<n(\xi)$. Otherwise there is a subset $\Phi'\subset\Phi(\xi)$ with $|\Phi'|=n(\xi)$, and then since h is multiplicative

$$\xi\in\bigcap\nolimits_{\phi\in\Phi'}h(\{\phi\})=h(\Phi')\subset g'(\Phi')\subset A_{|\Phi'|}=A_{n(\xi)},$$

contradicting the fact that $\xi\notin A_{n(\xi)}$. Hence $\Phi(\xi)\in\mathscr{P}_{\omega}(\Phi)$ and $\xi\in h(\Phi(\xi))\subset g(\Phi(\xi))$, *i.e.*, $\Phi(\xi)$ is satisfiable in $\langle\mathfrak{A}_{\xi},f_c(\xi)\rangle_{c\in C}$ for $\xi<\alpha$. We let $f\in\prod_{\xi<\alpha}A_{\xi}$ be such that

$$\langle\mathfrak{A}_{\xi},f_c(\xi)\rangle_{c\in C}\models_{\overline{f(\xi)}}\Phi(\xi)$$

for $\xi<\alpha$. Then

$$\{\xi<\alpha:\langle\mathfrak{A},f_c(\xi)\rangle_{c\in C}\models_{\overline{f(\xi)}}\Phi\}\supset\{\xi<\alpha:\phi\in\Phi(\xi)\}=h(\{\phi\})\in p$$

for $\phi\in\Phi$. Thus, by the theorem of Łoś,

$$\langle\mathfrak{A},c\rangle_{c\in C}\models_{\overline{f/p}}\Phi.$$

The following statement is a special case of this theorem.

If $|\rho|\leq\omega$, p is a countably incomplete ultrafilter on α, and $\{\mathfrak{A}_{\xi}:\xi<\alpha\}$ is a family of structures of type $\tau\in\omega^{\rho}$, then $\prod_{\xi<\alpha}\mathfrak{A}_{\xi}/p$ is ω^+-saturated.

This follows from the fact (proved in Lemma 9.8) that every (countably incomplete) ultrafilter is ω^+-good.

13.10. Corollary. *Let $\mathbb{K}$ be an elementary type with $\tau\in\omega^{\rho}$ and $|\rho|+\omega\leq\alpha$, and let $\mathfrak{A}\in\mathbb{K}$ with $|\mathfrak{A}|\leq 2^{\alpha}$. Then there is an ultrapower of $\mathfrak{A}$ which is α^+-saturated and of cardinality 2^{α}.*

Proof. Let p be a countably incomplete α^+-good ultrafilter on α (Theorem 10.4); then $\mathfrak{A}^{\alpha}/p$ is α^+-saturated by Theorem 13.9; since an α^+-good ultrafilter on α is regular (Corollary 9.10(b)), it follows from Theorem 12.22 that $2^{\alpha}=(2^{\alpha})^{\alpha}\geq|\mathfrak{A}^{\alpha}/p|=\omega^{\alpha}\geq 2^{\alpha}$.

13.11. Corollary. *Let $\mathbb{K}$ be an elementary type with $\tau\in\omega^{\rho}$ such that $|\rho|+\omega\leq\alpha$, and assume that $\alpha^+=2^{\alpha}$. The unique α^+-saturated structure of cardinality α^+ in $\mathbb{K}$ is representable as an ultrapower.*

Proof. This follows immediately from Corollary 13.10 and the fact that $\mathbb{K}$ has an infinite structure of cardinality (at most) α^+ (by properties (I) and $(\mathrm{VI})_{(|\rho|+\omega)^+}$).

13.12. Corollary. *Assume that* $\alpha^+ = 2^\alpha$, *and let* $\mathfrak{A}$ *and* $\mathfrak{B}$ *be structures of type* $\tau \in \omega^\rho$ *such that* $|\rho| + \omega \leq \alpha$, $|\mathfrak{A}| \leq \alpha^+$, *and* $|\mathfrak{B}| \leq \alpha^+$. *The following statements are equivalent.*

(a) $\mathfrak{A} \equiv \mathfrak{B}$;

(b) $\mathfrak{A}$ *and* $\mathfrak{B}$ *have isomorphic ultrapowers; and*

(c) *there is an ultrafilter* p *on* α *such that* $\mathfrak{A}^\alpha/p \cong \mathfrak{B}^\alpha/p$.

Proof. Obviously (c) implies (b), and (b) implies (a) by the Corollary to the theorem of Łoś. To prove that (a) implies (c), let p be a countably incomplete α^+-good ultrafilter on α, and use Corollary 13.11 if $\mathfrak{A}$ (and hence $\mathfrak{B}$) is infinite. If $\mathfrak{A}$ (and hence $\mathfrak{B}$) is finite, then (from the remark following Theorem 11.17) $\mathfrak{A} \cong \mathfrak{B}$ and (c) follows trivially; in fact $\mathfrak{A} \cong \mathfrak{A}^\alpha/p$ in such a case.

Significant parts of this corollary will be proved below without making use of any segment of the generalized continuum hypothesis (cf. Theorem 13.19, Remark 13.20, Corollary 13.21).

Definition. Let κ be an infinite cardinal. The structure $\mathfrak{P}(\kappa)$, of type $\tau: 1 \to \omega$ with $\tau(0) = 2$, is

$$\mathfrak{P}(\kappa) = \langle \mathscr{P}(\kappa), R(\kappa) \rangle,$$

where if $a, b \in \mathscr{P}(\kappa)$ then $\langle a, b \rangle \in R(\kappa)$ if and only if $a \subset b$.

13.13. Theorem *(Keisler). Let* α *and* β *be cardinals such that* $\beta > \omega$, *and let* p *be an ultrafilter on* α. *The following statements are equivalent.*

(a) *p is countably incomplete and β-good;*

(b) *if* $\rho < \beta$, $\tau \in \omega^\rho$, *and* $\langle \mathfrak{A}_\xi : \xi < \alpha \rangle$ *is a family of structures of type* τ, *then* $\prod_{\xi<\alpha} \mathfrak{A}_\xi/p$ *is* β*-saturated; and*

(c) *there is a cardinal* κ *such that*

$$\beta \leq \kappa^+, \quad \textit{and}$$

$$\mathfrak{P}(\kappa)^\alpha/p \quad \textit{is } \beta\textit{-saturated}.$$

Proof. (a) implies (b) by Theorem 13.9; (b) implies (c) in a trivial way; thus we only need prove that (c) implies (a). We assume that there is a cardinal κ such that

$$\beta \leq \kappa^+, \quad \text{and}$$

$$\mathfrak{P}(\kappa)^\alpha/p \quad \text{is } \beta\text{-saturated.}$$

For a structure $\mathfrak{A}$ of type $\tau: 1 \to \omega$ with $\tau(0) = 2$, and $a \in A$, we let ϕ_a be the formula

$$P_0(v_0, \bar{a}) \wedge \exists v_1 (P_0(v_1, v_0) \wedge \neg v_1 = v_0);$$

we note that $\phi_a \in \Phi_1(L_{\{a\}})$ and that the interpretation of $\phi_a(v_0)$ in $\langle \mathfrak{P}(\kappa), a \rangle$ is that v_0 is a non-empty subset of a.

We prove the following two statements.

(i) p is countably incomplete.

Since $\omega<\beta\leq\kappa^+$ we have $\kappa\geq\omega$, and thus there is a sequence $\{a_n:n<\omega\}\subset\mathscr{P}(\kappa)$ such that

$$a_0\supset a_1\supset\cdots\supset a_n\supset\cdots,$$
$$a_n\neq\emptyset\quad\text{for } n<\omega,\quad\text{and}$$
$$\bigcap\nolimits_{n<\omega}a_n=\emptyset.$$

Set $\Phi=\{\phi_{a_n}:n<\omega\}$; thus $\Phi\subset\Phi_1(L_{\{a_n:n<\omega\}})$, and Φ is finitely satisfiable in $\langle\mathfrak{P}(\kappa),a_n\rangle_{n<\omega}$. By the theorem of Łoś, Φ is finitely satisfiable in

$$\langle\mathfrak{P}(\kappa)^\alpha/p,\underline{a}_n/p\rangle_{n<\omega},$$

where $\underline{a}_{n,\xi}=a_n$ for $\xi<\alpha$, $n<\omega$. Since $\omega<\beta$, $\mathfrak{P}(\kappa)^\alpha/p$ is ω^+-saturated, and thus Φ is satisfiable in $\langle\mathfrak{P}(\kappa)^\alpha/p,\underline{a}_n/p\rangle_{n<\omega}$, *i.e.*, there is $b\in\mathscr{P}(\kappa)^\alpha$ such that

$$\langle\mathfrak{P}(\kappa)^\alpha/p,\underline{a}_n/p\rangle_{n<\omega}\models\phi_{a_n}[b/p]$$

for $n<\omega$. We define

$$A_n=\{\xi<\alpha:\langle\mathfrak{P}(\kappa),a_n\rangle\models\phi_{a_n}[b(\xi)]\}$$

for $n<\omega$. Then $A_n\in p$ for $n<\omega$ by the theorem of Łoś, but $A_n=\{\xi<\alpha:\emptyset\neq b(\xi)\subset a_n\}$ and hence $\bigcap_{n<\omega}A_n=\emptyset$; thus p is countably incomplete.

(ii) p is β-good.

Let $\gamma<\beta$ and let $g:\mathscr{P}_\omega(\gamma)\to p$ be a monotone function; we find a multiplicative function $h:\mathscr{P}_\omega(\gamma)\to p$ such that $h\subset g$. Using Lemma 9.7 we assume with no loss of generality that γ is infinite, and hence $|\mathscr{P}_\omega(\gamma)|=\gamma$. We choose a one-to-one onto function $t:\mathscr{P}_\omega(\gamma)\to\gamma$, and we define

$$e:\alpha\times\mathscr{P}_\omega(\gamma)\to\mathscr{P}(\gamma)$$

by the rule

$$e(\xi,F)=\{t(G):G\in\mathscr{P}_\omega(\gamma),F\subset G,\xi\in g(G)\}.$$

It is easy to verify that

$$e(\xi,F)=\emptyset\quad\text{if and only if }\xi\notin g(F),$$
$$e(\xi,F\cup G)=e(\xi,F)\cap e(\xi,G),\quad\text{and}$$
$$\{\xi<\alpha:t(F)\in e(\xi,F)\}=g(F)\in p$$

for $\xi<\alpha$ and $F,G\in\mathscr{P}_\omega(\gamma)$. We set

$$f_\zeta(\xi)=e(\xi,\{\zeta\})$$

for $\zeta<\gamma$ and $\xi<\alpha$; then $f_\zeta\in\mathscr{P}(\kappa)^\alpha$ for $\zeta<\gamma$. We set

$$\Phi=\{\phi_{f_\zeta}(v_0):\zeta<\gamma\},$$

so that $\Phi\subset\Phi_1((L_\tau)_{\{f_\zeta/p:\zeta<\gamma\}})$ and Φ is finitely satisfiable in $\langle\mathfrak{P}(\kappa)^\alpha/p, f_\zeta/p\rangle_{\zeta<\gamma}$. Since $\gamma<\beta$ and, by the assumptions of (c), $\mathfrak{P}(\kappa)^\alpha/p$ is β-saturated, it follows that Φ is satisfiable in $\langle\mathfrak{P}(\kappa)^\alpha/p, f_\zeta/p\rangle_{\zeta<\gamma}$ by some s/p, where $s\in\mathfrak{P}(\kappa)^\alpha$. We define

$$h:\mathscr{P}_\omega(\gamma)\to\mathscr{P}(\alpha)$$

by $h(F)=\{\xi<\alpha:\emptyset\neq s(\xi)\subset e(\xi,F)\}$. It is easy to verify that h is multiplicative and that $h\leq g$. Further if $\zeta<\gamma$, then $h(\{\zeta\})=\{\xi<\alpha:\emptyset\neq s(\xi)\subset e(\xi,\{\zeta\})\}\in p$, by the theorem of Łoś and the fact that $\langle\mathfrak{P}(\kappa)^\alpha/p, f_\zeta/p\rangle_{\zeta<\gamma}\models\phi_{f_\zeta}[s/p]$. Hence the values of h are elements of p, i.e.,

$$h[\mathscr{P}_\omega(\gamma)]\subset p.$$

The proof of the theorem is complete.

13.14. Theorem *(Keisler). Let α be an infinite cardinal.*

(a) *There are structures $\mathfrak{A}$ and $\mathfrak{B}$ of type $\tau\in\omega^{(\alpha^+)}$ such that*

$|A|=|B|=\alpha^+$,

$\mathfrak{A}\equiv\mathfrak{B}$, *and*

if p and q are ultrafilters on α then $\mathfrak{A}^\alpha/p$ and $\mathfrak{B}^\alpha/q$ are not isomorphic.

(b) *There is a structure $\mathfrak{A}$ of type $\tau\in\omega^{(2^\alpha)}$ such that*

$$|A|=\alpha,$$

and there are two countably incomplete α^+-good ultrafilters p and q on α such that

$$\mathfrak{A}^\alpha/p \text{ and } \mathfrak{A}^\alpha/q \text{ are not isomorphic.}$$

Proof. (a) We set

$$R_i=\{F\in\mathscr{P}_\omega(\alpha^+):i\in F\}$$

for $i<\alpha^+$, we let $\tau:\alpha^+\to\omega$ be the given by $\tau(i)=1$ for $i<\alpha^+$, and we define

$$\mathfrak{A}=\langle\mathscr{P}_\omega(\alpha^+),R_i\rangle_{i<\alpha^+},$$

a structure of type τ. Furthermore we set

$$\Phi=\{P_i(v_0):i<\alpha^+\},$$

so that $\Phi\subset\Phi_1(L_\tau)$ and Φ is finitely satisfiable in $\mathfrak{A}$. By the compactness theorem (11.7) and the downward Löwenheim-Skolem theorem (11.4), there is a structure $\mathfrak{B}$ such that

$$|\mathfrak{B}| = \alpha^+,$$
$$\mathfrak{B} \equiv \mathfrak{A}, \quad \text{and}$$
$$\Phi \text{ is satisfiable in } \mathfrak{B}.$$

We verify the following two statements.

(i) For any ultrafilter p on α, Φ is not satisfiable in $\mathfrak{A}^\alpha/p$.

Let p be any ultrafilter on α and let $f \in (\mathscr{P}_\omega(\alpha^+))^\alpha$. We note that the set

$$\{i < \alpha^+ : f(\xi) \in R_i\}$$

is finite for $\xi < \alpha$, and hence the set

$$S(f) = \{i < \alpha^+ : f(\xi) \in R_i \text{ for some } \xi < \alpha\}$$

has cardinality (at most) α and there is $i(f) \in \alpha^+ \setminus S(f)$; then $P_{i(f)}(v_0)$ is not satisfied in $\mathfrak{A}^\alpha/p$ by f/p.

(ii) For any ultrafilter q on α, Φ is satisfiable in $\mathfrak{B}^\alpha/q$.

This follows immediately from the theorem of Łoś.

Statement (a) now follows from (i) and (ii).

(b) Let $\langle R_i : i < 2^\alpha \rangle$ be a well-ordering of $\mathscr{P}(\alpha)$, let $\tau : 2^\alpha \to \omega$ be the type given by $\tau(i) = 1$ for $i < 2^\alpha$, and define

$$\mathfrak{A} = \langle \alpha, R_i \rangle_{i < 2^\alpha}.$$

Let p a countably incomplete, α^+-good ultrafilter on α. Then $|\mathfrak{A}^\alpha/p| \leq 2^\alpha$. (In fact $|\mathfrak{A}^\alpha/p| = 2^\alpha$ by Theorem 12.22, but this fact is not needed.) For $q \in \boldsymbol{\beta}(\alpha)$, set $\Phi_q = \{P_i(v_0) : R_i \in q\}$, and define

$$T(p) = \{q \in \boldsymbol{\beta}(\alpha) : q \text{ is countably incomplete and } \alpha^+\text{-good, and } \Phi_q \text{ is satisfiable in } \mathfrak{A}^\alpha/p\}.$$

We claim that $|T(p)| \leq 2^\alpha$. Indeed if $q \in T(p)$ there is $f_q \in \mathfrak{A}^\alpha$ such that Φ_q is satisfied by f_q/p, and the function $q \to f_q/p$ is one-to-one from $T(p)$ into $\mathfrak{A}^\alpha/p$; since $|\mathfrak{A}^\alpha/p| \leq 2^\alpha$, the claim follows. Since there are 2^{2^α} countably incomplete α^+-good ultrafilters on α (Theorem 10.6), there is a countably incomplete α^+-good ultrafilter on α such that $q \notin T(p)$. On the other hand, it is easy to see that

$$\mathfrak{A}^\alpha/q \models \phi[\mathrm{id}/q] \quad \text{for } \phi \in \Phi_q,$$

and thus $\mathfrak{A}^\alpha/p \not\equiv \mathfrak{A}^\alpha/q$.

Shelah's Characterization of Elementary Equivalence

Definition. Let α and κ be cardinals, $\mathscr{S}, \mathscr{H} \subset \kappa^\alpha$, and $\mathscr{F}$ a filter on α. Then $\mathscr{S}$ is a family of $\langle \kappa, \mathscr{H} \rangle$-*large oscillation* mod $\mathscr{F}$ if the following condition is satisfied:

if $\lambda<\kappa$, $\{f_i : i<\lambda\}$ are distinct elements of $\mathscr{S}$, and $\{\zeta_i : i<\lambda\}$ is a subset of κ,

if $n<\omega$, $\{g_k : k<n\}$ are distinct elements of $\mathscr{S}$ with $\{f_i : i<\lambda\} \cap \{g_k : k<n\}=\emptyset$, and $\{h_k : k<n\}$ is a subset of $\mathscr{H}$, and

if $A\in\mathscr{F}$,

then

$$\{\xi<\alpha : f_i(\xi)=\zeta_i \text{ for } i<\lambda\} \cap \{\xi<\alpha : g_k(\xi)=h_k(\xi) \text{ for } k<n\} \cap A \neq \emptyset .$$

13.15. Lemma. *Let α, β, κ and λ be cardinals such that $\lambda<\kappa\leq\beta$ and $\omega\leq\kappa$. Let $\mathscr{S}\subset\kappa^\alpha$, $\mathscr{H}\subset\lambda^\alpha$, and $\mathscr{F}$ a filter on α such that*

$\mathscr{F}$ is generated (as a filter) by at most β elements,

$|\mathscr{H}|\leq\beta$, and

$\mathscr{S}$ is of κ-large oscillation mod $\mathscr{F}$.

Then there is a family $\mathscr{S}'\subset\mathscr{S}$ such that

$|\mathscr{S}\setminus\mathscr{S}'|\leq\beta$, and

$\mathscr{S}'$ is of $\langle\kappa, \mathscr{H}\rangle$-large oscillation mod $\mathscr{F}$.

Proof. Let $\mathscr{F}$ be generated by the family $\{A_\rho : \rho<\beta\}$; since β is infinite we assume with no loss of generality that the family $\{A_\rho : \rho<\beta\}$ is closed under finite intersections. We consider two cases.

Case 1. $|\mathscr{H}|<\omega$. Since there is nothing to prove if $\mathscr{H}=\emptyset$, we assume $0<|\mathscr{H}|=n<\omega$, say $\mathscr{H}=\{h_0, \ldots, h_{n-1}\}$.

Let $\langle\theta_\eta : \eta<\lambda^n\rangle$ be a well-ordering of the set λ^n (of all ordered n-tuples of λ).

Suppose that there is no $\mathscr{S}'\subset\mathscr{S}$ satisfying the conclusion of the lemma. Then in particular $|\mathscr{S}|>\beta$. It is easy to define recursively the following for $\eta<\beta^+$:

a cardinal $\lambda_\eta<\kappa$,
a family $\{f_{\eta i} : i<\lambda_\eta\}$ of distinct elements of $\mathscr{S}$,
$\{\zeta_{\eta i} : i<\lambda_\eta\}\subset\kappa$,
a family $\{g_{\eta k} : k<n\}$ of distinct elements of $\mathscr{S}$, and
an ordinal $\rho_\eta<\beta$,

such that the family

$$\{f_{\eta i} : \eta<\beta^+, i<\lambda_\eta\} \cup \{g_{\eta k} : \eta<\beta^+, k<n\}$$

consists of distinct elements and

(*) $\{\xi<\alpha : f_{\eta i}(\xi)=\zeta_{\eta i} \text{ for } i<\lambda_\eta\} \cap \{\xi<\alpha : g_{\eta k}(\xi)=h_k(\xi) \text{ for } k<n\} \cap A_{\rho_\eta}=\emptyset$

for $\eta<\beta^+$.

We note that the function $\eta \to \langle \lambda_\eta, \rho_\eta \rangle$ is defined from β^+ into $\kappa \times \beta$ (a set of cardinality β). A simple pigeon-hole argument implies that there are a cardinal $\overline{\lambda} < \kappa$ and an ordinal $\overline{\rho} < \beta$ such that

$$|\{\eta < \beta^+ : \langle \lambda_\eta, \rho_\eta \rangle = \langle \overline{\lambda}, \overline{\rho} \rangle| = \beta^+ .$$

Thus, with no loss of generality, we assume that $\lambda_\eta = \overline{\lambda}$ and $\rho_\eta = \overline{\rho}$ for $\eta < \beta^+$. Then (*) becomes

(**) $\{\xi < \alpha : f_{\eta i}(\xi) = \zeta_{\eta i} \text{ for } i < \overline{\lambda}\} \cap \{\xi < \alpha : g_{\eta k}(\xi) = h_k(\xi) \text{ for } k < n\} \cap A_{\overline{\rho}} = \emptyset$

for $\eta < \beta^+$.

For the rest of the proof of part (a), we need these recursively defined objects only for ordinals $\eta < \lambda^n$. We define

$$B = \{\xi < \alpha : f_{\eta i}(\xi) = \zeta_{\eta i} \text{ for } i < \overline{\lambda} \text{ and } \eta < \lambda^n\}$$
$$\cap \{\xi < \alpha : g_{\eta k}(\xi) = \theta_{\eta k} \text{ for } k < n \text{ and } \eta < \lambda^n\} .$$

Since $\mathscr{S}$ is of κ-large oscillation, and $\overline{\lambda}, \lambda^n, n < \kappa$, and the family $\{f_{\eta i} : i < \overline{\lambda}, \eta < \lambda^n\} \cup \{g_{\eta k} : k < n, \eta < \lambda^n\}$ consists of distinct functions of $\mathscr{S}$, it follows that

$$B \cap A_{\overline{\rho}} \neq \emptyset .$$

On the other hand, let $\overline{\xi} \in B$; then $\overline{\xi} < \alpha$, and $\langle h_0(\overline{\xi}), \ldots, h_{n-1}(\overline{\xi}) \rangle \in \lambda^n$ and thus $\langle h_0(\overline{\xi}), \ldots, h_{n-1}(\overline{\xi}) \rangle = \theta_{\overline{\eta}}$ for some $\overline{\eta} < \lambda^n$. Thus $\overline{\xi}$ satisfies the relations

$$f_{\overline{\eta} i}(\overline{\xi}) = \zeta_{\overline{\eta} i} \quad \text{for } i < \overline{\lambda}, \quad \text{and}$$
$$g_{\overline{\eta} k}(\overline{\xi}) = \theta_{\overline{\eta} k} = h_k(\overline{\xi}) \quad \text{for } k < n.$$

Hence $\overline{\xi} \notin A_{\overline{\rho}}$ by (**), *i.e.*, $B \cap A_{\overline{\rho}} = \emptyset$. This contradiction completes the proof of *Case* 1.

Case 2. $\omega \leq |\mathscr{H}| \leq \beta$. For $\mathscr{G} \in \mathscr{P}_\omega(\mathscr{H})$ there is, by *Case* 1, $\mathscr{S}(\mathscr{G}) \subset \mathscr{S}$ such that

$$|\mathscr{S} \setminus \mathscr{S}(\mathscr{G})| \leq \beta, \quad \text{and}$$
$$\mathscr{S}(\mathscr{G}) \text{ is of } \langle \kappa, \mathscr{G} \rangle\text{-large oscillation } \bmod \mathscr{F} .$$

We set

$$\mathscr{S}' = \bigcap \{\mathscr{S}(\mathscr{G}) : \mathscr{G} \in \mathscr{P}_\omega(\mathscr{H})\} ;$$

it is clear that

$$|\mathscr{S} \setminus \mathscr{S}'| \leq \beta, \quad \text{and}$$
$$\mathscr{S}' \text{ is of } \langle \kappa, \mathscr{H} \rangle\text{-large oscillation } \bmod \mathscr{F} .$$

The proof of the lemma is complete.

We now introduce some notation that will be used only in the next lemma and in the proof of Shelah's theorem.

Let L be a language, $\mathfrak{A}=\langle A,\ldots\rangle$ a structure for L, and $\mathscr{F}$ a filter on a cardinal α. The reduced power is denoted, as usual, by $\mathfrak{A}^{\alpha}/\mathscr{F}$. For $C\subset A^{\alpha}$ we set

$$C/\mathscr{F}=\{h/\mathscr{F}:h\in C\},\quad\text{and}$$
$$C(\xi)=\{h(\xi):h\in C\}\quad\text{for }\xi<\alpha.$$

Recall that the symbol L_C denotes the language whose symbols are the symbols of L together with a constant symbol $\bar{c}$ for each $c\in C$; and $\Phi_1(L_C)$ denotes the set of formulas of L_C all of whose free variables are among v_0.

Let $\Phi\subset\Phi_1(L_C)$. We say that Φ is *$\mathscr{F}$-satisfiable (in $(\mathfrak{A}^{\alpha}/\mathscr{F})_{C/\mathscr{F}}$)* if there is $a\in A^{\alpha}$ such that

$$\{\xi<\alpha:\mathfrak{A}_{C(\xi)}\models\phi[a(\xi)]\}\in\mathscr{F}\quad\text{for }\phi\in\Phi;$$

and Φ is *finitely $\mathscr{F}$-satisfiable (in $(\mathfrak{A}^{\alpha}/\mathscr{F})_{C/\mathscr{F}}$)* if every finite subset of Φ is $\mathscr{F}$-satisfiable (in $(\mathfrak{A}^{\alpha}/\mathscr{F})_{C/\mathscr{F}}$).

The following lemma will be used in the recursion in Theorem 13.19, to insure β^{+}-saturation for all β such that $2^{\beta}\leq 2^{\alpha}$.

13.16. Lemma. *Let α,β,κ and λ be cardinals such that $\lambda<\kappa\leq\beta$ and $\omega\leq\kappa$. Let $\mathscr{S}\subset\kappa^{\alpha}$, $\mathscr{F}$ a filter on α, and $C\subset\lambda^{\alpha}$ such that*

$|\mathscr{S}|>\beta$,
$\mathscr{F}$ is generated (as a filter) by at most β elements, and
$\mathscr{S}$ is of κ-large oscillation mod $\mathscr{F}$.

Let $\mathfrak{A}=\langle\lambda,\ldots\rangle$ be a structure for a language L (whose universe is λ) and let $\Phi\subset\Phi_1(L_C)$ be such that

$|\Phi|\leq\beta$, and
Φ is finitely $\mathscr{F}$-satisfiable (in $(\mathfrak{A}^{\alpha}/\mathscr{F})_{C/\mathscr{F}}$).

Then there are a family $\mathscr{S}'\subset\mathscr{S}$ and a filter $\mathscr{F}'\supset\mathscr{F}$ such that

$|\mathscr{S}\setminus\mathscr{S}'|\leq\beta$,
$\mathscr{F}'$ is generated (as a filter) by at most β elements,
$\mathscr{S}'$ is of κ-large oscillation mod $\mathscr{F}'$, *and*
Φ is $\mathscr{F}'$-satisfiable (in $(\mathfrak{A}^{\alpha}/\mathscr{F}')_{C/\mathscr{F}'}$).

Proof. There is nothing to prove if Φ is finite, so we assume that Φ is infinite and, with no loss of generality, that Φ is closed under (finite) conjunctions.

For $\phi\in\Phi$ we define

$$A(\phi)=\{\xi<\alpha:\mathfrak{A}_{C(\xi)}\models\exists v_0\phi\}.$$

Then $A(\phi)\in\mathscr{F}$ because Φ is finitely $\mathscr{F}$-satisfiable in $(\mathfrak{A}^\alpha/\mathscr{F})_{C/\mathscr{F}}$. For $\phi\in\Phi$ we define a function

$$h_\phi:\alpha\to\lambda$$

by the rule:

$$\text{if } \xi\in A(\phi), \quad \text{then } h_\phi(\xi)\in\lambda \quad \text{and } \mathfrak{A}_{C(\xi)}\models\phi[h_\phi(\xi)];$$

and if $\xi\notin A(\phi)$, then $h_\phi(\xi)=0$. We set

$$\mathscr{H}=\{h_\phi:\phi\in\Phi\},$$

so that $\mathscr{H}\subset\lambda^\alpha$ and $|\mathscr{H}|\le\beta$.

By Lemma 13.15, there is $\mathscr{S}_0\subset\mathscr{S}$ such that

$$|\mathscr{S}\setminus\mathscr{S}_0|\le\beta, \quad \text{and}$$
$$\mathscr{S}_0 \text{ is of } \langle\kappa,\mathscr{H}\rangle\text{-large oscillation } \operatorname{mod}\mathscr{F}.$$

We note that $\mathscr{S}_0\neq\emptyset$, since $|\mathscr{S}|>\beta$ and $|\mathscr{S}\setminus\mathscr{S}_0|\le\beta$. We choose $g_0\in\mathscr{S}_0$, and we set

$$\mathscr{S}'=\mathscr{S}_0\setminus\{g_0\}.$$

Then $|\mathscr{S}\setminus\mathscr{S}'|\le\beta$, and $\mathscr{S}'\neq\emptyset$; we define a function

$$g_1:\alpha\to\lambda$$

by the rule

$$\begin{aligned} g_1(\xi)&=g_0(\xi) && \text{if } g_0(\xi)<\lambda\\ &=0 && \text{if } g_0(\xi)\ge\lambda, \end{aligned}$$

and we set

$$B(\phi)=\{\xi<\alpha:\mathfrak{A}_{C(\xi)}\models\phi[g_1(\xi)]\}.$$

We claim that

$$(*) \qquad A(\phi)\cap\{\xi<\alpha:g_0(\xi)=h_\phi(\xi)\}\subset B(\phi) \quad \text{for } \phi\in\Phi.$$

Indeed, let $\xi<\alpha$ be such that $\xi\in A(\phi)$ and $g_0(\xi)=h_\phi(\xi)$; then $g_0(\xi)<\lambda$ and hence $g_0(\xi)=g_1(\xi)$, so by the definition of h_ϕ,

$$A_{C(\xi)}\models\phi[g_1(\xi)]$$

and hence $\xi\in B(\phi)$; the claim is established. We define

$$\mathscr{F}'=\langle\mathscr{F}\cup\{B(\phi):\phi\in\Phi\}\rangle.$$

Then $\mathscr{F}'$ is generated by at most β elements, and Φ is $\mathscr{F}'$-satisfiable (in $(\mathfrak{A}^\alpha/\mathscr{F}')_{C/\mathscr{F}'}$) by $g_1/\mathscr{F}'$.

It remains to prove that

$$\mathscr{S}' \text{ is of } \kappa\text{-large oscillation } \operatorname{mod}\mathscr{F}'.$$

If this fails then there are a cardinal $\mu<\kappa$, a family $\{f_i: i<\mu\}$ of distinct elements of $\mathscr{S}'$, $\{\zeta_i: i<\mu\}\subset\kappa$, $A\in\mathscr{F}$, and (since Φ is closed under finite conjunctions) $\phi\in\Phi$ such that

$$(**)\qquad \{\xi<\alpha: f_i(\xi)=\zeta_i \text{ for } i<\mu\}\cap A\cap B(\phi)=\emptyset\,.$$

It follows from (*) and (**) that

$$\{\xi<\alpha: f_i(\xi)=\zeta_i \text{ for } i<\mu\}\cap\{\xi<\alpha: g_0(\xi)=h_\phi(\xi)\}\cap A\cap A(\phi)=\emptyset\,;$$

since $A(\phi)\in\mathscr{F}$ and $h_\phi\in\mathscr{H}$, this contradicts the fact that $\mathscr{S}_0$ is of $\langle\kappa,\mathscr{H}\rangle$-large oscillation $\operatorname{mod}\mathscr{F}$. The proof is complete.

The next lemma will be used for one step of the recursion in Theorem 13.19, to insure isomorphism between $\mathfrak{A}$ and $\mathfrak{B}$.

13.17. Lemma. *Let α, β and κ be infinite cardinals such that $\kappa\le\beta$. Let $\mathfrak{A}$ and $\mathfrak{B}$ be (similar) structures, of type $\tau_{\mathfrak{A}}:\rho_{\mathfrak{A}}\to\omega$, such that $|\rho_{\mathfrak{A}}|\le\alpha$, $\mathfrak{A}\equiv\mathfrak{B}$, and $|\mathfrak{A}|+|\mathfrak{B}|<\kappa$, and let $a\in A^\alpha$. Let $\mathscr{S}\subset\kappa^\alpha$, $\mathscr{F}$ a filter on α, $C\subset A^\alpha$, and let $h: C\to B^\alpha$ be such that*

$|C|\le\beta$;
$|\mathscr{S}|>\alpha+\beta$;
$\mathscr{F}$ is generated (as a filter) by at most $\alpha+\beta$ elements;
$\mathscr{S}$ is of κ-large oscillation $\operatorname{mod}\mathscr{F}$; *and*
if $\phi\in\Phi_0((L(\tau_{\mathfrak{A}}))_C$ then

(a) *either $\{\xi<\alpha:\mathfrak{A}_{C(\xi)}\models\phi\}\in\mathscr{F}$ or $\{\xi<\alpha:\mathfrak{A}_{C(\xi)}\models\neg\phi\}\in\mathscr{F}$, and*
(b) *$\{\xi<\alpha:\mathfrak{A}_{C(\xi)}\models\phi\}\in\mathscr{F}$ if and only if $\{\xi<\alpha:\mathfrak{B}_{(h[C])(\xi)}\models\phi\}\in\mathscr{F}$.*

Then there are a family $\mathscr{S}'\subset\mathscr{S}$, a filter $\mathscr{F}'\supset\mathscr{F}$, $C'\subset A^\alpha$, and $h': C'\to B^\alpha$, such that

$C'=C\cup\{a\}$, $\quad h\subset h'$;
$|\mathscr{S}\setminus\mathscr{S}'|\le\alpha+\beta$;
$\mathscr{F}'$ is generated (as a filter) by at most $\alpha+\beta$ elements;
$\mathscr{S}'$ is of κ-large oscillation $\operatorname{mod}\mathscr{F}'$; *and*
if $\phi\in\Phi_0((L(\tau_{\mathfrak{A}}))_{C'}$ then

(a) *either $\{\xi<\alpha:\mathfrak{A}_{C'(\xi)}\models\phi\}\in\mathscr{F}'$ or $\{\xi<\alpha:\mathfrak{A}_{C'(\xi)}\models\neg\phi\}\in\mathscr{F}'$, and*
(b) *$\{\xi<\alpha:\mathfrak{A}_{C'(\xi)}\models\phi\}\in\mathscr{F}'$ if and only if $\{\xi<\alpha:\mathfrak{B}_{(h'[C'])(\xi)}\models\phi\}\in\mathscr{F}'$.*

Proof. Since $|C|\le\beta$ and $|(L(\tau_{\mathfrak{A}}))_C|\le\alpha$, it follows that $|\Phi_1(L(\tau_{\mathfrak{A}})_C)|\le\alpha+\beta$. For $\phi\in\Phi_1(L(\tau_{\mathfrak{A}})_C)$, we set

$$A(\phi)=\{\xi<\alpha:\mathfrak{A}_{C(\xi)}\models\phi[a(\xi)]\}\,.$$

By a simple transfinite recursion, employing Lemma 10.1 at every stage, it follows that there are a family $\mathscr{S}_0 \subset \mathscr{S}$, and a filter $\mathscr{F}_0 \supset \mathscr{F}$, such that

$|\mathscr{S} \setminus \mathscr{S}_0| \leq \alpha + \beta$,
$\mathscr{F}_0$ is generated (as a filter) by at most $\alpha + \beta$ elements,
$\mathscr{S}_0$ is of κ-large oscillation $\bmod \mathscr{F}_0$, and
either $A(\phi) \in \mathscr{F}_0$ or $\alpha \setminus A(\phi) \in \mathscr{F}_0$ for $\phi \in \Phi_1(L(\tau_{\mathfrak{A}})_C)$.

Note that $|\mathscr{S}_0| > \alpha + \beta$. We set

$$\Phi = \{\phi \in \Phi_1(L(\tau_{\mathfrak{A}})_C) : A(\phi) \in \mathscr{F}_0\}.$$

Clearly $|\Phi| \leq \alpha + \beta$ and Φ is closed under finite conjunctions.

We claim that, for $\phi \in \Phi$,

$$\{\xi < \alpha : \mathfrak{A}_{C(\xi)} \models \exists v_0 \phi\} \in \mathscr{F}.$$

If the claim fails for some $\phi \in \Phi$, then since $\exists v_0 \phi \in \Phi_1(L(\tau_{\mathfrak{A}})_C)$ we have $\{\xi < \alpha : \mathfrak{A}_{C(\xi)} \models \neg v_0 \phi\} \in \mathscr{F} \subset \mathscr{F}_0$, and hence

$$A(\phi) \cap \{\xi < \alpha : \mathfrak{A}_{C(\xi)} \models \neg \exists v_0 \phi\} \in \mathscr{F}_0,$$

by sub-condition (a) of the assumptions. If ξ is an element of this last set, then $\mathfrak{A}_{C(\xi)} \models \phi[a(\xi)]$ and $\mathfrak{A}_{C(\xi)} \models \neg \exists v_0 \phi$, a contradiction.

It follows, by sub-condition (b) of the assumptions, that $\{\xi < \alpha : \mathfrak{B}_{(h[C])(\xi)} \models \exists v_0 \phi\} \in \mathscr{F} \subset \mathscr{F}_0$ for $\phi \in \Phi$, *i.e.*, since Φ is closed under finite conjunctions, that

$$\Phi \text{ is finitely } \mathscr{F}_0\text{-satisfiable in } (\mathfrak{B}^\alpha / \mathscr{F}_0)_{h[C]/\mathscr{F}_0}.$$

The conditions of Lemma 13.16 (with $\alpha, \alpha + \beta, \kappa, |\mathfrak{B}|, \mathscr{S}_0, \mathscr{F}_0, \mathfrak{B}, h[C]$, and Φ replacing $\alpha, \beta, \kappa, \lambda, \mathscr{S}, \mathscr{F}, \mathfrak{A}, C$ and Φ, respectively) are satisfied, and hence there are a family $\mathscr{S}' \subset \mathscr{S}_0$ and a filter $\mathscr{F}' \supset \mathscr{F}_0$ such that

$|\mathscr{S}_0 \setminus \mathscr{S}'| \leq \alpha + \beta$ (and hence $|\mathscr{S} \setminus \mathscr{S}'| \leq \alpha + \beta$),
$\mathscr{F}'$ is generated (as a filter) by at most $\alpha + \beta$ elements,
$\mathscr{S}'$ is of κ-large oscillation $\bmod \mathscr{F}'$, and
Φ is $\mathscr{F}'$-satisfiable (in $(\mathfrak{B}^\alpha / \mathscr{F}')_{h[C]/\mathscr{F}'}$).

By the last condition, there is $b \in B^\alpha$ such that

$$\{\xi < \alpha : \mathfrak{B}_{(h[C])(\xi)} \models \phi[b(\xi)]\} \in \mathscr{F}'.$$

We set $C' = C \cup \{a\}$, and we define

$$h' : C' \to B^\alpha$$

by $h'|C = h$ and $h(a) = b$. It is clear that these definitions satisfy the requirements of the lemma.

Let α be an infinite cardinal. We choose, and fix for the remainder of this section, a function (*i.e.*, a similarity type)

$$\tau_\alpha: \alpha \to \omega$$

such that $|\tau_\alpha^{-1}(\{n\})| = \alpha$ for $n < \omega$. If $\tau: \rho \to \omega$ is a similarity type, we say that τ is a *sub-type* of τ_α (denoted $\tau \preccurlyeq \tau_\alpha$) if there is a one-to-one, order-preserving function $e: \rho \to \alpha$ such that $\tau_\alpha \circ e = \tau$.

If $\tau: \rho \to \omega$ is a sub-type of τ_α then $\rho \le \alpha$; and for $\rho \le \alpha$ there are at most $\alpha^{|\rho|} \le \alpha^\alpha = 2^\alpha$ (one-to-one, order-preserving) functions $e: \rho \to \alpha$. Since every subset of α determines a sub-type of τ_α, it is clear that

$$\tau_\alpha \text{ has } 2^\alpha \text{ sub-types.}$$

We have the following simple reduction.

13.18. Lemma. *Let $\alpha^{\underline{\kappa}} = \alpha$ and $\omega \le \kappa$. For any structure $\mathfrak{A} = \langle A, R_i^{\mathfrak{A}} \rangle_{i < \rho_{\mathfrak{A}}}$ of type $\tau_{\mathfrak{A}}$, with $|\mathfrak{A}| < \kappa$, there is a structure $\mathfrak{A}^* = \langle \mathfrak{A}^*, R_j^{\mathfrak{A}^*} \rangle_{j < \rho_{\mathfrak{A}^*}}$ of type $\tau_{\mathfrak{A}}^*$ such that*

(i) $\tau_{\mathfrak{A}}^* \preccurlyeq \tau_\alpha$;

(ii) $A^* = A$;

(iii) *if p is an ultrafilter on α and β is a cardinal such that $\mathfrak{A}^{*\alpha}/p$ is β^+-saturated, then $\mathfrak{A}^\alpha/p$ is β^+-saturated; and*

(iv) *if $\mathfrak{B}$ is a structure with $|\mathfrak{B}| < \kappa$ and $\mathfrak{A} \equiv \mathfrak{B}$, and if p is an ultrafilter on α such that $\mathfrak{A}^{*\alpha}/p \cong \mathfrak{B}^{*\alpha}/p$, then $\mathfrak{A}^\alpha/p \cong \mathfrak{B}^\alpha/p$.*

Proof. Let $|\mathfrak{A}| = \lambda < \kappa$. The cardinality of the set of finitary relations on A together with the set of elements of A does not exceed $\lambda + \sum_{n<\omega} 2^{(\lambda^n)} \le \alpha^\lambda \le \alpha^{\underline{\kappa}} = \alpha$. We set

$$I_{\mathfrak{A}} = \{i < \rho_{\mathfrak{A}}: \langle \tau_{\mathfrak{A}}(i), R_i^{\mathfrak{A}} \rangle \neq \langle \tau_{\mathfrak{A}}(j), R_j^{\mathfrak{A}} \rangle \text{ for } j < i\}.$$

Thus $I_{\mathfrak{A}} \subset \rho_{\mathfrak{A}}$, and hence $I_{\mathfrak{A}}$ is well-ordered and $|I_{\mathfrak{A}}| \le \alpha \cdot \omega = \alpha$. Let $\rho_{\mathfrak{A}}^*$ be the unique ordinal which is order-isomorphic to $I_{\mathfrak{A}}$, and let $f_{\mathfrak{A}}: \rho_{\mathfrak{A}}^* \to I_{\mathfrak{A}}$ be the (unique) order-isomorphism. We note that $\rho_{\mathfrak{A}}^* \le \alpha$.

We define

$$e_{\mathfrak{A}}: I_{\mathfrak{A}} \to \alpha$$

recursively by the condition: $e_{\mathfrak{A}}(i)$ is the least $\xi < \alpha$ such that

$$\begin{aligned} &\tau_\alpha(\xi) = \tau_{\mathfrak{A}}(i) \quad \text{and} \\ &\xi > e_{\mathfrak{A}}(j) \quad \text{for } j < i \quad \text{and } j \in I_{\mathfrak{A}}. \end{aligned}$$

That there is such an ordinal $\xi < \alpha$ follows from the facts that $|\tau_\alpha^{-1}(\{n\})| = \alpha$ for $n < \omega$ and $|I_{\mathfrak{A}}| \le \alpha$. Clearly $e_{\mathfrak{A}}$ is one-to-one and order-preserving, and $\tau_\alpha \circ e_{\mathfrak{A}} = \tau_{\mathfrak{A}}$. We set

$$\tau_{\mathfrak{A}}^* = \tau_\alpha \circ e_\alpha \circ f_\alpha,$$

so that $\tau_{\mathfrak{A}}^*: \rho_{\mathfrak{A}}^* \to \omega$ is a similarity type. We define the structure

$$\mathfrak{A}^* = \langle A^*, R_j^{\mathfrak{A}^*} \rangle_{j < \rho_{\mathfrak{A}^*}}$$

by

$$A^* = A,$$
$$R_j^{\mathfrak{A}^*} = R_{f_{\mathfrak{A}}(j)}^{\mathfrak{A}} \quad \text{for } j < \rho_{\mathfrak{A}^*}.$$

A routine verification proves that, with the given definitions, the conclusions of the lemma are satisfied.

13.19. Theorem *(Shelah). Let $\alpha = \alpha^{\underline{\kappa}}$ and $\omega \leq \kappa$. Then there is a regular ultrafilter p on α such that*

(a) *if $\mathfrak{A}$ is a structure such that $|A| < \kappa$ and if $2^\beta \leq 2^\alpha$, then $\mathfrak{A}^\alpha/p$ is β^+-saturated; in particular, $\mathfrak{A}^\alpha/p$ is α^+-saturated; and,*

(b) *if $\mathfrak{A}$ and $\mathfrak{B}$ are two elementarily equivalent structures each of which is of cardinality less than κ, then $\mathfrak{A}^\alpha/p \cong \mathfrak{B}^\alpha/p$.*

Proof. Clearly it is sufficient to find a regular ultrafilter p on α that satisfies conditions (a) and (b) of the theorem for structures $\mathfrak{A}, \mathfrak{B}, \ldots$ whose universes are cardinals less than κ; and by Lemma 13.18 it is sufficient, in addition, to consider only structures $\mathfrak{A}$ and $\mathfrak{B}$ each of whose type is a sub-type of τ_α.

As remarked above, there are exactly 2^α sub-types of τ_α. For each type $\tau \in \omega^\rho$ with $\tau \preccurlyeq \tau_\alpha$ and $\lambda < \kappa$, the number of structures of type τ and cardinality λ is at most $\prod_{i<\rho}(\lambda + 2^{(\lambda^{\tau(i)})}) \leq (2^\lambda + \omega)^\alpha = 2^{\lambda \cdot \alpha} = 2^\alpha$. We denote by $\tau_{\mathfrak{A}}$ the type of a structure $\mathfrak{A}$. The set

$$\mathscr{A} = \{\mathfrak{A} : A \text{ is a cardinal less than } \kappa \text{ and } \tau_{\mathfrak{A}} \preccurlyeq \tau_\alpha\}$$

is then of cardinality 2^α. It follows that the set

$$\mathscr{B} = \{\langle \mathfrak{A}, \mathfrak{B} \rangle : \mathfrak{A}, \mathfrak{B} \in \mathscr{A} \text{ and } \mathfrak{A} \equiv \mathfrak{B}\}$$

is also of cardinality 2^α. Finally, the set

$$\mathscr{C} = \{\langle \mathfrak{A}, C, \Phi \rangle : \mathfrak{A} \in \mathscr{A}, C \in \mathscr{P}(A^\alpha), 2^{|C|} \leq 2^\alpha, \Phi \subset \Phi_1(L(\tau_{\mathfrak{A}})_C)\}$$

has cardinality 2^α, as is easy to see from the facts that $|A^\alpha| \leq 2^\alpha$ and

$$|\{C \in \mathscr{P}(2^\alpha) : 2^{|C|} \leq 2^\alpha\}| = 2^\alpha.$$

We set

$$\mathscr{D} = \mathscr{P}(\alpha) \cup \{\langle \mathfrak{A}, \mathfrak{B}, c \rangle : \langle \mathfrak{A}, \mathfrak{B} \rangle \in \mathscr{B}, c \in A^\alpha \cup B^\alpha\} \cup \mathscr{C},$$

and we have $|\mathscr{D}| = 2^\alpha$. Let $\langle d_\eta : \eta < 2^\alpha \rangle$ be a well-ordering of $\mathscr{D}$ such that every element of $\mathscr{C}$ is listed 2^α times.

By Theorem 3.16 there is a family $\mathscr{T} \subset \alpha^\alpha$ of functions of κ-large oscillation $(\mathrm{mod}\{\alpha\})$ such that $|\mathscr{T}| = 2^\alpha$. Let $\{B_\zeta : \zeta < \alpha\}$ be a family of subsets of α such that

$$\text{if } I \subset \alpha,\ |I| < \omega \quad \text{then } \textstyle\bigcap_{\zeta \in I} B_\zeta \neq \emptyset, \quad \text{and}$$
$$\text{if } I \subset \alpha,\ |I| \geq \omega \quad \text{then } \textstyle\bigcap_{\zeta \in I} B_\zeta = \emptyset .$$

We choose $g_0 \in \mathscr{T}$ and we set

$$A_\zeta = g_0^{-1}(B_\zeta) \quad \text{for } \zeta < \alpha .$$

It is clear that

$$\text{if } I \subset \alpha,\ |I| \geq \omega \quad \text{then } \textstyle\bigcap_{\zeta \in I} A_\zeta = \emptyset ,$$

and it is easy to verify that

$$\mathscr{T} \setminus \{g_0\} \quad \text{is of } \kappa\text{-large oscillation } \ \mathrm{mod}\langle\{A_\zeta : \zeta < \alpha\}\rangle .$$

We define for $f \in \mathscr{T} \setminus \{g_0\}$

$$f' : \alpha \to \kappa$$

by

$$\begin{aligned} f'(\xi) &= f(\xi) \quad && \text{if } f(\xi) < \kappa \\ &= 0 && \text{if } \kappa \leq f(\xi) < \alpha , \end{aligned}$$

and we set

$$\mathscr{S}_0 = \{f' : f \in \mathscr{T} \setminus \{g_0\}\}, \quad \text{and}$$
$$\mathscr{F}_0 = \langle\{A_\zeta : \zeta < \alpha\}\rangle .$$

It is clear that

$$|\mathscr{S}_0| = 2^\alpha, \quad \text{and}$$
$$\mathscr{S}_0 \text{ is of } \kappa\text{-large oscillation } \ \mathrm{mod}\, \mathscr{F}_0 .$$

We define families $\{\mathscr{S}_\eta : \eta < 2^\alpha\}$, filters $\{\mathscr{F}_\eta : \eta < 2^\alpha\}$, and functions $\{h_\eta^{\mathfrak{A},\mathfrak{B}} : \eta < 2^\alpha, \langle\mathfrak{A},\mathfrak{B}\rangle \in \mathscr{B}\}$, such that

(i) $\mathscr{S}_0$ and $\mathscr{F}_0$ are given as above, $h_0^{\mathfrak{A},\mathfrak{B}} = \emptyset$ for $\langle\mathfrak{A},\mathfrak{B}\rangle \in \mathscr{B}$;

(ii) $\mathscr{S}_\eta \subset \mathscr{S}$, $|\mathscr{S} \setminus \mathscr{S}_\eta| \leq \alpha + |\eta|$ for $\eta < 2^\alpha$;

(iii) $\mathscr{F}_\eta$ is a filter on α generated by at most $\alpha + |\eta|$ elements for $\eta < 2^\alpha$;

(iv) $h_\eta^{\mathfrak{A},\mathfrak{B}}$ is a function defined on a subset $\mathrm{dom}(h_\eta^{\mathfrak{A},\mathfrak{B}})$ of A^α, with $|\mathrm{dom}(h_\eta^{\mathfrak{A},\mathfrak{B}})| \leq |\eta|$, whose image $\mathrm{im}(h_\eta^{\mathfrak{A},\mathfrak{B}})$ is a subset of B^α for $\eta < 2^\alpha$, $\langle\mathfrak{A},\mathfrak{B}\rangle \in \mathscr{B}$;

(v) $\mathscr{S}_\zeta \subset \mathscr{S}_\eta$, $\mathscr{F}_\zeta \supset \mathscr{F}_\eta$, $h_\zeta^{\mathfrak{A},\mathfrak{B}} \supset h_\eta^{\mathfrak{A},\mathfrak{B}}$ for $\eta < \zeta < 2^\alpha$, $\langle\mathfrak{A},\mathfrak{B}\rangle \in \mathscr{B}$;

(vi) $\mathscr{S}_\eta = \bigcap_{\zeta < \eta} \mathscr{S}_\zeta$, $\mathscr{F}_\eta = \bigcup_{\zeta < \eta} \mathscr{F}_\zeta$, and $h_\eta^{\mathfrak{A},\mathfrak{B}} = \bigcup_{\zeta < \eta} h_\zeta^{\mathfrak{A},\mathfrak{B}}$ for non-zero limit ordinals $\eta < 2^\alpha$, $\langle\mathfrak{A},\mathfrak{B}\rangle \in \mathscr{B}$;

(vii) $\mathscr{S}_\eta$ is of κ-large oscillation $\mathrm{mod}\, \mathscr{F}_\eta$ for $\eta < 2^\alpha$;

(viii) if $d_\eta \in \mathscr{P}(\alpha)$, then either $d_\eta \in \mathscr{F}_{\eta+1}$ or $\alpha \setminus d_\eta \in \mathscr{F}_{\eta+1}$ for $\eta < 2^\alpha$;

(ix) if $d_\eta = \langle \mathfrak{A}, \mathfrak{B}, c\rangle$ with $\langle \mathfrak{A}, \mathfrak{B}\rangle \in \mathscr{B}$ and $c \in A^\alpha \cup B^\alpha$, then

(a) $\mathrm{dom}(h_{\eta+1}^{\mathfrak{A},\mathfrak{B}}) = \mathrm{dom}(h_\eta^{\mathfrak{A},\mathfrak{B}}) \cup \{c\}$ if $c \in A^\alpha$, and

$\mathrm{im}(h_{\eta+1}^{\mathfrak{A},\mathfrak{B}}) = \mathrm{im}(h_\eta^{\mathfrak{A},\mathfrak{B}}) \cup \{c\}$ if $c \in B^\alpha$;

(b) either $\{\xi < \alpha : \mathfrak{A}_{(\mathrm{dom}(h_{\eta+1}^{\mathfrak{A},\mathfrak{B}}))(\xi)} \models \phi\}$ or its complement is an element of $\mathscr{F}_{\eta+1}$; and

(c) $\{\xi < \alpha : \mathfrak{A}_{(\mathrm{dom}(h_{\eta+1}^{\mathfrak{A},\mathfrak{B}}))(\xi)} \models \phi\} \in \mathscr{F}_{\eta+1}$ if and only if $\{\xi < \alpha : \mathfrak{B}_{(\mathrm{im}(h_{\eta+1}^{\mathfrak{A},\mathfrak{B}}))(\xi)} \models \phi\} \in \mathscr{F}_{\eta+1}$ for $\eta < 2^\alpha$; and finally,

(x) if $d_\eta = \langle \mathfrak{A}, C, \Phi\rangle \in \mathscr{C}$ and

Φ is finitely $\mathscr{F}_\eta$-satisfiable (in $(\mathfrak{A}^\alpha/\mathscr{F}_\eta)_{C/\mathscr{F}_\eta}$), then

Φ is $\mathscr{F}_{\eta+1}$-satisfiable (in $(\mathfrak{A}^\alpha/\mathscr{F}_{\eta+1})_{C/\mathscr{F}_{\eta+1}}$) for $\eta < 2^\alpha$.

We define these objects by transfinite recursion. We define $\mathscr{S}_0, \mathscr{F}_0$ and $h_0^{\mathfrak{A},\mathfrak{B}} = \emptyset$ for $\langle \mathfrak{A}, \mathfrak{B}\rangle \in \mathscr{B}$ by condition (i). If η is a non-zero limit ordinal and $\eta < 2^\alpha$, we define $\mathscr{S}_\eta, \mathscr{F}_\eta$ and $h_\eta^{\mathfrak{A},\mathfrak{B}}$ for $\langle \mathfrak{A}, \mathfrak{B}\rangle$ by condition (vi). We now define $\mathscr{S}_{\eta+1}, \mathscr{F}_{\eta+1}$ and $h_{\eta+1}^{\mathfrak{A},\mathfrak{B}}$ for $\eta < 2^\alpha$, $\langle \mathfrak{A}, \mathfrak{B}\rangle \in \mathscr{B}$. There are five possibilities.

Case 1. $d_\eta \in \mathscr{P}(\alpha)$; we define $\mathscr{S}_{\eta+1}$ and $\mathscr{F}_{\eta+1}$ using Lemma 10.1 and the remark immediately following it, and we set $h_{\eta+1}^{\mathfrak{A},\mathfrak{B}} = h_\eta^{\mathfrak{A},\mathfrak{B}}$ for $\langle \mathfrak{A}, \mathfrak{B}\rangle \in \mathscr{B}$.

Case 2. $d_\eta = \langle \mathfrak{A}, \mathfrak{B}, a\rangle$, with $\langle \mathfrak{A}, \mathfrak{B}\rangle \in \mathscr{B}$ and $a \in A^\alpha$; we define $\mathscr{S}_{\eta+1}$, $\mathscr{F}_{\eta+1}$ and $h_{\eta+1}^{\mathfrak{A},\mathfrak{B}}$ using Lemma 13.17, and we set $h_{\eta+1}^{\mathfrak{A}',\mathfrak{B}'} = h_\eta^{\mathfrak{A}',\mathfrak{B}'}$ for $\langle \mathfrak{A}', \mathfrak{B}'\rangle \in \mathscr{B}$ and $\langle \mathfrak{A}', \mathfrak{B}'\rangle \neq \langle \mathfrak{A}, \mathfrak{B}\rangle$.

Case 3. $d_\eta = \langle \mathfrak{A}, \mathfrak{B}, b\rangle$, with $\langle \mathfrak{A}, \mathfrak{B}\rangle \in \mathscr{B}$ and $b \in B^\alpha$; we define $\mathscr{S}_{\eta+1}$, $\mathscr{F}_{\eta+1}$ and $h_{\eta+1}^{\mathfrak{A},\mathfrak{B}}$ using Lemma 13.17, with the obvious modification and the interchange of $\mathfrak{A}$ and $\mathfrak{B}$, and we set $h_{\eta+1}^{\mathfrak{A}',\mathfrak{B}'} = h_\eta^{\mathfrak{A}',\mathfrak{B}'}$ for $\langle \mathfrak{A}', \mathfrak{B}'\rangle \in \mathscr{B}$ and $\langle \mathfrak{A}', \mathfrak{B}'\rangle \neq \langle \mathfrak{A}, \mathfrak{B}\rangle$.

Case 4. $d_\eta = \langle \mathfrak{A}, C, \Phi\rangle \in \mathscr{C}$ and Φ is not finitely $\mathscr{F}_\eta$-satisfiable (in $(\mathfrak{A}^\alpha/\mathscr{F}_\eta)_{C/\mathscr{F}_\eta}$); we set $\mathscr{S}_{\eta+1} = \mathscr{S}_\eta$, $\mathscr{F}_{\eta+1} = \mathscr{F}_\eta$ and $h_{\eta+1}^{\mathfrak{A}',\mathfrak{B}'} = h_\eta^{\mathfrak{A}',\mathfrak{B}'}$ for $\langle \mathfrak{A}', \mathfrak{B}'\rangle \in \mathscr{B}$.

Case 5. $d_\eta = \langle \mathfrak{A}, C, \Phi\rangle \in \mathscr{C}$ and Φ is finitely $\mathscr{F}_\eta$-satisfiable (in $(\mathfrak{A}^\alpha/\mathscr{F}_\eta)_{C/\mathscr{F}_\eta}$); we define $\mathscr{S}_{\eta+1}$ and $\mathscr{F}_{\eta+1}$ using Lemma 13.16, and we set $h_{\eta+1}^{\mathfrak{A}',\mathfrak{B}'} = h_\eta^{\mathfrak{A}',\mathfrak{B}'}$ for $\langle \mathfrak{A}', \mathfrak{B}'\rangle \in \mathscr{B}$.

This completes the recursive definition, as required. We set

$$p = \bigcup\nolimits_{\eta < 2^\alpha} \mathscr{F}_\eta, \quad \text{and}$$
$$h^{\mathfrak{A},\mathfrak{B}} = \bigcup\nolimits_{\eta < 2^\alpha} h_\eta^{\mathfrak{A},\mathfrak{B}} \quad \text{for } \langle \mathfrak{A}, \mathfrak{B}\rangle \in \mathscr{B}.$$

By conditions (i), (iii), (v) and (viii), p is a regular ultrafilter on α; by (iv), (v) and (ix)(a), the domain of $h^{\mathfrak{A},\mathfrak{B}}$ is A^α and the image of $h^{\mathfrak{A},\mathfrak{B}}$ is B^α for $\langle \mathfrak{A}, \mathfrak{B}\rangle \in \mathscr{B}$; by (ix)(a) and the theorem of Łoś

$$(\mathfrak{A}^\alpha/p)_{A^\alpha/p} \models \phi \text{ if and only if } (\mathfrak{B}^\alpha/p)_{(h^{\mathfrak{A},\mathfrak{B}}[A^\alpha])/p} \models \phi$$

for $\langle \mathfrak{A}, \mathfrak{B} \rangle \in \mathscr{B}$ and $\phi \in \Phi_0(L(\tau_{\mathfrak{A}})_{A^\alpha})$. Thus $h^{\mathfrak{A},\mathfrak{B}}$ is an isomorphism and part (b) of the theorem is proved. Finally, from (x) and the facts that if $C \in \mathscr{P}(2^\alpha)$ and $2^{|C|} \leq 2^\alpha$ then $|C| < \mathrm{cf}(2^\alpha)$ (Corollary 1.21) and that every element of $\mathscr{C}$ has been listed 2^α times in the well-ordering of $\mathscr{D}$, it follows that $\mathfrak{A}^\alpha/p$ is β^+-saturated for $\mathfrak{A} \in \mathscr{A}$, for all β such that $2^\beta \leq 2^\alpha$.

The proof of the theorem is complete.

13.20. *Remark.* If $\alpha^\omega = \alpha$ then the ultrafilter p on α whose existence is proved for the pair $\alpha, \kappa = \omega^+$ in Theorem 13.19 is in fact countably incomplete and α^+-good (so that a part of Theorem 10.4 is proved here independently). Indeed p is countably incomplete since p is regular. Furthermore, according to Theorem 13.13(a)⇔(c), to prove that p is α^+-good it is enough to prove that the ultrapower

$$\mathfrak{P}(\alpha)^\alpha/p \ \text{ is } \alpha^+\text{-saturated.}$$

Since the cardinality of the language for the structure $\mathfrak{P}(\alpha)$ is countable, it follows from the Löwenheim-Skolem theorem (11.4) that there is a structure $\mathfrak{B}$ such that

$$\mathfrak{B} \equiv \mathfrak{P}(\alpha), \quad \text{and}$$
$$|\mathfrak{B}| = \omega\,.$$

According to Corollary 13.8 (with $\gamma = \alpha$, $\mathfrak{A}_\xi = \mathfrak{B}$, $\mathfrak{B}_\xi = \mathfrak{P}(\alpha)$ for $\xi < \alpha$) it is enough to prove that the ultrapower

$$\mathfrak{B}^\alpha/p \ \text{ is } \alpha^+\text{-saturated,}$$

a fact that follows from Theorem 13.19(a) (with $\kappa = \omega^+$).

13.21. Corollary. *Let $\mathfrak{A}$ and $\mathfrak{B}$ be two structures. Then $\mathfrak{A} \equiv \mathfrak{B}$ if and only if there is an ultrapower of $\mathfrak{A}$ isomorphic to an ultrapower of $\mathfrak{B}$. Furthermore, if $\mathfrak{A} \equiv \mathfrak{B}$ and $\lambda = |\mathfrak{A}| + |\mathfrak{B}| + \omega$, then there is an ultrafilter p on 2^λ such that*

$$\mathfrak{A}^{(2^\lambda)}/p \cong \mathfrak{B}^{(2^\lambda)}/p\,.$$

Proof. If there is an ultrapower of $\mathfrak{A}$ isomorphic to an ultrapower of $\mathfrak{B}$, then $\mathfrak{A} \equiv \mathfrak{B}$ by the theorem of Łoś. Now let $\mathfrak{A} \equiv \mathfrak{B}$ and $\lambda = |\mathfrak{A}| + |\mathfrak{B}| + \omega$. Then

$$(2^\lambda)^{\underline{\lambda^+}} = (2^\lambda)^\lambda = 2^\lambda$$

and thus the pair $\langle 2^\lambda, \lambda^+ \rangle$ satisfies the conditions on the pair $\langle \alpha, \kappa \rangle$ in Shelah's theorem. The corollary now follows from part (a) of Shelah's theorem.

Characterization of the Rudin-Keisler Order

Definition. Let p be an ultrafilter on α, q an ultrafilter on β, and let $e:\alpha\to\beta$ be such that $\bar{e}(p)=q$. For a structure $\mathfrak{A}$ we define

$$e_*:\mathfrak{A}^\beta/q\to\mathfrak{A}^\alpha/p$$

by $e_*(f/q)=f\circ e/p$.

13.22. Lemma. *With the notation of the above definition, e_* is a well-defined elementary embedding.*

Proof. To prove that e_* is well-defined let $f,f'\in\mathfrak{A}^\beta$ and $f/q=f'/q$. Then $\{\eta<\beta:f(\eta)=f'(\eta)\}\in q$, and hence

$$\{\zeta<\alpha:f\circ e(\zeta)=f'\circ e(\zeta)\}=e^{-1}(\{\eta<\beta:f(\eta)=f'(\eta)\})\in p,$$

since $\bar{e}(p)=q$; i.e., $f\circ e/p=f'\circ e/p$.

We next verify that e_* is an elementary embedding. Let $n<\omega$, $\phi=\phi(v_1,\dots,v_n)\in\Phi_n(L_{\mathfrak{A}})$, and let $g_1,\dots,g_n\in\mathfrak{A}^\beta$. Each of the next four statements is clearly equivalent to its successor.

$$\mathfrak{A}^\beta/q\models\phi(f_1/p,\dots,f_n/p);$$
$$\{\eta<\beta:\mathfrak{A}\models\phi(f_1(\eta),\dots,f_n(\eta))\}\in q=\bar{e}(p);$$
$$\{\xi<\alpha:\mathfrak{A}\models\phi(f_1\circ e(\xi),\dots,f_n\circ e(\xi))\}$$
$$=e^{-1}(\{\eta<\beta:\mathfrak{A}\models\phi(f_1(\eta),\dots,f_n(\eta))\})\in p;$$
$$\mathfrak{A}^\alpha/p\models\phi(f_1\circ e/p,\dots,f_n\circ e/p);\quad\text{and}$$
$$\mathfrak{A}^\alpha/p\models\phi(e_*(f_1/q),\dots,e_*(f_n/q)).$$

(Here the first and third equivalences use the theorem of Łoś.)

The proof of the lemma follows.

Definition. For a cardinal α we set

$$I=\alpha\cup\bigcup_{1\le n<\omega}\mathscr{P}(\alpha^n)$$

and we define the *complete structure on* α to be the structure

$$\mathfrak{C}(\alpha)=\langle\alpha,A\rangle_{A\in I}.$$

13.23. Lemma. *Let α be a cardinal, p and q ultrafilters on α and let*

$$f:\mathfrak{C}(\alpha)^\alpha/q\to\mathfrak{C}(\alpha)^\alpha/p$$

be an elementary embedding. Then there is $e\in\alpha^\alpha$ such that

$$\bar{e}(p)=q,\quad\textit{and}$$
$$e_*=f.$$

Furthermore if $e'\in\alpha^\alpha$ is such that $\bar{e}'(p)=q$ and $e'_*=f$, then $e\underset{p}{\equiv}e'$.

Proof. We denote by id the identity function on α. Thus $\mathrm{id}/q \in \mathfrak{C}(\alpha)^\alpha/q$ and $f(\mathrm{id}/q) \in \mathfrak{C}(\alpha)^\alpha/p$, so there is $e \in \alpha^\alpha$ such that $f(\mathrm{id}/q) = e/p$.

We first verify that $\bar{e}(p) = q$. For $A \in \mathscr{P}(\alpha)$ let R_A be the unique (unary) predicate of $L_{\mathfrak{C}(\alpha)}$ such that

$$R_A^{\mathfrak{C}(\alpha)} = A .$$

Now let $A \in q$; then $\{\xi < \alpha : \mathfrak{C}(\alpha) \models R_A[\xi]\} = A \in q$, and it follows from the theorem of Łoś that

$$\mathfrak{C}(\alpha)^\alpha/q \models R_A[\mathrm{id}/q] .$$

Since f is an elementary embedding and $f(\mathrm{id}/q) = e/p$ we have that

$$\mathfrak{C}(\alpha)^\alpha/p \models R_A[e/p] ,$$

and from the theorem of Łoś we have that

$$\{\eta < \alpha : \mathfrak{C}(\alpha) \models R_A[e(\eta)]\} = e^{-1}(A) \in p .$$

It follows that $\bar{e}(p) = q$.

We next prove that $e_* = f$. For $g \in \alpha^\alpha$ we are to prove that

$$e_*(g/q) = g \circ e/p = f(g/q) .$$

Since $g \subset \alpha^2$ there is a unique (binary) predicate R_g of $L_{\mathfrak{C}(\alpha)}$ such that

$$R_g^{\mathfrak{C}(\alpha)} = g .$$

Then $\{\xi < \alpha : \mathfrak{C}(\alpha) \models R_g[\mathrm{id}(\xi), g(\xi)]\} = \alpha \in q$, and it follows from the theorem of Łoś that

$$\mathfrak{C}(\alpha)^\alpha/q \models R_g[\mathrm{id}/q, g/q] .$$

Since f is an elementary embedding and $f(\mathrm{id}/q) = e/p$ we have that

$$\mathfrak{C}(\alpha)^\alpha/p \models R_g[e/p, f(g/q)] ,$$

and from the theorem of Łoś we have that

$$\{\xi < \alpha : \mathfrak{C}(\alpha) \models R_g[e(\xi), f(g/q)(\xi)]\} \in p .$$

Since $R_g^{\mathfrak{C}(\alpha)} = g$, it follows that

$$g \circ e \underset{p}{\equiv} f(g/q) ,$$

and hence that

$$f(g/q) = g \circ e/p = e_*(g/q) .$$

For the final statement of the lemma we note that $e/p = f(\mathrm{id}/q) = e'_*(\mathrm{id}/q) = e'/p$, *i.e.*, that $e \underset{p}{\equiv} e'$.

13.24. Theorem *(Blass). Let α be a cardinal and let p and q be ultrafilters on α. The following statements are equivalent.*

(a) $q \preccurlyeq p$.

(b) *For every structure $\mathfrak{A}$ there is an elementary embedding of $\mathfrak{A}^\alpha/q$ into $\mathfrak{A}^\alpha/p$.*

(c) *There is an elementary embedding of $\mathfrak{C}(\alpha)^\alpha/q$ into $\mathfrak{C}(\alpha)^\alpha/p$.*

Proof. (a)$\Rightarrow$(b). There is $e \in \alpha^\alpha$ such that $\bar{e}(p) = q$, and, by Lemma 13.22, $e_*: \mathfrak{A}^\alpha/q \to \mathfrak{A}^\alpha/p$ is an elementary embedding.

(b)$\Rightarrow$(c) is obvious.

(c)$\Rightarrow$(a). Let $f: \mathfrak{C}(\alpha)^\alpha/q \to \mathfrak{C}(\alpha)^\alpha/p$ be an elementary embedding. By Lemma 13.23 there is $e \in \alpha^\alpha$ such that $\bar{e}(p) = q$ and $e_* = f$. Thus $q \preccurlyeq p$.

Notes for § 13

The "infinitärcalcül" of du Bois-Reymond is treated by Hardy [10], who traces the historical development of the subject (including the contributions of du Bois-Reymond [77] (§ 6), Pincherle, and Hadamard [94] (§ 19)).

A variant of Lemma 13.1 is proved in Chang-Keisler [66] (Theorem 5.1.10 and Lemma 6.1.8); the result of Gillman-Jerison [60] (Theorem 12.7) is a forerunner. Theorem 13.2 and Corollary 13.3 are similar to Theorems 6.1.9 and 6.2.9 of Chang-Keisler [66]. Corollary 13.4 is essentially due to Keisler [63] (Theorem 3.10) (cf. Keisler [60] (Theorem 3)). The properties of limit ultraproducts are further examined in Keisler [65 d].

Lemma 13.5 is mentioned in Keisler [67 a] (before Theorem 1.2), and Theorem 13.6 is proved in Keisler [67 a] (Theorem 1.5). Theorem 13.7 and its Corollary 13.8 are proved in Keisler [67 a] (Theorem 2.1, Corollary 2.1 a).

Theorem 13.9 and Corollaries 13.10—13.12 are contained in Keisler [64 c] (Theorem 2.1 and Corollary 2.3), except that in 13.10 we have used Kunen's Theorem (10.4) and thus avoided the generalized continuum hypothesis. A part of Corollary 13.12 (namely the equivalence (a)$\Leftrightarrow$(b)) was proved earlier in Keisler [61] (Theorem 2.3).

Theorem 13.13 is due to Keisler [64 c] (Theorem 3.1). For related problems and results the reader is referred to Benda [72] and Shelah [72]. Theorem 13.14 (a) is Keisler [64 c] (Example 4.4) and Theorem 13.14 (b) is (a generalization of) Keisler [64 c] (Examples 4.5, 4.6).

The degree of saturation of ultraproducts modulo ultrafilters on ω is examined under Martin's axiom in Ellentuck-Rucker [72].

The results contained in 13.15—13.21 are due to Shelah [71]. Some remarks (especially Remark 13.20) are the result of some correspondence between S. Shelah and the authors (in letters of February, 1972 and January, 1973).

Prior to Shelah's results (13.15—13.21) there were attempts by various authors to eliminate GCH from Corollaries 13.10—13.12 (or from related model-theoretic results). Among these we note the following: the method of ultralimits in Kochen [61]; the method of limit ultrapowers (considered briefly in 13.1—13.4) of Keisler [63], [65c], [65d]; Boolean ultrapowers in Mansfield [71], [72]; the results of Galvin [67], [70]; and the adequate ultrafilters on special Boolean algebras of Negrepontis [70], [72a].

The ultraproduct characterization of the Rudin-Keisler order (13.22—13.24) is contained in the thesis of Blass [70] (§ 11).

The saturation of ultraproducts is a deep and far-reaching method in algebraic number theory; cf. the work of Ax-Kochen [65a], [65b], [66], Ax [67], [68] and Eršov [65].

§ 14. Topology of Spaces of Ultrafilters

The object of this section is the discussion and organization of the properties of the topology of certain subspaces of the Stone-Čech compactification of a cardinal. Stronger topologies on the space of uniform ultrafilters and their applications will be studied in § 15, while the characterization of the Gleason space of the space of uniform ultrafilters is already given in § 12.

A problem studied in detail (and in a sense unifying the contents of this section) concerns the C^*-embedding of open subsets of $U_\kappa(\alpha)$, with $\omega \leq \kappa \leq \alpha$. The results depend on the *type* of an open set G, *i.e.*, on the least cardinality of a family of open-and-closed subsets of $U_\kappa(\alpha)$ whose union is equal to G. Essentially these are consequences of certain special results about F_α-spaces which are given in 14.1—14.4. The more complete results are for the space $U(\alpha)$ of uniform ultrafilters on α when α is a regular cardinal, and they assume the segment $\alpha^+ = 2^\alpha$ of the generalized continuum hypothesis (Theorem 14.10).

The results about F_α-spaces just mentioned apply to the Stone space of the α-homogeneous-universal Boolean algebras (Theorem 14.6) as well. In general, assuming that $\alpha^+ = 2^\alpha$, there are some similarities (together with significant differences) between the space $U(\alpha)$ and the space S_{α^+}. We obtain as much information as is possible concerning the comparison of these spaces (14.12—14.22). The most satisfactory situation, reflecting the unusual saturating power of the Fréchet filter $\mathscr{P}^\omega(\omega)$ (see Notes), occurs when $\alpha = \omega$. In this case the Boolean algebra $\mathfrak{B}(\boldsymbol{\beta}(\omega)\setminus\omega)$ is ω^+-homogeneous-universal (Theorem 14.12), so if the continuum hypothesis $\omega^+ = 2^\omega$ is assumed then $\mathfrak{B}(\boldsymbol{\beta}(\omega)\setminus\omega)$ is isomorphic to $\mathfrak{C}_{\omega^+}$. A remarkable theorem of Hausdorff (Theorem 14.14) implies that if $\alpha > \omega^+$ then $\mathfrak{B}(\boldsymbol{\beta}(\alpha)\setminus\alpha)$ is not α-homogeneous-universal. This result can be analysed more closely with the following remarks: As we have seen in § 6, the α-homogeneous-universal property for Boolean algebras is closely related to condition (H_α); furthermore, condition (H_α) for $\mathfrak{B}(\boldsymbol{\beta}(\alpha)\setminus\alpha)$ is equivalent to the conjunction of the two following conditions: (a) S_α is an F_α-space (a separation type of condition); (b) the non-empty intersection of less than α open-and-closed subsets of S_α

has non-empty interior (a Baire category type of condition). It is condition (a) that fails, according to Hausdorff's theorem, for $\alpha > \omega^+$; on the other hand condition (b) holds for all cardinals $\alpha < 2^\omega$ in certain models of set theory (specifically, when Martin's axiom holds). For uncountable cardinals α the spaces $U(\alpha)$ and S_{α^+} are not homeomorphic, and in fact $\mathfrak{B}(U(\alpha))$ is not ω^{++}-homogeneous (nor even ω^+-homogeneous if $\mathrm{cf}(\alpha) > \omega$) (Theorem 14.19). According to the remarkable Theorem 14.22, however, if $\alpha^+ = 2^\alpha$ then there is an embedding of S_{α^+} into $U(\alpha)$; the proof uses the existence of countably incomplete, α^+-good ultrafilters on α, and the topological properties of S_{α^+} (as given in § 6).

Some of the results on the topology of $U_\kappa(\alpha)$ are applied to the space $SU(\alpha^+)$ of sub-uniform ultrafilters (*i.e.*, those which are α-uniform but not α^+-uniform) on α^+ when α is a regular cardinal. The main object is to find, for cardinals $\alpha = \alpha^{\underline{\alpha}}$, a topological property that distinguishes the space $\boldsymbol{\beta}(SU(\alpha^+)) \backslash SU(\alpha^+))$ from the space $U(\alpha^+)$; property Φ_α, defined earlier in § 10, fills this rôle (Theorem 14.27), and shows that the "infinitary" properties of $SU(\alpha^+)$ are more closely related to those of $U(\alpha)$ than to those of $U(\alpha^+)$. The consequence of this topological separation is that $\boldsymbol{\beta}(SU(\alpha^+)) \backslash SU(\alpha^+)$ is not homeomorphic to $U(\alpha^+)$, and in particular that $SU(\alpha^+)$ is not C^*-embedded in $U_\alpha(\alpha^+)$. As a by-product it is proved that if $\omega^+ = 2^\omega$ and $p \in \boldsymbol{\beta}(\omega) \backslash \omega$ then $\boldsymbol{\beta}(\omega) \backslash \omega \backslash \{p\}$ is not a normal space (Theorem 14.31). The results on sub-uniform ultrafilters are related to, and imply easily, the Aronszajn-Specker theorem: if $\alpha = \alpha^{\underline{\alpha}}$ then α^+ is not a ramifiable cardinal (Corollary 14.32). It seems possible that similar questions on the C^*-embedding of certain spaces of ultrafilters depend on (and imply) results in the partition calculus. A particularly simple such occurrence is given for weakly compact cardinals (Theorem 14.33). The similarities and comparison between $U(\alpha)$ and S_{α^+} are not exhausted by the results just described. The α-Baire category property of S_α (Corollary 6.16) has an analogue in Corollary 9.15(c) for the space $U(\alpha)$; the existence of P_α-points of S_α (Theorem 6.17) corresponds to the existence of P_α-points of $U(\alpha)$ (Corollary 9.15(d)); the existence of good ultrafilters on α (Theorem 9.13) has an analogue for Boolean algebras associated with the homogeneous-universal and special Boolean algebras (Theorem 10.19). Further similarities are implied by results in § 15.

In the last part of this section we examine some simple consequences of the results on the C^*-embedding of open subsets of $U_\kappa(\alpha)$ as they apply when we deal with measurable cardinals (14.34—14.40).

Certain Properties of F_α-Spaces

We recall that the type of an open set G (denoted $\tau(G)$) in a space X such that $\mathfrak{B}(X)$ is a base is the least cardinal β such that G is equal to the union of β open-and-closed subsets of G.

Definition. Let $\alpha \geq \omega$. A space X such that $\mathfrak{B}(X)$ is a base is an *F_α-space* if every open subset of X of type less then α is C^*-embedded.

We note that if G is an open subset of an F_α-space X with $\tau(G) < \alpha$, then G is itself an F_α-space. Indeed if H is open in G and the type of H (relative to G) is less than α then $\tau(H) < \alpha$ and hence H is C^*-embedded.

A space X is an *F-space* if every cozero set of X is C^*-embedded.

It is clear that a compact totally disconnected space is an F-space if and only if it is an F_{ω^+}-space.

In 14.1—14.4 below we prove some general results about F_α-spaces which provide information on the C^*-embedding of open subsets of type at most α.

14.1. Lemma. *Let α be an infinite, regular cardinal, let X be an F_α-space, and let G be the union of a family $\{C_\xi : \xi < \alpha\}$ of open-and-closed subsets of X. Then*

(a) *G is an F_α-space;*

(b) *if H is an open subset of G and $H \cap C_\xi$ is of type less than α in G for $\xi < \alpha$, then H is C^*-embedded in G; and*

(c) *if X is compact and zero-dimensional, then G is zero-dimensional.*

Proof. (b) We set $G_\xi = \bigcup_{\eta \leq \xi} C_\eta$ for $\xi < \alpha$. We note that for $\xi < \alpha$ the set G_ξ, because it is an open subset of X of type less than α, is an F_α-space. Let $h \in C^*(H)$ and set $h_\xi = h | H \cap G_\xi$ for $\xi < \alpha$. Let $\xi < \alpha$ and suppose that h_ζ has been extended to $g_\zeta \in C^*(G_\zeta)$ such that $\sup\{|g_\zeta(x)| : x \in G_\zeta\} \leq \sup\{|h(x)| : x \in H\}$ for $\zeta < \xi$, and that $g_\zeta \subset g_{\zeta'}$ for $\zeta < \zeta' < \xi$. The function $h_\xi \cup \bigcup_{\zeta < \xi} g_\zeta$ is well-defined and continuous on the set $(H \cap G_\xi) \cup \bigcup_{\zeta < \xi} G_\zeta$, an open set of type less than α in the F_α-space G_ξ; hence this function has an extension to a function $g_\xi \in C^*(G_\xi)$ such that

$$\sup\{|g_\xi(x)| : x \in G_\xi\} \leq \sup\{|h(x)| : x \in H\}.$$

Then $\bigcup_{\xi < \alpha} g_\xi$ is a continuous extension of h defined on the space G.

(a) if H is an open subset of G of type less than α, then, by part (b), H is C^*-embedded in G.

(c) We prove that if X is compact and zero-dimensional and Z_0 and Z_1 are disjoint zero-sets of G then there is a $\{0,1\}$-valued continuous function on G, which is equal to 0 on Z_0 and is equal to 1 on Z_1. We note that there are open sets A_0 and A_1 of G such that

$$\tau(A_0 \cap G_\xi) < \alpha, \quad \tau(A_1 \cap G_\xi) < \alpha \quad \text{for } \xi < \alpha,$$
$$Z_0 \subset A_0, \quad Z_1 \subset A_1, \quad \text{and}$$
$$A_0 \cap A_1 = \emptyset.$$

(Indeed, we take for A_0 and A_1 any disjoint cozero sets of G containing Z_0 and Z_1, respectively.) Now let $g \in C^*(A_0 \cup A_1)$ be the function which is equal to 0 on A_0 and is equal to 1 on A_1. By part (a) the set G_ξ is C^*-embedded in X, and hence G_ξ is zero-dimensional for $\xi < \alpha$. In the proof of (b), with $H = A_0 \cup A_1$, we add to the recursive assumptions that $\bigcup_{\zeta<\xi} g_\zeta$ is $\{0,1\}$-valued; and g_ξ is taken to be $\{0,1\}$-valued, as well. This completes the proof of the lemma.

Definition. Let $\alpha \geq \omega$, let X be a space such that $\mathfrak{B}(X)$ is a base, let G be a subset of X and let $p \in \overline{G} \setminus G$. Then *condition* $[p, G, \alpha]$ holds if there are an open-and-closed set V in X with $p \in V$, and an open set H such that $H \subset G$, $\tau(H) < \alpha$ and

$$(G \setminus H) \cap V \quad \text{has empty interior}.$$

This condition expresses the fact that locally at p the set G contains densely an open subset of type at most α.

14.2. Theorem. *Let α be an infinite regular cardinal, let X be an F_α-space, let G be an open subset of X such that*

$$\tau(G) \leq \alpha,$$

and let $p \in \overline{G} \setminus G$ be such that

the local weight of X at p is at most α.

Then G is C^*-embedded in $G \cup \{p\}$ if and only if condition $[p, G, \alpha]$ holds.

Proof. Assume that condition $[p, G, \alpha]$ holds. There are an open-and-closed neighborhood V of p and an open subset H of G, with $\tau(H) < \alpha$, such that $(G \setminus H) \cap V$ has empty interior. Let $f \in C^*(G)$ and consider $f|H$. Since X is an F_α-space the set H is C^*-embedded in X, so there is a unique continuous extension $g \in C^*(H \cup \{p\})$ of $f|H$. Then with $h = f \cup g$ we have $h \in C^*(G \cup \{p\})$, since $h|G = f$ and $h|V \cap (G \cup \{p\})$ are both continuous.

Conversely, let $G = \bigcup_{\xi<\alpha} C_\xi$ with C_ξ open-and-closed in X for $\xi < \alpha$, and let $\{V_\xi : \xi < \alpha\}$ be a base of neighborhoods at p, consisting of open-and-closed sets. We set $G_\xi = \bigcup_{\eta \leq \xi} C_\eta$ and we note that $\tau(G_\xi) < \alpha$ for $\xi < \alpha$.

We define sets A_ξ and B_ξ for $\xi<\alpha$ such that

(i) A_ξ and B_ξ are non-empty open-and-closed subsets of X,

(ii) $A_\xi\cap B_\xi=\emptyset$, and

(iii) $A_\xi\cup B_\xi\subset(G\setminus\bigcup_{\eta<\xi}(A_\eta\cup B_\eta\cup G_\eta))\cap V_\xi$.

We proceed by transfinite recursion. Let $\xi<\alpha$ and assume that A_η and B_η have been defined for $\eta<\xi$. If condition $[p,G,\alpha]$ fails then since the set

$$H_\xi=\bigcup_{\eta<\xi}(A_\eta\cup B_\eta\cup G_\eta)$$

is an open subset of G whose type in X is less than α, the set $(G\setminus H_\xi)\cap V_\xi$ has non-empty interior. Since $p\in\overline{G}\setminus G$, p cannot be an isolated element and thus the interior of $(G\setminus H_\xi)\cap V_\xi$ does not consist of a single element. Thus there are two non-empty, disjoint, open-and-closed sets A_ξ and B_ξ such that

$$A_\xi\cup B_\xi\subset(G\setminus H_\xi)\cap V_\xi\,.$$

The recursive definition of these sets is complete. We set $A=\bigcup_{\xi<\alpha}A_\xi$, $B=\bigcup_{\xi<\alpha}B_\xi$, and $C=A\cup B$. Clearly A, B and C are open subsets of X with $\tau(C)\leq\alpha$, and $A\cap B=\emptyset$ and $p\in\overline{A}\cap\overline{B}$. We claim that C is C^*-embedded in G. Indeed

$$C\cap C_\xi=\bigcup_{\eta\leq\xi}(A_\eta\cup B_\eta)\cap C_\xi\,,$$

so that $\tau(C\cap C_\xi)<\alpha$ for $\xi<\alpha$ and the claim follows from Lemma 14.1(b). Thus G is not C^*-embedded in $G\cup\{p\}$; in fact, there is a function $f\in C^*(G)$ such that $f=0$ on A and $f=1$ on B, and clearly f cannot be extended to a continuous function defined on $G\cup\{p\}$.

In the case of compact F_α-spaces we have the following simple description of condition $[p,G,\alpha]$.

14.3. Lemma. *Let $\alpha\geq\omega$, let X be a compact totally disconnected space, and let G be an open subset of X. Then G contains densely an open subset of type less than α if and only if condition $[p,G,\alpha]$ holds for $p\in\overline{G}\setminus G$.*

Proof. If G contains densely an open subset H with $\tau(H)<\alpha$, then condition $[p,G,\alpha]$ is satisfied for all $p\in\overline{G}\setminus G$ (with $V=X$ in the defining condition).

Suppose now that condition $[p,G,\alpha]$ holds for $p\in\overline{G}\setminus G$. Then for each such p there exist an open-and-closed neighborhood $V(p)$ of p and an open set $H(p)$ such that $H(p)\subset G$, $\tau(H(p))<\alpha$, and $(G\setminus H(p))\cap V(p)$ has empty interior. Since $\overline{G}\setminus G$ is compact and $\{V(p)\colon p\in\overline{G}\setminus G\}$ is an open cover of $\overline{G}\setminus G$, there are $n<\omega$ and $p_0,\dots,p_{n-1}\in\overline{G}\setminus G$ such that

$$V(p_0)\cup\cdots\cup V(p_{n-1})\supset\overline{G}\setminus G\,.$$

The set $K=\overline{G}\setminus(V(p_0)\cup\cdots\cup V(p_{n-1}))$ is compact and disjoint from $\overline{G}\setminus G$, so there is an open-and-closed set C of X such that

$$K\subset C\subset\overline{G} \quad \text{and}$$
$$C\cap(\overline{G}\setminus G)=\emptyset.$$

We set $H=C\cup H(p_0)\cup\cdots\cup H(p_{n-1})$; then H is open, H is a dense subset of G, and $\tau(H)<\alpha$.

14.4. Theorem. *Let α be an infinite regular cardinal, let X be a compact F_α-space, and let G be an open subset of X such that $\tau(G)\leq\alpha$. Then*

(a) *G is a zero-dimensional F_α-space.*

If in addition the local weight of X at each of its elements is at most α, then

(b) *G is C^*-embedded in X if and only if G contains densely an open subset of type less than α.*

Proof. (a) follows from Lemma 14.1; (b) follows from Theorem 14.2 and Lemma 14.3.

For the first of the two applications of Theorem 14.4, which concerns the Stone space of the homogeneous-universal Boolean algebras, we need the following fact.

14.5. Lemma. *Let $\alpha\geq\omega$, let $\mathfrak{B}$ be an α-homogeneous-universal Boolean algebra, let G be an open subset of $\mathsf{S}(\mathfrak{B})$ with $\tau(G)<\alpha$, and let $G\subsetneqq H\subset\overline{G}$. Then H is not open in $\mathsf{S}(\mathfrak{B})$.*

Proof. Set $\beta=\tau(G)$ and let $G=\bigcup_{\xi<\beta}U_\xi$, where U_ξ is open-and-closed in $\mathsf{S}(\mathfrak{B})$. Assume that there is an open set H for which $G\subsetneqq H\subset\overline{G}$, and let $p\in H\setminus G$. For every $\xi<\beta$ there is an open-and-closed set V_ξ of $\mathsf{S}(\mathfrak{B})$ such that

$$p\in V_\xi\subset H \quad \text{and}$$
$$U_\xi\cap V_\xi=\emptyset.$$

We set $V=\bigcap_{\xi<\beta}V_\xi$. Then $p\in V\subset H\setminus G$, so V is a non-empty, nowhere dense subset of $\mathsf{S}(\mathfrak{B})$ which is equal to the intersection of fewer than α open-and-closed sets. This is a contradiction to Lemma 6.5.

14.6. Theorem. *Let $\alpha=\alpha^{\underline{\alpha}}$ and let G be an open subspace of S_α. Then*

(a) *G is a zero-dimensional F_α-space;*

(b) *if $\tau(G)=\alpha$ and $p\in\overline{G}\setminus G$, then G is not C^*-embedded in $G\cup\{p\}$, and in fact there is a $\{0,1\}$-valued continuous function on G which cannot be continuously extended to $G\cup\{p\}$; and*

(c) *G is C^*-embedded in S_α if and only if $\tau(G)<\alpha$.*

Proof. (a) follows from Lemma 6.5 and Theorem 14.4; (b) follows from Theorem 14.2, Lemma 14.5 applied locally, and Lemma 2.23(b); and (c) follows from (a) and (b).

The Space of κ-Uniform Ultrafilters on α

14.7. Lemma. *Let $\omega \leq \kappa \leq \alpha$ and let G be an open subset of $U_\kappa(\alpha)$ of type $\tau(G) < \mathrm{cf}(\kappa)$. Then*

(a) *the closure $\bar{G}$ of G in $U_\kappa(\alpha)$ is open-and-closed;*

(b) *if H is an open subset of $U_\kappa(\alpha)$ such that $G \cap H = \emptyset$, then $\bar{G} \cap \bar{H} = \emptyset$; and*

(c) *G is C^*-embedded in $U_\kappa(\alpha)$.*

Proof. (a) Let $G = \bigcup_{\xi<\beta} \hat{A}_\xi$, where $A_\xi \in \mathscr{P}(\alpha)$ for $\xi < \beta$ and $\beta < \mathrm{cf}(\kappa)$. We set $A = \bigcup_{\xi<\beta} A_\xi$; it is clear that $\bar{G} \subset \hat{A}$. Suppose that $\bar{G} \neq \hat{A}$. Then there is $B \in \mathscr{P}(\alpha)$, with $|B| \geq \kappa$, such that $\bar{G} \cap \hat{B} = \emptyset$ and $\hat{B} \subset \hat{A}$. We assume with no loss of generality that $B \subset A$. Since $\hat{B} \cap \hat{A}_\xi = \emptyset$ we have $|B \cap A_\xi| < \kappa$ for $\xi < \beta$, and because $\beta < \mathrm{cf}(\kappa)$ we have

$$|B| = |B \cap A| = |B \cap \textstyle\bigcup_{\xi<\beta} A_\xi| = |\bigcup_{\xi<\beta} (B \cap A_\xi)| < \kappa\,,$$

a contradiction to the fact that $|B| \geq \kappa$.

(b) It is clear that $\bar{G} \cap H = \emptyset$. By part (a) above the set $\bar{G}$ is open-and-closed, and hence $\bar{G} \cap \bar{H} = \emptyset$.

(c) If $\mathrm{cf}(\kappa) = \omega$ then G is open-and-closed and hence C^*-embedded in $U_\kappa(\alpha)$. We assume that $\mathrm{cf}(\kappa) > \omega$. Let $\tau(G) = \beta < \mathrm{cf}(\kappa)$ and let $\{V_\xi : \xi < \beta\}$ be a family of open-and-closed subsets of $U_\kappa(\alpha)$ such that $G = \bigcup_{\xi<\beta} V_\xi$. We shall use Urysohn's extension theorem (Corollary 2.8). Let H and K be completely separated subsets of G, so there are subsets H_1 and K_1 of G such that

$$\begin{aligned}
&H \subset H_1 \quad \text{and} \quad K \subset K_1\,,\\
&H_1 \cap K_1 = \emptyset\,, \quad \text{and}\\
&H_1 \text{ and } K_1 \text{ are cozero-sets of } G.
\end{aligned}$$

For $\xi < \beta$ the set $H_1 \cap V_\xi$ is a cozero-set of V_ξ and hence $\tau(H_1 \cap V_\xi) \leq \omega$. Since $H_1 = \bigcup_{\xi<\beta}(H_1 \cap V_\xi)$ we have $\tau(H_1) \leq \omega \cdot \beta < \mathrm{cf}(\kappa)$, and from (b) above we have $\bar{H}_1 \cap \bar{K}_1 = \emptyset$ (the bar denoting closure in $U_\kappa(\alpha)$). The compactness of $U_\kappa(\alpha)$ implies that H_1 and K_1 are completely separated subsets of $U_\kappa(\alpha)$. Thus by Urysohn's extension theorem G is C^*-embedded in $U_\kappa(\alpha)$.

14.8. Lemma. *Let X be a space equal to the union of a compact subset K and a subset D consisting of isolated elements of X. Then X is a paracompact (and hence normal) space.*

Proof. Let $\mathscr{U}$ be an open cover of X. Since K is compact there is a finite subfamily $\mathscr{V}$ of $\mathscr{U}$ whose union contains K. Then

$$\mathscr{V}\cup\{\{p\}:p\in D\}$$

is a locally finite refinement of $\mathscr{U}$.

14.9. Theorem. *If* $\omega\le\kappa\le\alpha$ *then* $U_\kappa(\alpha)$ *is a compact* $F_{\mathrm{cf}(\kappa)^+}$*-space. In particular if* α *is an infinite regular cardinal then* $U_\alpha(\alpha^+)$ *and* $U(\alpha)$ *are compact* F_{α^+}*-spaces.*

Proof. If G is an open subset of $U_\kappa(\alpha)$ of type less than $\mathrm{cf}(\kappa)$ then G is C^*-embedded by Lemma 14.7(c). Thus we assume that G is an open subset of $U_\kappa(\alpha)$ such that $\tau(G)=\mathrm{cf}(\kappa)$. By Lemma 7.12(b) we have $G=\bigcup_{\xi<\mathrm{cf}(\kappa)}\hat{A}_\xi$ where $\{A_\xi:\xi<\mathrm{cf}(\kappa)\}$ is a family of subsets of α (and where $\hat{A}_\xi=\mathrm{cl}_{\beta(\alpha)}A_\xi\cap U_\kappa(\alpha)$ for $\xi<\mathrm{cf}(\kappa)$). Then G is equal to the union of a chain of length $\mathrm{cf}(\kappa)$ whose elements are open sets of type less than $\mathrm{cf}(\kappa)$; indeed we define

$$G_\xi=\bigcup_{\eta\le\xi}\hat{A}_\eta \quad \text{for } \xi<\mathrm{cf}(\kappa)$$

and we have that G_ξ is open, $\tau(G_\xi)<\mathrm{cf}(\kappa)$ and $G_\xi\subset G_\eta$ for $\xi<\eta<\mathrm{cf}(\kappa)$, and $G=\bigcup_{\xi<\mathrm{cf}(\kappa)}G_\xi$. It follows from Lemma 14.7(a) and (c) that $\bar{G}_\xi(=\mathrm{cl}_{U_\kappa(\alpha)}G_\xi)$ is open-and-closed in $U_\kappa(\alpha)$ and that G_ξ is C^*-embedded in $\bar{G}_\xi$ for $\xi<\mathrm{cf}(\kappa)$. We define $H=\bigcup_{\xi<\mathrm{cf}(\kappa)}\bar{G}_\xi$. Thus H is open in $U_\kappa(\alpha)$ and $\tau(H)\le\mathrm{cf}(\kappa)$ and G is dense in H.

We prove that G is C^*-embedded in H. Let $f\in C^*(G)$ and set $f_\xi=f|G_\xi$ for $\xi<\mathrm{cf}(\kappa)$. Let g_ξ be the unique Stone extension of f_ξ to $\bar{G}_\xi$. Since $f_\xi\subset f_\eta$ and these extensions are unique, we have $g_\xi\subset g_\eta$ for $\xi<\eta<\mathrm{cf}(\kappa)$. Set $g=\bigcup_{\xi<\mathrm{cf}(\kappa)}g_\xi$. Then g is continuous because $\bar{G}_\xi$ is open for $\xi<\mathrm{cf}(\kappa)$.

It is clear that $\bar{G}_\xi=(\bigcup_{\eta\le\xi}A_\eta)^\wedge$ for $\xi<\mathrm{cf}(\kappa)$ (cf. the proof of Lemma 14.7(a)). Set $B_\xi=\bigcup_{\eta\le\xi}A_\eta$ and $B=\bigcup_{\xi<\mathrm{cf}(\kappa)}B_\xi=\bigcup_{\xi<\mathrm{cf}(\kappa)}A_\xi$, so that $B_\xi\subset B_\eta$ for $\xi<\eta<\mathrm{cf}(\kappa)$. We claim that H is C^*-embedded in $B\cup H$. Let $f\in C^*(H)$ and set $f_\xi=f|\bar{G}_\xi$ for $\xi<\mathrm{cf}(\kappa)$. We proceed by transfinite recursion. By Lemma 14.8, the space $B_0\cup\bar{G}_0$ is normal; since $\bar{G}_0$ is a closed subset of $B_0\cup\bar{G}_0$ there is $g_0\in C^*(B_0\cup\bar{G}_0)$ such that $f_0\subset g_0$ and

$$\sup\{|g_0(x)|:x\in B_0\cup\bar{G}_0\}=\sup\{|f_0(x)|:x\in\bar{G}_0\}.$$

Let $\xi<\mathrm{cf}(\kappa)$ and suppose that for every $\eta<\xi$ we have defined g_η such that

$$g_\eta\in C^*(B_\eta\cup\bar{G}_\eta),$$
$$f_\eta\subset g_\eta,$$
$$\sup\{|g_\eta(x)|:x\in B_\eta\cup\bar{G}_\eta\}=\sup\{|f_\eta(x)|:x\in\bar{G}_\eta\}, \quad \text{and}$$
$$g_\zeta\subset g_\eta \quad \text{for } \zeta<\eta<\xi.$$

We define

$$g_\xi^* = ((\bigcup_{\eta<\xi} g_\eta) \cup f_\xi) : \bigcup_{\eta<\xi} (B_\eta \cup \overline{G}_\eta) \cup \overline{G}_\xi \to \mathbb{R}$$

(where $\mathbb{R}$ denotes, as usual, the space of real numbers). We verify that g_ξ^* is well-defined: the recursive assumption implies that $\bigcup_{\eta<\xi} g_\eta$ is well-defined on $\bigcup_{\eta<\xi}(B_\eta \cup \overline{G}_\eta)$, and f_ξ is defined on $\overline{G}_\xi$; further,

$$(\bigcup_{\eta<\xi}(B_\eta \cup \overline{G}_\eta)) \cap \overline{G}_\xi = \bigcup_{\eta<\xi} \overline{G}_\eta \subset H$$

and the functions $\bigcup_{\eta<\xi} g_\eta$ and f_ξ are equal to

$$f \mid (\bigcup_{\eta<\xi}(B_\eta \cup \overline{G}_\eta)) \cap \overline{G}_\xi$$

on this intersection. Thus g_ξ^* is well-defined. The function $\bigcup_{\eta<\xi} g_\eta$ is continuous because the sets $B_\eta \cup \overline{G}_\eta$ are open-and-closed for $\eta < \xi$. Finally, g_ξ^* is continuous, because the sets

$$\bigcup_{\eta<\xi}(B_\eta \cup \overline{G}_\eta) \quad \text{and } \overline{G}_\xi$$

are closed in $\bigcup_{\eta<\xi}(B_\eta \cup \overline{G}_\eta) \cup \overline{G}_\xi$.

We note that the space $B_\xi \cup \overline{G}_\xi$ is normal by Lemma 14.8 and that

$$\bigcup_{\eta<\xi}(B_\eta \cup \overline{G}_\eta) \cup \overline{G}_\xi$$

is a closed subset in $B_\xi \cup \overline{G}_\xi$. The recursive assumption implies that

$$\sup\{|g_\xi^*(x)| : x \in \bigcup_{\eta<\xi}(B_\eta \cup \overline{G}_\eta) \cup \overline{G}_\xi\} = \sup\{|f_\xi(x)| : x \in \overline{G}_\xi\},$$

so that g_ξ^* is bounded. Thus there is a continuous extension $g_\xi \in C^*(B_\xi \cup \overline{G}_\xi)$ such that

$$\sup\{|f_\xi(x)| : x \in \overline{G}_\xi\} = \sup\{|g_\xi(x)| : x \in B_\xi \cup \overline{G}_\xi\}.$$

This completes the recursive definition of g_ξ for $\xi < \mathrm{cf}(\kappa)$. We set $g = \bigcup_{\xi<\mathrm{cf}(\kappa)} g_\xi$. Since the sets $B_\xi \cup \overline{G}_\xi$ are open-and-closed for $\xi < \mathrm{cf}(\kappa)$, we have that g is continuous. Also

$$\sup\{|g(x)| : x \in B \cup H\} = \sup\{|f(x)| : x \in B\},$$

so g is bounded. This completes the proof of our claim that H is C^*-embedded in $B \cup H$.

We note finally that

$$B \subset B \cup H \subset \mathrm{cl}_{\boldsymbol{\beta}(\alpha)} B = \boldsymbol{\beta}(B),$$

so that $B \cup H$ is C^*-embedded in $\mathrm{cl}_{\boldsymbol{\beta}(\alpha)} B$. Since $\mathrm{cl}_{\boldsymbol{\beta}(\alpha)} B$ is open-and-closed in $\boldsymbol{\beta}(\alpha)$, it is C^*-embedded in $\boldsymbol{\beta}(\alpha)$ so by the transitivity of C^*-embedding, G is C^*-embedded in $\boldsymbol{\beta}(\alpha)$. Thus G is C^*-embedded in $U_\kappa(\alpha)$, and the proof is complete.

14.10. Theorem. *Let α be a regular cardinal such that $\alpha^+ = 2^\alpha$, and let G be an open subspace of $U(\alpha)$. Then*

(a) G is a zero-dimensional F_{α^+}-space;

(b) if $p \in \overline{G} \setminus G$, then G is C^*-embedded in $G \cup \{p\}$ if and only if condition $[p, G, \alpha^+]$ holds; and

(c) G is C^*-embedded in $U(\alpha)$ if and only if G contains densely an open subset of type (in $U(\alpha)$) at most α.

Proof. (a) follows from Theorems 14.4(a) (applied with α^+ replacing α) and 14.9; (b) follows from Theorems 14.2 and 14.9 and the fact that the local weight of $U(\alpha)$ at each of its elements is (at least) α^+ (Corollary 7.8(a)); (c) follows from Theorems 14.4(b) and 14.9 and Corollary 7.8(a).

Definitions. Let $\alpha \geq \omega$. A space X such that $\mathfrak{B}(X)$ is a base is *α-basically disconnected* if the closure of every open subset of X of type less than α is open-and-closed in X.

A *basically disconnected* space is an ω^+-basically disconnected space.

We note that a space X is extremally disconnected if and only if X is α-basically disconnected for all cardinals α.

It is clear that if X is a P_α-space such that $\mathfrak{B}(X)$ is a base then X is α-basically disconnected; and a compact α-basically disconnected space is an F_α-space, by Urysohn's extension theorem (Corollary 2.8) and Lemma 6.4.

14.11. Theorem. *Let $\omega \leq \kappa \leq \alpha$ and let G be an open subset of $U_\kappa(\alpha)$ with $\tau(G) = \mathrm{cf}(\kappa)$. The closure $\overline{G}$ in $U_\kappa(\alpha)$ is open-and-closed if and only if G contains densely an open subset of $U_\kappa(\alpha)$ of type less than $\mathrm{cf}(\kappa)$.*

In particular, $U_\kappa(\alpha)$ is not $\mathrm{cf}(\kappa)^+$-basically disconnected.

Proof. If G contains densely an open subset H of $U_\kappa(\alpha)$ and $\tau(H) < \mathrm{cf}(\kappa)$, then $\overline{G} = \overline{H}$ is open-and-closed by Lemma 14.7(a).

For the converse, suppose that G does not contain densely an open subset of type less than $\mathrm{cf}(\kappa)$, and let $G = \bigcup_{\xi < \mathrm{cf}(\kappa)} \hat{A}_\xi$ where $A_\xi \in \mathscr{P}(\alpha)$ for $\xi < \mathrm{cf}(\kappa)$. It is easy to define by recursion a family $\{G_\xi : \xi < \mathrm{cf}(\kappa)\}$ of open subsets of $U_\kappa(\alpha)$ such that

$$\begin{aligned} &\hat{A}_\xi \subset G_\xi \quad \text{for } \xi < \mathrm{cf}(\kappa), \\ &G_\xi \subset G_\zeta \quad \text{for } \xi < \zeta < \mathrm{cf}(\kappa), \\ &\tau(G_\xi) < \mathrm{cf}(\kappa) \quad \text{for } \xi < \mathrm{cf}(\kappa), \quad \text{and} \\ &\textstyle\bigcup_{\zeta < \xi} G_\zeta \quad \text{is not dense in } G_\xi \text{ for } \xi < \mathrm{cf}(\kappa). \end{aligned}$$

We set $H = \bigcup_{\xi < \mathrm{cf}(\kappa)} \overline{G}_\xi$ and we note that since by Lemma 14.7(a) $\overline{G}_\xi$ is open-and-closed we have that H is open in $U_\kappa(\alpha)$ and in fact $\tau(H) \leq \mathrm{cf}(\kappa)$. Since $G \subset H \subset \overline{G}$ we have $\overline{H} = \overline{G}$; hence it is enough to prove that $\overline{H}$ is not open.

By Lemma 7.12(b) there is $B_\xi \in \mathscr{P}(\alpha)$ such that $\bar{G}_\xi = \hat{B}_\xi$ for $\xi < \mathrm{cf}(\kappa)$, and if $\bar{H}$ is open there is $B \in \mathscr{P}(\alpha)$ such that $\bar{H} = \hat{B}$; we assume without loss of generality that $B_\xi \subset B$ for $\xi < \mathrm{cf}(\kappa)$. We claim that

$$|B_\xi \setminus \bigcup_{\eta<\xi} B_\eta| \geq \kappa \quad \text{for } \xi < \mathrm{cf}(\kappa).$$

Indeed from Lemma 14.7(a) we have

$$\emptyset \neq \bar{G}_\xi \setminus (\bigcup_{\eta<\xi} \bar{G}_\eta)^- = \hat{B}_\xi \setminus (\bigcup_{\eta<\xi} \hat{B}_\eta)^- = \hat{B}_\xi \setminus (\bigcup_{\eta<\xi} B_\eta)\hat{} = (B_\xi \setminus \bigcup_{\eta<\xi} B_\eta)\hat{},$$

and the claim follows from Lemma 7.12(c). Now let $\{\lambda_\xi : \xi < \mathrm{cf}(\kappa)\}$ be a family of cardinals such that

$$\kappa = \sum_{\xi<\mathrm{cf}(\kappa)} \lambda_\xi \quad \text{and}$$
$$\lambda_\xi < \kappa \quad \text{for } \xi < \mathrm{cf}(\kappa),$$

and for $\xi < \mathrm{cf}(\kappa)$ let $C_\xi \subset B_\xi \setminus \bigcup_{\eta<\xi} B_\eta$ with $|C_\xi| = \lambda_\xi$. We define

$$C = \bigcup_{\xi<\mathrm{cf}(\kappa)} C_\xi, \quad \text{and}$$
$$D = B \setminus C,$$

and we note that since $\hat{C} \cap \hat{D} = \emptyset$, $\hat{B} = \hat{C} \cup \hat{D}$, and $\hat{C} \neq \emptyset$ we have $\hat{B} \neq \hat{D}$. Since $B_\xi \setminus D = B_\xi \cap C = \bigcup_{\eta<\xi} C_\eta$ we have $|B_\xi \setminus D| < \kappa$ and hence $\hat{B}_\xi \subset \hat{D}$ for $\xi < \mathrm{cf}(\kappa)$; it follows that

$$H = \bigcup_{\xi<\mathrm{cf}(\kappa)} \bar{G}_\xi = \bigcup_{\xi<\mathrm{cf}(\kappa)} \hat{B}_\xi \subset \hat{D} \subset \hat{B} = \bar{H}$$

and hence $\bar{H} = \hat{D} = \hat{B}$, a contradiction.

For the final statement we note that since $\boldsymbol{S}(U_\kappa(\alpha)) \geq \alpha^+ \geq \mathrm{cf}(\kappa)^+$ there is a family $\{U_\xi : \xi < \mathrm{cf}(\kappa)\}$ of non-empty open-and-closed subsets of $U_\kappa(\alpha)$ such that

$$U_\xi \cap U_\zeta = \emptyset \quad \text{for } \xi < \zeta < \mathrm{cf}(\kappa).$$

We set $U = \bigcup_{\xi<\mathrm{cf}(\kappa)} U_\xi$; it is clear that if V is a dense and open subset of U then $\boldsymbol{\tau}(V) \geq \mathrm{cf}(\kappa)$. The proof is complete.

We recall from Corollary 12.10 that $\boldsymbol{S}(U(\alpha)) \geq \alpha^{++}$ for $\alpha \geq \omega$. It follows (as in the last paragraph of the proof above) that there is an open subset U of $U(\alpha)$ such that $\boldsymbol{\tau}(V) \geq \alpha^+$ for every dense and open subset V of U.

It follows from the theorem above and Lemma 14.7(a) that for cardinals α, β and κ with $\omega \leq \kappa \leq \alpha$ the space $U_\kappa(\alpha)$ is β-basically disconnected if and only if $\beta \leq \mathrm{cf}(\kappa)$. In particular $\boldsymbol{\beta}(\omega) \setminus \omega$ is a compact F-space which is not basically disconnected.

Spaces of Uniform Ultrafilters and Homogeneous-Universal Boolean Algebras

14.12. Theorem. *The Boolean algebra $\mathscr{P}(\omega)/\mathscr{P}^{\omega}(\omega)$ (which is isomorphic to $\mathfrak{B}(\boldsymbol{\beta}(\omega)\backslash\omega)$) is ω^+-homogeneous-universal (of cardinality 2^{ω}).*

Proof. We first prove that $\mathfrak{B}(\boldsymbol{\beta}(\omega)\backslash\omega)$ is ω^+-homogeneous. Let $\mathfrak{C}$ be a countable Boolean algebra and let ϕ and ψ be Boolean algebra embeddings of $\mathfrak{C}$ into $\mathfrak{B}(\boldsymbol{\beta}(\omega)\backslash\omega)$. We enumerate $\mathfrak{C}$, say $\{C_n : n<\omega\}$, and we set $U_n=\phi(C_n)$ and $V_n=\psi(C_n)$ for $n<\omega$. We define a sequence $\{A_n : n<\omega\}$ of subsets of ω such that

(i) if $C_n=\mathbb{0}$ then $A_n=\emptyset$,
(ii) $\hat{A}_n=U_n$,
(iii) if $C_n=C'_m$ then $A_n=\omega\backslash A_m$, and
(iv) if $C_n\leq C_m$ then $A_n\subset A_m$

for $n,m<\omega$. We proceed recursively. Let $n<\omega$ and suppose that A_m is defined for $m<n$. (We use here, as in § 2, the following notation: if b is an element of a Boolean algebra, then $1b=b$ and $-1b=b'$.) To define A_n we consider the 2^n expressions $\varepsilon_0 C_0\wedge\cdots\wedge\varepsilon_{n-1}C_{n-1}\wedge C_n$, where $\varepsilon\in\{-1,1\}^n$. If $\varepsilon_0 C_0\wedge\cdots\wedge\varepsilon_{n-1}C_{n-1}\wedge C_n\neq\mathbb{0}$, we choose a set $A_n(\varepsilon)$ such that

$$A_n(\varepsilon)\subset\varepsilon_0 A_0\cap\cdots\cap\varepsilon_{n-1}A_{n-1} \quad\text{and}$$
$$(A_n(\varepsilon))^{\wedge}=\varepsilon_0 U_0\cap\cdots\cap\varepsilon_{n-1}U_{n-1};$$

and if $\varepsilon_0 C_0\wedge\cdots\wedge\varepsilon_{n-1}C_{n-1}\wedge C_n=\mathbb{0}$, we set $A_n(\varepsilon)=\emptyset$. We set

$$A_n=\bigcup\{A_n(\varepsilon):\varepsilon\in\{-1,1\}^n\}.$$

It is easy to verify that A_n satisfies the recursive conditions. In a similar way we define a sequence $\{B_n : n<\omega\}$ of subsets of ω such that

(i) if $C_n=\mathbb{0}$ then $B_n=\emptyset$,
(ii) $\hat{B}_n=V_n$,
(iii) if $C_n=C'_m$ then $B_n=\omega\backslash B_m$, and
(iv) if $C_n\leq C_m$ then $B_n\subset B_m$.

We note that the following four conditions are equivalent for $n<\omega$ and $\varepsilon\in\{-1,1\}^n$:

$$|\varepsilon_0 A_0\cap\cdots\cap\varepsilon_{n-1}A_{n-1}|=\omega;$$
$$\varepsilon_0 A_0\cap\cdots\cap\varepsilon_{n-1}A_{n-1}\neq\emptyset;$$
$$|\varepsilon_0 B_0\cap\cdots\cap\varepsilon_{n-1}B_{n-1}|=\omega; \quad\text{and}$$
$$\varepsilon_0 B_0\cap\cdots\cap\varepsilon_{n-1}B_{n-1}\neq\emptyset.$$

Now for $k<\omega$ we define functions ε_k and ∂_k from ω to $\{-1,1\}$ as follows:

$$\begin{aligned} \varepsilon_k(n) &= \quad 1 \quad \text{if } n \in A_k \\ &= -1 \quad \text{if } n \notin A_k, \quad \text{and} \\ \partial_k(n) &= \quad 1 \quad \text{if } n \in B_k \\ &= -1 \quad \text{if } n \notin B_k. \end{aligned}$$

We define functions π and ρ from ω to ω such that

(i) $\pi|(n \cup \rho[n])$ is one-to-one onto $n \cup \pi[n]$,

(ii) $\rho|(n \cup \pi[n])$ is one-to-one onto $n \cup \rho[n]$,

(iii) $\rho|(n \cup \pi[n]) = (\pi|(n \cup \rho[n]))^{-1}$,

(iv) if $n \notin \rho[n]$ then $\pi(n) \in \varepsilon_0(n) B_0 \cap \cdots \cap \varepsilon_n(n) B_n$, and

(v) if $n \notin \pi[n+1]$ then $\rho(n) \in \partial_0(n) A_0 \cap \cdots \cap \partial_n(n) A_n$.

We proceed recursively. Let $n<\omega$ and suppose that $\pi(m)$ and $\rho(m)$ have been defined for $m<n$, satisfying conditions (i) to (v). If there is $k<n$ such that $n=\rho(k)$, we set $\pi(n)=k$; otherwise, we note that $n \in \varepsilon_0(n) A_0 \cap \cdots \cap \varepsilon_n(n) A_n$, so that $|\varepsilon_0(n) B_0 \cap \cdots \cap \varepsilon_n(n) B_n| = \omega$, and we choose

$$\pi(n) \in \varepsilon_0(n) B_0 \cap \cdots \cap \varepsilon_n(n) B_n \setminus (n \cup \pi[n]).$$

The definition of $\rho(n)$ is then given in a similar way. By (i) π is one-to-one and onto ω; by (ii), ρ is one-to-one and onto ω; by (iii), $\rho=\pi^{-1}$. We claim that

$$\pi[A_n \setminus (n \cup \rho[n])] \subset B_n \setminus (n \cup \pi[n]) \quad \text{for } n<\omega.$$

To verify this let $m \in A_n$ with $m \geq n$ and $m \neq \rho(k)$ for all $k<n$. There are two possibilities to consider.

Case 1. There is $k<m$ such that $m=\rho(k)$.
Since $m \notin \rho[n]$, we have $n \leq k$; and since, by (v),

$$m \in \partial_0(k) A_0 \cap \cdots \cap \partial_k(k) A_k,$$

we have $\partial_n(k)=1$. Thus $k=\pi(m) \in B_n$. And since $m \geq n$ and π is one-to-one, we have $\pi(m) \notin \pi[n]$. Thus

$$\pi(m) \in B_n \setminus (n \cup \pi[n]).$$

Case 2. $m \notin \rho[m]$.
Then $\pi(m) \in \varepsilon_0(m) B_0 \cap \cdots \cap \varepsilon_m(m) B_m$ by (iv). Since $m \in A_n$ we have $\varepsilon_n(m)=1$; and since also $m \geq n$ we have $\pi(m) \in B_n$. Because $m \notin \rho[n]$, it follows that $\pi(m) \notin n$; and $\pi(m) \notin \pi[n]$ because $m \geq n$. Thus

$$\pi(m) \in B_n \setminus (n \cup \pi[n]).$$

This completes the proof of the claim. In a similar way we show that $\rho[B_n \setminus (n \cup \pi[n])] \subset A_n \setminus (n \cup \rho[n])$ for $n < \omega$. It then follows that $\pi[A_n \setminus (n \cup \rho[n])] = B_n \setminus (n \cup \pi[n])$, and hence that $\bar{\pi}[\hat{A}_n] = \hat{B}_n$ for $n < \omega$. We define $h: \mathfrak{B}(\boldsymbol{\beta}(\omega) \setminus \omega) \to \mathfrak{B}(\boldsymbol{\beta}(\omega) \setminus \omega)$ by $h(U) = \bar{\pi}[U]$. Since π is a permutation of ω, h is a Boolean algebra automorphism of $\mathfrak{B}(\boldsymbol{\beta}(\omega) \setminus \omega)$; and $h(U_n) = V_n$ for $n < \omega$. Thus $h \circ \phi = \psi$, proving the ω^+-homogeneity of $\mathfrak{B}(\boldsymbol{\beta}(\omega) \setminus \omega)$.

We next prove that $\mathfrak{B}(\boldsymbol{\beta}(\omega) \setminus \omega)$ is ω^+-universal. In the presence of the property of ω^+-homogeneity, proved above, it follows from the general theory of Jónsson classes that it is enough to prove that $\mathfrak{B}(\boldsymbol{\beta}(\omega) \setminus \omega)$ is ω-universal (cf. the remark after Theorem 4.6). According to Theorem 2.29 (d) it is sufficient to prove that if X is a (totally disconnected) compact space with a countable base for its topology then there is a continuous function from $\boldsymbol{\beta}(\omega) \setminus \omega$ onto X. Let $\{d_n : n < \omega\}$ be a countable dense subset of X and let $\{A_n : n < \omega\}$ be a sequence of subsets of ω such that

$$|A_n| = \omega,$$
$$A_n \cap A_m = \emptyset \quad \text{if } n < m < \omega, \quad \text{and}$$
$$\bigcup_{n<\omega} A_n = \omega.$$

Let $f: \omega \to X$ be defined by $f[A_n] = \{d_n\}$ for $n < \omega$, and set $\phi = \bar{f} \,|\, \boldsymbol{\beta}(\omega) \setminus \omega$. Since $|A_n| = \omega$ we have $\hat{A}_n \neq \emptyset$ and $\phi[\hat{A}_n] = \{d_n\}$ for $n < \omega$. Thus the compact set $\phi[\boldsymbol{\beta}(\omega) \setminus \omega]$ is dense in X, so $\phi[\boldsymbol{\beta}(\omega) \setminus \omega] = X$. The proof of the theorem is now complete.

Remark. It follows from Theorem 14.12 and Lemma 6.3 that the Boolean algebra $\mathscr{P}(\omega)/\mathscr{P}^{\omega}(\omega)$ (which is isomorphic to $\mathfrak{B}(\boldsymbol{\beta}(\omega) \setminus \omega)$) satisfies condition (H_{ω^+}) (as defined in § 6). According to Lemma 6.5 condition (H_{ω^+}) for $\mathfrak{B}(\boldsymbol{\beta}(\omega) \setminus \omega)$ is equivalent to the following three properties:

(a) $\boldsymbol{\beta}(\omega) \setminus \omega$ has no isolated elements;

(b) every non-empty zero-set of $\boldsymbol{\beta}(\omega) \setminus \omega$ has non-empty interior; and

(c) $\boldsymbol{\beta}(\omega) \setminus \omega$ is an F-space.

A direct proof of these properties is contained in Lemma 14.16 and Remark 14.17 below.

Since $|\mathfrak{B}(\boldsymbol{\beta}(\omega) \setminus \omega)| = 2^{\omega}$ it follows that if $\omega^+ = 2^{\omega}$ and if X is a compact totally disconnected space with 2^{ω} open-and-closed sets, then X is homeomorphic to $\boldsymbol{\beta}(\omega) \setminus \omega$ if and only if X satisfies (the analogues of) conditions (a), (b) and (c) (cf. Corollary 14.15 below).

We examine the occurrence of spaces with these properties in Theorem 14.18 below.

Theorem 14.12 is the best possible in the sense that no strengthening of the property (H_{ω^+}) derived from its conclusion is possible. This is the content of Hausdorff's Theorem 14.14, for which the following material is preparatory. It is remarkable that this result can be proved without assuming the continuum hypothesis.

Definitions. The reduced product (or equivalently the quotient Boolean algebra) $\mathscr{P}(\omega)/\mathscr{P}^{\omega}(\omega)$ has a natural partial order given by

$$A/\mathscr{P}^{\omega}(\omega) \preccurlyeq B/\mathscr{P}^{\omega}(\omega) \quad \text{if } |A\setminus B| < \omega .$$

The induced binary relation $\preccurlyeq$ on $\mathscr{P}(\omega)$, defined by

$$A \preccurlyeq B \quad \text{if and only if } |A\setminus B| < \omega$$

for $A, B \in \mathscr{P}(\omega)$, is a pre-order (*i.e.*, it is a reflexive, transitive relation). We define for $A, B \in \mathscr{P}(\omega)$:

$$A \approx B \quad \text{if } A \preccurlyeq B \quad \text{and } B \preccurlyeq A , \quad \text{and}$$

$$A \prec B \quad \text{if } A \preccurlyeq B \quad \text{and it is false that } A \approx B .$$

If $\mathscr{A}, \mathscr{B} \subset \mathscr{P}(\omega)$ we define

$$\mathscr{A} \prec \mathscr{B} \quad \text{if } A \prec B \quad \text{for } A \in \mathscr{A} \quad \text{and } B \in \mathscr{B} ;$$

when $\mathscr{B} = \{B\}$ we write $\mathscr{A} \prec B$ instead of $\mathscr{A} \prec \mathscr{B}$, etc. If $A, B \in \mathscr{P}(\omega)$ and $A \preccurlyeq B$ we set

$$i(A,B) = \min\{k < \omega : A\setminus k \subset B\setminus k\} .$$

Let $\mathscr{A} \subset \mathscr{P}(\omega)$ and $B \in \mathscr{P}(\omega)$ and $\mathscr{A} \prec B$. Then B is *near* $\mathscr{A}$ if

$$|\{A \in \mathscr{A} : i(A,B) \leq k\}| < \omega \quad \text{for } k < \omega .$$

It is clear that if $|\mathscr{A}| > \omega$ and $\mathscr{A} \prec B$ then B is *not* near $\mathscr{A}$. The following simple remark will be used in the sequel:

if $\mathscr{A} \subset \mathscr{P}(\omega)$ and $B, C \in \mathscr{P}(\omega)$ and $\mathscr{A} \prec C \prec B$ and B is near $\mathscr{A}$, then C is near $\mathscr{A}$.

Indeed $\{A \in \mathscr{A} : i(A,C) \leq k\} \subset \{A \in \mathscr{A} : i(A,B) \leq k\}$ if $k \geq i(C,B)$.

14.13. Lemma. *Let $\mathscr{A}$ be a subset of $\mathscr{P}(\omega)$, and let $D \in \mathscr{P}(\omega)$ be such that*

$$|\mathscr{A}| \leq \omega ,$$

$\mathscr{A}$ is linearly ordered under $\preccurlyeq$,

$\mathscr{A} \prec D$, and

D is near $\{B \in \mathscr{A} : B \prec A\}$ for each $A \in \mathscr{A}$.

Then there is $C \in \mathscr{P}(\omega)$ such that

$$\mathscr{A} \prec C \prec D \quad \textit{and}$$

C is near $\mathscr{A}$.

Proof. (a) We assume first that $\mathscr{A}$ is a well-ordered set under $\preccurlyeq$ of type ω. Thus $\mathscr{A}=\{A_n : n<\omega\}$, with

$$A_0 \prec A_1 \prec \cdots \prec A_n \prec \cdots .$$

There is, from condition (H_{ω^+}) for $\mathscr{P}(\omega)/\mathscr{P}^{\omega}(\omega)$, a set $E\in\mathscr{P}(\omega)$ such that $\mathscr{A}\prec E\prec D$. We note that $|A_n\setminus(A_0\cup\cdots\cup A_{n-1})|=\omega$ for $n<\omega$, so there is

$$F_n \subset A_n\setminus(A_0\cup\cdots\cup A_{n-1}) \quad \text{such that}$$
$$|F_n|=n .$$

We set $C=E\setminus\bigcup_{n<\omega}F_n$. Then $C\preccurlyeq E\prec D$, and $\mathscr{A}\prec C$ since $|A_n\setminus C| =|(A_n\setminus E)\cup F_n|<\omega$ for $n<\omega$. Further $|A_n\setminus C|\geq n$ and hence $i(A_n, C)\geq n$, so that $\{A\in\mathscr{A} : i(A,C)\leq k\}\subset\{A_n\in\mathscr{A} : n\leq k\}$ for $k<\omega$; thus C is near $\mathscr{A}$.

(b) If $\mathscr{A}$ has a last element A_0, we let $C\in\mathscr{P}(\omega)$ be such that $A_0\prec C\prec D$. Then if $i(C,D)\leq k<\omega$ we have

$$\{A\in\mathscr{A} : i(A,C)\leq k\}\subset\{A\in\mathscr{A} : i(A,D)\leq k\}\cup\{A_0\},$$

and thus C is near $\mathscr{A}$.

(c) If D is near $\mathscr{A}$, we let $C\in\mathscr{P}(\omega)$ be such that $\mathscr{A}\prec C\prec D$. Such an element C exists, as we have remarked earlier, because the Boolean algebra $\mathscr{P}(\omega)/\mathscr{P}^{\omega}(\omega)$ satisfies condition (H_{ω^+}).

(d) Assume now that

D is not near $\mathscr{A}$, and

$\mathscr{A}$ does not have a last element.

We set $\mathscr{A}_k=\{A\in\mathscr{A} : i(A,D)\leq k\}$ for $k<\omega$. Because D is not near $\mathscr{A}$, there is $k_0<\omega$ such that $|\mathscr{A}_{k_0}|=\omega$, and therefore $|\mathscr{A}_k|=\omega$ for $k_0\leq k<\omega$. Because D is near every initial segment of $\mathscr{A}$, it follows that

$\mathscr{A}_k$ is cofinal in $\mathscr{A}$ under $\preccurlyeq$, and

$\mathscr{A}_k$ is of type ω under $\preccurlyeq$

for $k_0\leq k<\omega$. We define a sequence $\{C_k : k_0\leq k<\omega\}$ of subsets of ω such that

$$\mathscr{A}\prec\cdots\prec C_{k+1}\prec C_k\prec\cdots\prec C_{k_0+1}\prec C_{k_0}\prec D, \quad \text{and}$$
$$C_k \text{ is near } \mathscr{A}_k \text{ for } k_0\leq k<\omega .$$

We proceed recursively. By part (a) there is $C_{k_0}\in\mathscr{P}(\omega)$ such that

$$\mathscr{A}_{k_0}\prec C_{k_0}\prec D, \quad \text{and}$$
$$C_{k_0} \text{ is near } \mathscr{A}_{k_0}.$$

Since $\mathscr{A}_{k_0}$ is cofinal in $\mathscr{A}$, we have in fact $\mathscr{A} \prec C_{k_0}$. Let $k_0 < k < \omega$; by part (a) again there is $C_k \in \mathscr{P}(\omega)$ such that

$$\mathscr{A}_k \prec C_k \prec C_{k-1}, \quad \text{and}$$
$$C_k \text{ is near } \mathscr{A}_k.$$

Since $\mathscr{A}_k$ is cofinal in $\mathscr{A}$, we have in fact $\mathscr{A} \prec C_k$. This completes the recursive definition. We set $\mathscr{C} = \{C_k : k_0 \le k < \omega\}$; then $\mathscr{A} \prec \mathscr{C}$, so from condition (H_{ω^+}) for $\mathscr{P}(\omega)/\mathscr{P}^{\omega}(\omega)$ there is $C \in \mathscr{P}(\omega)$ such that $\mathscr{A} \prec C \prec \mathscr{C}$. Then $\mathscr{A} \prec C \prec D$. To verify that C is near $\mathscr{A}$, we note that if $k_0 \le k < \omega$ and $i(C, D) \le k$ then (because C_k and hence C is near $\mathscr{A}_k$) we have

$$\{A \in \mathscr{A} : i(A, C) \le k\} = \{A \in \mathscr{A}_k : i(A, C) \le k\},$$

so that $|\{A \in \mathscr{A} : i(A, C) \le k\}| < \omega$.

The proof of the lemma is complete.

14.14. Theorem *(Hausdorff). The Boolean algebra $\mathscr{P}(\omega)/\mathscr{P}^{\omega}(\omega)$ does not satisfy condition* $(\mathrm{H}_{\omega^{++}})$.

Proof. We define families $\{A_\xi : \xi < \omega^+\}$ and $\{B_\xi : \xi < \omega^+\}$ of subsets of ω such that

$$A_\xi \prec A_\zeta \quad \text{for } \xi < \zeta < \omega^+,$$
$$B_\zeta \prec B_\xi \quad \text{for } \xi < \zeta < \omega^+,$$
$$A_\xi \prec B_\zeta \quad \text{for } \xi, \zeta < \omega^+, \quad \text{and}$$
$$B_\xi \text{ is near } \{A_\zeta : \zeta < \xi\} \quad \text{for } \xi < \omega^+.$$

We proceed recursively. Let $A_0, B_0 \in \mathscr{P}(\omega)$ be such that $A_0 \prec B_0$. Let $0 < \xi < \omega^+$; because the Boolean algebra $\mathscr{P}(\omega)/\mathscr{P}^{\omega}(\omega)$ satisfies condition (H_{ω^+}), there is $C_\xi \in \mathscr{P}(\omega)$ such that

$$\{A_\zeta : \zeta < \xi\} \prec C_\xi \prec \{B_\zeta : \zeta < \xi\}.$$

We note that if $\zeta < \xi$ then the recursive assumptions imply that B_ζ, and therefore C_ξ, is near $\{A_\eta : \eta < \zeta\}$. It follows from Lemma 14.13 that there is $B_\xi \in \mathscr{P}(\omega)$ such that

$$\{A_\zeta : \zeta < \xi\} \prec B_\xi \prec C_\xi, \quad \text{and}$$
$$B_\xi \text{ is near } \{A_\zeta : \zeta < \xi\}.$$

From condition (H_{ω^+}) for $\mathscr{P}(\omega)/\mathscr{P}^{\omega}(\omega)$, there is $A_\xi \in \mathscr{P}(\omega)$ such that

$$\{A_\zeta : \zeta < \xi\} \prec A_\xi \prec B_\xi.$$

This completes the recursive definitions. We set

$$\mathscr{A} = \{A_\xi : \xi < \omega^+\} \quad \text{and} \quad \mathscr{B} = \{B_\xi : \xi < \omega^+\},$$

and we prove that condition $(\mathrm{H}_{\omega^{++}})$ fails for the families $\mathscr{A}$ and $\mathscr{B}$. For suppose that there is $C \in \mathscr{P}(\omega)$ such that $\mathscr{A} \prec C \prec \mathscr{B}$. Then there is $k < \omega$ such that

$$|\{\xi < \omega^+ : i(A_\xi, C) \leq k\}| = \omega^+ ,$$

and then there is $\eta < \omega^+$ such that

$$|\{\xi < \eta : i(A_\xi, C) \leq k\}| = \omega .$$

Then C is not near $\{A_\xi : \xi < \eta\}$, contradicting the fact that $\{A_\xi : \xi < \eta\} \prec C \prec B_\eta$. The proof is complete.

14.15. Corollary. *The following statements are equivalent.*

(a) $\omega^+ = 2^\omega$.

(b) *The Boolean algebra* $\mathscr{P}(\omega)/\mathscr{P}^\omega(\omega)$ (*which is isomorphic to* $\mathfrak{B}(\boldsymbol{\beta}(\omega)\setminus\omega)$) *is isomorphic to the* ω^+*-homogeneous-universal Boolean algebra* $\mathfrak{C}_{\omega^+}$ *of cardinality* ω^+.

(c) *The Boolean algebra* $\mathscr{P}(\omega)/\mathscr{P}^\omega(\omega)$ *is isomorphic to the* 2^ω*-homogeneous-universal Boolean algebra* $\mathfrak{C}_{2^\omega}$ *of cardinality* 2^ω.

Proof. That (a)⇒(b) follows from Theorem 14.12; that (b)⇒(c) is trivial; and that (c)⇒(a) follows from Theorems 6.12 and 14.14.

14.16. Lemma. *If X is a locally compact σ-compact space then $\boldsymbol{\beta}X\setminus X$ is an F-space.*

Proof. Let C be a cozero-set of $\boldsymbol{\beta}X\setminus X$. Then C is σ-compact, so $X \cup C$ is σ-compact and hence a normal space. Furthermore C is a closed subset of $X \cup C$ and hence according to the Tietze extension theorem C is C^*-embedded in $X \cup C$. Thus if $f \in C^*(C)$ there is $g \in C^*(X \cup C)$ such that $f \subset g$. Let $\overline{g}$ denote (as usual) the Stone extension of g defined on $\boldsymbol{\beta}X$; it is clear that $f \subset (\overline{g} | \boldsymbol{\beta}X\setminus X)$. Thus C is C^*-embedded in $\boldsymbol{\beta}X\setminus X$.

14.17. *Remark.* If X is a zero-dimensional, locally compact, σ-compact space then

(a) $\boldsymbol{\beta}X\setminus X$ is a compact, zero-dimensional space;

(b) $\boldsymbol{\beta}X\setminus X$ has no isolated elements; and

(c) every non-empty zero-set of $\boldsymbol{\beta}X\setminus X$ has non-empty interior in $\boldsymbol{\beta}X\setminus X$.

We verify these statements.

Proof of (a). Since X is locally compact, it follows that $\boldsymbol{\beta}X\setminus X$ is compact (Lemma 2.9(b)). Since X is zero-dimensional it follows from Corollary 2.26 that $\boldsymbol{\beta}X$, and hence also its closed subset $\boldsymbol{\beta}X\setminus X$, is zero-dimensional.

Proof of (b). This statement is a special case of Lemma 15.16(b); we give here a proof under the present assumptions.

If V is a non-empty open subset of $\boldsymbol{\beta}X\backslash X$ then there is, by Theorem 2.28, an open-and-closed non-compact subset A of X such that $\hat{A}(=\mathrm{cl}_{\boldsymbol{\beta}X}A\backslash X)\subset V$. Since A is a closed subset of X, it is then locally compact and σ-compact; and it is then easy to show that there is a family $\{A_n : n<\omega\}$ of non-empty, open-and-closed subsets of X such that

$$\bigcup\{A_n : n<\omega\}=A \quad \text{and}$$
$$A_n\cap A_m=\emptyset \quad \text{for } n<m<\omega .$$

We define

$$B_0=\bigcup\{A_n : n<\omega \text{ and } n \text{ even}\}, \quad \text{and}$$
$$B_1=\bigcup\{A_n : n<\omega \text{ and } n \text{ odd}\} .$$

Then $B_0\cap B_1=\emptyset$ and thus $\hat{B}_0\cap\hat{B}_1=\emptyset$. Since $\hat{B}_0\neq\emptyset$ and $\hat{B}_1\neq\emptyset$ and $\hat{B}_0\cup\hat{B}_1=\hat{A}$, we have $1<|\hat{A}|\leq|V|$. Thus $\boldsymbol{\beta}X\backslash X$ has no isolated elements.

Proof of (c). This statement is a special case of Theorem 15.18(a); we give here a proof under the present assumptions.

Let Z be a non-empty zero-set of $\boldsymbol{\beta}X\backslash X$. If Z is open then $\mathrm{int}_{\boldsymbol{\beta}X\backslash X}Z\neq\emptyset$, so we suppose that Z is not open. Then there is a sequence $\{V_n : n<\omega\}$ of open-and-closed subsets of $\boldsymbol{\beta}X\backslash X$ such that $V_{n+1}\subsetneqq V_n$ for $n<\omega$, and there is a sequence $\{A_n : n<\omega\}$ of open-and-closed subsets of X such that $V_n=(\mathrm{cl}_{\boldsymbol{\beta}X}A_n)\backslash X$ for $n<\omega$. We assume without loss of generality that $A_{n+1}\subset A_n$ for $n<\omega$, and that $\bigcap_{n<\omega}A_n=\emptyset$. For $n<\omega$ let K_n be a non-empty, open, compact subset of $A_n\backslash A_{n+1}$, and set $K=\bigcup_{n<\omega}K_n$. Then $\emptyset\neq\mathrm{int}_{\boldsymbol{\beta}X\backslash X}\,\mathrm{cl}_{\boldsymbol{\beta}X}K\subset Z$.

14.18. Theorem. *Assume that $\omega^+=2^\omega$. If X is a zero-dimensional space such that $|C(X)|=2^\omega$ and Z is a non-empty zero-set of $\boldsymbol{\beta}X$ such that $Z\subset\boldsymbol{\beta}X\backslash X$, then Z is homeomorphic to $\boldsymbol{\beta}(\omega)\backslash\omega$.*

Proof. We set $Y=\boldsymbol{\beta}X\backslash Z$. Then Y is a zero-dimensional locally compact σ-compact space such that $|C(Y)|=2^\omega$, and $Z=\boldsymbol{\beta}Y\backslash Y$. Thus Z is a zero-dimensional compact space without isolated elements such that every non-empty zero-set of Z has non-empty interior (Remark 14.17); furthermore Z is an F-space by Lemma 14.16, and clearly Z has 2^ω open-and-closed subsets. Hence according to the Remark following Theorem 14.12 the space Z is homeomorphic to $\boldsymbol{\beta}(\omega)\backslash\omega$.

14.19. Theorem. *If $\alpha>\omega$ then $U(\alpha)$ is not homeomorphic to the Stone space of an α^+-homogeneous Boolean algebra. In fact,*

(a) *if* $\mathrm{cf}(\alpha)>\omega$, *then the Boolean algebra $\mathscr{P}(\alpha)/\mathscr{P}^\alpha(\alpha)$ is not ω^+-homogeneous; and*

(b) *if* $\mathrm{cf}(\alpha)=\omega$, *then the Boolean algebra $\mathscr{P}(\alpha)/\mathscr{P}^\alpha(\alpha)$ is not ω^{++}-homogeneous.*

Proof. Since the Boolean algebras $\mathfrak{B}(U(\alpha))$ and $\mathscr{P}(\alpha)/\mathscr{P}^{\alpha}(\alpha)$ are isomorphic (Lemma 7.12) it suffices to prove (a) and (b). For an ordinal ξ we let $W(\xi)$ be the space whose underlying set is ξ and whose topology is the order topology determined by the well-ordering of ξ. We use only the spaces $W(\omega+1)$ and $W(\omega^+ +1)$, which are easily seen to be totally disconnected, compact spaces which have, respectively, ω and ω^+ open-and-closed subsets. We prove the following statements (1) and (2).

(1) If $\mathrm{cf}(\alpha)>\omega$, then there is a continuous onto function $f: U(\alpha) \to W(\omega+1)$ such that $f^{-1}(\omega)$ is dense in $U(\alpha)$; and

(2) if $\mathrm{cf}(\alpha)=\omega$ (and $\alpha>\omega$), then there is a continuous onto function $f: U(\alpha)\to W(\omega^+ +1)$ such that $f^{-1}(\omega^+)$ is dense in $U(\alpha)$.

Proof of (1). Let $\{A_n: n<\omega\}$ be a sequence of subsets of α such that

$$|A_n|=\alpha \quad \text{for } n<\omega,$$
$$A_n\cap A_m=\emptyset \quad \text{for } n<m<\omega, \quad \text{and}$$
$$\bigcup_{n<\omega} A_n=\alpha.$$

We set $\hat{A}_n=\mathrm{cl}_{\boldsymbol{\beta}(\alpha)} A_n \cap U(\alpha)$ for $n<\omega$, and with $G=\bigcup_{n<\omega}\hat{A}_n$ we see as follows that G is dense in $U(\alpha)$: if $A\in\mathscr{P}(\alpha)$ and $|A|=\alpha$, then $|A\cap A_n|=\alpha$ for some $n<\omega$ since $\mathrm{cf}(\alpha)>\omega$, so $\hat{A}\cap G\supset\hat{A}\cap\hat{A}_n\neq\emptyset$.

We define $h:\alpha\to\omega\subset W(\omega+1)$ by $h[A_n]=\{n\}$ for $n<\omega$, and we set $f=\bar{h}|U(\alpha)$. It is clear that $f[U(\alpha)]=W(\omega+1)$, and that $f^{-1}(\omega)\supset G$.

Proof of (2). Let $\{A_\xi: \xi<\omega^+\}$ be a family of subsets of α such that

$$A_\xi\subset A_\zeta \quad \text{for } \xi<\zeta<\omega^+,$$
$$|A_\xi\setminus\textstyle\bigcup_{\zeta<\xi} A_\zeta|=\alpha \quad \text{for } \xi<\omega^+, \quad \text{and}$$
$$\bigcup_{\xi<\omega^+} A_\xi=\alpha.$$

We set $\hat{A}_\xi=\mathrm{cl}_{\boldsymbol{\beta}(\alpha)} A_\xi\cap U(\alpha)$ for $\xi<\omega^+$, and with $G=\bigcup_{\xi<\omega^+}\hat{A}_\xi$ we claim that G is dense in $U(\alpha)$. Let $A\in\mathscr{P}(\alpha)$ with $|A|=\alpha$, and set $\beta_\xi=|A\cap A_\xi|$. Then

$$\beta_0\leq\beta_1\leq\cdots\leq\beta_\xi\leq\cdots\leq\alpha \quad \text{for } \xi<\omega^+, \quad \text{and}$$
$$\textstyle\sum_{\xi<\omega^+}\beta_\xi=\alpha$$

because $A=\bigcup_{\xi<\omega^+}(A\cap A_\xi)$. Since $\mathrm{cf}(\alpha)=\omega$ and $\omega<\alpha$, there is a sequence $\{\alpha_n: n<\omega\}$ of cardinals such that

$$\omega^+\leq\alpha_0\leq\alpha_1\leq\cdots\leq\alpha_n\leq\cdots,$$
$$\alpha_n<\alpha \quad \text{for } n<\omega, \quad \text{and}$$
$$\textstyle\sum_{n<\omega}\alpha_n=\alpha.$$

If there is $n<\omega$ such that $\beta_\xi\leq\alpha_n$ for $\xi<\omega^+$ then

$$\textstyle\sum_{\xi<\omega^+}\beta_\xi\leq\alpha_n\cdot\omega^+=\alpha_n<\alpha;$$

thus for every $n<\omega$, there is $\xi_n<\omega^+$ such that $\alpha_n<\beta_{\xi_n}$. We set $\xi=\sup\{\xi_n: n<\omega\}$. Then

$$\alpha=\sum_{n<\omega}\alpha_n\leq\sum_{n<\omega}\beta_{\xi_n}\leq\beta_\xi\leq\alpha,$$

so that $\beta_\xi=|A\cap A_\xi|=\alpha$. Thus $\hat{A}\cap G\supset\hat{A}\cap\hat{A}_\xi\neq\emptyset$, and G is dense in $U(\alpha)$ as claimed.

We define $h:\alpha\to\omega^+\subset W(\omega^++1)$ by the condition

$$(*)\qquad h[A_\xi\setminus\bigcup_{\zeta<\xi}A_\zeta]=\{\xi\}\quad\text{for } \xi<\omega^+,$$

and we set $f=\bar{h}|U(\alpha)$. Because $|A_\xi\setminus\bigcup_{\zeta<\xi}A_\zeta|=\alpha$ for $\xi<\omega^+$, it follows from (*) and the compactness of $U(\alpha)$ that $f[U(\alpha)]=W(\omega^++1)$ and that $f^{-1}(\omega^+)\supset G$.

Using (1) we prove (a) and using (2) we prove (b). Since the proofs are similar, we give only the proof of (b).

Let $\alpha=P_0\cup P_1$, with $|P_0|=|P_1|=\alpha$ and $P_0\cap P_1=\emptyset$. We choose any function $g:U(\alpha)\to W(\omega^++1)$ for which $g[\hat{P}_1]=\{\omega^+\}$, and $g[\hat{P}_0]=W(\omega^++1)$, and $g^{-1}(\omega^+)$ is dense in P_0. (That such a function exists is given by (2).) We now claim that with f defined as in (2) there is no auto-homeomorphism h of $U(\alpha)$ such that $f\circ h=g$. For $g[\hat{P}_1]=\{\omega^+\}$; and $\hat{P}_1$ (and hence $h[\hat{P}_1]$) is a non-empty, open subset of $U(\alpha)$, and the defining property of f (as in (2)) implies that $f[h[\hat{P}_1]]\neq\{\omega^+\}$. Statement (b) now follows from Stone's duality and the fact that $\mathscr{P}(\alpha)/\mathscr{P}^\alpha(\alpha)$ is isomorphic to $\mathfrak{B}(U(\alpha))$.

14.20. Lemma. *Let α and β be cardinal numbers, let $\mathscr{F}$ and $\mathscr{G}$ be filters on α and β respectively, and let $\mathfrak{A}$ be a structure. Then there is a (canonical) isomorphism between the structures*

$$(\mathfrak{A}^\alpha/\mathscr{F})^\beta/\mathscr{G}\quad\text{and}\quad\mathfrak{A}^{\alpha\times\beta}/t[\mathscr{G}\cdot\mathscr{F}].$$

Proof. We define a function

$$\phi:(\mathfrak{A}^\alpha/\mathscr{F})^\beta/\mathscr{G}\to\mathfrak{A}^{\alpha\times\beta}/t[\mathscr{G}\cdot\mathscr{F}]$$

as follows. For $a\in(\mathfrak{A}^\alpha/\mathscr{F})^\beta/\mathscr{G}$ choose $g=\langle g_\zeta:\zeta<\beta\rangle\in(\mathfrak{A}^\alpha/\mathscr{F})^\beta$ such that $a=g/\mathscr{G}$ and choose $f_\zeta\in\mathfrak{A}^\alpha$ such that $g_\zeta=f_\zeta/\mathscr{F}$ for $\zeta<\beta$ and define

$$h:\alpha\times\beta\to A$$

by $h(\xi,\zeta)=f_\zeta(\xi)$; then define $\phi(a)=h/t[\mathscr{G}\cdot\mathscr{F}]$. We verify the following statements.

(i) ϕ is a well-defined function.

Let $g'=\langle g'_\zeta:\zeta<\beta\rangle\in(\mathfrak{A}^\alpha/\mathscr{F})^\beta$ be such that $g'/\mathscr{G}=g/\mathscr{G}$ and let $f'_\zeta\in\mathfrak{A}^\alpha$ be such that $g'_\zeta=f'_\zeta/\mathscr{F}$ for $\zeta<\beta$. Then $\{\xi<\alpha: f_\zeta(\xi)=f'_\zeta(\xi)\}\in\mathscr{F}$ whenever $\zeta<\beta$ and $g_\zeta=g'_\zeta$. Then with $h'(\xi,\zeta)=f'_\zeta(\xi)$ for $\langle\xi,\zeta\rangle\in\alpha\times\beta$ we define

$$A=\{\langle\xi,\zeta\rangle\in\alpha\times\beta: h(\xi,\zeta)=h'(\xi,\zeta)\}.$$

Since

$$\{\zeta<\beta:\{\xi<\alpha:h(\xi,\zeta)=h'(\xi,\zeta)\}\in\mathscr{F}\}=\{\zeta<\beta:\{\xi<\alpha:f_\zeta(\xi)=f'_\zeta(\xi)\}\in\mathscr{F}\}$$
$$\supset\{\zeta<\beta:g'_\zeta=g_\zeta\}\in\mathscr{G},$$

we have $A\in t[\mathscr{G}\cdot\mathscr{F}]$. Thus ϕ is well-defined.

(ii) ϕ is a one-to-one function.
Let $g,g'\in(\mathfrak{A}^\alpha/\mathscr{F})^\beta$ with $g/\mathscr{G}\neq g'/\mathscr{G}$, so that

$$\{\zeta<\beta:g_\zeta\neq g'_\zeta\}\in\mathscr{G}.$$

We choose $f_\zeta\in\mathfrak{A}^\alpha$ and $f'_\zeta\in\mathfrak{A}^\alpha$ such that $g_\zeta=f_\zeta/\mathscr{F}$ and $g'_\zeta=f'_\zeta/\mathscr{F}$ for $\zeta<\beta$ and $g_\zeta\neq g'_\zeta$; then

$$\{\xi<\alpha:h(\xi,\zeta)=f_\zeta(\xi)\neq f'_\zeta(\xi)=h'(\xi,\zeta)\}\in\mathscr{F},$$

so that

$$\{\zeta<\beta:\{\xi<\alpha:h(\xi,\zeta)\neq h'(\xi,\zeta)\}\in\mathscr{F}\}\supset\{\zeta<\beta:g_\zeta\neq g'_\zeta\}\in\mathscr{G}.$$

It follows that

$$\{\langle\xi,\zeta\rangle\in\alpha\times\beta:h(\xi,\zeta)\neq h'(\xi,\zeta)\}\in t[\mathscr{G}\cdot\mathscr{F}],$$

and hence $\phi(g/\mathscr{G})\neq\phi(g'/\mathscr{G})$.

(iii) ϕ is an onto function.
Let $h\in\mathfrak{A}^{\alpha\times\beta}$. With $f_\zeta(\xi)=h(\xi,\zeta)$ for $\zeta<\beta$ and $\zeta<\alpha$, and $g=\langle f_\zeta/\mathscr{F}:\zeta<\beta\rangle$, it is clear that

$$\phi(g/\mathscr{G})=h/t[\mathscr{G}\cdot\mathscr{F}].$$

(iv) ϕ is an embedding of structures.
Let $\mathfrak{A}=\langle A,R_i^{\mathfrak{A}}\rangle_{i<\rho}$ be a structure of type $\tau\in\omega^\rho$. We set $\mathfrak{B}=\mathfrak{A}^\alpha/\mathscr{F}$, $\mathfrak{C}=\mathfrak{B}^\beta/\mathscr{G}$, and $\mathfrak{D}=\mathfrak{A}^{\alpha\times\beta}/t[\mathscr{G}\cdot\mathscr{F}]$. Thus ϕ is a function from $\mathfrak{C}$ to $\mathfrak{D}$.

We omit consideration of the simple case in which $i<\rho$ and $\tau(i)=0$. Let $i<\rho$ and $\tau(i)>0$. Let $\langle g_1/\mathscr{G},\dots,g_{\tau(i)}/\mathscr{G}\rangle\in R_i^{\mathfrak{C}}$, where $g_k=\langle g_{k,\zeta}:\zeta<\beta\rangle$ for $1\le k\le\tau(i)$. There are elements $f_{k,\zeta}\in\mathfrak{A}^\alpha$ such that $g_{k,\zeta}=f_{k,\zeta}/\mathscr{F}$ for $1\le k\le\tau(i)$ and $\zeta<\beta$, so with

$$G=\{\zeta<\beta:\langle g_{1,\zeta},\dots,g_{\tau(i),\zeta}\rangle\in R_i^{\mathfrak{B}}\}$$

we have $G\in\mathscr{G}$. Thus if $\zeta\in G$, then $\langle g_{1,\zeta},\dots,g_{\tau(i),\zeta}\rangle\in R_i^{\mathfrak{B}}$, and hence with

$$F_\zeta=\{\xi<\alpha:\langle f_{1,\zeta}(\xi),\dots,f_{\tau(i),\zeta}(\xi)\rangle\in R_i^{\mathfrak{A}}\}$$

we have $F_\zeta\in\mathscr{F}$.

We are to prove that $\langle h_1/t[\mathscr{G}\cdot\mathscr{F}],\dots,h_{\tau(i)}/t[\mathscr{G}\cdot\mathscr{F}]\rangle\in R_i^{\mathfrak{D}}$, *i.e.*, that

$$\{\langle\xi,\zeta\rangle\in\alpha\times\beta:\langle h_1(\xi,\zeta),\dots,h_{\tau(i)}(\xi,\zeta)\rangle\in R_i^{\mathfrak{A}}\}\in t[\mathscr{G}\cdot\mathscr{F}];$$

this is equivalent to the statement that

$$\{\zeta<\beta: F_\zeta\in\mathscr{F}\}\in\mathscr{G},$$

which follows from the relation

$$\{\zeta<\beta: F_\zeta\in\mathscr{F}\}\supset G\in\mathscr{G}$$

established above.

For the reverse inclusion let $\langle h_1/t[\mathscr{G}\cdot\mathscr{F}],\ldots,h_{\tau(i)}/t[\mathscr{G}\cdot\mathscr{F}]\rangle\in R_i^{\mathfrak{D}}$ and define $f_{k,\zeta}(\xi)=h_k(\xi,\zeta)$ for $1\le k\le\tau(i)$, $\zeta<\beta$, and $\xi<\alpha$, and $g_k=\langle f_{k,\zeta}/\mathscr{F}:\zeta<\beta\rangle$ for $1\le k\le\tau(i)$. As was noted in (iii) we have $\phi(g_k/\mathscr{G})=h_k/t[\mathscr{G}\cdot\mathscr{F}]$ for $1\le k\le\tau(i)$; and it follows from the definitions that $\langle g_1/\mathscr{G},\ldots,g_{\tau(i)}/\mathscr{G}\rangle\in R_i^{\mathfrak{C}}$.

14.21. Lemma. *If* $\omega\le\beta$ *and* $\mathfrak{B}$ *is an atomless, β-saturated Boolean algebra, then* $\mathfrak{B}$ *satisfies condition* (H_β).

Proof. Let $C_0\subset\mathfrak{B}\setminus\{\mathbb{1}\}$ and $C_1\subset\mathfrak{B}\setminus\{\mathbb{0}\}$ be such that

$|C_0|+|C_1|<\beta$,
$C_0<C_1$, and
C_0 is directed upwards and C_1 is directed downwards.

We must find $b\in\mathfrak{B}\setminus\{\mathbb{0},\mathbb{1}\}$ such that $C_0<b<C_1$. Set $C=C_0\cup C_1$, so that $|C|<\beta$. For $c\in C_0$ let ϕ_c be the formula $\bar{c}<v_0$; and for $d\in C_1$ let ϕ_d be the formula $v_0<\bar{d}$ (in $\Phi_1(L_C)$, where L is the language of the type of a Boolean algebra). We set $\Phi=\{\phi_c: c\in C\}$ and we claim that Φ is finitely satisfiable in the structure $\mathfrak{B}_C$.

To see this let $n,m<\omega$ and let $c_0,\ldots,c_{n-1}\in C_0$ and $d_0,\ldots,d_{m-1}\in C_1$. Since C_0 is directed upwards, there is $c\in C_0$ such that $c\ge c_0,\ldots,c_{n-1}$ if $n>0$ and $c=\mathbb{0}$ if $n=0$; and since C_1 is directed downwards, there is $d\in C_1$ such that $d\le d_0,\ldots,d_{m-1}$ if $m>0$ and $d=\mathbb{1}$ if $m=0$.

We have $c<d$, and since $\mathfrak{B}$ is atomless there is $b\in\mathfrak{B}\setminus\{\mathbb{0},\mathbb{1}\}$ such that $c<b<d$. This proves the claim.

Since $\mathfrak{B}$ is β-saturated the set Φ is satisfiable in $\mathfrak{B}_C$, say by $b\in\mathfrak{B}$. Then $c<b$ for $c\in C_0$ and $b<d$ for $d\in C_1$, *i.e.*, $C_0<b<C_1$. Thus $\mathfrak{B}$ satisfies property (H_β).

14.22. Theorem. *If* $\alpha^+=2^\alpha$ *then there is an embedding of* S_{α^+} *into* $U(\alpha)$.

Proof. It suffices to prove the dual statement concerning the corresponding Boolean algebras, *i.e.*, that the α^+-homogeneous-universal Boolean algebra $\mathfrak{C}_{\alpha^+}$ of cardinality $\alpha^+(=2^\alpha)$ is the homomorphic image of the Boolean algebra $\mathscr{P}(\alpha)/\mathscr{P}^\alpha(\alpha)$.

We note that the quotient Boolean algebra $\mathscr{P}(\omega)/\mathscr{P}^\omega(\omega)$, since it is isomorphic to $\mathfrak{B}(\boldsymbol{\beta}(\omega)\setminus\omega)$, is an atomless Boolean algebra of cardinality

2^ω (for by Lemma 7.16, $\boldsymbol{\beta}(\omega)\backslash\omega$ has no isolated elements). Furthermore $\mathscr{P}(\omega)/\mathscr{P}^\omega(\omega)$ is isomorphic to the reduced power $\mathbb{2}^\omega/\mathscr{P}^\omega(\omega)$, where $\mathbb{2}$ is the unique Boolean algebra consisting of two elements. From Theorem 10.4 there is a countably incomplete, α^+-good ultrafilter p on α.

The ultrapower $(\mathbb{2}^\omega/\mathscr{P}^\omega(\omega))^\alpha/p$ is of cardinality $(2^\omega)^\alpha = 2^\alpha = \alpha^+$ according to Theorem 12.22, since an α^+-good, countably incomplete ultrafilter is regular (Corollary 9.10(b)).

Furthermore the ultrapower $(\mathbb{2}^\omega/\mathscr{P}^\omega(\omega))^\alpha/p$ is α^+-saturated (from Theorem 13.9) and atomless (by the theorem of Łoś 11.5), so according to Lemma 14.21 it satisfies property (H_{α^+}); thus by Theorem 6.12 it is an α^+-homogeneous-universal Boolean algebra (of cardinality α^+). Thus the ultrapower in question is isomorphic to $\mathfrak{C}_{\alpha^+}$.

The ultrafilter p is regular (Corollary 9.10(b)) and hence uniform (Lemma 7.11(b)), so $\mathscr{P}^\alpha(\alpha) \subset p$ (Lemma 7.2(a)); it follows from Lemma 7.20(c) that $\mathscr{P}^\alpha(\alpha\times\omega) \subset p\cdot\mathscr{P}^\omega(\omega)$, and hence obviously we have

$$\mathscr{P}^\alpha(\omega\times\alpha) \subset t[p\cdot\mathscr{P}^\omega(\omega)]\,.$$

Furthermore according to Lemma 14.20 the ultrapower $(\mathbb{2}^\omega/\mathscr{P}^\omega(\omega))^\alpha/p$ is isomorphic to the reduced power $\mathbb{2}^{\omega\times\alpha}/t[p\cdot\mathscr{P}^\omega(\omega)]$. Thus there is a filter $\mathscr{F}$ on α such that $(\mathbb{2}^\omega/\mathscr{P}^\omega(\omega))^\alpha/p$ is isomorphic to the reduced power $\mathscr{P}(\alpha)/\mathscr{F}$. It is clear that $\mathscr{P}^\alpha(\alpha) \subset \mathscr{F}$, so that $\mathscr{P}(\alpha)/\mathscr{F}$ (and $\mathfrak{C}_{\alpha^+}$) is the homomorphic image of $\mathscr{P}(\alpha)/\mathscr{P}^\alpha(\alpha)$. This completes the proof of the theorem.

The Space of Sub-Uniform Ultrafilters

If α is an infinite cardinal and p is an ultrafilter on α^+ then p is a *sub-uniform* ultrafilter if p is α-uniform but not α^+-uniform. The set of sub-uniform ultrafilters on α^+ is denoted by $SU(\alpha^+)$.

The following few lemmas are needed for the proof of the main result on sub-uniform ultrafilters, given in Theorem 14.27.

14.23. Lemma. *Let $\alpha\geq\omega$ and let F be a non-empty subset of $\boldsymbol{\beta}(SU(\alpha^+))$ such that F is equal to the intersection of at most α open-and-closed subsets of $\boldsymbol{\beta}(SU(\alpha^+))$. Then $F\cap SU(\alpha^+)$ is non-empty.*

Proof. Let $F=\bigcap_{\eta<\alpha}F_\eta$, where F_η is open-and-closed in $\boldsymbol{\beta}(SU(\alpha^+))$, and for $\eta<\alpha$ set $U_\eta = SU(\alpha^+)\backslash F_\eta$. If $F\cap SU(\alpha^+)=\emptyset$, then $\{U_\eta : \eta<\alpha\}$ is an open cover of $SU(\alpha^+)$. Let $SU(\alpha^+)$ be equal to $\bigcup_{\xi<\alpha^+}S_\xi$, where

$S_\xi \subset S_\zeta$ for $\xi<\zeta<\alpha^+$,

S_ξ is open-and-closed in $SU(\alpha^+)$, and

S_ξ is homeomorphic to $U(\alpha)$

for $\xi<\alpha^+$. Let $\phi:\alpha^+\to\mathscr{P}_\omega(\alpha)$ be any function such that

$$\bigcup\{U_\eta:\eta\in\phi(\xi)\}\supset S_\xi \quad \text{for } \xi<\alpha^+.$$

Then there is $A\in\mathscr{P}_\omega(\alpha)$ such that

$$|\{\xi<\alpha^+:\phi(\xi)=A\}|=\alpha^+$$

and we have

$$\bigcup\{U_\eta:\eta\in A\}=SU(\alpha^+);$$

this contradicts the fact that F is non-empty.

14.24. Corollary. *Let $\alpha\geq\omega$ and for $\eta<\alpha$ let F_η be an open-and-closed set in $SU(\alpha^+)$. Then*

$$\mathrm{cl}_{\boldsymbol{\beta}(SU(\alpha^+))}(\bigcap_{\eta<\alpha}F_\eta)=\bigcap_{\eta<\alpha}\mathrm{cl}_{\boldsymbol{\beta}(SU(\alpha^+))}F_\eta.$$

Proof. We only need to prove that

$$\mathrm{cl}_{\boldsymbol{\beta}(SU(\alpha^+))}(\bigcap_{\eta<\alpha}F_\eta)\supset\bigcap_{\eta<\alpha}\mathrm{cl}_{\boldsymbol{\beta}(SU(\alpha^+))}F_\eta.$$

If $p\in\bigcap_{\eta<\alpha}\mathrm{cl}_{\boldsymbol{\beta}(SU(\alpha^+))}F_\eta\setminus\mathrm{cl}_{\boldsymbol{\beta}(SU(\alpha^+))}(\bigcap_{\eta<\alpha}F_\eta)$ then there is an open-and-closed subset N of $\boldsymbol{\beta}(SU(\alpha^+))$ such that $p\in N$ and

$$N\cap\mathrm{cl}_{\boldsymbol{\beta}(SU(\alpha^+))}(\bigcap_{\eta<\alpha}F_\eta)=\emptyset.$$

Then

$$F=N\cap\bigcap_{\eta<\alpha}\mathrm{cl}_{\boldsymbol{\beta}(SU(\alpha^+))}F_\eta$$

is the intersection of at most α open-and-closed subsets of $\boldsymbol{\beta}(SU(\alpha^+))$, and $p\in F$; but

$$F\cap SU(\alpha^+)\subset N\cap\bigcap_{\eta<\alpha}F_\eta=\emptyset,$$

contradicting Lemma 14.23.

We also use the following immediate consequence of earlier results.

14.25. Corollary. *If α is an infinite, regular cardinal then $SU(\alpha^+)$ is a zero-dimensional (locally compact) F_{α^+}-space.*

Proof. According to Theorem 14.9 the space $U_\alpha(\alpha^+)$ is an F_{α^+}-space. Thus Lemma 14.1 applies, with $U_\alpha(\alpha^+)$, $SU(\alpha^+)$ and α^+ replacing X, G and α respectively.

14.26. Lemma. *Let α be a regular infinite cardinal and let F be a closed subset of $\boldsymbol{\beta}(SU(\alpha^+))\setminus SU(\alpha^+)$ which is equal to the intersection of at most α open-and-closed subsets (of $\boldsymbol{\beta}(SU(\alpha^+))\setminus SU(\alpha^+)$). Then there is a closed subset W of $SU(\alpha^+)$ which is equal to the intersection of at most α open-and-closed subsets of $SU(\alpha^+)$ and such that*

$$\hat{W}\,(=\mathrm{cl}_{\boldsymbol{\beta}(SU(\alpha^+))}W\setminus SU(\alpha^+))=F.$$

Proof. Let $\{F_\eta : \eta < \alpha\}$ be a family of open-and-closed subsets of $\beta(SU(\alpha^+))\setminus SU(\alpha^+)$ such that $F = \bigcap_{\eta<\alpha} F_\eta$. Since, by Corollary 14.25, $SU(\alpha^+)$ is a zero-dimensional locally compact space, there is, by Theorem 2.28 (b), an open-and-closed set N_η of $SU(\alpha^+)$ such that $\hat{N}_\eta = F_\eta$ for $\eta < \alpha$. We set $W = \bigcap_{\eta<\alpha} N_\eta$; then W is equal to the intersection of at most α open-and-closed subsets of $SU(\alpha^+)$, and from Corollary 14.24 we have

$$\begin{aligned} \mathrm{cl}_{\beta(SU(\alpha^+))} W \setminus SU(\alpha^+) &= \mathrm{cl}_{\beta(SU(\alpha^+))}(\textstyle\bigcap_{\eta<\alpha} N_\eta) \setminus SU(\alpha^+) \\ &= \textstyle\bigcap_{\eta<\alpha} (\mathrm{cl}_{\beta(SU(\alpha^+))} N_\eta) \setminus SU(\alpha^+) \\ &= \textstyle\bigcap_{\eta<\alpha} \hat{N}_\eta = \bigcap_{\eta<\alpha} F_\eta = F . \end{aligned}$$

14.27. Theorem. *If* $\alpha = \alpha^{\underline{\alpha}}$ *then* $\beta(SU(\alpha^+))\setminus SU(\alpha^+)$ *has property* Φ_α.

Proof. Let $F = \bigcap_{\eta<\alpha} F_\eta$ be a closed, non-empty subset of $\beta(SU(\alpha^+))\setminus SU(\alpha^+)$, where for $\eta < \alpha$ the set F_η is open-and-closed in $\beta(SU(\alpha^+))\setminus SU(\alpha^+)$. Let $\{V_\eta : \eta < \alpha\}$ be a family of pairwise disjoint, non-empty open-and-closed subsets of $\beta(SU(\alpha^+))\setminus SU(\alpha^+)$. By Corollary 14.25 and Theorem 2.28 (b) there are open-and-closed sets N_η in $SU(\alpha^+)$ such that $\hat{N}_\eta = V_\eta$ for $\eta < \alpha$. Let $SU(\alpha^+)$ be equal to $\bigcup_{\xi<\alpha^+} S_\xi$, where

$$\begin{aligned} &S_\xi \subset S_\zeta \quad \text{for } \xi < \zeta < \alpha^+ , \\ &S_\xi \text{ is open-and-closed in } SU(\alpha^+) \text{ for } \xi < \alpha^+, \quad \text{and} \\ &S_\xi \text{ is homeomorphic to } U(\alpha) \text{ for } \xi < \alpha^+ . \end{aligned}$$

Since a subset of $SU(\alpha^+)$ is relatively compact (*i.e.*, has compact closure) in $SU(\alpha^+)$ if and only if it is a subset of S_ξ for some $\xi < \alpha^+$, there is for $\eta < \eta' < \alpha$ some $\xi(\eta, \eta')$ such that $N_\eta \cap N_{\eta'} \subset S_{\xi(\eta,\eta')}$. By taking ξ such that $\xi(\eta, \eta') < \xi < \alpha^+$ for all $\eta < \eta' < \alpha$, and by replacing N_η by $N_\eta \setminus S_\xi$, we may assume that the family $\{N_\eta : \eta < \alpha\}$ consists of pairwise disjoint sets. We assume also that $V_\eta \cap F = \emptyset$ for all $\eta < \alpha$, since otherwise any $p \in V_\eta \cap F$ (for some $\eta < \alpha$) will satisfy the conclusion of the theorem (by taking $N = V_\eta$).

By Lemma 14.26 there is a closed subset W of $SU(\alpha^+)$, the intersection of at most α open-and-closed subsets of $SU(\alpha^+)$, such that $\hat{W} = F$. Then $(W \cap N_\eta)^\wedge = \hat{W} \cap \hat{N}_\eta = F \cap V_\eta = \emptyset$ (using Corollary 14.24), so that $W \cap N_\eta \subset S_{\xi(\eta)}$ for some $\xi(\eta) < \alpha^+$; we may assume as above that $W \cap N_\eta = \emptyset$ for $\eta < \alpha$. We set

$$B = \{x \in W : x \text{ is not in the uniform closure of the family } \{N_\eta : \eta < \alpha\}\}.$$

Thus $x \in B$ if and only if there is some open-and-closed set $N(x)$ containing x such that

$$|\{\eta < \alpha : N(x) \cap N_\eta \neq \emptyset\}| < \alpha .$$

We claim now that B is not relatively compact in $SU(\alpha^+)$. If B is relatively compact then $B \subset S_\xi$ for some $\xi < \alpha^+$, so since $\hat{W} = F \neq \emptyset$ there is ζ with $\xi < \zeta < \alpha^+$ such that $(W \backslash B) \cap (S_\zeta \backslash S_\xi) \neq \emptyset$. Then $S_\zeta \backslash S_\xi$ is homeomorphic to $U(\alpha)$, and $W \backslash B$ is a non-empty closed subset which is the intersection of not more than α open-and-closed sets. But every element of $W \backslash B$ is in the uniform closure of the family $\{N_\eta \cap (S_\zeta \backslash S_\xi) : \eta < \alpha\}$. This contradicts Corollary 9.15(b), and establishes the claim that B is not relatively compact in $SU(\alpha^+)$.

By transfinite recursion, we can find a set $C = \{c_\xi : \xi < \alpha^+\}$, and open-and-compact subsets $N(c_\xi)$ of $SU(\alpha^+)$ for $\xi < \alpha^+$, such that

$$C \subset B,$$
$$|C| = \alpha^+,$$
$$\text{if } D \subset C, \quad \text{then } D \text{ is relatively compact if and only if } |D| \leq \alpha,$$
$$c_\xi \in N(c_\xi) \quad \text{for } \xi < \alpha^+,$$
$$N(c_\xi) \cap N(c_\zeta) = \emptyset \quad \text{for } \xi < \zeta < \alpha^+, \quad \text{and}$$
$$|\{\eta < \alpha : N(c_\xi) \cap N_\eta \neq \emptyset\}| < \alpha \quad \text{for } \xi < \alpha^+.$$

We set $A_\xi = \{\eta < \alpha : N(c_\xi) \cap N_\eta \neq \emptyset\}$, and let

$$\phi : \alpha^+ \to \mathcal{P}_\alpha(\alpha)$$

be defined by $\phi(\xi) = A_\xi$. Since $|\mathcal{P}_\alpha(\alpha)| = \alpha^{\underline{\alpha}} = \alpha$, there is $A \in \mathcal{P}_\alpha(\alpha)$ such that $A = A_\xi$ for α^+ many $\xi < \alpha^+$. We set $D = \{\xi < \alpha^+ : A = A_\xi\}$, and we define

$$G = \bigcup \{N_\eta : \eta \in \alpha \backslash A\} \cup \bigcup \{N(c_\xi) : \xi \in D\}.$$

It follows from Theorem 14.1, with $X = SU(\alpha^+)$, with G defined as above, and with α^+ replacing α, that G is C^*-embedded in $SU(\alpha^+)$. Let $f \in C^*(SU(\alpha^+))$ be such that

$$\begin{aligned} f(x) &= 1 \quad \text{if } x \in \bigcup \{N_\eta : \eta \in \alpha \backslash A\} \\ &= 0 \quad \text{if } x \in \bigcup \{N(c_\xi) : \xi \in D\}, \end{aligned}$$

and as usual let $\bar{f} \in C(\boldsymbol{\beta}(SU(\alpha^+))$ be the Stone extension of f to $\boldsymbol{\beta}(SU(\alpha^+))$. Let p be any element of $\boldsymbol{\beta}(SU(\alpha^+)) \backslash SU(\alpha^+)$ in the closure of $\{c_\xi : \xi \in D\}$. Then $p \in F$ because $C \subset B \subset W$ and $\hat{W} = F$; and $\bar{f}(p) = 0$. We set

$$N = \{q \in \boldsymbol{\beta}(SU(\alpha^+)) \backslash SU(\alpha^+) : \bar{f}(q) \neq 1\},$$

so that $p \in N$ and $N \cap V_\eta = \emptyset$ whenever $\eta \in \alpha \backslash A$. Thus

$$|\{\eta < \alpha : N \cap V_\eta \neq \emptyset\}| \leq |A| < \alpha$$

and the proof is complete.

14.28. Lemma. *If α and β are regular cardinals and $\omega \leq \beta < \alpha$, then $U(\alpha)$ does not have property Φ_β.*

Proof. Let $d = \{D_\eta : \eta < \beta\}$ be a family of subsets of α such that

$$|D_\eta| = \alpha \quad \text{for } \eta < \beta,$$
$$D_\eta \cap D_{\eta'} = \emptyset \quad \text{for } \eta < \eta' < \beta, \quad \text{and}$$
$$\bigcup_{\eta < \beta} D_\eta = \alpha.$$

Let $F = \bigcap_{\zeta < \beta} (\alpha \setminus \bigcup_{\eta < \zeta} D_\eta)\hat{}$ and $V_\eta = \hat{D}_\eta$ for $\eta < \beta$, (where $\hat{A} = (\mathrm{cl}_{\boldsymbol{\beta}(\alpha)} A) \cap U(\alpha))$ for $A \subset \alpha$). Then F is the intersection of at most β open-and-closed subsets of $U(\alpha)$, and $\{V_\eta : \eta < \beta\}$ is a family of β pairwise disjoint, non-empty, open-and-closed subsets of $U(\alpha)$. But it is clear that if $p \in F$ and N is any open subset of $U(\alpha)$ containing p, then

$$|\{\eta < \alpha : V_\eta \cap N \neq \emptyset\}| = \beta.$$

14.29. Corollary. *If $\alpha = \alpha^{\underline{\alpha}}$ then $\boldsymbol{\beta}(SU(\alpha^+)) \setminus SU(\alpha^+)$ is not homeomorphic to $U(\alpha^+)$. In particular $SU(\alpha^+)$ is not C^*-embedded in $\boldsymbol{\beta}(\alpha^+)$.*

Proof. By Theorem 14.27, $\boldsymbol{\beta}(SU(\alpha^+)) \setminus SU(\alpha^+)$ has property Φ_α, while by Lemma 14.28 (with α^+ and α replacing α and β respectively), $U(\alpha^+)$ does not have property Φ_α.

14.30. Corollary. *Let $\alpha = \alpha^{\underline{\alpha}}$.*

(a) *If $\alpha^+ \cup SU(\alpha^+) \subset X \subset \boldsymbol{\beta}(\alpha^+) \setminus U(\alpha^+)$ then X is not a normal space; and*

(b) *$SU(\alpha^+)$ is not a normal space.*

Proof. (a) follows from the fact that $SU(\alpha^+)$ is a closed subset of X which is not C^*-embedded in X.

(b) It follows from (a) that $\alpha^+ \cup SU(\alpha^+)$ is not a normal space. To prove (b) it is sufficient to prove that there is a closed subset of $SU(\alpha^+)$ which is homeomorphic to $\alpha^+ \cup SU(\alpha^+)$. Let $\phi : \alpha^+ \to \alpha^+$ be the unique one-to-one, order-preserving function such that

$$\phi[\alpha^+] = \{\xi < \alpha^+ : \mathrm{cf}(\xi) = \alpha\}.$$

We define

$$S_\xi = (\mathrm{cl}_{\boldsymbol{\beta}(\alpha^+)} \phi[\xi]) \cap SU(\alpha^+) \quad \text{for } \xi < \alpha^+$$

and we note that

$$SU(\alpha^+) = \bigcup_{\xi < \alpha^+} S_\xi,$$
S_ξ is homeomorphic to $U(\alpha)$ for $\xi < \alpha^+$,
S_ξ is open-and-closed in $SU(\alpha^+)$ for $\xi < \alpha^+$, and
$$S_\zeta \subsetneqq S_\xi \quad \text{for } \zeta < \xi < \alpha^+.$$

Let $\{H_\xi : \xi < \alpha^+\}$ be a family of non-empty open-and-closed subsets of $SU(\alpha^+)$ such that $H_\xi \subset S_{\xi+1} \backslash S_\xi$ for $\xi < \alpha^+$ and define $H = \bigcup_{\xi < \alpha^+} H_\xi$. It follows from Lemma 14.1(b) and Corollary 14.25 that H is an open subset of $SU(\alpha^+)$ which is C^*-embedded in $SU(\alpha^+)$. Let $a_\xi \in H_\xi$ for $\xi < \alpha^+$ and set $A = \{a_\xi : \xi < \alpha^+\}$. Then A is a C^*-embedded subset of $SU(\alpha^+)$ homeomorphic to α^+. It is clear that if $B \subset A$ and $|B| \leq \alpha$ then $\mathrm{cl}_{SU(\alpha^+)} B$ is compact and hence homeomorphic to $\boldsymbol{\beta}(B)$, and that $\mathrm{cl}_{SU(\alpha^+)} A = \bigcup \{\mathrm{cl}_{SU(\alpha^+)} B : B \subset A, |B| \leq \alpha\}$. It follows that $\mathrm{cl}_{SU(\alpha^+)} A$ is homeomorphic to $\alpha^+ \cup SU(\alpha^+)$. This completes the proof.

We remark that the argument given in part (b) of the above corollary proves that for any infinite regular cardinal α there is a closed subset of $SU(\alpha^+)$ homeomorphic to $\alpha^+ \cup SU(\alpha^+)$.

14.31. Theorem. *Assume $\omega^+ = 2^\omega$ and let $p \in \boldsymbol{\beta}(\omega) \backslash \omega$. Then*

(a) *$\boldsymbol{\beta}(\omega) \backslash \omega \backslash \{p\}$ is not C^*-embedded in $\boldsymbol{\beta}(\omega) \backslash \omega$; and*

(b) *$\boldsymbol{\beta}(\omega) \backslash \omega \backslash \{p\}$ is not a normal space.*

Proof. (a) If $\omega^+ = 2^\omega$ then $\boldsymbol{\beta}(\omega) \backslash \omega$ is homeomorphic to S_{ω^+} (Corollary 14.15) and hence from Theorem 14.6(c) (applied to ω^+) it follows that $\boldsymbol{\beta}(\omega) \backslash \omega \backslash \{p\}$ is not C^*-embedded in $\boldsymbol{\beta}(\omega) \backslash \omega$.

(b) There are two possibilities to consider.

Case 1. p is not a P-point of $\boldsymbol{\beta}(\omega) \backslash \omega$.
Then there is a zero-set Z of $\boldsymbol{\beta}(\omega) \backslash \omega$ such that $p \in BdZ$. We set $C = (\boldsymbol{\beta}(\omega) \backslash \omega) \backslash Z$; thus C is a cozero-set of $\boldsymbol{\beta}(\omega) \backslash \omega$ and since $\boldsymbol{\beta}(\omega) \backslash \omega$ is an F-space it follows that C is C^*-embedded in $\boldsymbol{\beta}(\omega) \backslash \omega$; hence BdZ is (homeomorphic to) $\boldsymbol{\beta} C \backslash C$. From Theorem 14.18 BdZ is homeomorphic to $\boldsymbol{\beta}(\omega) \backslash \omega$ and from part (a) above, $BdZ \backslash \{p\}$ is not C^*-embedded in BdZ. Thus there is $f \in C^*(BdZ \backslash \{p\})$ which has no continuous real-valued extension defined on BdZ.

Suppose now that $\boldsymbol{\beta}(\omega) \backslash \omega \backslash \{p\}$ is a normal space. Then since $BdZ \backslash \{p\}$ is a closed subset of $\boldsymbol{\beta}(\omega) \backslash \omega \backslash \{p\}$ it is C^*-embedded, so there is $g \in C^*(\boldsymbol{\beta}(\omega) \backslash \omega \backslash \{p\})$ such that $f \subset g$. Then $g|C \in C^*(C)$ and since C is C^*-embedded there is $h \in C(C \cup BdZ)$ such that $g|C \subset h$. Since C is dense in $C \cup BdZ$ and $h|C = g|C$ it follows that $h|C \cup (BdZ \backslash \{p\}) = g|C \cup (BdZ \backslash \{p\})$ and hence

$$f = g|BdZ \backslash \{p\} \subset h|BdZ,$$

contradicting the choice of f.

Case 2. p is a P-point of $\boldsymbol{\beta}(\omega) \backslash \omega$.
Then $\boldsymbol{\beta}(\omega) \backslash \omega \backslash \{p\}$ and $SU(\omega^+)$ are homeomorphic, since they are homeomorphic to Ω_{ω^+} (Theorem 6.19 and Corollary 6.20). Hence from Corollary 14.30, $\boldsymbol{\beta}(\omega) \backslash \omega \backslash \{p\}$ is not a normal space.

14.32. Corollary *(Aronszajn, Specker). If $\alpha = \alpha^{\underline{\alpha}}$ then α^+ is not a ramifiable cardinal.*

Proof. Since $SU(\alpha^+)$ is zero-dimensional (Corollary 14.25) and $SU(\alpha^+)$ is not C^*-embedded in $U_\alpha(\alpha^+)$, it follows from Lemma 2.23(b) and Theorem 2.27 that there is a $\{0,1\}$-valued continuous function f on $SU(\alpha^+)$ which cannot be extended to a $\{0,1\}$-valued continuous function on $U_\alpha(\alpha^+)$. For every $\xi<\alpha^+$ we choose a $\{0,1\}$-valued function $f_\xi : \xi \to \{0,1\}$ such that

$$\bar{f}_\xi | SU(\alpha^+) \cap \mathrm{cl}_{\boldsymbol{\beta}(\alpha^+)} \xi = f | SU(\alpha^+) \cap \mathrm{cl}_{\boldsymbol{\beta}(\alpha^+)} \xi .$$

Since $SU(\alpha^+) \cap \mathrm{cl}_{\boldsymbol{\beta}(\alpha^+)} \xi$ is a compact subset of $\alpha^+ \cup (SU(\alpha^+) \cap \mathrm{cl}_{\boldsymbol{\beta}(\alpha^+)} \xi)$, such a function f_ξ exists for $\xi<\alpha^+$. For $\xi<\alpha^+$ we set

$$\mathscr{A}_\xi = \{h \in \{0,1\}^\xi : h = f_\eta | \xi \text{ for some } \eta \text{ with } \xi \le \eta < \alpha^+\},$$

and we set $\mathfrak{A} = \bigcup_{\xi<\alpha^+} \mathscr{A}_\xi$. We make $\mathfrak{A}$ into a partially ordered set by setting $h \le h'$ if $h \subset h'$ (for h and h' in $\mathfrak{A}$). It is easy to verify that $\mathfrak{A}$ is a ramification system, that for $\xi<\alpha^+$ the set of elements in $\mathfrak{A}$ of order ξ is the set $\mathscr{A}_\xi$, and (since $f_\xi \in \mathscr{A}_\xi$) that $\mathscr{A}_\xi \neq \emptyset$ for $\xi<\alpha^+$. Furthermore if $h \in \mathscr{A}_\xi$, then $|\{\zeta<\xi : h(\zeta) \neq f_\xi(\zeta)\}| < \alpha$ (because $\bar{h} | SU(\alpha^+) \cap \mathrm{cl}_{\boldsymbol{\beta}(\alpha^+)} \xi = \bar{f}_\xi | SU(\alpha^+) \cap \mathrm{cl}_{\boldsymbol{\beta}(\alpha^+)} \xi$); it then follows that $|\mathscr{A}_\xi| \le \alpha^{\underline{\alpha}} = \alpha < \alpha^+$ for $\xi<\alpha^+$. And finally, $\mathfrak{A}$ does not contain any well-ordered subset of order-type α^+; if $\{h_\xi : \xi<\alpha^+\}$ is such a chain, then

$$h = \bigcup_{\xi<\alpha^+} h_\xi$$

is a well-defined, $\{0,1\}$-valued function on α^+ such that $f \subset \bar{h} | U_\alpha(\alpha^+)$, contrary to the choice of f. This completes the proof of the theorem.

For an infinite cardinal α we denote by $N(\alpha)$ the space of all ultrafilters p on α such that p is non-principal and non-uniform.

14.33. Theorem. *If α is an uncountable weakly compact cardinal, then $\boldsymbol{\beta}(N(\alpha)) = \boldsymbol{\beta}(\alpha) \backslash \alpha$.*

Proof. It is clear that $N(\alpha)$ is dense in $\boldsymbol{\beta}(\alpha) \backslash \alpha$. We claim that if Z_0 and Z_1 are disjoint zero-sets of $N(\alpha)$, then there is $A \in \mathscr{P}(\alpha)$ such that

$$Z_0 \subset \mathrm{cl}_{\boldsymbol{\beta}(\alpha)} A, \quad \text{and}$$
$$Z_1 \cap \mathrm{cl}_{\boldsymbol{\beta}(\alpha)} A = \emptyset .$$

Since $\mathrm{cl}_{\boldsymbol{\beta}(\alpha)} A$ is open-and-closed in $\boldsymbol{\beta}(\alpha)$, it follows from this claim that disjoint zero-sets of $N(\alpha)$ are completely separated in $\boldsymbol{\beta}(\alpha)$ (and hence in $\mathrm{cl}_{\boldsymbol{\beta}(\alpha)} N = \boldsymbol{\beta}(\alpha) \backslash \alpha$); thus the fact that $\boldsymbol{\beta}(N(\alpha)) = \boldsymbol{\beta}(\alpha) \backslash \alpha$ follows from Corollary 2.8.

For every $\xi<\alpha$, we note that $\mathrm{cl}_{\boldsymbol{\beta}(\alpha)}\xi$ (which we identify with $\boldsymbol{\beta}(\xi)$) is contained in $\alpha\cup N(\alpha)$, so that $\boldsymbol{\beta}(\xi)\backslash\xi=N(\alpha)\cap\boldsymbol{\beta}(\xi)$. Thus for $\xi<\alpha$ there is $A_\xi\in\mathscr{P}(\xi)$ such that

$$Z_0\cap\boldsymbol{\beta}(\xi)\subset\mathrm{cl}_{\boldsymbol{\beta}(\alpha)}A_\xi \quad\text{and}$$
$$Z_1\cap\boldsymbol{\beta}(\xi)\subset\mathrm{cl}_{\boldsymbol{\beta}(\alpha)}(\xi\backslash A_\xi)\,.$$

Now for $\xi<\alpha$ we define

$$\mathscr{A}_\xi=\{\langle\xi,A_\eta\cap\xi\rangle:\xi\leq\eta<\alpha\}\,,$$

and we note that

$$\mathscr{A}_\xi\neq\emptyset$$

since $\langle\xi,A_\xi\rangle\in\mathscr{A}_\xi$, and

$$|\mathscr{A}_\xi|<\alpha$$

since $\mathscr{A}_\xi\subset\{\xi\}\times\mathscr{P}(\xi)$ and $|\mathscr{P}(\xi)|=2^{|\xi|}<\alpha$ (using the fact that α is strongly inaccessible). We set

$$\mathfrak{A}=\bigcup_{\xi<\alpha}\mathscr{A}_\xi\,,$$

and we make $\mathfrak{A}$ into a partially ordered set by defining

$$\langle\xi.A_\eta\cap\xi\rangle\leq\langle\xi',A_{\eta'}\cap\xi'\rangle$$

if $\xi\leq\xi'$ and $A_{\eta'}\cap\xi=A_\eta\cap\xi$, where $\langle\xi,A_\eta\cap\xi\rangle$ and $\langle\xi',A_{\eta'}\cap\xi'\rangle$ are in $\mathfrak{A}$. It is easy to verify that $\mathfrak{A}$ is a ramification system and that the set of elements in $\mathfrak{A}$ of order ξ is the set $\mathscr{A}_\xi$ for $\xi<\alpha$. Since α is a weakly compact cardinal, it is ramifiable (cf. Theorem 8.23), so $\mathfrak{A}$ contains a well-ordered subset of order-type α. With no loss of generality we assume that this chain is of the form

$$\{\langle\xi,A_{\eta(\xi)}\cap\xi\rangle:\xi<\alpha\}$$

for some family $\{\eta(\xi):\xi<\alpha\}$ such that $\xi\leq\eta(\xi)<\alpha$ for $\xi<\alpha$. The set

$$A=\bigcup_{\xi<\alpha}(A_{\eta(\xi)}\cap\xi)$$

satisfies the requirements of the claim, and the proof is complete.

Relations to Measurable Cardinals

We recall for $\omega\leq\kappa\leq\alpha$ that $\Omega_\kappa(\alpha)$ denotes the set of κ-complete ultrafilters on α.

14.34. Lemma. *If $\omega\leq\kappa\leq\alpha$, then $\Omega_\kappa(\alpha)$ is a P_κ-space. If in addition κ is a measurable cardinal, then $\Omega_\kappa(\alpha)$ is a dense subspace of $U_\kappa(\alpha)$.*

Proof. Let $\lambda<\kappa$, let $\{U_\xi:\xi<\lambda\}$ be a family of open subsets of $\Omega_\kappa(\alpha)$, and let $p\in\bigcap_{\xi<\lambda}U_\xi$. For every $\xi<\lambda$ there is $A_\xi\in p$ such that $p\in\hat{A}_\xi\cap\Omega_\kappa(\alpha)\subset U_\xi$. Since $\lambda<\kappa$ and p is κ-complete we have $\bigcap_{\xi<\lambda}A_\xi\in p$, so that $p\in(\bigcap_{\xi<\lambda}A_\xi)^\wedge\cap\Omega_\kappa(\alpha)\subset\bigcap_{\xi<\lambda}U_\xi$. Thus $\Omega_\kappa(\alpha)$ is a P_κ-space.

Assume now in addition that κ is a measurable cardinal, and let U be a non-empty, open subspace of $U_\kappa(\alpha)$. Then there is $A\in\mathscr{P}(\alpha)$ such that $|A|=\kappa$ and $\hat{A}\subset U$, and since κ is measurable there is a κ-complete non-principal ultrafilter p on α such that $A\in p$. Then $p\in\hat{A}\cap\Omega_\kappa(\alpha)\subset U\cap\Omega_\kappa(\alpha)$, and we conclude that $\Omega_\kappa(\alpha)$ is a dense subspace of $U_\kappa(\alpha)$.

14.35. Lemma. *Let* $\omega\le\kappa\le\alpha$.

(a) *If* $\Omega_\kappa(\alpha)$ *is* κ*-compact then every open-and-closed subset of* $\Omega_\kappa(\alpha)$ *is equal to* $\hat{A}\cap\Omega_\kappa(\alpha)$ *for some* $A\in\mathscr{P}(\alpha)$; *and*

(b) *every open-and-closed subset of* $\Omega_\kappa(\alpha)$ *is equal to* $\hat{A}\cap\Omega_\kappa(\alpha)$ *for some* $A\in\mathscr{P}(\alpha)$ *if and only if* $\Omega_\kappa(\alpha)$ *is* C^**-embedded in* $U_\kappa(\alpha)$.

Proof. (a) Let U be an open-and-closed subset of $\Omega_\kappa(\alpha)$. For every $p\in U$ there is $A_p\in\mathscr{P}(\alpha)$ such that $p\in\hat{A}_p\cap\Omega_\kappa(\alpha)\subset U$, so since U is closed in $\Omega_\kappa(\alpha)$ (and is therefore itself κ-compact) there is $I\subset U$, with $|I|<\kappa$, such that $\bigcup\{\hat{A}_p\cap\Omega_\kappa(\alpha):p\in I\}=U$. We set

$$A=\bigcup_{p\in I}A_p,$$

and we claim that $U=\hat{A}\cap\Omega_\kappa(\alpha)$. Clearly

$$U=\bigcup\{\hat{A}_p\cap\Omega_\kappa(\alpha):p\in I\}\subset\hat{A}\cap\Omega_\kappa(\alpha);$$

if $q\in\hat{A}\cap\Omega_\kappa(\alpha)$, so that $\bigcup_{p\in I}A_p=A\in q$, then since q is κ-complete we have $A_p\in q$ for some $p\in I$ and then $q\in\hat{A}_p\cap\Omega_\kappa(\alpha)\subset U$.

(b) If $\kappa=\omega$ then $\Omega_\kappa(\alpha)=U_\kappa(\alpha)$ and if $\omega<\kappa$ then $\Omega_\kappa(\alpha)$ is a P_κ-space (by Lemma 14.34); hence $\Omega_\kappa(\alpha)$ is zero-dimensional. It is easy to verify, using Theorem 2.27 and Lemma 7.12(b), that the following statements are equivalent:

every open-and-closed subset of $\Omega_\kappa(\alpha)$ is equal to $\hat{A}\cap\Omega_\kappa(\alpha)$ for some $A\in\mathscr{P}(\alpha)$;

every open-and-closed subset of $\Omega_\kappa(\alpha)$ is equal to $U\cap\Omega_\kappa(\alpha)$ for some open-and-closed subset U of $\mathrm{cl}_{U_\kappa(\alpha)}\Omega_\kappa(\alpha)$;

every continuous $\{0,1\}$-valued function on $\Omega_\kappa(\alpha)$ has a continuous extension defined on $\mathrm{cl}_{U_\kappa(\alpha)}\Omega_\kappa(\alpha)$;

$\Omega_\kappa(\alpha)$ is C^*-embedded in $\mathrm{cl}_{U_\kappa(\alpha)}\Omega_\kappa(\alpha)$; and

$\Omega_\kappa(\alpha)$ is C^*-embedded in $U_\kappa(\alpha)$.

14.36. Lemma. *Let $\omega \leq \kappa \leq \alpha$ with κ a measurable cardinal, and let U be a dense, open subset of $U_\kappa(\alpha)$ such that $\tau(U) \leq \kappa$. Then there is an open subset V of $U_\kappa(\alpha)$ such that*

$$\tau(V) < \kappa \quad \textit{and}$$
$$\Omega_\kappa(\alpha) \subset V \subset U .$$

Proof. By Theorem 14.11 the set U contains densely an open subset V of $U_\kappa(\alpha)$ such that $\tau(V) < \kappa$. We show that $\Omega_\kappa(\alpha) \subset V$. If $V = \bigcup_{\xi < \lambda} V_\xi$, where $\lambda = \tau(V) < \kappa$ and each set V_ξ is open-and-closed in $U_\kappa(\alpha)$, then since $\Omega_\kappa(\alpha)$ is a P_κ-space (by Lemma 14.34) the set

$$\bigcup_{\xi < \lambda} (V_\xi \cap \Omega_\kappa(\alpha))$$

is closed in $\Omega_\kappa(\alpha)$. Since both V and $\Omega_\kappa(\alpha)$ are dense in $U_\kappa(\alpha)$ (by Lemma 14.34) and V is open, the set $V \cap \Omega_\kappa(\alpha)$, which is

$$\bigcup_{\xi < \lambda} (V_\xi \cap \Omega_\kappa(\alpha)) ,$$

is dense in $\Omega_\kappa(\alpha)$. Thus

$$V \cap \Omega_\kappa(\alpha) = \bigcup_{\xi < \lambda} (V_\xi \cap \Omega_\kappa(\alpha)) = \Omega_\kappa(\alpha)$$

and we have $\Omega_\kappa(\alpha) \subset V \subset U$.

14.37. Lemma. *If $\omega \leq \kappa \leq \alpha$, the following are equivalent.*

(a) *$\Omega_\kappa(\alpha)$ is κ-compact;*

(b) *if U is an open subset of $U_\kappa(\alpha)$ with $\Omega_\kappa(\alpha) \subset U$, then there is an open subset V of $U_\kappa(\alpha)$ such that*

$$\Omega_\kappa(\alpha) \subset V \subset U \quad \textit{and}$$
$$\tau(V) < \kappa .$$

If in addition κ is measurable then the following is also equivalent to (a) *and* (b).

(c) *If U is an open subset of $U_\kappa(\alpha)$ with $\Omega_\kappa(\alpha) \subset U$, then there is an open subset V of $U_\kappa(\alpha)$ such that*

$$V \textit{ is a dense subset of } U \textit{ and}$$
$$\tau(V) \leq \kappa .$$

Proof. (a)⇒(b). For every $p \in \Omega_\kappa(\alpha)$ there is $A_p \in \mathscr{P}(\alpha)$ such that $p \in \hat{A}_p \subset U$. Since $\Omega_\kappa(\alpha)$ is κ-compact there is $I \subset \Omega_\kappa(\alpha)$, with $|I| < \kappa$, such that

$$V = \bigcup \{\hat{A}_p : p \in I\} \supset \Omega_\kappa(\alpha) .$$

It is clear that $V \subset U$ and that $\tau(V) < \kappa$.

(b)⇒(a). Let $\{W_i : i \in I\}$ be an open cover of $\Omega_\kappa(\alpha)$, and for $i \in I$ let U_i be an open subset of $U_\kappa(\alpha)$ such that $W_i = U_i \cap \Omega_\kappa(\alpha)$. We set $U = \bigcup_{i \in I} U_i$. By condition (b) there is an open subset V of $U_\kappa(\alpha)$ such that

$$\Omega_\kappa(\alpha) \subset V \subset U \quad \text{and}$$
$$\tau(V) < \kappa .$$

Since V is κ-compact there is $J \subset I$, with $|J| < \kappa$, such that $\bigcup\{U_i : i \in J\} \supset V$. Then $\{W_i : i \in J\}$ is a cover of $\Omega_\kappa(\alpha)$, by elements of the given cover, of cardinality less than κ.

(b)⇒(c). Since $\Omega_\kappa(\alpha)$ is dense in $U_\kappa(\alpha)$ by Lemma 14.34, any set V as in (b) satisfies the conditions of (c).

(c)⇒(b). Let U be an open subset of $U_\kappa(\alpha)$ for which $\Omega_\kappa(\alpha) \subset U$. By condition (c) there is an open subset W of $U_\kappa(\alpha)$ such that

$$W \text{ is a dense subset of } U \text{ and}$$
$$\tau(W) \leq \kappa .$$

Since κ is measurable the set $\Omega_\kappa(\alpha)$, and hence also the sets U and W, are dense in $U_\kappa(\alpha)$ by Lemma 14.34. The result now follows from Lemma 14.36.

14.38. Theorem. *Let α be a measurable cardinal and assume that $\alpha^+ = 2^\alpha$. The following are equivalent.*

(a) *α is a strongly measurable cardinal;*

(b) *$\Omega_\alpha(\alpha)$ is α-compact;*

(c) *$\Omega_\alpha(\alpha)$ is C^*-embedded in $U(\alpha)$, i.e., $\beta(\Omega_\alpha(\alpha)) = U(\alpha)$; and*

(d) *each $\{0,1\}$-valued continuous function on $\Omega_\alpha(\alpha)$ has a continuous extension to a ($\{0,1\}$-valued) continuous function on $U(\alpha)$.*

Proof. The conditions are satisfied if $\alpha = \omega$, so we suppose that $\alpha > \omega$. The equivalence of (a) and (b) is clear from the definitions; (b)⇒(c) follows from Lemma 14.35, and (c)⇒(d) is obvious; and (d)⇒(c) follows from Lemma 2.23(c) and Theorem 2.27, since $\Omega_\alpha(\alpha)$ is a P_α-space which is dense in $U(\alpha)$ by Lemma 14.34. It remains to prove that (c)⇒(b). If (c) is satisfied and U is an open subset of $U(\alpha)$ such that $\Omega_\alpha(\alpha) \subset U$, then U is C^*-embedded in $U(\alpha)$ so by Theorem 14.10(c) U contains densely an open subset of type at most α. Thus condition (c) of Lemma 14.37 is satisfied with $\alpha = \kappa$, so $\Omega_\alpha(\alpha)$ is α-compact.

14.39. Theorem. *Let α be a measurable cardinal and let $p \in U(\alpha)$.*

(a) *If $p \notin \Omega_\alpha(\alpha)$, then $U(\alpha) \setminus \{p\}$ is C^*-embedded in $U(\alpha)$; and*

(b) *if $\alpha^+ = 2^\alpha$, then $U(\alpha) \setminus \{p\}$ is C^*-embedded in $U(\alpha)$ if and only if $p \notin \Omega_\alpha(\alpha)$.*

Proof. (a) Since $p \notin \Omega_\alpha(\alpha)$ there exist $\beta < \alpha$ and $\{A_\xi : \xi < \beta\} \subset p$ such that $\bigcap_{\xi<\beta} A_\xi \notin p$. With no loss of generality we assume that $\bigcap_{\xi<\beta} A_\xi = \emptyset$. We set

$$V = \bigcup_{\xi<\beta} (U(\alpha) \backslash \hat{A}_\xi)$$

(where $\hat{A}_\xi = \mathrm{cl}_{\beta(\alpha)} A_\xi \cap U(\alpha)$), so that V is a dense, open subset of $U(\alpha)$ with $\tau(V) \leq \beta < \alpha$; then from Theorem 14.9 it follows that V is C^*-embedded in $U(\alpha)$. Hence the intermediate space $U(\alpha)\backslash\{p\}$ is C^*-embedded in $U(\alpha)$.

(b) Conversely assume that $U(\alpha)\backslash\{p\}$ is C^*-embedded in $U(\alpha)$. From Theorem 14.10(c), there is an open, dense subset U of $U(\alpha)\backslash\{p\}$ for which $\tau(U) \leq \alpha$. Then $\Omega_\alpha(\alpha) \subset U$ by Lemma 14.36, so $p \notin \Omega_\alpha(\alpha)$.

14.40. Corollary. *Let α be a strongly measurable cardinal and assume that $\alpha^+ = 2^\alpha$. If $p \in \Omega_\alpha(\alpha)$ then $\Omega_\alpha(\alpha)\backslash\{p\}$ is not C^*-embedded in $\Omega_\alpha(\alpha)$.*

Proof. Assume that $p \in \Omega_\alpha(\alpha)$ and that $\Omega_\alpha(\alpha)\backslash\{p\}$ is C^*-embedded in $\Omega_\alpha(\alpha)$. We claim that, as a consequence, $U(\alpha)\backslash\{p\}$ is C^*-embedded in $U(\alpha)$, a contradiction to Theorem 14.39(b). Let f be a bounded, real-valued, continuous function on $U(\alpha)\backslash\{p\}$, and set $g = f\,|\,\Omega_\alpha(\alpha)\backslash\{p\}$. By our assumption g has a continuous extension $\bar{g}$ defined on $\Omega_\alpha(\alpha)$, and from Theorem 14.38 there is a continuous extension $\bar{f}$ of $\bar{g}$ defined on $U(\alpha)$. But from Lemma 14.34 $\Omega_\alpha(\alpha)$ is dense in $U(\alpha)$, so in fact $\bar{f}$ is an extension of f.

14.41. Theorem. *Let α be an infinite cardinal such that $\alpha^+ = 2^\alpha$ or $\alpha = 2^{\underline{\alpha}}$. Then there are $p \in U(\alpha)$ and a family $\{U_\zeta : \zeta < 2^\alpha\}$ of open subsets of $U(\alpha)$ such that*

$$U_\zeta \cap U_{\zeta'} = \emptyset \quad \text{for } \zeta < \zeta' < 2^\alpha \quad \text{and}$$
$$p \in (\mathrm{cl}_{\beta(\alpha)} U_\zeta) \backslash U_\zeta \quad \text{for } \zeta < 2^\alpha .$$

Proof. By (the Remark following) Lemma 12.19 in case $\alpha^+ = 2^\alpha$, and by Theorem 12.2 in case $\alpha = 2^{\underline{\alpha}}$, there is $\mathscr{B} \subset \mathscr{P}(\alpha)$ such that

$$|\mathscr{B}| = 2^\alpha,$$
$$|B| = \alpha \quad \text{for } B \in \mathscr{B}, \quad \text{and}$$
$$|B \cap C| < \alpha \quad \text{for } B, C \in \mathscr{B} \quad \text{with } B \neq C .$$

For $B \in \mathscr{B}$ it follows from Theorem 7.7(b) by taking $\mathscr{F} = \{B\}$ that there is $p_B \in U(\alpha)$ such that $|B \cap C| = \alpha$ for $C \in p_B$. We claim also that there is $p \in U(\alpha)$ such that

$$|\{B \in \mathscr{B} : p_B \in \hat{A}\}| = 2^\alpha \quad \text{for } A \in p .$$

Indeed if the claim fails then for $q\in U(\alpha)$ there is $A_q\in q$ such that $|\{B\in\mathscr{B}: p_B\in\hat{A}_q\}|<2^\alpha$, and since $U(\alpha)$ is compact there are $n<\omega$ and $\{q_k: k\leq n\}\subset U(\alpha)$ such that

$$\{p_B: B\in\mathscr{B}\}\subset U(\alpha)=\bigcup_{k\leq n}\hat{A}_{q_k}.$$

This contradicts the fact that the function $B\to p_B$ is one-to-one from $\mathscr{B}$ into $U(\alpha)$.

Let $\langle A_\xi:\xi<2^\alpha\rangle$ be a well-ordering of p. We claim that there is $\{B_\xi:\xi<2^\alpha\}\subset\mathscr{B}$ such that

$$|A_\xi\cap B_\xi|=\alpha \quad \text{for } \xi<2^\alpha \quad \text{and}$$
$$|B_\eta\cap B_\xi|<\alpha \quad \text{for } \eta<\xi<2^\alpha.$$

Indeed let $B_0\in\mathscr{B}$ with $|A_0\cap B_0|=\alpha$ and recursively, if $\xi<2^\alpha$ and B_η has been chosen for $\eta<\xi$, use the facts that $|\xi|<\alpha$ and $|\{B\in\mathscr{B}: |A_\xi\cap B|=\alpha\}|=2^\alpha$ to choose $B_\xi\in\mathscr{B}$ such that

$$|A_\xi\cap B_\xi|=\alpha, \quad \text{and}$$
$$B_\xi\neq B_\eta \quad \text{for } \eta<\xi.$$

Since $|A_\xi\cap B_\xi|=\alpha$ for $\xi<\alpha$ there is (by Lemma 12.19 in case $\alpha^+=2^\alpha$, and by Theorem 12.2 in case $\alpha=2^{\underline{\alpha}}$) a family $\{C_\zeta(\xi):\zeta<2^\alpha\}$ of subsets of $A_\xi\cap B_\xi$ such that

$$|C_\zeta(\xi)|=\alpha \quad \text{for } \zeta<2^\alpha \quad \text{and}$$
$$|C_\zeta(\xi)\cap C_{\zeta'}(\xi)|<\alpha \quad \text{for } \zeta<\zeta'<2^\alpha.$$

We suppose without loss of generality that $C_\zeta(\xi)\notin p$ for $\zeta<2^\alpha$.

We set $U_\zeta=\bigcup\{C_\zeta(\xi)\hat{}:\xi<2^\alpha\}$ for $\zeta<2^\alpha$; here $\hat{C}$ denotes $(\mathrm{cl}_{\beta(\alpha)}C)\cap U(\alpha)$ for $C\in\mathscr{P}(\alpha)$. We verify that

$$U_\zeta\cap U_{\zeta'}=\emptyset \quad \text{for } \zeta<\zeta'<2^\alpha, \quad \text{and}$$
$$p\in(\mathrm{cl}_{\beta(\alpha)}U_\zeta)\backslash U_\zeta \quad \text{for } \zeta<2^\alpha.$$

Let $\zeta<\zeta'<2^\alpha$ and $\xi\leq\xi'<2^\alpha$. If $\xi=\xi'$ then since

$$|C_\zeta(\xi)\cap C_{\zeta'}(\xi')|<\alpha$$

we have $C_\zeta(\xi)\hat{}\cap C_{\zeta'}(\xi')\hat{}=(C_\zeta(\xi)\cap C_{\zeta'}(\xi'))\hat{}=\emptyset$ by Theorem 2.24 (c) (i) and the fact that $\hat{C}=\emptyset$ for $C\in\mathscr{P}_\alpha(\alpha)$; if $\xi\neq\xi'$ then since

$$C_\zeta(\xi)\cap C_\zeta(\xi')\subset B_\xi\cap B_{\xi'} \quad \text{and } |B_\xi\cap B_{\xi'}|<\alpha$$

we again have $C_\zeta(\xi)\hat{}\cap C_{\zeta'}(\xi')\hat{}=(C_\zeta(\xi)\cap C_{\zeta'}(\xi'))\hat{}=\emptyset$. Thus $U_\zeta\cap U_{\zeta'}=\emptyset$.

Since $C_\zeta(\xi)\notin p$ for $\xi<2^\alpha$ we have

$$p\notin\bigcup\{C_\zeta(\xi)\hat{}:\xi<2^\alpha\}=U_\zeta.$$

Finally for $\xi < 2^\alpha$ we have $C_\zeta(\xi)\hat{} \neq \emptyset$ and

$$\hat{A}_\xi \cap U_\zeta \supset \hat{A}_\xi \cap C_\zeta(\xi)\hat{} = C_\zeta(\xi)\hat{}\,;$$

it follows that $p \in \mathrm{cl}_{\boldsymbol{\beta}(\alpha)} U_\zeta$.

The proof is complete.

14.42. Theorem. *Let α be a measurable cardinal such that $\alpha^+ = 2^\alpha$ and let $p \in \Omega_\alpha(\alpha)$. Then there is a family $\{U_\zeta : \zeta < \alpha^+\}$ of open subsets of $U(\alpha)$ such that*

$$U_\zeta \cap U_{\zeta'} = \emptyset \quad \textit{for } \zeta < \zeta' < \alpha^+ \quad \textit{and}$$
$$p \in (\mathrm{cl}_{\boldsymbol{\beta}(\alpha)} U_\zeta) \setminus U_\zeta \quad \textit{for } \zeta < \alpha^+.$$

Proof. Let $p = \{A_\xi : \xi < \alpha^+\}$. It is enough to prove that there is $\{B_\xi : \xi < \alpha^+\}$ such that

$$B_\xi \subset A_\xi \quad \text{for } \xi < \alpha^+,$$
$$|B_\xi| = \alpha \quad \text{for } \xi < \alpha^+,$$
$$B_\xi \notin p \quad \text{for } \xi < \alpha^+, \quad \text{and}$$
$$|B_\xi \cap B_{\xi'}| < \alpha \quad \text{for } \xi < \xi' < \alpha^+.$$

For if $\{B_\xi : \xi < \alpha^+\}$ is such a family then by Theorem 12.2 there is $\{C_\zeta(\xi) : \zeta < \alpha^+\} \subset \mathscr{P}(B_\xi)$ such that

$$|C_\zeta(\xi)| = \alpha \quad \text{for } \zeta < \alpha^+ \quad \text{and}$$
$$|C_\zeta(\xi) \cap C_{\zeta'}(\xi)| < \alpha \quad \text{for } \zeta < \zeta' < \alpha^+\,;$$

we define $U_\zeta = \bigcup \{C_\zeta(\xi)\hat{} : \xi < \alpha^+\}$, where $\hat{C}$ denotes $(\mathrm{cl}_{\boldsymbol{\beta}(\alpha)} C) \cap U(\alpha)$ for $C \in \mathscr{P}(\alpha)$, and just as in the proof of Theorem 14.41 we verify that

$$U_\zeta \cap U_{\zeta'} = \emptyset \quad \text{for } \zeta < \zeta' < \alpha^+ \quad \text{and}$$
$$p \in (\mathrm{cl}_{\boldsymbol{\beta}(\alpha)} U_\zeta) \setminus U_\zeta \quad \text{for } \zeta < \alpha^+.$$

We do not repeat these verifications.

To find the required family $\{B_\xi : \xi < \alpha^+\}$ we choose $B_0 \subset A_0$ so that $|B_0| = \alpha$ and $B_0 \notin p$ and we proceed by recursion. Let $\xi < \alpha^+$ and suppose that B_η has been defined for $\eta < \xi$. We set $I = \{\eta < \xi : |A_\xi \cap B_\eta| = \alpha\}$ and we consider two cases.

Case 1. $|I| < \alpha$. Since $B_\eta \notin p$ for $\eta \in I$ and p is an α-complete ultrafilter we have

$$\bigcap \{A_\xi \setminus B_\eta : \eta \in I\} \in p\,.$$

Hence there is $B_\xi \subset \bigcap \{A_\xi \setminus B_\eta : \eta \in I\}$ such that $|B_\xi| = \alpha$ and $B_\xi \notin p$. It is clear that $B_\xi \cap B_\eta = \emptyset$ for $\eta < \xi$ with $\eta \in I$, and that $|B_\xi \cap B_\eta| < \alpha$ for $\eta < \xi$ with $\eta \notin I$.

Case 2. $|I|=\alpha$. Set $I=\{\eta(i):i<\alpha\}$. Since α is regular by the remark following Theorem 8.30 and

$$|A_\xi\cap B_{\eta(i)}|=\alpha \quad \text{for } i<\alpha \quad \text{and}$$
$$|B_{\eta(i)}\cap B_{\eta(i')}|<\alpha \quad \text{for } i'<i<\alpha\,,$$

there is

$$\rho(i)\in(A_\xi\cap B_{\eta(i)})\setminus\bigcup_{i'<i}B_{\eta(i')} \quad \text{for } i<\alpha\,.$$

We set

$$C=\{\rho(i):i<\alpha\}$$

and we choose $B_\xi\subset C$ so that

$$|B_\xi|=\alpha \quad \text{and}$$
$$B_\xi\notin p\,;$$

such a set B_ξ exists because $|C|=\alpha$. If $\eta<\xi$ and $\eta\notin I$ then since $B_\xi\subset A_\xi$ and $|A_\xi\cap B_\eta|<\alpha$ we have $|B_\xi\cap B_\eta|<\alpha$; if $\eta<\xi$ and $\eta\in I$ there is $i<\alpha$ such that $\eta=\eta(i)$ and we have

$$|B_\xi\cap B_\eta| = |B_\xi\cap B_{\eta(i)}| \le |i|<\alpha\,.$$

The required family $\{B_\xi:\xi<\alpha^+\}$ has been defined and the proof is complete.

We note the following consequence of Theorems 14.41 and 14.42. There is $p\in\boldsymbol{\beta}(\omega)\setminus\omega$ and a family $\{U_\zeta:\zeta<2^\omega\}$ of pairwise disjoint open subsets of $\boldsymbol{\beta}(\omega)\setminus\omega$ such that $p\in(\mathrm{cl}_{\boldsymbol{\beta}(\omega)}U_\zeta)\setminus U_\zeta$ for $\zeta<2^\omega$; and if $\omega^+=2^\omega$ then for every $p\in\boldsymbol{\beta}(\omega)\setminus\omega$ there is such a family $\{U_\zeta:\zeta<2^\omega\}$.

Notes for § 14

The results 14.1—14.11 are closely related to some results in Fine-Gillman [60] and Negrepontis [68], [73]. Specifically, Lemma 14.1 generalizes Fine-Gillman [60] (Theorem 4.1), Negrepontis [68] (Theorem 4.4), and Negrepontis [73] (Theorem 2.6); Theorem 14.2 generalizes Fine-Gillman [60] (Theorem 4.3 and Corollary 4.4) and Negrepontis [68] (Theorem 6.1); and Theorem 14.4 generalizes Fine-Gillman [60] (Theorem 4.6).

Theorem 14.6 is stated in Negrepontis [69a] (Theorems 1.7 and 3.7). Theorem 14.9 generalizes Negrepontis [68] (Theorem 3.1) and Negrepontis [73] (Theorem 2.5). Theorem 14.10 is proved in Negrepontis [68] (Theorem 6.1 and Corollaries 6.2, 6.3). Theorem 14.11 generalizes Negrepontis [68] (Theorem 3.2) and the well-known (proof of the) fact that

$\boldsymbol{\beta}(\omega)\setminus\omega$ is not basically disconnected. As Engelking ([72] (p. 91)) remarks, the fact that $\boldsymbol{\beta}(\omega)\setminus\omega$ is not basically disconnected follows immediately from the fact, proved by Hausdorff [36 b] (Erster Einschaltungssatz), that there is no "ω-limit" in $\mathscr{P}(\omega)/\mathscr{P}^{\omega}(\omega)$.

Assuming the continuum hypothesis, the ω^+-homogeneity-universality of $\mathscr{P}(\omega)/\mathscr{P}^{\omega}(\omega)$ is proved by Keisler [66]. It follows from Jónsson-Olin [68] (Theorem 1) that $\mathscr{P}(\omega)/\mathscr{P}^{\omega}(\omega)$ is ω^+-saturated, but without assuming the continuum hypothesis this result cannot be used to conclude that $\mathscr{P}(\omega)/\mathscr{P}^{\omega}(\omega)$ is ω^+-homogeneous-universal (cf. the conditions of Theorem 11.17 above and Jónsson-Olin [68] (Lemma 3)). The ω^+-homogeneity of $\mathscr{P}(\omega)/\mathscr{P}^{\omega}(\omega)$ in theorem 14.12 is implicit in Steiner-Steiner [70] (Theorem 4), where it is proved using a Schröder-Bernstein type of theorem due to von Neumann [35] (pp. 11—12) (whose proof has been successively simplified by Halmos [68] and Yorke [69]). The direct proof given here is new.

The remark following Theorem 14.12 has been established by Hausdorff [36 b] in set-theoretic language. More specifically, part (b) corresponds to Hausdorff's statement that there is no ω-limit in $\mathscr{P}(\omega)/\mathscr{P}^{\omega}(\omega)$; and part (c) to Hausdorff's statement that there is no (ω,ω^*)-gap in $\mathscr{P}(\omega)/\mathscr{P}^{\omega}(\omega)$. The set $\mathscr{P}(\omega)/\mathscr{P}^{\omega}(\omega)$ with the pre-order $\preccurlyeq$ has been investigated by Hausdorff [36 b], Rothberger [41], [48], and Lusin [47]. Novák [53] translated these results in the topology of $\boldsymbol{\beta}(\omega)\setminus\omega$. There is an informative discussion of the relation of these papers to the topology of $\boldsymbol{\beta}(\omega)\setminus\omega$ in Engelking [72]. Hechler [72] (Lemma 2.1 and Theorem 3.2) proves various consistency results concerning the length of Hausdorff's gaps that are related to the remark following Theorem 14.12.

Lemma 14.13 and Theorem 14.14 are proved by Hausdorff [36 b] (pp. 245—247). As is pointed out there (footnote, p. 244), the proof of this result was essentially known to Hausdorff in 1909.

The ideas related to the proof of Theorem 14.15 (a)$\Rightarrow$(b) (together with the remark following Theorem 14.12) were considered by W. Rudin [56], Parovičenko [63] (Theorem 1), Keisler [66], and Hoffman and Ramsay [65] (Proposition 2). Lemma 14.16 was originally proved by Gillman-Henriksen [56] (Theorem 2.7) and Gillman-Jerison [60] (Theorem 14.27). The proof given here appears in Negrepontis [67 b] (Remark 3.3). Theorem 14.18 appears in Comfort-Negrepontis [68] (Theorem 1.1).

Theorem 14.19 is an amplification of a remark by Keisler [66] (p. 131). Lemma 14.20 is given in Kochen [61] (Theorem 6.7) and Frayne-Morel-Scott [62] (Theorem 1.10). We note that Lemma 14.21 together with the results of § 6 can be used to prove the completeness and model-completeness of the theory of atomless Boolean algebras. Theorem 14.22 is a special case of (the dual form) of a result by Shelah [*] (Theorem 3.8) whose proof has been simplified by Keisler (cf. Benda [*] (discussion after Proposition 8)) and independently by one of the authors (S. N.). The proof given here follows the simplified version.

The results 14.23—14.30 appear in Negrepontis [73]. For $\alpha=\omega$ the statement that $SU(\omega^+)$ is not C^*-embedded in $\boldsymbol{\beta}(\omega^+)$ (part of Corollary 14.29) and Corollary 14.30 were proved by Warren [70] (Theorem 2 and Corollary 2), [72]. Theorem 14.31(a) is proved in Fine-Gillman [60] (Theorem 4.6) and Comfort-Negrepontis [68] (Corollary 2.4), where it follows from a stronger property (Comfort-Negrepontis [68] (Theorem 2.3)). Theorem 14.31(b) for non-P-points is due to Gillman (oral communication to the authors); and a proof appears in Comfort-Negrepontis [68] (Corollary 2.6). Theorem 14.31(b) for P-points is due to Warren [70] (Theorem 1), [72] and Rajagopalan [72]; these and related results are presented in the text of Walker [74].

The deduction of Corollary 14.32 from Theorem 14.27 appears in Negrepontis [73] (Corollary 3.6). The original proof of Corollary 14.32 is given (as mentioned in the Notes for § 8) in Kurepa [36] and Specker [51]. (Cf. also Erdős-Tarski [66] (footnote 10, p. 76) concerning Monk's observation.) Theorem 14.33 is given in Negrepontis [72b], [*].

The results 14.34—14.40 were proved in Comfort-Negrepontis [70].

Some of the results of § 14 and § 15 on the topology of spaces of ultrafilters are described in Negrepontis [***].

The case $\alpha=2^{\mathfrak{s}}$ of Theorem 14.41 and the case $\alpha=\omega$ of Theorem 14.42 are from Hindman [69]. Earlier Pierce [67] (Corollary 21.3) had shown that if $\omega^+=2^\omega$ there are $p\in\boldsymbol{\beta}(\omega)\backslash\omega$ and pairwise disjoint open subsets U_0, U_1, and U_2 of $\boldsymbol{\beta}(\omega)\backslash\omega$ such that $p\in(\mathrm{cl}_{\boldsymbol{\beta}(\omega)}U_n)\backslash U_n$ for $n\leq 2$.

It has been shown by Comfort and Hindman [73] and Prikry [*] that if $\alpha^+=2^\alpha$ and α is regular, then the following statements are equivalent for $p=\{A_\xi:\xi<2^\alpha\}\in U(\alpha)$: (a) p is in the boundary of two disjoint open subsets of $U(\alpha)$; (b) p is in the boundary of 2^α disjoint open subsets of $U(\alpha)$; (c) there is an α-almost disjoint family $\{B_\xi:\xi<2^\alpha\}$ such that $B_\xi\subset A_\xi$ for $\xi<2^\alpha$. More recently, Prikry [**] has shown that (c) holds for all $p\in U(\alpha)$ (assuming $\alpha^+=2^\alpha$ and α regular); on the other hand it follows from the result of Baumgartner [70] mentioned in the Notes for § 12 that there is a model of set theory in which condition (c) fails for all $p\in U(\omega^+)$.

§ 15. Spaces Homeomorphic to $(2^\alpha)_\alpha$

We begin this section with an abstract study, independent of previous material, of the spaces $(2^\alpha)_\alpha$ for regular infinite α. These Cantor-like spaces have a canonical structure and their characterization with an economical list of properties (Theorem 15.3) suggests the possibility of their frequent "natural" occurrence. The property of $\mathscr{B}$-compactness in the characterization theorem suggests that if X is a *compact* space satisfying certain natural cardinality conditions then the associated space X_α is homeomorphic to $(2^\alpha)_\alpha$. That this is indeed the case is a simple and particularly useful result (Theorem 15.9). A special case occurs as follows: assuming that 2^α is a regular cardinal and that no uniform ultrafilter on α has a filter base of cardinality less than 2^α, there is a homeomorphism between $U(\alpha)_{2^\alpha}$ and $(2^{(2^\alpha)})_{2^\alpha}$ (Theorem 15.14(a)). This homeomorphism forms the crucial link relating the general results about the spaces $(2^\alpha)_\alpha$ considered thus far to the study of ultrafilters. It implies that the topological properties of $(2^{(2^\alpha)})_{2^\alpha}$ may be studied and applied profitably to the space $U(\alpha)_{2^\alpha}$.

The property examined in detail is the α-Baire category property of $(2^\alpha)_\alpha$: the intersection of at most α open and dense subsets of $(2^\alpha)_\alpha$ is dense in $(2^\alpha)_\alpha$. Such an "α-residual" set (as it will be called for the purposes of this introduction) is generally found (with some care necessary for the weakly compact cardinals α) to be again a space homeomorphic to $(2^\alpha)_\alpha$ or to $(\alpha^\alpha)_\alpha$ (Theorem 15.8). We remark here that since, as we have seen in § 8, the space $(2^\alpha)_\alpha$ is α-compact if and only if α is weakly compact, it is not surprising that some of the statements of the present section (Theorems 15.3, 15.8) depend on whether α is weakly compact.

The stability of the class of spaces homeomorphic to $(2^\alpha)_\alpha$ when taking α-residual sets (expressed in Theorem 15.8) has remarkable consequences when applied to the space $U(\alpha)$. As was noted in the introduction of § 10, each of the two most significant classes of ultrafilters, namely the set $G(\alpha)$ of α^+-good ultrafilters, and the set $RK(\alpha)$ of uniform ultrafilters on α that are Rudin-Keisler minimal in $\tau[U(\alpha)]$, forms an α^+-residual set in $U(\alpha)_{\alpha^+}$ (assuming the segment $\alpha^+=2^\alpha$ of the generalized continuum hypothesis); therefore each of the sets $G(\alpha)_{\alpha^+}$,

$RK(\alpha)_{\alpha^+}$, and $(G(\alpha)\cap RK(\alpha))_{\alpha^+}$ is homeomorphic to $(2^{(\alpha^+)})_{\alpha^+}$ (Theorem 15.14(b)); in particular each of these sets is non-empty and alternative proofs of the existence of such ultrafilters are thus implied by these results.

We observe here yet another occurrence of the familiar but always surprising phenomenon of Banach that is associated (almost always, one is tempted to say...) with arguments of Baire category type: the extremal (perhaps, pathological) property (in our case, α^+-goodness and Rudin-Keisler minimality of ultrafilters) is in fact typical, and is shared by a "majority" (in the topological sense) of the elements under consideration.

The same general situation occurs not only for $U(\alpha)_{\alpha^+}$ but for the "growth" spaces of the form $\beta X\backslash X$ for other (suitably restricted) spaces X. Thus if $\omega^+=2^\omega$ and if X is locally compact, non-compact, σ-compact, and $|C(X)|\leq 2^\omega$, then $(\beta X\backslash X)_{\omega^+}$ is homeomorphic to $(2^{(\omega^+)})_{\omega^+}$ (Theorem 15.20). For the results on the growth $\beta X\backslash X$ this statement is the fundamental application of Theorem 15.9 opening the way to the exploitation of the topological properties of $(2^{(\omega^+)})_{\omega^+}$. Again the two most interesting classes of elements of $\beta X\backslash X$ emerge as ω^+-residual sets. Thus, under the above assumptions on X, the set of P_{ω^+}-points of $\beta X\backslash X$ is an ω^+-residual set of $\beta X\backslash X$, and if (in addition) X is a metric space without isolated elements then the set of remote points of βX is also an ω^+-residual set of $\beta X\backslash X$ (Theorem 15.20).

The dependence of these results on the continuum hypothesis is not absolute; some of them can be established assuming only Martin's axiom. (See Notes.)

The applications of Theorems 15.3 and 15.4 are limited to the systematic use of the Baire category property of these spaces mentioned in Theorem 15.8. However, the spaces $(2^\alpha)_\alpha$ and $(\alpha^\alpha)_\alpha$ form the higher cardinal analogues of the Cantor set and the space of irrationals respectively and thus they serve as the fundamental spaces for the analogues to higher cardinalities of a large part of the classical descriptive set theory (sometimes depending on the size of the cardinal α). We consider it quite possible that these higher cardinality analogues of the classical descriptive set theory, when used in the corresponding spaces of ultrafilters, can produce similarly useful results.

The Topological Characterization of $(2^\alpha)_\alpha$

15.1. Lemma. *Let X and Y be spaces, let $\mathscr{A}$ and $\mathscr{B}$ be subbases of X and Y respectively, and suppose that there is a one-to-one function ϕ*

from $\mathscr{A}$ onto $\mathscr{B}$ such that

$$\bigcap \mathscr{C} = \emptyset \quad \textit{if and only if} \quad \bigcap \phi[\mathscr{C}] = \emptyset$$

for $\mathscr{C} \subset \mathscr{A}$. Then X is homeomorphic to Y.

Proof. We claim first that if $p \in X$ then $\bigcap \{\phi(A): A \in \mathscr{A}$ and $p \in A\}$ contains a single element of Y. By the assumption on ϕ this intersection is not empty. If it contains two distinct elements q_0 and q_1 then there is $B \in \mathscr{B}$ such that $q_0 \in B$ and $q_1 \notin B$, and then

$$B \notin \{\phi(A): A \in \mathscr{A} \text{ and } p \in A\}$$

so $\phi^{-1}(\{B\}) \notin \{A: A \in \mathscr{A}$ and $p \in A\}$. Thus

$$\phi^{-1}(\{B\}) \cap \bigcap \{A: A \in \mathscr{A} \text{ and } p \in A\} = \emptyset,$$

and we have

$$q_0 \in B \cap \bigcap \{\phi(A): A \in \mathscr{A} \text{ and } p \in A\} = \emptyset,$$

a contradiction.

We define $f: X \to Y$ by letting $f(p)$ be the unique element of $\bigcap \{\phi(A): A \in \mathscr{A}$ and $p \in A\}$ for $p \in X$. We verify the following statements.

(i) f is one-to-one and onto.

There is a function $g: Y \to X$ defined symmetrically to f and for which we have

$$\begin{aligned} \{f \circ g(q)\} &= \bigcap \{\phi(A): A \in \mathscr{A} \text{ and } g(q) \in A\} \\ &\subset \bigcap \{\phi(A): q \in \phi(A)\} \\ &= \bigcap \{B: B \in \mathscr{B} \text{ and } q \in B\} = \{q\}, \end{aligned}$$

for $q \in Y$. It follows that f is onto, and symmetrically that f is one-to-one.

(ii) f is a homeomorphism.

Since ϕ and f are one-to-one it suffices to prove that $\phi(A) = f[A]$. It is clear from the definition of f that $\phi(A) \supset f[A]$. By symmetry $A \supset f^{-1}[\phi(A)]$, *i.e.*, $f[A] \supset \phi(A)$.

The proof of the lemma is complete.

We introduce some definitions and notation.

If $\mathscr{B}$ is a family of closed non-empty subsets of a space X, we say that X is *$\mathscr{B}$-compact* if whenever $\emptyset \neq \mathscr{C} \subset \mathscr{B}$ and $\mathscr{C}$ has the finite intersection property (*i.e.*, every non-empty finite subset of $\mathscr{C}$ has non-empty intersection), then $\bigcap \mathscr{C} \neq \emptyset$. Thus X is compact if and only if X is $\mathscr{F}^*$-compact, where $\mathscr{F}^*$ is the family of all non-empty closed subsets of X.

An *open partition* $\mathscr{C}$ of a space X is a family of open pairwise disjoint non-empty subsets of X whose union is X. If $\mathscr{B}$ and $\mathscr{C}$ are open partitions of X, we say that $\mathscr{B}$ is a refinement of $\mathscr{C}$ (denoted $\mathscr{C} \prec \mathscr{B}$)

if every element of $\mathscr{B}$ is a subset of a (necessarily unique) element of $\mathscr{C}$. If $\alpha \geq \omega$, if X is a P_α-space, if $\{\mathscr{B}_i : i \in I\}$ is a family of open partitions of X and if $|I| < \alpha$, we let $\bigwedge_{i\in I}\mathscr{B}_i$ be the family of all non-empty subsets of X of the form $\bigcap_{i\in I} B_i$, where $B_i \in \mathscr{B}_i$ for $i \in I$. It is clear that $\bigwedge_{i\in I}\mathscr{B}_i$ is an open partition of X, and that (despite the misleading notation) we have $\mathscr{B}_j \prec \bigwedge_{i\in I}\mathscr{B}_i$ for all $j \in I$. We note that if $I = \emptyset$ then $\bigwedge_{i\in I}\mathscr{B}_i = \{X\}$.

For an infinite (regular) cardinal α the space $(2^\alpha)_\alpha$ has a *canonical* base $\mathscr{E}$ for its topology such that

$(2^\alpha)_\alpha$ is $\mathscr{E}$-compact,

$\mathscr{E}$ is a ramification system under reverse set inclusion,

the set $\mathscr{E}_\xi$ of elements of $\mathscr{E}$ of order ξ is an open partition of $(2^\alpha)_\alpha$ for $\xi < \alpha$,

$\mathscr{E}_\xi = \bigwedge_{\zeta<\xi}\mathscr{E}_\zeta$ for limit ordinals $\xi < \alpha$, and

if $E \in \mathscr{E}_\xi$ then E has *exactly two* immediate successors (in $\mathscr{E}_{\xi+1}$) for $\xi < \alpha$.

In fact, set $E(s) = \{t \in 2^\alpha : t|\xi = s\}$ for $s \in 2^\xi$ and set

$$\mathscr{E}_\xi = \{E(s) : s \in 2^\xi\} \quad \text{for } \xi < \alpha, \quad \text{and}$$
$$\mathscr{E} = \bigcup_{\xi<\alpha}\mathscr{E}_\xi .$$

It is clear that $|\mathscr{E}_\xi| = 2^{|\xi|}$ for $\xi < \alpha$, so from Lemma 1.22(f) we have

$$w((2^\alpha)_\alpha) \leq \sum_{\xi<\alpha} 2^{|\xi|} = 2^{\underline{\alpha}} .$$

Let $\mathscr{C}$ be a family of open subsets of X such that $\mathscr{C}$ is a ramification system of order α under reverse set inclusion and let $\mathscr{C}_\xi$ denote the set of elements of $\mathscr{C}$ of order $\xi < \alpha$. If $C \in \mathscr{C}_\xi$ and $\eta < \alpha$ we set

$$\mathscr{S}(C,\eta) = \{D \in \mathscr{C}_{\xi+\eta} : D \subset C\} .$$

In particular $\mathscr{S}(C,1)$ denotes the set of immediate successors of C (in the system $\mathscr{C}$). We note that if $\mathscr{C}_\xi$ is an open partition of X for $\xi < \alpha$ then $\mathscr{S}(C,\eta)$ is an open partition of C for $C \in \mathscr{C}$ and $\eta < \alpha$.

15.2. Lemma. *Let α be an infinite, regular cardinal and let X be a P_α-space without isolated elements. Suppose that X has an α-subbase $\mathscr{B}$ such that*

X is $\mathscr{B}$-compact,

$\mathscr{B} = \bigcup_{\xi<\alpha}\mathscr{B}_\xi$, and

$\mathscr{B}_\xi$ is an open partition of X for $\xi < \alpha$.

Then there is a base $\mathscr{C}$ for X such that

X is $\mathscr{C}$-compact,

$\mathscr{C}$ is a ramification system of order α with respect to reverse set inclusion,

the set $\mathscr{C}_\xi$ of elements of $\mathscr{C}$ of order ξ is an open partition of X, and

$\mathscr{C}_\xi = \bigwedge_{\zeta<\xi}\mathscr{C}_\zeta$ for limit ordinals $\xi<\alpha$.

Further, the space $(2^\alpha)_\alpha$ is the continuous image of X.

Proof. We set

$$\mathscr{A}_\xi = \bigwedge_{\zeta<\xi}\mathscr{B}_\zeta \quad \text{for } \xi<\alpha, \quad \text{and}$$
$$\mathscr{C} = \bigcup_{\xi<\alpha}\mathscr{A}_\xi .$$

We claim that if $A=\bigcap_{\zeta<\xi}B_\zeta\in\mathscr{A}_\xi$, where $\xi<\alpha$ and $B_\zeta\in\mathscr{B}_\zeta$ for $\zeta<\xi$, then

$$(*) \qquad \{B\in\mathscr{C}: B\supset A\} = \{\textstyle\bigcap_{\zeta<\eta}B_\zeta: \eta\leq\xi\}.$$

Indeed, if $\eta\leq\xi$ then $\bigcap_{\zeta<\eta}B_\zeta\supset\bigcap_{\zeta<\xi}B_\zeta=A$. For the reverse inclusion let $B\in\mathscr{C}$ and $B\supset A$. Since $B\in\mathscr{C}$, we have $B=\bigcap_{\zeta<\eta}B'_\zeta\neq\emptyset$, where $\eta<\alpha$ and $B'_\zeta\in\mathscr{B}_\zeta$ for $\zeta<\eta$. If $B'_\zeta\neq B_\zeta$ for some $\zeta<\eta$ and $\zeta<\xi$, then since $B_\zeta, B'_\zeta\in\mathscr{B}_\zeta$ and $\mathscr{B}_\zeta$ is an open partition of X we have $B_\zeta\cap B'_\zeta=\emptyset$ and hence $A=B\cap A=\emptyset$, a contradiction. Thus $B_\zeta=B'_\zeta$ when $\zeta<\eta$ and $\zeta<\xi$. Now if $\eta>\xi$ then

$$A=\bigcap_{\zeta<\xi}B_\zeta=\bigcap_{\zeta<\xi}B'_\zeta\supset\bigcap_{\zeta<\eta}B'_\zeta=B$$

so that $B=A=\bigcap_{\zeta<\xi}B_\zeta$; and if $\eta\leq\xi$ then

$$B=\bigcap_{\zeta<\eta}B'_\zeta=\bigcap_{\zeta<\eta}B_\zeta$$

as required.

We now verify the following statements.

(i) $\mathscr{C}$ is a base for X.

Let V be a neighborhood of $p\in X$. There are $\beta<\alpha$, ordinals $\xi(i)<\alpha$ for $i<\beta$, and $B_{\xi(i)}\in\mathscr{B}_{\xi(i)}$, such that

$$p\in\bigcap_{i<\beta}B_{\xi(i)}\subset V.$$

Set $\xi=\sup_{i<\beta}\xi(i)$; then $\xi<\alpha$ since α is regular, and it is clear that there is $A\in\mathscr{C}$ such that

$$p\in A\subset\bigcap_{i<\beta}B_{\xi(i)}.$$

(ii) X is $\mathscr{C}$-compact.

Let $\{A^{(i)}: i\in I\}$ be a family in $\mathscr{C}$ with the finite intersection property. By property (*) this family is a chain under reverse set inclusion. Hence there are $\xi\le\alpha$, and $B_\zeta\in\mathscr{B}_\zeta$ for $\zeta<\xi$, and $\zeta(i)<\xi$ such that

$$A^{(i)}=\bigcap\nolimits_{\zeta<\zeta(i)} B_\zeta \quad \text{for } i\in I\,.$$

Then $\{B_\zeta:\zeta<\xi\}$ has the finite intersection property, and since X is $\mathscr{B}$-compact we have

$$\emptyset \neq \bigcap\nolimits_{\zeta<\xi} B_\zeta \subset \bigcap\nolimits_{i\in I}\bigcap\nolimits_{\zeta<\zeta(i)} B_\zeta = \bigcap\nolimits_{i\in I} A^{(i)}\,.$$

(iii) $\mathscr{C}$ is a ramification system of order α under reverse set inclusion. That $\mathscr{C}$ is a ramification system follows directly from property (*). It is clear that the order of $\mathscr{C}$ is at most α. Since X is a P_α-space without isolated elements and $\mathscr{C}$ is a base for X by (i) above, the order of $\mathscr{C}$ cannot be less than α.

(iv) The set $\mathscr{C}_\xi$ of elements of $\mathscr{C}$ of order ξ is an open partition of X. Let $p\in X$. The set $\{A\in\mathscr{C}: p\in A\}$ is a well-ordered set under reverse set inclusion since $\mathscr{C}$ is a ramification system, and its order-type is α since X is a P_α-space without isolated elements. Hence there is $A\in\mathscr{C}$ such that $p\in A$ and $A\in\mathscr{C}_\xi$. This proves that $\mathscr{C}_\xi$ is an open cover of X. It is clear from the definition of $\mathscr{C}$ and the fact that each $\mathscr{B}_\zeta$ is a partition that distinct elements of $\mathscr{C}$ are either disjoint or comparable. Thus $\mathscr{C}_\xi$ is a partition of X.

(v) $\mathscr{C}_\xi=\bigwedge_{\zeta<\xi}\mathscr{C}_\zeta$ for limit ordinals $\xi<\alpha$.
If $C\in\mathscr{C}_\xi$ then there is an order-isomorphism ϕ from ξ to the set of predecessors of C in $\mathscr{C}$. Because $C\in\mathscr{C}$, there are $\eta<\alpha$ and $B_i\in\mathscr{B}_i$ for $i<\eta$ such that $C=\bigcap_{i<\eta} B_i$. By property (*) there is $\eta(\zeta)<\eta$ such that

$$\phi(\zeta)=\bigcap\nolimits_{i<\eta(\zeta)} B_i \quad \text{for } \zeta<\xi\,.$$

It is clear that

$$\eta(\zeta)<\eta(\zeta') \quad \text{if } \zeta<\zeta'<\xi\,,$$
$$\phi(\zeta)\in\mathscr{C}_\zeta \quad \text{for } \zeta<\xi, \quad \text{and}$$
$$C=\bigcap\nolimits_{\zeta<\xi}\phi(\zeta)\in\bigwedge\nolimits_{\zeta<\xi}\mathscr{C}_\zeta\,.$$

Conversely let $C\in\bigwedge_{\zeta<\xi}\mathscr{C}_\zeta$; thus $C=\bigcap_{\zeta<\xi} C_\zeta$ where $C_\zeta\in\mathscr{C}_\zeta$ for $\zeta<\xi$. Since the order of C_ζ is ζ, and

$$C_{\zeta'}\subsetneqq C_\zeta \quad \text{if } \zeta<\zeta'<\xi\,,$$

it follows that the order of C is ξ, *i.e.*, that $C\in\mathscr{C}_\xi$.

The proof of the existence of $\mathscr{C}$ with the required properties is complete. It remains to prove that there is a continuous function from X onto $(2^\alpha)_\alpha$. Let $\mathscr{E}$ be the canonical base for $(2^\alpha)_\alpha$. We define a function $\phi:\mathscr{C}\to\mathscr{E}$

recursively, as follows. We let $\phi_0:\mathscr{C}_0\to\mathscr{E}_0$ be given by $\phi_0(X)=(2^\alpha)_\alpha$. If ξ is a non-zero limit ordinal, $\xi<\alpha$, and if $\phi_\zeta:\mathscr{C}_\zeta\to\mathscr{E}_\zeta$ has been defined for $\zeta<\xi$, we define $\phi_\xi:\mathscr{C}_\xi\to\mathscr{E}_\xi$ by

$$\phi_\xi(\bigcap_{\zeta<\xi}C_\zeta)=\bigcap_{\zeta<\xi}\phi_\zeta(C_\zeta),$$

where $C_\zeta\in\mathscr{C}_\zeta$ for $\zeta<\xi$ and $\bigcap_{\zeta<\xi}C_\zeta\neq\emptyset$.

We now define $\phi_{\xi+1}$ for $\xi<\alpha$. Let $C\in\mathscr{C}_\xi$; then $|\mathscr{S}(C,1)|\geq 2$, and thus there are sets $\mathscr{S}_0(C,1)$ and $\mathscr{S}_1(C,1)$ such that

$$\begin{aligned}&\mathscr{S}(C,1)=\mathscr{S}_0(C,1)\cup\mathscr{S}_1(C,1),\\&\mathscr{S}_0(C,1)\neq\emptyset,\quad \mathscr{S}_1(C,1)\neq\emptyset,\ \text{ and}\\&\mathscr{S}_0(C,1)\cap\mathscr{S}_1(C,1)=\emptyset.\end{aligned}$$

Let $E_{C,0}$ and $E_{C,1}$ be the two immediate successors of $\phi(C)$ in $\mathscr{E}_{\xi+1}$. We define $\phi_{\xi+1}$ by the conditions

$$\begin{aligned}&\phi_{\xi+1}[\mathscr{S}_0(C,1)]=\{E_{C,0}\},\quad\text{and}\\&\phi_{\xi+1}[\mathscr{S}_1(C,1)]=\{E_{C,1}\}.\end{aligned}$$

Finally, we set $\phi=\bigcup_{\xi<\alpha}\phi_\xi:\mathscr{C}\to\mathscr{E}$. It is clear that ϕ satisfies the following conditions:

$$\begin{aligned}&\phi[\mathscr{C}_\xi]=\mathscr{E}_\xi \quad\text{for } \xi<\alpha,\quad\text{and}\\&\text{if } C,D\in\mathscr{C}\quad\text{and } C\subset D\quad\text{then } \phi(C)\subset\phi(D).\end{aligned}$$

We define $f:X\to(2^\alpha)_\alpha$ by the condition

$$\{f(p)\}=\bigcap\{\phi(C):C\in\mathscr{C}\text{ and } p\in C\}$$

for $p\in C$, and we verify the following statements.

(i) f is well-defined.

If $p\in X$ then $\{C\in\mathscr{C}:p\in C\}$ is a (well-ordered) chain of $\mathscr{C}$ and hence $\{\phi(C):C\in\mathscr{C}$ and $p\in C\}$ is a chain of $\mathscr{E}$; further $\{C\in\mathscr{C}:p\in C\}\cap\mathscr{C}_\xi\neq\emptyset$ and hence $\{\phi(C):C\in\mathscr{C}$ and $p\in C\}\cap\mathscr{E}_\xi\neq\emptyset$ for $\xi<\alpha$. Thus the set $\bigcap\{\phi(C):C\in\mathscr{C}$ and $p\in C\}$ consists of a single element of $(2^\alpha)_\alpha$, *i.e.*, f is well-defined.

(ii) f is an onto function.

Let $q\in(2^\alpha)_\alpha$. We define a family $\{C_\xi:\xi<\alpha\}$ such that

$$\begin{aligned}&C_\xi\in\mathscr{C}_\xi\quad\text{for } \xi<\alpha,\\&\{C_\xi:\xi<\alpha\}\ \text{is a chain in } \mathscr{C},\ \text{and}\\&q\in\phi(C_\xi)\quad\text{for } \xi<\alpha.\end{aligned}$$

We proceed recursively. Let $C_0=X$. If ξ is a non-zero limit ordinal and $\xi<\alpha$ we set $C_\xi=\bigcap_{\zeta<\xi}C_\zeta$. Now for $\xi<\alpha$ we define $C_{\xi+1}$. Since

$q\in\phi(C_\xi)\in\mathscr{E}_\xi$ and $q\in E_{\xi+1}$ for a unique $E_{\xi+1}\in\mathscr{E}_{\xi+1}$ we have $E_{\xi+1}\subset\phi(C_\xi)$. The properties of ϕ imply that there is $C_{\xi+1}\in\mathscr{C}_{\xi+1}$ such that $\phi(C_{\xi+1}) = E_{\xi+1}$ and $C_{\xi+1}\subset C_\xi$. This completes the recursive definition. We define $p\in X$ by the rule $\{p\}=\bigcap_{\xi<\alpha}C_\xi$. It is clear that $f(p)=q$.

(iii) f is continuous.
Let $p\in X$, $E\in\mathscr{E}$, and $f(p)\in E$. We must prove that there is $C\in\mathscr{C}$ such that $p\in C$ and $f[C]\subset E$. Let $\xi<\alpha$ be such that $E\in\mathscr{E}_\xi$ and choose $C\in\mathscr{C}_\xi$ such that $\phi(C)=E$. We claim that $f[C]\subset\phi(C)$. Indeed let $q\in C$. Then $\{f(q)\}=\bigcap\{\phi(D): D\in\mathscr{C}$ and $q\in D\}\subset\phi(C)$.

The proof of the lemma is complete.

15.3. Theorem. *Let α be an infinite regular cardinal. Let X be a P_α-space without isolated elements and with an α-subbase $\mathscr{A}$ such that*

$$|\mathscr{A}|\le 2^{\underline{\alpha}},$$
$$X \text{ is } \mathscr{A}\text{-compact},$$
$$\mathscr{A}=\bigcup\nolimits_{\xi<\alpha}\mathscr{A}_\xi, \quad \text{and}$$
$$\mathscr{A}_\xi \text{ is an open partition of } X \text{ for } \xi<\alpha.$$

(a) *If α is not a weakly compact cardinal, then X is homeomorphic to $(2^\alpha)_\alpha$.*

(b) *If α is a weakly compact cardinal, then X is homeomorphic to $(2^\alpha)_\alpha$ if and only if*

$$|\mathscr{A}_\xi|<\alpha \quad \text{for } \xi<\alpha.$$

Proof. From Lemma 15.2 there is a base $\mathscr{C}$ for the topology of X such that

(i) X is $\mathscr{C}$-compact,

(ii) $\mathscr{C}$ is a ramification system of order α with respect to reverse set inclusion,

(iii) the set $\mathscr{C}_\xi$ of elements of $\mathscr{C}$ of order ξ is an open partition of X of cardinality at most $2^{\underline{\alpha}}$ for $\xi<\alpha$, and

(iv) $\mathscr{C}_\xi=\bigwedge_{\zeta<\xi}\mathscr{C}_\zeta$ for limit ordinals $\xi<\alpha$.

We note from (ii) and (iii) (or from (iv)) that we have $\mathscr{C}_0=\{X\}$; and from (ii) that every element of $\mathscr{C}$ has *at least two* immediate successors. The additional fact (not given by Lemma 15.2) that $|\mathscr{C}_\xi|\le 2^{\underline{\alpha}}$ for $\xi<\alpha$ follows from the method of proof of Lemma 15.2 and the equality $(2^{\underline{\alpha}})^{\underline{\alpha}}=2^{\underline{\alpha}}$ which according to Lemma 1.26 holds for infinite regular cardinals α. We note as a consequence that if $C\in\mathscr{C}$ and $\eta<\alpha$ then

$$2^{|\eta|}\le|\mathscr{S}(C,\eta)|\le 2^{\underline{\alpha}}.$$

We set

$$\alpha_{C,\eta}=|\mathscr{S}(C,\eta)| \quad \text{for } \eta<\alpha.$$

Proof of (a). We define a base $\mathscr{B}$ for the topology of X such that

$\mathscr{B} \subset \mathscr{C}$,

$\mathscr{B}$ satisfies conditions (i), (ii), (iv), and

(v) if $B \in \mathscr{B}$ then B has exactly $2^{\underline{\alpha}}$ immediate successors (in the system $\mathscr{B}$).

We proceed recursively. Let $\mathscr{B}_0 = \{X\}$. If ξ is a non-zero limit ordinal and $\xi < \alpha$ we set $\mathscr{B}_\xi = \bigwedge_{\zeta < \xi} \mathscr{B}_\zeta$. Now for $\xi < \alpha$ we define $\mathscr{B}_{\xi+1}$. It is clear from the recursive assumptions that $|\mathscr{B}_\xi| = 2^{\underline{\alpha}}$ if $\xi > 0$. There are two (partially overlapping) cases to consider.

Case 1. α is not strongly inaccessible. Then there is a cardinal $\beta < \alpha$ such that $\alpha \leq 2^\beta$, and

$$\alpha \leq 2^\beta \leq \alpha_{B,\beta} \leq 2^{\underline{\alpha}}$$

for $B \in \mathscr{B}_\xi$. Let $\phi_{B,\beta}$ be a function from $\alpha_{B,\beta}$ onto α and let $\langle C_{B,\beta,i} : i < \alpha_{B,\beta} \rangle$ be a well-ordering of $\mathscr{S}(B,\beta)$ for $B \in \mathscr{B}_\xi$. We define

$$\mathscr{B}_{\xi+1} = \bigcup \{\mathscr{S}(C_{B,\beta,i}, \phi_{B,\beta}(i)) : B \in \mathscr{B}_\xi \text{ and } i < \alpha_{B,\beta}\}.$$

It is easy to verify that $\mathscr{B}_{\xi+1}$ satisfies the recursive conditions.

Case 2. $\alpha = \alpha^{\underline{\alpha}}$ and α is not a weakly compact cardinal. Then by Theorem 8.23 the space $(2^\alpha)_\alpha$ is not α-compact. Since according to Lemma 15.2 the space $(2^\alpha)_\alpha$ is the continuous image of X, it follows that X is not α-compact. In fact every non-empty, open-and-closed subset of X is easily seen to satisfy the conditions of the theorem and hence also the conditions of Lemma 15.2, so in particular if $C \in \mathscr{C}$ then C is not α-compact. Since $\mathscr{C}$ is a base for the topology of X, for every $C \in \mathscr{C}$ there is an open cover $\mathscr{U}_C$ of C such that

$\mathscr{U}_C \subset \mathscr{C}$, and

no subcover of $\mathscr{U}_C$ has cardinality less than α.

Since $\alpha = \alpha^{\underline{\alpha}}$, we have $|\mathscr{U}_C| = \alpha$. We define

$$\mathscr{V}_C = \{B \in \mathscr{U}_C : \text{there is no } B' \in \mathscr{U}_C \text{ such that } B \subsetneqq B'\},$$

and we prove that $\mathscr{V}_C$ is an open partition of C. Let $p \in C$; since $\mathscr{C}$ is a ramification system under reverse inclusion the family $\{B \in \mathscr{U}_C : p \in B\}$ is a well-ordered chain in $\mathscr{C}$ and thus there is a least element B_0 of this family. It is clear that $B_0 \in \mathscr{V}_C$ and thus $\mathscr{V}_C$ is a cover of C. Further if $B, B' \in \mathscr{V}_C$ and $B \neq B'$ then the inclusions $B \subset B'$ and $B' \subset B$ both fail (from the definition of $\mathscr{V}_C$) and therefore $B \cap B' = \emptyset$.

We note from the fact that $\mathscr{V}_C$ is a subcover of $\mathscr{U}_C$ that $|\mathscr{V}_C| = \alpha$. We define

$$\mathscr{B}_{\xi+1} = \bigcup \{\mathscr{V}_B : B \in \mathscr{B}_\xi\}.$$

It is clear, since $\alpha=2^{\underline{\alpha}}$, that $\mathscr{B}_{\xi+1}$ satisfies the recursive conditions.

The recursive definition of the family $\{\mathscr{B}_\xi:\xi<\alpha\}$ is now complete in both cases. We set $\mathscr{B}=\bigcup_{\xi<\alpha}\mathscr{B}_\xi$.

It follows easily from Lemma 15.1 that any two spaces X and Y with bases $\mathscr{B}$ and $\mathscr{D}$, respectively, satisfying (the analogues of) conditions (i), (ii), (iii) and (iv) are homeomorphic. The space $(2^\alpha)_\alpha$ clearly satisfies the conditions of the present theorem. Thus X is homeomorphic to $(2^\alpha)_\alpha$. The proof of part (a) is complete.

Proof of (b). Assume that $|\mathscr{A}_\xi|=\alpha$ for some $\xi<\alpha$. Then X has an open partition of X of cardinality α and clearly it is not α-compact. Thus X is not homeomorphic to the α-compact space $(2^\alpha)_\alpha$.

For the converse implication assume that $|\mathscr{A}_\xi|<\alpha$ for $\xi<\alpha$. It is clear from the definition of $\mathscr{C}$ in Lemma 15.2 and the fact that α is strongly inaccessible that $|\mathscr{C}_\xi|<\alpha$ for $\xi<\alpha$.

If $\alpha=\omega$ then X is a totally disconnected space without isolated elements and with a countable base for its topology. Since X is $\mathscr{C}$-compact and $\mathscr{C}$ is a ramification system of order ω it follows from D. König's theorem (Corollary 8.5) that X is compact. Hence X is homeomorphic to the Cantor set $2^\omega(=(2^\omega)_\omega)$ by Theorem 6.12(b).

If $\alpha>\omega$ and X and X' are spaces satisfying the conditions of the present theorem then as noted above there are bases $\mathscr{C}$ and $\mathscr{C}'$ for X and X' respectively satisfying conditions (i) to (iv) above and such that

$$|\mathscr{C}_\xi|<\alpha \quad \text{and } |\mathscr{C}'_\xi|<\alpha \quad \text{for } \xi<\alpha\,.$$

We define bases $\mathscr{B}$ and $\mathscr{B}'$ for X and X', respectively, such that

$$\mathscr{B}\subset\mathscr{C} \quad \text{and } \mathscr{B}'\subset\mathscr{C}',$$

$\mathscr{B}$ and $\mathscr{B}'$ satisfy (the analogues of) conditions (i), (ii), (iv) above, and

$$|\mathscr{B}_\xi|<\alpha \quad \text{and } |\mathscr{B}'_\xi|<\alpha \quad \text{for } \xi<\alpha\,;$$

and we define also a function $\phi:\mathscr{B}\to\mathscr{B}'$ such that

ϕ is one-to-one and onto,

$\phi[\mathscr{B}_\xi]=\mathscr{B}'_\xi$ for $\xi<\alpha$, and

if $B_0, B_1\in\mathscr{B}$ and $B_0\subset B_1$ then $\phi(B_0)\subset\phi(B_1)$.

We proceed recursively. Set $\mathscr{B}_0=\{X\}$, $\mathscr{B}'_0=\{X'\}$, and $\phi_0:\mathscr{B}_0\to\mathscr{B}'_0$ the unique function. If ξ is a non-zero limit ordinal and $\xi<\alpha$ we set

$$\mathscr{B}_\xi=\bigwedge\nolimits_{\zeta<\xi}\mathscr{B}_\zeta,$$
$$\mathscr{B}'_\xi=\bigwedge\nolimits_{\zeta<\xi}\mathscr{B}'_\zeta, \quad \text{and}$$
$$\phi_\xi\left(\bigcap\nolimits_{\zeta<\xi}B_\zeta\right)=\bigcap\nolimits_{\zeta<\xi}\phi_\xi(B_\zeta) \quad \text{if } B_\zeta\in\mathscr{B}_\zeta \quad \text{for } \zeta<\xi \quad \text{and } \bigcap\nolimits_{\zeta<\xi}B_\zeta\neq\emptyset\,.$$

We let $\xi<\alpha$ and we define $\mathscr{B}_{\xi+1}, \mathscr{B}'_{\xi+1}$, and $\phi_{\xi+1}$. The recursive assumptions clearly imply that

$$|\mathscr{B}_\xi| = |\mathscr{B}'_\xi| < \alpha .$$

For $B \in \mathscr{B}_\xi$ we set $B' = \phi_\xi(B)$ and for $n<\omega$ we define recursively ordinals $\eta_{B,n}<\alpha$ and $\eta_{B',n}<\alpha$ such that

$$\begin{aligned}&\eta_{B,0}=1,\\ &\eta_{B,n}\le\eta_{B',n}\le\eta_{B,n+1}, \quad \text{and}\\ &2\le|\mathscr{S}(B,\eta_{B,n})|\le|\mathscr{S}(B',\eta_{B',n})|\le|\mathscr{S}(B,\eta_{B,n+1})|.\end{aligned}$$

We set

$$\eta_B = \sup_{n<\omega}\eta_{B,n} = \sup_{n<\omega}\eta_{B',n},$$

and it is clear that

$$|\mathscr{S}(B,\eta_B)| = |\mathscr{S}(B',\eta_B)| < \alpha$$

for $B \in \mathscr{B}_\xi$. We let $\phi_{B,\xi+1}$ be any one-to-one function from $\mathscr{S}(B,\eta_B)$ onto $\mathscr{S}(B',\eta_B)$ and we define

$$\begin{aligned}\mathscr{B}_{\xi+1} &= \bigcup\{\mathscr{S}(B,\eta_B): B\in\mathscr{B}_\xi\},\\ \mathscr{B}'_{\xi+1} &= \bigcup\{\mathscr{S}(B',\eta_B): B\in\mathscr{B}_\xi\}, \quad \text{and}\\ \phi_{\xi+1} &= \bigcup\{\phi_{B,\xi+1}: B\in\mathscr{B}_\xi\}.\end{aligned}$$

The recursive definitions are complete. Finally we set

$$\begin{aligned}\mathscr{B} &= \textstyle\bigcup_{\xi<\alpha}\mathscr{B}_\xi,\\ \mathscr{B}' &= \textstyle\bigcup_{\xi<\alpha}\mathscr{B}'_\xi, \quad \text{and}\\ \phi &= \textstyle\bigcup_{\xi<\alpha}\phi_\xi.\end{aligned}$$

It follows from Lemma 15.1 that X and X' are homeomorphic, and in particular that X is homeomorphic to $(2^\alpha)_\alpha$.

The proof of the theorem is complete.

The Baire Category Properties of $(2^\alpha)_\alpha$

The following characterization of the space $(\alpha^\alpha)_\alpha$ is not as economical as the one given in Theorem 15.3 for $(2^\alpha)_\alpha$; it is an auxiliary result towards the proof of Theorem 15.8.

15.4. Lemma. *Let α be an infinite regular cardinal and let X be a (P_α-) space (without isolated elements) having a base $\mathscr{B}$ for its topology such that*

(i) *X is $\mathscr{B}$-compact,*

(ii) *$\mathscr{B}$ is a ramification system of order α under reverse set inclusion,*

(iii) *the set* $\mathscr{B}_\xi$ *of elements of order* ξ *is an open partition of* X *of cardinality at most* $2^{\underline{\alpha}}$ *for* $\xi<\alpha$,

(iv) $\mathscr{B}_\xi = \bigwedge_{\zeta<\xi}\mathscr{B}_\zeta$ *for limit ordinals* $\xi<\alpha$, *and*

(v) *if* $B\in\mathscr{B}$ *then* B *has at least* α *immediate successors (with respect to the partial order of the ramification system* $\mathscr{B}$*).*

Then X *is homeomorphic to* $(\alpha^\alpha)_\alpha$.

Proof. We define a base $\mathscr{C}$ for X such that

$$\mathscr{C}\subset\mathscr{B},$$

$\mathscr{C}$ satisfies (i), (ii), (iii), (iv), and

(vi) if $C\in\mathscr{C}$ then C has *exactly* $2^{\underline{\alpha}}$ immediate successors (with respect to the order of the ramification system $\mathscr{C}$).

We proceed recursively. Set $\mathscr{C}_0=\mathscr{B}_0=\{X\}$. If ξ is a non-zero limit ordinal and $\xi<\alpha$, we set

$$\mathscr{C}_\xi = \bigwedge_{\zeta<\xi}\mathscr{C}_\zeta .$$

We let $\xi<\alpha$ and we define $\mathscr{C}_{\xi+1}$. Clearly from the recursive assumptions we have $|\mathscr{C}_\xi|=2^{\underline{\alpha}}$ if $\xi>0$, and for every $C\in\mathscr{C}_\xi$ we have $\alpha\le|\mathscr{S}(C,1)|\le 2^{\underline{\alpha}}$. Let $\langle B_{C,i}: i<|\mathscr{S}(C,1)|\rangle$ be a well-ordering of $\mathscr{S}(C,1)$ and let $\phi_C: |\mathscr{S}(C,1)|\to\alpha$ be an onto function. We define

$$\mathscr{C}_{\xi+1} = \bigcup\{\mathscr{S}(B_{C,i},\phi_C(i)): C\in\mathscr{C}_\xi, i<|\mathscr{S}(C,1)|\} .$$

The recursive definition is complete.

It follows easily from Lemma 15.1 that any two spaces X and Y with bases $\mathscr{C}$ and $\mathscr{D}$ respectively satisfying (the analogues of) conditions (i), (ii), (iii), (iv), and (vi) are homeomorphic.

The space $(\alpha^\alpha)_\alpha$ has a *canonical* base $\mathscr{E}'$ for its topology satisfying conditions (i) to (v) of the present lemma, and in fact

> the set $\mathscr{E}'_\xi$ of elements of order ξ has cardinality $\alpha^{|\xi|}$ for $\xi<\alpha$, and
> if $E\in\mathscr{E}'$ then E has *exactly* α immediate successors.

In fact set $E'(s)=\{t\in\alpha^\alpha: t|\xi=s\}$ for $s\in\alpha^\xi$ and set

$$\mathscr{E}'_\xi = \{E'(s): s\in\alpha^\xi\} \quad \text{for } \xi<\alpha, \quad \text{and}$$
$$\mathscr{E}' = \bigcup_{\xi<\alpha}\mathscr{E}'_\xi .$$

This completes the proof of the lemma.

15.5. Theorem. *Let* α *be an infinite regular cardinal.*

(a) *If* $\alpha\le\beta_\xi\le 2^{\underline{\alpha}}$ *for* $\xi<\alpha$ *then the space* $(\prod_{\xi<\alpha}\beta_\xi)_\alpha$ *is homeomorphic to* $(\alpha^\alpha)_\alpha$.

In particular, if $\alpha\le\beta\le 2^{\underline{\alpha}}$ *then* $(\beta^\alpha)_\alpha$ *is homeomorphic to* $(\alpha^\alpha)_\alpha$*; thus* $((2^{\underline{\alpha}})^\alpha)_\alpha$ *is homeomorphic to* $(\alpha^\alpha)_\alpha$.

(b) *If* $2 \leq \beta_\xi < \alpha$ *for* $\xi < \alpha$ *then the space* $(\prod_{\xi<\alpha} \beta_\xi)_\alpha$ *is homeomorphic to* $(2^\alpha)_\alpha$.

In particular, if $2 \leq \beta < \alpha$ *then* $(\beta^\alpha)_\alpha$ *is homeomorphic to* $(2^\alpha)_\alpha$.

(c) *If* α *is not weakly compact and if* $2 \leq \beta_\xi \leq 2^{\underline{\alpha}}$ *for* $\xi < \alpha$ *then the space* $(\prod_{\xi<\alpha} \beta_\xi)_\alpha$ *is homeomorphic to* $(2^\alpha)_\alpha$ *(and also to* $(\alpha^\alpha)_\alpha$*).*

(d) *If* $1 \leq \beta_\xi < \alpha$ *for* $\xi < \alpha$ *then the space* $(\prod_{\xi<\alpha} 2^{\beta_\xi})_\alpha$ *is homeomorphic to* $(2^\alpha)_\alpha$.

In particular, if $1 \leq \beta < \alpha$ *then* $((2^\beta)^\alpha)_\alpha$ *is homeomorphic to* $(2^\alpha)_\alpha$.

Proof. (a) follows from Lemma 15.4.

(b) follows from Theorem 15.3(a) and (b).

(c) follows from Theorem 15.3(a).

(d) If $2^{\beta_\xi} < \alpha$ for $\xi < \alpha$, then the statement follows from part (b). If $2^{\beta_\xi} \geq \alpha$ for some $\xi < \alpha$ then α is not strongly inaccessible, so in particular α is not weakly compact; then since $2^{\beta_\xi} \leq 2^{\underline{\alpha}}$ for $\xi < \alpha$ the statement follows from part (c).

15.6. Corollary. *Let* α *be an infinite, regular cardinal. The following statements are equivalent.*

(a) $\alpha = 2^{\underline{\alpha}}$;

(b) *if* $\beta > \alpha$ *then* $(2^\alpha)_\alpha$ *is not homeomorphic to* $(\beta^\alpha)_\alpha$*; and*

(c) *if* $\beta > \alpha$ *then* $(\alpha^\alpha)_\alpha$ *is not homeomorphic to* $(\beta^\alpha)_\alpha$.

Proof. (a)⇒(b). Assume $\alpha = 2^{\underline{\alpha}}$ and $\beta > \alpha$. Then $(2^\alpha)_\alpha$ is α^+-compact and hence β-compact, while $(\beta^\alpha)_\alpha$ is not β-compact.

(b)⇒(a). If $\alpha < 2^{\underline{\alpha}}$ then there is $\gamma < \alpha$ such that $2^\gamma > \alpha$. By Theorem 15.5(d) we have that $((2^\gamma)^\alpha)_\alpha$ is homeomorphic to $(2^\alpha)_\alpha$.

(a)⇒(c). Assume $\alpha = 2^{\underline{\alpha}}$ and $\beta > \alpha$. Then $\alpha = \alpha^{\underline{\alpha}}$ by Theorem 1.27(a) so $(\alpha^\alpha)_\alpha$ is α^+-compact and thus β-compact; but $(\beta^\alpha)_\alpha$ is not β-compact.

(c)⇒(a). This follows from Theorem 15.5(a).

15.7. Corollary. *Let* α *be an infinite regular cardinal. Then* α *is weakly compact if and only if* $(2^\alpha)_\alpha$ *is not homeomorphic to* $(\alpha^\alpha)_\alpha$.

Proof. If $(\alpha^\alpha)_\alpha$ is homeomorphic to $(2^\alpha)_\alpha$ then $(2^\alpha)_\alpha$ is not α-compact and hence α is not weakly compact by Theorem 8.23. Conversely if α is not weakly compact then $(2^\alpha)_\alpha$ is homeomorphic to $(\alpha^\alpha)_\alpha$ by Theorem 15.3(a).

We recall that a space X is said to have the *α-Baire category property* if the intersection of a family of at most α open and dense subsets of X is dense.

The following notation will be used below. If $\mathscr{B}$ is a family of non-empty open subsets of X and G is a dense subset of X then we set

$$\mathscr{B} \upharpoonright G = \{B \cap G : B \in \mathscr{B}\}.$$

15.8. Theorem. *Let α be an infinite regular cardinal.*

(a) *The space $(2^\alpha)_\alpha$ has the α-Baire category property.*

(b) *If α is not weakly compact, the intersection of any family of at most α open and dense subsets of $(2^\alpha)_\alpha$ is homeomorphic to $(2^\alpha)_\alpha$.*

(c) *The intersection G of any family at most α open subsets of $(2^\alpha)_\alpha$ such that G and $(2^\alpha)_\alpha \backslash G$ are both dense in $(2^\alpha)_\alpha$ is homeomorphic to $(\alpha^\alpha)_\alpha$.*

Proof. (a) Let $G = \bigcap_{\xi<\alpha} G_\xi$, where G_ξ is open and dense in $(2^\alpha)_\alpha$ for $\xi<\alpha$. It suffices to prove that $E \cap G \neq \emptyset$ for all $E \in \mathscr{E}$, where $\mathscr{E}$ denotes the canonical base of $(2^\alpha)_\alpha$. We define a family $\{E_\xi : \xi<\alpha\}$ such that

$$\begin{aligned}
&E_0 = E,\\
&E_\xi \in \mathscr{E} \quad \text{for } \xi<\alpha,\\
&E_{\xi+1} \subset E_\xi \cap G_\xi \quad \text{for } \xi<\alpha, \quad \text{and}\\
&E_\xi = \textstyle\bigcap_{\zeta<\xi} E_\zeta \quad \text{for non-zero limit ordinals } \xi<\alpha.
\end{aligned}$$

We proceed recursively. We define $E_0 = E$; if ξ is a non-zero limit ordinal and $\xi<\alpha$ we set $E_\xi = \bigcap_{\zeta<\xi} E_\zeta$. The properties of the canonical base $\mathscr{E}$ imply that $E_\xi \in \mathscr{E}$. Now let $\xi<\alpha$. We define $E_{\xi+1}$. Since $E_\xi \in \mathscr{E}$ and E_ξ is open and non-empty and G_ξ is open and dense in $(2^\alpha)_\alpha$, it follows that $E_\xi \cap G_\xi$ is open and non-empty. The fact that $\mathscr{E}$ is a base implies that there is $E_{\xi+1} \in \mathscr{E}$ such that $E_{\xi+1} \subset E_\xi \cap G_\xi$. This completes the recursive definition. Since $(2^\alpha)_\alpha$ is $\mathscr{E}$-compact, we have

$$\emptyset \neq \textstyle\bigcap_{\xi<\alpha} E_\xi = \bigcap_{\xi<\alpha} E_{\xi+1} \subset E \cap \bigcap_{\xi<\alpha} G_\xi = E \cap G.$$

(b) Let $G = \bigcap_{\xi<\alpha} G_\xi$, where $\{G_\xi : \xi<\alpha\}$ is a family of open and dense subsets of $(2^\alpha)_\alpha$. Because of part (a) we assume with no loss of generality that $G_\xi \subset G_\zeta$ for $\zeta<\xi<\alpha$.

We define families $\mathscr{B}_\xi$ for $\xi<\alpha$ such that

$$\begin{aligned}
&\mathscr{B}_\xi \subset \textstyle\bigcup_{\xi\le\zeta<\alpha} \mathscr{E}_\zeta, \quad \text{and}\\
&\mathscr{B}_\xi \text{ is an open partition of } G_\xi.
\end{aligned}$$

Let $\mathscr{U}_\xi$ be an open cover of G_ξ such that

$$\mathscr{U}_\xi \subset \textstyle\bigcup_{\xi\le\zeta<\alpha} \mathscr{E}_\zeta.$$

Set $\mathscr{B}_\xi = \{E \in \mathscr{U}_\xi : \text{there is no } E' \in \mathscr{U}_\xi \text{ such that } E \subsetneqq E'\}$. We prove that $\mathscr{B}_\xi$ is an open partition of G_ξ for $\xi<\alpha$. Indeed let $p \in G_\xi$; since $\mathscr{E}$ is a ramification system under reverse inclusion, the family $\{E \in \mathscr{U}_\xi : p \in E\}$ is a well-ordered chain in $\mathscr{E}$ and thus there is a least element E_0 of this family. It is clear that $E_0 \in \mathscr{B}_\xi$ and thus $\mathscr{B}_\xi$ is a cover of G_ξ. Further if $E, E' \in \mathscr{B}_\xi$ and $E \neq E'$ then the inclusions $E \subset E'$ and $E' \subset E$ both fail (from the definition of $\mathscr{B}_\xi$) and therefore $E \cap E' = \emptyset$.

We set $\mathscr{B} = \bigcup_{\xi<\alpha}(\mathscr{B}_\xi \upharpoonright G)$ and we prove that $\mathscr{B}$ satisfies the conditions of Theorem 15.3(a). It suffices to verify the following statements.

(i) $\mathscr{B}$ is a base for the topology of G.
Let $p \in G$ and let U be a neighborhood of p in $(2^\alpha)_\alpha$. There are $\xi < \alpha$ and $E \in \mathscr{E}_\xi$ such that $p \in E \subset U$. Let B_ξ be the unique element of $\mathscr{B}_\xi$ such that $p \in B_\xi$. Since $\mathscr{B}_\xi \subset \bigcup_{\xi \le \zeta < \alpha} \mathscr{E}_\zeta$ it follows that $B_\xi \subset E$.

(ii) G is $\mathscr{B}$-compact.
Let $\mathscr{C}$ be a subfamily of $\mathscr{B}$ with the finite intersection property. Let $\beta = |\mathscr{C}|$ and let $\langle C_i : i < \beta \rangle$ be a well-ordering of $\mathscr{C}$ such that $C_i \neq C_j$ if $i < j < \beta$. There is $B_i \in \bigcup_{\xi<\alpha} \mathscr{B}_\xi \subset \mathscr{E}$ such that $C_i = B_i \cap G$ for $i < \beta$; since $\{B_i : i < \beta\}$ has the finite intersection property we have $\bigcap_{i<\beta} B_i \neq \emptyset$. There are two possibilities to consider.

Case 1. $\beta < \alpha$. Then $\bigcap_{i<\beta} B_i$ is open and non-empty and $\bigcap_{i<\beta} C_i = \bigcap_{i<\beta} B_i \cap G \neq \emptyset$.

Case 2. $\beta = \alpha$. Then $B_i \in \mathscr{B}_{\xi(i)}$ for some $\xi(i) < \alpha$ and $|\{\xi(i) : i < \beta\}| = \alpha$ and hence the set $\{\xi(i) : i < \beta\}$ is cofinal in α. It follows that $\bigcap_{i<\beta} B_i \subset \bigcap_{i<\beta} G_{\xi(i)} = \bigcap_{\xi<\alpha} G_\xi = G$ and thus again $\bigcap_{i<\beta} C_i = \bigcap_{i<\beta} B_i \neq \emptyset$.

The proof of part (b) is complete.

(c) Let $G = \bigcap_{\xi<\alpha} G_\xi$, where $\{G_\xi : \xi < \alpha\}$ is a family of open and dense subsets of $(2^\alpha)_\alpha$ and $(2^\alpha)_\alpha \setminus G$ is dense in $(2^\alpha)_\alpha$. Again because of part (a) we assume in addition with no loss of generality that $G_\xi \subset G_\zeta$ for $\zeta < \xi < \alpha$.

We define sets H_ξ and families $\mathscr{B}_\xi$ for $\xi < \alpha$ such that

(i) $H_0 = G_0$ and $\mathscr{B}_0 = \{G_0\}$;
(ii) H_ξ is open and dense in $(2^\alpha)_\alpha$ for $\xi < \alpha$;
(iii) $H_\xi \subset H_\zeta$ for $\zeta < \xi < \alpha$;
(iv) $H_\xi = \bigcap_{\zeta<\xi} H_\zeta$ for non-zero limit ordinals $\xi < \alpha$;
(v) $G \subset H_{\xi+1} \subset G_\xi$ for $\xi < \alpha$;
(vi) $\mathscr{B}_\xi \subset \bigcup_{\xi \le \zeta < \alpha} \mathscr{E}_\zeta$ for $0 < \xi < \alpha$;
(vii) $\mathscr{B}_\xi$ is an open partition of H_ξ for $\xi < \alpha$;
(viii) $\mathscr{B}_\zeta \upharpoonright H_\xi \prec \mathscr{B}_\xi$ for $\zeta < \xi < \alpha$;
(ix) $\mathscr{B}_\xi = \bigwedge_{\zeta<\xi} (\mathscr{B}_\zeta \upharpoonright H_\xi)$ for non-zero limit ordinals $\xi < \alpha$; and
(x) if $B \in \mathscr{B}_\zeta$ and $\zeta < \xi < \alpha$ then $H_\xi \cap B$ is a *proper* subset of B.

We proceed recursively. We let H_0 and $\mathscr{B}_0$ be defined by (i).

If ξ is a non-zero limit ordinal and $\xi < \alpha$ we define H_ξ and $\mathscr{B}_\xi$ by (iv) and (ix), respectively. We verify the (relevant) recursive conditions for ξ. For condition (ii) we note that H_ξ is open because X is a P_α-space, and H_ξ is dense by part (a); (iii) is obvious. For (vi) let $B \in \mathscr{B}_\xi$. Then $B = \bigcap_{\zeta<\xi} (B_\zeta \cap H_\xi)$ where $B_\zeta \in \mathscr{B}_\zeta \subset \bigcup_{\zeta \le \eta < \alpha} \mathscr{E}_\eta$ for $0 < \zeta < \xi$. We note that

$$\begin{aligned} B &= \textstyle\bigcap_{\zeta<\xi} (B_\zeta \cap H_\xi) = \bigcap_{\zeta<\xi} B_\zeta \cap H_\xi = \bigcap_{\zeta<\xi} B_\zeta \cap \bigcap_{\zeta<\xi} H_\zeta \\ &= \textstyle\bigcap_{\zeta<\xi} (B_\zeta \cap H_\zeta) = \bigcap_{\zeta<\xi} B_\zeta \end{aligned}$$

and hence, since the non-empty, decreasing intersection of a family of less than α elements of the canonical base $\mathscr{E}$ is in $\mathscr{E}$, we have that $B \in \mathscr{E}$; it is clear that in fact $B \in \bigcup_{\xi \le \zeta < \alpha} \mathscr{E}_\zeta$. Condition (vii) follows from the fact that X is a P_α-space; (viii) is clear; and for (x) we note that if $B \in \mathscr{B}_\zeta$ and $\zeta < \xi < \alpha$ then $\zeta + 1 < \xi$ and the recursive assumptions imply that

$$B \cap H_\xi \subset B \cap H_{\zeta+1} \subsetneqq B .$$

Now let $\xi < \alpha$. We define $H_{\xi+1}$ and $\mathscr{B}_{\xi+1}$. We note that $G_\xi \cap H_\xi$ is open and dense in $(2^\alpha)_\alpha$ and contains G. Because $(2^\alpha)_\alpha \backslash G$ is dense in $(2^\alpha)_\alpha$ it follows that for $B \in \mathscr{B}_\xi \restriction (G_\xi \cap H_\xi)$ there is ξ_B such that

$$\xi < \xi_B < \alpha, \quad \text{and}$$
$$B \cap G_{\xi_B} \text{ is a } \textit{proper} \text{ subset of } B.$$

We define

$$H_{\xi+1} = \bigcup \{B \cap G_{\xi_B} : B \in \mathscr{B}_\xi \restriction (G_\xi \cap H_\xi)\} .$$

Let $\mathscr{U}_{\xi+1}$ be an open cover of $H_{\xi+1}$ such that

$$\mathscr{U}_{\xi+1} \subset \bigcup_{\xi < \zeta < \alpha} \mathscr{E}_\zeta \quad \text{and}$$
$$\mathscr{B}_\xi \restriction H_{\xi+1} \prec \mathscr{U}_{\xi+1} .$$

We define

$$\mathscr{B}_{\xi+1} = \{E \in \mathscr{U}_{\xi+1} : \text{there is no } E' \in \mathscr{U}_{\xi+1} \text{ such that } E \subsetneqq E'\} .$$

The only recursive condition not obviously satisfied is (vii), and this is proved as was the corresponding statement in part (b); the details are omitted. This completes the recursive definitions.

We define $\mathscr{B} = \bigcup_{\xi < \alpha} (\mathscr{B}_\xi \restriction G)$. As in part (b), we prove that

$$\mathscr{B} \text{ is a base for } G, \text{ and}$$
$$G \text{ is } \mathscr{B}\text{-compact.}$$

Further we claim that if $\xi < \alpha$ and $C \in \mathscr{B}_\xi \restriction G$ then C has *at least* α immediate successors in $\mathscr{B}_{\xi+1} \restriction G$. There is a unique $B \in \mathscr{B}_\xi$ such that $C = B \cap G$, and by (x) the set $H_{\xi+1} \cap B$ is a proper subset of B. Because $H_{\xi+1} \cap B$ is open and dense in B and $\mathscr{B}_{\xi+1}$ is an open partition of $H_{\xi+1}$ refining $\mathscr{B}_\xi \restriction H_{\xi+1}$, it follows that

$$|\{D \in \mathscr{B}_{\xi+1} : D \subset H_{\xi+1} \cap B\}| \ge \alpha .$$

For otherwise, since X is a P_α-space, the set

$$\bigcup \{D \in \mathscr{B}_{\xi+1} : D \subset H_{\xi+1} \cap B\}$$

would be a closed subset of B, contradicting the fact that it is a proper and dense subset of B.

Then $D\cap C\neq\emptyset$ for $D\in\mathscr{B}_{\xi+1}$ with $D\subset H_{\xi+1}\cap B$, because each such set D is open and non-empty and C is dense in $H_{\xi+1}\cap B$ (since G is dense in $(2^\alpha)_\alpha$). Thus C has at least α immediate successors in $\mathscr{B}_{\xi+1}\restriction G$.

The conditions of Lemma 15.4 are satisfied, so G is homeomorphic to $(\alpha^\alpha)_\alpha$.

The proof of the theorem is complete.

We remark that the assumption that the cardinal α is not weakly compact in Theorem 15.8(b) is essential; it is clear that the statement fails for $\alpha=\omega$ (and in fact for every weakly compact cardinal α).

Spaces of Ultrafilters Homeomorphic to Spaces $(2^\alpha)_\alpha$

15.9. Theorem. *Let α be an infinite regular cardinal and let X be a compact space. If*

(a) *the weight of X is (at most) α,*
(b) *the local weight of X at every element of X is (at least) α, and*
(c) *X is totally disconnected if $\alpha=\omega$,*

then X_α is homeomorphic to $(2^\alpha)_\alpha$.

Conversely if $\alpha=\alpha^{\underline{\alpha}}$ and X_α is homeomorphic to $(2^\alpha)_\alpha$ then X satisfies conditions (a), (b) *and* (c).

Proof. If $\alpha=\omega$ we let $\{B_n:n<\omega\}$ be a base for X consisting of open-and-closed sets, and we set $\mathscr{A}_n=\{B_n,X\setminus B_n\}$ for $n<\omega$ and $\mathscr{A}=\bigcup_{n<\omega}\mathscr{A}_n$. We note that $|\mathscr{A}_n|\leq 2<\omega$ for $n<\omega$.

If $\alpha>\omega$ we let $\{B_\xi:\xi<\alpha\}$ be a family of zero-sets of X such that $\{\operatorname{int}B_\xi:\xi<\alpha\}$ is a base for X (where $\operatorname{int}B_\xi$ denotes the interior of B_ξ in X). We have used the fact, proved in Lemma 2.12, that every base for X contains a subfamily which is itself a base for X and of cardinality $\boldsymbol{w}(X)$. It is clear that $\{B_\xi:\xi<\alpha\}$ is an α-subbase for X_α. For $\xi<\alpha$ let f_ξ be a continuous real-valued function defined on X such that $B_\xi=\{x\in X:f_\xi(x)=0\}$. We set

$$\mathscr{A}_\xi=\{f_\xi^{-1}(\{t\}):t\in f_\xi[X]\}$$

for $\xi<\alpha$, and

$$\mathscr{A}=\bigcup_{\xi<\alpha}\mathscr{A}_\xi.$$

Then the elements of $\mathscr{A}$ are open-and-closed subsets of X_α (since a zero-set is closed in X and thus in X_α, and is a G_δ-set in X and thus an open set in X_α). Further $\mathscr{A}_\xi$ is an open partition of X_α and $|\mathscr{A}_\xi|\leq 2^\omega\leq 2^{\underline{\alpha}}$ for $\xi<\alpha$. Since $\{B_\xi:\xi<\alpha\}\subset\mathscr{A}$ the family $\mathscr{A}$ is an α-subbase of X_α. By condition (b) the space X_α is a P_α-space without isolated elements. Since X is compact and $\mathscr{A}$ consists of closed, non-empty subsets of X, it follows that X_α is $\mathscr{A}$-compact.

If α is not weakly compact then $|\mathscr{A}_\xi| \leq 2^{\underline{\alpha}}$ for $\xi < \alpha$ and by Theorem 15.3(a) the space X_α is homeomorphic to $(2^\alpha)_\alpha$. If α is weakly compact, then either $\alpha = \omega$ in which case $|\mathscr{A}_\xi| < \omega$ for $\xi < \alpha$, or $\alpha > \omega$ in which case $|\mathscr{A}_\xi| \leq 2^\omega < \alpha$ for $\xi < \alpha$. It follows from Theorem 15.3(b) that X_α is homeomorphic to $(2^\alpha)_\alpha$.

For the (partial) converse assume that X is compact, $\alpha = \alpha^{\underline{\alpha}}$, and X_α is homeomorphic to $(2^\alpha)_\alpha$. It is clear that X satisfies conditions (b) and (c). Condition (a) is obvious if $\alpha = \omega$, so we assume $\alpha > \omega$. Condition (a) will follow from the existence of an embedding of X into the power space $[0,1]^\alpha$, since then we will have $w(X) \leq w([0,1]^\alpha) = \alpha$.

Since X_α is homeomorphic to $(2^\alpha)_\alpha$ we have

$$w(X_\alpha) = w((2^\alpha)_\alpha) \leq 2^{\underline{\alpha}} = \alpha .$$

Thus from Lemma 2.12 and the discussion in § 1 following the definition of the space X_α it follows that there is a base $\mathscr{A}$ for X_α such that

$|\mathscr{A}| = \alpha$, and

for $A \in \mathscr{A}$ there is a set I_A and for $i \in I_A$ a zero-set Z_i of X such that $A = \bigcap_{i \in I_A} Z_i$.

We assume without loss of generality that $I_{A_0} \cap I_{A_1} = \emptyset$ for $A_0, A_1 \in \mathscr{A}$ with $A_0 \neq A_1$. We define

$$I = \bigcup \{I_A : A \in \mathscr{A}\} \quad \text{and } \mathscr{B} = \{Z_i : i \in I\},$$

for $i \in I$ we choose a continuous function $f_i: X \to [0,1]$ such that $Z_i = f_i^{-1}(\{0\})$, and we set

$$\mathscr{S} = \{f_i : i \in I\} .$$

Then $|\mathscr{S}| \leq |I| = \alpha$ and we claim that $\mathscr{S}$ separates elements of X. Indeed if $p, q \in X$ and $p \neq q$ there is $A = \bigcap_{i \in I_A} Z_i \in \mathscr{A}$ such that $p \in A$ and $q \notin A$, and hence there is $i \in I_A$ such that $q \notin Z_i$; we have $f_i(p) = 0$ and $f_i(q) \neq 0$. It follows from Lemma 2.10(c) and the compactness of X that there is an embedding of X into $[0,1]^\alpha$. The proof is complete.

We do not know if condition (a) of the above theorem can be replaced by the weaker condition

(a′) $w(X) \leq 2^\alpha$

(it being understood that condition (b) is sharpened to the condition that the local weight of X at each of its elements is exactly α).

We recall from § 5 that $\Lambda(\alpha)$ denotes the set $\{0,1\}^\alpha$ ordered lexicographically, and also the space whose underlying set is $\Lambda(\alpha)$ and whose (order) topology is determined by the requirement that the open intervals form a base for the topology; we recall from § 6 that if X is a space then $P_\alpha(X)$ denotes the set of P_α-points in X.

15.10. Corollary. (a) *If* $\alpha=\alpha^{\underline{\alpha}}$ *then* $\Lambda(\alpha)_\alpha$ *is homeomorphic to* $(2^\alpha)_\alpha$.

(b) $P_\omega(\Lambda(\omega))(=\Lambda(\omega))$ is homeomorphic to 2^ω.

(c) *If* $\alpha=\alpha^{\underline{\alpha}}>\omega$ *then* $P_\alpha(\Lambda(\alpha))$ *is homeomorphic to* $(\alpha^\alpha)_\alpha$.

Proof. (a) From the remarks at the end of § 5 the space $\Lambda(\alpha)$ is a compact, totally disconnected space whose weight is (at most) $2^{\underline{\alpha}}=\alpha$ and whose local weight at each element is (exactly) α. The result follows from Theorem 15.9.

(b) is a special case of (a).

(c) We note that α is regular by Theorem 1.27(a), $\Lambda(\alpha)_\alpha$ is homeomorphic to $(2^\alpha)_\alpha$ by part (a), $(2^\alpha)_\alpha$ has the α-Baire category property by Theorem 15.8(a), and if $p\in 2^\alpha$ then $\{p\}$ is not an open subset of $(2^\alpha)_\alpha$ (and hence $\Lambda(\alpha)_\alpha\backslash\{p\}$ is a dense, open subset of $\Lambda(\alpha)_\alpha$). Since

$$P_\alpha(\Lambda(\alpha))=\bigcap\{\Lambda(\alpha)_\alpha\backslash\{p\}:p\in\Lambda(\alpha)\backslash P_\alpha(\Lambda(\alpha))\}\,,$$

and

$$|\Lambda(\alpha)\backslash P_\alpha(\Lambda(\alpha))|=2^{\underline{\alpha}}=\alpha$$

by Remark (d) at the end of § 5, it follows that $P_\alpha(\Lambda(\alpha))$ is dense in $\Lambda(\alpha)_\alpha$ and is equal to the intersection of α open subsets of $\Lambda(\alpha)_\alpha$. Furthermore the set $\Lambda(\alpha)_\alpha\backslash P_\alpha(\Lambda(\alpha))$ is dense in $\Lambda(\alpha)_\alpha$, since the non-empty intersection of a family of less than α open subsets of $\Lambda(\alpha)$ contains an open interval of $\Lambda(\alpha)$ (and hence contains non-P_α-points). Thus $P_\alpha(\Lambda(\alpha))$ is homeomorphic to $(\alpha^\alpha)_\alpha$ by Theorem 15.8(c).

15.11. Corollary. (a) *If* $\alpha=\alpha^{\underline{\alpha}}$ *then* $(\mathsf{S}_\alpha)_\alpha$ *is homeomorphic to* $(2^\alpha)_\alpha$.

(b) $P_\omega(\mathsf{S}_\omega)=P_\omega(2^\omega)=2^\omega$.

(c) *If* $\alpha=\alpha^{\underline{\alpha}}>\omega$ *then* $P_\alpha(\mathsf{S}_\alpha)$ *is homeomorphic to* $(\alpha^\alpha)_\alpha$.

Proof. (a) The space S_α is compact, totally disconnected of weight α and of local weight α at each of its elements (cf. Lemma 6.5).

(b) We have noted in Remark (1) preceding Theorem 6.15 that S_ω is (homeomorphic to) 2^ω.

(c) By part (a) the space $(\mathsf{S}_\alpha)_\alpha$ is homeomorphic to $(2^\alpha)_\alpha$. We note that the identity function $(\mathsf{S}_\alpha)_\alpha\to\mathsf{S}_\alpha$ is a homeomorphism when it is restricted to the subspace $P_\alpha(\mathsf{S}_\alpha)$. Furthermore $(\mathsf{S}_\alpha)_\alpha\backslash P_\alpha(\mathsf{S}_\alpha)$ is dense in $(\mathsf{S}_\alpha)_\alpha$ and $P_\alpha(\mathsf{S}_\alpha)$ is the intersection of (at most) α open subsets of S_α (and hence of $(\mathsf{S}_\alpha)_\alpha$), each of which is dense in S_α, and in fact in $(\mathsf{S}_\alpha)_\alpha$. Indeed

$$P_\alpha(\mathsf{S}_\alpha)=\bigcap\{\mathsf{S}_\alpha\backslash BdU: U \text{ is a subset of } \mathsf{S}_\alpha \text{ equal to the union of less than } \alpha \text{ open-and-closed subsets of } \mathsf{S}_\alpha\}$$

(where BdU denotes the boundary of U). The result now follows from Theorem 15.8(c).

The following two corollaries are simple consequences of the earlier Lemmas 9.11, 9.12, respectively.

15.12. Corollary. *Let α be an infinite cardinal, let $f \in \alpha^\alpha$ and set*

$$D_f = \bigcup \{\hat{A} : A \in \mathscr{P}(\alpha) \text{ and either } f|A \text{ is one-to-one or } |f[A]| < \alpha\}.$$

Then D_f is an open and dense subset of $(U(\alpha))_{\alpha^+}$.

Proof. Clearly D_f is open in $U(\alpha)$, hence in $(U(\alpha))_{\alpha^+}$. That D_f is a dense subset of $(U(\alpha))_{\alpha^+}$ is a direct consequence of Lemma 9.11.

If $\phi : \mathscr{P}_\omega(\alpha) \to \mathscr{P}(\alpha)$ is a monotone function, we set

$$U_\phi = \bigcap \{\phi(F)^\wedge : F \in \mathscr{P}_\omega(\alpha)\}$$

(where $\hat{A}$ denotes $\mathrm{cl}_{\beta(\alpha)} A \cap U(\alpha)$ for $A \in \mathscr{P}(\alpha)$).

15.13. Corollary. *Let α be an infinite cardinal, let $\phi : \mathscr{P}_\omega(\alpha) \to \mathscr{P}(\alpha)$ be a monotone function, and set*

$$E_\phi = ((U(\alpha))_{\alpha^+} \setminus U_\phi) \cup \bigcup \{U_\psi : \psi \text{ is a multiplicative function, } \psi : \mathscr{P}_\omega(\alpha) \to \mathscr{P}(\alpha), \text{ and } \psi \leq \phi\}.$$

Then E_ϕ is an open and dense subset of $(U(\alpha))_{\alpha^+}$.

Proof. Each U_ψ is open-and-closed in $(U(\alpha))_{\alpha^+}$, because it is equal to the intersection of at most α open-and-closed subsets of $U(\alpha)$; hence E_ϕ is open in $(U(\alpha))_{\alpha^+}$.

To prove that E_ϕ is dense in $(U(\alpha))_{\alpha^+}$, let $\{V_\xi : \xi < \alpha\}$ be a family of open-and-closed subsets of $U(\alpha)$ such that $V = \bigcap_{\xi<\alpha} V_\xi \neq \emptyset$; we prove that $V \cap E_\phi \neq \emptyset$. Let $A_\xi \in \mathscr{P}(\alpha)$ be such that $\hat{A}_\xi = V_\xi$ for $\xi < \alpha$. Then the family $\mathscr{F} = \{A_\xi : \xi < \alpha\}$ is a family of subsets of α of cardinality at most α and with the uniform finite intersection property. There are two possibilities to consider.

Case 1. $\mathscr{F} \cup \phi[\mathscr{P}_\omega(\alpha)]$ has the uniform finite intersection property.

By Lemma 9.12, there is a multiplicative function $\psi : \mathscr{P}_\omega(\alpha) \to \mathscr{P}(\alpha)$ such that $\psi \leq \phi$ and $\mathscr{F} \cup \psi[\mathscr{P}_\omega(\alpha)]$ has the uniform finite intersection property. Then

$$\emptyset \neq V \cap U_\psi \subset V \cap E_\phi.$$

Case 2. $\mathscr{F} \cup \phi[\mathscr{P}_\omega(\alpha)]$ does not have the uniform finite intersection property.

Then $V \cap U_\phi = \emptyset$ and $V \subset (U(\alpha))_{\alpha^+} \setminus U_\phi \subset E_\phi$.

We denote by $RK(\alpha)$ the set of $p \in U(\alpha)$ such that $\tau(p)$ is minimal in $\tau[U(\alpha)]$ in the Rudin-Keisler order, and by $G(\alpha)$ the set of $p \in U(\alpha)$ such that p is α^+-good (cf. definitions in § 9).

The following theorem is the main result of the present section concerning ultrafilters.

15.14. Theorem. *Let α be an infinite cardinal.*

(a) *If 2^α is a regular cardinal and no uniform ultrafilter on α has a filter base of cardinality less than 2^α, then*

$$(U(\alpha))_{2^\alpha} \text{ is homeomorphic to } (2^{(2^\alpha)})_{2^\alpha}.$$

(b) *If $\alpha^+ = 2^\alpha$, then each of the spaces $(U(\alpha))_{\alpha^+}$, $(G()_{\alpha^+}$, $(RK(\alpha))_{\alpha^+}$, $(G(\alpha) \cap RK(\alpha))_{\alpha^+}$ is homeomorphic to $(2^{(\alpha^+)})_{\alpha^+}$.*

Proof. (a) is a direct application of Theorem 15.9 (with α replaced by 2^α).

(b) The statement about $(U(\alpha))_{\alpha^+}$ follows from part (a), since α^+ is regular and no uniform ultrafilter on α has a filter base of cardinality at most α (Corollary 7.8(a)).

We verify that

(i) $(RK(\alpha))_{\alpha^+} = \bigcap\{D_f : f \in \alpha^\alpha\}$, and

(ii) $(G(\alpha))_{\alpha^+} = \bigcap\{E_\phi : \phi$ monotone function, $\phi : \mathscr{P}_\omega(\alpha) \to \mathscr{P}(\alpha)\}$ where the sets D_f, E_ϕ are defined in Corollaries 15.12, 15.13, respectively.

To prove (i), let p be a uniform ultrafilter on α whose type is Rudin-Keisler minimal in $\tau[U(\alpha)]$, and let $f \in \alpha^\alpha$. Then there is $A \in p$ such that either $f|A$ is one-to-one or $|f[A]| < \alpha$; hence $p \in \hat{A} \subset D_f$. Conversely, let $p \in \bigcap\{D_f : f \in \alpha^\alpha\}$ and let $f \in \alpha^\alpha$. Since $p \in D_f$ we have $p \in \hat{A}$ for some $A \in \mathscr{P}(\alpha)$ such that either $f|A$ is one-to-one or $|f[A]| < \alpha$. Hence by Lemma 9.4 the type of p is Rudin-Keisler minimal in $\tau[U(\alpha)]$.

To prove (ii) let $p \in (G(\alpha))_{\alpha^+}$ and let ϕ be a monotone function, $\phi : \mathscr{P}_\omega(\alpha) \to \mathscr{P}(\alpha)$. If $\phi(F) \notin p$ for some $F \in \mathscr{P}_\omega(\alpha)$, then $\alpha \backslash \phi(F) \in p$ and hence

$$p \in (U(\alpha))_{\alpha^+} \backslash \phi(F) \subset E_\phi .$$

If $\phi(F) \in p$ for $F \in \mathscr{P}_\omega(\alpha)$ then, because p is α^+-good, there is a multiplicative function $\psi : \mathscr{P}_\omega(\alpha) \to p$ such that $\psi \leq \phi$. Then $p \in \psi(F)$ for $F \in \mathscr{P}_\omega(\alpha)$; *i.e.*,

$$p \in U_\psi \subset E_\phi .$$

Thus, in any case, we have $p \in E_\phi$ for every monotone function ϕ. Conversely, let $p \in \bigcap\{E_\phi : \phi$ monotone, $\phi : \mathscr{P}_\omega(\alpha) \to \mathscr{P}(\alpha)\}$ and let $\phi : \mathscr{P}_\omega(\alpha) \to p$ be a monotone function. Then $p \in E_\phi$ and also $p \in U_\phi$; hence there is a multiplicative function $\psi : \mathscr{P}_\omega(\alpha) \to \mathscr{P}(\alpha)$ such that $\psi \leq \phi$ and $p \in U_\psi$. Then $\psi(F) \in p$ for $F \in \mathscr{P}_\omega(\alpha)$, *i.e.*, $\psi : \mathscr{P}_\omega(\alpha) \to p$. This proves that $p \in (G(\alpha))_{\alpha^+}$.

The remaining statements of part (b) now follow from the fact (proved in (a)) that $(U(\alpha))_{\alpha^+}$ is homeomorphic to $(2^{(\alpha^+)})_{\alpha^+}$, together with Theorem 15.8(b), Lemmas 15.12 and 15.13 and statements (i) and (ii) just proved. Alternatively, since each of the sets $(RK(\alpha))_{\alpha^+}$ and $(G(\alpha))_{\alpha^+}$ has dense complement in $(U(\alpha))_{\alpha^+}$, Theorem 15.8(c) may be used in place of Theorem 15.8(b).

The proof of the theorem is complete.

Applications to the Growth Spaces $\beta X \setminus X$

We recall that a space is σ-compact if it is the union of a sequence of compact sets.

Definitions. A space X is *realcompact* if for every $p \in \beta X \setminus X$ there is a zero-set Z of βX such that $p \in Z \subset \beta X \setminus X$.

A space X is *pseudocompact* if there is no unbounded, real-valued, continuous function on X.

15.15. Theorem. *Let X be a space and consider the following conditions on X.*

(a) *X is a non-compact, σ-compact space;*

(b) *X is a non-compact, realcompact space;*

(c) *X is not pseudocompact;*

(d) *there is a subset N of X such that N is homeomorphic to ω and is C-embedded in X;*

(e) *there is a subset N of X such that N is homeomorphic to ω and a closed, C^*-embedded subset of X; and*

(f) $|\beta X \setminus X| \geq 2^{2^\omega}$.

Then (a)$\Rightarrow$(b)$\Rightarrow$(c)$\Leftrightarrow$(d)$\Rightarrow$(e)$\Rightarrow$(f).

Proof. (a)$\Rightarrow$(b). Let X be a σ-compact space and let $p \in \beta X \setminus X$. There is a sequence $\{K_n : n < \omega\}$ of compact subsets of X such that $X = \bigcup_{n<\omega} K_n$. There is a zero-set Z_n of βX such that $p \in Z_n$ and $Z_n \cap K_n = \emptyset$ for $n < \omega$. We define $Z = \bigcap_{n<\omega} Z_n$. Then Z is a zero-set of βX by Lemma 2.1(a), and clearly $p \in Z \subset \beta X \setminus X$.

(b)$\Rightarrow$(c). Let X be a non-compact, realcompact space. There are $p \in \beta X \setminus X$ and a zero-set Z of βX such that $p \in Z \subset \beta X \setminus X$. There is $f \in C(\beta X)$ such that

$$Z = \{x \in \beta X : f(x) = 0\}.$$

We define $g: X \to \mathbb{R}$ by $g(x) = 1/f(x)$; then $g \in C(X)$ and g is unbounded and thus X is not pseudocompact.

(c)$\Rightarrow$(d). Let $h \in C(X)$ and h be unbounded; then $h[X]$ is an unbounded subset of $\mathbb{R}$ so there is a sequence $\{t_n : n < \omega\} \subset h[X]$ such that

$$|t_n - t_m| \geq 1 \quad \text{if } m < n < \omega.$$

It is clear that $\{t_n : n < \omega\}$ is homeomorphic to ω and a closed subset of $\mathbb{R}$. Let $p_n \in h^{-1}(\{t_n\})$ for $n < \omega$ and define

$$N = \{p_n : n < \omega\}.$$

It is clear that N is homeomorphic to ω. To prove that N is C-embedded in X let $f \in C(N) = \mathbb{R}^N$. We define

$$\tilde{f} : \{t_n : n < \omega\} \to \mathbb{R}$$

by $\tilde{f}(t_n)=f(p_n)$ for $n<\omega$. Since a closed subset of $\mathbb{R}$ is C-embedded it follows that there is $\tilde{g}\in C(\mathbb{R})$ such that $\tilde{f}\subset\tilde{g}$. We set $g=\tilde{g}\circ h$. Then $g\in C(X)$, and

$$g(p_n)=\tilde{g}(h(p_n))=\tilde{g}(t_n)=\tilde{f}(t_n)=f(p_n) \quad \text{for } n<\omega,$$

i.e., $f\subset g$.

(d)⇒(c). This is clear.

(d)⇒(e). Let N be a subset of X satisfying the conditions of (d). Then N is C^*-embedded in X. Suppose that N is not closed in X and let $p\in \text{cl}_X N\setminus N$. Let $\langle p_n : n<\omega\rangle$ be a faithful enumeration of N and define $f: N\to\mathbb{R}$ by $f(p_n)=n$ for $n<\omega$. Since f assumes arbitrarily large values on every neighborhood of p in X, it is clear that f cannot be extended to a continuous, real-valued function defined at p. This contradicts the assumption that N is C-embedded in X.

(e)⇒(f). We note that $\boldsymbol{\beta}N\setminus N$ is homeomorphic to $\boldsymbol{\beta}(\omega)\setminus\omega$ and hence of cardinality $2^{2^{\omega}}$ (Corollary 7.4). Since N is closed in X, we have that $\text{cl}_{\boldsymbol{\beta}X} N\setminus N\subset\boldsymbol{\beta}X\setminus X$. The result will follow if it is proved that $\text{cl}_{\boldsymbol{\beta}X} N$ is homeomorphic to $\boldsymbol{\beta}N$. It is clear that $\text{cl}_{\boldsymbol{\beta}X} N$ is compact and contains N as a dense subspace. Furthermore if $f\in C^*(N)$ then since N is C^*-embedded in X there is $g\in C^*(X)$ such that $f\subset g$. Let $\bar{g}$ denote the (unique) Stone extension of g defined on $\boldsymbol{\beta}X$. Then $f\subset\bar{g}|\text{cl}_{\boldsymbol{\beta}X} N$, so the uniqueness theorem for the Stone-Čech compactification (Theorem 2.6) implies that $\text{cl}_{\boldsymbol{\beta}X} N$ is homeomorphic to $\boldsymbol{\beta}N$.

15.16. Lemma. *Let X be a realcompact space. Then*

(a) *if $p\in\boldsymbol{\beta}X\setminus X$ then $\{p\}$ is not a zero-set of $\boldsymbol{\beta}X\setminus X$;*

(b) *$\boldsymbol{\beta}X\setminus X$ has no isolated elements; and*

(c) *the local weight of $\boldsymbol{\beta}X\setminus X$ at each of its elements is uncountable.*

Proof. (a) Let $p\in\boldsymbol{\beta}X\setminus X$. Since X is realcompact there is a zero-set Z of $\boldsymbol{\beta}X$ such that $p\in Z\subset\boldsymbol{\beta}X\setminus X$. If $\{p\}$ is a zero-set of $\boldsymbol{\beta}X\setminus X$ then certainly $\{p\}$ is a zero-set of Z and thus $Z\setminus\{p\}$ is σ-compact. Because Z is a zero-set of $\boldsymbol{\beta}X$ the space $\boldsymbol{\beta}X\setminus Z$ is σ-compact (Lemma 2.1(c)). Hence the space $Y=\boldsymbol{\beta}X\setminus\{p\}=(\boldsymbol{\beta}X\setminus Z)\cup(Z\setminus\{p\})$ is a σ-compact space. It is clear further that Y is non-compact and (since $X\subset Y\subset\boldsymbol{\beta}X$) that Y is C^*-embedded in $\boldsymbol{\beta}X$ and that $\boldsymbol{\beta}Y=\boldsymbol{\beta}X$ and $\boldsymbol{\beta}Y\setminus Y=\{p\}$. This contradicts the implication (a)⇒(f) of Theorem 15.15.

Statements (b) and (c) follow from (a).

15.17. Lemma. *Let X be a locally compact space, let F be a nowhere dense subset of X, and let Z be a non-empty zero-set of $\boldsymbol{\beta}X$ such that $Z\subset\boldsymbol{\beta}X\setminus X$. Then*

$$\text{int}_{\boldsymbol{\beta}X\setminus X}(Z\setminus\text{cl}_{\boldsymbol{\beta}X} F)\neq\emptyset.$$

Proof. There is $f\in C^*(\boldsymbol{\beta}X)$ such that

$$f(p)\geq 0 \quad \text{for } p\in\boldsymbol{\beta}X \quad \text{and}$$
$$Z=f^{-1}(\{0\}).$$

Since $Z\neq\emptyset$ we have $\inf\{f(x):x\in X\}=0$ and hence there is a sequence $\{t_n:n<\omega\}$ of real numbers such that

$$0<t_{n+1}<t_n \quad \text{and } t_n\in f[X] \quad \text{for } n<\omega, \quad \text{and}$$
$$\inf\{t_n:n<\omega\}=0.$$

Let $x_n\in X\cap f^{-1}(\{t_n\})$ for $n<\omega$, let $\{s_n:n<\omega\}$ be a sequence of real numbers such that

$$s_{n+1}<t_n<s_n \quad \text{for } n<\omega,$$

and let K_n be a compact neighborhood in X of x_n such that

$$f[K_n]\subset(s_{n+1},s_n).$$

Since F is nowhere dense in X there are for $n<\omega$ an element p_n of $\operatorname{int}_X(K_n\backslash F)$ and a continuous function $g_n:X\to[0,1]$ such that

$$g_n(p_n)=1 \quad \text{and}$$
$$g_n=0 \quad \text{on } F\cup(X\backslash\operatorname{int}_X K_n).$$

We note that every element of X has a neighborhood that meets at most one element of the sequence $\{K_n:n<\omega\}$. Thus the function $g:X\to[0,1]$ given by $g(x)=\sum_{n<\omega}g_n(x)$ is a well-defined continuous function.

We set $C=\{p\in\boldsymbol{\beta}X\backslash X:\overline{g}(p)\neq 0\}$, where $\overline{g}$ is the continuous Stone extension of g. Then C is open in $\boldsymbol{\beta}X\backslash X$, and C is non-empty because $\{p_n:n<\omega\}$ is not compact and

$$\operatorname{cl}_{\boldsymbol{\beta}X}\{p_n:n<\omega\}\backslash X\subset C.$$

Since $g=0$ on F we have $\overline{g}=0$ on $\operatorname{cl}_{\boldsymbol{\beta}X}F$; hence $C\cap\operatorname{cl}_{\boldsymbol{\beta}X}F=\emptyset$. Furthermore if $p\in C$ then every neighborhood of p has non-empty intersection with infinitely many elements of the family $\{K_n:n<\omega\}$, so that $f(p)=0$. It follows that $C\subset Z$. We have shown that

$$\emptyset\neq C\subset\operatorname{int}_{\boldsymbol{\beta}X\backslash X}(Z\backslash\operatorname{cl}_{\boldsymbol{\beta}X}F).$$

15.18. Theorem. *Let X be a locally compact, realcompact space.*

(a) *Every non-empty zero-set of $\boldsymbol{\beta}X\backslash X$ contains a non-empty zero-set of $\boldsymbol{\beta}X$;*

(b) *if Z is a zero-set of $\boldsymbol{\beta}X\backslash X$ then*

$$\operatorname{cl}_{\boldsymbol{\beta}X\backslash X}\operatorname{int}_{\boldsymbol{\beta}X\backslash X}Z=Z$$

(i.e., Z is a regular-closed subset of $\beta X \setminus X$);

(c) *$\beta X \setminus X$ has the ω^+-Baire category property; and*

(d) *if F is a (closed) nowhere dense subset of X then $\mathrm{cl}_{\beta X} F \setminus X$ is nowhere dense in $\beta X \setminus X$.*

Proof. (a) Let Z be a non-empty zero-set of $\beta X \setminus X$, choose $p \in Z$ and let Z_0 be a zero-set of βX such that $p \in Z_0 \subset \beta X \setminus X$. Since $\beta X \setminus X$ is compact (by Lemma 2.9(b)) it is C^*-embedded in βX so there is a zero-set Z_1 of βX such that $Z = Z_1 \cap (\beta X \setminus X)$. Then $Z_0 \cap Z_1$ is a non-empty zero-set of βX, and $Z_0 \cap Z_1 \subset Z$.

(b) If $p \in Z \setminus (\mathrm{cl}_{\beta X \setminus X} \mathrm{int}_{\beta X \setminus X} Z)$ then there is a zero-set Z_0 of $\beta X \setminus X$ such that

$$p \in Z_0 \quad \text{and}$$
$$Z_0 \cap \mathrm{cl}_{\beta X \setminus X} \mathrm{int}_{\beta X \setminus X} Z = \emptyset .$$

Since $Z_0 \cap Z$ is a non-empty zero-set of $\beta X \setminus X$ there is by part (a) a non-empty zero-set W of βX such that $W \subset Z_0 \cap Z$, and from Lemma 15.17 (with Z and F replaced by W and $\emptyset$, respectively) we have

$$\emptyset \neq \mathrm{int}_{\beta X \setminus X} W \subset W \subset Z_0 \cap Z = \emptyset ,$$

a contradiction.

(c) The space $\beta X \setminus X$ is compact by Lemma 2.9(b) and from part (b) it follows that if $\{V_n : n < \omega\}$ is a sequence of open subsets of $\beta X \setminus X$ such that $\bigcap_{n<\omega} V_n \neq \emptyset$ then $\mathrm{int}_{\beta X \setminus X}(\bigcap_{n<\omega} V_n) \neq \emptyset$. According to Theorem 6.15, $\beta X \setminus X$ has the ω^+-Baire category property.

(d) If $\mathrm{int}_{\beta X \setminus X}(\mathrm{cl}_{\beta X} F \setminus X) \neq \emptyset$ then there is a non-empty zero-set Z of $\beta X \setminus X$ such that

$$Z \subset (\mathrm{cl}_{\beta X} F) \setminus X ,$$

and by part (a) there is no loss in generality in assuming that Z is a zero-set of βX. We apply Lemma 15.17 to obtain a contradiction.

The following simple result supplies us with a class of spaces for which the next theorem (15.20) applies.

15.19. Lemma. *If X is a σ-compact space and $w(X) \leq 2^{\omega}$, then $|C(X)| \leq 2^{\omega}$.*

Proof. (a) Assume first that X is a compact space of weight at most 2^{ω}. From Corollary 2.11 there is an embedding e of X into $[0,1]^{(2^{\omega})}$. Since X is compact the space $e[X]$ is C-embedded in $[0,1]^{(2^{\omega})}$. According to Corollary 3.18 the space $[0,1]^{(2^{\omega})}$ is separable (*i.e.*, its density character is ω) and hence clearly $|C([0,1]^{(2^{\omega})})| = 2^{\omega}$. Thus

$$|C(X)| \leq |C([0,1]^{(2^{\omega})})| = 2^{\omega} .$$

We remark that if $X \neq \emptyset$ then in fact $|C(X)| = 2^{\omega}$.

(b) Let $\{K_n : n<\omega\}$ be a sequence of compact subsets of X whose union is X. We define a function

$$\phi : C(X) \to \prod_{n<\omega} C(K_n)$$

by $\phi(f) = \langle f|K_n : n<\omega\rangle$. Then ϕ is a one-to-one function from $C(X)$ into $\prod_{n<\omega} C(K_n)$ and from part (a) we have

$$|C(X)| \le \prod_{n<\omega} |C(K_n)| \le (2^\omega)^\omega = 2^\omega .$$

For a space X we recall that $\mathscr{Z}(X)$ denotes the family of all zero-sets of X.

15.20. Theorem. *Assume that $\omega^+ = 2^\omega$ and that X is a non-compact, locally compact, realcompact space such that $|C(X)| \le 2^\omega$. Then each of the spaces*

$$(\boldsymbol{\beta} X \setminus X)_{\omega^+} \quad \textit{and} \quad P_{\omega^+}(\boldsymbol{\beta} X \setminus X)$$

is homeomorphic to $(2^{(\omega^+)})_{\omega^+}$.

Proof. We note first that $\boldsymbol{\beta} X \setminus X$ is closed and hence C^*-embedded in $\boldsymbol{\beta} X$, so that

$$(*) \quad w(\boldsymbol{\beta} X \setminus X) \le |\mathscr{Z}(\boldsymbol{\beta} X \setminus X)| \le |\mathscr{Z}(\boldsymbol{\beta} X)| \le |C(\boldsymbol{\beta} X)| \le |C(X)| \le 2^\omega = \omega^+ .$$

To prove that $(\boldsymbol{\beta} X \setminus X)_{\omega^+}$ is homeomorphic to the space $(2^{(\omega^+)})_{\omega^+}$ we make use of Theorem 15.9. Indeed ω^+ is regular; $\boldsymbol{\beta} X \setminus X$ is compact; it follows from (*) that $w(\boldsymbol{\beta} X \setminus X) \le \omega^+$; and if $p \in \boldsymbol{\beta} X \setminus X$ then the local weight of $\boldsymbol{\beta} X \setminus X$ at p is at least ω^+ according to Lemma 15.16(c).

To establish the second homeomorphism we note that

$$P_{\omega^+}(\boldsymbol{\beta} X \setminus X) = \bigcap \{\boldsymbol{\beta} X \setminus X \setminus Bd\, Z : Z \in \mathscr{Z}(\boldsymbol{\beta} X \setminus X)\}$$

and that the restriction of the identity function $(\boldsymbol{\beta} X \setminus X)_{\omega^+} \to \boldsymbol{\beta} X \setminus X$ to $P_{\omega^+}(\boldsymbol{\beta} X \setminus X)$ is a homeomorphism. We make use of Theorem 15.8(b) and the homeomorphism already established between $(\boldsymbol{\beta} X \setminus X)_{\omega^+}$ and $(2^{(\omega^+)})_{\omega^+}$: since ω^+ is not weakly compact and $|\mathscr{Z}(\boldsymbol{\beta} X \setminus X)| \le \omega^+$ by (*), we only have to verify that

$$(\boldsymbol{\beta} X \setminus X \setminus Bd\, Z)_{\omega^+} \quad \text{is dense in } (\boldsymbol{\beta} X \setminus X)_{\omega^+} \quad \text{for } Z \in \mathscr{Z}(\boldsymbol{\beta} X \setminus X) .$$

The family $\mathscr{Z}(\boldsymbol{\beta} X \setminus X)$ is a base for the topology of $(\boldsymbol{\beta} X \setminus X)_{\omega^+}$, and thus it is sufficient to verify that

$$(\boldsymbol{\beta} X \setminus X \setminus Bd\, Z) \cap W \neq \emptyset \quad \text{for } W \in \mathscr{Z}(\boldsymbol{\beta} X \setminus X) \text{ and } W \neq \emptyset .$$

If this relation fails for some $W, Z \in \mathscr{Z}(\boldsymbol{\beta} X \setminus X)$ then according to Theorem 15.18(a) we would have

$$\emptyset \neq \operatorname{int}_{\boldsymbol{\beta} X \setminus X} W \subset W \subset Bd\, Z \quad \text{and} \quad \operatorname{int}_{\boldsymbol{\beta} X \setminus X} Bd\, Z = \emptyset ,$$

a contradiction. The proof of the theorem is complete.

Definition. If X is a space and $p \in \beta X \setminus X$, then p is a *remote point* of $\beta X \setminus X$ if p is not in the closure (in βX) of any discrete subset of X. The set of remote points of $\beta X \setminus X$ is denoted $R(\beta X \setminus X)$.

15.21. Theorem. *Assume that $\omega^+ = 2^{\omega}$ and that X is a non-compact, locally compact, separable metric space without isolated elements. Then each of the spaces*

$$(R(\beta X \setminus X))_{\omega^+} \quad \text{and} \quad R(\beta X \setminus X) \cap P_{\omega^+}(\beta X \setminus X)$$

is homeomorphic to $(2^{(\omega^+)})_{\omega^+}$.

Proof. It is clear that X is σ-compact and that $|C(X)| \leq 2^{\omega}$. Thus from Theorem 15.20 and Lemma 15.15 it follows that $(\beta X \setminus X)_{\omega^+}$ is homeomorphic to $(2^{(\omega^+)})_{\omega^+}$.

For the first homeomorphism we note that

$$R(\beta X \setminus X) = \bigcap \{\beta X \setminus X \setminus \mathrm{cl}_{\beta X} D : D \text{ is a discrete subset of } X\}.$$

We make use of Theorem 15.8(b) and the homeomorphism between $(\beta X \setminus X)_{\omega^+}$ and $(2^{(\omega^+)})_{\omega^+}$. If D is a discrete subset of X then D is countable (since every subspace of a separable metric space is separable); thus the cardinal of the family of all discrete subsets of X is at most $2^{\omega} = \omega^+$. We claim that if D is a discrete subspace of X then $\mathrm{cl}_X D$ is nowhere dense in X. If there is a non-empty, open subset V of X such that $V \subset \mathrm{cl}_X D$, then $V \cap D$ is both dense in V and locally compact (because it is discrete). Then $V \cap D$ is open in V by Lemma 2.9(a), and hence is open in X. This contradicts the assumption that X does not have isolated elements. From the facts that X is σ-compact and that the intersection of countably many zero-sets of βX is a zero-set of βX (Lemma 2.1), it follows that X is realcompact. Thus from Lemma 15.18(c) it follows that if D is a discrete subset of X then $\mathrm{cl}_{\beta X} D \setminus X$ is nowhere dense in $\beta X \setminus X$. We verify that

$$(\beta X \setminus X \setminus \mathrm{cl}_{\beta X} D)_{\omega^+} \quad \text{is dense in} \quad (\beta X \setminus X)_{\omega^+}$$

for every discrete set D of X. The family $\mathscr{Z}(\beta X \setminus X)$ is a base for $(\beta X \setminus X)_{\omega^+}$, so it is sufficient to verify that

$$(\beta X \setminus X \setminus \mathrm{cl}_{\beta X} D) \cap Z \neq \emptyset \quad \text{for } Z \in \mathscr{Z}(\beta X \setminus X) \quad \text{and} \quad Z \neq \emptyset.$$

Indeed if this relation fails for some $Z \in \mathscr{Z}(\beta X \setminus X)$ then according to Theorem 15.18(a) we would have

$$\emptyset \neq \mathrm{int}_{\beta X \setminus X} Z \subset Z \subset \mathrm{cl}_{\beta X} D \quad \text{and} \quad \mathrm{int}_{\beta X \setminus X} \mathrm{cl}_{\beta X} D = \emptyset,$$

a contradiction.

The second homeomorphism follows from a combination of the proofs of the second part of Theorem 15.20 and the proof just given; the details are omitted.

Notes for § 15

The principal references for this section are the papers by Hung-Negrepontis [73], [74], which we follow closely.

Lemma 15.1 is noted in Hung [72] (Lemma 3.4) and Hung-Negrepontis [74] (Lemma 2.1). Lemma 15.2 and Theorem 15.3 appear in Hung-Negrepontis [74] (Lemma 2.2 and Theorem 2.3, respectively). Special cases of Lemma 15.2 and Theorem 15.3 were proved in Hung [72] (Chapter III).

Lemma 15.4 is given in Hung-Negrepontis [74] (Lemma 3.1). Theorem 15.5 is partly proved in Hung [72] (Theorem 3.19) and in its present form in Hung-Negrepontis [74] (Theorem 3.2). Theorem 15.5 and Corollaries 15.6, 15.7 are given in Hung-Negrepontis [74] (Theorem 3.2, and Corollaries 3.3, 3.4, respectively). Theorem 15.8(a) is due to Sikorski [50] (§ 4, Theorem (xv)); Theorem 15.8(c) for $\alpha=\omega$ appears in Kuratowski [66] (§ 36, II, Theorem 3) and is attributed there to Mazurkiewicz [18], Brouwer [17], and Alexandrov-Urysohn [27]. Theorem 15.8, in its present form, appears in Hung-Negrepontis [74] (Theorem 3.5). Theorem 15.9 is proved for $\alpha=\omega^+=2^{\omega}$ in Comfort-Negrepontis [68] (Theorem 3.2) and in general in Hung [72] (Theorem 3.13) and Hung-Negrepontis [74] (Theorem 4.1). Corollaries 15.10 and 15.11 are stated for $\alpha=\omega^+=2^{\omega}$ in Parovičenko [63] and proved in Comfort-Negrepontis [68] (§ 3); their general case is stated in Negrepontis [69 a] (Theorem 3.8) and proved in Hung [72] (Corollaries 3.15 and 3.17, respectively) and in Hung-Negrepontis [74] (Corollaries 4.2 and 4.3, respectively). Corollaries 15.12 and 15.13 are proved in Blass [70] (Theorems 8.3 and 8.1, respectively) and in Hung-Negrepontis [74] (Corollaries 4.5 and 4.7, respectively).

Theorem 15.14 is proved in Hung-Negrepontis [74] (Theorem 4.8). The fact that if $\alpha^+=2^{\alpha}$ then $(U(\alpha))_{\alpha^+}$ is homeomorphic to $(2^{(\alpha^+)})_{\alpha^+}$ is proved in Hung [72] (Corollary 3.14); and the fact that $(U(\alpha))_{\alpha^+}$ has the α^+-Baire category property, a consequence of Theorems 15.8(a) and 15.14(b), has been proved directly by Blass [70] (Theorem 7.8). A Baire category type of proof of the existence of α^+-good and uniformly selective ultrafilters on α, assuming $\alpha^+=2^{\alpha}$, was first given by Blass [70] (Corollary 8.8).

Realcompact and pseudocompact spaces are studied in Gillman-Jerison [60], where all implications of Theorem 15.15 are noted (*e.g.*, cf. Gillman-Jerison [60] (Corollary 1.21) for the equivalence (c)⇔(d) of Theorem 15.15).

Lemma 15.17 for $F=\emptyset$ and Theorem 15.18(a), (b) are essentially proved in Fine-Gillman [60] (Lemma 3.1). The proof of Theorem

15.18(d) was known to one of the authors (S. N.) since 1963 (unpublished); its proof is given in Hung-Negrepontis [74] (Lemma 4.14). A generalization is given in Woods [71 b] (Lemma 2.6). Lemma 15.19 is a special case of Comfort-Hager [70] (Theorem 2.2); it is also given in Hung-Negrepontis [74] (Lemma 4.10).

Theorem 15.20 appears in Hung-Negrepontis [74] (Theorem 4.15). Most of Theorem 15.20 has been proved in Comfort-Negrepontis [68]. Specifically, Comfort-Negrepontis [68] (Corollary 3.3) states that $(\beta X\setminus X)_{\omega^+}$ is homeomorphic to $(2^{(\omega^+)})_{\omega^+}$; and Comfort-Negrepontis [68] (Theorem 3.5) states that $P_{\omega^+}(\beta X\setminus X)$ is also homeomorphic to $(2^{(\omega^+)})_{\omega^+}$ under certain assumptions on X which are stronger than those of Theorem 15.20. The existence of remote points of $\beta\mathbb{R}\setminus\mathbb{R}$ was proved, assuming the continuum hypothesis, by Fine and Gillman [62]. A Baire-category type of proof of the existence of P_{ω^+}-points and of remote points of $\beta X\setminus X$ was first given by one of the authors (S. N.) in 1963 (unpublished). For more information on remote points the reader is referred to Plank [69], Woods [71 a], [71 b]. Theorem 15.21 is proved in Hung-Negrepontis [74] (Theorem 4.16).

Finally, we mention that most of Theorems 15.20 and 15.21 can be proved assuming Martin's axiom instead of the continuum hypothesis. In fact we have the following result (stated in Hung-Negrepontis [74], (Corollary 4.17)):

Theorem. *Assume Martin's axiom.*

(a) *If X is a locally compact, σ-compact, non-compact metric space and $A\subset\beta X\setminus X$ with $|A|\le 2^\omega$ then each of the spaces*

$$(\beta X\setminus X)_{2^\omega}, \quad P_{2^\omega}(\beta X\setminus X)\setminus A, \quad (RK(\omega))_{2^\omega}$$

is homeomorphic to $(2^{(2^\omega)})_{2^\omega}$.

(b) *If in addition to the properties of X given in* (a) *X does not have isolated elements, then each of the spaces*

$$(R(\beta X\setminus X))_{2^\omega}, \quad R(\beta X\setminus X)\cap P_{2^\omega}(\beta X\setminus X)$$

is homeomorphic to $(2^{(2^\omega)})_{2^\omega}$.

The following results should be contrasted: if $\omega^+=2^\omega$ then $\beta(\omega)\setminus\omega$ is homeomorphic to the Stone space of the 2^ω-homogeneous-universal Boolean algebra S_{2^ω} of cardinality 2^ω (Theorem 14.15). However if $\omega^+<2^\omega=(2^\omega)^{\underline{2^\omega}}$ then $\beta(\omega)\setminus\omega$ is definitely not homeomorphic to S_{2^ω} according to the result of Hausdorff (Theorem 14.14), even though if Martin's axiom holds as well, then part (a) of the above theorem implies that $(\beta(\omega)\setminus\omega)_{2^\omega}$ *is* homeomorphic to $(\mathsf{S}_{2^\omega})_{2^\omega}$ (since both spaces are homeomorphic to $(2^{(2^\omega)})_{2^\omega}$).

§ 16. Ultrafilters on ω

In this final section, perhaps the least complete of all, we collect a number of results about ultrafilters on ω whose proofs do not seem to generalize in any obvious way to proofs of results about ultrafilters on arbitrary cardinals. The first sub-section (16.1—16.11) analyses the similarities and the differences between the notion of a Rudin-Keisler minimal ultrafilter and the (more general) notion of a P-point. The results here depend on the continuum hypothesis; among them are these: for every P-point p of $\boldsymbol{\beta}(\omega)\backslash\omega$ there is a P-point q whose type is strictly greater in the Rudin-Keisler order than the type of p (Theorem 16.1); and, there is a P-point p of $\boldsymbol{\beta}(\omega)\backslash\omega$ such that $|\boldsymbol{\phi}^{-1}(\{\langle p,p\rangle\})|=2^{2^\omega}$ (where $\boldsymbol{\phi}:\boldsymbol{\beta}(\omega\times\omega)\to\boldsymbol{\beta}(\omega)\times\boldsymbol{\beta}(\omega)$ is the canonical function) (Theorem 16.11), while as we know $|\boldsymbol{\phi}^{-1}(\{\langle q,q\rangle\})|=3$ for every Rudin-Keisler minimal ultrafilter q (Theorem 9.6). From either of these results it follows that there are P-points of $\boldsymbol{\beta}(\omega)\backslash\omega$ that are not Rudin-Keisler minimal (in $\boldsymbol{\beta}(\omega)\backslash\omega$). We note that the difficulty in generalizing the results of this sub-section is mainly due to the inductive method used in the proof of existence of ultrafilters with specific properties.

In the second sub-section we introduce the Rudin-Frolík order on $\tau[\boldsymbol{\beta}(\omega)]$ and describe its most elementary properties (16.12—16.17). The difficulty in generalizing the properties and results on the Rudin-Frolík order to arbitrary cardinals lies with the special character of Lemma 16.15. Various structural theorems concerning this order, given by Booth, Louveau, and others, are not treated.

The final and most satisfactory sub-section deals with a general method (Theorem 16.18) that proves the non-homogeneity of the space of uniform ultrafilters (Theorem 16.22), and other results. In contrast to the results of the first sub-section, the methods here are designed to avoid making use of the continuum hypothesis.

The Rudin-Keisler Order on $\boldsymbol{\beta}(\omega)$ and the Canonical Function $\boldsymbol{\phi}: \boldsymbol{\beta}(\omega \times \omega) \to \boldsymbol{\beta}(\omega) \times \boldsymbol{\beta}(\omega)$

16.1. Theorem. *Assume $\omega^+ = 2^\omega$, and let p be a P-point of $\boldsymbol{\beta}(\omega)\setminus\omega$. Then there is $q \in \boldsymbol{\beta}(\omega)\setminus\omega$ such that*

$$q \text{ is a P-point of } \boldsymbol{\beta}(\omega)\setminus\omega, \text{ and}$$
$$\tau(p) \prec \tau(q).$$

Proof. Let $f: \omega \to \omega$ be a function such that

$$|f^{-1}(\{n\})| = n \quad \text{for } n < \omega,$$

and set

$$\mathscr{A} = \{A \in \mathscr{P}(\omega): \sup\{|f^{-1}(\{n\}) \cap A| : n < \omega\} < \omega\}.$$

It is clear that $\mathscr{P}_\omega(\omega) \subset \mathscr{A}$, that $\mathscr{A}$ is closed under finite unions, that if $A \in \mathscr{A}$ and $B \subset A$ then $B \in \mathscr{A}$, and that

$$(\omega\setminus A) \cap f^{-1}(C) \neq \emptyset \quad \text{for } A \in \mathscr{A}, \quad C \in p.$$

Let $\langle S_\xi : \xi < \omega^+ \rangle$ and $\langle d_\xi : \xi < \omega^+ \rangle$ be well-orderings of $\mathscr{P}(\omega)$ and the set of ω-partitions of ω respectively, and for $\xi < \omega^+$ let $\langle d_{\xi,n} : n < \omega \rangle$ be a well-ordering of d_ξ. For $\xi < \omega^+$ we define a family $\mathscr{F}_\xi$ of subsets of ω such that

(i) $\mathscr{F}_0 = \{\omega\}$,
(ii) $\mathscr{F}_\xi$ is a countable filter base on ω for $\xi < \omega^+$,
(iii) $\mathscr{F}_\xi \subset \mathscr{F}_\zeta$ for $\xi < \zeta < \omega^+$,
(iv) $\mathscr{F}_\xi = \bigcup_{\zeta < \xi} \mathscr{F}_\zeta$ for non-zero limit ordinals $\xi < \omega^+$,
(v) $S_\xi \in \mathscr{F}_{\xi+1}$ or $\omega\setminus S_\xi \in \mathscr{F}_{\xi+1}$ for $\xi < \omega^+$,
(vi) there is $B_\xi \in \mathscr{F}_{\xi+1}$ such that

$$|\{n < \omega : |d_{\xi,n} \cap B_\xi| = \omega\}| < \omega$$

for $\xi < \omega^+$, and
(vii) $(\omega\setminus A) \cap f^{-1}(C) \cap F \neq \emptyset$ for $A \in \mathscr{A}$, $C \in p$, $F \in \mathscr{F}_\xi$ for $\xi < \omega^+$.

We proceed by transfinite recursion. We set $\mathscr{F}_0 = \{\omega\}$ and if ξ is a non-zero limit ordinal we set $\mathscr{F}_\xi = \bigcup_{\zeta < \xi} \mathscr{F}_\zeta$.

Now for $\xi < \omega^+$ we define B_ξ and $\mathscr{F}_{\xi+1}$.

Case 1. There is $n < \omega$ such that if $A \in \mathscr{A}$, $C \in p$ and $F \in \mathscr{F}_\xi$, then

$$(\omega\setminus A) \cap f^{-1}(C) \cap F \cap (d_{\xi,0} \cup \cdots \cup d_{\xi,n}) \neq \emptyset.$$

We then define

$$B_\xi = d_{\xi,0} \cup \cdots \cup d_{\xi,n}$$

and we note that

$$|\{m<\omega:|d_{\xi,m}\cap B_\xi|=\omega\}|=n+1<\omega.$$

It is clear that the family

$$\{\omega\setminus A:A\in\mathscr{A}\}\cup\{f^{-1}(C):C\in p\}\cup\mathscr{F}_\xi\cup\{B_\xi\}$$

has the finite intersection property, and hence either the family

$$\{\omega\setminus A:A\in\mathscr{A}\}\cup\{f^{-1}(C):C\in p\}\cup\mathscr{F}_\xi\cup\{B_\xi\}\cup\{S_\xi\}$$

has the finite intersection property or the family

$$\{\omega\setminus A:A\in\mathscr{A}\}\cup\{f^{-1}(C):C\in p\}\cup\mathscr{F}_\xi\cup\{B_\xi\}\cup\{\omega\setminus S_\xi\}$$

has the finite intersection property; in the former case we define $\mathscr{F}_{\xi+1}$ to be the (countable) filter base generated by

$$\mathscr{F}_\xi\cup\{B_\xi\}\cup\{S_\xi\}$$

and in the latter case we define $\mathscr{F}_{\xi+1}$ to be the (countable) filter base generated by

$$\mathscr{F}_\xi\cup\{B_\xi\}\cup\{\omega\setminus S_\xi\}.$$

Case 2. For every $n<\omega$ there are $A_n\in\mathscr{A}$, $C_n\in p$ and $F_n\in\mathscr{F}_\xi$ such that

$$(\omega\setminus A_n)\cap f^{-1}(C_n)\cap F_n\cap(d_{\xi,0}\cup\cdots\cup d_{\xi,n})=\emptyset.$$

Since $\mathscr{A}$ is closed under finite unions, p is a filter and $\mathscr{F}_\xi$ is a countable filter base, we assume with no loss of generality that

$A_n\subset A_{n+1}$, $C_{n+1}\subset C_n$ and $F_{n+1}\subset F_n$ for $n<\omega$, and $\{F_n:n<\omega\}$ is a filter base for $\mathscr{F}_\xi$.

We set

$$D_n=\{m<\omega:|f^{-1}(\{m\})\cap F_n\setminus(d_{\xi,0}\cup\cdots\cup d_{\xi,n})|>n\}\quad\text{for } n<\omega$$

and we claim that $D_n\in p$ for $n<\omega$. To prove this we first set

$$E_n=f^{-1}(\omega\setminus D_n)\cap F_n\setminus(d_{\xi,0}\cup\cdots\cup d_{\xi,n})$$

and we claim that $E_n\in\mathscr{A}$. Indeed if $m\in\omega\setminus D_n$ then

$$|f^{-1}(\{m\})\cap E_n|\leq|f^{-1}(\{m\})\cap F_n\setminus(d_{\xi,0}\cup\cdots\cup d_{\xi,n})|\leq n$$

and if $m\in D_n$ then $|f^{-1}(\{m\})\cap E_n|=0$; hence $E_n\in\mathscr{A}$. Since

$$f^{-1}(C_n)\cap F_n\cap(d_{\xi,0}\cup\cdots\cup d_{\xi,n})\subset A_n\in\mathscr{A}$$

we have $f^{-1}(C_n)\cap F_n\cap(d_{\xi,0}\cup\cdots\cup d_{\xi,n})\in\mathscr{A}$, and hence if $D_n\notin p$ we have $C_n\cap(\omega\setminus D_n)\in p$. Thus from

$$f^{-1}(C_n\cap(\omega\setminus D_n))\cap F_n\subset f^{-1}(C_n)\cap F_n\cap(d_{\xi,0}\cup\cdots\cup d_{\xi,n})\cup E_n\in\mathscr{A}$$

we have

$$f^{-1}(C_n \cap (\omega \setminus D_n)) \cap F_n \in \mathscr{A},$$

so that $f^{-1}(C_n \cap (\omega \setminus D_n))$ intersects its own complement by the recursive assumption (vii). This contradiction proves that $D_n \in p$ for $n < \omega$. We note, using the fact that $F_{n+1} \subset F_n$ for $n < \omega$, that $D_{n+1} \subset D_n$ for $n < \omega$.

Since p is a P-point of $\beta(\omega) \setminus \omega$ and $D_n \in p$ for $n < \omega$ there is (by Lemma 9.14(c)) $D \in p$ such that $|D \setminus D_n| < \omega$ for $n < \omega$; we assume without loss of generality that $D \subset D_0$. We now define the set B_ξ by a subsidiary (finite) recursion.

For $m \in D \cap D_0 = D$ we have $0 < |f^{-1}(\{m\}) \cap F_0| \leq m$ and since

$$f^{-1}(\{m\}) \cap F_0 = \bigcup_{k<\omega} (f^{-1}(\{m\}) \cap F_0 \cap d_{\xi,k})$$

there is a largest $k(m,0) < \omega$ such that

$$f^{-1}(\{m\}) \cap F_0 \cap d_{\xi,k(m,0)} \neq \emptyset.$$

We choose $s_{m,0} \in f^{-1}(\{m\}) \cap F_0 \cap d_{\xi,k(m,0)}$.

For $m \in D \cap D_n$ with $1 \leq n < \omega$ we have

$$n < |f^{-1}(\{m\}) \cap F_n| \leq m$$

and hence $f^{-1}(\{m\}) \cap F_n \setminus \{s_{m,0}, \ldots, s_{m,n-1}\}$ is a finite, non-empty set. Thus there is a largest $k(m,n) < \omega$ such that

$$(f^{-1}(\{m\}) \cap F_n \setminus \{s_{m,0}, \ldots, s_{m,n-1}\}) \cap d_{\xi,k(m,n)} \neq \emptyset;$$

we choose $s_{m,n}$ to be an element of this set. We finally define

$$B_\xi = \{s_{m,n} : m \in D \cap D_n, m < \omega, n < \omega\}$$

and we prove statements (a) and (b).

(a) $|d_{\xi,k} \cap B_\xi| < \omega$ for $k < \omega$.

Before proving (a) we make some simple remarks. We note first that since $D_{n+1} \subset D_n$ and $F_{n+1} \subset F_n$ for $n < \omega$ we have $k(m,n+1) \leq k(m,n)$ for $m \in D \cap D_{n+1}$. We note next that $\bigcap_{n<\omega} D_n = \emptyset$ (since $|f^{-1}(\{n\})| = n$), and thus for $m \in D$ there is a largest $h(m) < \omega$ such that $m \in D \cap D_{h(m)}$; clearly $0 \leq h(m) < m$ for $m \in D$. Finally we note that if $m \in D \cap D_n$ then $n < k(m,n)$. Indeed

$$n < |f^{-1}(\{m\}) \cap F_n \setminus (d_{\xi,0} \cup \cdots \cup d_{\xi,n})|$$

and hence

$$f^{-1}(\{m\}) \cap F_n \setminus \{s_{m,0}, \ldots, s_{m,n-1}\} \text{ is not a subset of } d_{\xi,0} \cup \cdots \cup d_{\xi,n},$$

i.e., $n < k(m,n)$.

We fix k and prove that $|d_{\xi,k}\cap B_\xi|<\omega$. If $s_{m,j}\in d_{\xi,k}\cap B_\xi$ and $m\in D\cap D_k$ then from $j\leq h(m)$ we have

$$k\leq h(m)<k(m,h(m))\leq k(m,j)$$

and hence $s_{m,j}\notin d_{\xi,k}$. Since further $j\leq h(m)$ for $s_{m,j}\in B_\xi$ we conclude that

$$d_{\xi,k}\cap B_\xi\subset\{s_{m,j}\in B_\xi: m\in D\setminus D_k \text{ and } j\leq h(m)\}\,.$$

Since $|D\setminus D_k|<\omega$ we have $|d_{\xi,k}\cap B_\xi|<\omega$.

(b) The family $\{\omega\setminus A: A\in\mathscr{A}\}\cup\{f^{-1}(C): C\in p\}\cup\mathscr{F}_\xi\cup\{B_\xi\}$ has the finite intersection property.

Let $A\in\mathscr{A}$, $C\in p$ and $F\in\mathscr{F}_\xi$. We choose $n_0, k<\omega$ such that

$$F_{n_0}\subset F \quad \text{and}$$
$$|f^{-1}(\{m\})\cap A|\leq k \quad \text{for } m\in C\cap D\cap D_{n_0+k}\,.$$

Since

$$s_{m,n_0}, s_{m,n_0+1},\ldots, s_{m,n_0+k}\in f^{-1}(C\cap D\cap D_{n_0})\cap F_{n_0}\cap B_\xi$$

and at most k of these elements are elements of A there is j such that

$$n_0\leq j\leq n_0+k \quad \text{and}$$
$$s_{m,j}\in(\omega\setminus A)\cap f^{-1}(C\cap D\cap D_{n_0})\cap F_{n_0}\cap B_\xi$$
$$\subset(\omega\setminus A)\cap f^{-1}(C)\cap F\cap B_\xi\,.$$

We now proceed as in *Case* 1. It follows from (b) that either the family

$$\{\omega\setminus A: A\in\mathscr{A}\}\cup\{f^{-1}(C): C\in p\}\cup\mathscr{F}_\xi\cup\{B_\xi\}\cup\{S_\xi\}$$

has the finite intersection property or the family

$$\{\omega\setminus A: A\in\mathscr{A}\}\cup\{f^{-1}(C): C\in p\}\cup\mathscr{F}_\xi\cup\{B_\xi\}\cup\{\omega\setminus S_\xi\}$$

has the finite intersection property; in the former case we define $\mathscr{F}_{\xi+1}$ to be the (countable) filter base generated by

$$\mathscr{F}_\xi\cup\{B_\xi\}\cup\{S_\xi\}$$

and in the latter case we define $\mathscr{F}_\xi$ to be the (countable) filter base generated by

$$\mathscr{F}_\xi\cup\{B_\xi\}\cup\{\omega\setminus S_\xi\}\,.$$

This completes the definition of the family $\mathscr{F}_\xi$ for all $\xi<\omega^+$; it is clear that the recursive assumptions (i)—(vii) are satisfied.

Finally we define

$$q=\bigcup\nolimits_{\xi<\omega^+}\mathscr{F}_\xi\,.$$

Then q is an ultrafilter on ω by (i), (ii) and (v), and $q\in\boldsymbol{\beta}(\omega)\setminus\omega$ by (vii) and the fact that $\mathscr{P}_\omega(\omega)\subset\mathscr{A}$; further q is a P-point of $\boldsymbol{\beta}(\omega)\setminus\omega$ by (vi), and $\bar{f}(q)=p$ (and hence $\tau(p)\leqslant\tau(q)$) by (vii) and Lemma 7.19(a); if $A\in\mathscr{P}(\omega)$ and $f|A$ is one-to-one then $A\in\mathscr{A}$ and hence $A\notin q$ by (vii), so from Theorem 9.2(b) we have $\tau(q)\neq\tau(p)$. The proof is complete.

16.2. Corollary. *Assume $\omega^+=2^\omega$. There are P-points of $\boldsymbol{\beta}(\omega)\backslash\omega$ whose types are not minimal in the Rudin-Keisler order of $\tau[\boldsymbol{\beta}(\omega)\backslash\omega]$.*

Proof. This follows from Theorem 16.1 and Theorem 9.13 (or, Corollary 14.15).

In the following theorem we set $U=\{\langle m,n\rangle\in\omega\times\omega: m<n\}$, $L=\{\langle m,n\rangle\in\omega\times\omega: n<m\}$ and $\Delta=\{\langle n,n\rangle: n<\omega\}$. Also as in § 7, for cardinals α and β we denote by $\boldsymbol{\phi}$ the (unique) continuous function from $\boldsymbol{\beta}(\alpha\times\beta)$ to $\boldsymbol{\beta}(\alpha)\times\boldsymbol{\beta}(\beta)$ such that $\boldsymbol{\phi}(\xi,\zeta)=\langle\xi,\zeta\rangle$ for $\langle\xi,\zeta\rangle\in\alpha\times\beta$.

16.3. Theorem. *Assume $\omega^+=2^\omega$. There are $p,q\in\boldsymbol{\beta}(\omega)\backslash\omega$ such that*

$\tau(p)$ and $\tau(q)$ are minimal in the Rudin-Keisler order of $\tau[\boldsymbol{\beta}(\omega)\backslash\omega]$, and $|\boldsymbol{\phi}^{-1}(\{\langle p,q\rangle\})|=2$.

Proof. Let $\langle S_\xi:\xi<\omega^+\rangle$ be a well-ordering of the set $\mathscr{P}(\omega\times\omega)\cup\omega^\omega$ (made disjoint by some simple device). We define families $\{\mathscr{F}_\xi:\xi<\omega^+\}$ and $\{\mathscr{G}_\xi:\xi<\omega^+\}$ of subsets of ω such that

(i) $\mathscr{F}_0=\mathscr{G}_0=\mathscr{P}^\omega(\omega)$;

(ii) $\mathscr{F}_\xi$ and $\mathscr{G}_\xi$ are countable filter bases for $\xi<\omega^+$;

(iii) $\mathscr{F}_\xi\subset\mathscr{F}_\zeta$ and $\mathscr{G}_\xi\subset\mathscr{G}_\zeta$ for $\xi<\zeta<\omega^+$;

(iv) $\mathscr{F}_\xi=\bigcup_{\zeta<\xi}\mathscr{F}_\zeta$ and $\mathscr{G}_\xi=\bigcup_{\zeta<\xi}\mathscr{G}_\zeta$ for non-zero limit ordinals $\xi<\omega^+$;

(v) if $S_\xi\in\omega^\omega$ then there are $A_\xi\in\mathscr{F}_{\xi+1}$ and $B_\xi\in\mathscr{G}_{\xi+1}$ such that

$S_\xi|A_\xi$ is either one-to-one or constant, and
$S_\xi|B_\xi$ is either one-to-one or constant, for $\xi<\omega^+$; and

(vi) if $S_\xi\in\mathscr{P}(\omega\times\omega)$ then
(a) there are $A_\xi\in\mathscr{F}_{\xi+1}$ and $B_\xi\in\mathscr{G}_{\xi+1}$ such that either

$$(A_\xi\times B_\xi)\cap(U\cup\Delta)\subset S_\xi \quad\text{or}$$
$$(A_\xi\times B_\xi)\cap(U\cup\Delta)\subset(\omega\times\omega)\backslash S_\xi, \quad\text{and}$$

(b) there are $\tilde{A}_\xi\in\mathscr{F}_{\xi+1}$ and $\tilde{B}_\xi\in\mathscr{G}_{\xi+1}$ such that either

$$(\tilde{A}_\xi\times\tilde{B}_\xi)\cap(L\cup\Delta)\subset S_\xi \quad\text{or}$$
$$(\tilde{A}_\xi\times\tilde{B}_\xi)\cap(L\cup\Delta)\subset(\omega\times\omega)\backslash S_\xi .$$

We proceed by transfinite recursion. We set

$$\mathscr{F}_0=\mathscr{G}_0=\mathscr{P}^\omega(\omega),$$

and we define $\mathscr{F}_\xi$ and $\mathscr{G}_\xi$ by (iv) for non-zero limit ordinals $\xi<\omega^+$. We now define $\mathscr{F}_{\xi+1}$ and $\mathscr{G}_{\xi+1}$ for $\xi<\omega^+$. There are two cases to consider.

Case 1. $S_\xi\in\omega^\omega$. We note that $\{S_\xi^{-1}(\{m\}):m<\omega\}$ is a partition of ω. It follows easily from Lemma 9.11 that there is $A_\xi\subset\omega$ such that

$$\begin{aligned}&\mathscr{F}_\xi\cup\{A_\xi\} \text{ has the finite intersection property, and}\\ &S_\xi|A_\xi \text{ is either one-to-one or constant.}\end{aligned}$$

Similarly there is $B_\xi\subset\omega$ such that

$$\begin{aligned}&\mathscr{G}_\xi\cup\{B_\xi\} \text{ has the finite intersection property, and}\\ &S_\xi|B_\xi \text{ is either one-to-one or constant.}\end{aligned}$$

We define $\mathscr{F}_{\xi+1}$ and $\mathscr{G}_{\xi+1}$ to be the (countable) filter bases generated by $\mathscr{F}_\xi\cup\{A_\xi\}$ and $\mathscr{G}_\xi\cup\{B_\xi\}$ respectively.

Case 2. $S_\xi\in\mathscr{P}(\omega\times\omega)$. Let $\langle F_n:n<\omega\rangle$ and $\langle G_n:n<\omega\rangle$ be well-orderings of $\mathscr{F}_\xi$ and $\mathscr{G}_\xi$ respectively, and set

$$T_\xi=\{m<\omega:\pi_1[S_\xi\cap(\{m\}\times\omega)]\in\langle\mathscr{G}_\xi\rangle\}\,.$$

We distinguish two sub-cases.

Case 2a. $\mathscr{F}_\xi\cup\{T_\xi\}$ has the finite intersection property. By a subsidiary recursion we define $a_n, b_n<\omega$ for $n<\omega$. Let $a_0\in F_0\cap T_\xi$; since $\pi_1[S_\xi\cap(\{a_0\}\times\omega)]\in\langle\mathscr{G}_\xi\rangle$ we have

$$\pi_1[S_\xi\cap(\{a_0\}\times\omega)]\cap G_0\neq\emptyset$$

and we choose b_0 to be an element of this non-empty set. Now let $1\leq n<\omega$ and suppose that a_k, b_k have been defined for $k<n$; we define a_n and b_n. Since $\mathscr{F}_\xi\supset\mathscr{P}^\omega(\omega)$ we have $|F_n\cap T_\xi|=\omega$ and hence there is $a_n\in F_n\cap T_\xi$ such that

$$b_k<a_n \quad \text{for } k<n\,.$$

Since $\pi_1[S_\xi\cap(\{a_k\}\times\omega)]\in\langle\mathscr{G}_\xi\rangle$ for $k\leq n$ we have

$$\Big(\bigcap\nolimits_{k\leq n}\pi_1[S_\xi\cap(\{a_k\}\times\omega)]\Big)\cap G_n\neq\emptyset$$

and we choose b_n to be an element of this non-empty set. This completes the subsidiary recursion. We set

$$A_\xi=\{a_n:n<\omega\} \quad \text{and} \quad B_\xi=\{b_n:n<\omega\}\,,$$

we note that the two families $\mathscr{F}_\xi\cup\{A_\xi\}$, $\mathscr{G}_\xi\cup\{B_\xi\}$ have the finite intersection property, and we define $\hat{\mathscr{F}}_{\xi+1}$ and $\hat{\mathscr{G}}_{\xi+1}$ to be the (countable) filter bases generated by $\mathscr{F}_\xi\cup\{A_\xi\}$ and $\mathscr{G}_\xi\cup\{B_\xi\}$ respectively.

We claim that $(A_\xi \times B_\xi) \cap (U \cup \Delta) \subset S_\xi$. Indeed if

$$\langle a_n, b_m \rangle \in (A_\xi \times B_\xi) \cap (U \cup \Delta)$$

then (since $b_k < a_n$ for $k < n$) we have $n \leq m$ and hence

$$b_m \in \pi_1 [S_\xi \cap (\{a_n\} \times \omega)],$$

i.e., $\langle a_n, b_m \rangle \in S_\xi$.

We now let $\langle \hat{F}_n : n < \omega \rangle$ and $\langle \hat{G}_n : n < \omega \rangle$ be well-orderings of $\hat{\mathscr{F}}_{\xi+1}$ and $\hat{\mathscr{G}}_{\xi+1}$ respectively, and we set

$$U_\xi = \{n < \omega : \pi_0 [S_\xi \cap (\omega \times \{n\})] \in \hat{\mathscr{F}}_{\xi+1}.$$

We distinguish two sub-cases.

Sub-case 2a(i). $\hat{\mathscr{G}}_{\xi+1} \cup \{U_\xi\}$ has the finite intersection property. We then use the argument given under *Case* 2a so far to define $\tilde{A}_\xi, \tilde{B}_\xi \in \mathscr{P}(\omega)$ such that

$\hat{\mathscr{F}}_{\xi+1} \cup \{\tilde{A}_\xi\}$ and $\hat{\mathscr{G}}_{\xi+1} \cup \{\tilde{B}_\xi\}$ have the finite intersection property, and $(\tilde{A}_\xi \times \tilde{B}_\xi) \cap (L \cup \Delta) \subset S_\xi$;

we then define $\mathscr{F}_{\xi+1}$ and $\mathscr{G}_{\xi+1}$ to be the (countable) filter bases generated by $\hat{\mathscr{F}}_{\xi+1} \cup \{\tilde{A}_\xi\}$ and $\hat{\mathscr{G}}_{\xi+1} \cup \{\tilde{B}_\xi\}$ respectively. We omit the details of the argument.

Sub-case 2a(ii). $\hat{\mathscr{G}}_{\xi+1} \cup \{U_\xi\}$ does not have the finite intersection property. Then $\hat{\mathscr{G}}_{\xi+1} \cup \{\omega \backslash U_\xi\}$ has the finite intersection property and we use the argument above to define $\tilde{A}_\xi, \tilde{B}_\xi \in \mathscr{P}(\omega)$ so that

$\hat{\mathscr{F}}_{\xi+1} \cup \{\tilde{A}_\xi\}$ and $\hat{\mathscr{G}}_{\xi+1} \cup \{\tilde{B}_\xi\}$ have the finite intersection property, and $(\tilde{A}_\xi \times \tilde{B}_\xi) \cap (L \cup \Delta) \subset \omega \backslash S_\xi$;

we then define $\mathscr{F}_{\xi+1}$ and $\mathscr{G}_{\xi+1}$ to be the (countable) filter bases generated by $\hat{\mathscr{F}}_{\xi+1} \cup \{\tilde{A}_\xi\}$ and $\hat{\mathscr{G}}_{\xi+1} \cup \{\tilde{B}_\xi\}$ respectively.

The discussion of *Case* 2a is complete.

Case 2b. $\mathscr{F}_\xi \cup \{T_\xi\}$ does not have the finite intersection property. Then $\mathscr{F}_\xi \cup \{\omega \backslash T_\xi\}$ has the finite intersection property and we use the argument above to define $A_\xi, B_\xi \in \mathscr{P}(\omega)$ such that the two families $\mathscr{F}_\xi \cup \{A_\xi\}$, $\mathscr{G}_\xi \cup \{B_\xi\}$ have the finite intersection property and

$$(A_\xi \times B_\xi) \cap (U \cup \Delta) \subset \omega \backslash S_\xi.$$

Sub-cases 2b(i) and 2b(ii) are then considered as with *Case* 2a. We omit the details.

We note from (ii) and (iii) that $\bigcup_{\xi < \omega^+} \mathscr{F}_\xi$ and $\bigcup_{\xi < \omega^+} \mathscr{G}_\xi$ are filter bases on ω, and hence there are $p, q \in \boldsymbol{\beta}(\omega)$ such that

$$\bigcup_{\xi < \omega^+} \mathscr{F}_\xi \subset p \quad \text{and} \quad \bigcup_{\xi < \omega^+} \mathscr{G}_\xi \subset q.$$

It follows from (i) that $p,q\in\boldsymbol{\beta}(\omega)\backslash\omega$ and from (v) (and Lemma 9.4) that $\tau(p)$ and $\tau(q)$ are minimal in the Rudin-Keisler order of $\tau[\boldsymbol{\beta}(\omega)\backslash\omega]$. Since $p,q\in\boldsymbol{\beta}(\omega)\backslash\omega$ we have from Lemma 7.22(a) that $U\in p\cdot q$ and $L\in\bar{t}(q\cdot p)$, hence $|\boldsymbol{\phi}^{-1}(\{\langle p,q\rangle\})|\geq 2$. Furthermore from (vi) we have

$$\{r\in\boldsymbol{\beta}(\omega\times\omega): p\times q\cup\{U\cup\Delta\}\subset r\}=\{p\cdot q\} \quad \text{and}$$
$$\{r\in\boldsymbol{\beta}(\omega\times\omega): p\times q\cup\{L\cup\Delta\}\subset r\}=\{\bar{t}(q\cdot p)\}\,.$$

Thus if $\boldsymbol{\phi}(r)=\langle p,q\rangle$ with $r\neq p\cdot q$ then $U\cup\Delta\notin r$, so that $L\in r$ and hence $r=\bar{t}(q\cdot p)$. It follows that

$$\{p\cdot q,\bar{t}(q\cdot p)\}=\boldsymbol{\phi}^{-1}(\{\langle p,q\rangle\})\,.$$

The proof is complete.

16.4. Theorem. *If* $p,q\in\boldsymbol{\beta}(\omega)\backslash\omega$ *and* $|\boldsymbol{\phi}^{-1}(\{\langle p,q\rangle\})|=2$ *then* p *and* q *are P-points of* $\boldsymbol{\beta}(\omega)\backslash\omega$.

Proof. It is enough to show that if $\{A_n:n<\omega\}\subset p$ then there is $C\in p$ such that $|C\backslash A_n|<\omega$ for $n<\omega$. We assume without loss of generality that $A_{n+1}\subset A_n$ for $n<\omega$. We set

$$B=\{\langle m,n\rangle\in\omega\times\omega: m\in A_n\}\cap L$$

and we claim that $\langle p,q\rangle\in\mathrm{cl}_{\boldsymbol{\beta}(\omega)\times\boldsymbol{\beta}(\omega)}B$. Indeed if $C\in p$ and $D\in q$ then for $n\in D$ there is $m\in A_n\cap C$ such that $n<m$, and then

$$\langle m,n\rangle\in B\cap(C\times D)\,;$$

the claim is proved. We claim next that

$$\langle p,q\rangle\notin\mathrm{cl}_{\boldsymbol{\beta}(\omega)\times\boldsymbol{\beta}(\omega)}(L\backslash B)\,.$$

If the claim fails then since $B\cap(L\backslash B)=\emptyset$ there are $r,s\in\boldsymbol{\phi}^{-1}(\{\langle p,q\rangle\})$ such that

$$r\in\mathrm{cl}_{\boldsymbol{\beta}(\omega\times\omega)}B \quad \text{and} \quad s\in\mathrm{cl}_{\boldsymbol{\beta}(\omega\times\omega)}(L\backslash B)\,,$$

and since $p\cdot q\in(\mathrm{cl}_{\boldsymbol{\beta}(\omega\times\omega)}U)\cap\boldsymbol{\phi}^{-1}(\{\langle p,q\rangle\})$ (by Lemmas 7.21(c) and 7.22(a)) we have

$$\{p\cdot q,r,s\}\subset\boldsymbol{\phi}^{-1}(\{\langle p,q\rangle\}) \quad \text{and hence } |\boldsymbol{\phi}^{-1}(\{\langle p,q\rangle\})|\geq 3\,;$$

this contradicts the assumption $|\boldsymbol{\phi}^{-1}(\{\langle p,q\rangle\})|=2$.

It follows that there are $C\in p$ and $D\in q$ such that $(L\backslash B)\cap(C\times D)=\emptyset$. We claim that $|C\backslash A_n|<\omega$ for $n<\omega$. Indeed for $n<\omega$ there is $k(n)<\omega$ such that $n<k(n)\in D$, and if $m\in C\backslash A_n$ and $k(n)<m$ then since $\langle m,k(n)\rangle\in(C\times D)\cap L$ we have $\langle m,k(n)\rangle\in B$ and hence $m\in A_{k(n)}\subset A_n$; it follows that

$$C\backslash A_n\subset\{m<\omega: m\leq k(n)\} \quad \text{for } n<\omega\,,$$

and hence $|C \setminus A_n| < \omega$ for $n < \omega$.

The proof is complete.

We recall that if p and q are ultrafilters on β and α respectively then p,q and $p \cdot q$ are (identified with) ultrafilters on $\beta \times \alpha$. We note that if p is a principal ultrafilter then there is a one-to-one function $f: \beta \times \alpha \to \beta \times \alpha$ such that $\bar{f}(p \cdot q) = q$, and hence $p \cdot q \approx q$ by Theorem 9.2(b); that if q is a principal ultrafilter then $p \cdot q \approx p$; and that if $p \approx p'$ and $q \approx q'$, then $p \cdot q \approx p' \cdot q'$.

16.5. Lemma. *Let α and β be cardinals with $\alpha \geq \omega$ and let $e: \beta \to \boldsymbol{\beta}(\alpha)$ be a one-to-one function such that $e[\beta]$ is a strongly discrete subset of $\boldsymbol{\beta}(\alpha)$. If $p \in \boldsymbol{\beta}(\beta)$, $q \in \boldsymbol{\beta}(\alpha)$, and $e(\zeta) \approx q$ for $\zeta < \beta$, then $\bar{e}(p) \approx p \cdot q$.*

Proof. If q is a principal ultrafilter then $e[\beta] \subset \alpha$, and since e is a one-to-one function we have

$$\bar{e}(p) \approx p \approx p \cdot q$$

by Theorem 9.2(b) and the remarks preceding this lemma. We thus assume that q is non-principal, so that $e(\zeta) \in \boldsymbol{\beta}(\alpha) \setminus \alpha$ for $\zeta < \beta$. Since $e[\beta]$ is strongly discrete there is a family $\{A_\zeta : \zeta < \beta\}$ of subsets of α such that

$$A_\zeta \cap A_\eta = \emptyset \quad \text{for } \zeta < \eta < \beta,$$
$$\textstyle\bigcup_{\zeta < \beta} A_\zeta = \alpha, \quad \text{and}$$
$$A_\zeta \in e(\zeta) \quad \text{for } \zeta < \beta.$$

We assume without loss of generality that

$$|A_\zeta| = \alpha \quad \text{for } \zeta < \beta.$$

Since $e(\zeta) \approx q \approx \zeta \cdot q$ for $\zeta < \beta$, there is a one-to-one function

$$f_\zeta : \pi_0^{-1}(\{\zeta\}) \to A_\zeta$$

such that $\bar{f}_\zeta(\zeta \cdot q) = e(\zeta)$. We set

$$f = \textstyle\bigcup_{\zeta < \beta} f_\zeta : \beta \times \alpha \to \alpha$$

and we note that f is a one-to-one function such that

$$\bar{f}(\zeta \cdot q) = e(\zeta) = e \circ \bar{\pi}_0(\zeta \cdot q) \quad \text{for } \zeta < \beta.$$

Since $p \cdot q \in \mathrm{cl}_{\boldsymbol{\beta}(\beta \times \alpha)}\{\zeta \cdot q : \zeta < \beta\}$ we have

$$\bar{f}(p \cdot q) = \bar{e} \circ \bar{\pi}_0(p \cdot q) = \bar{e}(p)$$

and hence (by Theorem 9.2(b)) $\bar{e}(p) \approx p \cdot q$, as required.

16.6. Corollary. *If $p, q \in \boldsymbol{\beta}(\omega) \setminus \omega$ then there are two countable discrete subsets of $\boldsymbol{\beta}(\omega) \setminus \omega$ whose boundaries contain $p \cdot q$ and $\bar{t}(q \cdot p)$ respectively.*

Proof. There are one-to-one functions $e, f: \omega \to \boldsymbol{\beta}(\omega) \setminus \omega$ such that

$$e[\omega] \text{ and } f[\omega] \text{ are (strongly) discrete subsets of } \boldsymbol{\beta}(\omega), \text{ and}$$
$$e(n) \approx q \quad \text{and} \quad f(n) \approx p \quad \text{for } n < \omega.$$

Then $\bar{e}(p) \approx p \cdot q$ and $\bar{f}(q) \approx q \cdot p \approx \bar{t}(q \cdot p)$ by Lemma 16.5, and $\bar{e}(p)$ and $\bar{f}(q)$ are in the boundaries of $e[\omega]$ and $f[\omega]$, respectively.

16.7. Lemma. *Let $f \in \omega^\omega$ and let p be a P-point of $\boldsymbol{\beta}(\omega) \setminus \omega$. If $\bar{f}(p) \in \boldsymbol{\beta}(\omega) \setminus \omega$ then $\bar{f}(p)$ is a P-point of $\boldsymbol{\beta}(\omega) \setminus \omega$.*

Proof. If $\{A_n : n < \omega\} \subset \bar{f}(p)$ then since $f^{-1}(A_n) \in p$ for $n < \omega$ and p is a P-point of $\boldsymbol{\beta}(\omega) \setminus \omega$, there is $B \in p$ such that $|B \setminus f^{-1}(A_n)| < \omega$ for $n < \omega$. Then $f[B] \in \bar{f}(p)$ and $|f[B] \setminus A_n| < \omega$ for $n < \omega$. Hence $\bar{f}(p)$ is a P-point of $\boldsymbol{\beta}(\omega) \setminus \omega$.

16.8. Theorem. *Let $p, q \in \boldsymbol{\beta}(\omega) \setminus \omega$ such that*

$\tau(p)$ and $\tau(q)$ are minimal in the Rudin-Keisler order of $\tau[\boldsymbol{\beta}(\omega) \setminus \omega]$ and $|\boldsymbol{\phi}^{-1}(\{\langle p, q \rangle\})| = 2$.

Then $\{\tau(p), \tau(q)\}$ does not have a supremum in $\tau[\boldsymbol{\beta}(\omega) \setminus \omega]$ and does not have an upper bound in $\tau[P_{\omega^+}(\boldsymbol{\beta}(\omega) \setminus \omega)]$.

Thus, assuming $\omega^+ = 2^\omega$, $\tau[\boldsymbol{\beta}(\omega) \setminus \omega]$ is not an upper semi-lattice and $\tau[P_{\omega^+}(\boldsymbol{\beta}(\omega) \setminus \omega)]$ is not directed upwards in the Rudin-Keisler order.

Proof. We establish the following three statements:

(a) $\tau(p)$ and $\tau(q)$ are not comparable in the Rudin-Keisler order;

(b) $\tau(p \cdot q) \neq \tau(\bar{t}(q \cdot p))$; and

(c) both $\tau(p \cdot q)$ and $\tau(\bar{t}(q \cdot p))$ are immediate successors of $\tau(p)$ and of $\tau(q)$ in the Rudin-Keisler order.

Proof of (a). Suppose there is $f: \omega \to \omega$ such that $\bar{f}(p) = q$. We define $g: \omega \to \omega \times \omega$ by $g(m) = \langle m, f(m) \rangle$. Since g is one-to-one we have $\bar{g}(p) \approx p$ by Theorem 9.2(b), and since $\bar{g}(p) \supset p \times q$ we have $\bar{g}(p) \in \boldsymbol{\phi}^{-1}(\{\langle p, q \rangle\})$ by Lemma 7.21(c). Since $\bar{g}(p) \approx p \in P_{\omega^+}(\boldsymbol{\beta}(\omega) \setminus \omega)$ we have $\bar{g}(p) \in P_{\omega^+}(\boldsymbol{\beta}(\omega) \setminus \omega)$; and

$$p \cdot q,\ \bar{t}(q \cdot p) \notin P_{\omega^+}(\boldsymbol{\beta}(\omega) \setminus \omega)$$

by Corollary 16.6. Thus $|\boldsymbol{\phi}^{-1}(\{\langle p, q \rangle\})| \geq 3$ by Lemmas 7.21 and 7.22, contradicting the assumption $|\boldsymbol{\phi}^{-1}(\{\langle p, q \rangle\})| = 2$.

Proof of (b). By Lemma 16.5 there are one-to-one functions $e, f: \omega \to \boldsymbol{\beta}(\omega \times \omega)$ such that

$$e[\omega] \text{ and } f[\omega] \text{ are (strongly) discrete subsets of } \boldsymbol{\beta}(\omega \times \omega),$$
$$e(n) \approx q \quad \text{and} \quad f(m) \approx p \quad \text{for } n, m < \omega, \quad \text{and}$$
$$\bar{e}(p) = p \cdot q \quad \text{and} \quad \bar{f}(q) = q \cdot p.$$

Suppose now that $\tau(\bar{t}(q\cdot p))=\tau(p\cdot q)$. Then there is a permutation π of $\omega\times\omega$ such that $\bar{\pi}(\bar{t}(q\cdot p))=p\cdot q$; we note that $\pi\circ t$ is a permutation of $\omega\times\omega$ and hence

$$\bar{\pi}(\bar{t}(f(m)))\approx f(m)\approx p \quad \text{for } m<\omega\,.$$

We note also that $e[\omega]\cap(\bar{\pi}\circ\bar{t}\circ f)[\omega]=\emptyset$, since otherwise there are $n,m<\omega$ such that

$$q\approx e(n)=(\bar{\pi}\circ\bar{t}\circ f)(m)\approx f(m)\approx p$$

contrary to statement (a) above. Since $p,q\in P_{\omega^+}(\boldsymbol{\beta}(\omega)\backslash\omega)$ and $e[\omega]$ and $(\bar{\pi}\circ\bar{t}\circ f)[\omega]$ are disjoint (strongly) discrete subsets of $\boldsymbol{\beta}(\omega\times\omega)$ there are $\{A_n:n<\omega\}\subset\mathscr{P}(\omega\times\omega)$ and $\{B_m:m<\omega\}\subset\mathscr{P}(\omega\times\omega)$ such that

$A_n\in e(n)$ and $B_m\in(\bar{\pi}\circ\bar{t}\circ f)(m)$ for $m,n<\omega$,
$A_n\cap A_m=A_n\cap B_m=A_m\cap B_n=B_n\cap B_m=\emptyset$ for $m<n<\omega$, and
$(\bigcup_{n<\omega}A_n)\cup(\bigcup_{m<\omega}B_m)=\omega\times\omega\,.$

We set

$$\begin{aligned} h(i,j)&=0 \quad \text{if } \langle i,j\rangle\in \textstyle\bigcup_{n<\omega}A_n\\ &=1 \quad \text{if } \langle i,j\rangle\in\omega\times\omega\backslash\textstyle\bigcup_{n<\omega}A_n\,. \end{aligned}$$

Since

$$p\cdot q\in \mathrm{cl}_{\boldsymbol{\beta}(\omega\times\omega)}e[\omega]\subset \mathrm{cl}_{\boldsymbol{\beta}(\omega\times\omega)}h^{-1}(\{0\}) \quad\text{and}$$
$$p\cdot q=\bar{\pi}\circ\bar{t}(q\cdot p)\in\mathrm{cl}_{\boldsymbol{\beta}(\omega\times\omega)}(\bar{\pi}\circ\bar{t})[f[\omega]]\subset\mathrm{cl}_{\boldsymbol{\beta}(\omega\times\omega)}h^{-1}(\{1\})$$

we have $\bar{h}(p\cdot q)=0$ and $\bar{h}(p\cdot q)=1$, a contradiction.

Proof of (c). We note that $p,q\leqslant p\cdot q$ by Lemma 7.21(b); furthermore $\tau(p)\neq\tau(p\cdot q)$ and $\tau(q)\neq\tau(p\cdot q)$ since $p,q\in P_{\omega^+}(\boldsymbol{\beta}(\omega)\backslash\omega)$ and (by Corollary 16.6) $p\cdot q$ is not a P-point of $\boldsymbol{\beta}(\omega)\backslash\omega$. Hence to prove that $\tau(p\cdot q)$ is an immediate successor of $\tau(p)$ and $\tau(q)$ it is enough to show that if $0\prec r\prec p\cdot q$ then $r\approx p$ or $r\approx q$.

Since $r\prec p\cdot q$ there is $f:\omega\times\omega\to\omega$ such that $\bar{f}(p\cdot q)=r$. We set

$$f_m(n)=f(m,n) \quad \text{for } m,n<\omega$$

and for each $m<\omega$ we note that since $\bar{f}_m(q)\leqslant q$ we have

$$\bar{f}_m(q)\approx q \quad \text{or } \bar{f}_m(q)<\omega\,.$$

We set $A=\{m<\omega:\bar{f}_m(q)<\omega\}$ and we consider two cases.

Case 1. $A\in p$. We define $g:\omega\to\omega$ by

$$\begin{aligned} g(m)&=\bar{f}_m(q) \quad \text{if } m\in A\\ &=0 \qquad\quad \text{if } m\notin A \end{aligned}$$

and we set

$$C=\{\langle m,n\rangle\in\omega\times\omega: m\in A, f(m,n)=g(m)\}\,.$$

Since $C \in p \cdot q$ and

$$g \circ \pi_0(m,n) = f(m,n) \quad \text{for } \langle m,n \rangle \in C$$

we have

$$r = \bar{f}(p \cdot q) = \bar{g} \circ \bar{\pi}_0(p \cdot q) = \bar{g}(p)$$

(by Lemma 7.21(b)) and hence $r \preccurlyeq p$. Since $0 \prec r$ and $\tau(p)$ is minimal in the Rudin-Keisler order of $\tau[\boldsymbol{\beta}(\omega) \setminus \omega]$, it follows that $p \approx r$.

Case 2. $A \notin p$. Since $\tau(q)$ is minimal in the Rudin-Keisler order of $\tau[\boldsymbol{\beta}(\omega) \setminus \omega]$ we have $\bar{f}_m(q) \approx q$ for $m \in \omega \setminus A$ and hence

$$\omega \setminus A = \{m < \omega : \bar{f}_m(q) \approx q\} \in p .$$

We define $e: \omega \to \boldsymbol{\beta}(\omega \times \omega)$ by $e(m) = m \cdot q$ and we note from the definition of $p \cdot q$ that $\bar{e}(p) = p \cdot q$. We define

$$k = |\{\bar{f}_m(q) : m \in \omega \setminus A\}| ,$$

we let $\langle q_i : i < k \rangle$ be a well-ordering of $\{\bar{f}_m(q) : m \in \omega \setminus A\}$ such that $q_i \neq q_j$ for $i < j < k$, and we define $g: \omega \to k$ and $d: k \to \boldsymbol{\beta}(\omega)$ by

$$\begin{aligned} g(m) &= i \quad \text{if } m \in \omega \setminus A, \quad \bar{f}_m(q) = q_i \\ &= 0 \quad \text{if } m \in A \end{aligned}$$

and

$$d(i) = q_i \quad \text{for } i < k .$$

If $m \in \omega \setminus A$ and $\bar{f}_m(q) = q_i$ then

$$d(g(m)) = q_i = \bar{f}_m(q) = \bar{f}(m \cdot q) = \bar{f}(e(m)) ;$$

hence $r = \bar{f}(p \cdot q) = \bar{f}(\bar{e}(p)) = \bar{d}(\bar{g}(p))$. Since d is a one-to-one function such that

$$\begin{aligned} &d[k] \text{ is a (strongly) discrete subset of } \boldsymbol{\beta}(\omega) \text{ and} \\ &d(i) = q_i \approx q \quad \text{for } i < k \end{aligned}$$

we have

$$\bar{g}(p) \cdot q \approx \bar{d}(\bar{g}(p)) = r$$

by Lemma 16.5. If $\bar{g}(p) \in \boldsymbol{\beta}(\omega) \setminus \omega$ then $\bar{g}(p) \approx p$ and $r \approx p \cdot q$, contradicting the assumption $r \prec p \cdot q$; hence $\bar{g}(p)$ is a principal ultrafilter and we have

$$r \approx \bar{g}(p) \cdot q \approx q ,$$

as required. The proof that $\tau(p \cdot q)$ is an immediate successor of $\tau(p)$ and of $\tau(q)$ is complete. By interchanging p and q it follows that $\tau(q \cdot p)$, which is $\tau(\bar{t}(q \cdot p))$, is an immediate successor of $\tau(q)$ and $\tau(p)$. The proof of (c) is complete.

It follows immediately from (a), (b) and (c) that $\{\tau(p), \tau(q)\}$ does not have a supremum in the Rudin-Keisler order of $\tau[\boldsymbol{\beta}(\omega) \setminus \omega]$.

Now let $p \preccurlyeq r$ and $q \preccurlyeq r$; we show that $r \notin P_{\omega^+}(\boldsymbol{\beta}(\omega)\setminus\omega)$. There are $f,g:\omega\to\omega$ such that $\overline{f}(r)=p$ and $\overline{g}(r)=q$. We define $h:\omega\to\omega\times\omega$ by

$$h(n)=\langle f(n),g(n)\rangle \quad \text{for } n<\omega\,;$$

clearly $\overline{h}(r)\supset p\times q$, so from Lemma 7.21(c) and the assumption $|\boldsymbol{\phi}^{-1}(\{\langle p,q\rangle\})|=2$ it follows that $\overline{h}(r)=p\cdot q$ or $\overline{h}(r)=\overline{t}(q\cdot p)$. Hence either $p\cdot q\preccurlyeq r$ or $\overline{t}(q\cdot p)\preccurlyeq r$, and if $r\in P_{\omega^+}(\boldsymbol{\beta}(\omega)\setminus\omega)$ then from Lemma 16.7 it follows that $p\cdot q\in P_{\omega^+}(\boldsymbol{\beta}(\omega)\setminus\omega)$ or $\overline{t}(q\cdot p)\in P_{\omega^+}(\boldsymbol{\beta}(\omega)\setminus\omega)$; this contradicts Corollary 16.6.

The final statement of the theorem follows from what has been proved above and the fact (given in Theorem 16.3) that if $\omega^+=2^\omega$ then there are $p,q\in\boldsymbol{\beta}(\omega)\setminus\omega$ such that

$\tau(p)$ and $\tau(q)$ are minimal in the Rudin-Keisler order of $\tau[\boldsymbol{\beta}(\omega)\setminus\omega]$, and $|\boldsymbol{\phi}^{-1}(\{\langle p,q\rangle\})|=2$.

A consequence of Theorem 16.8 and its proof is the fact that, assuming $\omega^+=2^\omega$, the type of a non-principal ultrafilter on ω is not determined by the set of minimal types below it. Indeed in the notation of Theorem 16.8 we have $\tau(p\cdot q)\neq\tau(\overline{t}(q\cdot p))$ and both $\tau(p\cdot q)$ and $\tau(\overline{t}(q\cdot p))$ have $\{\tau(p),\tau(q)\}$ as the set of Rudin-Keisler minimal predecessors.

16.9. Theorem. *Let $\alpha\geq\omega$ and $p\in U(\alpha)$. Then there is $q\in\boldsymbol{\beta}(\alpha)$ such that $|\boldsymbol{\phi}^{-1}(\{\langle p,q\rangle\})|\geq 2^\alpha$.*

Proof. By Corollary 3.18 there is $f:\alpha\to\alpha^{(2^\alpha)}$ such that $f[\alpha]$ is dense in $\alpha^{(2^\alpha)}$. We define

$$U_{\eta,\eta'}(A)=\{r\in\alpha^{(2^\alpha)}: r_\eta<r_{\eta'} \text{ and } r_\eta,r_{\eta'}\in A\} \quad \text{for } \eta<\eta'<2^\alpha \text{ and } A\in p,$$

and we note that $\{U_{\eta,\eta'}(A):\eta<\eta'<2^\alpha, A\in p\}$ is a family of open subsets of $\alpha^{(2^\alpha)}$ with the finite intersection property. Since $f[\alpha]$ is dense in $\alpha^{(2^\alpha)}$ the family

$$\{f^{-1}(U_{\eta,\eta'}(A)):\eta<\eta'<2^\alpha, A\in p\}$$

has the finite intersection property and hence there is $q\in\boldsymbol{\beta}(\alpha)$ such that

$$f^{-1}(U_{\eta,\eta'}(A))\in q \quad \text{for } \eta<\eta'<2^\alpha,\ A\in p\,.$$

For $\eta<2^\alpha$ we denote as usual by π_η the projection function from $\alpha^{(2^\alpha)}$ to α defined by $\pi_\eta(r)=r_\eta$ and (for this proof only) we denote by $\tilde{\pi}_0$ and $\tilde{\pi}_1$ the projection functions from $\alpha\times\alpha$ to α defined by

$$\tilde{\pi}_0(\xi,\zeta)=\xi \quad \text{and} \quad \tilde{\pi}_1(\xi,\zeta)=\zeta\,.$$

Since $\pi_\eta[U_{\eta,\eta'}(A)]=A$ for $\eta<\eta'<2^\alpha$ and $A\in p$ we have

$$A\in(\pi_\eta\circ f)[q]=\overline{\pi_\eta\circ f}(q) \quad \text{for } \eta<2^\alpha \quad \text{and } A\in p$$

by Lemma 7.19(a) and hence

$$\overline{\pi_\eta\circ f}(q)=p \quad \text{for } \eta<2^\alpha .$$

We define $g_\eta:\alpha\to\alpha\times\alpha$ by the rule

$$g_\eta(\xi)=\langle(\pi_\eta\circ f)(\xi),\xi\rangle$$

for $\eta<2^\alpha$ and we denote by $\overline{g_\eta}$ the Stone extension of g_η from $\boldsymbol{\beta}(\alpha)$ into $\boldsymbol{\beta}(\alpha\times\alpha)$. Since $\tilde{\pi}_0\circ g_\eta=\pi_\eta\circ f$ and $\tilde{\pi}_1\circ g_\eta$ is the identity function on α we have

$$\overline{\tilde{\pi}_0\circ g_\eta}(q)=\overline{\pi_\eta\circ f}(q)=p \quad \text{and } \overline{\tilde{\pi}_1\circ g_\eta}(q)=q$$

and hence $p\times q\subset\overline{g_\eta}(q)$ for $\eta<2^\alpha$ by Lemma 7.21. Thus

$$\overline{g_\eta}(q)\in\boldsymbol{\phi}^{-1}(\{\langle p,q\rangle\}) \quad \text{for } \eta<2^\alpha$$

and to complete the proof it is enough to show that $\overline{g_\eta}(q)\neq\overline{g_{\eta'}}(q)$ for $\eta<\eta'<2^\alpha$.

We note that $f^{-1}(U_{\eta,\eta'}(\alpha))\in q$ and hence $g_\eta[f^{-1}(U_{\eta,\eta'}(\alpha))]\in g_\eta[q]=\overline{g_\eta}(q)$ and $g_{\eta'}[f^{-1}(U_{\eta,\eta'}(\alpha))]\in g_{\eta'}[q]=\overline{g_{\eta'}}(q)$ by Lemma 7.19(a). Now let $\xi,\xi'\in f^{-1}(U_{\eta,\eta'}(\alpha))$. If $\xi=\xi'$ then

$$\tilde{\pi}_0(g_\eta(\xi))=f(\xi)_\eta<f(\xi)_{\eta'}=\tilde{\pi}_0(g_{\eta'}(\xi))$$

and hence $g_\eta(\xi)\neq g_{\eta'}(\xi)$; and if $\xi\neq\xi'$ then

$$\tilde{\pi}_1(g_\eta(\xi))=\xi\neq\xi'=\tilde{\pi}_1(g_{\eta'}(\xi'))$$

and hence $g_\eta(\xi)\neq g_{\eta'}(\xi')$. Thus

$$g_\eta[f^{-1}(U_{\eta,\eta'}(\alpha))]\cap g_{\eta'}[f^{-1}(U_{\eta,\eta'}(\alpha))]=\emptyset$$

and $\overline{g_\eta}(q)\neq\overline{g_{\eta'}}(q)$ for $\eta<\eta'<2^\alpha$. The proof is complete.

We note that if $p,q\in\boldsymbol{\beta}(\omega)$ and $|\boldsymbol{\phi}^{-1}(\{\langle p,q\rangle\})|\geq\omega$ then in fact $|\boldsymbol{\phi}^{-1}(\{\langle p,q\rangle\})|=2^{2^\omega}$. For the compact set $\boldsymbol{\phi}^{-1}(\{\langle p,q\rangle\})$ contains an infinite discrete subset by Lemma 3.3(b), and since this set is C^*-embedded in $\boldsymbol{\beta}(\omega\times\omega)$ (see for example Lemma 14.16 and Lemma 16.15(b) below) we have $|\boldsymbol{\phi}^{-1}(\{\langle p,q\rangle\})|\geq|\boldsymbol{\beta}(\omega)|=2^{2^\omega}$ by Corollary 7.4.

Definition. Let $d=\langle d_n:n<\omega\rangle$ be an ω-partition of $\omega\times\omega$ and let $A\subset\omega$. Then A is *d-thick* if for $n<\omega$, $\tau\in\omega^n$ and for every equality $A=\bigcup_{i<n}A_i$ there is $i<n$ such that

$$(A_i\times A_i)\cap d_{\tau(i)}\neq\emptyset .$$

The following statements are clear from the definition.

Let d be an ω-partition of $\omega\times\omega$ and let $A, B\subset\omega$.

(a) If A is d-thick and $A\subset B$, then B is d-thick; and

(b) if neither A nor B is d-thick, then $A\cup B$ is not d-thick.

16.10. Lemma. *Let d be an ω-partition of $\omega\times\omega$ and let $A\subset\omega$. If A is d-thick then for $n<\omega$ and $\tau\in\omega^n$ there is $G\in\mathscr{P}_\omega(A)$ such that for every equality $G=\bigcup_{i<n}G_i$ there is $i<n$ such that*

$$(G_i\times G_i)\cap d_{\tau(i)}\neq\emptyset .$$

Proof. Let $\mathscr{A}$ be the family of ordered n-tuples $\langle A_0, A_1, \ldots, A_{n-1}\rangle$ such that

$A_i\in\mathscr{P}_\omega(A)$ for $i<n$, and

if $G\in\mathscr{P}_\omega(A)$ with $\bigcup_{i<n}A_i\subset G$, then there is an equality $G=\bigcup_{i<n}G_i$ such that

$$A_i\subset G_i \quad \text{for } i<n, \quad \text{and}$$
$$(G_i\times G_i)\cap d_{\tau(i)}=\emptyset \quad \text{for } i<n.$$

We assume that the lemma is false, so that the n-tuple $\langle\emptyset,\emptyset,\ldots,\emptyset\rangle$ is an element of $\mathscr{A}$.

We note that $|A|=\omega$, and hence

$$\bigcup_{i<n}A_i\neq A \quad \text{for } \langle A_0, A_1, \ldots, A_{n-1}\rangle\in\mathscr{A} .$$

We define a relation $\le$ on $\mathscr{A}$ by the rule

$$\langle A_0, A_1, \ldots, A_{n-1}\rangle<\langle B_0, B_1, \ldots, B_{n-1}\rangle$$

if

$$A_i\subset B_i \quad \text{for } i<n \quad \text{and}$$
$$\min(\omega\setminus\bigcup_{i<n}A_i)\in\bigcup_{i<n}B_i .$$

It is clear that $\langle\mathscr{A},\le\rangle$ is a ramification system. For

$$a=\langle A_0, A_1, \ldots, A_{n-1}\rangle\in\mathscr{A}$$

we denote (as in § 8) by $\mathscr{S}(a)$ the set of immediate successors of a, and for $k<\omega$ we set

$$\mathscr{A}_k=\{a\in\mathscr{A}: \text{the order of } a \text{ is } k\} .$$

Since $|\mathscr{A}_0|=1$ and $|\mathscr{S}(a)|\le n$ for $a\in\mathscr{A}$ we have

$$|\mathscr{A}_k|<\omega \quad \text{for } k<\omega .$$

We claim that the ramification system $\langle\mathscr{A},\le\rangle$ has order ω; it is enough to prove that $\mathscr{S}(a)\neq\emptyset$ for $a\in\mathscr{A}$.

Let $a=\langle A_0,A_1,\dots,A_{n-1}\rangle\in\mathscr{A}$ with $\mathscr{S}(a)=\emptyset$, set $m=\min(A\setminus\bigcup_{i<n}A_i)$, and for $j<n$ define

$$b(j)=\langle B_0(j),B_1(j),\dots,B_{n-1}(j)\rangle$$

by the rule

$$\begin{aligned} B_i(j)&=A_i && \text{if } i\neq j\\ &=A_i\cup\{m\} && \text{if } i=j\,. \end{aligned}$$

Then $b(j)\notin\mathscr{A}$ for $j<n$ so there is $G(j)\in\mathscr{P}_\omega(A)$ such that

$\bigcup_{i<n}B_i(j)\subset G(j)$, and

for every equality $G(j)=\bigcup_{i<n}G_i(j)$ with $B_i(j)\subset G_i(j)$ for $i<n$, there is $i<n$ such that

$$(G_i(j)\times G_i(j))\cap d_{\tau(i)}\neq\emptyset\,.$$

We set $G=\bigcup_{j<n}G(j)$ and we note that since

$$G\in\mathscr{P}_\omega(A)\quad\text{and}\quad\bigcup_{i<n}A_i\subset G$$

there is an equality $G=\bigcup_{i<n}G_i$ such that

$$\begin{aligned} &A_i\subset G_i \quad \text{for } i<n, \quad\text{and}\\ &(G_i\times G_i)\cap d_{\tau(i)}=\emptyset \quad \text{for } i<n\,. \end{aligned}$$

There is $\bar{j}<n$ such that $m\in G_{\bar{j}}$, and since

$$\begin{aligned} &G(\bar{j})=\bigcup_{i<n}(G(\bar{j})\cap G_i) \quad\text{and}\\ &B_i(\bar{j})\subset G(\bar{j})\cap G_i \quad \text{for } i<n \end{aligned}$$

there is $\bar{i}<n$ such that

$$((G(\bar{j})\cap G_{\bar{i}})\times(G(\bar{j})\cap G_{\bar{i}}))\cap d_{\tau(\bar{i})}\neq\emptyset\,.$$

This contradiction completes the proof that the ramification system $\langle\mathscr{A},\leq\rangle$ has order ω. By D. König's Theorem (Corollary 8.5) there is

$$a(k)=\langle A_0(k),A_1(k),\dots,A_{n-1}(k)\rangle\in\mathscr{A}_k$$

such that $a(k)<a(k+1)$ for $k<\omega$. We set

$$S(k)=\bigcup_{i<n}A_i(k)$$

and we note that if $A=\{m_k:k<\omega\}$ with $m_k<m_{k+1}$ for $k<\omega$ then $m_k\in S(k+1)$; thus

$$A=\bigcup_{k<\omega}S(k)\,.$$

Finally we define

$$A_i=\bigcup_{k<\omega}A_i(k)\quad\text{for } i<n.$$

Then $A=\bigcup_{i<n}A_i$ and

$$(A_i\times A_i)\cap d_{\tau(i)}=\emptyset \quad \text{for } i<n,$$

contrary to the assumption that A is d-thick. The proof is complete.

16.11. Theorem. *Assume $\omega^+=2^\omega$. There is a P-point p of $\boldsymbol{\beta}(\omega)\backslash\omega$ such that*

$$|\boldsymbol{\phi}^{-1}(\{\langle p,p\rangle\})|=2^{2^\omega}.$$

Proof. By the remark following the proof of Theorem 16.9 it is enough to prove that there is a P-point p of $\boldsymbol{\beta}(\omega)\backslash\omega$ such that $|\boldsymbol{\phi}^{-1}(\{\langle p,p\rangle\}|\geq\omega$.

Let $d=\langle d_n:n<\omega\rangle$ be an ω-partition of $\omega\times\omega$ and $\langle S_\xi:\xi<\omega^+\rangle$ a well-ordering of $\mathscr{P}(\omega)$. For $\xi<\omega^+$ we define $A_\xi,B_\xi\in\mathscr{P}(\omega)$ and a family $\mathscr{F}_\xi$ of subsets of ω such that

(i) $\mathscr{F}_0=\{\omega\}$;
(ii) $\mathscr{F}_\xi\subset\mathscr{F}_\zeta$ for $\xi<\zeta<\omega^+$;
(iii) $\mathscr{F}_\xi=\bigcup_{\zeta<\xi}\mathscr{F}_\zeta$ for non-zero limit ordinals $\xi<\omega^+$;
(iv) $\mathscr{F}_\xi$ is a countable filter base for $\xi<\omega^+$;
(v) $A_\xi=S_\xi$ or $A_\xi=\omega\backslash S_\xi$ for $\xi<\omega^+$;
(vi) $|B_\xi\backslash A|<\omega$ for $A\in\mathscr{F}_\xi$ and $\xi<\omega^+$; and
(vii) A is d-thick for $A\in\mathscr{F}_\xi$ and $\xi<\omega^+$.

We proceed by transfinite recursion. We set $\mathscr{F}_0=\{\omega\}$ and for non-zero limit ordinals $\xi<\omega^+$ we set $\mathscr{F}_\xi=\bigcup_{\zeta<\xi}\mathscr{F}_\zeta$. Now for $\xi<\omega^+$ we define A_ξ, B_ξ and $\mathscr{F}_{\xi+1}$.

If $A\cap S_\xi$ is d-thick for all $A\in\mathscr{F}_\xi$ we set $A_\xi=S_\xi$; otherwise we set $A_\xi=\omega\backslash S_\xi$. We claim that $A\cap A_\xi$ is d-thick for all $A\in\mathscr{F}_\xi$. If the claim is false then $A_\xi=\omega\backslash S_\xi$ and there are $A,B\in\mathscr{F}_\xi$ such that

$$\text{neither } A\cap S_\xi \text{ nor } B\cap(\omega\backslash S_\xi) \text{ is } d\text{-thick}$$

and hence

$$\text{neither } (A\cap B)\cap S_\xi \text{ nor } (A\cap B)\cap(\omega\backslash S_\xi) \text{ is } d\text{-thick};$$

but then $A\cap B$ is not d-thick, in contradiction to (iv) and (vii).

Let $\langle F_m:m<\omega\rangle$ be a filter base for the filter generated by $\mathscr{F}_\xi\cup\{A_\xi\}$ such that

$$F_{m+1}\subset F_m \quad \text{for } m<\omega$$

and let $\langle\tau_k:k<\omega\rangle$ be a well-ordering of the set $\bigcup_{n<\omega}\omega^n$; we define $\rho:\omega\to\omega$ by the rule

$$\rho(k)=n \quad \text{if } \tau_k\in\omega^n.$$

By the claim proved above we have

$$F_m \text{ is } d\text{-thick for } m<\omega.$$

Hence by Lemma 16.10 for $m,k<\omega$ there is $G_{m,k}\in\mathscr{P}_\omega(F_m)$ such that for every equality $G_{m,k}=\bigcup_{i<\rho(k)}G_i$ there is $i<\rho(k)$ such that

$$(G_i\times G_i)\cap d_{\tau_k(i)}\neq\emptyset.$$

We define

$$B_\xi=\bigcup_{m<\omega}(\bigcup_{k\leq m}G_{m,k}),$$

and we note that since

$$G_{m,k}\subset F_m\subset F_n \quad \text{for } n\leq m<\omega$$

we have $B_\xi\backslash F_n\subset\bigcup_{m<n}\bigcup_{k\leq m}G_{m,k}$ and hence $|B_\xi\backslash F_n|<\omega$ for $n<\omega$.

We define $\mathscr{F}_{\xi+1}$ to be the (countable) filter base generated by $\mathscr{F}_\xi\cup\{A_\xi,B_\xi\}$ and we claim that

$$A \text{ is } d\text{-thick for } A\in\mathscr{F}_{\xi+1}.$$

It is enough to show that

$$F_m\cap B_\xi \text{ is } d\text{-thick for } m<\omega.$$

Let $m,k<\omega$ and let

$$F_m\cap B_\xi=\bigcup_{i<\rho(k)}A_i.$$

We define

$$\bar{m}=\max\{m,\rho(k)\} \quad \text{and}$$
$$G_i=A_i\cap G_{\bar{m},k} \quad \text{for } i<\rho(k).$$

Since $\bigcup_{i<\rho(k)}G_i=G_{\bar{m},k}$ there is $i<\rho(k)$ such that

$$\emptyset\neq(G_i\times G_i)\cap d_{\tau_k(i)}\subset(A_i\times A_i)\cap d_{\tau_k(i)}.$$

The proof that

$$A \text{ is } d\text{-thick for } A\in\mathscr{F}_{\xi+1}$$

is complete.

Finally we define

$$p=\bigcup_{\xi<\omega^+}\mathscr{F}_\xi.$$

It is clear from (ii), (iv) and (v) that p is an ultrafilter on ω, and from (vi) that p is a P-point of $\boldsymbol{\beta}(\omega)\backslash\omega$. Since A is d-thick for $A\in p$ we have

$$(A\times A)\cap d_n\neq\emptyset \quad \text{for } A\in p, \quad n<\omega$$

so for $n<\omega$ there is $q_n\in\boldsymbol{\beta}(\omega\times\omega)$ such that

$$p\times p\subset q_n, \quad \text{and}$$
$$d_n\in q_n.$$

For $n<m<\omega$ we have $d_n \cap d_m = \emptyset$ and hence $q_n \neq q_m$. Hence

$$|\boldsymbol{\phi}^{-1}(\{\langle p,p\rangle\})| \geq \omega$$

by Lemma 7.21(c), as required. The proof is complete.

Remarks. (a) We showed in Corollary 16.2, assuming $\omega^+ = 2^\omega$, that there are P-points of $\boldsymbol{\beta}(\omega)\backslash\omega$ whose type is not minimal in the Rudin-Keisler order of $\tau[\boldsymbol{\beta}(\omega)\backslash\omega]$. We note that Theorem 16.11 gives an alternate proof of this statement. Indeed from Theorem 9.6 for $p \in U(\alpha)$ we have $\tau(p)$ minimal in $\tau[\boldsymbol{\beta}(\alpha)\backslash\alpha]$ if and only if $|\boldsymbol{\phi}^{-1}(\{\langle p,p\rangle\})| = 3$, and hence the ultrafilter p of Theorem 16.11 is not of minimal type in $\tau[\boldsymbol{\beta}(\omega)\backslash\omega]$.

(b) We note that assuming $\omega^+ = 2^\omega$ there are P-points p, q of $\boldsymbol{\beta}(\omega)\backslash\omega$ such that $p \neq q$ and $|\boldsymbol{\phi}^{-1}(\{\langle p,q\rangle\})| = 2^{2^\omega}$. Let p be the ultrafilter defined in Theorem 16.11, let π be a permutation of ω such that $\overline{\pi}(p) \neq p$, set $q = \overline{\pi}(p)$ and define $h: \omega\times\omega \to \omega\times\omega$ by the rule $h(m,n) = \langle m, \pi(n)\rangle$. Then

$$(\pi_0 \circ h)(m,n) = \pi_0(m,n) \quad \text{and} \quad (\pi_1 \circ h)(m,n) = (\pi\circ\pi_1)(m,n)$$

for $\langle m,n\rangle \in \omega\times\omega$,

so for $r \in \boldsymbol{\phi}^{-1}(\{\langle p,p\rangle\})$ we have

$$\overline{\pi}_0(\overline{h}(r)) = \overline{\pi}_0(r) = p \quad \text{and} \quad \overline{\pi}_1(\overline{h}(r)) = \overline{\pi}(\overline{\pi}_1(r)) = \overline{\pi}(p) = q$$

by Lemma 7.21(c). Thus $\overline{h}[\boldsymbol{\phi}^{-1}(\{\langle p,p\rangle\})] \subset \phi^{-1}(\{\langle p,q\rangle\})$, and since $\overline{h}$ is a homeomorphism of $\boldsymbol{\beta}(\omega\times\omega)$ we have $|\boldsymbol{\phi}^{-1}(\{\langle p,q\rangle\})| = 2^{2^\omega}$.

The Rudin-Frolík Order

We define here the Rudin-Frolík partial order $\sqsubseteq$ on $\tau[\boldsymbol{\beta}(\omega)]$, a partial order stronger than the Rudin-Keisler partial order $\preccurlyeq$, and we examine its most elementary properties. Since the properties of the Rudin-Frolík order depend on Lemma 16.15(a) below, which does not appear to generalize to cardinals higher than ω, we restrict the definition of the Rudin-Frolík order to types of ultrafilters on ω.

We recall from § 9 that a subset D of $\boldsymbol{\beta}(\alpha)$ is strongly discrete if there is a family $\{A_d : d \in D\}$ of subsets of α such that

$$A_d \cap A_{d'} = \emptyset \quad \text{for } d, d' \in D,\ d \neq d', \quad \text{and}$$
$$A_d \in d \quad \text{for } d \in D.$$

It is easy to see that a countable subset of $\boldsymbol{\beta}(\omega)$ is strongly discrete if and only if it is discrete.

Definition. The *Rudin-Frolík pre-order* $\sqsubseteq$ on $\boldsymbol{\beta}(\omega)$ is the binary relation given by

$p \sqsubseteq q$ if there is $f:\omega \to \boldsymbol{\beta}(\omega)$ such that $f[\omega]$ is a (strongly) discrete subset of $\boldsymbol{\beta}(\omega)$, and $\bar{f}(p)=q$.

It is clear that

$$\text{if } p \approx p',\ q \approx q' \text{ and } p \sqsubseteq q, \text{ then } p' \sqsubseteq q' \text{ for } p, p', q, q' \in \boldsymbol{\beta}(\omega).$$

The *Rudin-Frolík partial order* (also denoted by $\sqsubseteq$) on $\tau[\boldsymbol{\beta}(\omega)]$ is the binary relation on $\tau[\beta(\omega)]$ defined by

$\sigma \sqsubseteq \tau$ if there are $p, q \in \boldsymbol{\beta}(\omega)$ such that $\tau(p)=\sigma$, $\tau(q)=\tau$, and $p \sqsubseteq q$.

We verify that $\sqsubseteq$ is a well-defined partial order on $\tau[\boldsymbol{\beta}(\omega)]$.

16.12. Lemma. (a) *If* $p, q \in \boldsymbol{\beta}(\omega) \setminus \omega$ *and* $p \sqsubseteq q$ *then* $p \preccurlyeq q$;
(b) $\sqsubseteq$ *is a partial order on* $\tau[\boldsymbol{\beta}(\omega)]$.

Proof. (a) There is a one-to-one function $f:\omega \to \boldsymbol{\beta}(\omega)$ such that

$f[\omega]$ is a (strongly) discrete subset of $\boldsymbol{\beta}(\omega)$, and $\bar{f}(p)=q$.

We show that there is $g \in \omega^\omega$ such that $\bar{g}(q)=p$. Indeed let $\{A_n : n<\omega\}$ be a family of pairwise disjoint subsets of ω such that

$$f(n) \in \mathrm{cl}_{\boldsymbol{\beta}(\omega)} A_n \quad \text{for } n<\omega \quad \text{and}$$
$$\textstyle\bigcup_{n<\omega} A_n = \omega,$$

and define $g:\omega \to \omega$ by the rule

$$g[A_n]=\{n\};$$

since $\bar{g}(f(n))=n$ for $n<\omega$ we have

$$p=\bar{g}(\bar{f}(p))=\bar{g}(q),$$

as required.

(b) Since $\sqsubseteq$ is clearly a well-defined, reflexive relation on $\tau[\boldsymbol{\beta}(\omega)]$, and is anti-symmetric by part (a) and Corollary 9.3, it is enough to prove that $\sqsubseteq$ is transitive on $\boldsymbol{\beta}(\omega)$.

Let $p, q, r \in \boldsymbol{\beta}(\omega)$ with $p \sqsubseteq q$ and $q \sqsubseteq r$; there are one-to-one functions $f, g:\omega \to \boldsymbol{\beta}(\omega)$ such that

$f[\omega]$ and $g[\omega]$ are (strongly) discrete subsets of $\boldsymbol{\beta}(\omega)$, and
$\bar{f}(p)=q, \quad \bar{g}(q)=r.$

We set $h=\bar{g}\circ f$. Then

$$\bar{h}(p)=\bar{g}(\bar{f}(p))=\bar{g}(q)=r,$$

and h is a one-to-one function since both f and the homeomorphic embedding $\overline{g}$ are one-to-one functions. We verify that $h[\omega]$ is a strongly discrete subset of $\boldsymbol{\beta}(\omega)$. There are $\{A_n : n<\omega\}$ and $\{B_n : n<\omega\}$, families of subsets of ω, such that

$$A_n \cap A_m = B_n \cap B_m = \emptyset \quad \text{for } n<m<\omega, \quad \text{and}$$
$$f(n) \in \mathrm{cl}_{\boldsymbol{\beta}(\omega)} A_n \quad \text{and } g(n) \in \mathrm{cl}_{\boldsymbol{\beta}(\omega)} B_n \quad \text{for } n<\omega.$$

We set

$$C_n = \bigcup \{B_k : k \in A_n\} \quad \text{for } n<\omega$$

and we note that $C_n \cap C_m = \emptyset$ for $n<m<\omega$. Further for $k \in A_n$ we have $B_k \subset C_n$ and hence

$$g(k) \in \mathrm{cl}_{\boldsymbol{\beta}(\omega)} B_k \subset \mathrm{cl}_{\boldsymbol{\beta}(\omega)} C_n;$$

thus $g[A_n] \subset \mathrm{cl}_{\boldsymbol{\beta}(\omega)} C_n$ and we have

$$h(n) = \overline{g}(f(n)) \in \overline{g}[\mathrm{cl}_{\boldsymbol{\beta}(\omega)} A_n] \subset \mathrm{cl}_{\boldsymbol{\beta}(\omega)} C_n$$

for $n<\omega$. The proof is complete.

16.13. Lemma. *Let* $p \in \boldsymbol{\beta}(\omega)$. *Then*
(a) $|\{\tau(q) \in \tau[\boldsymbol{\beta}(\omega)] : \tau(q) \sqsubseteq \tau(p)\}| \le 2^\omega$; *and*
(b) $|\{\tau(q) \in \tau[\boldsymbol{\beta}(\omega)] : \tau(p) \sqsubseteq \tau(q)\}| = 2^{2^\omega}$.

Proof. (a) follows from statement (b) preceding Lemma 9.1 and Lemma 16.12(a).

(b) Let $\{A_n : n<\omega\}$ be an ω-partition of ω, let $\langle p_{n,\xi} : \xi < 2^{2^\omega}\rangle$ be a well-ordering of $\hat{A}_n$ such that $p_{n,\xi} \neq p_{n,\zeta}$ for $\xi<\zeta<2^{2^\omega}$, and for $\xi < 2^{2^\omega}$ let $f_\xi : \omega \to \boldsymbol{\beta}(\omega)$ be defined by the rule

$$f_\xi(n) = p_{n,\xi} \quad \text{for } n<\omega.$$

Now let $p \in \boldsymbol{\beta}(\omega)$. It is clear that $p \sqsubseteq \overline{f}_\xi(p)$ for $\xi < 2^{2^\omega}$; further since

$$f_\xi[\omega] \cap f_\zeta[\omega] = \emptyset \quad \text{and}$$
$$f_\xi[\omega] \cup f_\zeta[\omega] \text{ is } C^*\text{-embedded in } \boldsymbol{\beta}(\omega)\backslash\omega$$

for $\xi<\zeta<2^{2^\omega}$

we have $\mathrm{cl}_{\boldsymbol{\beta}(\omega)} f_\xi[\omega] \cap \mathrm{cl}_{\boldsymbol{\beta}(\omega)} f_\zeta[\omega] = \emptyset$ and hence

$$\overline{f}_\xi(p) \neq \overline{f}_\zeta(p) \quad \text{for } \xi<\zeta<2^{2^\omega}.$$

It follows that $|\{q \in \boldsymbol{\beta}(\omega) : p \sqsubseteq q\}| = 2^{2^\omega}$. Since $|\tau(q)| \le 2^\omega$ for $q \in \boldsymbol{\beta}(\omega)$, we have

$$|\{\tau(q) \in \tau[\boldsymbol{\beta}(\omega)] : \tau(p) \sqsubseteq \tau(q)\}| = 2^{2^\omega}.$$

We remark that (in the notation of the paragraph above) we have $p \sqsubset \bar{f}_\xi(p)$ for $\xi < 2^{2^\omega}$. Indeed if there is $\xi < 2^{2^\omega}$ such that $p \approx \bar{f}_\xi(p)$ then there is a permutation π of ω such that

$$(\bar{\pi} \circ \bar{f}_\xi)(p) = \bar{\pi}(\bar{f}_\xi(p)) = p,$$

and since $\bar{\pi} \circ \bar{f}_\xi : \boldsymbol{\beta}(\omega) \to \boldsymbol{\beta}(\omega)$ is a homeomorphic embedding we have

$$\emptyset = \{n < \omega : (\bar{\pi} \circ \bar{f}_\xi)(n) = n\} \in p.$$

Thus $p \not\approx \bar{f}_\xi(p)$, and hence $p \sqsubset \bar{f}_\xi(p)$, for $\xi < 2^{2^\omega}$.

16.14. Lemma. *If $p \in \boldsymbol{\beta}(\omega) \backslash \omega$ then the following statements are equivalent.*

(a) *$\tau(p)$ is minimal in the Rudin-Frolík order of $\tau[\boldsymbol{\beta}(\omega) \backslash \omega]$;*

(b) *there is no countable discrete subset A of $\boldsymbol{\beta}(\omega) \backslash \omega$ such that $p \in (\operatorname{cl} A) \backslash A$.*

Proof. (a)⇒(b). If (b) fails there is a one-to-one function $f : \omega \to \boldsymbol{\beta}(\omega) \backslash \omega$ such that

$$f[\omega] \text{ is discrete and}$$
$$p \in (\operatorname{cl} f[\omega]) \backslash f[\omega].$$

We choose $q \in \bar{f}^{-1}(\{p\})$ and we note that since $q \sqsubseteq p$ and $q \in \boldsymbol{\beta}(\omega) \backslash \omega$ we have $\tau(p) = \tau(q)$ by (a); hence there is a permutation π of ω such that $\bar{\pi}(p) = q$. We define

$$e = \bar{f} \circ \bar{\pi} : \boldsymbol{\beta}(\omega) \to \boldsymbol{\beta}(\omega)$$

and we note that e is a homeomorphic embedding of $\boldsymbol{\beta}(\omega)$ into $\boldsymbol{\beta}(\omega)$ such that

$$e(p) = \bar{f}(\bar{\pi}(p)) = \bar{f}(q) = p$$

(and such that $e[\omega]$ is a strongly discrete subset of $\boldsymbol{\beta}(\omega) \backslash \omega$)). It follows from Corollary 9.17(b) that

$$\{n < \omega : e(n) = n\} \in p,$$

contrary to the fact that

$$e(n) = f(\bar{\pi}(n)) \in \bar{f}[\omega] \subset \boldsymbol{\beta}(\omega) \backslash \omega \quad \text{for } n < \omega.$$

(b)⇒(a). If $q \in \boldsymbol{\beta}(\omega) \backslash \omega$ and $\tau(q) \sqsubset \tau(p)$ then there is a one-to-one function $f : \omega \to \boldsymbol{\beta}(\omega)$ such that

$$f[\omega] \text{ is a (strongly) discrete subset of } \boldsymbol{\beta}(\omega), \text{ and}$$
$$\bar{f}(q) = p.$$

We define

$$A = \{n < \omega : f(n) < \omega\} \quad \text{and } B = \omega \backslash A.$$

If $A \in p$ then it follows easily from Theorem 9.2(b) that

$$p = \bar{f}(q) \approx q$$

and hence $\tau(p) = \tau(q)$, a contradiction. Hence $B \in q$ and $f[B]$ is a countable discrete subset of $\boldsymbol{\beta}(\omega) \setminus \omega$ such that

$$p \in (\operatorname{cl} f[B]) \setminus f[B].$$

16.15. Lemma. *Let X be a space.*

(a) *If A and B are countable subsets of X such that $A \cap \operatorname{cl} B = B \cap \operatorname{cl} A = \emptyset$, then there are disjoint cozero-sets U and V of X such that $A \subset U$ and $B \subset V$.*

(b) *If X is an F-space, $C \subset X$ and $|C| \leq \omega$, then C is C^*-embedded in X.*

Proof. (a) Let $A = \{p_n : n < \omega\}$ and $B = \{q_n : n < \omega\}$. We define families $\{U_n : n < \omega\}$ and $\{V_n : n < \omega\}$ of cozero-sets of X such that

(i) $p_n \in U_n$ and $q_n \in V_n$ for $n < \omega$;

(ii) $\operatorname{cl} U_n \cap (\bigcup_{k<n} \operatorname{cl} V_k) = \emptyset$ and $\operatorname{cl} U_n \cap B = \emptyset$ for $n < \omega$; and

(iii) $\operatorname{cl} V_n \cap (\bigcup_{k \leq n} \operatorname{cl} U_k) = \emptyset$ and $\operatorname{cl} V_n \cap A = \emptyset$ for $n < \omega$.

We proceed by finite recursion. Let U_0 be a cozero-set of X such that $p_0 \in U_0$ and $\operatorname{cl} U_0 \cap B = \emptyset$, and let V_0 be a cozero-set of X such that $q_0 \in V_0$, $\operatorname{cl} V_0 \cap \operatorname{cl} U_0 = \emptyset$ and $\operatorname{cl} V_0 \cap A = \emptyset$. If $0 < n < \omega$ and U_k, V_k have been defined for $k < n$, we choose cozero-sets U_n, V_n of X so that (i), (ii) and (iii) are satisfied. Finally we set

$$U = \bigcup_{n<\omega} U_n \quad \text{and} \quad V = \bigcup_{n<\omega} V_n.$$

Then U and V are cozero-sets of X by Lemma 2.1(a), and it is obvious that $A \subset U$, $B \subset V$, and $U \cap V = \emptyset$.

(b) By Urysohn's extension theorem (Corollary 2.8) it is enough to show that if A and B are subsets of C which are completely separated in C, then A and B are completely separated in X. Since $A \cap \operatorname{cl} B = B \cap \operatorname{cl} A = \emptyset$ there are by part (a) disjoint cozero-sets U and V of X such that $A \subset U$ and $B \subset V$. We define $f : U \cup V \to [0,1]$ by

$$\begin{aligned} f(p) &= 0 \quad \text{for } p \in U \\ &= 1 \quad \text{for } p \in V. \end{aligned}$$

Since $U \cup V$ is a cozero-set and $f \in C^*(U \cup V)$ there is $g \in C^*(X)$ such that $f \subset g$. Since

$$A \subset U \subset g^{-1}(\{0\}) \quad \text{and} \quad B \subset V \subset g^{-1}(\{1\})$$

the sets A and B are completely separated in X.

16.16. Theorem. *Let* $p \in \boldsymbol{\beta}(\omega)$. *The set*

$$P(\tau(p)) = \{\tau(q) : q \in \boldsymbol{\beta}(\omega),\ \tau(q) \sqsubseteq \tau(p)\}$$

is linearly ordered in the Rudin-Frolík order of $\tau[\boldsymbol{\beta}(\omega)]$.

Proof. It is enough to show that if

$$q, r \in \boldsymbol{\beta}(\omega) \quad \text{and} \quad q, r \sqsubseteq p$$

then $q \sqsubseteq r$ or $r \sqsubseteq q$. There are one-to-one functions $f, g : \omega \to \boldsymbol{\beta}(\omega)$ such that

$f[\omega]$ and $g[\omega]$ are (strongly) discrete subsets of $\boldsymbol{\beta}(\omega)$, and

$\bar{f}(q) = p$ and $\bar{g}(r) = p$.

We note that $p \in \operatorname{cl} f[\omega] \cap \operatorname{cl} g[\omega]$ and we claim that

$$p \in \operatorname{cl}(f[\omega] \cap \operatorname{cl} g[\omega]) \cup \operatorname{cl}(g[\omega] \cap \operatorname{cl} f[\omega]).$$

We define

$$A = f[\omega] \setminus \operatorname{cl} g[\omega] \quad \text{and} \quad B = g[\omega] \setminus \operatorname{cl} f[\omega],$$

and we note that by Lemma 16.15(a) there are disjoint cozero-sets U and V of $\boldsymbol{\beta}(\omega)$ such that

$$A \subset U \quad \text{and} \quad B \subset V;$$

it is clear that $\boldsymbol{\beta}(\omega)$ is an F-space, and hence there is $h \in C^*(\boldsymbol{\beta}(\omega))$ such that

$$\begin{aligned} h(x) &= 0 \quad \text{if } x \in U \\ &= 1 \quad \text{if } x \in V. \end{aligned}$$

If the claim is false we have $p \in (\operatorname{cl} A) \cap (\operatorname{cl} B)$ and hence

$$h(p) = 0 \quad \text{and} \quad h(p) = 1;$$

this contradiction establishes the claim.

We assume without loss of generality that

$$p \in \operatorname{cl}(f[\omega] \cap \operatorname{cl} g[\omega])$$

(and we show below that $q \sqsubseteq r$). We assume also, choosing $\{p_n : n < \omega\} \subset \operatorname{cl} g[\omega]$ such that

$\{p_n : n < \omega\}$ is a (strongly) discrete subset of $\boldsymbol{\beta}(\omega)$ and

$\{p_n : n < \omega\} \cap f[\omega] = \emptyset$

and then replacing $f(n)$ by p_n if $f(n) \notin \operatorname{cl} g[\omega]$, that $f[\omega] \subset \operatorname{cl} g[\omega]$. Since $\bar{g} : \boldsymbol{\beta}(\omega) \to \boldsymbol{\beta}(\omega)$ is a homeomorphic embedding the function

$$\bar{g}^{-1} : \bar{g}[\boldsymbol{\beta}(\omega)] \to \boldsymbol{\beta}(\omega)$$

is a homeomorphism. We set

$$k = \bar{g}^{-1} \circ f : \omega \to \boldsymbol{\beta}(\omega).$$

Since $f[\omega]$ is a strongly discrete subset of $\boldsymbol{\beta}(\omega)$ there is $\{A_n : n < \omega\} \subset \mathscr{P}(\omega)$ such that

$$f(n) \in \mathrm{cl}_{\boldsymbol{\beta}(\omega)} A_n \quad \text{for } n < \omega \quad \text{and}$$
$$A_n \cap A_m = \emptyset \quad \text{for } n < m < \omega,$$

and since $(\mathrm{cl}_{\boldsymbol{\beta}(\omega)} A_n) \cap \bar{g}[\boldsymbol{\beta}(\omega)]$ is a non-empty, open-and-closed subset of $\bar{g}[\boldsymbol{\beta}(\omega)]$ there is by Lemma 7.12(b) $\{B_n : n < \omega\} \subset \mathscr{P}(\bar{g}[\omega])$ such that

$$\mathrm{cl}_{\bar{g}[\boldsymbol{\beta}(\omega)]} B_n = (\mathrm{cl}_{\boldsymbol{\beta}(\omega)} A_n) \cap \bar{g}[\boldsymbol{\beta}(\omega)] \quad \text{for } n < \omega;$$

we assume without loss of generality that

$$B_n \cap B_m = \emptyset \quad \text{for } n < m < \omega.$$

Finally we set

$$C_n = g^{-1}(B_n) \quad \text{for } n < \omega.$$

Then

$$C_n \cap C_m = \emptyset \quad \text{for } n < m < \omega,$$

and since

$$k(n) = \bar{g}^{-1}(f(n)) \in \bar{g}^{-1}(\mathrm{cl}_{\bar{g}[\boldsymbol{\beta}(\omega)]} B_n) = \mathrm{cl}_{\boldsymbol{\beta}(\omega)} \bar{g}^{-1}(B_n) = \mathrm{cl}_{\boldsymbol{\beta}(\omega)} C_n \quad \text{for } n < \omega,$$

the function $k : \omega \to \boldsymbol{\beta}(\omega)$ is a one-to-one function such that

$$k[\omega] \text{ is a strongly discrete subset of } \boldsymbol{\beta}(\omega);$$

further $\bar{k}(q) = \bar{g}^{-1}(\bar{f}(q)) = \bar{g}^{-1}(p) = r$. Thus $q \sqsubseteq r$. The proof is complete.

It follows from Theorem 16.16 that $\langle \tau[\boldsymbol{\beta}(\omega)], \sqsubseteq \rangle$ is not directed upwards; we recall from Corollary 10.10(a) that $\langle \tau[\boldsymbol{\beta}(\alpha)], \preccurlyeq \rangle$ is directed upwards for $\alpha \geq \omega$.

Remark. For $p \in \boldsymbol{\beta}(\omega) \setminus \omega$ we thus have the following implications:

$\tau(p)$ is minimal in the Rudin-Keisler order of $\tau[\boldsymbol{\beta}(\omega) \setminus \omega] \Rightarrow p$ is a P-point of $\boldsymbol{\beta}(\omega) \setminus \omega \Rightarrow p$ is not in the boundary of a countable subset of $\boldsymbol{\beta}(\omega) \setminus \omega \Rightarrow p$ is not in the boundary of a countable discrete subset of $\boldsymbol{\beta}(\omega) \setminus \omega \Leftrightarrow \tau(p)$ is minimal in the Rudin-Frolík order of $\tau[\boldsymbol{\beta}(\omega) \setminus \omega]$.

Assuming $\omega^+ = 2^\omega$ none of the direct implications can be reversed. Indeed, the first implication cannot be reversed because of Corollary 16.2; and the second because of Theorem 16.17 below. The fact that the third implication cannot be reversed will not be proved here (cf. Notes).

For use in the following theorem we remark that if p is a P-point of $\boldsymbol{\beta}(\omega)\backslash\omega$ then, assuming $\omega^+=2^\omega$, there is a family $\{A_\eta:\eta<\omega^+\}$ of elements of p such that

$$|A_\zeta\backslash A_\eta|<\omega \quad \text{for } \eta<\zeta<\omega^+,$$
$$|A_\eta\backslash A_\zeta|=\omega \quad \text{for } \eta<\zeta<\omega^+, \quad \text{and}$$
$$\{\hat{A}_\eta:\eta<\omega^+\} \quad \text{is a local base in } \boldsymbol{\beta}(\omega)\backslash\omega \text{ at } p.$$

Indeed let $\langle S_\eta:\eta<\omega^+\rangle$ be a well-ordering of p, set $A_0=S_0$ and recursively, if $0<\zeta<\omega^+$ and A_η has been chosen for $\eta<\zeta$, let B_ζ be an element of p (as is given by Remark 14.17(c)) such that

$$\hat{B}_\zeta\subset(\textstyle\bigcap_{\eta<\zeta}\hat{A}_\eta)\cap\hat{S}_\zeta;$$

there is $A_\zeta\in p$ such that

$$A_\zeta\subset B_\zeta \quad \text{and} \quad |B_\zeta\backslash A_\zeta|=\omega,$$

and the family $\{A_\eta:\eta<\omega^+\}$ is as required.

16.17. Theorem. *Assume $\omega^+=2^\omega$. There is $p\in\boldsymbol{\beta}(\omega)\backslash\omega$ such that*

p is not a P-point of $\boldsymbol{\beta}(\omega)\backslash\omega$ and

p is not in the boundary of any countable subset of $\boldsymbol{\beta}(\omega)\backslash\omega$.

Proof. Let $\langle S_\xi:\xi<\omega^+\rangle$ be a well-ordering of $\mathscr{P}(\omega)$ and let f be a function from ω to ω such that $|f^{-1}(\{n\})|=\omega$ for $n<\omega$. For every non-zero limit ordinal $\xi<\omega^+$ let $\langle\xi(i):i<\omega\rangle$ be an indexing of ξ by ω, i.e., let $\xi(i)$ be such that

$$\xi=\{\eta:\eta<\xi\}=\{\xi(i):i<\omega\};$$

and for $\xi<\omega^+$ set $(\xi+1)(i)=\xi$ for $i<\omega$.

For $\xi,\eta<\omega^+$ and $n<\omega$ we define sets M_ξ and $A(\xi,n,\eta)$, and elements $p_{\xi,n}$ of $\boldsymbol{\beta}(\omega)\backslash\omega$, such that

(i) $M_0=\omega$ and $A(0,n,0)=f^{-1}(\{n\})$ for $n<\omega$;

(ii) if $\xi,\eta<\omega^+$ and $n<\omega$, then $A(\xi,n,\eta)\subset A(0,n,0)$, $|A(\xi,n,\eta)|=\omega$, and $|M_\xi|=\omega$;

(iii) if $\eta<\xi<\omega^+$ and $n<\omega$, then $|A(\xi,n,0)\cap A(\eta,n,\xi+1)|<\omega$;

(iv) if $\xi<\omega^+$ and $n<\omega$, then there is $\zeta<\xi$ such that $|A(\xi,n,0)\backslash A(\zeta,n,\xi)|<\omega$;

(v) if $\eta<\xi<\omega^+$ and $n<\omega$ and $|A(\xi,n,0)\cap A(\eta,n,0)|=\omega$, then $|A(\xi,n,0)\backslash A(\eta,n,0)|<\omega$;

(vi) if $\xi<\omega^+$ and $n<\omega$, then $p_{\xi,n}$ is a P-point of $\boldsymbol{\beta}(\omega)\backslash\omega$;

(vii) if $\xi<\omega^+$ and $n<\omega$, then $\{\hat{A}(\xi,n,\eta):\eta<\omega^+\}$ is a local base in $\boldsymbol{\beta}(\omega)\backslash\omega$ at $p_{\xi,n}$, and

$$|A(\xi,n,\zeta)\backslash A(\xi,n,\eta)|<\omega \quad \text{and} \quad |A(\xi,n,\eta)\backslash A(\xi,n,\zeta)|=\omega \quad \text{for } \eta<\zeta<\omega^+;$$

(viii) if $\xi<\omega^+$ and $M_\xi=\{n_i(\xi):i<\omega\}$ with

$$n_0(\xi)<n_1(\xi)<\cdots<n_i(\xi)<n_{i+1}(\xi)<\cdots,$$

and if $n=n_i(\xi)$, then

$$n\in M_{\xi(0)}\cap M_{\xi(1)}\cap\cdots\cap M_{\xi(i)} \quad \text{and}$$

$$|A(\xi(0),n,0)\cap A(\xi(1),n,0)\cap\cdots\cap A(\xi(i),n,0)\cap A(\xi,n,0)|=\omega.$$

We proceed by transfinite recursion, using (i) to define M_0 and $A(0,n,0)$ for $n<\omega$. Since $|A(0,n,0)|=\omega$ there are by Theorem 6.17, Corollary 14.15, and the remark preceding this theorem a P-point $p_{0,n}$ of $\boldsymbol{\beta}(\omega)\backslash\omega$ such that $A(0,n,0)\in p_{0,n}$ and a local base $\{A(0,n,\eta):\eta<\omega^+\}$ in $\boldsymbol{\beta}(\omega)\backslash\omega$ at $p_{0,n}$ satisfying (vii) (with $\xi=0$).

Now let ξ be a non-zero limit ordinal such that $\xi<\omega^+$. We define M_ξ, $p_{\xi,n}$ for $n<\omega$, and $A(\xi,n,\eta)$ for $\eta<\omega^+$.

Choose $n_0(\xi)\in M_{\xi(0)}$ and if $i<\omega$ and $\{n_j(\xi):j\leq i\}$ have been chosen with

$$n_j(\xi)\in M_{\xi(0)}\cap M_{\xi(1)}\cap\cdots\cap M_{\xi(j)}$$

and with $n_{j-1}(\xi)<n_j(\xi)$ for $0<j\leq i$, let

$$n_{i+1}(\xi)\in M_{\xi(0)}\cap M_{\xi(1)}\cap\cdots\cap M_{\xi(i+1)}$$

with $n_i(\xi)<n_{i-1}(\xi)$ and with

$$|A(\xi(0),n_{i+1}(\xi),0)\cap A(\xi(1),n_{i+1}(\xi),0)\cap\cdots\cap A(\xi(i+1),n_{i+1}(\xi),0)|=\omega\,.$$

This defines $n_i(\xi)$ for $i<\omega$. We set

$$M_\xi=\{n_i(\xi):i<\omega\}\,.$$

If $n\in\omega\backslash M_\xi$ we set

$$A(\xi,n,0)=A(0,n,\xi)\backslash A(0,n,\xi+1)$$

and if $n\in M_\xi$, say $n=n_i(\xi)$, we set

$$\eta=\max\{\xi(j):j\leq i\}$$

and

$$A(\xi,n,0)=A(\eta,n,\xi)\backslash A(\eta,n,\xi+1)\,.$$

Since in each case $|A(\xi,n,0)|=\omega$ by (vii) (with η and ξ in place of ξ and ζ respectively) there are by Theorem 6.17 and Corollary 14.15 a P-point $p_{\xi,n}$ of $\boldsymbol{\beta}(\omega)\backslash\omega$ such that $A(\xi,n,0)\in p_{\xi,n}$ and a local base $\{A(\xi,n,\eta):\eta<\omega^+\}$ in $\boldsymbol{\beta}(\omega)\backslash\omega$ at $p_{\xi,n}$ satisfying (vii).

Now let $\xi<\omega^+$. We define $M_{\xi+1}$ and $p_{\xi+1,n}$ for $n<\omega$, and $A(\xi+1,n,\eta)$ for $\eta<\omega^+$.

For $n\in M_\xi$ we have either

$$|S_\xi\cap(A(\xi,n,\xi)\setminus A(\xi,n,\xi+1))|=\omega$$

or

$$|(\omega\setminus S_\xi)\cap(A(\xi,n,\xi)\setminus A(\xi,n,\xi+1)|=\omega\,.$$

We set

$$M'_\xi=\{n\in M_\xi:|S_\xi\cap(A(\xi,n,\xi)\setminus A(\xi,n,\xi+1))|=\omega\}\,,$$

and we set

$$\begin{aligned} M_{\xi+1}&=M'_\xi \qquad\text{if } |M'_\xi|=\omega\,,\\ &=M_\xi\setminus M'_\xi \quad \text{if } |M'_\xi|<\omega\,; \end{aligned}$$

Then in any event we have

$$M_{\xi+1}\subset M_\xi \quad\text{and}\quad |M_{\xi+1}|=\omega\,.$$

We define

$$A(\xi+1,n,0)=S_\xi\cap(A(\xi,n,\xi)\setminus A(\xi,n,\xi+1)) \quad\text{if } n\in M_{\xi+1} \quad\text{and}$$

$M_{\xi+1}=M'_\xi$,

$$A(\xi+1,n,0)=(\omega\setminus S_\xi)\cap(A(\xi,n,\xi)\setminus A(\xi,n,\xi+1)) \quad\text{if } n\in M_{\xi+1}$$

and $M_{\xi+1}=M_\xi\setminus M'_\xi$, and

$$A(\xi+1,n,0)=A(\xi,n,\xi)\setminus A(\xi,n,\xi+1) \quad\text{if } n\notin M_{\xi+1}\,;$$

then in any event we have

$$A(\xi+1,n,0)\subset A(\xi,n,\xi) \quad\text{and}\quad |A(\xi+1,n,0)|=\omega\,.$$

Again there are by Theorem 6.17 and Corollary 14.15 a P-point $p_{\xi+1,n}$ of $\boldsymbol{\beta}(\omega)\setminus\omega$ such that $A(\xi+1,n,0)\in p_{\xi+1,n}$ and a local base $\{A(\xi+1,n,\eta):\eta<\omega^+\}$ in $\boldsymbol{\beta}(\omega)\setminus\omega$ at $p_{\xi+1,n}$ satisfying (vii).

We have defined M_ξ, $p_{\xi,n}$, and $A(\xi,n,\eta)$ for each non-zero limit ordinal $\xi<\omega^+$ and $n<\omega$ and $\eta<\omega^+$, and we have defined $M_{\xi+1}$, $p_{\xi+1,n}$ and $A(\xi+1,n,\eta)$ for $\xi<\omega^+$, $n<\omega$ and $\eta<\omega^+$. The verification of the inductive hypotheses, tedious but elementary, is omitted here and left to the reader.

We define

$$E(\xi)=\bigcup\{A(\xi,n,0):n\in M_\xi\}$$

for $\xi<\omega^+$, and we set

$$p=\{E\subset\omega:\text{there is } \xi<\omega^+ \text{ such that } E(\xi)\subset E\}\,.$$

for $k<\omega$ and $q_m\in Q(1)$. (Such an ordinal ζ_{i+1} exists because q_m is a P-point, $q_m\neq p_{\eta,k}$ for $\eta\leq\zeta_i$ and $k<\omega$, and

$$|\zeta_i\times\omega\times Q(1)|\leq\omega.)$$

Now set $\xi=\sup\{\zeta_i:i<\omega\}$ and $U(1)=\hat{E}(\xi)$. Then $p\in U(1)$. Suppose there is $q_m\in Q(1)$ such that $q_m\in U(1)$. There is $k<\omega$ such that $f^{-1}(\{k\})\in q_m$, and then

$$f^{-1}(\{k\})\cap E(\xi)\in q_m\,.$$

Since $f^{-1}(\{k\})\cap E(\xi)=\emptyset$ if $k\notin M_\xi$ we have $k\in M_\xi$, so

$$A(\xi,k,0)=f^{-1}(\{k\})\cap E(\xi)\in q_m\,.$$

By (iv) there is $i<\omega$ such that

$$|A(\xi,k,0)\setminus A(\zeta_i,k,\xi)|<\omega\,,$$

so $A(\zeta_i,k,\xi)\in q_m$. But $\zeta_{i+1}<\xi$, so

$$|A(\zeta_i,k,\xi)\setminus A(\zeta_i,k,\zeta_{i+1})|<\omega$$

by (vii) and therefore $A(\zeta_i,k,\zeta_{i+1})\in q_m$. This contradiction shows that $q_m\notin U(1)$. We conclude that $U(1)\cap Q(1)=\emptyset$.

If $Q(2)=\emptyset$ we may take $U(2)=\boldsymbol{\beta}(\omega)\setminus\omega$, so we assume $Q(2)\neq\emptyset$. Since $q_m\neq p$ for $m<\omega$ there is $\xi_m<\omega^+$ such that $E(\xi_m)\notin q_m$; we choose the sequence $\{\xi_m:m<\omega\}$ so that $\xi_m\neq\xi_{m'}$ for $m<m'<\omega$.

Let ξ be a limit ordinal such that $\xi<\omega^+$ and $\xi>\xi_m$ for $m<\omega$. Since $\xi=\{\xi(i):i<\omega\}$ there is $i_m<\omega$ such that $\xi_m=\xi(i_m)$. For $q_m\in Q(2)$ we set

$$W_m=(E(\xi)\setminus E(\xi(i_m)))\setminus f^{-1}(\{0,1,\ldots,n_{i_m}(\xi)\})\,,$$

and we define

$$W=E(\xi)\setminus\bigcup\{W_m:q_m\in U(2)\}$$

and $U(2)=\hat{W}$. We must show that $U(2)\cap Q(2)=\emptyset$ and that $p\in U(2)$. Suppose there is $q_m\in Q(2)$ such that $q_m\in U(2)$. Then $E(\xi)\in q_m$, so

$$E(\xi)\setminus E(\xi_m)=E(\xi)\setminus E(\xi(i_m))\in q_m\,.$$

Since $f^{-1}(\{k\})\notin q_m$ for $k<\omega$ we have $W_m\in q_m$ and hence $W\notin q_m$. Thus $q_m\notin U(2)$. We conclude that $U(2)\cap Q(2)=\emptyset$.

To show $p\in U(2)$ it will suffice to show that

$$\bigcup\{W_m:q_m\in Q(2)\}\notin p\,.$$

We show first that if $n<\omega$ then

(a) $|W_m\cap f^{-1}(\{n\})|<\omega$ for $q_m\in Q(2)$, and

(b) $|\{m<\omega:q_m\in Q(2)$ and $W_m\cap f^{-1}(\{n\})\neq\emptyset\}|<\omega$.

If $n\notin M_\xi$ then

$$W_m\cap f^{-1}(\{n\})\subset E(\xi)\cap f^{-1}(\{n\})=\emptyset$$

for $q_m \in Q(2)$ so (a) and (b) hold. We assume in what follows that $n \in M_\xi$, say $n = n_i(\xi)$.

If $n = n_i(\xi) \le n_{i_m}(\xi)$ then

$$W_m \cap f^{-1}(\{n\}) = W_m \cap f^{-1}(\{n_i(\xi)\}) = \emptyset .$$

If $n = n_i(\xi) > n_{i_m}(\xi)$ then $i > i_m$ and since by (viii)

$$n \in M_{\xi(0)} \cap M_{\xi(1)} \cap \cdots \cap M_{\xi(i)} \quad \text{and}$$
$$|A(\xi(0), n, 0) \cap A(\xi(1), n, 0) \cap \cdots \cap A(\xi(i), n, 0) \cap A(\xi, n, 0)| = \omega$$

we have in particular $n \in M_{\xi(i_m)}$ and

$$|A(\xi, n, 0) \cap A(\xi(i_m), n, 0)| = \omega .$$

Since $\xi(i_m) < \xi$ we have

$$|A(\xi, n, 0) \setminus A(\xi(i_m), n, 0)| < \omega$$

by (v), so

$$|W_m \cap f^{-1}(\{n\})| = |A(\xi, n, 0) \setminus (\xi(i_m), n, 0)| < \omega .$$

Thus statement (a) is proved.

To prove (b) (with $n = n_i(\xi) \in M_\xi$) we note first that there are at most finitely many $m < \omega$ such that $q_m \in Q(2)$ and $n_{i_m}(\xi) < n$. For if $m < m' < \omega$ then

$$\xi(i_m) = \xi_m \neq \xi_{m'} = \xi(i_{m'}) ,$$

so $i_m \neq i_{m'}$ and $n_{i_m}(\xi) \neq n_{i_{m'}}(\xi)$. Let $\bar{m}$ be the largest $m < \omega$ such that $q_m \in Q(2)$ and $n_{i_m}(\xi) < n$ (or, if no such m exist, set $\bar{m} = 0$). Then for $m > \bar{m}$ with $q_m \in Q(2)$ we have $n_{i_m}(\xi) \ge n$ and then

$$W_m \cap f^{-1}(\{n\}) = \emptyset .$$

Thus statement (b) is proved.

Since (a) and (b) hold for $n < \omega$ we have

$$|[\bigcup \{W_m : q_m \in Q(2)\}] \cap f^{-1}(\{n\})| < \omega$$

for $n < \omega$. From (*) it follows that

$$\bigcup \{W_m : q_m \in Q(2)\} \notin p ,$$

so $p \in U(2)$. The proof is complete.

Non-Homogeneity of Certain Spaces

16.18. Theorem. *Let X be a compact space, $\mathscr{B}$ a family of non-empty open subsets of X, $\alpha \geq \omega$, and $\phi: \alpha \to \mathscr{B}$, such that*

$$|\mathscr{B}|^{\alpha} < 2^{2^{\alpha}},$$
$$h[B] \in \mathscr{B} \quad \text{for every auto-homeomorphism } h \text{ of } X \text{ and } B \in \mathscr{B},$$
$$\phi(\xi) \cap \phi(\zeta) = \emptyset \quad \text{for } \xi < \zeta < \alpha, \quad \text{and}$$
$$\textstyle\bigcup_{\xi<\alpha} \phi(\xi) \text{ is } C^*\text{-embedded in } X.$$

Then X is not point-homogeneous.

Proof. We set

$$\Phi = \{\psi \in \mathscr{B}^{\alpha} : \psi(\xi) \cap \psi(\zeta) = \emptyset \text{ for } \xi < \zeta < \alpha \text{ and } \textstyle\bigcup_{\xi<\alpha} \psi(\xi) \text{ is } C^*\text{-embedded in } X\},$$

and for $p \in X$ and $\psi \in \Phi$ we define

$$\begin{aligned} q(p,\psi) &= \{\{\xi < \alpha : \psi(\xi) \cap U \neq \emptyset\} : U \text{ is a neighborhood of } p\} \\ &\qquad\qquad \text{if } p \in \mathrm{cl}_X(\textstyle\bigcup_{\xi<\alpha} \psi(\xi)) \\ &= 0 \qquad\quad \text{if } p \notin \mathrm{cl}_X(\textstyle\bigcup_{\xi<\alpha} \psi(\xi)). \end{aligned}$$

We claim that $q(p,\psi) \in \boldsymbol{\beta}(\alpha)$ for $p \in X$, $\psi \in \Phi$.

If $p \notin \mathrm{cl}_X(\bigcup_{\xi<\alpha} \psi(\xi))$ then $q(p,\psi) = 0 \in \alpha \subset \boldsymbol{\beta}(\alpha)$, so we assume that $p \in \mathrm{cl}_X(\bigcup_{\xi<\alpha} \psi(\xi))$. It is enough to prove

(i) $q(p,\psi)$ has the finite intersection property, and

(ii) if $C \subset \alpha$ then $C \in q(p,\psi)$ or $\alpha \backslash C \in q(p,\psi)$.

If $A, B \in q(p,\psi)$ there are neighborhoods U and V of p such that

$$A = \{\xi < \alpha : \psi(\xi) \cap U \neq \emptyset\} \quad \text{and} \quad B = \{\xi < \alpha : \psi(\xi) \cap V \neq \emptyset\},$$

and since $U \cap V$ is a neighborhood of p and $p \in \mathrm{cl}_X(\bigcup_{\xi<\alpha} \psi(\xi))$ we have

$$A \cap B \supset \{\xi < \alpha : \psi(\xi) \cap (U \cap V) \neq \emptyset\} \neq \emptyset;$$

statement (i) follows. To prove (ii) let $C \subset \alpha$ and set

$$N_C = \bigcup \{\psi(\xi) : \xi \in C\} \quad \text{and} \quad N_{\alpha\backslash C} = \bigcup \{\psi(\xi) : \xi \in \alpha \backslash C\}.$$

Since N_C and $N_{\alpha\backslash C}$ are disjoint, open-and-closed subsets of $\bigcup_{\xi<\alpha} \psi(\xi)$ and $\bigcup_{\xi<\alpha} \psi(\xi)$ is C^*-embedded in X, there is $f \in C^*(X)$ such that

$$f(x) = 0 \quad \text{for } x \in N_C \quad \text{and}$$
$$f(x) = 1 \quad \text{for } x \in N_{\alpha\backslash C}.$$

We set

$$U = \{x \in X : f(x) \neq 1\} \quad \text{and} \quad V = \{x \in X : f(x) \neq 0\}.$$

Since U or V is a neighborhood of p and

$$C=\{\xi<\alpha:\psi(\xi)\cap U\neq\emptyset\} \quad\text{and}\quad \alpha\backslash C=\{\xi<\alpha:\psi(\xi)\cap V\neq\emptyset\},$$

we have $C\in q(p,\psi)$ or $\alpha\backslash C\in q(p,\psi)$. The proof of statement (ii) is complete.

Now for $p\in X$ we define $A(p)=\{q(p,\psi):\psi\in\Phi\}$. We note that if $p,p'\in X$ and h is an auto-homeomorphism of X such that $h(p)=p'$, then for every neighborhood U of p and $\xi<\alpha$ such that $\psi(\xi)\cap U\neq\emptyset$ we have

$$h[U] \text{ is a neighborhood of } p' \text{ and}$$
$$h[\psi(\xi)]\cap h[U]\neq\emptyset;$$

it follows that $A(p)\subset A(p')$. Similarly (since h^{-1} is an auto-homeomorphism of X) we have $A(p')\subset A(p)$. Thus if $p,p'\in X$ and there is an auto-homeomorphism h of X such that $h(p)=p'$, then $A(p)=A(p')$.

To complete the proof it is enough to fix $p\in X$ and to show that there is $p'\in X$ such that $A(p)\neq A(p')$. Since

$$|A(p)|\leq|\Phi|\leq|\mathscr{B}^{\alpha}|<2^{2^{\alpha}}=|\boldsymbol{\beta}(\alpha)|$$

by Corollary 7.4 there is $q\in\boldsymbol{\beta}(\alpha)\backslash A(p)$. For $\xi<\alpha$ we choose $f(\xi)\in\phi(\xi)$ and we note that $f[\alpha]$ is C^*-embedded in X. We define

$$p'=\bar{f}(q)$$

and we claim that $q=q(p',\phi)$. Indeed for every neighborhood U of p' we have

$$\{\xi<\alpha:f(\xi)\in U\}\subset\{\xi<\alpha:\phi(\xi)\cap U\neq\emptyset\},$$

and $\{\xi<\alpha:f(\xi)\in U\}\in q$ since $\bar{f}$ is an embedding of $\boldsymbol{\beta}(\alpha)$ into X; thus every element of the ultrafilter $q(p',\phi)$ contains an element of the ultrafilter q and we have $q(p',\phi)=q$ by Lemma 2.3(c). It follows that

$$q\in A(p')\backslash A(p),$$

and the proof is complete.

Before the theorem just proved can be used to show that a given compact space X is not point-homogeneous, it is necessary to find a family $\mathscr{B}$ which is not very large and which is invariant under auto-homeomorphisms of X. This may be difficult, but in practice there are certain natural candidates for $\mathscr{B}$. One is the family of non-empty cozero-sets of X and another, in case X is zero-dimensional, is the family $\mathfrak{B}(X)\backslash\{\emptyset\}$ of non-empty open-and-closed subsets of X. Since the number of cozero-sets of X does not exceed $|C(X)|$, the following result is useful in this connection.

We recall from § 2 and § 3 that $\boldsymbol{pw}(X)$ and $\boldsymbol{S}(X)$ denote respectively the pseudoweight and the Souslin number of the space X.

16.19. Lemma. *If X is an infinite space, then*

$$|C(X)| \leq (\boldsymbol{pw}(X))^{\underset{\smile}{\boldsymbol{S}(X)}}.$$

Proof. Let $\mathscr{B}$ be a pseudobase for X such that $|\mathscr{B}| = \boldsymbol{pw}(X)$ and set

$$\Phi = \{\mathscr{S} : \mathscr{S} \subset \mathscr{B} \text{ and } \mathscr{S} \text{ is a cellular family of } X\}.$$

We note that $\emptyset \in \Phi$. Since $\boldsymbol{S}(X) \leq (\boldsymbol{pw}(X))^+$ and $|\mathscr{S}| < \boldsymbol{S}(X)$ for $\mathscr{S} \in \Phi$ we have

$$|\Phi| \leq (\boldsymbol{pw}(X))^{\underset{\smile}{\boldsymbol{S}(X)}}$$

by Lemma 1.22(c). Since $\boldsymbol{S}(X)$ is a regular, uncountable cardinal by Theorem 3.5 we have

$$|\Phi|^\omega \leq ((\boldsymbol{pw}(X))^{\underset{\smile}{\boldsymbol{S}(X)}})^\omega = (\boldsymbol{pw}(X))^{\underset{\smile}{\boldsymbol{S}(X)}}$$

by Lemma 1.26. To complete the proof it is enough to show that $|C(X)| \leq |\Phi|^\omega$.

Let $\mathbb{Q}$ denote the set of rational numbers and for

$$\langle f, r \rangle \in C(X) \times \mathbb{Q}$$

let $\mathscr{S}\langle f, r \rangle$ be an element of Φ such that

$$\bigcup \mathscr{S}\langle f, r \rangle \text{ is dense in } \{x \in X : f(x) < r\}.$$

(We note that $\mathscr{S}\langle f, r \rangle = \emptyset$ if $r \leq f[X]$.) We define

$$\phi : C(X) \to \Phi^{\mathbb{Q}}$$

by the rule

$$\phi(f)_r = \mathscr{S}\langle f, r \rangle$$

and we claim that ϕ is a one-to-one function. Indeed if $f, g \in C(X)$ and there is $p \in X$ such that $f(p) < g(p)$ then there is $r \in \mathbb{Q}$ such that $f(p) < r < g(p)$ and we have

$$p \in \mathrm{cl}_X(\bigcup \mathscr{S}\langle f, r \rangle) \setminus \mathrm{cl}_X(\bigcup \mathscr{S}\langle g, r \rangle).$$

Thus ϕ is a one-to-one function and we have

$$|C(X)| \leq |\Phi^{\mathbb{Q}}| = |\Phi|^\omega.$$

16.20. Corollary. *Let $\alpha \geq \omega$ and let X be a compact space such that*

$$(\boldsymbol{pw}(X))^{\underset{\smile}{\boldsymbol{S}(X)}} < 2^{2^\alpha},$$

and there is a family $\{\phi(\xi):\xi<\alpha\}$ *of non-empty cozero-sets of* X *such that* $\phi(\xi)\bigcap\phi(\zeta)=\emptyset$ *for* $\xi<\zeta<\alpha$ *and* $\bigcup_{\xi<\alpha}\phi(\xi)$ *is* C^**-embedded in* X.

Then X *is not point-homogeneous.*

Proof. Let $\mathscr{B}$ be the family of non-empty cozero-sets of X. Then

$$|\mathscr{B}|\leq|C(X)|\leq(\boldsymbol{pw}(X))^{\underset{\smile}{\boldsymbol{S}(X)}}$$

by Lemma 16.19 and since $\alpha<\boldsymbol{S}(X)$ we have

$$|\mathscr{B}|^{\alpha}\leq((\boldsymbol{pw}(X))^{\underset{\smile}{\boldsymbol{S}(X)}})^{\alpha}=(\boldsymbol{pw}(X))^{\underset{\smile}{\boldsymbol{S}(X)}}<2^{2^{\alpha}}$$

by Lemma 1.26. Since $h[B]\in\mathscr{B}$ for every auto-homeomorphism h of X and $B\in\mathscr{B}$, the space X is not point-homogeneous by Theorem 16.18.

Remark. For F_{α^+}-spaces (defined in § 14) the last condition of the corollary above may be replaced by the condition $\alpha<\boldsymbol{S}(X)$ and we have the following statement:

Let X *be a compact space and suppose there is an infinite cardinal* α *such that*

X is an F_{α^+}-space (for example, an extremally disconnected space),
$(\boldsymbol{pw}(X))^{\underset{\smile}{\boldsymbol{S}(X)}}<2^{2^{\alpha}}$, *and*
$\alpha<\boldsymbol{S}(X)$.

Then X *is not point-homogeneous.*

16.21. Theorem. *Let* X *be an infinite, compact* F*-space such that* $\boldsymbol{w}(X)\leq 2^{\omega}$. *Then* X *is not point-homogeneous.*

Proof. Let $\mathscr{B}$ be the family of non-empty cozero-sets of X. By Lemma 3.3(b) there is $\{\phi(n):n<\omega\}\subset\mathscr{B}$ such that

$$\phi(n)\cap\phi(m)=\emptyset\quad\text{for } n<m<\omega,$$

and since $\bigcup_{n<\omega}\phi(n)$ is a cozero-set of X by Lemma 2.1(a) this set is C^*-embedded in X. Since $|\mathscr{B}|\leq 2^{\omega}$ by Lemma 15.19, the statement follows from Theorem 16.18.

It follows from Lemma 14.16 that $\boldsymbol{\beta}(\omega)\backslash\omega$ is a (compact) F-space. It follows from Theorem 16.21 that $\boldsymbol{\beta}(\omega)\backslash\omega$ is not point-homogeneous. (We remark that, assuming $\omega^+=2^{\omega}$, the non-homogeneity of $\boldsymbol{\beta}(\omega)\backslash\omega$ is a immediate consequence of the existence of P-points; cf. Corollary 6.18.) We generalize this statement.

16.22. Theorem. *If* $\alpha \geq \omega$, *then the space* $U(\alpha)$ *is not point-homogeneous.*

Proof. We set $\mathfrak{B} = \mathfrak{B}(U(\alpha))$ and for $A \in \mathscr{P}(\alpha)$ we set $\hat{A} = \mathrm{cl}_{\boldsymbol{\beta}(\alpha)} A \cap U(\alpha)$. From Lemma 7.12(b) for every $B \in \mathfrak{B}$ there is $A \in \mathscr{P}(\alpha)$ such that $B = \hat{A}$, and hence we have

$$|\mathfrak{B}|^{\alpha} \leq |\mathscr{P}(\alpha)|^{\alpha} = (2^{\alpha})^{\alpha} = 2^{\alpha} < 2^{2^{\alpha}};$$

since $h[B] \in \mathfrak{B}$ for every auto-homeomorphism h of $U(\alpha)$ and $B \in \mathfrak{B}$, it is (by Theorem 16.18) enough to prove that there is a family $\{A_{\xi}: \xi < \alpha\} \subset \mathscr{P}(\alpha)$ such that

(i) $\hat{A}_{\xi} \neq \emptyset$ for $\xi < \alpha$,

(ii) $\hat{A}_{\xi} \cap \hat{A}_{\zeta} = \emptyset$ for $\xi < \zeta < \alpha$, and

(iii) $\bigcup_{\xi < \alpha} \hat{A}_{\xi}$ is C^*-embedded in $U(\alpha)$.

Let $\{A_{\xi}: \xi < \alpha\}$ be an α-partition of α, *i.e.*, a family of subsets of α such that

$$|A_{\xi}| = \alpha \quad \text{for } \xi < \alpha,$$
$$A_{\xi} \cap A_{\zeta} = \emptyset \quad \text{for } \xi < \zeta < \alpha, \quad \text{and}$$
$$\textstyle\bigcup_{\xi < \alpha} A_{\xi} = \alpha.$$

Properties (i) and (ii) follow from parts (c) and (d)(i) respectively of Lemma 7.12. To verify (iii) let $f \in C^*(\bigcup_{\xi < \alpha} \hat{A}_{\xi})$, for $\xi < \alpha$ let g_{ξ} be an element of $C^*(A_{\xi} \cup \hat{A}_{\xi})$ such that

$$f \,|\, \hat{A}_{\xi} \subset g_{\xi} \quad \text{and}$$
$$\sup\{|g_{\xi}(\eta)|: \eta \in A_{\xi}\} = \sup\{|f(p)|: p \in \hat{A}_{\xi}\},$$

set $g = \bigcup_{\xi < \alpha} g_{\xi}$ and denote by $\overline{g}$ the (unique) element of $C^*(\boldsymbol{\beta}(\alpha))$ such that $g \subset \overline{g}$. Then

$$f \subset \overline{g} \,|\, U(\alpha) \in C^*(U(\alpha)).$$

The proof is complete.

In Theorem 16.25 below we show that if X is not pseudocompact then $\boldsymbol{\beta} X \backslash X$ is not point-homogeneous.

For a space K and $p \in K$ we set

$$B_K(p) = \{q \in \boldsymbol{\beta}(\omega) \backslash \omega : \text{there is a homeomorphism } h \text{ from } \boldsymbol{\beta}(\omega) \text{ into } K \text{ such that } h(q) = p\}.$$

16.23. Lemma. *Let X be a space, let N be a discrete, C-embedded subset of X such that $|N| = \omega$, and set $K = (\mathrm{cl}_{\boldsymbol{\beta} X} N) \backslash X$.*

(a) *If h is a homeomorphism of $\boldsymbol{\beta}(\omega)$ into $\boldsymbol{\beta} X \backslash X$, then $K \cap \mathrm{cl}_{\boldsymbol{\beta} X}(h[\omega] \backslash K) = \emptyset$.*

(b) *$B_K(p) = B_{\boldsymbol{\beta} X \backslash X}(p)$ for $p \in K$.*

Proof. (a) Since N is closed in X we have

$$(\mathrm{cl}_{\beta X} N)\cap(h[\omega]\setminus K)=\mathrm{cl}_{\beta X}(h[\omega]\setminus K)\cap N=\emptyset$$

and hence by Lemma 16.15(a) there are disjoint cozero-sets U and V of βX such that $N\subset U$ and $h[\omega]\setminus K\subset V$. There is $f\in C^*(\beta X)$ such that

$$f(p)>0 \quad \text{for } p\in U \quad \text{and}$$
$$f(p)=0 \quad \text{for } p\in \beta X\setminus U\,;$$

we denote by g the function defined on N by

$$g(p)=1/f(p) \quad \text{for } p\in N$$

and we note that since $g\in C(N)$ and N is C-embedded in X there is $h\in C(X)$ such that $g\subset h$. We set $j=h\cdot(f|X)$ and for $p\in X$ we define

$$\begin{aligned} k(p)&=1 && \text{if } j(p)>1\\ &=j(p) && \text{if } 0\le j(p)\le 1\\ &=0 && \text{if } j(p)<0\,. \end{aligned}$$

Then $k\in C^*(X)$ and we have

$$k(p)=j(p)=1 \quad \text{for } p\in N \quad \text{and}$$
$$k(p)=j(p)=0 \quad \text{for } p\in V\cap X\,.$$

Let $\overline{k}$ denote the Stone extension of k. Then

$$\begin{aligned} \overline{k}(p)&=1 && \text{for } p\in \mathrm{cl}_{\beta X} N\\ &=0 && \text{for } p\in \mathrm{cl}_{\beta X}(V\cap X)\,. \end{aligned}$$

Since $K\subset \mathrm{cl}_{\beta X} N$ and $(h[\omega]\setminus K)\subset \mathrm{cl}_{\beta X}(V\cap X)$ we have

$$K\cap \mathrm{cl}_{\beta X}(h[\omega]\setminus K)=\emptyset\,.$$

(b) Clearly $B_K(p)\subset B_{\beta X\setminus X}(p)$ for $p\in K$. Let $q\in B_{\beta X\setminus X}(p)$ with $p\in K$ and let h be a homeomorphism of $\beta(\omega)$ into $\beta X\setminus X$ such that $h(q)=p$. Clearly there is $g:\omega\to K$ such that

g is a one-to-one function,

$g[\omega]$ is discrete, and

$g(n)=h(n)$ for $n<\omega$ and $h(n)\in K$.

Since K is an F-space (by Lemma 14.16) the set $g[\omega]$ is C^*-embedded in K (by Lemma 16.15(b)) and hence the Stone extension $\overline{g}:\beta(\omega)\to K$ is a homeomorphism of $\beta(\omega)$ into K. We set

$$A=\{n<\omega: h(n)\in K\}\,.$$

Since $p \notin \mathrm{cl}_{\beta X} h[\omega \setminus A]$ (by part (a)) we have $p \in \mathrm{cl}_{\beta X} h[A]$ and hence $q \in \mathrm{cl}_{\beta(\omega)} A$. Since $g(n) = h(n)$ for $n \in A$ we have

$$\bar{g}(q) = h(q) = p\,,$$

and therefore $q \in B_K(p)$. The proof that $B_K(p) = B_{\beta X \setminus X}(p)$ is complete.

16.24. Lemma. *If* $p \in \beta(\omega) \setminus \omega$, *then* $|B_{\beta(\omega) \setminus \omega}(p)| \leq 2^\omega$.

Proof. We set $\hat{A} = (\mathrm{cl}_{\beta(\omega)} A) \setminus \omega$ for $A \subset \omega$. Let $\mathscr{D}$ denote the set of partitions

$$d = \{d_n : n < \omega\}$$

of ω for which there is a homeomorphism h from $\beta(\omega)$ into $\beta(\omega) \setminus \omega$ such that

$$h(n) \in \hat{d}_n \quad \text{for } n < \omega\,,$$
$$p \notin h[\omega]\,, \quad \text{and}$$
$$p \in \mathrm{cl}_{\beta(\omega)} h[\omega]\,,$$

and for $d \in \mathscr{D}$ let h_d be such a homeomorphism. For $d \in \mathscr{D}$ there is a (unique) element q_d of $\beta(\omega) \setminus \omega$ such that $h_d(q_d) = p$. Since $|\mathscr{D}| \leq 2^\omega$, to complete the proof it is enough to show that if $q \in B_{\beta(\omega) \setminus \omega}(p)$ then there is $d \in \mathscr{D}$ such that $q = q_d$.

Let $q \in B_{\beta(\omega) \setminus \omega}(p)$ and let h be a homeomorphism of $\beta(\omega)$ into $\beta(\omega) \setminus \omega$ such that $h(q) = p$. Since $p \notin h[\omega]$ and h is a one-to-one function there is a partition $d = \{d_n : n < \omega\}$ of ω such that $h(n) \in d_n$; since $p \in \mathrm{cl}_{\beta(\omega)} h[\omega]$ we have $d \in \mathscr{D}$. We set

$$A = \{n < \omega : h(n) = h_d(n)\}$$

and we claim that $p \in \mathrm{cl}_{\beta(\omega)} h[A]$. If $n \in \omega \setminus A$ then

$$d_n \cap (h[\omega \setminus A] \cup h_d[\omega \setminus A]) = \{h(n), h_d(n)\}\,,$$

and thus the set $h[\omega \setminus A] \cup h_d[\omega \setminus A]$ is discrete. It follows from Lemma 16.15(b) that there is $f \in C^*(\beta(\omega) \setminus \omega)$ such that $f(r) = 0$ for $r \in h[\omega \setminus A]$ and $f(r) = 1$ for $r \in h_d[\omega \setminus A]$. If

$$p \notin \mathrm{cl}_{\beta(\omega)} h[A] = \mathrm{cl}_{\beta(\omega)} h_d[A]$$

then

$$p \in \mathrm{cl}_{\beta(\omega)} h[\omega \setminus A] \cap \mathrm{cl}_{\beta(\omega)} h_d[\omega \setminus A]$$

and hence $f(p) = 0$ and $f(p) = 1$. This contradiction shows that $p \in \mathrm{cl}_{\beta(\omega)} h[A]$; hence $q \in \mathrm{cl}_{\beta(\omega)} A$. Since $h(n) = h_d(n)$ for $n \in A$ we have

$$h_d(q) = h(q) = p$$

and thus $q = q_d$. The proof is complete.

16.25. Theorem. *If X is a non-pseudocompact space, then $\boldsymbol{\beta} X\backslash X$ is not point-homogeneous.*

Proof. By Theorem 15.15(c)⇔(d) there is a discrete, C-embedded subset N of X such that $|N|=\omega$; we set $K=(\mathrm{cl}_{\boldsymbol{\beta} X} N)\backslash X$. We note that if $p,p'\in\boldsymbol{\beta} X\backslash X$ and there is an auto-homeomorphism k of $\boldsymbol{\beta} X\backslash X$ such that $k(p)=p'$, then $B_{\boldsymbol{\beta} X\backslash X}(p)=B_{\boldsymbol{\beta} X\backslash X}(p')$. Indeed if $q\in B_{\boldsymbol{\beta} X\backslash X}(p)$ and h is a homeomorphism of $\boldsymbol{\beta}(\omega)$ into $\boldsymbol{\beta} X\backslash X$ such that $h(q)=p$, then $k\circ h(q)=p'$; thus $B_{\boldsymbol{\beta} X\backslash X}(p)\subset B_{\boldsymbol{\beta} X\backslash X}(p')$, and the reverse inclusion is proved similarly.

Now let $p\in K$. Since K is homeomorphic with $\boldsymbol{\beta}(\omega)\backslash\omega$ we have $|B_K(p)|\leq 2^{\omega}$ by Lemma 16.24 and hence there is $q\in(\boldsymbol{\beta}(\omega)\backslash\omega)\backslash K$. Let h be a homeomorphism of $\boldsymbol{\beta}(\omega)$ into K and set $p'=h(q)$. Then $q\in B_K(p')$ and from Lemma 16.23(b) we have

$$B_{\boldsymbol{\beta} X\backslash X}(p)=B_K(p)\neq B_K(p')=B_{\boldsymbol{\beta} X\backslash X}(p')\,.$$

The proof is complete.

Notes for § 16

Theorem 16.1 appears in M. E. Rudin [71] (III.C). Booth [69] (Theorem 1.11) proves Corollary 16.2. Theorem 16.3 is due to Blass [70] (Theorem 20.2, Corollary 20.3). Using similar methods, Blass also proves that $|\boldsymbol{\phi}^{-1}(\{\langle p,q\rangle\})|$ can be any finite non-zero cardinal (for suitable p, q) (Blass [70] (Corollary 20.15)). That $|\boldsymbol{\phi}^{-1}(\{\langle p,q\rangle\})|$ can also have the value $2^{2^{\omega}}$ was first noted by Blass [70] (Corollary 20.15). Assuming CH the stronger statement given in Theorem 16.11 (together with Lemma 16.10) is due to Hindman [73]. Lemma 16.5 is due to Blass [70] (Proposition 16.9), and Theorem 16.8 appears in Blass [70] (§ 21). Other results on the Rudin-Keisler order on $\tau[\boldsymbol{\beta}(\omega)]$, using CH or Martin's axiom, have been given by Blass [70], [73]. Various properties of ultrafilters on ω in terms of partitions, similar to the properties of a P-point and of a Rudin-Keisler minimal ultrafilter, have been studied by Pitt [71].

Pitt [71] (Theorem 2.12) and Mathias [72] prove, assuming $\omega^{+}=2^{\omega}$, that there is $p\in\boldsymbol{\beta}(\omega)\backslash\omega$ such that $\{q\in\boldsymbol{\beta}(\omega)\backslash\omega\colon q\preccurlyeq p$ and q is a P-point$\}=\emptyset$.

Frolík [67b] implicitly has defined the Rudin-Frolík partial (pre-) order (where $p\sqsubseteq q$ if, in his terminology, *p produces q*) and he established Lemma 16.13 there. The verification that $\sqsubseteq$ is a partial order on $\tau[\boldsymbol{\beta}(\omega)]$ is given in M. E. Rudin [71] (II.A). Lemma 16.15(a) is proved in Gillman-Jerison [60] (Problem 3B.4), and Theorem 16.16 is proved in M. E. Rudin [71] (II.B), using M. E. Rudin [66] (Lemma 2). The long Theorem 16.17 was announced by Kunen [70b] and proved, by a method that is different from Kunen's, in M. E. Rudin [71] (IV. C. 2.); the proof in the text follows that of M. E. Rudin. Kunen [70b] has also announced, assuming CH, the existence of $p \in \boldsymbol{\beta}(\omega) \setminus \omega$ such that p is in the boundary of a countable subset of $\boldsymbol{\beta}(\omega) \setminus \omega$, but p is not in the boundary of a countable discrete subset of $\boldsymbol{\beta}(\omega) \setminus \omega$.

The thesis of Booth [69], [70] contains results on the Rudin-Frolík order. Thus, it is proved that $\langle \tau[\boldsymbol{\beta}(\omega)], \sqsubseteq \rangle$ is not well-founded (Booth [69] (Theorem 2.12)), and in fact contains a chain of order type $(\omega^+)^*$ (*i.e.* ω^+ with its usual order reversed), and a chain order-isomorphic to the real numbers (Booth [69] (Corollaries 2.27 and 2.28 respectively)). Both of these results are consequences of certain embeddings of the ultraproduct $\langle \omega, \varepsilon \rangle^\omega / p$ into $\langle \tau[\boldsymbol{\beta}(\omega)], \sqsubseteq \rangle$ (Booth [69] (Theorem 2.26)). Booth [69] (Theorem 2.43) contains also the following result due to Kunen: $\langle \tau[\boldsymbol{\beta}(\omega)], \sqsubseteq \rangle^\omega / p$ is order-isomorphic to $\{\tau(q) \colon \tau(p) \sqsubseteq \tau(q)\}$.

Louveau [72] (especially § 3) also studies the Rudin-Frolík order, together with a classification of ultrafilters on ω that has properties analogous to the Rudin-Frolík order. Assuming CH he proves that there is an ultrafilter p on ω such that $\{\tau(p) \colon \tau(q) \sqsubset \tau(p)\}$ is order-isomorphic to ω^* (*i.e.* ω with its usual order reversed) (Louveau [72] (Théorème 32, Corollaire 34)). Solomon [73] has independently proved the weaker result: assuming CH there is $p \in \boldsymbol{\beta}(\omega) \setminus \omega$ such that $\{\tau(q) \colon \tau(q) \sqsubseteq \tau(p)\}$ has cardinality ω. Cf. also Steiner-Steiner [71].

Using the continuum hypothesis, W. Rudin [56] proved the existence of P-points in $\boldsymbol{\beta}(\omega) \setminus \omega$, thus proving that $\boldsymbol{\beta}(\omega) \setminus \omega$ is not point-homogeneous; he showed also (Theorem 4.5 and page 633), assuming the continuum hypothesis, that if X is locally compact and not pseudocompact, then $\boldsymbol{\beta} X \setminus X$ is not point-homogeneous. The method of W. Rudin was extended by Isiwata [57] who proved, assuming the continuum hypothesis, that if X is not pseudocompact then $\boldsymbol{\beta} X \setminus X$ is not point-homogeneous.

Frolík [67a], [67b] introduced the set of types of ultrafilters of ω and, via his "summation" procedure and "producing" relation, the Rudin-Frolík order on $\mathbb{T}(\boldsymbol{\beta}(\omega))$. With these techniques Frolík [67a], [67b] proved, without assuming the continuum hypothesis, that $\boldsymbol{\beta}(\omega) \setminus \omega$ is not point-homogeneous. Lemma 16.24 is 1.8 of Frolík [67b], and Lemma 16.23 (a) and Theorem 16.25 are from Frolík [67c] (Lemma and Theorem 1 respectively).

Efimov [68] (Theorem 1) proved that if X is an infinite, compact, extremally disconnected space such that $\boldsymbol{S}(X) = \alpha^+$ and $\boldsymbol{pw}(X) \leq 2^\alpha$, then X is not point-homogeneous; Corollary 16.20 and the statement following it improve this result. Arhangel'skiĭ [67] proved Theorem 16.21 for extremally disconnected spaces X. For related results cf. Frolík [67d].

The statement that $U(\alpha)$ is not point-homogeneous is proved, assuming $\alpha^+ = 2^\alpha$, in Negrepontis [69c]; the proof given here of Theorem 16.22 is new. Lemma 16.19 improves (since it deals with the Souslin number instead of the cellular number) one of the inequalities given by Comfort and Hager [70].

Bibliography

Adler, Andrew
[71] The cardinality of ultrapowers—an example. Proc. Amer. Math. Soc. **28**, 311—312 (1971).

Alexandrov, P. and Urysohn, P.
[27] Über nulldimensionale Punktmengen. Math. Ann. **98**, 89—106 (1927).

Alling, Norman L.
[61] A characterization of abelian η_α-groups in terms of their natural valuation. Proc. Nat. Acad. Sci. U.S.A. **47**, 711—713 (1961).
[62] On the existence of real-closed fields that are η_α-sets of power $\aleph_\alpha$. Trans. Amer. Math. Soc. **103**, 341—352 (1962).

Arhangel'skiĭ, A. V.
[67] An extremally disconnected bicompactum of weight $\mathfrak{c}$ is inhomogeneous. Dokl. Akad. Nauk SSSR **175**, 751—754 (1967) [Russian]. (English translation: Soviet Math. Dokl. **8**, 897—900 (1967).)

Artin, Emil and Schreier, Otto
[26] Algebraische Konstruktion reeller Körper. Abh. Math. Sem. Univ. Hamburg **5**, 85—99 (1926).

Ax, James
[67] Solving diophantine problems modulo every prime. Ann. of Math. (2) **85**, 161—183 (1967).
[68] The elementary theory of finite fields. Ann. of Math. (2) **88**, 239—271 (1968).

Ax, James and Kochen, Simon
[65a] Diophantine problems over local fields I. Amer. J. Math. **87**, 605—630 (1965).
[65b] Diophantine problems over local fields II. A complete set of axioms for p-adic number theory. Amer. J. Math. **87**, 631—648 (1965).
[66] Diophantine problems over local fields III. Decidable fields. Ann. of Math. (2) **83**, 437—456 (1966).

Bachmann, Heinz
[67] Transfinite Zahlen. 2. Aufl. Ergebnisse der Mathematik und ihrer Grenzgebiete, Bd. 1. Berlin-Heidelberg-New York: Springer 1967.

Baer, R.
[40] Abelian groups that are direct summands of every containing abelian group. Bull. Amer. Math. Soc. **46**, 800—806 (1940).

Balcar, B. and Vopěnka, P.
[72] On systems of almost disjoint sets. Bull. Acad. Polon. Sci. Sér. Sci. Math. Astronom. Phys. **20**, 421—424 (1972).

Banach, S.
[29] Sur les fonctionnelles linéaires I, II. Studia Math. **1**, 211—216; 223—239 (1929).

Banach, Stefan and Kuratowski, Casimir
[29] Sur une généralization du problème de la mesure. Fund. Math. **14**, 127—131 (1929).

Baumgartner, James E.
[70] Results and independence proofs in combinatorial set theory. Doctoral dissertation, University of California (Berkeley) 1970.

Bell, J. L. and Slomson, A. B.
[69] Models and ultraproducts: An introduction. Amsterdam-London: North-Holland 1969.

Benda, Miroslav
[72] On reduced products and filters. Ann. Math. Logic **4**, 1—29 (1972).
[*] On saturated reduced products, manuscript.

Blass, Andreas Raphael
[70] Orderings of ultrafilters. Doctoral dissertation, Harvard University 1970.
[73] The Rudin-Keisler ordering of *P*-points. Trans. Amer. Math. Soc. **179**, 145—166 (1973).

Bockstein, M.
[48] Un théorème de séparabilité pour les produits topologiques. Fund. Math. **35**, 242—246 (1948).

Booth, David D.
[69] Countably indexed ultrafilters. Doctoral dissertation, University of Wisconsin 1969.
[70] Ultrafilters on a countable set. Ann. Math. Logic **2**, 1—24 (1970).

Bourbaki, N. and Dieudonné, J.
[39] Note de teratopologie II. Revue Scientifique **77**, 180—181 (1939).

Brouwer, L. E. J.
[17] On linear inner limiting sets. Amsterdam Akad. Wetensch. Proc. Sect. Sci. **20**, 1191 (1917).
[24] Beweis, daß jede volle Funktion gleichmäßig stetig ist. Amsterdam Akad. Wetensch. Proc. Sect. Sci. **27**, 189—193 (1924).

Bukovský, Lev
[68] ∇-model and distributivity in Boolean algebras. Comment. Math. Univ. Carolinae **9**, 595—612 (1968).

Całczyńska-Karłowicz, M. [= Karłowicz] (see also R. Engelking —)
[64] Theorem on families of finte sets. Bull. Acad. Polon. Sci. Sér. Sci. Math. Astronom. Phys. **12**, 87—89 (1964).

Cantor, Georg
[78] Ein Beitrag zur Mannigfaltigkeitslehre. J. Reine Angew. Math. (Crelle) **84**, 242—258 (1878).
[83] Über unendliche, lineare Punktmannigfaltigkeiten, 5. Math. Ann. **21**, 545—591 (1883).
[90] Über eine elementare Frage der Mannigfaltigkeitslehre. Jber. Deutsch. Math.-Verein. **1**, 75—78 (1890—91).
[95] Beiträge zur Begründung der transfiniten Mengen. Math. Ann. **46**, 481—512 (1895).
[62] Gesammelte Abhandlungen mathematischen und philosophischen Inhalts. Herausgegeben von Ernst Zermelo. Hildesheim: Georg Olms Verlagsbuchhandlung 1962.

Cartan, H.
[37a] Théorie des filtres. C. R. Acad. Sci. Paris **205**, 595—598 (1937).
[37b] Filtres et ultrafiltres. C. R. Acad. Sci. Paris **205**, 777—779 (1937).

Čech, Eduard
[37] On bicompact spaces. Ann. of Math. (2) **38**, 823—844 (1937).

Chang, C. C.
[67a] Descendingly incomplete ultrafilters. Trans. Amer. Math. Soc. **126**, 108—118 (1967).
[67b] Ultraproducts and other methods of constructing models. In: Sets, Models, and Recursion Theory. Proc. Summer School Math. Logic and Tenth Logic Colloq., Leicester, 1965, edited by J. Crossley, pp. 85—121. Amsterdam-London: North-Holland 1967.

Chang, Chen Chung and Keisler, H. Jerome
[62] Applications of ultraproducts of pairs of cardinals to the theory of models. Pacific J. Math. **12**, 835—845 (1962).
[66] Continuous Model Theory. Annals of Mathematics Studies, no. 58. Princeton, N. J.: Princeton University Press 1966.
[73] Model Theory. Amsterdam-London: North-Holland 1973.

Chang, C. C. and Morel, Anne C.
[58] On closure under direct product. J. Symbolic Logic **23**, 149—154 (1958).

Choquet, Gustave
[68a] Construction d'ultrafiltres sur $\mathbb{N}$. Bull. Sci. Math. (2) **92**, 41—48 (1968).
[68b] Deux classes remarquables d'ultrafiltres sur $\mathbb{N}$. Bull. Sci. Math. (2) **92**, 143—153 (1968).

Cohen, Paul J.
[63] The independence of the continuum hypothesis I. Proc. Nat. Acad. Sci. U.S.A. **50**, 1143—1148 (1963).
[64] The independence of the continuum hypothesis II. Proc. Nat. Acad. Sci. U.S.A. **51**, 105—110 (1964).
[66] Set Theory and the Continuum Hypothesis. New York: Benjamin 1966.

Comfort, W. W. and Hager, Anthony W.
[70] Estimates for the number of real-valued continuous functions. Trans. Amer. Math. Soc. **150**, 619—631 (1970).

Comfort, W. Wistar and Hindman, Neil B.
[73] Almost disjoint refining families for uniform ultrafilters. Notices Amer. Math. Soc. **20**, A-644 (1973).

Comfort, W. W. and Negrepontis, S.
[68] Homeomorphs of three subspaces of $\beta\mathbb{N}\setminus\mathbb{N}$. Math. Z. **107**, 53—58 (1968).
[70] Some topological properties associated with measurable cardinals, Fund. Math. **69**, 191—205 (1970).
[72a] On families of large oscillation. Fund. Math. **75**, 275—290 (1972).
[72b] Continuous functions on products with strong topologies. In: General Topology and Its Relations to Modern Analysis and Algebra III. Proc. Third Prague Topological Sympos., Prague, 1971, edited by J. Novák, pp. 89—92. Prague: Academia 1972.
[*] Topological products and intersection systems, manuscript.

Corson, H. H.
[59] Normality in subsets of product spaces. Amer. J. Math. **81**, 785—796 (1959).

Corson, H. H. and Isbell, J. R.
[60] Some properties of strong uniformities. Quart. J. Math. Oxford Ser. (2) **11**, 17—33 (1960).

Craig, William
[57] Linear Reasoning. A new form of the Herbrand-Gentzen theorem. J. Symbolic Logic **22**, 250—268 (1957).

Čudnovskiĭ, G. V. and Čudnovskiĭ, D. V.
[71] Regular and descending incomplete ultrafilters. Dokl. Akad. Nauk SSSR **198**, 779—782 (1971) [Russian]. (English translation: Soviet Math. Dokl. **12**, 901 – 905 (1971).)

Cuesta, N.
[47] Notas sobre unos trabajos de Sierpiński. Rev. Mat. Hisp.-Amer. **4**, 130—131 (1947); **9**, 132 (1949).

Daigneault, Aubert
[64] On automorphisms of polyadic algebras. Trans. Amer. Math. Soc. **112**, 84—130 (1964).

Davies, Roy O.
[67] An intersection theorem of Erdős and Rado. Proc. Cambridge Philos. Soc. **63**, 995—996 (1967).

Dickmann, M. A.
[70] Model theory of infinitary languages. Vols. I, II, Lecture Notes Series, No. 20. Matematisk Institut, Aarhus Universitet, Aarhus 1970.

Dieudonné, J. (see also N. Bourbaki)
[48] Sur le théorème de Lebesgue-Nikodym, III. Ann. Univ. Grenoble. Sect. Sci. Math. Phys. (N.S.) **23**, 25—53 (1948).
[51] Sur le théorème de Lebesgue-Nikodym, IV. J. Indian Math. Soc. (N.S.) **15**, 77—86 (1951).

du Bois-Reymond, P.
[77] Über die Paradoxen des Infinitärcalcüls. Math. Ann. **11**, 149—167 (1877).

Dwinger, Ph. and Yaqub, F. M.
[63] Generalized free products of Boolean algebras with an amalgamated subalgebra. Nederl. Akad. Wetensch. Proc. Ser. **A 66** = Indag. Math. **25**, 225—231 (1963).

Easton, William Bigelow
[64] Powers of regular cardinals. Doctoral dissertation, Princeton University 1964.
[70] Powers of the regular cardinals. Ann. Math. Logic **1**, 139—178 (1970).

Efimov, B.
[68] Extremally disconnected bicompacta of continuum π-weight. Dokl. Akad. Nauk SSSR **183**, 511—514 (1968) [Russian]. (English translation: Soviet Math. Dokl. **9**, 1404—1407 (1968).)

Eklof, P. C.
[73] The structure of ultraproducts of abelian groups. Pacific J. Math. **47**, 67—79 (1973).

Ellentuck, Erik and Rucker, R. V. B.
[72] Martin's axiom and saturated models. Proc. Amer. Math. Soc. **34**, 243—249 (1972).

Engelking, R.
[65] Cartesian products and dyadic spaces. Fund. Math. **57**, 287—304 (1965).
[72] Hausdorff's gaps and limits and compactifications. In: Theory of Sets and Topology, edited by Asser *et al.*, pp. 89—94. Berlin: VEB Deutscher Verlag der Wissenschaften 1972.

Engelking, R. and Karłowicz, M.
[65] Some theorems of set theory and their topological consequences. Fund. Math. **57**, 275—285 (1965).

Erdős, P.
[42] Some set-theoretical properties of graphs. Univ. Nac. Tucumán. Revista A. **3**, 363—367 (1942).
[50] Some remarks on set theory. Proc. Amer. Math. Soc. **1**, 133—137 (1950).
[62] Review of Kurepa [59b], [39]; Math. Reviews **23**A, 444 (1962).

Erdős, P., Gillman, L., and Henriksen, M.
[55] An isomorphism theorem for real-closed fields. Ann. of Math. (2) **61**, 542—554 (1955).

Erdős, P. and Hajnal, A.
[58] On the structure of set-mappings. Acta Math. Acad. Sci. Hungar. **9**, 111—131 (1958).
[62] Some remarks concerning our paper "On the structure of set-mappings".—Non-existence of a two-valued σ-measure for the first uncountable inaccessible cardinal. Acta Math. Acad. Sci. Hungar. **13**, 223—226 (1962).
[71] Unsolved problems in set theory. In: Axiomatic Set Theory. Proc. Sympos. Pure Math., Vol. XIII, Part I, Univ. California, Los Angeles, 1967, edited by D. S. Scott, pp. 17—48. Providence, R. I.: American Mathematical Society 1971.

Erdős, P., Hajnal, A., and Rado, R.
[65] Partition relations for cardinal numbers. Acta Math. Acad. Sci. Hungar. **16**, 93—196 (1965).

Erdős, Paul and Hechler, Stephen H.
[*] On maximal almost-disjoint families over singular cardinals, manuscript.

Erdős, P. and Rado, R.
[56] A partition calculus in set theory. Bull. Amer. Math. Soc. **62**, 427—489 (1956).
[60] Intersection theorems for systems of sets. J. London Math. Soc. **35**, 85—90 (1960).
[69] Intersection theorems for systems of sets (II). J. London Math. Soc. **44**, 467—479 (1969).

Erdős, P. and Tarski, Alfred
[43] On families of mutually exclusive sets. Ann. of Math. (2) **44**, 315—329 (1943).
[61] On some problems involving inaccessible cardinals. In: Essays on the Foundations of Mathematics, edited by Y. Bar-Hillel, pp. 50—82. Jerusalem: The Magnes Press 1961.

Erdős, P. and Ulam, S.
[68] On equations with sets as unknowns. Proc. Nat. Acad. Sci. U.S.A. **60**, 1189—1195 (1968).

Eršov, Ju. L.
[65] On the elementary theory of maximal normed fields. Dokl. Akad. Nauk SSSR **165**, 21—23 (1965) [Russian]. (English translation: Soviet Math. Dokl. **6**, 1390—1393 (1965).)

Esenin-Vol'pin, A. S.
[49] On the existence of a universal bicompactum of arbitrary weight. Dokl. Akad. Nauk SSSR **68**, 649—652 (1949) [Russian].

Fichtenholz, G. and Kantorovitch, L.
[35] Sur les opérations linéaires dans l'espace des fonctions bornées. Studia Math. **5**, 69—98 (1935).

Fine, N. J. and Gillman, L.
[60] Extension of continuous functions in $\beta\mathbb{N}$. Bull. Amer. Math. Soc. **66**, 376—381 (1960).
[62] Remote points in $\beta\mathbb{R}$. Proc. Amer. Math. Soc. **13**, 29—36 (1962).

Fleischer, Isidore
[66] A note on universal homogeneous models. Math. Scand. **19**, 183—184 (1966).

Fraenkel, Abraham A.
[22] Zu den Grundlagen der Cantor-Zermeloschen Mengenlehre. Math. Ann. **86**, 230—237 (1922).
[53] Abstract Set Theory. Amsterdam-London: North-Holland 1953.

Fraïssé, Roland Jean
[54] Sur l'extension aux relations de quelques propriétés des ordres. Ann. Sci. École Norm. Sup. (3) **71**, 363—388 (1954).

Frayne, T., Morel, A. C., and Scott, D. S.
[62] Reduced direct product. Fund. Math. **51**, 195—228 (1962); **53**, 117 (1963).

Frolík, Zdeněk
[67a] Types of ultrafilters on countable sets. In: General Topology and its Relations to Modern Analysis and Algebra II. Proc. Second Prague Topological Sympos., Prague, 1966, edited by J. Novák, pp. 142—143. Prague: Academia Prague 1967.
[67b] Sums of ultrafilters. Bull. Amer. Math. Soc. **73**, 87—91 (1967).
[67c] Non-homogeneity of $\beta P \setminus P$. Comment. Math. Univ. Carolinae **8**, 705—709 (1967).
[67d] Homogeneity problems for extremally disconnected spaces. Comment. Math. Univ. Carolinae **8**, 757—763 (1967).
[68a] Fixed points of maps of $\beta\mathbb{N}$. Bull. Amer. Math. Soc. **74**, 187—191 (1968).
[68b] Fixed points of maps of extremally disconnected spaces and complete Boolean algebras. Bull. Acad. Polon. Sci. Sér. Sci. Math. Astronom. Phys. **16**, 269—275 (1968).

Fuhrken, Gebhard
[65] Languages with added quantifier "there exist at least $\aleph_\alpha$". In: The Theory of Models. Proc. 1963 Internat. Sympos. Berkeley, edited by Addison *et al.*, pp. 121—131. Amsterdam-London: North-Holland 1965.

Gaifman, H.
[64] Infinite Boolean polynomials I. Fund. Math. **54**, 229—250 (1964); **57**, 117 (1965).

Galvin, Fred
[67] Reduced products, Horn sentences, and decision problems. Bull. Amer. Math. Soc. **73**, 59—64 (1967).
[70] Horn sentences. Ann. Math. Logic **1**, 389—422 (1970).

Gillman, L. (see also P. Erdős, —, M. Henriksen; N. J. Fine —)
[55] Remarque sur les ensembles η_α. C. R. Acad. Sci. Paris **241**, 12—13 (1955).
[56] Some remarks on η_α-sets. Fund. Math. **43**, 77—82 (1956).

Gillman, Leonard and Henriksen, Melvin
[56] Rings of continuous functions in which every finitely generated ideal is principal. Trans. Amer. Math. Soc. **82**, 366—391 (1956).

Gillman, Leonard and Jerison, Meyer
[60] Rings of Continuous Functions. Princeton: Van Nostrand 1960.

Gleason, Andrew M.
[58] Projective topological spaces. Illinois J. Math. **2**, 482—489 (1958).

Gödel, Kurt
[38] The consistency of the axiom of choice and of the generalized continuum hypothesis. Proc. Nat. Acad. Sci. U.S.A. **24**, 556—557 (1938).
[39] Consistency-proof for the generalized continuum-hypothesis. Proc. Nat. Acad. Sci. U.S.A. **25**, 220—224 (1939).
[40] The Consistency of the Continuum Hypothesis. Annals of Mathematics Studies, no. 3. Princeton: Princeton University Press 1940.
[47] What is Cantor's continuum problem? Amer. Math. Monthly **54**, 515—525 (1947).

Goodner, Dwight B.
[50] Projections in normed linear spaces. Trans. Amer. Math. Soc. **69**, 89—108 (1950).

Hadamard, J.
[94] Sur les caractères de convergence des séries à termes positifs et sur les fonctions indéfiniment croissantes. Acta Math. **18**, 319 — 336 (1894); 421.

Hager, Anthony W. (see also W. W. Comfort —)
[71] The projective resolution of a compact space. Proc. Amer. Math. Soc. **28**, 262—266 (1971).

Hahn, H.
[27] Über lineare Gleichungssysteme in linearen Räumen. J. Reine Angew. Math. **157**, 214—229 (1927).

Hájek, Petr
[71] Sets, semisets, models. In: Axiomatic Set Theory. Proc. Sympos. Pure Math., Vol. XIII, Part I, Univ. California, Los Angeles, 1967, edited by D. S. Scott, pp. 67—81. Providence, R. I.: American Mathematical Society 1971.

Hajnal, A. (see also P. Erdős —; P. Erdős, —, R. Rado)
[61] Proof of a conjecture of S. Ruziewicz. Fund. Math. **50**, 123—128 (1961).
[64] Remarks on a theorem by W. P. Hanf. Fund. Math. **54**, 109—113 (1964).
[69] Ulam matrices for inaccessible cardinals. Bull. Acad. Polon. Sci. Sér. Math. Astronom. Phys. **17**, 683—688 (1969).

Hajnal, A. and Juhász, I.
[72] On disjoint representation of ultrafilters. In: Theory of Sets and Topology, edited by Asser *et al.*, pp. 215—219. Berlin: VEB Deutscher Verlag der Wissenschaften 1972.

Hales, A. W.
[64] On the non-existence of free complete Boolean algebras. Fund. Math. **54**, 45—66 (1964).

Halmos, Paul R.
[59] The representation of monadic Boolean algebras. Duke Math. J. **26**, 447—454 (1959).
[68] Permutations of sequences and the Schröder-Bernstein theorem. Proc. Amer. Math. Soc. **19**, 509—510 (1968).

Hanf, W. P.
[63] Some fundamental problems concerning languages with infinitely long expressions. Doctoral dissertation, University of California (Berkeley) 1963.
[64a] Incompactness in languages with infinitely long expressions. Fund. Math. **53**, 309—324 (1964).
[64b] On a problem of Erdős and Tarski. Fund. Math. **53**, 325—334 (1964).

Hanf, W. P. and Scott, Dana
[61] Classifying inaccessible cardinals. Notices Amer. Math. Soc. **8**, 445 (1961).

Haratomi, Keitaro
[31] Über höherstufige Separabilität und Kompaktheit (erster Teil). Jap. J. Math. **8**, 113—141 (1931).

Hardy, G. H.
[10] Orders of infinity: The "Infinitärcalcül" of Paul du Bois-Reymond. Cambridge Tracts in Mathematics & Mathematical Physics, no. 12. Cambridge: Cambridge University Press 1910.

Hausdorff, Felix
[04] Der Potenzbegriff in der Mengenlehre. Jber. Deutsch. Math.-Verein. **13**, 569—571 (1904).
[08] Grundzüge einer Theorie der geordneten Mengen. Math. Ann. **65**, 435—505 (1908).
[14] Grundzüge der Mengenlehre. Leipzig: Veit 1914.
[36a] Über zwei Sätze von G. Fichtenholz und L. Kantorovitch. Studia Math **6**, 18—19 (1936).
[36b] Summen von $\aleph_1$ Mengen. Fund. Math. **26**, 241—255 (1936).

Hechler, Stephen H. (see also Paul Erdős —)
[71] Classifying almost-disjoint families with applications to $\beta\mathbb{N}\setminus\mathbb{N}$. Israel J. Math. **10**, 413—432 (1971).
[72] Short complete nested sequences in $\beta\mathbb{N}\setminus\mathbb{N}$ and small almost-disjoint families. General Topology and App. **2**, 139—149 (1972).

Hedrlín, Z.
[66] An application of Ramsey's theorem to the topological products. Bull. Acad. Polon. Sci. Sér. Sci. Math. Astronom. Phys. **14**, 25—26 (1966).

Henriksen, Melvin (see Leonard Gillman —)

Hewitt, Edwin
[43] A problem of set-theoretic topology. Duke Math. J. **10**, 309—333 (1943).
[46] A remark on density characters. Bull. Amer. Math. Soc. **52**, 641—643 (1946).
[48] Rings of real-valued continuous functions I. Trans. Amer. Math. Soc. **64**, 54—99 (1948).

Hindman, Neil (see also W. Wistar Comfort —)
[69] On the existence of $\mathfrak{c}$-points in $\beta\mathbb{N}\setminus\mathbb{N}$. Proc. Amer. Math. Soc. **21**, 277—280 (1969).
[73] Preimages of points under the natural map from $\beta(\mathbb{N}\times\mathbb{N})$ to $\beta\mathbb{N}\times\beta\mathbb{N}$. Proc. Amer. Math. Soc. **37**, 603—608 (1973).

Hirschfeld, Joram (see Moshé Machover —)

Hoffman, Kenneth and Ramsay, Arlan
[65] Algebras of bounded sequences. Pacific J. Math. **15**, 1239—1248 (1965).

Hrbáček, Karel
[67] A note on generalized Souslin's problem. Comment. Math. Univ. Carolinae **8**, 307—310 (1967).

Hung, Henry Hin-Lai
[72] The amalgamation property for G-metric spaces and homeomorphs of the space $(2^\alpha)_\alpha$. Doctoral dissertation, McGill University 1972.

Hung, H. H. and Negrepontis, S.
[73] Spaces homeomorphic to $(2^\alpha)_\alpha$. Bull. Amer. Math. Soc. **79**, 143—146 (1973).
[74] Spaces homeomorphic to $(2^\alpha)_\alpha$ II. Trans. Amer. Math. Soc. **188**, 1—30 (1974).

Ionescu Tulcea, A. and Ionescu Tulcea, C.
[69] Topics in the Theory of Lifting. Ergebnisse der Mathematik und ihrer Grenzgebiete, Bd. 48. Berlin-Heidelberg-New York: Springer 1969.

Isbell, J. R. (see H. H. Corson —)

Isiwata, Takesi
[57] A generalization of Rudin's theorem for the homogeneity problem. Sci. Rep. Tokyo Kyoiku Daigaku Sect. **A 5**, 300—303 (1957).

Jech, Thomas J. [Tomáš Jech]
[67] Non-probability of Souslin's hypothesis. Comment. Math. Univ. Carolinae **8**, 291—305 (1967).
[71a] Trees. J. Symbolic Logic **36**, 1—14 (1971).
[71b] Lectures in set theory with particular emphasis on the method of forcing. Lecture Notes in Mathematics, Vol. 217. Berlin-Heidelberg-New York: Springer 1971.

Jerison, Meyer (see Leonard Gillman —)

Jónsson, Bjàrni
[56] Universal relational systems. Math. Scand. **4**, 193—208 (1956).
[60] Homogeneous universal relational systems. Math. Scand. **8**, 137—142 (1960).
[65] Extension of relational structures. In: The Theory of Models. Proc. 1963 Internat. Sympos. Berkeley, edited by Addison *et al.*, pp. 146—157. Amsterdam-London: North-Holland 1965.

Jónsson, Bjàrni and Olin, Philip
[68] Almost direct products and saturation. Compositio Math. **20**, 125—132 (1968).

Juhász, I. (see also A. Hajnal —)
[67] Remarks on a theorem of B. Pospíšil. Comment. Math. Univ. Carolinae **8**, 231—247 (1967) [Russian].
[69] On the character of points in $\beta\mathbb{N}_m$. In: Contributions to Extension Theory of Topological Structures. Proc. 1967 Berlin Sympos., edited by J. Flachsmeyer *et al.*, pp. 139—140. Berlin: VEB Deutscher Verlag der Wissenschaften 1969.
[70] Martin's axiom solves Ponomarev's problem. Bull. Acad. Polon. Sci. Sér. Sci. Math. Astronom. Phys. **18**, 71—74 (1970).
[71] Cardinal functions in topology. Mathematical Centre Tracts **34**, Mathematisch Centrum Amsterdam 1971.

Kakutani, Shizuo (see Masahiro Nakamura —)

Kantorovitch, L. (see G. Fichtenholz —)

Katětov, M.
[67] A theorem on mappings. Comment. Math. Univ. Carolinae **8**, 431—433 (1967).

Keisler, H. Jerome (see also Chen Chung Chang —)
[60] On the class of limit ultrapowers of a relational system. Notices Amer. Math. Soc. **7**, 878—879 (1960).
[61] Ultraproducts and elementary classes. Nederl. Akad. Wetensch. Proc. Ser. **A 64** = Indag. Math. **23**, 477—495 (1961).
[62] Some applications of the theory of models to set theory. In: Logic, Methodology and Philosophy of Science. Proc. 1960 Internat. Congr., edited by Nagel *et al.*, pp. 80—86. Stanford: Stanford University Press 1962.
[63] Limit ultrapowers. Trans. Amer. Math. Soc. **107**, 382—408 (1963).
[64a] On cardinalities of ultraproducts. Bull. Amer. Math. Soc. **70**, 644—647 (1964).
[64b] Good ideals in fields of sets. Ann. of Math. (2) **79** 338—359 (1964).
[64c) Ultraproducts and saturated models. Nederl. Akad. Wetensch. Proc. Ser. **A 67** = Indag. Math. **26**, 178—186 (1964).
[65a] A survey of ultraproducts. In: Logic, Methodology and Philosophy of Science. Proc. 1964 Internat. Congr., edited by Y. Bar-Hillel, pp. 112—126. Amsterdam-London: North-Holland 1965.
[65b] Ideals with prescribed degree of goodness. Ann. of Math. (2) **81**, 112—116 (1965).
[65c] Reduced products and Horn classes. Trans. Amer. Math. Soc. **117**, 307—328 (1965).
[65d] Limit ultraproducts. J. Symbolic Logic **30**, 212—234 (1965).

[66] Universal homogeneous Boolean algebras. Mich. J. Math. **13**, 129—132 (1966).
[67a] Ultraproducts which are not saturated. J. Symbolic Logic **32**, 23—46 (1967).
[67b] Mimeographed Lecture Notes. University of California (Los Angeles) 1967.
[67c] Ultraproducts of finite sets. J. Symbolic Logic **32**, 47—57 (1967).
[71] Model Theory for Infinitary Logic. Amsterdam-London: North-Holland 1971.

Keisler, H. Jerome and Prikry, Karel
[74] A result concerning cardinalities of ultraproducts. J. Symbolic Logic **39**, 43—48 (1974).

Keisler, H. J. and Tarski, A.
[64] From accessible to inaccessible cardinals. Fund. Math. **53**, 225—308 (1964); **57**, 119 (1965).

Kelley, J. L.
[52] Banach spaces with the extension property. Trans. Amer. Math. Soc. **72**, 323—326 (1952).

Ketonen, Jussi Antero
[71] Everything you wanted to know about ultrafilters—but were afraid to ask. Doctoral dissertation, University of Wisconsin 1971.
[72] Strong compactness and other cardinal sins. Ann. Math. Logic **5**, 47—76 (1972).
[73] Ultrafilters over measurable cardinals. Fund. Math. **77**, 257—269 (1973).

Kleene, S. C.
[55] Hierarchies of number theoretic predicates. Bull. Amer. Math. Soc. **61**, 193—213 (1955).

Kleinberg, E. M.
[73] Infinitary Combinatorics. In: 1971 Cambridge Summer School in Mathematical Logic. Lecture Notes in Mathematics, Vol. 337, pp. 361—418. Berlin-Heidelberg-New York: Springer 1973.

Kochen, Simon (see also James Ax —)
[58] Ultrafiltered products and arithmetical extensions. Doctoral dissertation, Princeton University 1958.
[61] Ultraproducts in the theory of models. Ann. of Math. (2) **74**, 221—261 (1961).

König, Denes
[27] Über eine Schlußweise aus dem Endlichen ins Unendliche. Acta Litt. Sci. Szeged **3**, 121—130 (1927).

König, J.
[04] Zum Kontinuumproblem. Math. Ann. **60**, 177—180 (1904).

Kripke, Saul A.
[67] An extension of a theorem of Gaifman-Hales-Solovay. Fund. Math. **61**, 29—32 (1967).

Kunen, Kenneth
[68] Inaccessibility properties of cardinals. Doctoral dissertation, Stanford University 1968.
[70a] Some applications of iterated ultrapowers in set theory. Ann. of Math. Logic **2**, 179—227 (1970).
[70b] On the compactification of the integers. Notices Amer. Math. Soc. **17**, 299 (1970).
[72] Ultrafilters and independent sets. Trans. Amer. Math. Soc. **172**, 299—306 (1972).

Kunen, Kenneth and Prikry, Karel
[71] On descendingly incomplete ultrafilters. J. Symbolic Logic **36**, 650—652 (1971).

Kuratowski, Casimir (see also Stefan Banach —)
[47] Sur l'extension de deux théorèmes topologiques à la théorie des ensembles. Fund. Math. **34**, 34—38 (1947).

Kuratowski, K. and Mostowski, A.
[68] Set Theory. Amsterdam-London: North-Holland 1968.

Kurepa, Georges
[36] Ensembles linéaires et une classe de tableaux ramifiés (tableaux ramifiés de M. Aronszajn). Publ. Math. Univ. Belgrade **6**, 129—160 (1936).
[39] Sur la puissance des ensembles partiellement ordonnés. C. R. Soc. Sci. Varsovie **32**, 62—67 (1939).
[50] La condition de Souslin et une propriété caractéristique des nombres réels. C. R. Acad. Sci. Paris **231**, 1113—1114 (1950).
[52] Sur une propriété caractéristique du continu linéaire et le problème de Suslin. Acad. Serbe Sci. Publ. Inst. Math. **4**, 97—108 (1952).
[59a] On the cardinal number of ordered sets and of symmetrical structures in dependence on the cardinal numbers of its chains and antichains [Serbo-Croatian summary]. Glasnik Mat.-Fiz. Astronom. Društvo Mat. Fiz. Hrvatske Ser. II **14**, 183—203 (1959).
[59b] Sur la puissance des ensembles partiellement ordonnés [Serbo-Croatian summary]. Glasnik Mat.-Fiz. Astronom. Društvo Mat. Fiz. Hrvatske Ser. II **14**, 205—211 (1959).
[62] The cartesian multiplication and the cellularity numbers. Publ. Inst. Math. (Beograd) (N.S.) **2**, 121—139 (1962).

Lázár, D.
[36] On a problem in the theory of aggregates. Compositio Math. **3**, 304 (1936).

Łoš, Jerzy
[55] Quelques remarques, théorèmes et problèmes sur les classes définissables d'algèbres. In: Mathematical Interpretation of Formal Systems, edited by Th. Skolem *et al.*, pp. 98—113. Amsterdam-London: North-Holland 1955.

Louveau, Alain
[72] Ultrafiltres sur ℕ, et dérivation séquentielle. Bull. Sci. Math. (2) **96**, 353—382 (1972).

Lusin, N.
[43] Sur les parties de la suite naturelle des nombres entiers. Dokl. Akad. Nauk SSSR **40**, 175—178 (1943).
[47] On subsets of the series of natural numbers [Russian; English summary]. Izv. Akad. Nauk SSSR Ser. Mat. **11**, 403—410 (1947).

Luxemburg, W. A. J.
[64] Non-Standard Analysis. Lectures on A. Robinson's theory of infinitesimals and infinitely large numbers, Second edition. California Institute of Technology, Pasadena, Calif. 1964.

Luxemburg, W. A. J. (Editor)
[69] Applications of Model Theory to Algebra, Analysis, and Probability. New York: Holt, Rinehart and Winston 1969.

Luxemburg, W. A. J. and Robinson, A. (Editors)
[72] Contributions to Non-Standard Analysis. Proc. (1970) Oberwolfach Symposium on Non-Standard Analysis. Amsterdam-London: North-Holland 1972.

Machover, Moshé and Hirschfeld, Joram
[69] Lectures on Non-Standard Analysis. Lecture Notes in Mathematics, Vol. 94. Berlin-Heidelberg-New York: Springer 1969.

MacNeille, H. M.
[37] Partially ordered sets. Trans. Amer. Math. Soc. **42**, 416—460 (1937).

Mansfield, Richard
[71] The theory of Boolean ultrapowers. Ann. Math. Logic **2**, 297—323 (1971).
[72] Horn classes and reduced direct products. Trans. Amer. Math. Soc. **172**, 279—286 (1972).

Marczewski, Edward [= Szpilrajn]
[41] Remarque sur les produits cartésiens d'espaces topologiques. C. R. (Doklady) Acad. Sci. URSS (N.S.) **31**, 525—527 (1941).
[47] Séparabilité et multiplication cartésienne des espaces topologiques. Fund. Math. **34**, 127—143 (1947).

Marek, W.
[64] On families of sets. Bull. Acad. Polon. Sci. Sér. Sci. Math. Astronom. Phys. **12**, 443—448 (1964).

Martin, D. A. and Solovay, R. M.
[69] A basis theorem for Σ_3^1 sets of reals. Ann. of Math. (2) **89**, 138—159 (1969).
[70] Internal Cohen extensions. Ann. Math. Logic **2**, 143—178 (1970).

Mathias, A. R. D.
[72] Solution of problems of Choquet and Puritz. In: Proc. Conference in Math. Logic, London 1970. Lecture Notes in Mathematics, Vol. 255, pp. 204—210. Berlin-Heidelberg-New York: Springer 1972.

Mazur, S.
[52] On continuous mappings in product spaces. Fund. Math. **39**, 229—238 (1952).

Mazurkiewicz, S.
[18] Teorja zbiorów G_δ. Wektor 1918.

McDowell, Robert H.
[58] Extension of functions from dense subspaces. Duke Math. J. **25**, 297—304 (1958).

Michael, E.
[62] A note on intersections. Proc. Amer. Math. Soc. **13**, 281—283; 1000 (1962).

Mirimanoff, D.
[17] Les antinomies de Russell et de Burali-Forti et le problème fondamental de la théorie des ensembles. Enseignement Math. **19**, 37—52 (1917).

Mokobodzki, Gabriel
[69] Ultrafiltres rapides sur $\mathbb{N}$. Construction d'une densité relative de deux potentiels comparables. Séminaire de Théorie du Potentiel, dirigé par M. Brelot, G. Choquet et J. Deny, 12ᵉ année, 1967/68, exposé 12. Secrétariat Mathématique, Paris 1969.

Monk, D. and Scott, D.
[64] Additions to some results of Erdős and Tarski. Fund. Math. **53**, 335—343 (1964).

Morel, A. C. (see C. C. Chang —; T. Frayne, —, D. S. Scott)

Morel, A. C., Scott, D., and Tarski, A.
[58] Reduced products and the compactness theorem. Notices Amer. Math. Soc. **5**, 674—675 (1958).

Morley, Michael
[68a] Partitions and Models. In: Proc. Summer School in Logic, Leeds 1967. Lecture Notes in Mathematics, Vol. 70, pp. 109—158. Berlin-Heidelberg-New York: Springer 1968.
[68b] Review of Tomáš Jech, Non-provability of Souslin's hypothesis. Math. Reviews **35**, 1210 (1968).

Morley, Michael and Vaught, Robert
[62] Homogeneous universal models. Math. Scand. **11**, 37—57 (1962).

Mostowski, A. (see also K. Kuratowski —)
[69] Constructible Sets with Applications. Studies in Logic and the Foundations of Mathematics. Amsterdam-London: North-Holland; Warsaw: PWN—Polish Scientific Publishers 1969.

Mrówka, S.
[59] On the potency of subsets of $\beta\mathbb{N}$. Colloq. Math. **7**, 23—25 (1959).

Nachbin, Leopoldo
[50] A theorem of Hahn-Banach type for linear transformations. Trans. Amer. Math. Soc. **68**, 28—46 (1950).

Nakamura, Masahiro and Kakutani, Shizuo
[43] Banach limits and the Čech compactification of a countable discrete set. Proc. Imp. Acad. Tokyo **19**, 224—229 (1943).

Negrepontis, S. (see also W. W. Comfort —; H. H. Hung —)
[67a] On a theorem of Hoffman and Ramsay. Pacific J. Math. **20**, 281—282 (1967).
[67b] Absolute Baire sets. Proc. Amer. Math. Soc. **18**, 691—694 (1967).
[68] Extension of continuous functions in βD. Nederl. Akad. Wetensch. Proc. Ser. **A 71** = Indag. Math. **30**, 393—400 (1968).
[69a] The Stone space of the saturated Boolean algebras. Trans. Amer. Math. Soc. **141**, 515—527 (1969).
[69b] The Stone space of the saturated Boolean algebras. In: Proc. Internat. Sympos. on Topology and its Applications, Herceg-Novi, Aug., 1968, edited by D. R. Kurepa, pp. 265—268. Savez Društava Matematičara, Fizičara I Astronoma, Belgrade 1969.
[69c] The existence of certain uniform ultrafilters. Ann. of Math. (2) **90**, 23—32 (1969).
[70] Adequate ultrafilters of special Boolean algebras. Notices Amer. Math. Soc. **17**, 832—833 (1970).
[71] A property of the saturated Boolean algebras. Nederl. Akad. Wetensch. Proc. Ser. **A 74** = Indag. Math. **33**, 117—120 (1971).
[72a] Adequate ultrafilters of special Boolean algebras. Trans. Amer. Math. Soc. **174**, 345—367 (1972).
[72b] Ramification systems and spaces of ultrafilters. In: General Topology and Its Relations to Modern Analysis and Algebra, III. Proc. Third Prague Topological Sympos., Prague, 1971, edited by J. Novák, p. 333. Prague: Academia 1972.
[73] The growth of sub-uniform ultrafilters. Trans. Amer. Math. Soc. **175**, 155—165 (1973).
[*] Ultrafilters on weakly compact cardinals, manuscript.
[**] Note, manuscript.
[***] Recent results on the topology of spaces of ultrafilters. In: Conference on Rings of Continuous Functions (Rome, Nov. 1973), Symposia Mathematica. London-New York: Academic Press, in preparation.

Neumann, J. von
[23] Zur Einführung der transfiniten Zahlen. Acta Litt. Sci. Szeged **1**, 199—208 (1923).
[28] Die Axiomatisierung der Mengenlehre. Math. Z. **27**, 669—752 (1928).
[35] Charakterisierung des Spektrums eines Integraloperators. Paris: Hermann 1935.

Noble, N. and Ulmer, M.
[72] Factoring functions on Cartesian products. Trans. Amer. Math. Soc. **163**, 329—340 (1972).

Novák, J.
[53] On some problems of Luzin concerning the subsets of natural numbers. Czechoslovak Math. J. **3**, 385—395 (1953).

Olin, Philip (see Bjàrni Jónsson —)

Parovičenko, I. I.
[57] On the theory of spaces not satisfying the first countability axiom. Study Notes of Kishinev State University **29**, 15—24 (1957) [Russian].
[59] On the branching problem. Study Notes of Kishinev State University **39**, 189—194 (1959) [Russian].
[63] A universal bicompact of weight $\aleph$. Dokl. Akad. Nauk SSSR **150**, 36—39 (1963) [Russian]. (English translation: Soviet Math. Dokl. **4**, 592—595 (1963).)
[67] The branching hypothesis and the correlation between local weight and power to topological spaces. Dokl. Akad. Nauk SSSR **174**, 30—32 (1967) [Russian]. (English translation: Soviet Math. Dokl. **8**, 589—592 (1967).)
[68] Topological equivalents of the branching hypothesis. Dokl. Akad. Nauk SSSR **183**, 1258—1261 (1968) [Russian]. (English translation: Soviet Math. Dokl. **9**, 1544—1547 (1968).)

Piccard, Sophie
[37] Sur un problème de M. Ruziewicz de le théorie des relations. Fund. Math. **29**, 5—9 (1937).

Pierce, R. S.
[57] Distributivity in Boolean algebras. Pacific J. Math. **7**, 988—992 (1957).
[58a] A note on complete Boolean algebras. Proc. Amer. Math. Soc. **9**, 892—896 (1958).
[58b] Distributivity and the normal completion of Boolean algebras. Pacific J. Math. **8**, 133—140 (1958).
[61] Rings of integer-valued continuous functions. Trans. Amer. Math. Soc. **100**, 371—394 (1961).
[67] Modules over commutative regular rings. Mem. Amer. Math. Soc., no. **70**. Providence, R. I.: American Mathematical Society 1967.

Pitt, R. A.
[71] The classification of ultrafilters on $\mathbb{N}$. Doctoral dissertation, University of Leicester 1971.

Plank, Donald
[69] On a class of subalgebras of $C(X)$ with applications to $\beta X \backslash X$. Fund. Math. **64**, 41—54 (1969).

Pondiczery, E. S.
[44] Power problems in abstract spaces. Duke Math. J. **11**, 835—837 (1944).

Pospíšil, Bedřich
[37] Remark on bicompact spaces. Ann. of Math. (2) **38**, 845—846 (1937).
[39] On bicompact spaces. Publ. Fac. Sci. Univ. Masaryk **270**, 3—16 (1939).

Prikry, Karel (see also H. Jerome Keisler —; Kenneth Kunen —)
[68] Changing measurable cardinals into accessible cardinals. Doctoral dissertation, University of California (Berkeley) 1968.
[70] On a problem of Gillman and Keisler. Ann. of Math. Logic **2**, 179—187 (1970).
[73] On descendingly complete ultrafilters. In: 1971 Cambridge Summer School in Mathematical Logic. Lecture Notes in Mathematics, Vol. 337, pp. 459—488. Berlin-Heidelberg-New York: Springer 1973.
[*] Ultrafilters and almost disjoint sets. General Topology and Appl., to appear.
[**] Ultrafilters and almost disjoint sets II, manuscript.

Rado, R. (see P. Erdős, A. Hajnal, —; P. Erdős, —)

Rainwater, John
[59] A note on projective resolutions. Proc. Amer. Math. Soc. **10**, 734—735 (1959).

Rajagopalan, M.
[72] $\beta\mathbb{N}\setminus\mathbb{N}\setminus\{p\}$ is not normal. Journal of Indian Math. Soc. **36**, 173—176 (1972).

Ramsay, Arlan (see Kenneth Hoffman —)

Ramsey, F. P.
[30] On a problem of formal logic. Proc. London Math. Soc. (2) **30**, 264—286 (1930).

Riesz, F.
[09] Stetigkeitsbegriff und abstrakte Mengenlehre. In: Atti del IV Congresso Internaz. Mat., Roma, 1908, edited by G. Castelnuovo, Vol. II, pp. 18—24. Roma: Tipografia della R. Accademia dei Lincei 1909.

Robinson, Abraham (see also W. A. J. Luxemburg —)
[56a] A result on consistency and its applications to the theory of definition. Nederl. Akad. Wetensch. Proc. Ser. **A 59** = Indag. Math. **18**, 47—58 (1956).
[56b] Complete Theories. Amsterdam-London: North-Holland 1956.
[65] Introduction to Model Theory and to the Metamathematics of Algebra, Second edition. Amsterdam-London: North-Holland 1965.
[66] Non-Standard Analysis. Amsterdam-London: North-Holland 1966.

Ross, K. A. and Stone, A. H.
[64] Products of separable spaces. Amer. Math. Monthly **71**, 398—403 (1964).

Rothberger, Fritz
[41] Sur les familles indénombrables de suites de nombres naturels et les problèmes concernant la propriété *C*. Proc. Cambridge Philos. Soc. **37**, 109—126 (1941).
[48] On some problems of Hausdorff and of Sierpiński. Fund. Math. **35**, 29—46 (1948).

Rowbottom, Frederick
[64] Some strong axioms of infinity incompatible with the axiom of constructibility. Doctoral dissertation, University of Wisconsin 1964.

Rubin, Herman and Rubin, Jean E.
[63] Equivalents of the Axiom of Choice. Amsterdam-London: North-Holland 1963.

Rucker, R. V. B. (see Erik Ellentuck —)

Rudin, Mary Ellen
[66] Types of ultrafilters. In: Topology Seminar, Wisconsin, 1965. Annals of Mathematics Studies no. 60, pp. 147—151. Princeton: Princeton Univ. Press 1966.
[69] Souslin's conjecture. Amer. Math. Monthly **76**, 1113—1119 (1969).
[71] Partial orders on the types of $\beta\mathbb{N}$. Trans. Amer. Math. Soc. **155**, 353—362 (1971).

Rudin, Walter
[56] Homogeneity problems in the theory of Čech compactifications. Duke Math. J. **23**, 409—419; 633 (1956).

Šanin, N. A. [= Shanin]
[46a] A theorem from the general theory of sets. C. R. (Doklady) Acad. Sci. URSS **53**, 399—400 (1946).
[46b] On intersection of open subsets in the product of topological spaces. C. R. (Doklady) Acad. Sci. URSS **53**, 499—501 (1946).
[48] On the product of topological spaces. Trudy Mat. Inst. Akad. Nauk SSSR **24**, 1—112 (1948) [Russian].

Schoenflies, A. (and H. Hahn)
[13] Entwickelung der Mengenlehre und ihrer Anwendungen, erste Hälfte: Allgemeine Theorie der unendlichen Mengen und Theorie der Punktmengen. Leipzig und Berlin: A. Schoenflies 1913.

Schreier, Otto (see Emil Artin)

Scott, Dana (see also T. Frayne, A. C. Morel —; W. P. Hanf —; D. Monk —; A. C. Morel, —, A. Tarski)
[61] Measurable cardinals and constructible sets. Bull. Acad. Polon. Sci. Sér. Sci. Math. Astronom. Phys. **9**, 521—524 (1961).

Semadeni, Z.
[64] Periods of measurable functions and the Stone-Čech compactification. Amer. Math. Monthly **71**, 891—893 (1964).

Sevee, Denis
[72] On generalized Jónsson classes. Master's thesis, McGill University 1972.

Shelah, Saharon
[70] On the cardinality of ultraproduct of finite sets. J. Symbolic Logic **35**, 83—84 (1970).
[71] Every two elementarily equivalent models have isomorphic ultrapowers. Israel J. Math. **10**, 224—233 (1971).
[72] Saturation of ultrapowers and Keisler's order. Ann. Math. Logic **4**, 75—114 (1972).
[*] Generalization of theorems on ultraproducts to reduced product, manuscript.

Sierpiński, Wacław
[22] Sur un problème concernant les sous-ensembles croissants du continu. Fund. Math. **3**, 109—112 (1922).
[28] Sur une décomposition d'ensembles. Monatsh. Math. Phys. **35**, 239—242 (1928).
[37] Sur les suites transfinies finalement disjoints. Fund. Math. **28**, 115—119 (1937).
[38a] Sur la décomposition des ensembles en sous-ensembles presque disjoints. Mathematica (Cluj) **14**, 15—17 (1938).
[38b] Fonctions additives non complètement additives et fonctions non mesurables. Fund. Math. **30**, 96—99 (1938).
[40] Sur une propriété des ensembles ordonnés. Pont. Acad. Sci. Acta **4**, 207—208 (1940).
[45] Sur les espaces métriques universels. Fund. Math. **33**, 123—136 (1945).
[49] Sur une propriété des ensembles ordonnés. Fund. Math. **36**, 56—67 (1949).
[56] Hypothèse du continu. Second edition. New York: Chelsea 1956.
[65] Cardinal and ordinal numbers. Second revised edition. Monografie Matematyczne, Vol. 34. Warsaw: Panstwowe Wydawnictwo Naukowe 1965.

Sierpiński, W. and Tarski, A.
[30] Sur une propriété caractéristique des nombres inaccessibles. Fund. Math. **15**, 292—300 (1930).

Sikorski, Roman
[48] A theorem on extension of homomorphisms. Ann. Soc. Polon. Math. **21**, 332—335 (1948).
[50] Cartesian products of Boolean algebras. Fund. Math. **37**, 25—54 (1950).
[63] On extensions and products of Boolean algebras. Fund. Math. **53**, 99—116 (1963).
[64] Boolean Algebras. Second edition. Ergebnisse der Mathematik und ihrer Grenzgebiete, Bd. 25. Berlin-Heidelberg-New York: Springer 1964.

Silver, Jack H.
[66] Some applications of model theory in set theory. Doctoral dissertation, University of California (Berkeley) 1966.
[70] A large cardinal in the constructible universe. Fund. Math. **69**, 93—100 (1970).

Simpson, Stephen G.
[70] Model-theoretic proof of a partition theorem. Notices Amer. Math. Soc. **17**, 964 (1970).

Skolem, Th.
[22] Einige Bemerkungen zur axiomatischen Begründung der Mengenlehre, in wissenschaftlichen Vorträgen gehalten auf dem fünften Kongreß der Skandinavischen Mathematiker in Helsingfors, pp. 217—232. Helsingfors 1922.
[34] Über die Nicht-Charakterisierbarkeit der Zahlenreihe mittels endlich oder abzählbar unendlich vieler Aussagen mit ausschließlich Zahlenvariablen. Fund. Math. **23**, 150—161 (1934).
[55] Peano's axioms and models of arithmetic. In: Mathematical Interpretations of Formal Systems, edited by Skolem *et al.*, pp. 1—14. Amsterdam-London: North-Holland 1955.

Slomson, A. B. (see J. L. Bell —)

Smith, E. C., Jr. and Tarski, Alfred
[57] Higher degrees of distributivity and completeness in Boolen algebras. Trans. Amer. Math. Soc. **84**, 230—257 (1957).

Solomon, R. C.
[73] A type of $\beta\mathbb{N}$ with $\aleph_0$ relative types. Fund. Math. **79**, 209—212 (1973).

Solovay, Robert M. (see also D. A. Martin —)
[66] New proof of a theorem of Gaifman and Hales. Bull. Amer. Math. Soc. **72**, 282—284 (1966).
[70] A model of set-theory in which every set of reals is Lebesgue-measurable. Ann. of Math. (2) **92**, 1—56 (1970).
[71] Real-valued measurable cardinals. In: Axiomatic Set Theory. Proc. Sympos. Pure Math., Vol. XIII, Part I, Univ. California, Los Angeles, 1967, edited by D. S. Scott, pp. 397—428. Providence, R. I.: American Mathematical Society 1971.

Solovay, R. M. and Tennenbaum, S.
[71] Iterated Cohen extensions and Souslin's problem. Ann. of Math. (2) **94**, 201—245 (1971).

Souslin, M.
[20] Problème 3. Fund. Math. **1**, 223 (1920).

Specker, E.
[51] Sur un problème de Sikorski. Colloq. Math. **2**, 9—12 (1951).

Stavi, Jonathan
[73] Extensions of Kripke's embedding theorem. Doctoral dissertation, The Hebrew University of Jerusalem 1973.

Steiner, A. K. and Steiner, E. F.
[70] Graph closures and metric compactifications of $\mathbb{N}$. Proc. Amer. Math. Soc. **25**, 593—597 (1970).
[71] Relative types of points of $\beta\mathbb{N}\setminus\mathbb{N}$. Trans. Amer. Math. Soc. **160**, 279—286 (1971).

Štěpánek, Petr and Vopěnka, Petr
[67] Decomposition of metric spaces into nowhere dense sets. Comment. Math. Univ. Carolinae **8**, 387—404; 567—568 (1967).

Stone, A. H. (see K. A. Ross —)

Stone, M. H.
[34] Boolen algebras and their application to topology. Proc. Nat. Acad. Sci. U.S.A. **20**, 197—202 (1934).

[36] The representation theorem for Boolean algebras. Trans. Amer. Math. Soc. **40**, 37—111 (1936).
[37a] Applications of the theory of Boolean rings to general topology. Trans. Amer. Math. Soc. **41**, 375—481 (1937).
[37b] Algebraic characterizations of special Boolean rings. Fund. Math. **29**, 223—303 (1937).

Szpilrajn, Edward [see Edward Marczewski]

Tarski, Alfred (see also P. Erdős —; H. J. Keisler —; A. C. Morel, D. Scott —; W. Sierpiński —; E. C. Smith, Jr. —)
[25] Quelques théorèmes sur les alephs. Fund. Math. **7**, 1—14 (1925).
[28] Sur la décomposition des ensembles en sous-ensembles presque disjoints. Fund. Math. **12**, 188—205 (1928).
[30a] Une contribution à la théorie de la mesure. Fund. Math. **15**, 42—50 (1930).
[30b] Sur les classes d'ensembles closes par rapport à certaines opérations élémentaires. Fund. Math. **16**, 181—304 (1930).
[36] Der Wahrheitsbegriff in den formalisierten Sprachen. Studia Philos. **1**, 261—404 (1936).
[38a] Über unerreichbare Kardinalzahlen. Fund. Math. **30**, 68—89 (1938).
[38b] Drei Überdeckungssätze der allgemeinen Mengenlehre. Fund. Math. **30**, 132—155 (1938).
[39] Ideale in vollständigen Mengenkörpern I. Fund. Math. **32**, 45—63 (1939).
[62] Some problems and results relevant to the foundations of set theory. In: Logic, Methodology and Philosophy of Science. Proc. 1960 Internat. Congr., edited by Nagel *et al.*, pp. 125—135. Stanford: Stanford University Press 1962.

Tarski, Alfred and Vaught, Robert L.
[57] Arithmetical extensions of relational systems. Compositio Math. **13**, 81—102 (1957).

Tennenbaum, S. (see also R. M. Solovay —)
[68] Souslin's problem. Proc. Nat. Acad. Sci. U.S.A. **59**, 60—63 (1968).

Tychonoff, A.
[30] Über die topologische Erweiterung von Räumen. Math. Ann. **102**, 544—561 (1930).

Ulam, Stanislaw (see also P. Erdős —)
[29] Concerning functions of sets. Fund. Math. **14**, 231—233 (1929).
[30] Zur Maßtheorie in der allgemeinen Mengenlehre. Fund. Math. **16**, 140—150 (1930).

Ulmer, Milton Don (see also N. Noble —)
[70] Functions on product spaces and C-embedded Σ-spaces. Doctoral dissertation, Wesleyan University 1970.

Urysohn, Paul (see also P. Alexandrov —)
[25] Sur un espace métrique universel. C. R. Acad. Sci. Paris **180**, 803—806 (1925).
[27] Sur un espace métrique universel. Bull. Sci. Math. (2) **51**, 43—64 (1927).

Vaught, Robert (see Michael Morley —; Alfred Tarski —)

Vopěnka, Petr (see also B. Balcar —; Petr Štěpánek —)
[65] Properties of ∇-model. Bull. Acad. Polon. Sci. Sér. Sci. Math. Astronom. Phys. **13**, 441—444 (1965).
[71] The theory of semisets. In: Proc. 1970 Internat. Congr., Nice, Vol. 1, pp. 255—260. Paris: Gauthier-Villars 1971.

Walker, Russell C.
[74] The Stone-Čech compactification. Ergebnisse der Mathematik und ihrer Grenzgebiete, Bd. 83. Berlin-Heidelberg-New York: Springer 1974.

Warren, Nancy McMaster
[70] Extending continuous functions in Stone-Čech compactifications of discrete spaces and in zero-dimensional spaces. Doctoral dissertation, University of Wisconsin 1970.
[72] Properties of the Stone-Čech compactifications of discrete spaces. Proc. Amer. Math. Soc. **33**, 599—606 (1972).

Woods, R. Grant
[71a] A Boolean algebra of regular closed subsets of $\beta X \setminus X$. Trans. Amer. Math. Soc. **154**, 23—36 (1971).
[71b] Co-absolutes of remainders of Stone-Čech compactifications. Pacific J. Math. **37**, 545—560 (1971).

Yaqub, F. M. (see Ph. Dwinger —)

Yorke, James A.
[69] Permutations and two sequences with the same cluster set. Proc. Amer. Math. Soc. **20**, 606 (1969).

Zermelo, E.
[08] Untersuchungen über die Grundlagen der Mengenlehre I. Math. Ann. **65**, 261—281 (1908).

Index of Symbols

Subject Index

Die Grundlehren der mathematischen Wissenschaften in Einzeldarstellungen mit besonderer Berücksichtigung der Anwendungsgebiete

Eine Auswahl

---

23. Pasch: Vorlesungen über neuere Geometrie. In Vorbereitung
41. Steinitz: Vorlesungen über die Theorie der Polyeder. In Vorbereitung
45. Alexandroff/Hopf: Topologie. Band 1
46. Nevanlinna: Eindeutige analytische Funktionen
63. Eichler: Quadratische Formen und orthogonale Gruppen
102. Nevanlinna/Nevanlinna: Absolute Analysis
114. Mac Lane: Homology
127. Hermes: Enumerability, Decidability, Computability
131. Hirzebruch: Topological Methods in Algebraic Geometry
135. Handbook for Automatic Computation. Vol. 1/Part a: Rutishauser: Description of ALGOL 60
136. Greub: Multilinear Algebra
137. Handbook for Automatic Computation. Vol. 1/Part b: Grau/Hill/Langmaack: Translation of ALGOL 60
138. Hahn: Stability of Motion
139. Mathematische Hilfsmittel des Ingenieurs. 1. Teil
140. Mathematische Hilfsmittel des Ingenieurs. 2. Teil
141. Mathematische Hilfsmittel des Ingenieurs. 3. Teil
142. Mathematische Hilfsmittel des Ingenieurs. 4. Teil
143. Schur/Grunsky: Vorlesungen über Invariantentheorie
144. Weil: Basic Number Theory
145. Butzer/Berens: Semi-Groups of Operators and Approximation
146. Treves: Locally Convex Spaces and Linear Partial Differential Equations
147. Lamotke: Semisimpliziale algebraische Topologie
148. Chandrasekharan: Introduction to Analytic Number Theory
149. Sario/Oikawa: Capacity Functions
150. Iosifescu/Theodorescu: Random Processes and Learning
151. Mandl: Analytical Treatment of One-dimensional Markov Processes
152. Hewitt/Ross: Abstract Harmonic Analysis. Vol. 2: Structure and Analysis for Compact Groups. Analysis on Locally Compact Abelian Groups
153. Federer: Geometric Measure Theory
154. Singer: Bases in Banach Spaces I
155. Müller: Foundations of the Mathematical Theory of Electromagnetic Waves
156. van der Waerden: Mathematical Statistics
157. Prohorov/Rozanov: Probability Theory
158. Constantinescu/Cornea: Potential Theory on Harmonic Spaces
159. Köthe: Topological Vector Spaces I
160. Agrest/Maksimov: Theory of Incomplete Cylindrical Functions and their Applications
161. Bhatia/Szegö: Stability Theory of Dynamical Systems
162. Nevanlinna: Analytic Functions
163. Stoer/Witzgall: Convexity and Optimization in Finite Dimensions I
164. Sario/Nakai: Classification Theory of Riemann Surfaces
165. Mitrinović/Vasić: Analytic Inequalities

166. Grothendieck/Dieudonné: Eléments de Géométrie Algébrique I
167. Chandrasekharan: Arithmetical Functions
168. Palamodov: Linear Differential Operators with Constant Coefficients
169. Rademacher: Topics in Analytic Number Theory
170. Lions: Optimal Control of Systems Governed by Partial Differential Equations
171. Singer: Best Approximation in Normed Linear Spaces by Elements of Linear Subspaces
172. Bühlmann: Mathematical Methods in Risk Theory
173. Maeda/Maeda: Theory of Symmetric Lattices
174. Stiefel/Scheifele: Linear and Regular Celestial Mechanics. Perturbed Two-body Motion—Numerical Methods—Canonical Theory
175. Larsen: An Introduction to the Theory of Multipliers
176. Grauert/Remmert: Analytische Stellenalgebren
177. Flügge: Practical Quantum Mechanics I
178. Flügge: Practical Quantum Mechanics II
179. Giraud: Cohomologie non abélienne
180. Landkof: Foundations of Modern Potential Theory
181. Lions/Magenes: Non-Homogeneous Boundary Value Problems and Applications I
182. Lions/Magenes: Non-Homogeneous Boundary Value Problems and Applications II
183. Lions/Magenes: Non-Homogeneous Boundary Value Problems and Applications III
184. Rosenblatt: Markov Processes. Structure and Asymptotic Behavior
185. Rubinowicz: Sommerfeldsche Polynommethode
186. Handbook for Automatic Computation. Vol. 2. Wilkinson/Reinsch: Linear Algebra
187. Siegel/Moser: Lectures on Celestial Mechanics
188. Warner: Harmonic Analysis on Semi-Simple Lie Groups I
189. Warner: Harmonic Analysis on Semi-Simple Lie Groups II
190. Faith: Algebra: Rings, Modules, and Categories I
192. Mal'cev: Algebraic Systems
193. Pólya/Szegö: Problems and Theorems in Analysis I
194. Igusa: Theta Functions
195. Berberian: Baer $*$-Rings
196. Athreya/Ney: Branching Processes
197. Benz: Vorlesungen über Geometrie der Algebren
198. Gaal: Linear Analysis and Representation Theory
200. Dold: Lectures on Algebraic Topology
202. Schmetterer: Introduction to Mathematical Statistics
203. Schoeneberg: Elliptic Modular Functions
204. Popov: Hyperstability of Control Systems
206. André: Homologie des algèbres commutatives
207. Donoghue: Monotone Matrix Functions and Analytic Continuation
208. Lacey: The Isometric Theory of Classical Banach Spaces
209. Ringel: Map Color Theorem
210. Gihman/Skorohod: The Theory of Stochastic Processes I
211. Comfort/Negrepontis: The Theory of Ultrafilters